WILDLIFE MANAGEMENT AND CONSERVATION

WILDLIFE MANAGEMENT AND CONSERVATION

Contemporary Principles and Practices

Second Edition

EDITED BY Paul R. Krausman and James W. Cain III

Published in Association with The Wildlife Society

JOHNS HOPKINS UNIVERSITY PRESS | BALTIMORE

Johns Hopkins University Press
2715 North Charles Street
Baltimore, Maryland 21218-4363
www.press.jhu.edu

Library of Congress Cataloging-in-Publication Data

Names: Krausman, Paul R., 1946– editor. | Cain, James W., editor.
Title: Wildlife management and conservation : contemporary principles
    and practices / edited by Paul R. Krausman and James W. Cain III.
Description: Second edition. | Baltimore : Johns Hopkins University
    Press, 2022. | Published in Association with The Wildlife Society. |
    Includes bibliographical references and index.
Identifiers: LCCN 2021038985 | ISBN 9781421443966 (hardcover) |
    ISBN 9781421443973 (ebook)
Subjects: LCSH: Wildlife management—North America. | Wildlife
    conservation—North America.
Classification: LCC SK361 .W49 2022 | DDC 639.9097—dc23
LC record available at https://lccn.loc.gov/2021038985

A catalog record for this book is available from the British Library.

Special discounts are available for bulk purchases of this book. For more information, please contact Special Sales at specialsales@jh.edu.

CONTENTS

CONTRIBUTORS

C. Jane Anderson
611 Virginia Drive
Round Rock, TX 78664, USA

Bart M. Ballard
Caesar Kleberg Wildlife Research
Institute
Texas A&M–Kingsville
MSC 218, 700 University Boulevard
Kingsville, TX 78363, USA

Brent D. Bibles
Distance Education Graduate Faculty
Unity College
Unity, ME 04988, USA

John A. Bissonette
Department of Wildlands Resources
Quinney College of Natural Resources
Utah State University
Logan, UT 84322-5230, USA

Clint W. Boal
US Geological Survey Texas Coopera-
tive Fish and Wildlife Research Unit
Texas Tech University
Lubbock, TX 79409, USA

Scott T. Boyle
Department of Biology
New Mexico State University
PO Box 30001, MSC 3AF
Las Cruces, NM 88003, USA

Leonard A. Brennan
Caesar Kleberg Wildlife Research
Institute
Texas A&M–Kingsville
MSC 218, 700 University Boulevard
Kingsville, TX 78363, USA

Robert D. Brown
109 Cape Cod Drive
Cary, NC 27511, USA

James W. Cain III
US Geological Survey
New Mexico Cooperative Fish and
Wildlife Research Unit
Department of Fish, Wildlife, and
Conservation Ecology
New Mexico State University
PO Box 30003, MSC 4901
Las Cruces, NM 88003, USA

Tyler A. Campbell
East Foundation
200 Concord Plaza Drive, Suite 410
San Antonio, TX 78216, USA

Michael J. Cherry
Caesar Kleberg Wildlife Research
Institute
Texas A&M–Kingsville
MSC 218, 700 University Boulevard
Kingsville, TX 78363, USA

Michael R. Conover
Department of Wildland Resources
Utah State University
Logan, UT 84322-5230, USA

Daniel J. Decker
Department of Natural Resources and
the Environment
Cornell University
Ithaca, NY 14853-3001, USA

Randall W. DeYoung
Caesar Kleberg Wildlife Research
Institute
Texas A&M–Kingsville
MSC 218, 700 University Boulevard
Kingsville, TX 78363, USA

Jonathan B. Dinkins
Department of Animal and Rangeland
Sciences
Oregon State University
206 Withycombe Hall
Corvallis, OR 97331, USA

W. Sue Fairbanks
Department of Natural Resource
Ecology and Management
Oklahoma State University
008C Agricultural Hall
Stillwater, OK 74078, USA

James B. Grand
US Geological Survey Cooperative
Research Units
1444 Beri Barfield Drive
Dadeville, AL 36853, USA

Michael J. Haney
3169 E 3500 N
Twin Falls, ID 83301, USA

James R. Heffelfinger
5219 West Bobwhite Way
Tucson, AZ 85742, USA

Scott E. Henke
Caesar Kleberg Wildlife Research
Institute
Texas A&M–Kingsville
MSC 218, 700 University Boulevard
Kingsville, TX 78363, USA

Fidel Hernandez
Caesar Kleberg Wildlife Research
Institute
Texas A&M–Kingsville
MSC 218, 700 University Boulevard
Kingsville, TX 78363, USA

David G. Hewitt
Caesar Kleberg Wildlife Research
Institute
Texas A&M–Kingsville
MSC 218, 700 University Boulevard
Kingsville, TX 78363, USA

Christopher L. Hoving
Michigan Department of Natural
Resources
Constitution Hall
525 West Allegan Street
PO Box 30444
Lansing, MI 48909, USA

David A. Jessup
Wildlife Health Center
School of Veterinary Medicine
University of California–Davis
Davis, CA 95616, USA

Heather E. Johnson
USGS Alaska Science Center
4210 University Drive
Anchorage, AK 99508, USA

John L. Koprowski
Haub School of Environment and
Natural Resources
201 Bim Kendall House
804 E Fremont Street
Laramie, WY 82072, USA

Paul R. Krausman
263 Camino Los Abuelos
Santa Fe, NM 87508, USA

William P. Kuvlesky Jr.
Caesar Kleberg Wildlife Research
Institute
Texas A&M–Kingsville
MSC 218, 700 University Boulevard
Kingsville, TX 78363, USA

Roel R. Lopez
Texas A&M Natural Resources
Institute
1919 Oakwell Farms Parkway
Suite 100
San Antonio, TX 78218, USA

R. William Mannan
School of Natural Resources and the
Environment
University of Arizona
Tucson, AZ 85721, USA

Melissa J. Merrick
School of Natural Resources and the
Environment
University of Arizona
Tucson, AZ 85721, USA

L. Scott Mills
University of Montana
University Hall, Room 116
Missoula, MT 59812, USA

Michael S. Mitchell
USGS Montana Cooperative Wildlife
Research Unit
University of Montana
205 Natural Science Building
Missoula, MT 59812, USA

Michael L. Morrison
Department of Rangeland, Wildlife
and Fisheries Management
Texas A&M University
College Station, TX 77832-2138, USA

Anna M. Muñoz
US Fish and Wildlife Service
134 S. Union Boulevard, Suite 400
Lakewood, CO 80228, USA

John F. Organ
US Geological Survey
Cooperative Fish and Wildlife
Research Units
12201 Sunrise Valley Drive
Reston, VA 20192, USA

Katherine L. Parker
Ecosystem Science and Management
University of Northern British
Columbia
Prince George, BC V2N 4Z9, Canada

William F. Porter†
Michigan State University

Shawn J. Riley
Department of Fisheries and Wildlife
Michigan State University
480 Wilson Road, 2D Natural
Resources Building
East Lansing, MI, 48824, USA

Steven S. Rosenstock
2700 Woodlands
Building 300-308
Flagstaff, AZ 86001, USA

Michael C. Runge
US Geological Survey
12100 Beech Forest Road
Laurel, MD 20708, USA

Susan P. Rupp
Enviroscapes Ecological Consulting,
LLC
13117 Bluebird Road
Gravette, AR 72736, USA

William F. Siemer
Department of Natural Resources and
the Environment
Cornell University
Fernow Hall
Ithaca, NY 14853-3001, USA

Robert J. Steidl
School of Natural Resources and the
Environment
University of Arizona
Tucson, AZ 85721, USA

Kelley M. Stewart
Department of Natural Resources and
Environmental Science
University of Nevada, Reno
Reno, NV 89557, USA

John M. Tomeček
Department of Rangeland, Wildlife
and Fisheries Management
Texas A&M University
College Station, TX 77832-2138, USA

PREFACE

Wildlife management and conservation are at a crossroads. Wildlife habitat is being altered at an unprecedented rate because of human influences. Efficient and effective management is important for the future of wildlife habitats, but it can only be accomplished by incorporating all stakeholders into the process. This textbook examines animal species, their habitat, and how human management influences them both—the wildlife management triad. These three components are central to wildlife management and form the core of the profession. Unfortunately, the human side of wildlife management has not received as much attention as wildlife species or their habitats, but it must be recognized for effective management. Wildlife in North America belongs to the public, which must be considered in management decisions. Simply discussing animals and their habitats will not address the complexity of wildlife management in the 21st century, especially with increasing human demands for landscape use. Habitat for wildlife is decreasing, and human use of those lands is facing serious challenges. We need leaders in wildlife management and conservation who understand the biology of wildlife species and how landscapes can be managed to ensure their survival and long-term viability. This second edition of *Wildlife Management and Conservation* considers the ways wildlife is managed and explores how management is only successful when the animal, habitat, and people are equally considered. It emphasizes the importance of structured decision-making and planning in the process of wildlife management. This edition includes a new chapter on plant–animal interactions, updated revisions to all chapters, additional bibliographies, and a glossary.

This volume is for future wildlife leaders, written by current leaders in the field who are members of The Wildlife Society (TWS), the professional society for wildlife biologists. No other course is as important for undergraduate wildlife biologists as the one that addresses the basics of management and conservation. As such, this text is endorsed by TWS and written by TWS professionals across the nation.

Designed for use in various wildlife curricula, this edition can serve as a standalone text for programs that do not have classes in all of the topics covered, or it can be used as an introductory text for students in programs that have a complete suite of wildlife classes. It also contains a solid review of the profession for graduate students and practicing professionals.

This volume consists of 20 chapters, beginning with definitions of the wildlife profession and history of wildlife conservation in North America (Chapters 1–3). We then discuss the human dimensions of wildlife management (Chapter 4) and include that important dynamic throughout the text. Chapter 5 addresses the importance of structured decision-making, a critical component in management. Following those chapters is an exploration of the biological component of management (Chapters 6–15), habitat and wildlife restoration (Chapters 16 and 17), and how climate change influences wildlife populations (Chapter 18). The last two chapters address conservation planning and include case studies of wildlife population management. Each chapter introduces readers to some of the many key personalities in the profession through short biographies.

Future leaders in the wildlife profession must use numerous tools to be effective, relevant, and current. One of those tools is a solid education related to wildlife biology, habitat, and the human dimensions that dictate what we do. *Wildlife Management and Conservation* has been written to assist in that effort.

ACKNOWLEDGMENTS

A work of this scope is possible only because of the combined efforts of numerous researchers and managers who have dedicated their professional lives to the management of wildlife. Our work environments with The Wildlife Society and the US Geological Survey, New Mexico Cooperative Fish and Wildlife Research Unit, were also instrumental toward the completion of this book. We appreciate the support of The Wildlife Society; the US Geological Survey, New Mexico Cooperative Fish and Wildlife Research Unit; the Department of Fish, Wildlife and Conservation Ecology, New Mexico State University; the Wildlife Health Center, School of Veterinary Medicine, University of California, Davis; University of Montana; and all the other organizations that supported authors while revising chapters.

We also continue to thank numerous colleagues who willingly agreed to review book chapters for both editions, including Bill Bartush, Merav Ben-David, John A. Bissonette, Vernon C. Bleich, Bill Block, R. Terry Bowyer, Regan Brown, Melanie Bucci, Jamal N. Butler, Colin M. Callahan, Casey J. Cardinal, Kevin Crooks, Ashley D'Antonio, Thomas Decker, Steven De Stefano, T. Donovan, Adria Elskus, Colin Gillin, James B. Grand, Kevin Gutzwiler, Matthew Kauffman, Bryan M. Kluever, Amy J. Kuenzi, Cole J. Lamoreaux, Wilson Laney, Ryan L. Lokteff, Mark Madison, Monika E. Maier, Jason P. Marshal, Heather A. Matheson, James E. Miller, Michael L. Morrison, James D. Nichols, Janet L. Rachlow, Ron Regan, Gary W. Roemer, Steve Running, Dana Sanchez, M. Schrage, Sarah Sells, James H. Shaw, William W. Shaw, Lisa A. Shipley, Tom Smith, Donald E. Spalinger, Jennifer Szymanski, Mark C. Wallace, Terry Walshe, Gary C. White, Tamara L. Wright, Kyle W. Young, numerous students in wildlife classes (including N. Butler, C. Callahan, C. Cardinal, A. D'Antonio, B. Kluever, C. Lamoreaux, R. Lokteff, M. Maier, T. Wright, and K. Young), several anonymous referees, and authors of the first edition, including Marta A. Jarzyna and Benjimin Zuckerberg. Scott Bonar, Daniel Edge, Selma Glascock, Kevin Hurley, Terry Johnson, James Earl Kennamer, Winifred Kessler, and Ray Lee provided information for various chapters and biographies.

We appreciate the cooperation of all organizations and individuals (e.g., P. Budy, S. Leavitt, and K. Mock) that supported the production of this work, including the Caesar Kleberg Wildlife Research Institute, the Boone & Crockett Program in Wildlife Conservation, and the Michigan Department of Natural Resources through the Partnership for Ecosystem Research and Management, and provided a sounding board for some chapters (T. Donovan and K. Gutzwiler). We again thank Mary Bissonette and family for putting up with John A. Bissonette, the household curmudgeon, while he prepared his chapter. We are also thankful for the excellent assistance we received from Johns Hopkins University Press, especially from Tiffany Gasbarrini and Ezra Rodriguez, copyeditor Ashleigh McKown, Robert Brown, and indexer Devon Thomas.

Anyone attempting a work of this scope knows that long hours are required at home and outside the office, and we are grateful to our wives and families—Carol Lee Krausman, Ellie Cain, Muriel Cain, and Logan Cain—for tolerating the days, evenings, and weekends we spent working on this project. Their support was instrumental in seeing it to completion. We thank all of the budding professionals who used the first edition of the book and those who will use the text to advance to leadership in the profession. After all, we wrote this book for you.

Finally, we dedicate this second edition with our personal appreciation to Jim Cain, Jane Cain, Amanda and Gabriel Wegner, current and future wildlifers, those who share our love of the wild, and those who inspire the same in others.

WILDLIFE MANAGEMENT AND CONSERVATION

1

DEFINING WILDLIFE AND WILDLIFE MANAGEMENT

PAUL R. KRAUSMAN

INTRODUCTION

Wildlife is such a part of society that most people have an idea of what wildlife is, but wildlife has many definitions. To be able to discuss wildlife management throughout this text in a consistent manner, it is important to have a common definition. To that end, this chapter defines wildlife, makes distinctions between active and inactive wildlife management, introduces the goals of management, and concludes with classifications of wildlife that were initiated by Leopold (1933) and added to by others (International Union for Conservation of Nature [IUCN] 2012).

WHAT IS WILDLIFE?

Initially, wildlife was considered game—animals that were hunted. Aldo Leopold defined game management as "the art of making land produce sustained annual crops of wild game for recreational use" (Leopold 1933:3). Contemporary use of the term *game* is declining, but state fish and game agencies commonly designate game based on the legislature of the state in which the animal resides. The word can have a narrow, precise, legislated meaning like the one that first appeared in British law in the Qualification Act of 1389 (McKelvie 1985). The Qualification Act stated that one had to be a landowner worth at least 40 shillings or a clergyman earning at least 10 pounds per year to legally take game (Lueck 1989). In England, two classes of animals were designated based on common law of property: *domitae naturae* (domestic animals) and *ferae naturae* (wild animals). For *domitae naturae*, absolute property rights were in place (i.e., the animals belonged to the owners even when they strayed from their property), but *ferae naturae* did not have owners until they were captured (Lueck 1989). The use of the term game found its way to North America, and numerous anomalies have arisen. For example, mourning doves (*Zenaida macroura*) are classified as songbirds (i.e., nongame) in parts of the eastern United States but are hunted as game in many western states. In addition, many species are not classified as game, but some states require a game license to collect them (e.g., lizards, snakes). The law in the United

States followed a different path than that in England. The fundamental legal doctrine regarding wildlife in the United States implies that states—and in some cases the federal government—have authority over wildlife, not landowners. Wildlife is not universally defined; it changes with the viewpoint of the user (Caughley and Sinclair 1994).

In general, the wildlife profession considers wildlife to be free-living, wild animals (excluding feral or exotic species) of major significance to humans. This definition includes the associated plants and lower animals (e.g., microorganisms) because habitats that support wildlife have to be considered. Species and their habitats are interlocked and cannot be considered separately.

For several decades, the profession of wildlife management concentrated on game species and their habitats. By the 1970s and 1980s, however, there was increased emphasis on consideration of the interests of all citizens when making management decisions. Because wildlife is classified as belonging to the public in the United States and Canada, it makes sense that human opinions be considered. A holistic view considers wildlife a triad of the animal, its habitat, and people, and the interactions between them (Giles 1978; Fig. 1.1). The animal component considers all aspects of biology, ecology, behavior, genetics, physiology, and life history characteristics and other important factors related to the species. The habitat component considers vegetation, soils, weather, topography, and other relationships within the ecological community. Human dimensions—i.e., "how people value wildlife, how they want wildlife to be managed, and how they affect or are affected by wildlife and wildlife management decisions" (Decker et al. 2001:3)—consider all anthropogenic influences on wildlife and their habitats. The human dimension is often the most important aspect of the triad because humans dictate how species will be managed, what intrusions into their habitat are acceptable, and how management will be funded (see Chapter 4). Each part of the triad involves different education and expertise, but all three parts are necessary to be efficient.

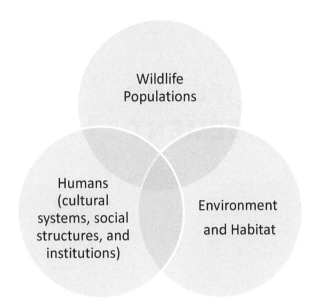

Figure 1.1. Wildlife management triad

One definition of wildlife is provided in the Endangered Species Act: "The term *fish* or *wildlife* means any member of the animal kingdom, including without limitation any mammal, fish, bird, (including any migratory, non-migratory, or endangered bird for which protection is also afforded by treaty or other international agreement), amphibian, reptile, mollusk, crustacean, arthropod or other invertebrate, and includes any part, products, egg, or off-spring thereof, or the dead body parts or parts thereof" (US Department of Commerce, National Oceanic and Atmospheric Administration, National Marine Fisheries Service, Enforcement Division 1979:4). This definition does not take into account important considerations of wildlife (e.g., whether it is free-living, restricted by fences, or includes human dimensions).

The Wildlife Society (the professional society for wildlifers) also defines wildlife as free-ranging animals of major significance to man. Until the 1970s, wildlife was synonymous with animals that were hunted, but in the past four decades most free-living animals have become significant to humans, especially with the nation's emphasis on the conservation and management of biodiversity. In the wildlife profession, wildlife is often restricted to terrestrial and aquatic vertebrates other than fish because of a long political history.

THE POLITICAL DISCIPLINE OF WILDLIFE MANAGEMENT

Many wildlife departments in universities are part of, or associated with, departments of agriculture because they begin with an agricultural focus. Some of the earliest efforts to manage wildlife in the United States were established in the US Department of Agriculture's Division of Entomology, which was funded by the American Ornithologists' Union. The Division of Entomology was established to determine the status of bird distributions and their migrations. This group was then transferred to the Division of Economic Ornithology and Mammalogy in 1885; their main function was to determine bird distributions and the damage they caused to agricultural crops. This agricultural base was expanded to address the relationship of all wildlife and agriculture and was placed in the Bureau of Biological Survey in 1896. In 1940, a political decision by President Franklin Delano Roosevelt created the Bureau of Wildlife, which combined the Bureau of Biological Survey (which addressed birds and mammals) and the Bureau of Fisheries. Fisheries biologists, however, did not believe the newly formed bureau adequately represented them under this broad title, and the bureau's name was changed to the Bureau of Sport Fish and Wildlife. It was later changed to the present name, the US Fish and Wildlife Service. The name change implied that fish were to be treated differently from other wildlife, resulting in different disciplines and societies: one for fisheries and one for other terrestrial vertebrates. Thus wildlife and fisheries management evolved into different disciplines.

Bureaucracies are rarely permanent, and in the late 1990s, Secretary of the Interior Bruce Babbitt attempted to create a freestanding organization that would combine all of the research conducted in the US Department of the Interior into a single organization called the National Biological Survey. Babbitt's reason was that the purpose of science was to understand the mechanisms of ecosystems and to fit humans into the resources available. The evolution of the agencies that manage wildlife at the national level has expanded from understanding bird distributions and their influence on agricultural crops (i.e., Division of Entomology) to placing humans in ecosystems (i.e., National Biological Survey). These are huge changes over a short period; however, the changes reflect how public attitudes evolve and how conservation organizations are formed and named. The National Biological Survey was eventually renamed the Biological Resources Division within the US Geological Survey. As the political faces change at the federal level, so will the names of those divisions that are responsible for management. As wildlife populations are dynamic, so are the organizations that contain the people that manage them (see Chapter 2).

Regardless of the organizations that manage our wildlife, or the various definitions of wildlife, for the purposes of this volume we consider wildlife as free-ranging, undomesticated animals in natural environments. Animals that are kept on property owned by private landowners with a barrier (most often a fence) are not free-ranging and are not considered wildlife by most wildlifers. Leopold recognized this when he described his theorems expressing the relationships between game and humans: "1. The denser the human population, the more intense the system of game management needed to supply the same proportion of people with hunting. 2. The recreational value of a head of game is inverse to the artificiality of its origin, and hence in a broad way to the intensiveness of the system of game management which produced it. 3. A proper game policy seeks a happy medium between the intensity of management necessary to maintain a game supply and

JOHN RAYMOND MORGART (1951–2009)

John Raymond Morgart is the perfect example of a professional wildlifer. Wildlife was his lifelong passion, and he served as the ideal role model for field biologists, working for the integrity of wildlife, their habitats, and the profession. He received BS and MS degrees from Arizona State University and a PhD from the University of Arizona in 1990. He worked for the US Bureau of Land Management and Bureau of Reclamation but spent most of his career in Alaska, Arizona, and New Mexico with the US Fish and Wildlife Service.

During his time with the Fish and Wildlife Service, his roles as team leader for the Sonoran pronghorn and the Mexican gray wolf teams were instrumental in the recovery of both species. By involving multiple agencies and individuals, he was able to form coalitions that worked diligently to improve the status of Sonoran pronghorn, and the success of the mammal today is in large part due to his leadership.

Morgart was active in local and national activities through The Wildlife Society (TWS), served as an associate editor of the *Journal of Wildlife Management*, and received the Jim McDonough Award from TWS.

Ethical, honest, compassionate, and friendly, Morgart loved to learn. He had little tolerance for ignorance or laziness. An avid hunter, outdoorsman, and naturalist, Morgart was the definition of a true wildlifer, one with drive, dedication, and devotion. His legacy is similar to that of other devoted professionals, whose achievements serve as a pathway to the successful conservation of wildlife.

that which would deteriorate its quality or recreational value" (Leopold 1933:394). Not everyone will agree, and there is certainly controversy about raising wildlife behind wire (Knox 2011). Animals in captivity, however, regardless of the size of the enclosure, are often intensively managed. They are provided supplemental food, enclosed, often genetically manipulated (e.g., for large antlers), and subject to game ranching, which is developing into a separate discipline akin to animal husbandry. Those animals are not wildlife, and their management is not wildlife management. Knox (2011:45) states it well when he asks, "is shooting a privately-owned, half tame, semi-domesticated, supplementally-fed, genetically-engineered buck standing in a bait pile inside a pen the future of deer management?"

Caughley and Sinclair (1994) stated that defining wildlife management as the management of wildlife populations may be too restrictive because it could exclude the human aspect of the wildlife management triad. The authors of this text agree that the human aspect of wildlife management is important because management of wildlife incorporates managing human behavior through human education, extension, law enforcement, and administration, among many other related issues. Human dimensions today are recognized as part of the core practice of manipulating or protecting wildlife populations to achieve a goal (see Chapter 4). Biologists and managers who understand animal ecology can make sound recommendations to those developing policy to achieve public goals, but to manage populations effectively requires a combination of biological and sociological strategies. As Decker et al. articulate in Chapter 4, a contemporary, comprehensive, and inclusive definition of wildlife management is guidance of

decision-making processes and implementation of practices to influence interactions between people, wildlife, and wildlife habitats and among people about wildlife to achieve effects desired by stakeholders.

This book focuses on both, because a sound understanding of wildlife management is central to understanding the biology and habitat of the animal and the human dimensions associated with both. Leopold (1933:394) emphasized this philosophy by strongly encouraging management to consider the public in their activities (i.e., his fifth theorem, "only the landowner can practice game management cheaply").

For decades, wildlife management in the United States used the client model, which paid more attention to licensed hunters and anglers who paid for the services of management. It was not until Decker et al. (1996) called for a philosophical shift from the client model to the stakeholder model, which more closely followed Leopold's ideas for wildlife management to incorporate all wildlife and people. The stakeholder model of wildlife management involved hunters and anglers but also included anyone who had a vested interest in a wildlife issue, program, action, or decision leading to an action (Decker et al. 1996). As the wildlife profession discusses the North American Model of Wildlife Conservation (Mahoney and Geist 2019; Chapter 3), more of the public will need to be included in management of our natural resources, and the stakeholder model goes a long way toward that goal.

ACTIVE VERSUS INACTIVE MANAGEMENT

There are numerous ways wildlife is managed, but they all imply stewardship and can be classified into two broad categories: active management and inactive management. Active management does something to the population—such as increasing or decreasing its size—in a direct manner through strategies like translocations or hunting, respectively. Populations can also be actively managed by altering the habitat to the benefit or detriment of a population. If population numbers are too low for the goal of management agencies, other active management can be incorporated, such as predator control to minimize neonatal mortality or habitat improvement to provide required cover for neonates from predators. These efforts represent active approaches to management. Other populations may not be actively managed, like those in national parks. In such situations, management activities minimize external influences on populations and habitat, which often involves management of humans and not animals. Still, other populations may be so poorly understood that no action is taken because managers do not have enough information to make management decisions. Thus the management is inactive. It is important to understand that all populations are managed either actively or inactively. For some populations, no active management occurs, but they are still managed because the decision to do nothing is a management decision; in these cases, management is often referred to as passive management or nonmanagement. There are yet other situations where populations (i.e., small vertebrates, insects) are essentially invisible to agencies because a decision to actively or passively manage is not even contemplated.

THE GOALS OF WILDLIFE MANAGEMENT

Most management goals for wildlife can be categorized into one of four options (Caughley 1977). When biologists are working with endangered species or declining populations, their goal is often to increase population numbers. For example, biologists have been working for years to increase the population of Sonoran pronghorn (*Antilocapra americana sonoriensis*) in southwest Arizona and northern Mexico (Krausman et al. 2005b) through active (Hervert et al. 2005) and inactive management (Krausman et al. 2005a). Indeed, most management plans for endangered populations aim to increase the population.

In other situations, the population may be too high, and the management goal is to reduce the population. White-tailed deer (*Odocoileus virginianus*) in suburban and urban areas throughout the eastern United States have increased, and the public wants numbers reduced to minimize deer–vehicle collisions, disease, damage to vegetation, and threats to children (Krausman et al. 2011). Similar situations occur with elk (*Cervus canadensis*) in Arizona and Montana, American black bears (*Ursus americanus*) in national parks and housing areas (Merkle et al. 2011), and in other situations where wildlife poses a problem to the public, whether the concern is real or perceived. In some cases, even endangered species may be too abundant. A 2011 controversy with the endangered gray wolf (*Canis lupus*) created nationwide concern because some individuals in some states (Montana, Idaho, Wyoming) believe that they

have too many wolves and need to reduce their numbers, while others claim more wolves are needed for the establishment of viable populations. In most of these situations the overall management goal is to reduce the population.

The third option is to manage the population for a sustained yield, as in the case of game animals throughout the United States. Hunting can be a tool for management, can provide game and recreation for hunters, and is a common management goal (see Chapter 9).

The final option is to do nothing except monitor the population, as is done with some bats, reptiles, amphibians, small mammals, and other wildlife in national parks. At a minimum, management goals should include keeping track of the species under supervision. Unfortunately, monitoring is often time-consuming and expensive and is not even considered until there is a problem that must be addressed.

The four basic options to increase populations, decrease populations, harvest for sustained yield, or to only monitor the population are available to wildlife biologists and will vary depending on the goals of the organization. After goals are established in conjunction with the public, the manager can determine the appropriate action and how that action can be achieved in the best possible manner. The establishment of goals is a value judgment that is neither right nor wrong, good nor bad, but how the goals are achieved requires technical skills and decisions to ensure success.

THE WILDLIFE BEING MANAGED

All wildlife is managed actively or passively; even if there is no active management, the decision to do nothing is a management decision. Wildlife has been classified numerous ways, however, and the way it is classified often dictates how it will be managed. There is a danger to classification schemes because, by setting up a classification criterion, some species will not be considered if they do not fall into a particular classification scheme. For example, funds set aside for threatened species cannot be used for other animals that are not threatened regardless of the need. Classification of flora and fauna, however, is an ongoing process and will continue to be. The question is, Can we describe all the species before they become extinct? Estimates of the number of species on earth are as high as 30 million, yet fewer than 1.7 million have been formally described since Linnaeus initiated the binomial system in 1753 (Wilson 1985). There are ≥5.5 million insects (Stork et al. 2015); 374,000 plants, including algae and fungi (Christenhusz and Byng 2016); 47,000 vertebrates; and the rest is made up of assorted invertebrates and microorganisms (Wilson 1985). Of the approximately 9 million eukaryotic species globally, of which 2.2 million are marine, 86% of the existing species on earth and 91% of those in the oceans still await description (Mora et al. 2011). Wildlifers concentrate their efforts on the ecology of vertebrates and thus address only a small portion of the world's fauna. Leopold (1933) established some of the earlier classifications of types of wildlife that are managed, but other groupings are more recent (e.g., rare and

WINIFRED B. KESSLER (1944–)

Winifred "Wini" B. Kessler is unquestionably one of the most influential wildlife biologists and conservationists in the field today. Her career, which has bridged academics and government in two countries, the United States and Canada, began when positions within the wildlife field were dominated by men. Wini's desire to obtain a field biologist position at such a time when positions were prevailingly open only to men spurred her to complete by age 27 undergraduate and master's degrees from the University of California, Berkeley, and a doctorate from Texas A&M. She was then qualified as a certified wildlife biologist by The Wildlife Society (1979). With these outstanding credentials and a steadfast determination to open doors, Wini was one of the first women selected for nearly every position she sought. Her academic career consisted of teaching positions at the University of Idaho, Utah State University, and the University of Northern British Columbia, where she built and chaired the Forestry Program. In 1997 she was named British Columbia Academic of the Year. Wini also served 21 years with the US Forest Service in the positions of Alaska regional ecologist, national wildlife ecologist, and Alaska regional director of wildlife, fisheries, ecology, watershed, and subsistence management. In addition to her professional positions, she has dedicated many years of effort to nonprofit organizations, including the Boone & Crockett Club, where she was the first female professional member to be accepted into the club. She held positions on the board of directors of the Habitat Conservation Trust Foundation, Ducks Unlimited Canada, and Canadian Wildlife Federation. She has been an active, longtime member of The Wildlife Society, where she was elected northwest section representative to council and served as associate editor for The Wildlife Society Bulletin for six years. Her dedication and efforts to The Wildlife Society led her to be selected as president of The Wildlife Society and a Wildlife Society fellow, and she was also honored with the profession's highest award, the coveted Aldo Leopold Memorial Award. Wini is an avid writer and has contributed many articles to scientific and conservation-based journals. She was a coeditor of the 2018 textbook *North American Wildlife Policy and Law*. Aside from her many professional accomplishments, perhaps one of the most important contributions Wini has made to the field of wildlife science and conservation has been that of a role model, mentor, and friend to many young biologists—both women and men. She has encouraged women to seek professional positions within conservation and academic organizations and supported their nominations to committees and boards. Her mentorship and advice have helped guide many young biologists into leadership positions, thus strengthening the field of wildlife science and conservation.

endangered, urban, park, critically endangered, vulnerable) and are described below.

Farm Species

Farm species are wild species that do not migrate, can reside and reproduce on farms, and are a suitable by-product of farming. Typical farm species include quail, pheasants, squirrels, rabbits, raccoons (*Procyon lotor*), and Virginia opossums (*Didelphis virginiana*).

Forest and Range Species

Forest and range species are also sedentary but are compatible with forestry or livestock operations and are a suitable by-product of such. Typical examples are wild turkeys (*Meleagris gallopavo*), deer, elk, raccoons, foxes (*Vulpes* spp.), grouse, pronghorn, and collared peccaries (*Pecari tajacu*).

Wilderness Species

Wilderness species are those that are harmful to, or harmed by, economic land use and require special reserves of forests to be preserved as their habitat. Grizzly bears (*Ursus arctos*

horribilis), cougars (*Puma concolor*), mountain goats (*Oreamnos americanus*), and bighorn sheep (*Ovis canadensis*) are true wilderness species. Others can include pronghorn, elk, deer, and moose (*Alces alces*).

Migratory Species

Migratory species are those that normally leave the land on which they were raised in the course of their seasonal movements. Ducks, geese, and swans are excellent examples, but elk, deer, bats, and other mammals can also be placed into this category.

Furbearers

Furbearers are species that can be produced and marketed because of the commercial value of their pelts. Coyotes (*Canis latrans*), bobcats (*Lynx rufus*), muskrats (*Ondatra zibethicus*), otters (*Lontra canadensis*), and beavers (*Castor canadensis*) are just a few of those typically in this category.

Predators

Predators are animals that kill other species or are considered dangerous to livestock. Obvious examples are mountain lions, bobcats, coyotes, and feral cats and dogs.

Threatened Species

Threatened species are species of native fish and wildlife that are threatened with extinction. They are classified as such by the US secretary of the interior whenever their existence is endangered; because their habitat faces destruction, drastic modification, or severe curtailment; or the species itself faces exploitation, disease, predation, or other factors. In such cases, their very survival requires assistance.

Urban Wildlife

More recently, emphasis has been placed on species that find habitat in towns, villages, and cities, and a relatively new discipline that studies habitat relationships with wildlife in cities is developing; these species are defined as urban wildlife (Adams 2016, Kobilinsky 2020). At one time, pigeons and rats were the species of cities, but deer, raccoons, falcons, and coyotes are other examples of wild species that are increasingly adapting to urban life.

Park Wildlife

Park wildlife is another classification for those species that exist in parks, which includes all those mentioned above plus others. The category has developed because of the confined management that occurs within human-created boundaries and the philosophy of many parks (i.e., no harvest, limited management of wildlife, enhanced human management; Kerlinger et al. 2013).

Unfortunately, because of declining populations, decreased range, limited mature individuals, and a high probability of extinction within years, other categories of wildlife are being used and have been defined by the International Union for the Conservation of Nature (IUCN 2012).

Extinct

A taxon is extinct when there is no reasonable doubt that the last individual has died (e.g., saber-toothed cat [*Smilodon fatalis*], dodo [*Raphus cucullatus*]). A taxon is presumed extinct when exhaustive surveys in known or expected habitat, at appropriate times (diurnal, seasonal, annual), throughout its historic range have failed to record an individual.

Extinct in the Wild

A taxon is extinct in the wild when it is known only to survive in cultivation, in captivity or as a naturalized population (or populations) well outside the past range (e.g., Père David's deer [*Elaphurus davidianus*], Socorro dove [*Zenaida graysoni*]). A taxon is presumed extinct in the wild when exhaustive surveys in known or expected habitat, at appropriate times (diurnal, seasonal, annual), throughout its historic range have failed to record an individual.

Critically Endangered

A taxon is critically endangered when the best available evidence indicates that the population will be reduced in size >90% over 10 years or three generations, have a reduction of the geographical range to <100 km², and a low number of mature individuals. It is therefore considered to be facing an extremely high risk (>50%) of extinction in the wild within 10 years or three generations (e.g., Dama gazelle [*Nanger dama*], California condor [*Gymnopys californianus*]).

Endangered

A taxon is endangered when the best available evidence indicates that the population will be reduced by ≥70% over 10 years or three generations, the range will be <5,000 km², with <2,500 mature individuals, and the probability of the population has a very high risk of extinction in the wild (≥20% within 20 years or five generations; e.g., African wild ass [*Equus africanus*], Banggai crow [*Covus unicolor*]).

Vulnerable

A taxon is vulnerable when the best available evidence indicates that the population is reduced by >50% over 10 years or three generations, reduced in geographical range to <20,000 km², has <10,000 individuals in the population, and is therefore considered to be facing a high risk of extinction (>10% in 100 years) in the wild (e.g., Thorold's deer [*Cervus albirostris*], blue-eyed cockatoo [*Cacatua ophthalmica*]).

Near Threatened

A taxon is near threatened when it has been evaluated against the criteria for critically endangered, endangered, and vulnerable and does not qualify for any of those categories now but is close to qualifying for or is likely to qualify for one of those categories in the near future (e.g., white rhinoceros [*Ceratotherium simum*], gentoo penguin [*Pygoscelis papua*]).

Least Concern

A taxon is least concern when it has been evaluated against the criteria for other categories and does not qualify for critically endangered, endangered, vulnerable, or near threatened. Widespread and abundant taxa are included in this category (e.g., giraffe [*Giraffa giraffa*], snowy owl [*Bubo scandiacus*]).

Data Deficient

A taxon is data deficient when there is inadequate information to make a direct, or indirect, assessment of its risk of extinction based on its distribution or population status. A taxon in this category may be well studied and its biology well known, but appropriate data on abundance or distribution are lacking (i.e., silver dik-dik [*Madoqua piacentinii*], Mayr's swiftlet [*Aerodramus orientalis*]). Data deficient is therefore not a category of threat. Listing of taxa in this category indicates that more information is required and acknowledges the possibility that future research will show that threatened classification is appropriate. It is important to make positive use of whatever data are available. In many cases, great care should be exercised in choosing between data deficient and a threatened status. If the range of a taxon is suspected to be relatively circumscribed, and a considerable period has elapsed since the last record of the taxon, threatened status may well be justified.

Regardless of how humans classify wildlife, the basic goals for management are the same: influence the population to increase or decrease, manage for harvest, or simply monitor the population (Caughley 1977). The classification of wildlife being managed is not static, but the definitions provided are those used in the past and in contemporary management and conservation.

SUMMARY

Wildlife includes free-ranging, undomesticated animals in natural environments. Wildlife is not restricted to animals but includes their interrelationships with their habitats and with humans. Fish are usually considered separately from other animals because of political divisions and history. The goals of management are to increase populations, decrease populations, harvest the populations for sustained yield, or just monitor populations. Management can be active (i.e., influences populations directly) or inactive (i.e., no intentional management). There are numerous categories of wildlife, including farm, forest and range, wilderness, migratory, furbearers, predators, rare, endangered, urban, park, and a series of classifications for extinct and declining species (i.e., extinct, extinct in the wild, critically endangered, endangered, vulnerable, near threatened), species of least concern, and those that do not have enough data to accurately classify.

Literature Cited

Adams, C. E. 2016. Urban wildlife management. CRC Press, Boca Raton, Florida, USA.

Caughley, G. 1977. Analysis of vertebrate populations. John Wiley and Sons, New York, New York, USA.

Caughley, G., and A. R. E. Sinclair. 1994. Wildlife ecology and management. Blackwell Scientific, Boston, Massachusetts, USA.

Christenhusz, M. J. M, and J. W. Byng. 2016. The number of known plants species in the world and its annual increase. Phytotaxa 261:201–217.

Decker, D. J., T. L. Brown, and W. F. Siemer. 2001. Evolution of people–wildlife relations. Pages 3–22 in D. J. Decker, T. L. Brown, and W. F. Siemer, editors. Human dimensions of wildlife in North America. The Wildlife Society, Bethesda, Maryland, USA.

Decker, D. J., C. C. Krueger, R. A. Baer Jr., B. A. Knuth, and M. E. Richmond. 1996. From clients to stakeholders: a philosophical shift for fish and wildlife management. Human Dimensions of Wildlife 1:70–82.

Giles, R. H., Jr. 1978. Wildlife management. W. H. Freeman, San Francisco, California, USA.

Hervert, J. J., J. L. Bright, R. H. Henry, L. A. Piest, and M. T. Brown. 2005. Home-range and habitat-use patterns of Sonoran pronghorn in Arizona. Wildlife Society Bulletin 33:8–15.

International Union for Conservation of Nature (IUCN). 2012. IUCN Red List categories and criteria. Version 3.1., 2nd edition. Colchester Print Group, Gland, Switzerland, and Cambridge, UK.

Kerlinger, P., J. Burger, H. K. Cordell, D. J. Decker, D. N. Cole, P. Landres, E. N. Smith, J. Brett, R. Larsan, T. O'Shea, and S. Temple. 2013. Wildlife and recreationists: coexistence through management and research. Island Press, Washington, DC, USA.

Knox, W. M. 2011. The antler religion. Wildlife Society Bulletin 35:45–48.

Kobilinsky, D. 2020. Wild cities. Wildlife Professional 14:18–26.

Krausman, P. R., L. K. Harris, S. K. Haas, K. K. G. Koenen, P. Devers, D. Bunting, and M. Barb. 2005a. Sonoran pronghorn habitat use on landscapes disturbed by military activities. Wildlife Society Bulletin 33:16–23.

Krausman, P. R., J. R. Morgart, L. K. Harris, C. S. O'Brien, J. W. Cain III, and S. S. Rosenstock. 2005b. Introduction: management for the survival of Sonoran pronghorn in the United States. Wildlife Society Bulletin 33:5–7.

Krausman, P. R., S. M. Smith, J. Derbridge, and J. Merkle. 2011. The cumulative effects of suburban and exurban influences on wildlife. Pages 135–191 in P. R. Krausman, and L. K. Harris, editors. Cumulative effects in wildlife management: impact mitigation. CRC Press, Boca Raton, Florida, USA.

Leopold, A. 1933. Game management. Charles Scribner's Sons, New York, New York, USA.

Lueck, D. 1989. The economic nature of wildlife law. Journal of Legal Studies 18:291–324.

Mahoney, S. P., and V. Geist. 2019. The North American model of wildlife conservation. Johns Hopkins University Press, Baltimore, Maryland, USA.

McKelvie, C. L. 1985. A future for game? George Allen and Unwin, London, UK.

Merkle, J. A., P. R. Krausman, N. J. Decesare, and J. J. Jonkel. 2011. Predicting spatial distribution of human–black bear interactions in urban areas. Journal of Wildlife Management 75:1121–1127.

Mora, C., D. P. Tittensor, S. Adl, A. G. B. Simpson, and B. Worm. 2011. How many species are there in the earth and in the ocean? PLoS Biology e1001127.

Stork, N. E., J. McBroom, C. Gely, and A. J. Hamilton. 2015. New approaches narrow global species estimates for beetles, insects, and terrestrial arthropods. Proceedings of the National Academy of Sciences 112:7519–7523.

US Department of Commerce, National Oceanic and Atmospheric Administration, National Marine Fisheries Service, Enforcement Division. 1979. Endangered Species Act of 1973. Stock number 003-020-00150-5. US Government Printing Office, Washington, DC, USA.

Wilson, E. O. 1985. The biological diversity crisis. BioScience 35:700–706.

2 THE HISTORY OF WILDLIFE CONSERVATION IN NORTH AMERICA

ROBERT D. BROWN

INTRODUCTION

The three North American countries of Mexico, the United States, and Canada comprise a total of almost 22 million km². The border between Mexico and the United States is 3,115 km long, and that between the United States and Canada is 8,891 km long. These lands have an amazing diversity of wildlife and their habitats, from the Sonoran Desert and the jungles of the Yucatan to the boreal forest of Canada and the arctic tundra of Alaska. Although to a degree our cultures have merged over time in terms of languages, foods, and even peoples, our approaches to conservation have not been the same. This chapter explores the differences and similarities of our approaches, as well as the biological, geological, and political reasons for them. We have had and will continue to have much to learn from each other. For the purposes of this chapter, I apply the term *North America* to include what are now Canada, the United States, and Mexico, and use *Native Americans* for the Indigenous peoples of those countries.

NATIVE AMERICANS AND WILDLIFE

Archeologists and anthropologists disagree over the time of arrival of the first inhabitants to North America, variously dating it between 80,000 years ago and as recently as 10,000 years ago, with most settling on their appearance ranging from about 14,000 to 16,000 years ago (Barnes 2019, Waters 2019). One could say their arrival signaled the end of the "natural" landscape and biota of our three countries. Most agree that the Paleo-Indians came across a continental ice bridge from Siberia and eventually worked their way east in North America and 7,000 miles south by land or by boat to South America (Krech 1999, Wade 2019), although some settlers of Canada may have come from Polynesia or across the North Atlantic from Europe (Taber and Payne 2003). Estimates of Native American population size prior to European exploration vary from 500,000 to 5 million in what is now the United States to as many as 100 million in both North and South America, with widely varying densities (Krech 1999, Mann 2011). Populations are believed to have doubled about every 20 years. Eventually, population densities were largest in Mexico and smallest in Canada (Druschka 2003, Simonian 1995). They spoke more than 2,000 languages (Barnes 2019).

The first organized civilization in Mexico was the Mayan, which began about 2600 BC. At their peak, about AD 250–900, they built cities of tens of thousands of people with great stone monuments and had an organized political system headed by kings. They and the Olmecs, whose cultures overlapped, continue to be known for their amazing progress in astronomy, mathematics, and agriculture. They cleared forests and built canals, reservoirs, and roads to develop extensive agricultural systems based mostly on corn (maize), squash, and beans. They supplemented their diets by harvesting wild plants, game, and fish to the extent of depleting those populations and causing soil erosion. The Aztecs followed the Mayan culture about 400 years later. Though they made great progress in many areas as well, their civilization lasted only about 300 years, until the arrival of Europeans in the late 1400s (Simonian 1995).

Many Native Americans in what is now the United States—including the Sioux, Arapaho, Blackfoot, Cheyenne, Cree, Kiawah, Mandan, Pawnee, and Shawnee, as well as the Huron of Canada—were hunter-gatherers, living off the meat of some large (but often small) mammals, fish, birds, and the nuts, berries, and fruits they gathered. Some Native American societies planted vast terraced and irrigated agricultural crops, again mostly of squash, corn (maize), and beans, though some also raised cotton and tobacco and even had tree farms (Trefethen 1975, Krech 1999, Mann 2011). Some made alcohol from fermented corn, honey cactus, or persimmons (Barnes 2019).

Historians also differ about the abundance of wildlife on our continent prior to the arrival of Europeans. Some writers speak of vast numbers of deer (*Odocoileus* spp.), elk (*Cervus canadensis*), moose (*Alces alces*), and beavers (*Castor canadensis*) in eastern forests; polar bears (*Ursus maritimus*), muskoxen (*Ovibos moschatus*), moose, and caribou (*Rangifer* spp.) in the Northwest; and millions of buffaloes (*Bison bison*) and prong-

22,000 years before the arrival of Europeans. Some call the idea of an abundance of wildlife the "pristine myth" (Denevan 2003:5) and note that English settlers had a difficult time finding game to hunt, and that early Spanish explorers do not even mention seeing bison.

Over these many centuries, 35 genera of mammals and 10 genera of birds disappeared (Krech 1999). This was especially true of the large megafauna, such as mammoths (*Mammuthus primigenius*), mastodons (*Mammut americanum*), giant sloths (Mylodoute), camels (Dromedary), and horses (Equidae; Krech 1999). Surviving species were sometimes extirpated from localized areas. It is unclear whether predation by humans, increasing global temperatures, changing vegetation, or other factors caused the decline of these species.

Native Americans hunted mammals for food and perhaps attempted to protect their crops from wildlife depredation. They could be effective hunters. The Mississippians (AD 750–1500) of the US Southeast used spears with flint tips, bows and arrows, and the atlatl (Barnes 2019). They diverted water for irrigation and cleared forests for crops, firewood, and construction materials. They terraced land and built dams. They built hundreds, if not thousands, of mounds for burials and temples. For more than 10,000 years, Indigenous peoples of the entire continent set fires, perhaps to clear underbrush for easier hunting, to improve wildlife habitat, to clear land for crops, to enhance forage production, and to herd wild animals. Thomas Morton (Mann 2011:285) reported in 1637 that eastern Native Americans carried flints "to set fire of the country in all places where they come." The surveyor Peter Fidler of Alberta reported constant burning by Indigenous people. He said, "these fires burning off the old grass in the ensuing spring and summer makes excellent sweet feed for horses and buffalo" (Mann 2011:287). They also used torches to hunt at night. Certainly, the use of fire by Indigenous peoples could be called the first attempts at game management in North America. The extent of this burning and its effect on habitat are debated. Some scholars believe Native Americans may have decimated large game around their population centers, which were unevenly distributed. Naturally, the use of fire varied by tribe and by ecosystem. Some fires may have been used as a weapon against other tribes or later European immigrants. Other times, fire probably got away, causing substantial damage to the local biome. It is also believed that in some areas, including what is now Yellowstone National Park, occasional winter die-offs of game animals and localized overharvesting led to near starvation of the Native Americans (Kay 1998, Krech 1999, Mann 2011).

Scholars also debate how Native Americans valued and used wildlife. Because game was generally abundant, at least in certain areas at certain times, it was not always used wisely. At one point it was believed there were as many as 30–50 million bison, ranging from Canada to Mexico and Oregon to Georgia. The bison has been called "the tribe's department store . . . a builder's emporium, furniture mart, drugstore and supermarket rolled into one" (Krech 1999:128) (Fig. 2.1). Some Native Americans used the entire carcass of a bison,

Figure 2.1. Many Native American tribes depended on the bison for their sustenance

horns (*Antilocapra americana*) along with thousands of grizzly bears (*Ursus arctos horribilis*) and elk in the West.

Although the human population of North America increased prior to European arrival, some Native American populations disappeared, including the Anasazi in the 13th to 15th centuries in the Southwest, the Cahokian in Missouri in the 12th to 14th centuries, and the Hohokam in the 15th century. The Hohokam of Arizona built more than 725 km of canals to irrigate nearly 2 million km² of farmland. They were known to trade with other cultures buffalo hides, deerskins, exotic feathers, and marine shells. In fact, trade among Indigenous tribes was from Mexico to Canada and from the West to the East Coast (Barnes 2019). Unfortunately, droughts, depletion of soil quality from growing crops or grazing animals, lack of firewood, warfare, pre-European diseases, or a combination of factors may have caused the disappearance or scattering of these populations (Krech 1999).

Archeologists examining animal and bird bones found in campfire sites believe that Native Americans may have had a significant effect on North American wildlife in the 10,000–

including the meat, hide, entrails, bones, and hooves. Others killed only female bison or took only their tongues or fetuses, which were considered delicacies. The practice of running herds of bison over cliffs was exceptionally wasteful, though some tribes stored carcasses underwater for months to eventually consume them as "green bison soup." Horatio Jones, who was captured by the Algonquians in 1782, spoke of his captors feasting on passenger pigeons (*Extopistes migratoris*; Mann 2011). With such abundant wildlife, there may not have been a concept of waste. Eventually, the Inuit of Alaska and Canada nearly exterminated caribou, musk oxen, and mountain sheep (*Ovis canadensis*), but the Algonquians rationed their take of beaver, and the Selawiks controlled their tribe's take of fish (Martin 2002).

The practice of localized overexploitation may also have been supported by some Native American belief systems, such as those that believed bison were supernatural and emanated from underground, or that dead bison would be reincarnated (Krech 1999), or that their own ancestors were wild animals (Simonian 1995). In fact, wild animals were important influences on the spiritual culture of most tribes from the Mayans and Aztecs to today's North American tribes. Many rituals revolved around animals, such as the buffalo, deer, and eagle dances of Plains Indians, the snake-antelope ceremony of the Hopi, or the coyote creation stories of the Navaho and Zuni tribes. Some tribes had hunting taboos, like not killing young animals or not wounding animals. In what is now the United States and Mexico, only dogs (*Canis familiaris*) were domesticated; in Canada, however, caribou (reindeer) were domesticated more than 3,000 years ago (Hughes 1996, Taber and Payne 2003, Harkin and Lewis 2007).

So the question remains, Were Indigenous peoples, prior to the arrival of Europeans, conservationists? (See Hughes 1996; Harkin and Lewis 2007.) Again, opinions differ. William Jacobs, in Hughes (1996:137), stated, "After having studied the mass of evidence in the biological, physical and social sciences, I am convinced that Indians were indeed conservators. They were America's first ecologists." Conversely, Paul Martin in Krech (1999:29) stated, "The destruction of fauna, if not habitat, was far greater before Columbus than at any time since." Krech (1999:122) stated, "By the time Europeans arrived, North America was a manipulated continent. Indians had long since altered the landscape by burning or clearing woodland for farming and food. Despite European images of untouched Eden, this nation was not cultural, not virgin, not anthropogenic, not primeval, and nowhere is this more evident than in the Indian use of fire."

The effect of Native Americans on wildlife can only be guessed from the reports of early explorers. Spanish explorer Cabeza de Vaca traveled from Florida to Mexico across the southern and southwestern United States from 1527 to 1537. He saw no alligators (*Alligator mississippieusis*), wolves (*Canis lupus*), or coyotes (*Canis latrans*) in Florida and no peccaries (Tayassuidae), grizzly bears, beaver, eagles (Acciptridae), wild turkeys (*Meleagris gallopavo*), or passenger pigeons on his trip. He did report seeing buffalo on the northern reaches of his ex-

ploration (Mahoney and Geist 2019). Hernando de Soto, who in 1541 was the first European to see the Mississippi River, also saw no bison. Mann (2011:28) stated, "Rather than the thick, unbroken monumental swath of trees imagined by Thoreau, the great Eastern forest was an ecological kaleidoscope of garden plots, blackberry rambles, pine barrens, and spacious groves of chestnut hickory and oak." He said that John Smith could have ridden his horse through Virginia forests, and that the eastern forests looked more like English parks, with wide spaces between trees, presumably from the Indian use of fire.

ARRIVAL OF THE EUROPEANS

The history of wildlife conservation in the United States, Canada, and Mexico is discussed largely in the context of European migration. Some explorers, such as the Spaniards, came to the Americas looking for gold and silver. Others, like the French, English, Portuguese, Italians, Dutch, and Russians, came to escape an overcrowded Europe, looking for land and economic opportunity. Many came to escape religious and political persecution. Although initial settlements had difficult times (about 90% of the Jamestown settlers perished), once established here, the emigrants saw unlimited land, timber, and wildlife with little or no government or laws to restrict their ambitions. As a result, the attitude of Americans, primarily in the United States, has been one of individual independence and the right to life, liberty, and the pursuit of happiness. These attitudes differed from those in Europe and eventually from other countries influenced by European traditions, including Canada and Mexico (Trefethen 1975, Colpitts 2002, Lopez-Hoffman et al. 2009). To this day, Americans bristle at attempts by local or federal government to restrict their personal freedom and independence. Nonetheless, the North American attitude toward wildlife gradually evolved from one of unrestricted abundance and harvest to one of regulated management and equitable access. We now call this the North American Model of Wildlife Conservation (Geist 1995, Geist et al. 2001), although its implementation has varied somewhat across Canada, the United States, and Mexico.

Once European explorers arrived, their diseases quickly reduced the population of Native Americans. The English, Spanish, and French, having lived closely with their livestock and having survived numerous plagues themselves, brought to the New World smallpox, typhus, bubonic plague, cholera, typhoid, influenza, measles, mumps, yellow fever, common cold, diphtheria, chickenpox, scarlet fever, pneumonia, dysentery, whooping cough, and tropical malaria. The explorers also brought hundreds of cattle, sheep, horses, and pigs, all with their own diseases such as anthrax and bovine tuberculosis. Exploration and the following settlements of immigrants were vast, wide, and surprisingly rapid. European countries were competing with each other to discover and claim these new lands.

Even the early explorers were many and extensive in their travels. Although Columbus actually landed in Hispaniola in 1492, John Cabot explored Newfoundland in 1497–1498 and was followed in other parts of North America by Ponce de Leon (1523), Italian Giovanni da Verrazzano (1523–1524), Jacque Cartier (1534–1541), Cabeza de Vaca (1536), Hernando de Soto (1539–1543), Coronado (1540–1542), and Sir Francis Drake and Sir Walter Raleigh (1578). In the following century, Frenchman Samuel de Champlain founded Quebec in 1608. The English established Jamestown in 1607 and started importing slaves in 1619. In the late 1600s the Spanish introduced horses, eventually changing the culture of the Plains Indians. The Spanish founded St. Augustine, Florida, in 1565 and Santa Fe, New Mexico, in 1610. In 1741 Chirikov, a Russian, explored Alaska, to be followed by Juan Perez in 1774 and Captain James Cook in 1778. Cook traded with the Youquot tribe for sea otter (*Enhydra lutris*) furs, which he sold for a profit in China. In the late 1700s the Russians established fur trading posts from Alaska to northern California as Spanish Franciscans built missions in Florida and California (Barnes 2019).

Diseases spread quickly among the Indigenous peoples, often ahead of the explorers. Native American runners and traders might have spread the word about the visitors to other villages, or captives might have escaped and returned home, taking the diseases with them. Some explorers came upon entire villages of the sick and dying or already depopulated. Importantly, Europeans also contributed to slavery and introduced alcohol to the Indigenous peoples, and eventually genocide (Dunbar-Ortiz 2019).

As early as 1581, King Philip of Spain reported to his supreme court that one-third of Latin American Indians had been wiped out (Galeano 1997). Hernan Cortez, with 450 Spanish soldiers and some Indigenous allies, destroyed the Aztec Empire in 1521 when he captured Tenochtitlan, a city of 140,000, which was already weakened by disease. Smallpox subsequently decimated the Aztec population of 5–6 million. One estimate suggested that the US population of about 3.8 million Native Americans was reduced by 74% to 1 million within 50 years of the European arrival. Later, diseases spread by individual trappers, explorers, and traders would increase that decline of Native Americans to more than 90% of their pre-European population (Denevan 2003). Others estimate the total population of North American Indigenous peoples of 70 million was reduced to 3.5 million within 150 years, or from about 1500 to 1650 (Galeano 1997). Whatever the numbers, the effect was clearly devastating. As historian Francis Jennings stated (Dunbar-Ortiz 2019:46), it was a myth that "America was a virgin land, or wilderness, inhabited by a non-people called savages . . . they did not settle a virgin land. They invaded and displaced a resident population."

It should be briefly noted that the first non-natives to arrive in North American were the Vikings from Scandinavia. After populating Iceland in 870 and Greenland about 980, the Vikings, led by Leif Erickson, son of Eric the Red, landed in Newfoundland about the year AD 1000. Three small villages were established on Baffin Island and the Labrador Coast. After two or three years, violence with the Indigenous Inuits led the Vikings to abandon their villages. This was 500 years before the voyages of Columbus (Ginger 2020).

WILDLIFE REBOUND AND THEN DECLINE

As the Native American population dwindled, the wildlife populations and their habitats rebounded in those areas where wildlife had been under harvest pressure. Logs from the Lewis and Clark expedition of 1805 stated that "the whole face of the country was covered with buffalo, elk and antelopes; deer were also abundant" (Krech 1999:74). As settlers arrived in the eastern United States and Canada, they cleared land for farming and cut forests for shipbuilding. Eastern tribes, such as the Sioux, migrated westward. At the same time, new land-clearing reduced habitat for wildlife. That habitat loss combined with hunting for subsistence, market hunting (harvesting wildlife for commercial meat, fur, and feather markets), and trapping for furs had a marked effect on eastern wildlife populations. In the Southwest, domestic livestock introduced by the Spaniards competed for forage with grazing wildlife. The introduction of Native Americans to horses changed their lifestyle and improved their hunting success, but not to the extent that it had much effect on the wildlife populations.

To some extent, Europeans made modest efforts to conserve the New World's resources. From the 14th to 16th centuries, Spanish kings enacted laws for Mexico that prohibited the use of iron traps for bears, boars (Suidae), or deer (1348), the use of poison for fishing (1435), killing mourning doves (*Zenaida macroura*; 1465), and hunting with hounds over snow (1516). Other laws limited timber cutting (1559), required reforestation (1763), and banned animal sacrifices. Even though Mexican kings put in place a few similar rules as early as the 13th century, the laws were rarely followed, and eventually the Spanish conquest cost Mexico 50% to 75% of its forests and much of its wildlife (Simonian 1995).

In Canada, after Inuits turned away Leif Ericksson from Newfoundland, explorers came from Portugal, Italy, France, Holland, England, and, on the West Coast, Russia. Fisheries were the first resource to be exploited. As the British, Dutch, and French moved inland, they settled in the south of Canada, where winters were less severe. In 1728 the British appointed a Surveyor General of His Majesty's Woods and claimed Nova Scotia's forests for shipbuilding. In 1763 King George gave Canada's aboriginal people well-defined property rights for subsistence hunting. The extraction of natural resources in this part of the continent by the British, however, eventually helped spark the American Revolution (Taber and Payne 2003).

The growth of the European settler population from emigrants and their offspring was astounding, and human population growth had an increasing effect on wildlife and habitats. Emigration began in earnest after about 100 years of exploration in the 1400s and 1500s by the French, English, Spanish, and Dutch. By 1638 there were about 30,000 whites or Anglos in North America. That population grew to 1.3 million emigrants, their offspring, and slaves by 1700 and to 3.9 million by 1790. By 1850, the population of the United States was over 23 million Anglos, with more than 4 million additional slaves (Taylor 2003).

In addition to clearing land for farming and grazing, introducing domestic livestock, and hunting wildlife for food and markets, the new settlers discovered the value of exporting fur and feathers as a source of cash for the purchase of the other goods they needed. As early as 1607, Captain John Smith reported that the French were shipping 25,000 beaver pelts per year to Europe, and by 1650 most beavers were eliminated from the entire East Coast. The exploitation of furbearers in the Northeast and Canada was by the French and by England's Hudson's Bay Company. In the Pacific Northwest, the Russian-American Fur Company took seals (Phocidae) and sea otters, and by 1768 it had extirpated the Steller's sea cow (*Hydrodamalis gigas*). Bird populations suffered from being taken for meat and for their plumage, which was used for ladies' hats in Europe. Deer and turkey populations also declined, again largely because of market hunting. In 1748 alone, South Carolina shipped 160,000 deer pelts to England. Much of the fur came from trade with Indigenous peoples. They traded pelts for guns, knives, blankets, axes, beads, clothes, and eyeglasses. This brought wealth to some but not all tribes, and it made the tribes dependent on the fur companies for food (Barnes 2019). As wildlife populations declined, the settlers at first blamed it on predators. In 1630 the Massachusetts Bay Colony offered a bounty of one shilling for each wolf killed. When the deer did not rebound, the city of Portsmouth, Rhode Island, enacted the first closed season on deer hunting in 1646. This was only the beginning of American's game management efforts, as well as North America's never-ending conflict over our attitude toward predators (Trefethen 1975).

As the nation expanded, the new settlers moved westward, where land and wildlife seemed free and unlimited. An estimated 25 million or more buffaloes and 10 million pronghorns resided in the West at their peak, although those numbers are obviously hard to confirm. Land purchases and other types of land acquisitions, such as the Louisiana Purchase, the Mexican-American War, the Oregon Compromise, the Gadsden Purchase of parts of Arizona and New Mexico, and the purchase of Alaska fulfilled the concept of "manifest destiny" and led to an America that spanned from coast to coast. Settlers followed, clearing more land and harvesting more wildlife. President Thomas Jefferson sent the famed Lewis and Clark Expedition from St. Louis, Missouri, to the Pacific Northwest in 1804. Although they encountered isolated stretches with little game, they found abundant herds of buffalo and deer, as well as grizzly bears and prairie dog (*Cynomys ludovicianus*) towns larger than 2.6 km². By the early 1800s, trading posts had been established across the West, again encouraging Native Americans to harvest game for their hides and to provide meat for settlers and trappers.

In Spain, parliament declared in 1813 that all lands in Mexico would be privatized, denying Indigenous people access to public land and guaranteeing land ownership for the wealthy. This policy helped to spark the Mexican revolution of 1810–1821, though there were few subsequent reforms in land tenure (Simonian 1995, Valdez et al. 2006). Canada, however, was under the English policy, whereby most land was federal

Figure 2.2. Market hunting nearly extirpated bison from North America

or owned by the Crown but managed by the provinces. The Crown managed land in the territories. In the United States in 1776, land not privately owned became property of the states. As states joined the Union, much state land was ceded to the federal government to pay for the revolution and in return for federal services (Taber and Payne 2003). Wildlife was believed to be superabundant and something to be used for food, fur, and income, or to be eradicated as pests (Dunlap 1988).

In 1830 Congress passed the Indian Removal Act, removing all title to lands from Indian tribes and moving many Native Americans to reservations in Oklahoma. By then, most of the beavers were gone from the continent, and silk had replaced beaver for the manufacture of men's hats. This collapse of the beaver hat market saved the beaver from total extirpation. Former beaver trappers became buffalo hunters. In 1833 alone, the American Fur Company shipped 43,000 buffalo hides, most received in trade from Native Americans. Buffalo meat was also used for camp towns and for crews building railroads to the West. There was wanton wastefulness, with many buffalo again being killed solely for their hides or for their tongues. In 1845 the Hudson Bay Company shipped 4,300 buffalo tongues to England. By the mid-1840s a noticeable decline in buffalo numbers was already evident in the United States and in Canada (Fig. 2.2), though some Canadians tried unsuccessfully to domesticate bison and to breed them with domestic cattle (Trefethen 1975, Krech 1999, Colpitts 2002).

The California Gold Rush of 1848 and later the Klondike Gold Rush of 1896 brought thousands of settlers westward, as did the Mormon migration to Salt Lake City. The population of the United States nearly doubled from 17 to 32 million between 1840 and 1860. The Civil War, from 1861 to 1865, slowed the western expansion and market hunting, but also any thoughts of conservation. It also reduced the US population by an estimated 600,000–750,000, more than all of the other US wars combined. But the homestead laws of 1862 allowed anyone to mark out 259 ha of land for private ownership at no cost if they would live on it for five years. This of course led to the final Indian Wars, as settlers crowded Native Americans out of their traditional hunting grounds and onto reservations in Oklahoma and the Dakotas (Trefethen 1975, Mackie 2000).

MARKET HUNTING LEADS TO CONTROL EFFORTS

The late 1860s and '70s saw the final diminution of the bison herds. In 1871 Colonel R. I. Dodge reported a single herd in Colorado being 80.5 km wide and 32.1 km long, with an estimated 4 million head. The influx of hunters and railroads made the shipment of hides, meat, and tongues very profitable. The famous "Buffalo Bill" Cody, a hunter for the railroad, once killed 69 in a single day and 4,240 in an 18-month period; other hunters were actually more successful. The annual kill in 1865 was 1 million; by 1871, it was 5 million. In 1864 Idaho imposed the first closed season on bison hunting, and Colorado and Kansas followed in 1875. In 1876, however, the annihilation of General George Armstrong Custer and 276 soldiers of the Seventh Cavalry at the infamous Battle of the Little Big Horn outraged the public and sealed the fate of both the Native Americans and the bison. Congress passed a bill that would have stopped the slaughter of the bison, but President Grant vetoed it. The US Army knew that extermination of the bison would lead to control of Native Americans. The last commercial shipment of buffalo hides was in 1884 (Krech 1999). It is estimated that between 1868 and 1881, more than 31 million bison were killed (Taber and Payne 2003). A complete census in 1886 reported that there were only 540 bison left in the entire United States, mostly in the Yellowstone area of Montana (Trefethen 1975).

The disappearance of the bison was the turning point for wildlife conservation in the United States and Canada. It became one of the rallying cries for those concerned about the future of wildlife in the United States. Another was the demise of the passenger pigeon. Today it is hard to imagine the abundance of this bird. In 1806 Alexander Wilson, an ornithologist for whom the Wilson Society is named, recorded a flight 1.6 km wide and 64.4 km long, estimated to be more than 2 billion birds. In 1813 James Audubon, famous artist and ornithologist, observed in one day a flight 88.5 km long that held an estimated 1 billion birds—and the flight continued two more days. These flocks often devoured agricultural crops, and market hunters—using cannon-like "punt guns," nets, and even clubs—decimated these flocks, as well as populations of ducks (Anatidae), swans (*Cygnus* spp.), and geese (*Anser* spp.) for the meat market in America and Europe. The last passenger pigeon died at the Cincinnati Zoo in 1914 (Trefethen 1975).

A CONSERVATION MOVEMENT BEGINS

The concept of conservation was clearly different across the three countries of North America. Mexicans had a utilitarian view of forests, and wildlife captured little economic interest. In the early 1800s, Mexico forbade foreigners from harvesting furbearers, passed a forestry act, protected sea otter pups, and licensed timber harvesting. Unfortunately, the laws were rarely heeded by the public or by government officials. The Treaty of Hidalgo, which ended the Mexican–American War in 1848, and the Gadsden Purchase of 1854 voided all previous Mexican laws in those territories. After the French intervention in Mexico (1862–1867), additional laws were passed to protect wildlife, except "ferocious and dangerous animals," but they, too, were ignored (Simonian 1995). The federal department of fish and game was established in Mexico in 1894 (Valdez et al. 2006).

In Canada, and to some extent in the United States, the concept of wildlife was a cultural attitude shared with England. Subsistence hunting was accepted for settlers, but for the upper class, the hunting of game was associated with sport. Upper-class immigrants and Canadians and Americans who traveled to England were familiar with the pursuit of game on English estates. The evolution of hunters as conservationists ran parallel in the United States and Canada, though it developed more slowly in Canada because of its smaller human population and vast forested habitat unsuitable for farming (Taber and Payne 2003).

Most early settlers in the United States had little time for sport hunting, however, and wildlife was viewed as a source for sustenance and profit. Naturalist publications, such as *Nature* (1836) by Ralph Waldo Emerson or *Walden* (1854) by Henry David Thoreau, had little effect until generations later. Nonetheless, the concept of wildlife as a "public trust," so different from European experience, became codified law. In 1842 the US Supreme Court case *Martin v. Waddell* denied a landowner's claim to exclude others from taking oysters from some mudflats in New Jersey. The judge quoted the English Magna Carta of 1215 and codified the concept that in the United States, wildlife and fish belong to all the people, and stewardship of those fauna is entrusted to the individual states. This guaranteed the food supply at the time, although it continued to apply as wildlife became valued for sport, aesthetics, and for spiritual and cultural reasons. This decision, and others that followed, was the legal basis for the North American Model of Wildlife Conservation (Geist 1995; Geist et al. 2001; Organ et al. 2010, 2012)

It was not until the nation became more prosperous that sport hunters would become the impetus for early conservation efforts. Wealthy, mostly eastern, landowners and businessmen who no longer had to hunt for subsistence formed clubs of like-minded friends to promote comradeship, a kinship with the pioneer spirit, and ethical hunting practices. The first sportsmen's club in the United States was the Carroll's Island Club, formed in 1832 near Baltimore, Maryland, largely for waterfowl hunting. In 1844 the New York Sportsmen's Club was formed, which drafted model game laws that recommended closed hunting seasons on woodcock (*Scolopax minor*), quail (Odontophoridae), deer, and trout (Salmoninae) fishing. Orange and Rockland Counties in New York passed these laws in 1848. Many club members were attorneys, and they personally sued violators to encourage compliance with the law.

In 1859 British Columbia passed ordinances that limited mammal and bird hunting and forbade the sales of some wild game meat to secure their food supply (Colpitts 2002). Eventually, hundreds of local sportsmen's clubs were formed across both countries, and similar game laws were passed. In addition to game limits and seasons, some states outlawed the use of hunting dogs and night hunting with lights, and others banned the use of traps, snares, and pitfalls, as well as poisoning wildlife to protect crops, all of which were common at the time. The laws were hard to enforce, however, and in 1852 Maine became the first state to employ a game warden (Trefethen 1975).

PRESERVING LANDS FOR PUBLIC USE

During the 1870s much of North America was still being explored and surveyed. In 1870, a group of explorers pleaded with the governor of Montana that the Yellowstone area was too beautiful to exploit for profit and that it should be held in trust for the entire public. The concept of a public, national park was first raised by a painter, George Catlin, who in 1832 proposed "a nation's park, containing man and beast, in all the wild and freshness of their natural beauty" (Adams 2004:77). In 1872 President Ulysses S. Grant established the first national park in the United States, Yellowstone, encompassing 8,671 km². The first national park in North America was Chapultepec, established in what is now Mexico City in the 16th century (Simonian 1995). The Yellowstone law had little effect, however, and the US Army was later dispatched to guard the park from squatters and poachers for more than 30 years. The protection of this land was particularly important, as it signified the concept of a national asset in the public interest, and it began a tradition of setting aside lands for public parks, forests, and wildlife refuges. Rather than develop a "Tragedy of the Commons" (Hardin 1968), where European public lands were overutilized to their detriment, Americans agreed to share their commons fairly and democratically: "In no other nation was nature seen as such an essential part of national identity" (Wellock 2007:17). Waterways were eventually included. In 1872 the US Supreme Court case *Massachusetts v. Holyoke Water Power Company* decided that the company could not build a dam without a fish ladder. The court stated that the use of a river "may be regulated as public rights, subject to legislative control" (Wellock 2007:20). In 1871 Congress created the US Commission on Fish and Fisheries, one of our first federal conservation agencies. The primary purpose of this agency was to establish fish hatcheries, but again the move was significant in that it signaled the importance of federal management of wildlife and fisheries over that of states (Wellock 2007).

Canadians deplored the loss of the great auk (*Pinguinus impennis*), bison, Labrador ducks (*Camptorhynchus labradorius*), and the devastation of eastern sea bird colonies. In 1887 parliament established the first "Dominion Park" in Canada, first called the Rocky Mountain Park and later known as Banff National Park. Parliament also established the first national wildlife refuge at Lost Mountain Lake in Saskatchewan. Numerous provincial parks were established, as were Glacier, Yoho, Jasper, and Waterton Lakes National Parks. The Canadian Pacific Railway developed these protected areas with hotels and other amenities to entice tourism (Burnett 2003, Adams 2004). Eventually, the 1911 Dominion Forest Reserves and Parks Act standardized management of Canada's parks (Adams 2004).

THE EFFECT OF CLUBS AND THE MEDIA ON CONSERVATION

One cannot understate the influence of the formation of sportsmen's clubs, conservation organizations, scientific societies, and the print media on the North American conservation movement. The early apostle for hunting and conservation in America as a fair chase sport evoking manhood, character, and virtue was William Henry Herbert, who wrote under the pseudonym Frank Forester (Dunlap 1988). Magazines such as the *American Sportsman* (1871), *Forest and Stream* (1873), *Field and Stream* (1874), and the *American Angler* (1881) informed readers of the bounty and the plight of western wildlife. Public attitudes changed as they saw some species of wildlife disappearing because of market hunting. They came to realize that natural resources in America were not unlimited, and that conservation efforts should be used. In some circles, Charles Darwin's *On the Origin of Species* (1859) and *The Descent of Man* (1871) led people to realize that man had a genetic relationship with animals, and that animals might be able to reason and suffer (Dunlap 1988).

The American Association for the Advancement of Science (AAAS) in 1881 pressured the US secretary of agriculture to form the Forestry Division, which later became the US Forest Service. In 1891 the American Forestry Association and the AAAS convinced Congress to establish national forest reserves to ensure a future timber supply for the nation. By then, the US General Land Office and the US Geological Survey had been formed to survey and keep track of federal lands (Trefethen 1975).

Gifford Pinchot, a friend of Teddy Roosevelt's, was named director of the US Forestry Division in 1898. Pinchot was trained in Europe, as there were no forestry courses in American universities at that time, but he understood that forests could be used for timber, wildlife, and watershed, and conserved through sustained yield management. In fact, he and his staff coined the term *conservation*, derived from the British *conservator*. He led a survey of American forests with the AAAS, and in 1896 President Grover Cleveland added more than 8.5 million ha to the forest reserves and established Grand Canyon and Mount Rainier as national parks. In 1904 Pinchot became the first director of the US Forest Service and was widely known as "the father of American forestry." Although he, too, was part of the eastern elite, Pinchot strongly believed in equity in access to America's resources. He stated, "natural resources must be developed and preserved for the benefit of the many, not just the profit of the few" (Krech 1999:25). About the same time, he formed the Society of American Foresters, an organization composed of forestry professionals. In 1905 they held the first American Forest Conference to bring together managers, educators, and scientists (Trefethen 1975).

During the 1870s and later, dozens of additional hunting, conservation, and scientific organizations were formed, including the League of American Sportsmen, the American Ornithologists' Union, the Camp Fire Club, the New York Zoological Society, the Audubon Society, and the American Bison Society. These groups, along with local sportsmen's clubs, lobbied for stricter laws to stop market hunting, to ban unethical sport hunting, and to begin game restoration efforts. They recognized that states had difficulty enforcing their game laws. In 1900 Congress passed the Lacey Game and Wild Birds Preservation and Disposition Act, the first national legislation for wildlife conservation. This law made it a federal offense to transport wild game across state borders if taken illegally. It also strictly controlled the importation of exotic species. This strengthened state game laws, and it helped stop the trade in plume and feathers, as well as the poaching and smuggling of wildlife meat products (Trefethen 1975).

In addition, sportsmen's clubs started wildlife restoration efforts. An interesting result was the importation in 1881 of 28 ring-neck pheasants (*Phasianus colchicus*) to Oregon from Shanghai, China. Game bird farms blossomed, raising birds in captivity and releasing them into the wild to be hunted. Pheasant hunting is now popular through much of the United States, and few Americans recognize this bird as an exotic species. Destructive logging techniques during that period ac-

tually led to better deer habitat, as more diverse vegetation replaced the mature forests. In 1878, a sportsmen's club in Vermont was the first to trap and restock deer. Other states followed (Trefethen 1975).

Pressure from these sportsmen's, conservation, and scientific organizations also affected natural resource education. The Land Grant Act, signed by President Abraham Lincoln in the middle of the Civil War in 1862, established agricultural and technical colleges in all states, making higher education in agriculture and engineering affordable to the sons and daughters of farmers and mechanics. Later legislation established the Agricultural Experiment Station System and the Cooperative Extension Service with joint funding from the state, federal, and county governments. In 1898 Cornell University began offering courses in forestry, and in 1900 Yale University established the School of Forestry, the first of its kind. Within three years, forestry schools opened at the universities of Maine, Michigan, Minnesota, and at Michigan State. Wildlife was not yet a scientific or management discipline of its own, as most studies of biota were still of a taxonomic sort found in botany and zoology programs (Trefethen 1975).

THE ROOSEVELT ERA

Born into a wealthy New York family, weak and asthmatic as a child, Theodore Roosevelt rose to become our country's most successful conservationist. Homeschooled as a child, he read voraciously about natural history, learned taxidermy, exer-

THEODORE ROOSEVELT (1858–1919)

A young, affluent politician named Theodore Roosevelt aided the Yellowstone Park effort in the United States. Roosevelt was a 23-year-old Harvard graduate and member of the New York State legislature when in 1883 he went on his first buffalo hunting trip in North Dakota. He would go on to become an avid and adventurous outdoorsman and hunter. The following year his wife and mother died a few hours apart, and he returned to North Dakota for three years to reflect on his life and then to purchase and establish the Elkhorn Ranch. Roosevelt became a cowboy and a hunter who loved trophies but who felt a near-spiritual kinship with nature. He detested the decline of the buffalo (Wilson 2009).

Roosevelt's enthusiasm for conservation was born in the Dakota Badlands during those years, and he became the most active president and leader in the history of North American conservation. In 1887 he gathered a group of influential American hunters in New York to form the Boone & Crockett Club, with a mission of preserving the big game of North America. One of the members, an outspoken editor named George Bird Grinnell, wrote numerous articles about the plight of Yellowstone in his magazine *Forest and Stream*. In 1894 President Grover Cleveland signed the Yellowstone Protection Act, making the park the first wildlife refuge in the United States, and provided it with guaranteed funding and administration. Another early conservationist, John Muir, advocated the establishment of Yosemite National Park and other parks in California. He later formed the Sierra Club in 1890. Muir was a preservationist who believed public lands should not be used for timber harvesting or hunting. Roosevelt, the conservationist, believed lands for public enjoyment should also be used for multiple purposes, including resource extraction. The philosophical argument over preservation—or strictly limiting access to public lands—versus conservation—to include recreation, grazing, logging, and hunting on public lands—continues to this day in the United States (Trefethen 1975).

cised himself to physical fitness, and took hunting and camping trips to Maine and other states. After graduating from Harvard in 1880, at 23 he became New York's youngest state legislator. He wrote of his hunting trips to the Dakota Territory in his book *Hunting Trips of a Ranchman*. After the death of his mother from yellow fever and his wife from childbirth on the same day, he returned to the Dakotas for 16 months, purchasing the Elkhorn Ranch. He remarried in New York in 1886 and produced a plethora of books on hunting, ranching, and western lore. He was an active member of the Explorers Club, the Camp Fire Club, and he personally formed the Boone & Crockett Club and the NY Zoological Society, which established the Bronx Zoo.

In 1888 Roosevelt accepted a position on the federal Civil Service Commission. That was followed in 1885 at age 35 as New York City police commissioner, and in 1897 as assistant secretary of the Navy. He resigned in 1898 to form the Rough Riders and lead the decisive battle of the Spanish-American War at San Juan Hill in Cuba, for which he was posthumously awarded the Medal of Honor by President Bill Clinton in 2001. He returned from the war to become governor of New York and then vice president under President William McKinley. He became the nation's youngest president at age 42 upon McKinley's assassination in 1901. It was on a hunting trip to Mississippi in 1902 that he refused to shoot a captive bear, which subsequently became known as "Teddy's Bear," an icon for toy manufacturers. During his eight years in office, Roosevelt established 51 wildlife refuges, 5 national parks, and 16 national monuments, protecting 59,893,475 ha in addition to enacting significant conservation legislation, building the Panama Canal, and mediating the Russo-Japanese War, for which he was awarded the Nobel Prize. In 1908, after his presidency, he took a three-month hunting trip to Africa, collecting dozens of animals for the Smithsonian collection. In 1908 he ran again for president against Taft and Wilson as a third-party candidate of the Progressive or "Bull Moose" Party. At one point he delivered an hour-long speech with a bullet in his lung from an assassination attempt. After losing the election to Wilson, he embarked on a three-month trip down the River of Doubt, a tributary of the Amazon, where he was seriously injured and contracted malaria. Roosevelt died in his sleep at his Sagamore Hill, New York, estate in 1919, at age 60. Roosevelt's legacy includes 26 books, more than 1,000 magazine articles, and thousands of letters and speeches. He is remembered as one of our country's greatest historians, biographers, statesmen, hunters, naturalists, and orators.

When Theodore Roosevelt became president in 1901, a new era of wildlife conservation began. He once stated, "there can be no greater issue than conservation in this country" (Wilson 2009:i). Roosevelt was advised by Gifford Pinchot, George Bird Grinnell, and to some extent John Muir. Roosevelt formed the Agriculture Department's Division of Economic Ornithology and Mammalogy, which soon became the Bureau of Biological Survey, and tasked it with surveying the nation's biota. It was initially a research organization but was given policing powers with the passage of the Lacey Act. Major John Wesley

Powell, the well-known explorer of Grand Canyon, became director of the Geological Survey. He supported the Reclamation Act of 1902, which authorized more than 30 federally funded reservoir and dam projects. Although this legislation was not directly associated with wildlife, it had a major effect on future wildlife populations. It recognized the authority of the US government over the rights of states in national natural resources issues, though it led to the development of significant irrigated farming in the West (Trefethen 1975).

Although Roosevelt and John Muir were friends, their relationship was tested over the first major political battle over land preservation. After the tragic San Francisco earthquake and fire of 1906, it was clear that San Francisco needed a dependable water supply. A proposal was made to dam part of the Tuolumne River in California, which would flood part of Yosemite. Muir strenuously opposed the Hetch Hetchy Dam project, which Pinchot and Roosevelt supported. The dam was completed in 1914, and Muir died a year later. The battle over this use of a public resource split the conservationists from the preservationists. Eventually, however, the argument let to the formation of the US National Park Service (NPS) in 1916 under the auspices of the Department of the Interior (Cronon 2003).

Theodore Roosevelt believed in a strong role for the federal government in protecting as much land as possible for public use. In 1903 Pelican Island, Florida, became the first unit of the National Wildlife Refuge System, which now encompasses more than 60.7 million ha (Fig. 2.3). Roosevelt used the National Antiquities Act of 1906 to declare scenic lands and wildlife habitat as parks and as forest reserves. In 1908 he added additional wildlife refuges in Alaska, Florida, Nebraska, and Oregon. That same year, Roosevelt appointed the National Conservation Commission, which was tasked with inventorying the national forests, waters, and minerals and with recommending management strategies. In all, Roosevelt set aside 60 million ha during his presidency, more than 20,000 ha for each

Figure 2.3. Pelican Island was the first national wildlife refuge. Courtesy of US Fish and Wildlife Service

day he was in office, including 16 national monuments, 51 wildlife refuges, and 5 national parks (Wellock 2007, Brinkley 2009).

CONSERVATION EFFORTS CONTINUE

Congress was relatively slow to address the problem of market hunting waterfowl. The Weeks-McLean Migratory Bird Act, proposed in 1904 by Pennsylvania Congressman George Shiras III, asserted the federal government's power over states' rights and let the secretary of agriculture set hunting seasons and limits on migratory game birds. This was opposed for nine years as a states' rights issue, but it finally passed in 1913. Three years later the United States signed a treaty with Great Britain for the Protection of Migratory Birds in the United States and Canada. Mexico was invited to sign the treaty, but it was in revolution at the time. Mexico later signed the treaty in 1936 (Valdez et al. 2006). This landmark legislation was the first between countries to protect wildlife. Similarly, after a decade of debate, Congress passed the Weeks Act of 1911, which authorized the federal government to purchase lands to protect streamflow and eventually led to the creation of the eastern national forests (Lewis 2011).

The Canadian wilderness lifestyle had evolved from that of hunter-trappers to a more agrarian type about the time of the confederation of the provinces in 1870. With it came a new ethic of wildlife conservation. The formation of the Historical Society of Manitoba (1879) was followed by the formation of the Winnipeg Game Preservation League (1882) and the Macleod Game Protective Association (1889). But as the railroads and developers enticed western emigration, they exaggerated the abundance of wild game available for sport hunting. Taxidermy exhibits went eastward to advertise the size and diversity of game in western Canada and Alaska. As the railroad, mining, and logging crews used more wildlife as a food source, conflicts with sport-hunting tourists became inevitable. In the 1890s, sportsmen successfully lobbied for controls on sport and meat hunting, including subsistence hunting by Indigenous Canadians. In 1909 the Canadian Commission on Conservation was formed to mirror Theodore Roosevelt's efforts in the conservation clubs formed in the United States, such as the Camp Fire Club, the Boone & Crockett Club, and the Audubon Society, and their individual members influenced Canadian legislatures. The 1911 Canadian Game Act prohibited the shooting and selling of 26 types of game species (Colpitts 2002).

Roosevelt, Pinchot, and Muir influenced the attitudes of Canadians as well. In 1906 game protection and management acts were enacted in Alberta, New Brunswick, Prince Edward Island, and Saskatchewan; British Columbia followed in 1913. *Life Histories of North American Mammals*, by Seton and Roberts (1909), further enhanced public interest in protecting wildlife. US and Canadian biologists and policy makers collaborated a great deal, and in 1909 parliament formed the 32-member Commission for the Conservation of Natural Resources, which later developed the National Parks Act, the Northwest Territories Game Act, and the Migratory Birds Convention Act. In 1911 Howard Douglas, superintendent of Rocky Mountain National Park, purchased 703 bison in Montana and released them in the park. In 1918 Hoyes Lloyd, supervisor of wildlife protection, authorized the Royal Canadian Mounted Police (RCMP) to enforce game laws (Burnett 2003). Two activist officials—Gordon Hewitt, dominion entomologist with the Canadian Department of Agriculture and author of *The Conservation of the Wild Life in Canada* (1921), and James Harkin, commissioner of parks—were instrumental in keeping wildlife protection under the parks department rather than under the forestry department. They were also involved in lobbying for the Migratory Bird Treaty. Hewitt and Harkin were supported politically by Clifford Sifton, the minister of the interior, and by Prime Minister Wilfred Lauier (Taber and Payne 2003).

In Mexico, Miguel Angel de Quevedo, an engineer, architect, and founder of Mexico City's arboretum, was an advocate for conservation of Mexico's forests. He spoke at the North American Wildlife Conference in 1909, where he met Roosevelt and Pinchot. He promoted legislation that would preserve forests for watersheds, recreation, and scenic values, but the revolution undid all of his efforts. He later headed the effort to form the Mexican Forestry Society and headed the Commission for Protection of Wild Birds (Simonian 1995).

By 1910 every state in the United States had some sort of commission for the protection of wild game and fisheries. The National Association of Game Wardens and Commissioners became the International Association of Game, Fish and Conservation Commissioners (now known as the Association of Fish and Wildlife Agencies). But funding was still a problem. In 1913 Pennsylvania was the first state to issue hunting licenses. When the sale of these licenses brought in more than $300,000 the next year, many other states followed suit. These funds paid for wildlife restoration efforts, enforcement of game laws, and predator control. Predators interfered with livestock operations, and hunters also still believed that predators limited game abundance. Many states paid bounties for wolves, mountain lions (*Puma concolor*), foxes (*Vulpes* spp.), coyotes, bobcats (*Lynx rufus*), hawks (Accipitridae), owls (Strigiformes), and even eagles (Trefethen 1975).

Debates continued in the state houses and at the national level over the concepts of preservation versus conservation. Dr. William T. Hornaday, superintendent of the National Zoo in Washington, DC, and head of the National Bison Society, published *Our Vanishing Wildlife* in 1913 and led his followers to oppose hunting. He was an elitist who once remarked, "all members of the lower classes of southern Europe are a dangerous menace to our wild life," and that protection was needed from "Italians, negroes and others who shoot songbirds for food" (Wellock 2007:87). He was also opposed to federal funding of waterfowl restoration through hunting fees. Hunters and conservationists created the organizations listed above plus new ones like the Izaak Walton League, Forests and Wild Life, and American Wild Fowlers (later to become Ducks Unlimited). Scientists formed the Ecological Society of America in 1914 and American Society of Mammalogists in 1919. Both societies were to then publish scientific journals. Wildlife science began to move beyond taxonomy. The scientific societies had a close relationship with the Biological Survey, and together they developed better methods of censusing wildlife and studying diets, cover requirements, and disease issues in the national parks and refuges. As ecology developed as a discipline, the concepts of plant succession, niche, community scales, trophic levels, and food chains were developed and debated in the United States and abroad (Trefethen 1975).

Preservation of public land continued, with the formation of national parks in the United States at Glacier (1910), Lassen (1916), Denali (1917), and Grand Canyon (1919). In 1917 Mexico's President Carranza declared Desierto de los Leones as the country's first modern national park. The US National Park Service was eager to sustain wildlife in their parks for visitors to see, a much different mission than that of the US Forest Service, which was to preserve a sustainable timber supply. The NPS hired scientists to study the natural resources they managed. Efforts by the NPS were instrumental in preserving trumpeter swans (*Cygnus buccinator*), grizzly bears, bighorn sheep, and wild burros (*Equus asinus*). Still, however, there was little funding at the national level to support wildlife research (Simonian 1995).

In the early 1900s, America's assault on predators continued on private and public lands, as it was believed they harmed domestic livestock and wildlife populations. In the Kaibab Plateau of Arizona, a herd of 3,000 deer on about 405,000 ha was thought to be in decline. The government hired hunters and trappers to kill 120 bobcats, 11 wolves, 674 mountain lions, and more than 3,000 coyotes. By 1924 the deer herd had grown to 100,000. During that year's severe winter, 60,000 deer died of starvation. This episode was later to become a national example of the perils of predator control as a means of managing wildlife populations (Trefethen 1975).

In 1937 the scientists, with the help of the conservation groups, lobbied the US Congress to pass the Pittman-Robertson Federal Aid in Wildlife Restoration Act (P-R Act). This act excised an 11% tax on all hunting weapons and ammunition. The act provided nearly $3 million in its first year. The funds are collected by the federal government and distributed to the states based on the number of hunting licenses they sell, their population, and their land area. The act provides 75% of the funding, which must be matched by 25% from the states. The funds are used for wildlife restoration projects, research, and education, and they cannot be used for any other purpose. This is still one of the major sources of funding for wildlife research in the United States (Trefethen 1975).

WILDLIFE RESEARCH BEGINS TO SUPPORT MANAGEMENT

The 1930s were the beginning of wildlife management research in the United States, led by Aldo Leopold. Although trained as a forester at Yale, Leopold became "the father of wildlife management" in the United States. Working for the US Forest Service in New Mexico, later for their Forest Products Lab in Wisconsin, and then as a consultant for the Sporting Arms and Ammunition Manufacturer's Institute, he conducted the first intensive analysis of wildlife populations in the Midwest. In 1933 he became the first professor of game management at the University of Wisconsin and published *Game Management*, the first book of its kind in America. In it he said, "we have the scientist, but not his science, employed as an instrument of game conservation" (Leopold 1933:20). Leopold was well known for his many essays on land ethics, and he developed a series of wildlife management principles: every species has a defined set of habitat requirements that sets it apart from all other species, all animal and plant biota are interconnected, and the habitat has a seasonal carrying capacity that should not be exceeded. He said the tools of wildlife management were "the ax, the plow, the cow, fire and the gun" (Leopold 1933:332).

The stock market crash of 1929 and the dust bowl era of the early 1930s were disastrous for American wildlife. Pressure from livestock growers and the public pushed the Biological Survey toward predator control rather than wildlife research as its primary mission. They formed the Division of Predator and Rodent Control (later to become Animal Damage Control and then Wildlife Services under the Department of Agriculture). Much of the federal funding for other wildlife-related programs was cut at the same time waterfowl and game bird habitat was disappearing in the Midwest. Wildlife conservation and restoration needed more funding, and in 1934 Congress passed the Migratory Bird Hunting Stamp Act (similarly adopted in Canada in 1966). Because waterfowl are migratory across state boundaries, the federal government limits their seasons and bag, and hunters must purchase a duck stamp to affix to their state hunting licenses. These funds have now protected more than 2.5 million ha of waterfowl habitat (Ducks Unlimited 2021).

FRANKLIN DELANO ROOSEVELT AND A NEW DEAL FOR WILDLIFE

The presidents who followed Theodore Roosevelt were not nearly as supportive of conservation. President Taft (1909–1913) thought Roosevelt had gone too far, and he opposed habitat preservation and even forest fire control in Alaska. He fired Pinchot and Interior Secretary James Garfield (the former president's son) and replaced them with men more amenable to dealing favorably with developers and industrialists. Deals were cut to sell federal coalfields and to encourage development of railroads, copper mines, and steamship companies in Alaska. Presidents Wilson (1913–1921), Harding (1921–1923), Coolidge (1923–1929), and Hoover (1929–1933) showed little interest in conservation, though the latter two were fly fishers (Brinkley 2011).

Franklin D. Roosevelt, who was elected to office in 1933, commonly referred to himself as a tree farmer (Brinkley 2011) and made conservation a national jobs program. The Civilian Conservation Corps (CCC) employed more than 3 million

JAY NORWOOD "DING" DARLING (1876–1962)

Born in Norwood, Michigan, and raised in Sioux City, Iowa, Jay Darling intended to pursue a career in medicine. But at Beloit College in Wisconsin, he began newspaper writing, cartooning, and signing his name "Ding," a contraction of his last name. His professional newspaper career began in 1900, with posts at the *Sioux City Journal*, *Des Moines Register and Leader*, the *New York Globe*, and the *New York Herald Tribune*. Though he eventually moved back to Des Moines, the *Tribune* published his political cartoons from 1917 to 1949. His cartoons were wildly popular and syndicated in 130 papers across the country. They covered many topics, but as an avid hunter and fisher, Ding was especially concerned about pollution and loss of wildlife habitat. Ding received the Pulitzer Prize for editorial cartooning in 1924 and 1943, and he was voted the best cartoonist in America by a group of leading editors in 1934.

Despite his inexperience, Ding was appointed head of the US Biological Survey (later to become the US Fish and Wildlife Service) in 1934 by President Franklin D. Roosevelt. During his one-year tenure at the bureau, Ding raised $17 million for wildlife habitat restoration. The idea of the Duck Stamp was his, and he drew its first design. The funds he raised led to the expansion of the National Wildlife Refuge System. He established the Migratory Bird Commission to bring together hunters and other conservationists, later founded the National Wildlife Federation, and the first North American Wildlife Conference was his idea. He often vacationed on Sanibel Island, Florida, and there he convinced his neighbors to purchase land that President Harry Truman named a national wildlife refuge in 1945. In 1967 the refuge was named the J. N. Ding Darling National Wildlife Refuge.

The Cooperative Wildlife Research Unit Program established research units of 2 to 4 biologists at ten land grant universities, with funding supplied by the Bureau of Biological Survey, the universities, the American Wildlife Institute (later the Wildlife Management Institute), and state fish and game agencies. These units provided research on practical wildlife problems, taught university courses, and helped train thousands of biologists. There are now 40 such units in the country. The Wildlife Management Institute receives its funding from the sporting arms and ammunition industry, and it lobbies for conservation laws and conducts professional reviews of university wildlife programs and state wildlife agencies. It also annually convenes the North American Wildlife and Natural Resources Conservation Conference of scientists, state and federal agency personnel, and nongovernmental organizations (NGOs). The first meeting, held in 1935, produced more information in its proceedings than had ever before been accumulated in one volume on wildlife in North America. The North American Wildlife Federation later became the National Wildlife Federation, now a private organization with more than 6 million members, and the North American Wildlife Institute later became the North American Wildlife Foundation, a semiprivate granting agency of the federal government (Trefethen 1975).

out-of-work factory workers and farmers. Between 1933 and 1942, the CCC built 800 new parks, 193,000 km of roads, and 1,147 fire lookout towers; reforested more 809,000 ha with 2 billion trees; and restored thousands of hectares of wetlands for waterfowl breeding. Not all of this construction was ecologically sound, but it popularized the concept of conservation to the American working classes (Wellock 2007). President Roosevelt hired Jay "Ding" Darling, a political cartoonist critical of the New Deal, to head the Bureau of Biological Survey. He was an enthusiastic leader, and he enhanced the morale of the bureau. He saw the need to develop wildlife research programs and to enhance educational programs to produce wildlife managers and biologists. In 1934 he urged Congress and private organizations to establish the Cooperative Wildlife Research Unit Program, the American Wildlife Institute, the North American Wildlife Federation, the North American Wildlife Institute, and the North American Wildlife and Natural Resources Conference (Trefethen 1975).

Other accomplishments during the New Deal were the passage of the Fish and Wildlife Coordination Act, to force federal agencies to communicate with each other; the establishment of the Soil Erosion Service, later to be called the Soil Conservation Service and now called the Natural Resources Conservation Service; and the Taylor Grazing Act, which increased management of public lands and charged fees for restoration of more than 32 million ha (Wellock 2007). These improvements had a powerful, positive effect on wildlife in North America, restoring deer to the Northeast and waterfowl to the Midwest. Other initiatives, such as the Tennessee Valley Authority (TVA), had mixed effects. The TVA built 16 dams and brought electricity to thousands, but it destroyed many streams and wildlife habitat in doing so. In 1930, a committee headed by Aldo Leopold developed the Model Game Law, recommending that states set up wildlife commissions of volunteers, appointed by governors, for staggered terms. That eliminated the problem of turnover of state wildlife agency personnel caused by new governors firing supervisors after each election. The P-R Act prohibited the governors from redirecting its funds to other uses (Trefethen 1975).

The Wildlife Society, an organization of professional wildlife biologists, was formed in 1936, and the first issue of the *Journal of Wildlife Management* was published the next year. The American Fisheries Society and the Society for Range Management also represented professionals and published scientific journals. Just before World War II, the Bureau of Fisheries in the Department of Commerce and the Biological Survey of the Department of Agriculture were merged into the US Fish and Wildlife Service (USFWS) under the secretary of the Department of the Interior, a significant development, as governmental leaders recognized that fish and wildlife were more than crops to be grown and harvested (Trefethen 1975).

Things did not always go so smoothly in Canada. Subsistence hunting and fishing by Indigenous people have been, and continue to be, an issue. In 1949 Newfoundland joined the Canadian confederation and challenged the Migratory Bird Act, hoping to reduce or eliminate hunting restrictions. The political debate continued into the 1950s, until new laws allowed needy residents to hunt for food but also outlawed the sale of game meat (Taber and Payne 2003). Unlike Canada, Mexico did not recognize Indigenous communities or their rights to land or hunting and fishing privileges. In 1936 Mexico signed the Protection of Migratory Birds and Animals Act, which established hunting regulations and wildlife refuges, and outlawed transportation of live or dead game, much like the Lacey Act in the United States (Lopez-Hoffman et al. 2009).

In 1935 Mexico's President Cardenas (1934–1940) appointed Angel de Quevedo to head the Department of Forestry, Fish and Game. Cardenas had made land reform a major platform of his election campaign. During his administration, Cardenas and Quevedo created 40 national parks, plus additional wildlife refuges, agricultural cooperatives, irrigation programs, and dams. This amounted to the largest land reform in Mexico's history. In 1935 the United States–Mexico International Parks Commission was formed. Unfortunately, a strong hunter-conservationist lobby never formed in Mexico, largely owing to lack of interest and the lack of money spent on sport hunting. Quevedo's initiatives were strongly opposed by agricultural interests. Game laws continued to be poorly enforced, and Mexico's grizzly bear and wolf populations were nearly extirpated. Finally, bowing to agricultural interests, President Cardenas abolished the Department of Forestry, Fish and Game in 1940 and accused Quevedo of being antirevolutionary. Quevedo died in 1946 (Simonian 1995).

WORLD WAR II AND THE 1950s

As with the Civil War and World War I, conservation efforts took a backseat during World War II. After the war, however, returning servicemen bought hunting licenses by the thousands. Sales in the United States increased from $7 million per year before the war to $12 million by 1947. States benefited from the influx of hunting license fees and the P-R Act funds they generated, as did federal agencies from the sale of duck stamps. Unused P-R Act funds had accumulated during the war and were now available for larger projects. Large restocking efforts resulted for deer, pronghorns, elk, mountain goats (*Oreamnos americanus*) and sheep, bears, beavers, and turkeys. By then, most land grant universities had wildlife departments, and the GI Bill allowed returning military personnel to pursue wildlife biology as a profession. Aldo Leopold died in 1948, and his final essays were published in a book titled *A Sand County Almanac* (Leopold 1949), which became required reading for Americans interested in conservation for generations to come (Trefethen 1975).

During the 1950s, support for conservation took a downturn in the United States as the federal government, concerned over military threats in Europe, redirected funds toward increasing the size and power of America's military. Funding for fish, wildlife, park, and forest programs diminished, while logging, grazing, and oil and gas exploration leases were granted on public lands. In 1950 concern over funding for restoration of freshwater fisheries led to the Dingell-Johnson Federal Aid in Fisheries Restoration Act (D-J Act). The act was funded by an excise tax on fishing equipment and boats, and it functioned much as the P-R Act did for wildlife. The Magnuson Act was passed to separate commercial fisheries from the USFWS and to pass it to a new US Fisheries Commission.

Americans during the 1950s flocked by the millions to our national parks, wildlife refuges, seashores, and forests, establishing recreation as an integral component of the natural resource value of public lands. In response to the decrease in federal land acquisition, The Nature Conservancy was formed in 1951 to preserve lands the government could not afford. It is now the wealthiest conservation organization in the nation, protecting more than 50 million ha worldwide. Bird-watching became a popular pastime with the continued publication and updates of Roger Tory Peterson's *A Field Guide to Birds* (1934). Disney Studios, with the feature film *Bambi* and television documentaries like *The Living Desert*, had a huge effect on the public's attitude toward nature (Trefethen 1975).

In 1951 Mexico revised the Federal Game Law, which was first passed in 1940. The new law mirrored much of the US and Canadian laws. It specified that wildlife was public property with the federal government as legal custodian, required hunters to belong to hunting clubs, defined which poisons could be used to kill wild animals, and forbade the commercialization and exportation of game animals (Valdez et al. 2006).

But the new prosperity and growth of the 1950s brought more commercial development of land for housing and concentrated farming and livestock operations. Farming efficiency was helped by mechanization as well as liberal use of pesticides and herbicides. One of the most notorious was dichlorodiphenyltrichloroethane (DDT). DDT was an effective pesticide, but it was released before being adequately tested. As early as 1946, it was found to be lethal to crustaceans like crabs (Brachyura) and crayfish (Astacoidea). As insects became impervious to DDT, other chemicals such as chlordane, dieldren, aldrin, and methoxychlor went on the market. Wildlife

RACHEL LOUISE CARSON (1907–1964)

Rachel Carson was a nationally known scientist and author well before she published *Silent Spring*. Born in Springdale, Pennsylvania, Carson graduated from the Pennsylvania College for Women (now Chatham College) in 1925, studied at the Woods Hole Marine Biological Laboratory, and obtained an MA in zoology from Johns Hopkins University in 1932. She began her professional career writing radio scripts for the US Bureau of Fisheries in 1936 and eventually became editor-in-chief for all US Fish and Wildlife Service publications. She wrote scientific articles and pamphlets, and published popular articles in the *Atlantic Monthly*, the *New Yorker*, and the *Baltimore Sun*. Her books *Under the Sea-Wind* (1941), *The Sea around Us* (1952), and *The Edge of the Sea* (1955) brought her fame as a naturalist and author. She was awarded a National Book Award, the John Burroughs Medal, the Henry Grier Bryant Gold Medal, the New York Zoological Society Gold Medal, and a Simon Guggenheim Fellowship.

Carson resigned from federal service in 1952 to concentrate on writing, and later she changed her focus to the misuse of pesticides. In 1962 her book *Silent Spring* was published in three installments in the *New Yorker* and was a book-of-the-month club selection. She challenged agricultural practices and lack of government oversight as she linked the health of the natural world to that of humans. Though she called only for prudent use of DDT rather than a total ban, she was vigorously attacked by the chemical industry. Her testimony before Congress in 1963 eventually led to the formation of the Environmental Defense Fund in 1967, the formation of the Environmental Protection Agency in 1970, and the passage of the federal Insecticide, Fungicide and Rodenticide Act in 1972, which banned the use of DDT in the United States. Prior to her death from cancer in 1964, she was awarded the National Audubon Society Medal and was inducted into the American Academy of Arts and Letters. She was posthumously awarded the Presidential Medal of Freedom by President Jimmy Carter (Nielsen 2017).

biologists first reported dying songbirds, followed by brown pelicans (*Pelecanus occidentalis*), ospreys (*Pandion haliaetus*), and bald eagles. Researchers at the USFWS Patuxent Wildlife Research Center in Laurel, Maryland, fed these chemicals to wildlife and proved that they were toxic (Gottlieb 2003).

WILDLIFE BECOME PART OF THE ENVIRONMENT

In 1962 Rachel Carson, a former USFWS editor, researcher, and internationally successful writer, published *Silent Spring*, a book that documented the effect of these chemicals on the environment, especially on wildlife. She spoke to the issue of science being ignored in public conservation policy and corrupted by commercial interests. She stated that the increasing use of pesticides was indicative of "an era dominated by industry, in which the right to make money, at whatever costs to others, is seldom challenged" (Carson 1962:67). She predicted a future of spring seasons when no birds would be heard. No book before or after has had as great an effect on arousing the American public's awareness of environmental concerns. It set the stage for the environmental movement for the next three decades. In Canada, the book and the public outcry that followed led to the formation of the Wildlife Toxicology Division of the Canadian Wildlife Service. Chemical companies tried to discredit Carson, but President Kennedy defended her and made protecting the environment part of his political plat-

form, though DDT was not actually banned until 1974. Kennedy also established the Land and Water Conservation Fund to acquire land for scenic, recreational, and public values, and he signed the Sikes Act of 1960, which required the Department of Defense to prepare natural resources management plans for all 12 million ha of military bases. In 1964 President Lyndon B. Johnson signed the Wilderness Act, which incorporated roadless and untrammeled lands of 2,023 ha or larger into wilderness areas. This legislation removed them from the National Forest Inventory, placing them into national parks and wildlife refuges (Cronon 2003). Next, Johnson signed the Wild and Scenic Rivers Act, setting aside seven major rivers for recreational and conservation purposes (Trefethen 1975).

NIXON: THE RELUCTANT ENVIRONMENTALIST

Concerns over pesticide toxins eventually led to public dissatisfaction over predator control techniques. Common methods for killing predators included the use of sodium floroacetate (compound 1080) and coyote getters, explosive shells filled with cyanide pellets, as well as the use of strychnine and thallium poisons. These devices and poisons killed all manner of small mammals, hawks, and eagles. Under the Nixon administration, an executive order prohibited the use of chemical poisons on all public lands. In response to public outcry over the potential loss of our national bird, the bald eagle, and other species, Congress passed the Endangered Species Con-

Figure 2.4. The bald eagle was one of the first listings on the endangered species list. Courtesy of US Fish and Wildlife Service

servation Act in 1969, but it had little in the way of enforcement power or funding. Congress followed with a tougher Endangered Species Act (ESA) in 1973. It defined and divided species and subspecies into threatened versus endangered. It eliminated all commercial traffic in live, dead, parts, or products from endangered species, and it funded research on why species were becoming threatened and how to recover them. The act defined "critical habitat," included harassing wildlife to be a "taking," set substantial fines, and required teams of scientists and managers to develop recovery plans for each endangered species. Early endangered species included wolves, whooping cranes (*Grus americana*), Key deer (*O.v. clavium*), Sonoran pronghorns (*Antilocapra americana sonoriensis*), peregrine falcons (*Falco peregrinus*), bald eagles (Fig. 2.4), alligators, Kirtland's warblers (*Dendroica kirtlandii*), and California condors (*Gymnogyps californianus*). Captive breeding programs were often begun as part of the restoration projects. Swift foxes (*Vulpes velox*), black-footed ferrets (*Mustela nigripes*), and California condors are examples of translocation and breeding and release projects. The USFWS and the National Marine Fisheries Service were given authority to enforce the act. In 1978 the US Supreme Court case of *Tennessee Valley Authority v. Hill* delayed the federal Tellico Dam project to protect the snail darter (*Percina tanasi*), a tiny endangered fish. The court ruled that "the plain intent of the statute was to halt and reverse the trend towards extinction, whatever the cost" (Lueck 2008:154). The ESA is often touted as the most significant conservation legislation ever enacted in the United States. It continues to be the basis and the source of much funding for wildlife research, though the funding is far below what is needed for this enterprise.

Surprisingly to many, the Nixon administration was one of the most environmentally progressive of all American presidencies. In addition to the ESA, Nixon signed the Marine Mammal Protection Act, the National Environmental Policy Act (NEPA), the Clean Air Act, and the Clean Water Act, and he established the Environmental Protection Agency (EPA).

This legislation, taken together, had a monumental effect on guiding what could and could not be done with America's natural resources, and it clearly established the environment as an issue of national importance. The NEPA led to the Council on Environmental Quality, which requires environmental impact statements on all government projects. It also requires open hearings for public input, thus making environmental decision-making an open and democratic process. The EPA is now one of the major funding sources for wildlife research, in addition to being an environmental enforcement and research agency. In 1969 the United States signed the Convention on International Trade in Endangered Flora and Fauna Species Act (CITES), making it illegal to import or export items made from endangered species.

Nixon's impetus for signing all of this environmental and conservation legislation was strictly political. Public opposition to the war in Vietnam—in the form of street demonstrations and even riots—was growing during his administration. Simultaneously, the public had lost faith in the government to protect them from protracted war, environmental pollution, and nuclear proliferation. Nixon inherited Lyndon Johnson's unpopularity over the war, but Johnson had enjoyed the popular Secretary of the Interior Stuart Udall, and Lady Bird Johnson had made beautifying America by planting flowers a priority. Public concerns over the environment were heightened by television ads of Iron Eyes Cody, an actor who depicted an American Indian shedding a tear over highway trash. In addition, population growth became a concern in 1968 with the publication of Paul Ehrlich's book *The Population Bomb* (his coauthor and wife, Anne, was not credited). On 22 April 1970, the nation participated in the first Earth Day, a project suggested by Senator Gaylord Nelson, in which 2,000 universities, 10,000 schools, and more than 20 million people voiced their frustrations and suggested solutions to the nation's environmental problems. During this time there were legal pressures on the federal government, as well. The Environmental Defense Fund was formed in 1967, to be followed by the Natural Resources Defense Council and the Sierra Club Legal Defense Fund, all of which commonly sued the government. Though Nixon did not initiate any of the environmental or conservation legislation he signed, he was so unpopular over the war (and especially his ordering of the invasion of Cambodia in 1970) that he saw environmentalists as perhaps the only friends he had (Flippen 2003).

During this period, numerous challenges to the Migratory Bird Act continued in Canada. Canadian Wildlife Service Director David Munro ordered the RCMP to allow Indigenous people to hunt on their reservations and on unoccupied Crown land, but legal guarantees of traditional hunting rights were not legislated until the 1980s (Burnett 2003). From the 1940s to the 1970s, Mexico's human population more than doubled, bringing massive pollution problems. The Green Revolution, led by future Nobel Prize winner Norman Borlaug and funded by the Rockefeller Foundation, increased Mexico's crop productivity by 22%, but it required heavy use of fertilizers, pesticides, and herbicides and led to clearing forests and draining wetlands. In the 1950s, Enrique Beltran, who held a PhD in zoology from Columbia University and who served as Mexico's undersecretary of forestry and fauna, was director of the Institute de Recursos Naturales Renovables and sponsored wildlife and fisheries surveys, research, and conferences. Unfortunately, he opposed Leopold's land ethic, believing natural resources should be exploited primarily for their economic value. In fact, "between 1940 and 1970, Mexican governments abandoned conservation altogether" (Simonian 1995:131). Others might disagree considering the Mexican legislature's revision of the game laws in 1952 (Valdez et al. 2006).

In 1971 the United Nations Educational, Scientific, and Cultural Organization initiated the Man in the Biosphere Program for scientific study and the protection of biodiversity. Biosphere reserves consist of a core protected area surrounded by a managed use area encircled by zones of cooperation. All of these involve local people with attention to their cultural and economic needs to find a balance between nature and mankind. Mexico established their first two biosphere reserves in Durango in 1974, and eventually 85% of Mexico's protected areas fell under the biosphere reserve program. The World Wildlife Fund and Conservation International were instrumental in establishing the programs. Success varied, but the program contributed to the recovery of the Bolson tortoise (*Gopherus flavomarginatus*), Kemp's Ridley turtle (*Lepidochelys kempii*), and the Mexican crocodile (*Crocodylus moreletii*; Simonian 1995).

THE REAGAN ERA

Ronald Reagan, the former governor of California, became president in 1980, and he appointed James Watt as his secretary of interior. They sensed attitudes had changed in the previous decade. The conservation movement, interested in protecting scenic places and other resources, was largely outpaced by the environmental movement, interested in pollution, clean air, and clean water. The Earth First movement, some of which included environmental terrorism like burning logging equipment and spiking trees with nails, radicalized the environmental movement and led to a fracturing of mainstream and radical environmental groups. The Sagebrush Rebellion or wise use movement of western farmers and ranchers called for private control of public lands (Wellock 2007). The Reagan administration sought to dismantle the ESA via the Taskforce for Regulatory Reform. David Stockman, head of the federal Office of Management and Budget, recommended selling off national parks and forests to balance the federal budget. Morale in the federal agencies was low during this period, yet there was great support for established legislation from the memberships of the Sierra Club, Audubon Society, and The Wilderness Society, all of which had doubled their memberships during the previous decade.

Despite being reelected in 1984, Reagan was not able to dismantle the federal agencies or sell off public lands, and he eventually retreated from these efforts. In 1980 the Alaska National Interest Lands Conservation Act expanded the National Wildlife Refuge System by 21.4 million ha. Twenty-nine new wildlife refuges were added to the system, 200 plants and animals were added to the Endangered and Threatened Species List, and the ESA and Clean Water Act were reauthorized during Reagan's eight years in office (Pope 2003). A Fish and Wildlife Conservation Act was passed, to do for nongame species what the P-R and D-J Acts had done for game species. Unfortunately, no funding was provided. Funding for federal agencies and universities continued to lag for the next two decades. Nonetheless, conservation groups found a new source of revenue for preserving wildlife habitat and for restoration projects.

The federal Farm Bill establishes the authority and funding for the Department of Agriculture and its many crop subsidy and price support programs. It is renewed every five years. During the 1980s and '90s, conservation became more prominent each time the bill was renewed. Programs such as the Conservation Reserve Program, the Wetlands Reserve Program, the Wildlife Habitat Incentive Program, the Grasslands Reserve Program, and the Environmental Quality Incentives Program, all managed by the Natural Resources Conservation Service, provided private landowners with payments to take land out of production or to make conservation improvements. Most states developed similar programs (Wellock 2007).

THE 21ST CENTURY BECKONS

During the Clinton administration, Secretary of the Interior Bruce Babbitt felt that there was too much duplication among the scientific research projects by the many agencies of that department, so in 1996 he reassigned all Department of Interior scientists to a new organization called the National Biological Survey and later renamed it the Biological Resource Division of the US Geological Survey. Although that may have improved federal efficiency, the field resource managers of the NPS, the Bureau of Land Management, the USFWS, and others felt they had lost control over the research they needed. Subsequently, these agencies formed a coalition with university researchers to conduct their needed research, and additional educational and outreach programs. The organization, known as the Cooperative Ecosystem Studies Units, is composed of 270 universities organized into 17 regions and allows funding from more than a dozen federal agencies to flow to the academic researchers (Brown 2010).

But as we entered the 21st century, funding for nongame species was still inadequate. A concerted effort to correct the shortfall was developed and supported by more than 3,000 conservation organizations. Termed the Conservation and Reinvestment Act (CARA), it would have used a tax on outdoor recreational items like canoes, binoculars, and birdseed to provide $350 million a year to the states in much the same manner as the P-R and D-J Acts. The national effort to support this legislation was called Teaming with Wildlife. A few equipment manufacturers supported the endeavor, but many others strenuously opposed it. A compromise was reached in 2001 with "CARA-lite," now called the State Wildlife Grants

Program, which provides $50 million a year for the USFWS budget. The funds are awarded competitively and require a match of 25% to 50% from the states. Importantly, it required each state to develop a comprehensive wildlife conservation plan; plans were completed in 2005. This funding has increased each year, and some of the money is set aside specifically for Native American tribes (Brown 2010).

In Mexico, the end of the 20th century included the administrations of Miguel de la Madrid (1982–1988), Jose Lopez Portillo (1982–1988), and Carlos Salinas de Gortari (1988–1994). Public concerns over pollution and loss of forests and wildlife led to the formation of new nongovernmental organizations (NGOs), including Pronotura (1981), the Mexican Ecologists Movement (1981), and Biocenosis (1982). Under Portillo and Madrid, pollution continued unabated, largely owing to the oil boom and poor control of PEMEX, the state-owned oil company. Illegal trade in jaguars (*Panthera onca*), pumas, ocelots (*Leopardus pardalis*), Mexican parrots (*Amazona viridigenalis*), and other birds continued. Mexico refused to sign the CITES agreement. President Salinas, however, passed the General Law on Ecological Balance and Environmental Protection. In 1989, when it was determined that Mexican slaughterhouses had processed 35,000 Kemp's Ridley turtles (*Lepidochelys kempii*), Salinas banned the practice. In 1991 he signed CITES. The NGOs helped persuade the government to join Partners in Flight (1990) and the United States–Canada Waterfowl Management Plan (1994) to form the Sonoran and Rio Grande Joint Ventures. The North American Free Trade Agreement of 1994 required Mexico to establish the Commission on Environmental Protection, but its success has been hard to assess (Simonian 1995). In 2000 Mexico again passed a revised General Wildlife Law, said to be the most comprehensive wildlife legislation ever enacted in Mexico (Valdez et al. 2006:275). It covered a plethora of issues, including species at risk, ethics, landowner incentives, sustainable use, exotic species, reintroductions and translocations, cooperation among agencies, Indigenous peoples, critical habitat, wildlife scientific research, and law enforcement (Valdez et al. 2006).

In 1993 Canada overhauled its Migratory Bird Conservation Act and passed the Wild Animal and Plant Protection and Regulation of International and Interprovincial Trade Act. This consolidated 40 years of wildlife protection policy. In 2002 parliament passed the Species at Risk Act, modeled after the ESA (Gallardo 2018, Taber and Payne 2003).

CONSERVATION SUCCESSES AND STRUGGLES IN THE 21ST CENTURY

In 2008 Barack Obama was elected president of the United States. Many of his conservation initiatives were by executive order and often opposed by the Republican legislature. Over his two terms in office, he often used the Antiquities Act of 1906, which did not require additional congressional approval. He expanded the marine sanctuaries in the Pacific and Atlantic and strengthened laws concerning illegal marine fishing and sea-

food mislabeling. Through a variety of orders and programs, he added 222 million ha to the United States' protected lands, nearly double the record of Theodore Roosevelt. He added 22 new national parks, 6 national monuments, and 4 conservation areas. His administration declared the bison as the "national mammal" and initiated the National Landscape Conservation System, modernized planning of the National Forests, expanded the Arctic National Wildlife Refuge by 4.9 million ha, withdrew millions of hectares of public land from mining and oil and gas claims, and introduced the Sage Grouse Initiative, the Comprehensive Everglades Restoration Program, the Great Lakes Restoration Initiative, the Gulf Coast Restoration Task Force, and a number of other environmental and conservation programs. Some of these, however, were not without controversy. The 5,500-ha Bears Ear National Monument in Utah was proposed by the Bears Ear Intertribal Coalition and approved by President Obama after the 2016 election. Republicans opposed it, and President Donald Trump reduced the size of the protected area by 85% (Northcott 2017). Likewise, the 7,610-ha Grand Staircase-Escalante National Monument in Utah, approved by President Clinton in 1996, was reduced by 47% by the Trump administration in 2017. Most of these changes were done by executive order, and time will tell if the Biden administration will reverse some or many of these changes. On the same day as he approved the Bears Ear Monument, President Obama approved the 121,000-ha Gold Butte National Monument in Nevada. The Paiute Indians recommended this monument, but it was opposed because it was adjacent to land owned by Cliven Bundy (Department of the Interior 2016).

Cliven Bundy was a rancher who had been grazing his cattle on federal lands for nearly 20 years without paying the required fees when federal agents arrived to confiscate his herd in 2014. By that time, he owed the government over $1 million in fees and fines. Hundreds of Tea Party supporters and armed militiamen rallied to his side. The government eventually backed down and returned his cattle. In 2016 Bundy's son, Ammon Bundy, led an armed occupation of the Malheur National Wildlife Refuge in Oregon, claiming that the federal government did not have a right to the land. After about a month, 27 militants were arrested, and one person was killed. These confrontations were part of the Sagebrush Rebellion, a term for the opposition of western ranchers to federal ownership of grazing lands and the fees charged by the Bureau of Land Management (National Public Radio 2020).

As of 2016, there were 566 recognized Native American tribes in the United States on 300 reservations managing over 22 million ha of lands in the lower 48 states and 18 million ha in Alaska (Czech 1995). The Navaho holdings are the largest, with 2.9 million ha. These tribes and their lands are all considered sovereign nations, and each has its own form of government. The Consultation and Coordination with Indian Tribal Governments Act of 2000 requires the US government to consult with tribes over any policies that might affect them. The lands on reservations are a fragmented mix of tribal, federal, and private holdings, mostly held in trust by the Bureau of Indian Affairs (Harkin and Lewis 2007, Blackwell 2018). Over the years, US federal courts have decided both for and against Native American rights concerning wildlife. Most of these conflicts have been over recreational and subsistence hunting and fishing, including the hunting of polar bears, seals, and gray whales (*Eschrichtius robustus*) by Inuit tribes, and the capture and collection of feathers of bald and golden (*Aquila chrysactos*) eagles (Hoagland and Voirin 2018). As early as 1971, the Alaska Native Claims Act gave 12% of the state's land and $1 billion for Native Americans. The 1980 Alaska National Interest Conservation Claims Act allowed for subsistence hunting for both native and non-native Alaskans. This concept was strengthened in 1990 with the Federal Subsistence Management Program, which established a Federal Subsistence Board to oversee this hunting and fishing (Hughes 1996, Kessler 1998). Many tribes also obtain substantial income from guided trophy game hunts as well as selling of fish.

Native Americans were major players in the opposition to oil and gas pipelines in the United States and Canada. In 2016 the Standing Rock and Cheyenne River Sioux opposed the $3.7 billion, 1770-km-long Dakota Access pipeline over fears it would contaminate their water supply and damage their hunting and fishing areas. Hundreds of Native Americans and their supporters physically blocked the construction, and the protest won international attention when police used water cannons during freezing weather and attack dogs to disperse the crowd. Nearly 500 people were arrested, 300 were injured, and 1 was killed. The pipeline was completed in 2017, but a federal court shut it down in 2020 (BBC News 2016). Subsequently, North Dakota and three other states criminalized protests against oil and gas pipelines (Gurley 2020). In 2020 Duke Energy and Dominion Energy proposed the Atlantic Coast Pipeline. The Lumbee, Coharie, and Haliwa-Sponi tribes joined environmental and conservation groups, basing their opposition on environmental issues. The pipeline would cross 51 waterways and the Appalachian Trail. The plan was abandoned in 2020, though the concerns of the tribes were only marginally considered because they were not federally recognized tribes (Gurley 2020).

In Canada, the huge land area and relatively small population has made government–First Nation interactions at least seem less controversial. The Constitution Act of 1982 recognized specific categories of Indigenous people and declared that their rights and cultures must be considered and respected in natural resource decisions. The 2015 Truth and Reconciliation Commission listed the physical, cultural, and biological effect of genocide and provided reserves to the Indigenous people, though that land is still owned by the Crown (Muir and Booth 2018).

In addition to reducing the size of some national monuments, the Trump administration, elected in 2016, attacked environmental and conservation policies with vigor. In 2020, near the end of President Trump's term, a list of 100 environmental and conservation rules being reversed or weak-

ened was published. Some had been reversed, some were in progress of being reversed, and the Supreme Court had overturned some. They included 27 air pollution and emission rules, including the Clean Air Act and efforts to reduce air pollution in national parks and wildlife refuges, as well as the withdrawal from the Paris Climate Agreement; 19 drilling and extraction rules, including voiding the ban on drilling in the Arctic National Wildlife Refuge; 13 infrastructure and planning rules, including lifting the freeze on coal leases on public lands and approval of logging in the Tongrass National Forest; 12 wildlife rules, including weakening of the Endangered Species Act, the National Environmental Policy Act, the US-Mexico-Canada Treaty on Protection of Migratory Birds, and sage grouse (*Centrocercus urophasianus*), marine mammal, sea turtle (Testudines), and bluefin tuna (*Thunnus thynnus*) protection; 11 water pollution rules, including weakening of the Clean Water Act; and 18 toxic substance and safety rules. President Trump's effort to remove the Yellowstone grizzly bear from the Endangered Species List was reversed by a federal judge. Many if not most of these rollbacks are being contested in court by environmental and conservation groups and Indian tribes. Some of the reversals were overturned and were being appealed by Trump administration attorneys (Malakoff 2018, Popovich et al. 2020). As of early 2021, as the new Biden administration assumes office, the status of many of these weakened rules is unclear.

Nonetheless, in late 2020, a significant piece of legislation was passed by both houses of Congress and signed by President Trump. The Great American Outdoors Act will provide $9.5 billion over five years to address infrastructure projects at national parks and wildlife refuges managed by the National Park Service, Bureau of Land Management, US Fish and Wildlife Service, and the US Forest Service. The Land and Water Conservation Fund will receive $900 million in perpetuity, of which $15 million will be allocated for increasing public access for hunting, fishing, and other outdoor recreation. This fund, which is largely used for land acquisition, was established in 1965 with a cap of $900 million per year, but that goal was rarely met. Previously all of the funds came from fees and royalties from offshore oil and gas drilling, motorboat fuel, and sale of excess federal real estate. Over $22 billion have been diverted to other uses over the years. The Great American Outdoors Initiative was an Obama administration proposal in 2010 (Council on Environmental Quality 2012). The Pitman-Robertson Modernization Act of 2020 expanded the uses of the funds to include marketing and recruiting of hunters and recreational shooters to slow or reverse the decline in hunting participation in the United States (Congress.gov 2020*a*). Currently under consideration, the Recovering America's Wildlife Act would provide $1.4 billion per year in additional funds to states and tribes for recovering fish and wildlife species of greatest concern (Congress.gov 2020*b*). Also, the 21st Century Conservation Corps Act, which is a partial response to the recession caused by the COVID-19 pandemic, would provide over $40 billion to employ veterans and other young people on service projects to address conservation, forestry, recreational, and infrastructure needs of our national parks and forests (Wyden 2020). Finally, the American Conservation Enhancement Act would form local, state, and federal partnerships to restore aquatic habitats for recreational fishing, allow lead fishing tackle for five more years, and provide funds for the health of the Chesapeake Bay (Barrasso 2020).

CONTINUING PROBLEMS AND NEW OPPORTUNITIES

The US government owns or controls 148 million ha, or about 16% of the total landmass of the country; states own or control another 79 million ha (Lubowski et al. 2006). The United States, Canada, and Mexico now have just about all the public land they will ever have, and our wildlife resources are limited and in most cases declining. The human population continues to grow, using more food, goods, and services. Forests and farmlands are being converted for commercial uses at alarming rates, removing wildlife habitat and fragmenting what is left. Water quality is a concern, and water quantity is restricting development and wildlife habitat in some areas. The infrastructure in our national parks and monuments is eroding, and decades of fire suppression along with global warming have led to devastating fires in national parks and forests. Overpopulation of some species, game ranching, and international travel have led to outbreaks of wildlife–domestic and animal–human diseases in our wild mammal and bird populations, including the West Nile virus, avian influenza, chronic wasting disease, brucellosis, and tuberculosis. Today, Canada has lost more than 16 million ha of forests and habitat to the mountain pine beetle (*Dendroctonus ponderosae*), Colorado and Wyoming have lost 1.5 million ha, and there is no end in sight. It is estimated that 80% of Mexico's landmass suffers from erosion and over 40% of Mexico's population lives in poverty, often turning to wildlife for sustenance (Valdez et al. 2006).

Figure 2.5. Wolf releases in the Yellowstone area have been remarkably successful but not without controversy. Courtesy of US Fish and Wildlife Service

Our economies thrive on energy, and there are increased political efforts to expand oil and gas exploration in our national parks and refuges and off our coasts. The process of fracking to release natural gas deposits, and extraction of oil from sand and shale, may be environmentally unviable. Hunting, although still generally supported by the nonhunting public, is declining as a recreational activity, although trophy wildlife hunting has brought substantial financial revenues to parts of all three countries. Animal rights and welfare groups, such as the People for the Ethical Treatment of Animals, have organized public support to ban trapping in some states and hunting of carnivores such as bears and mountain lions. Human–wildlife conflicts have increased, including millions of deer killed on our highways and mountain lion and bear attacks on humans each year. There are still political threats to the future of the Endangered Species Act, and some hunting groups favor the private ownership of wildlife, even to the extent of allowing artificial insemination of wild deer and cloning of lucrative species. Arguments over listing polar bears or delisting wolves and grizzly bears continue. Enforcement of wildlife laws is often inadequate, as state and federal agencies have not had the funds to hire new employees. Climate change threatens the world's ecosystems, and yet we are slow to respond to the threat of global warming. Funding for wildlife research in the United States in 2020 remained at about the 1980s level, as the wars in Iraq and Afghanistan continued to absorb much of the federal budget. The 2008 recession hit all three countries, affecting funding for wildlife as well as conservation and environmental efforts. Then the 2020–2021 COVID-19 pandemic and concomitant recession occurred, and all three countries increased their national debts by massive amounts.

Mexican gray wolves (*Canis lupus baileyi*), lesser long-nosed bats (*Leptonycteris yerbabuenae*), black-tailed prairie dogs, and even Monarch butterflies (*Danaus plesippus*) share habitat across the border between the United States and Mexico, yet collaboration on recovery efforts is stymied by immigration and drug-smuggling issues. Smugglers of people and drugs (and the law enforcement officers sent to interdict them) disrupt wildlife movement and damage habitat. The 2006 US Secure Fence Act called for 1,127 km of 3.1-m fences separating Mexico and the United States, further restricting movement of wildlife. The Trump administration's plans to build a 9-m-tall wall along the entire US-Mexico border threatened wildlife movement of both predators and prey.

Conversely, the public considers the environment, natural resources, and wildlife to be important to their quality of life. Recent books by Richard Louv, *Last Child in the Woods* (2006) and *The Nature Principle* (2011), have raised public awareness of the importance of both children and adults connecting with nature if we are to remain psychologically well in a complex and technological world. Other books, like Jane Goodall's *Seeds of Hope* (2014), Ray and Ulrike Hilborn's *Ocean Recovery* (2019), and E. O. Wilson's *Half-Earth: Our Planet's Fight for Life* (2016), offer plans for saving wildlife and habitats around the world. Television shows like those on the Discovery, Smithsonian, History, and National Geographic channels, including popular hunting and fishing shows and their counterparts on YouTube and other social media, have opened many eyes to the wonderment of nature. Funding for scientific research, such as through the National Science Foundation and the National Institutes of Health, continues to increase, though slowly, in 2020. New technologies like satellite radio-tracking, acoustic tracking of fish, geographic information system mapping, the use of drones for animal and habitat censusing, DNA genetic analysis, and eDNA survey techniques provide new tools for the study of conservation science. The development by the USFWS of the Safe Harbor concept, wherein landowners are not penalized if their managed lands attract endangered species, has provided flexibility to farmers and ranchers. Federal fishery management agencies have shifted from single-species management to ecosystem-based fishery management, which includes fish, birds, turtles, marine mammals, and reptiles (Sanchiricoa et al. 2007). Some species once at risk of disappearing—bison, timber and gray wolves (Fig. 2.5), bald eagles, peregrine falcons, American alligators, grizzly bears, ospreys, sea turtles, and eastern brown pelicans—have made spectacular comebacks. There are more deer and turkeys in the United States now than ever before. Canada reached its goal of 12% of its land in wilderness areas by 2000, and there are now numerous collaborative agreements across Canada, the United States, and Mexico to protect wildlife, from CITES to the North American Waterfowl Management Plan to the Black-Tailed Prairie Dog Conservation Action Plan. Private landowners, looking toward hunting leases and ecotourism as means of income, see the value of conserving their wildlife and wildlands, and state and federal agencies acknowledge the importance and value of private lands for the conservation of game and nongame wildlife (Benson et al. 1999). The *Conservation Directory* (2017), published annually by the National Wildlife Federation, lists more than 1,600 private conservation organizations in the United States and Canada. Mexico now has more than 200 conservation NGOs (Valdez et al. 2006). Hundreds of private land trusts and local governments raise money to purchase important habitat land for greenways or permanent conservation easements to restrict commercial development. Slowly, the concept of paying private landowners for "ecosystems services" is developing in the United States. Renewable energy, in terms of wind, solar, and ocean-wave power, is becoming affordable and popular. Initial experiments with ridding the oceans of trash have begun. A fledgling carbon-trading market has already developed, and some entities are beginning to pay landowners for their watersheds and endangered species habitat.

On 20 January 2021, a new presidential administration took office in the United States. Within the first two weeks, President Joe Biden appointed a new secretary of the interior, new directors of the EPA and USFWS, and many other administrators and their deputies. Many of those positions must be approved by the Senate. President Biden issued a flurry of executive orders that canceled many of President Trump's executive orders, especially those concerning oil and gas drill-

ing and others affecting climate change initiatives. There are indications that there will be major initiatives concerning climate change, environmental justice, green energy, and other environmental and conservation topics in the future of this administration. One must be cautious, however, to recognize that rule-making by executive order may last for only one presidential administration. The next administration may rescind any or all such orders.

LESSONS LEARNED

The North American Model of Wildlife Conservation, based on the public trust doctrine of public ownership of wildlife, has a checkered history, but in general it has served society well in the United States and Canada. Clearly, wildlife in North America does not know international boundaries any more than they know state or provincial ones. Our three countries share thousands of kilometers of common borders, and if we are to conserve our shared wildlife, we need to understand each other and share at least some common goals.

The North American model holds that wildlife is a public and international resource managed by policies based on sound science. Under this model, the killing of wildlife should only occur for legitimate purposes, the allocation of wildlife is by law, markets for game are eliminated, and the democracy of hunting is standard (Organ et al. 2012). Though wildlife is considered an international resource, even in North America, there should be no commercial markets for wildlife, and wild birds and animals should only be killed for legitimate reasons. Hunting has been an essential component of this model, but democracy means that there is equitable access to game (The Wildlife Society 2010).

Unfortunately, few modern-day Americans, Canadians, or Mexicans know of or understand this history. Wildlife conservation is a political issue with multiple stakeholders. Conservation issues—such as drilling for oil in the Arctic National Wildlife Refuge, control of urban deer or feral cats, or shooting wolves—become highly emotional in our societies. As our population becomes more urbanized and children spend less time outdoors, the public, though supportive of conservation, becomes less knowledgeable and less able to make informed decisions. It is thus critical that we in the wildlife conservation profession provide public and private educational programs to inform our citizens about the choices before us. As Richard Stengle, editor of *Time* magazine, stated in 2006, "being an American is not based on a common ancestry, a common religion, even a common culture—it's based on accepting an uncommon set of ideas. And if we don't understand those ideas, we don't value them; and if we don't value them, we don't protect them" (Posewitz 2010:32).

SUMMARY

The history of wildlife conservation in Canada, the United States, and Mexico is a history of human migration, changing values, and associated behavior. Native Americans, having arrived about 10,000–22,000 years ago, exploited wildlife for their food, clothing, shelter, tools, weapons, and even fuel. Wildlife populations were sometimes diminished because of hunting, more so around areas of human concentration with large and elaborate agricultural developments. Upon the arrival of Europeans, Native American populations crashed, mostly because of diseases brought by immigrants and their livestock. Wildlife populations rebounded, only to be soon reduced significantly owing to market hunting and trapping for meat, fur, and feathers. As early as the 16th century, Spanish kings enacted laws in Mexico to limit harvests and to protect animals, birds, and forests.

Those laws and similar ones enacted in Canada and the United States in later centuries had limited effect. It was not until the late 1800s, when more affluent North Americans noted the extirpation of some wildlife species and near loss of others, that effective laws were passed and enforced. Once game meat was replaced by domestic meat production and felt, cotton, silk, and wool replaced fur and feathers, hunting became a sport rather than a commercial enterprise. Under the leadership of President Theodore Roosevelt, many influential individuals, sportsmen's groups, the media, and conservation groups rallied to protect and manage land and wildlife in the United States with laws and refuges. Canada followed with similar laws, but with a vast land area and relatively small human population, the stress of human effects on wildlife populations was only moderate. Mexico's wildlife populations suffered from greater human effect and weaker enforcement of wildlife protection laws. This may be partially due to Mexico's tumultuous political history, its lack of economic development as compared to its northern neighbors, and a continuing attitude among the public that natural resources need to be exploited for the survival of its people.

Overall, the history of wildlife conservation in North America is a time line of the evolution of what we now call the North American Model of Wildlife Conservation. It is largely based on our European ancestry, with imbued values of public ownership of wildlife—management by policies based on science and protected by laws enforced by the state and federal government. As our human population continues to increase, and as our consumption of goods per capita continues to rise, only time will tell if these values can be sustained.

Literature Cited

Adams, W. M. 2004. Against extinction: the story of conservation. Earthscan, London, UK.

Barnes, I. 2019. The historical atlas of Native Americans. Chartwell Books, New York, New York, USA.

Barrasso, J. 2020. America's conservation enhancement act. https://www.congress.gov/bill/116th-congress/senate-bill/3051.

BBC News. 2016. The Dakota access pipeline. https://www.bbc.com/news/world-us-canada-37322266.

Benson, D. E., R. Shelton, and D. W. Steinbach. 1999. Wildlife stewardship and recreation on private lands. Texas A&M University Press, College Station, USA.

Blackwell, J. 2018. Policy and laws relating to tribal wildlife management. Pages 392–399 in B. D. Leopold, W. B. Kessler, and J. L. Cummins, editors. North American wildlife policy and law. Boone & Crockett Club, Missoula, Montana, USA.

Brinkley, D. 2009. The wilderness warrior: Theodore Roosevelt and the crusade for America. Harper Collins, New York, New York, USA.

Brinkley, D. 2011. The quiet world: saving Alaska's wilderness kingdom, 1879–1960. Harper Collins, New York, New York, USA.

Brown, R. D. 2010. A conservation timeline: milestones of the model's evolution. Wildlife Professional 4:28–31.

Burnett, J. A. 2003. A passion for wildlife: the history of the Canadian Wildlife Service. University of British Columbia Press, Vancouver, Canada.

Carson, R. 1941. Under the sea-wind. Penguin Books, London, UK.

Carson, R. 1955. The edge of the sea. Mariner Books, Boston, Massachusetts, USA.

Carson, R. 1961. The sea around us. Open Road, New York, New York, USA.

Carson, R. 1962. Silent spring. Houghton Mifflin, Boston, Massachusetts, USA.

Colpitts, G. 2002. Game in the garden: a human history of wildlife in western Canada to 1940. University of British Columbia Press, Vancouver, Canada.

Congress.gov. 2020a. The P-R modernization act. https://www.congress.gov/bill/116th-congress/house-bill/877.

Congress.gov. 2020b. The recovering America's wildlife act. https://www.congress.gov/bill/116th-congress/house-bill/3742.

Council on Environmental Quality. 2012. America's great outdoors initiative. https://obamawhitehouse.archives.gov/admoinistration/eop/ceq/initiatives/ago.

Cronon, W. 2003. The trouble with wilderness; or, getting back to the wrong nature. Pages 213–243 in L. S. Warren, editor. American environmental history. Blackwell, Malden, Massachusetts, USA.

Czech, B. 1995. American Indians and wildlife conservation. Wildlife Society Bulletin 23(4):568–573.

Darwin, C. 1859. The origin of the species. P. F. Collier and Sons, New York, New York, USA.

Darwin, C. 1871. The descent of man, and selection in relation to sex. P. Appleton, New York, New York, USA.

Denevan, W. 2003. The pristine myth: the landscape of the Americas in 1492. Pages 5–42 in L. S. Warren, editor. American environmental history. Blackwell, Malden, Massachusetts, USA.

Department of Interior. 2016. Fact sheet: the Obama administration's conservation record. https://www.doi.gov/news/upload/obama-administration-fact-shet-on-conservation.pdf.

Druschka, K. 2003. Canada's forests: a history. Forest History Society Issues Series. McGill-Queen's University Press, Montreal, Quebec, Canada.

Ducks Unlimited. 2021. Great value in duck stamps. www.ducks.org.

Dunbar-Ortiz, R. 2019. An Indigenous people's history of the United States. Beacon Press, Boston, Massachusetts, USA.

Dunlap, T. R. 1988. Saving America's wildlife. Princeton University Press, Princeton, New Jersey, USA.

Ehrlich, P. 1968. The population bomb. Ballantine Books, New York, New York, USA.

Emerson, R. W. 1836. Nature. James Monroe, Boston, Massachusetts, USA.

Flippen, J. B. 2003. Richard Nixon and the triumph of environmentalism. Pages 272–297 in L. S. Warren, editor. American environmental history. Blackwell, Malden, Massachusetts, USA.

Galeano, E. 1997. Open veins of Latin America. Monthly Review Press, New York, New York, USA.

Gallardo, J. C. 2018. Wildlife policy and law in Mexico. Pages 343–358 *in* B. D. Leopold, W. B. Kessler, and J. L. Cummins, editors. North American wildlife policy and law. Boone & Crockett Club, Missoula, Montana. USA.

Geist, V. 1995. North American policies of wildlife conservation. Pages 75–129 *in* V. Geist and I. McTaggert-Cowan, editors. Wildlife conservation policy. Detselig Enterprises, Calgary, Alberta, Canada.

Geist, V., S. Mahoney, and J. Organ. 2001. Why hunting has defined the North American model of wildlife conservation. Transactions of the North American Wildlife and Natural Resources Conference 66:175–185.

Ginger, C. 2020. Everything you need to know about the Vikings. Future Publishing, Bournemouth, Dorset, UK.

Goodall, J. 2014. Seeds of hope. Grand Central Publishing, New York, New York, USA.

Gottlieb, R. 2003. Reconstructing environmentalism: complex movements, diverse roots. Pages 245–270 *in* L. S. Warren, editor. American environmental history. Blackwell, Malden, Massachusetts, USA.

Gurley, G. 2020. Native Americans hail oil and gas pipeline decisions. https://prospect.org/environment/native-americans-halt-oil-and-gas-pipeline-decisions.

Hardin, G. 1968. The tragedy of the commons. Science 162:1243–1248.

Harkin, M. E., and D. R. Lewis. 2007. Native Americans and the environment: perspectives on the ecological Indian. University of Nebraska Press, Lincoln, USA.

Hewitt, G. 1921. The conservation of the wild life in Canada. C. Scribner's Sons, New York, New York, USA.

Hilborn, R., and U. Hilborn. 2019. Ocean recovery. Oxford University Press, Oxford, UK.

Hoagland, S. J., and C. Voirin. 2018. Relationships of Indigenous peoples to natural resources. Pages 103–113 *in* B. D. Leopold, W. B. Kessler, and J. L. Cummins, editors. North American wildlife policy and law. Boone & Crockett Club, Missoula, Montana, USA.

Hornaday, W. T. 1913. Our vanishing wildlife: its extermination and preservation. Kissinger Legacy Reprints, Whitefish, Montana, USA.

Hughes, J. D. 1996. North American Indian ecology. 2nd ed. Texas Western Press, University of Texas, El Paso, USA.

Kay, C. E. 1998. Are ecosystems structured from the top-down or bottom-up: a new look at an old debate. Wildlife Society Bulletin 23:484–498.

Kessler, S. 1998. Use of natural resources for subsistence in Alaska: concepts, policy and law. Pages 116–121 *in* B. D. Leopold, W. B. Kessler and J. L. Cummins, editors. North American wildlife policy and law. Boone & Crockett Club, Missoula, Montana, USA.

Krech, S., III. 1999. The ecological Indian: myth and history. W. W. Norton, New York, New York, USA.

Leopold, A. 1933. Game management. University of Wisconsin Press, Madison, USA.

Leopold, A. 1949. A Sand County almanac. Oxford University Press, New York, New York, USA.

Lewis, J. G. 2011. The Weeks Act at 100: the "organic act" of the eastern national forests. Forest Landowner 69:22–27.

Lopez-Hoffman, L., E. D. McGovern, R. G. Varady, and K. W. Flessa. 2009. Conservation of shared environments: learning from the US and Mexico. University of Arizona Press, Tucson, USA.

Louv, R. 2006. Last child in the woods: saving our children from nature-deficit disorder. Algonquin Books, Chapel Hill, North Carolina, USA.

Louv, R. 2011. The nature principle. Algonquin Books, Chapel Hill, North Carolina, USA.

Lubowski, R. N., M. Vesterby, S. Bucholtz, A. Baez, and M. Roberts. 2006. Major uses of land in the United States, 2002. USDA Economic Information Bulletin. May:1–54.

Lueck, D. 2008. Wildlife: sustainability and management. Pages 133–174 *in* R. A. Sedjo, editor. Perspectives on sustainable resources in America. Resources for the Future Press, Washington, DC, USA.

Mackie, R. J. 2000. History of management of large mammals in North America. Pages 292–320 *in* S. Demarais and P. R. Krausman, editors. Ecology and management of large mammals in North America. Prentice-Hall, Upper Saddle River, New Jersey, USA.

Mahoney, S. P., and V. Geist. 2019. The North American model of wildlife conservation. Johns Hopkins University Press, Baltimore, Maryland, USA.

Malakoff, D. 2018. Trump's new oceans policy washes away Obama's emphasis on conservation and climate. https://www.sciencemag.org/news/2018/06/trump-s.

Mann, C. C. 2011. 1491: new revelations of the Americas before Columbus. Vintage Books, New York, New York, USA.

Martin, P. S. 2002. Prehistoric extinctions. Pages 1–27 *in* C. E. Kay and R. T. Simmons, editors. Wilderness and political ecology—aboriginal influences and the original state of nature. University of Utah Press, Salt Lake City, USA.

Muir, B. R., and A. L. Booth. 2018. Musckat helped make the world: priority rights of aboriginal peoples to wildlife in Canada. Pages 311–329 *in* B. D. Leopold, W. B. Kessler, and J. L. Cummins, editors. North American wildlife policy and law. Boone & Crockett Club, Missoula, Montana, USA.

National Public Radio. 2020. Ammon Bundy is arrested and wheeled out of the Idaho statehouse—again. https://www.npr.org/2020/08/25/9060469.

National Wildlife Federation. 2017. Conservation directory. Island Press, Reston, Virginia, USA.

Nielsen, L. A., 2017. Nature's allies: eight conservationists who changed our world. Island Press, Washington, DC, USA.

Northcott, C. 2017. Obama's historic conservation legacy beats Teddy Roosevelt. https://www.bbc.com/news/world-us-canada-38311093.

Organ, J. F., V. Geist, S. P. Mahoney, S. Williams, P. R. Krausman, et al. 2012. The North American model of wildlife conservation. Wildlife Society Technical Review 12-04. The Wildlife Society, Bethesda, Maryland, USA.

Organ, J. F., S. P. Mahoney, and V. Geist. 2010. Born in the hands of hunters: the North American model of wildlife conservation. Wildlife Professional 4:22–27.

Peterson, R. T. 1934. A field guide to birds. Houghton Mifflin, Boston, Massachusetts, USA.

Pope, C. 2003. The politics of plunder. Pages 325–327 *in* L. S. Warren, editor. American environmental history. Blackwell, Malden, Massachusetts, USA. [Reprinted from Boyer, P. 1990. Reagan as president: contemporary views of the man, his politics and his policies. Ivan R. Dee, Chicago, Illinois, USA.]

Popovich, N., L. Albeck-Ripka, and K. Pierre-Louis. 2020. The Trump administration is reversing 100 environmental rules. Here's the full list. https://nytimes.com/interactive/2020/15july2020/.

Posewitz, J. 2010. The hunter's ethic: the past, the peril and the future. Wildlife Professional 4:32–34.

Sanchiricoa, J. N., M. D. Smith, and D. W. Lipton. 2007. An empirical approach to ecosystem-based fishery management. Ecological economics 64:586-596.

Seton, E. T., and C. G. D. Roberts. 1909. Life histories of North American mammals: an account of the mammals of Manitoba. Volume 1. Grass eaters. Constable, London, UK.

Simonian, L. 1995. Defending the land of the jaguar: a history of conservation in Mexico. University of Texas Press, Austin, USA.

Taber, R. D., and N. F. Payne. 2003. Wildlife conservation and human welfare: a US and Canadian perspective. Krieger, Malabar, Florida, USA.

Taylor, A. 2003. Wasty ways: stories of American settlement. Pages 102–124 in L. S. Warren, editor. American environmental history. Blackwell, Malden, Massachusetts, USA.

The Wildlife Society. 2010. The public trust doctrine: implications for wildlife management and conservation in the US and Canada. Technical Review 10-01. Bethesda, Maryland, USA.

Thoreau, H. D. 1854. Walden. Tichnor and Fields, Boston, Massachusetts, USA.

Trefethen, J. B. 1975. An American crusade for wildlife. Boone & Crockett Club, Missoula, Montana, USA.

Valdez, R., J. C. Guzman-Aranda, F. J. Abarca, L. A. Tarango-Arambula, and F. C. Sanchex. 2006. Wildlife conservation and management in Mexico. Wildlife Society Bulletin. 34:270–282.

Wade, L. 2019. Ancient site in Idaho implies first Americans came by sea. Science 365:848–849.

Waters, M. R. 2019. Late Pleistocene exploration and settlement of the Americas by modern humans. Science 365:east5447. doi:10.1126/science.aat5447.

Wellock, T. R. 2007. Preserving the nation: the conservation and environmental movements, 1870–2000. Harlan Davidson, Wheeling, Illinois, USA.

Wilson, E. O. 2016. Half-earth: our planet's fight for life. W. W. Norton, New York, New York, USA.

Wilson, R. L. 2009. Theodore Roosevelt, hunter-conservationist. Boone & Crockett Club, Missoula, Montana, USA.

Wyden, R. 2020. The 21st century conservation corps for our health and our jobs act. https://www.wyden.senate.gov/imo/media/doc/21stcenturyconservationcorpsforourhealth andourjobsactof2020.

3

THE WILDLIFE PROFESSIONAL

JOHN F. ORGAN

INTRODUCTION

Wildlife management is a young profession. Leopold (1933) identified precursors involving controls on hunting that date back to biblical times. Indeed, one might construe the first human attempts to domesticate wild animals circa 10,000 BP to be the original precursor of wildlife management (Diamond 1997). Wildlife management in its current form in North America arose out of a unique set of social and environmental circumstances. The wildlife conservation movement in North America began in earnest in the mid-19th century, when organized sport hunters paved the way for restrictive laws and regulations designed to eliminate commercial market hunting, a by-product of the Industrial Revolution, and to curtail pot hunting (i.e., the rural practice of shooting purely for food; Reiger 1975, Trefethen 1975).

Until around 1905 the dominant paradigm was to perpetuate wildlife through restrictive laws and regulations, predator control, protected areas, and restocking. Science as a tool in wildlife management was a novel concept at the onset of the 20th century. Early naturalists discovered, cataloged, and described wildlife. President Thomas Jefferson, in commissioning the Lewis and Clark Expedition, instructed them to describe "the animals of the country generally and especially those not known in the U.S.; the remains and accounts of any which may be deemed rare or extinct" (Lewis and Clark 2002:xxix). The Division of Economic Ornithology and Mammalogy was formed in the US Department of Agriculture in 1885, and its first chief, C. Hart Merriam, dedicated its resources to the discovery and cataloging of birds and mammals of the western United States.

Many naturalists realized that species were not like planets and geologic strata; they were different because civilizations can destroy them. George Bird Grinnell, a Yale-educated naturalist, accompanied George Armstrong Custer on his first expedition into the Black Hills in present-day South Dakota in 1874. Nearly 10 years later, Harvard-educated naturalist Theodore Roosevelt would embark on his own expedition to the Badlands of present-day North Dakota and publish a book about his experiences (Roosevelt 1885). Grinnell produced a review of the book (Grinnell 1885) that contained some criticisms, prompting a meeting between them. That meeting spawned a realization that exemplified the emerging paradigm shift among naturalists who recognized that civilization's conquest was leading to the extirpation of species.

Later, as president of the United States, Roosevelt articulated the need for scientifically based wildlife management in what Leopold termed the Roosevelt doctrine (Leopold 1933). In short, the Roosevelt doctrine states:

1. All outdoor resources are one integral whole.
2. Conservation through wise use is a public responsibility, and private ownership is a public trust.
3. Science is the proper tool for discharging that responsibility.

Important milestones in the application of science to managing wildlife occurred in the 1920s with Herbert Stoddard's work on northern bobwhite (*Colinus virginianus*) in Georgia and Aldo Leopold's game surveys (Leopold 1931), funded by the Sporting Arms and Ammunition Manufacturers Institute. These science-based investigations led to another realization: restrictive laws, predator control, refuges, and stocking were not enough to stem the decline of wildlife. A program of active restoration was needed. Leopold chaired a committee of leading conservationists who published *The American Game Policy* (1930), which called for advancements including the establishment of a wildlife management profession. At that time, there were no university programs teaching wildlife management or any organizations promoting professional standards. Within 10 years, the first university programs would be established at the University of Wisconsin and the University of Michigan, the first textbook on the topic published (Leopold 1933), and the first professional scientific society for wildlife biologists established—The Wildlife Society (TWS), established in 1937.

At its inception as a profession, wildlife management was considered an art practiced by people with scientific training. During the early formative years of the wildlife management profession, three subdisciplines were recognized: game research, game administration, and game keeping (King 1938). This narrow scope was reflective of the dynamics that surrounded the movement to refocus wildlife conservation on restoration programs. These subdisciplines were considered interrelated, yet many believed they could not be combined into a single undergraduate curriculum. Leopold (1939) described a wildlife professional as an individual with an intense conviction of the need for and usefulness of science as a tool for the accomplishment of conservation; the ability to diagnose the landscape to discern and predict trends in its biotic community and to modify them where necessary in the interest of conservation; knowledge of plants, animals, soil, and water; and familiarity with other professions and their influence and effect on the landscape. Wildlife professionals in these early years were primarily public employees, and the lack of private employment and private lands management was viewed as a weakness of the profession (Leopold 1940). Additionally, it was presumed that the wildlife professional was male, Caucasian, from a rural background, and a hunter.

During the 1950s, a second generation of wildlife professionals emerged, and the concept of the wildlife professional showed signs of maturation. McCabe (1954) redefined wildlife managers as wildlife ecologists and stated unequivocally that they are scientists and not artisans. He further stated that a wildlife professional must have an ethical code and sense of aesthetic values toward conservation as a whole in addition to knowledge and skills acquired from academic training. Murie (1954) expanded upon the notion of professional ethics and deemed it a responsibility of every member of the wildlife profession. Murie framed his argument in the context of ethical thinking that was ongoing in society at that time. The post–World War II era in America saw a new social consciousness brought about by recent genocide, nuclear proliferation, communist expansion, and a realization that the country was not as insulated from global conflict as once thought. Murie viewed this social consciousness as an effort by people to try to understand their proper place in nature, and he felt the highest calling of the wildlife professional was to contribute to that understanding. Leopold predicted that the fusion of wildlife biology and social sciences would be the outstanding accomplishment of the 20th century (Meine 1988); Murie viewed people to be on equal footing with animals, plants, soil, and water as the fundamental knowledge base and realm of responsibility for the wildlife professional.

During the 1960s, an increasing awareness of environmental issues arose in American society. A number of factors contributed to this, including the publication of Rachel Carson's *Silent Spring* (1962) and advances in media that brought issues of pollution and animal exploitation into people's homes (Organ et al. 1998). The number of nongovernmental wildlife organizations expanded from 56 in 1945 to more than 300 by the mid-1970s and to more than 400 by the 1980s (Dunlap 1988). In 1973 the Wildlife Management Institute published the North American Wildlife Policy in response to these social and environmental changes (Allen 1973). The policy called for greater federal oversight and broader environmental pro-

grams. Sweeping federal legislation enacted during the 1970s mandated clean air, clean water, protection of endangered species, and public input on federal actions that affect the environment. Many state fish and wildlife agencies became divisions or subunits of larger environmental agencies. The ranks of wildlife professionals were no longer confined to wildlife biologists, wildlife researchers, wildlife educators, wildlife administrators, wildlife managers, and wildlife law enforcement officers. The expanding profession now included within its ranks wildlife damage management specialists, wildlife toxicologists, wildlife pathologists, land-use planners, geographic information system analysts, statisticians, wetland scientists, community ecologists, wildlife veterinarians, and other practitioners. Human dimensions of wildlife management emerged as a formal discipline, with sociologists, resource economists, political scientists, and cultural anthropologists contributing to wildlife management (Decker et al. 2012).

In the 21st century the term *wildlife professional* might be appropriately applied to professionals ranging from traditional wildlife biologists to filmmakers. Additionally, in contrast to the early years of the profession, the human diversity of the field has begun to broaden. Increased participation by a wider range of people from different backgrounds (e.g., gender, sexual orientation, race, ethnicity, culture, socioeconomic status, and others) promises to add valuable talent and insights as wildlife professionals are pressed to manage in evermore challenging landscapes and situations. Recent years have seen the number of women entering the workforce equal or exceed men, and with most Americans now living in urban areas, the proportion of wildlife professionals coming from cities has increased substantially. Ethnic and racial diversity within the wildlife profession increased also, although it remains far less than society in general. Data for members of TWS (Table 3.1) indicate that women and minorities are underrepresented, although a substantial proportion of members did not identify their gender or ethnicity, and response bias is unknown. Gender and ethnicity within the Wildlife Biologist Series (GS-0486) for US Fish and Wildlife Service (USFWS) employees from

Table 3.1. Change in the demographic composition of The Wildlife Society membership as self-reported by members, 2014 and 2020 (in percentages)

Demographic	2014	2020
African American	0.4	0.7
Asian	1.4	1.7
Caucasian	93.7	90.1
Hispanic	1.9	4.0
Native American / Indigenous	0.8	0.9
Other	1.9	2.6
Female	32.0	41.9
Male	68.0	58.1

Source: Data courtesy of C. Kovach, The Wildlife Society.

Note: Differences in total members and members not reporting between the two time periods prevent accurate estimation of percentage of change within the different groups.

Table 3.2. Diversity trends within the Wildlife Biologist Series (GS-0486) in the US Fish and Wildlife Service, 1999–2020 (in percentages)

Demographic	1999	2011	2020
Black or African American	1.2	0.5	1.0
Hispanic	4.1	5.2	4.2
Asian	1.4	1.7	1.9
Native American	1.7	1.7	1.9
Native Hawaiian or other Pacific Islander			0
Two or more races			0.2
White			90.9
Women	27.7	35.5	41.5
Minority women	2.5	3.5	3.6
Total minority	8.5	9.9	9.1

Source: US Fish and Wildlife Service, Management Directive (MD) 715, https://www.fws.gov/odiwm/MD-715.html.

1999 to 2020 have not been representative of the American population from 1999 to 2011 (Table 3.2; see https://www.census.gov/quickfacts/fact/table/US/PST045219). These data represent less than 10% of the total USFWS workforce, and other series, such as Fish and Wildlife Biologist, Refuge Manager, and Biological Technician, are not presented. Declines in some categories could represent movement into other series. Nevertheless, these data suggest that women and minorities are underrepresented as wildlife biologists. The wildlife profession in its current state, as indicated above, is broad and encompasses many disciplines. The focus hereafter will be on the core discipline (i.e., wildlife biology) and the educational requirements, ethical responsibilities, and employment opportunities associated with it. Part of being a professional involves maintaining currency with science, policy, and ethical standards. Membership in professional scientific societies is an indicator of one's professionalism.

THE WILDLIFE BIOLOGIST

Wildlife biology by its nature is an integrative discipline. It is a mixture of whole-animal biology, ecology, zoology, botany, genetics, policy, and social science with grounding in life, physical, and quantitative sciences. Background in forestry, range science, marine biology, or other disciplines might be necessary depending on the species or landscape one focuses on. The Wildlife Society has established educational requirements for becoming a certified wildlife biologist. These requirements are fluid and subject to change; current information can be found at www.wildlife.org. Coursework requirements are nested within six core areas:

1. Biological sciences. Thirty-six semester credit hours in biological sciences are required and must include courses in the following subcategories:
    a. Wildlife management (6 hours). Wildlife management courses focus on the principles and techniques of managing wildlife. Traditional wildlife biology curricula typically have a course dedicated to principles of wildlife management and one dedicated to wildlife management techniques. A principles course covers population dynamics theory (e.g., population growth rates, *r* versus *K* species, compensatory versus additive mortality), habitat management principles (e.g., structural requirements, species/area dynamics, successional patterns, island biogeography), history, policies, laws, and other components that form the scientific basis for wildlife management. A techniques course will introduce students to the tools used to manage wildlife and how to use them. This can include capture devices, chemical immobilization, radio telemetry, geographic positioning systems, habitat mapping, survey and monitoring techniques, and other field and laboratory efforts. Courses in conservation biology can be considered wildlife management if they contain a specific focus on management and decision-making.
    b. Wildlife biology (6 hours). Wildlife biology courses focus on the biology and behavior of birds, mammals, reptiles, or amphibians. These courses should provide the student with an understanding of species biology and habitat requirements sufficient to provide a basic knowledge of management needs. At least one course in this category must deal solely with the science of mammalogy, ornithology, or herpetology.
    c. Ecology (3 hours). A course in general animal or plant ecology will meet this requirement, which provides the wildlife biologist with an understanding of the interrelationships within and among ecological systems.
    d. Zoology (9 hours). Courses in taxonomy, biology, behavior, physiology, anatomy, and natural history of vertebrates and invertebrates will meet this requirement. A wildlife biologist should have a basic understanding of animal biology, systematics, and evolutionary mechanisms.
    e. Botany (9 hours). Courses in general botany, plant genetics, plant morphology, plant physiology, dendrology, or plant taxonomy will meet this requirement. A background in botany is essential for the wildlife biologist to understand and manage habitat.
2. Physical sciences (9 hours). Courses in the physical sciences can include chemistry, physics, geology, or soil science, with at least two disciplines represented. A background in physical sciences is important for the wildlife biologist to understand ecosystem processes and physiological mechanisms.
3. Quantitative sciences (9 hours). Study in the quantitative sciences must include:
    a. Basic statistics (3 hours). A course in basic statistics is essential for a wildlife biologist. It would be rare to go through a career in wildlife biology without having to collect and analyze data, and even rarer to get an advanced degree in wildlife biology without doing so.

Without this grounding, the wildlife biologist will be ill equipped to be a scientist.

b. Quantitative sciences (6 hours). This can include courses in calculus, biometry, advanced algebra, systems analysis, mathematical modeling, sampling, and computer science. For the wildlife biologist who desires a career in research, expertise in quantitative sciences will pay dividends. White (2001) provides a compelling argument for the importance of quantitative science in the training of a wildlife biologist.

4. Humanities and social sciences (9 hours). This includes courses such as economics, sociology, psychology, political science, government, history, literature, or foreign language. Murie (1954:293) cautioned that "our training in universities should be such that we do not come out pretty good technicians but philosophical illiterates." Study in humanities is essential for development of critical thinking ability, a characteristic that will serve one well in any profession. Social science background has become a necessity in the wildlife profession (Organ et al. 2006). Many wildlife agencies have social scientists on their staff in recognition of this need. The wildlife profession, in striving to maintain relevancy within a changing society (Jacobson et al. 2010), increasingly applies social science, with equal rigor as it does biological science, toward understanding wildlife stakeholder attitudes, values, normative behaviors, and preferences. Decision-making processes that integrate biological and social sciences are becoming more rigorous and transparent (Riley et al. 2003, Williams et al. 2009).

5. Communications (12 hours). These courses are designed to improve communication skills. They can include courses in English composition, technical writing, journalism, public speaking, and media. Effective communication is important in every aspect of a wildlife professional's work, but wildlife agencies are increasingly looking for individuals with skills who can assist them in communicating with a broadening group of stakeholders through diverse information streams.

6. Policy, administration, and law (6 hours). These courses focus on natural resource policy, administration, wildlife or environmental law, or natural resource or land-use planning. The wildlife professional should have an understanding of the legal bedrock for wildlife conservation. In the United States, this is provided for in the public trust doctrine (Geist and Organ 2004), the US Constitution, and various state constitutions and laws (Batcheller et al. 2010). The wildlife professional should be aware of how law and policy differ, and how each is developed.

Graduate education has become a virtual necessity for the student desiring employment as a wildlife biologist. The diverse and specialized needs in the field of wildlife conservation make it difficult for an individual with an undergraduate degree to compete, even in entry-level positions, with those who have advanced degrees and specialized training. During the course of undergraduate studies, the student should assess his or her interests and desires, and seek graduate education that will afford the skills and experience necessary for entering the workforce and embarking on a career directed toward those goals.

Conservation biology emerged as a discipline around 1978 and, by 1986 the Society for Conservation Biology had formed (Hunter 1996). Conservation biology is the applied science of maintaining the earth's biological diversity. As such, it is a multidisciplinary science that ideally integrates biology, ecology, physical sciences, and social sciences (economics, sociology, anthropology, political science) in efforts to sustain natural ecosystems and biological processes (Meffe and Carroll 1994). Conservation biology principles are an essential part of a wildlife professional's knowledge base.

PROFESSIONAL BEHAVIOR AND THE WILDLIFE PROFESSION'S CODE OF ETHICS

McCabe (1954) and Murie (1954) articulated the importance of professional ethics in the wildlife profession. TWS developed a code of ethics for its members (see https://wildlife.org/wp-content/uploads/2017/07/Code-of-Ethics-May-2017.pdf). The code demands that wildlife professionals:

1. Uphold the dignity and integrity of the wildlife profession. They shall endeavor to avoid even the suspicion of dishonesty, fraud, deceit, misrepresentation, or unprofessional demeanor.

2. Refrain from plagiarism in verbal or written communications and shall give credit to the works and ideas of others.

3. Refrain from fabrication, falsification, or suppression of results, and shall not deliberately misrepresent research findings, or otherwise commit scientific fraud.

4. Exercise high standards in the care and use of live vertebrate animals used for research, in accordance with accepted professional guidelines for the respective classes of animals under study.

5. Protect the rights and welfare of human subjects used in research and obtain the informed consent of those individuals, in accordance with approved professional guidelines for human subjects.

6. Be mindful of their responsibility to society, and seek to meet the needs of all people when seeking advice in wildlife-related matters. They shall studiously avoid discrimination in any form, or the abuse of professional authority for personal satisfaction.

7. Recognize and inform clients or employers of the wildlife professional's prime responsibility to the public interest, conservation of the wildlife resource, and the environment. They shall exercise professional judgment, and avoid actions or omissions that may compromise these broad responsibilities. They shall cooperate fully

with other professionals in the best interest of the wild-life resource.

8. Provide maximum possible effort in the best interest of each client or employer, regardless of the degree of remuneration.

9. Accept employment to perform professional services only in areas of their own competence, and consistent with the code of ethics. They shall seek to refer clients or employers to other natural resource professionals when the expertise of such professionals shall best serve the interests of the public, wildlife, and the client or employer.

10. Maintain a confidential relationship between profession-als and clients or employers except when specifically authorized by the client or employer or required by due process of law or the code of ethics to disclose pertinent information. They shall not use such confidences to their personal advantage or to the advantage of other parties, nor shall they permit personal interests or other client or employer relationships to interfere with their professional judgment.

11. Refrain from advertising in a self-laudatory manner—beyond statements intended to inform prospective clients or employers of one's qualifications—or in a manner detrimental to fellow professionals and the wild-life resource. They shall clearly distinguish among facts, hypotheses, and opinions. They shall provide profes-sional advice and guidance only when qualified to do so by training and experience.

12. Refuse compensation or rewards of any kind intended to influence their professional judgment or advice or to secure preferential treatment. They shall not permit a person who recommends or employs them, directly or indirectly, to regulate or impair their professional judg-ment. They shall not accept compensation for the same professional services from any source other than the cli-ent or employer without prior consent of all the clients or employers involved.

13. Avoid performing professional services for any client or employer when such service is judged to be contrary to the code of ethics or detrimental to the well-being of wildlife resources and their environments. If a wildlife professional believes that his or her employment activi-ties conflict with the code of ethics, that person shall advise the client or employer of such conflict.

14. Advise against an action by a client or employer that violates any statute or regulation.

The Wildlife Society has provisions for enforcing this code among its members, and violations could result in censure or suspension from membership in the society. An ethics board appointed by the society president will review allegations of ethical misconduct and will follow a process outlined in the society's bylaws in determining whether a violation oc-curred.

EMPLOYMENT OPPORTUNITIES FOR THE WILDLIFE BIOLOGIST

Employment within the wildlife profession is skewed toward the public sector, but private sector employment today repre-sents a large percentage of available jobs. In recent decades, the growth of private lands, hunting-for-fee facilities, envi-ronmental consulting firms, wildlife organizations, and other private endeavors have broadened the employment field. In 2020, 23.8% of TWS members were employed in the private sector (i.e., consulting firm, nonprofit organization, corpora-tion, self-employed). Another 36.7% of TWS members were employed by federal, state, or provincial authorities; 25.8% were employed by universities (Table 3.3).

Students and others desiring careers as wildlife biologists must be prepared for competition. Students should take ad-vantage of opportunities to volunteer on research projects within their academic department. These opportunities will help develop applied skills that can enhance résumés and make an impression on potential future employers or references. Summer seasonal work opportunities are a valuable means to develop skills and to make contacts within agencies and orga-nizations that could ultimately lead to full-time employment. Seasonal and full-time employment opportunities can be found on several Internet sites. Federal job opportunities (sea-sonal and full-time) are listed at www.usajobs.gov. Texas A&M University's Department of Wildlife and Fisheries Sciences maintains an extensive employment website at wfscjobs.tamu .edu, and TWS has a job board at https://careers.wildlife.org/. As mentioned above, the wildlife profession has broadened beyond the traditional wildlife biologist, largely in response to increased demands from society. Specialists within the wildlife profession include sociologists who address the hu-man dimensions of wildlife conservation (Decker et al. 2012), communication experts who can work with stakeholders and convert complex science and policy into plain language (Bonar 2007), range scientists and managers who can address habi-tat issues of rangeland wildlife (Fulbright and Ortega-Santos 2013). Outlined below is an overview of the major entities

Table 3.3. Employment category data for members of The Wild-life Society as self-reported by 77% of members, 2020

Employment Category	Percentage
University	25.8
Federal	17.2
State or province	19.5
Consulting firm	9.3
Nonprofit organization	8.0
Retired	7.5
Corporation	3.1
Tribal	0.0
Self-employed	3.4
Other	6.5

Source: Data courtesy of C. Kovach, The Wildlife Society.

ROBERT A. McCABE (1914–1995)

At a time when specialization became vogue in the wildlife profession, Robert McCabe was asked to describe his specialty. Without hesitation, he responded that he was a generalist, a matter-of-fact reply that typified him.

Bob McCabe was a product of impoverished South Milwaukee, Wisconsin. Of Germanic and Irish heritage, he took advantage of the few opportunities available during the Depression era. He proved to be a gifted athlete but only an average student. With the same drive that earned him athletic achievements, McCabe sought to escape his hardscrabble neighborhood. His determination led him to pursue higher education at Carroll College in Waukesha, Wisconsin, where he discovered three passions. First was Marie Stanfield, his wife of 54 years. Second was rabbit hunting. Third was an abiding curiosity in nature that fostered his academic enthusiasm.

McCabe went on to attend graduate school at the University of Wisconsin–Madison. Applying for admission in 1939 to zoological study in game conservation, he was directed to "see Leopold on the Ag campus." He did so reluctantly, but it was a fortuitous, momentous encounter. Aldo Leopold's graduate student, protégé, assistant, and occasional hunting companion, McCabe eventually succeeded him as chair of the Department of Wildlife Management. More important was the deep friendship they shared—the inspiration for McCabe's career and a daily guidepost until his passing in 1995. In his copy of *A Sand County Almanac*, Leopold's wife poignantly inscribed: "A son by affection to Aldo."

During his 27 years as department chair, McCabe dedicated himself to enhancing its reputation of academic excellence by accepting only the most promising graduate students; hiring faculty of high intelligence, character, and outdoor experience; and by brooking no challenge, institutional or otherwise, to the Leopold legacy. Although an administrator, advisor, and teacher, McCabe relished being a researcher. The most enduring of his work concerned the alder flycatcher, subject of his book *The Little Green Bird* (1991), published several years after his cathartically crafted biography *Aldo Leopold: The Professor* (1987).

McCabe was an inveterate, indefatigable pursuer of upland game birds, particularly ruffed grouse and woodcock. He was a very good shot, nonpareil as a hunter. Time afield with him was a treat for his sons and his few other, favored hunting companions. A Fulbright scholar, advisor on international wildlife matters (Africa, Canada, Ireland, Russia), recipient of honorary degrees (Carroll College and University of Dublin), TWS president, conferee of the Aldo Leopold Medal (1986) and other honors, McCabe most cherished his time with family, especially at Rusty Rock, their farm in Iowa County, Wisconsin.

To some, he was seen as a taciturn, even stern taskmaster. He insisted on quality effort in all endeavors and was unimpressed with those he deemed prima donnas and pretenders. To friends and family, Bob McCabe was sensitive, fun, and funny. And he was as unstintingly devoted to them as he was to his profession.

---

that employ wildlife biologists in the United States. Additional job opportunities are detailed in Henke and Krausman (2017).

Federal Government

The legal public trust authority for most species of wildlife rests with the states; the federal government has primary legal authority for species captured within one of three clauses of the US Constitution: the property clause (wildlife on federally owned land such as national wildlife refuges), the commerce clause (wildlife affected by interstate commerce such as trade in endangered species), and the supremacy clause, which makes federal treaties supreme over any other law of the land (e.g., the Migratory Bird Treaty of 1916).

US Department of Agriculture (USDA)

The USDA contains several bureaus with responsibilities for fish and wildlife management. The USDA has tremendous influence over management of private agricultural lands and public lands that collectively represent significant wildlife habitat.

US Forest Service (USFS)

The USFS manages a nationwide network of 78 million ha of forests and grasslands. Its mission is to sustain the health, diversity, and productivity of the nation's forests and grasslands to meet the needs of present and future generations. The USFS has a Watershed, Fish, Wildlife, Air, and Rare Plants Program that coordinates management activities and research for fish, wildlife, and endangered species nationwide. National forests are managed for multiple uses, and hunting and fishing are primary uses, with seasons and limits typically adhering to state regulations.

Animal and Plant Health Inspection Service: Wildlife Services

Wildlife Services provides leadership in managing human–wildlife conflicts. The agency has state-based field staff and major research facilities that develop tools and techniques for managing wildlife conflicts. Wildlife damage management has matured as a major subdiscipline within the wildlife profession, and Wildlife Services provides leadership in this dimension.

Animal and Plant Health Inspection Service: Veterinary Services

Veterinary Services works to protect and improve the health, quality, and marketability of our nation's animals, animal products, and veterinary biologics by preventing, controlling, and eliminating animal diseases, and monitoring and promoting animal health and productivity. Veterinary Services addresses wildlife zoonoses, such as avian influenza, that have potential to affect domestic agricultural animals.

Natural Resources Conservation Service (NRCS)

The NRCS works with landowners through conservation planning and assistance designed to benefit the soil, water, air, plants, and animals that result in productive lands and healthy ecosystems. Funding for much of NRCS activities comes through the federal Farm Bill.

Department of Commerce

The Department of Commerce, through the National Oceanic and Atmospheric Administration (NOAA) and its bureau NOAA Fisheries, has federal authority over certain species of marine organisms and regulates commerce in marine fisheries. Under the Marine Mammal Protection Act of 1972, NOAA has authority over the order Cetacea and seals and sea lions. NOAA supports regional fisheries management councils and commissions that work collaboratively to regulate commerce in marine fisheries.

Department of Defense

The Department of Defense employs wildlife professionals on many of its installations, where they coordinate efforts to manage wildlife populations on their extensive landholdings, including hunting programs and endangered species recovery efforts. Wildlife professionals also provide guidance on the effects of proposed field activities and exercises on wildlife resources. Information can be found at http://www.nmfwa.org/.

US Army Corps of Engineers (USACE)

The USACE has jurisdictional responsibility for portions of the federal Clean Water Act. It employs wildlife professionals to implement portions of the regulatory program that oversees impacts to navigable waterways, including wetlands, by land and water development activities.

US Department of the Interior (DOI)

The DOI has several bureaus with management responsibility for fish and wildlife. Similar to the USDA, the DOI has influence over a significant portion of public and private lands and some regulatory authority over land and water development activities.

US Fish and Wildlife Service (USFWS)

The USFWS is the largest fish and wildlife conservation agency in the world. Its mission is to work with others to conserve, protect, and enhance fish, wildlife, plants, and their habitats for the continuing benefit of the American people. The USFWS manages a network of national wildlife refuges, fish hatcheries, fishery resource stations, law enforcement offices, and other field stations that provide a multitude of conservation services, including endangered species recovery activities, wetland protection, environmental contaminants, and private landowner conservation assistance (Fig. 3.1). The USFWS also has an international affairs office that assists conservation efforts abroad and oversees activities in the United States, and its Division of Wildlife and Sport Fish Restoration provides federal aid to state fish and wildlife agencies.

Bureau of Land Management (BLM)

The mission of the BLM is to sustain the health, diversity, and productivity of public lands for the use and enjoyment of present and future generations. The BLM's Fish, Wildlife, and Plant Conservation Program oversees management activities on lands under its control.

National Park Service (NPS)

The NPS manages a network of national parks across the United States. The NPS's Biological Resource Management

Figure 3.1. US Fish and Wildlife Service wildlife biologist bands an endangered red-cockaded woodpecker. Courtesy of US Fish and Wildlife Service

Division provides policy, planning, and operational support to NPS personnel concerning the management of native vegetation and wildlife resources, the control of nonnative species, and the biological restoration of disturbed ecosystems. It formulates biological resource policy recommendations and conducts legislative, regulatory, and environmental reviews related to biological resource protection. It assists the field in vegetation and wildlife resource management activities.

US Geological Survey (USGS)

The USGS's Ecosystem Mission Area provides science support to other DOI bureaus (e.g., USFWS) and partners (e.g., universities). It has a network of ecological and climate science centers and cooperative fish and wildlife research units that specialize in applied fish and wildlife research.

State Government

There are 56 states, territories, and insular areas (including the District of Columbia) within the United States, and each has an agency or bureau dedicated to the conservation and management of fish and wildlife resources. Most species of wildlife, except for those reserved by the US Constitution (i.e., fall within federal oversight) and its amendments (e.g., migratory birds, marine mammals, endangered species), are under the legal jurisdiction of the states. State fish and wildlife agencies are typically governed by a board or commission, or by a politically appointed director. Many state fish and wildlife bureaus are nested within larger environmental or natural resource agencies; a few, such as the Pennsylvania Game Commission and Idaho Fish and Game, are independent agencies. State fish and wildlife agencies have broad responsibilities that range from research, management, and law enforcement to environmental review, education, policy, and administration.

Historically, these agencies focused on game species, but recent decades have seen an increase in staffing and effort for endangered species and wildlife diversity programs (Fig. 3.2). The development of state wildlife action plans in 2005 and funding from the State Wildlife Grants Program have enhanced the capacity of state fish and wildlife agencies to broaden programs (https://www1.usgs.gov/csas/swap/). Major funding for state fish and wildlife agencies comes from hunting license revenues and the Pittman-Robertson Wildlife Restoration Program, with funding derived from federal excise taxes on firearms, ammunition, handguns, and archery equipment (http://wsfrprograms.fws.gov/).

Wildlife biologists can find employment in other state government agencies, too, including state conservation and recreation agencies focused on parks and forests, and transportation and utility departments where biologists assist in planning and mitigation of adverse effects of roads and other infrastructure on wildlife.

Tribal Nations

Federally recognized Indian tribes within the lower 48 states have jurisdiction over a reservation land base of more than 21 million ha (130,759 km²). Alaskan Native lands comprise

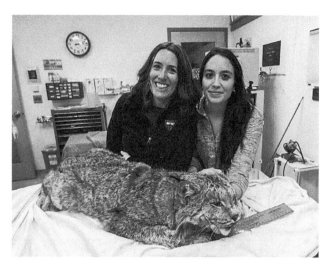

Figure 3.2. State wildlife biologist Jennifer Vashon of the Maine Department of Inland Fisheries and Wildlife and wildlife conservation geneticist Tanya Lama of the Massachusetts Cooperative Fish and Wildlife Research Unit with live Canada lynx LIC47, the reference animal used to sequence the Canada lynx genome. Genomic tools can greatly assist in understanding population and landscape dynamics, and in the development of conservation strategies. Courtesy of Bill Byrne, Massachusetts Division of Fisheries and Wildlife

another 18 million ha. Some tribes control resources outside of reservations owing to federal court decisions and voluntary cooperative agreements that allow a comanagement status between tribes and states. These lands are called ceded and usual and accustomed areas, and they equal over 15 million ha. In these areas, tribes maintain comanagement jurisdiction for fisheries and wildlife management and use. Tribal lands coupled with the ceded and usual and accustomed areas total a natural resource base of over 226,314 km², containing more than 295,420 ha of lakes and impoundments, and more than 16,093 km of streams and rivers. This land combined would constitute the fifth-largest state in the United States. Many tribal nations employ wildlife professionals to manage their natural resources. Additional information can be found at http://www.nafws.org/.

Private Sector

Private sector employment for wildlife biologists is broad and diversified, in contrast to the early years of the profession. Land management firms and private individuals hire wildlife biologists to manage wildlife for hunting, to restore rare species, and to ensure extractive uses comply with environmental regulations. Many environmental consulting firms provide services to private individuals and corporations in planning and environmental review and compliance (Fig. 3.3). In the southeastern United States, a number of private wildlife consulting firms specialize in developing land management programs for private landholdings to enhance hunting opportunities.

There are numerous nongovernmental wildlife organizations that focus on advocacy, policy, science, or some combi-

Figure 3.3. Private consultant wildlife biologists work with individuals and corporations to ensure land management activities minimize adverse impacts to wildlife habitat. Courtesy of Jon Haufler

Figure 3.4. Wildlife veterinarian Miryam Quevedo of San Marcos University vaccinates a dog in a remote village in the Peruvian Amazon as part of a surveillance effort for wildlife diseases that uses pets as sentinels. Courtesy of Jesus Lescano

nation of the three. Depending upon the mission of the organization, a wildlife biologist employed in this realm could be immersed in legal court actions, working the halls of federal or state capitol buildings, or designing and implementing field research investigations. The larger national (and international) organizations include professional scientific societies such as TWS, professional organizations such as the Association of Fish and Wildlife Agencies and the Wildlife Management Institute, and member-based groups such as the Boone & Crockett Club, Defenders of Wildlife, Ducks Unlimited, National Wildlife Federation, National Wild Turkey Federation, National Audubon Society, Pheasants Forever, Quail Unlimited, Quality Deer Management Association, Rocky Mountain Elk Foundation, Safari Club International, The Nature Conservancy, Whitetails Unlimited, and the Wildlife Conservation Society. Many other groups operate at the regional and local levels.

In some parts of the United States, leasing lands for hunting has become big business. In many cases, wildlife biologists are employed to manage populations and habitats on these properties to optimize opportunity and quality of hunting for big game, upland game birds, and waterfowl. Within the wildlife profession, concern has been raised over the implication of this trend relative to privatizing wildlife (Organ and Batcheller 2009, Batcheller et al. 2010).

Wildlife veterinary work is a distinct subdiscipline within the wildlife profession (Fig. 3.4). Amendments to the Animal Welfare Act promulgated in 1985 require veterinarians with expertise on wildlife for certain activities regulated under the act. Many federal and state conservation agencies, zoos, and private facilities employ wildlife veterinarians. Many zoos are active in captive breeding programs for endangered wildlife and employ wildlife veterinary professionals. The field of wildlife veterinary science has become established at many leading veterinary schools. The American Association of Wildlife Veterinarians (http://www.aawv.net/) provides a forum for the advocacy of veterinary medicine within the wildlife conservation field. Conservation medicine is gaining traction as an area of inquiry and management necessary for addressing contemporary conservation challenges. The emergence of white-nose syndrome in bats, chronic wasting disease in ungulates, avian influenza, and other diseases suggests there will be many opportunities for wildlife professionals interested in pursuing careers in wildlife veterinary science and conservation medicine.

Universities

Universities represent a major pillar within the wildlife profession (Gill 1996). The National Association of University Fisheries and Wildlife Programs (http://naufwp.org/index.html) lists 63 member colleges and universities in the United States, and this does not include many programs at smaller state colleges and private institutions. As a rule, faculty positions require a PhD at a minimum. Many programs maintain a professional technical staff to assist in field and laboratory research.

The Cooperative Fish and Wildlife Research Units Program (https://www1.usgs.gov/coopunits//) was established in 1935 to enhance graduate education in fisheries and wildlife sciences and to facilitate research between natural resource agencies and universities on topics of mutual concern. Today, there are 41 cooperative research units in 38 states. Each unit is a partnership among the USGS, a state natural resource agency, a host university, and the Wildlife Management Institute. Staffed by federal personnel, cooperative research units conduct research on renewable natural resource questions, participate in the education of graduate students, provide technical assistance and consultation on natural resource issues, and provide continuing education for natural resource professionals.

The Cooperative Research and Extension Services (https://www.usda.gov/topics/rural/cooperative-research-and-extension-services) of the USDA supports a cooperative extension system with programs at land-grant universities, providing funding to advance agriculture, the environment, human health, and communities by supporting research, education, and technical assistance.

SUMMARY

The wildlife professional of today is part of a diverse network of men and women who have expertise in a variety of disciplines and who work in virtually every employment sector. What was once a profession focused on restoration of game species by government biologists is now much too broad to characterize in simple terms. Individuals interested in a career as a wildlife professional would be well advised to focus their undergraduate education on meeting requirements for becoming a certified wildlife biologist, and then focus their postgraduate education on a specialty field, whether it be wildlife population or habitat management, human dimensions, genomics (Fig. 3.2), spatial ecology, law, policy, medicine, or otherwise. An individual with solid educational grounding and a demonstrated commitment to the profession as evidenced by participation in professional scientific societies will be better able to compete for professional wildlife jobs.

Literature Cited

Allen, D. L. 1973. Report of the committee on North American wildlife policy. Wildlife Society Bulletin 1:73–92.

Batcheller, G. R., M. C. Bambery, L. Bies, T. Decker, S. Dyke, et al. 2010. The public trust doctrine: implications for wildlife management and conservation in the United States and Canada. Technical Review 10-1. The Wildlife Society, Bethesda, Maryland, USA.

Bonar, S. A. 2007. The conservation professional's guide to working with people. Island Press, Washington, DC, USA.

Carson, R. 1962. Silent spring. Houghton Mifflin, Boston, Massachusetts, USA.

Decker, D. J., S. J. Riley, and W. E. Siemer. 2012. Human dimensions of wildlife management. 2nd ed. Johns Hopkins University Press, Baltimore, Maryland, USA.

Diamond, J. M. 1997. Guns, germs, and steel: the fate of human societies. Norton, New York, New York, USA.

Dunlap, T. R. 1988. Saving America's wildlife. Princeton University Press, Princeton, New Jersey, USA.

Fulbright, T. E., and J. A. Ortega-Santos. 2013. White-tailed deer habitat: ecology and management on rangelands. Texas A&M University Press, College Station, Texas, USA.

Geist, V., and J. F. Organ. 2004. The public trust foundation of the North American model of wildlife conservation. Northeast Wildlife 58:49–56.

Gill, R. B. 1996. The wildlife professional subculture: the case of the crazy aunt. Human Dimensions of Wildlife 1:60–69.

Grinnell, G. B. 1885. New publications: hunting trips of a ranchman. Forest and Stream 24:450–451.

Henke, S. E., and P. R. Krausman. 2017. Becoming a wildlife professional. Johns Hopkins University Press, Baltimore, Maryland, USA.

Hunter, M. L., Jr. 1996. Fundamentals of conservation biology. Blackwell Science, Cambridge, Massachusetts, USA.

Jacobson, C. A., J. F. Organ, D. J. Decker, G. R. Batcheller, and L. Carpenter. 2010. A conservation institution for the 21st century: implications for state wildlife agencies. Journal of Wildlife Management 74:203–209.

King, R. T. 1938. What constitutes training in wildlife management. Transactions of the North American Wildlife and Natural Resources Conference 3:548–557.

Leopold, A. 1930. Report to the American game conference on an American game policy. Transactions of the American Game Conference 17:281–283.

Leopold, A. 1931. Report on a game survey of the north central states. Sporting Arms and Manufacturers' Institute. Madison, Wisconsin, USA.

Leopold, A. 1933. Game management. Charles Scribner's Sons, New York, New York, USA.

Leopold, A. 1939. Academic and professional training in wildlife work. Journal of Wildlife Management 3:156–161.

Leopold, A. 1940. The state of the profession. Journal of Wildlife Management 4:343–346.

Lewis, M., and W. Clark. 2002. The journals of Lewis and Clark. National Geographic Society, Washington, DC, USA.

McCabe, R. A. 1954. Training for wildlife management. Journal of Wildlife Management 18:145–149.

Meffe, G. K., and C. R. Carroll. 1994. Principles of conservation biology. Sinauer, Sunderland, Massachusetts, USA.

Meine, C. 1988. Aldo Leopold: his life and work. University of Wisconsin Press, Madison, USA.

Murie, O. J. 1954. Ethics in wildlife management. Journal of Wildlife Management 18:289–293.

Organ, J. F., and G. R. Batcheller. 2009. Reviving the public trust doctrine as a foundation for wildlife management in North America. Pages 161–171 in M. J. Manfredo, J. J. Vaske, P. J. Brown, D. J. Decker, and E. A. Duke, editors. Wildlife and society: the science of human dimensions. Island Press, Washington, DC, USA.

Organ, J. F., D. J. Decker, L. H. Carpenter, W. F. Siemer, and S. R. Riley. 2006. Thinking like a manager: reflections on wildlife management. Wildlife Management Institute, Washington, DC, USA.

Organ, J. F., R. M. Muth, J. E. Dizard, S. J. Williamson, and T. A. Decker. 1998. Fair chase and humane treatment: balancing the ethics of hunting and trapping. Transactions of the North American Wildlife and Natural Resources Conference 63:528–543.

Reiger, J. F. 1975. American sportsmen and the origins of conservation. Winchester, New York, New York, USA.

Riley, S. J., W. F. Siemer, D. J. Decker, L. H. Carpenter, J. F. Organ, and L. Berchielli. 2003. Adaptive impact management: an integrative approach to wildlife management. Human Dimensions of Wildlife 8:81–95.

Roosevelt, T. 1885. Hunting trips of a ranchman. G. P. Putnam's Sons, New York, New York, USA.

Trefethen, J. B. 1975. An American crusade for wildlife. Winchester, New York, New York, USA.

White, G. C. 2001. Why take calculus? rigor in wildlife management. Wildlife Society Bulletin 29:380–386.

Williams, B. K., R. C. Szaro, and C. D. Shapiro. 2009. Adaptive management: the US Department of the Interior technical guide. Adaptive Management Working Group, US Department of the Interior, Washington, DC, USA.

HUMAN DIMENSIONS OF WILDLIFE MANAGEMENT

DANIEL J. DECKER, SHAWN J. RILEY, AND WILLIAM F. SIEMER

INTRODUCTION

Wildlife management is a set of decision-making and implementation activities intent on influencing several key, interdependent elements: humans, wildlife populations, environments and habitats, and their interactions (see Fig. 1.1). From a wildlife management standpoint, these elements together with the natural and social processes operating within and between them create a social-ecological system (Berkes and Folke 1998) that we refer to as the wildlife management system. Management of any part of such a system requires relevant information about the three main elements and their interactions. Such knowledge is needed to support sound decisions that are the essence of wildlife management—identifying meaningful and achievable objectives and then selecting actions likely to achieve those objectives. Reflecting the vital roles of decision-informing and decision-making, wildlife management can be defined as the guidance of decision-making processes and implementation of practices to influence interactions between people, wildlife, and wildlife habitats, and among people about wildlife, to achieve outcomes valued by stakeholders (Riley et al. 2002).

This chapter focuses on the people element or human dimensions of wildlife management, which include traits of individuals, groups, organizations, social structures, economic activity, cultural systems, communities, and institutions within the wildlife management system. Human dimensions of wildlife management can be described as discovering and applying insight about how humans value wildlife, how humans want wildlife to be managed, and how humans affect or are affected by wildlife and outcomes of wildlife management.

The emphasis of wildlife management is similar to any management activity: to turn complexity, information, and specialized activity into value-producing performance (Margretta 2002). Wildlife management is aimed at production of value for and defined by society, where value, or benefits, is measured by the outcomes experienced by stakeholders (e.g., positive effects created or negative effects reduced; values associated with biodiversity, recreation, and economic activity). Human dimensions efforts tackle the fundamentally human

purpose of wildlife management (Decker et al. 2001), focusing on understanding and addressing the reasons for people's interest in the management of wildlife and wildlife habitat.

Management decision processes and social science research that supports management are key foci of this chapter because no matter where you are working in the world, conservation is accomplished through management within a framework of decision-making referred to as governance (Armitage et al. 2012, Rudolph et al. 2012, Decker et al. 2016). Good governance of wildlife resources necessitates knowledge about affected people and an understanding of their interactions with wildlife, with other people about wildlife and with the environment (wildlife habitat), and of the overall social-ecological systems within which those interactions occur.

Wildlife managers of an earlier era learned that despite best efforts to apply their understanding of wildlife habitats and populations to management, objectives seldom could be achieved without considering people's interests in wildlife and preferences for management. Aldo Leopold voiced the need to attend to humans in the management equation in 1943 when he wrote, "The real problem of wildlife management is not how we shall handle the animals . . . the real problem is one of human management" (Flader 1974:188). More recently, wildlife managers also learned that public participation in decisions about management objectives and methods results in more durable and effective public wildlife management. Thus human dimensions of wildlife management emerged as a field of social science research and application focused on improving managers' understanding of the broad social and cultural ideals (e.g., social norms, ethical tenets), societal aspirations (e.g., about good governance), and specific traits of individuals and groups (e.g., motivations to participate in wildlife-related activities) that influence management (Decker et al. 2008, 2012a).

Although human dimensions research and application initially focused mainly on game species, the field broadened topically and geographically so that today it includes all species, ecosystems, and interests in wildlife (e.g., hunting, viewing, conflict mitigation, biodiversity, ecosystem services). Human dimensions research includes several social science disciplines, some of which we discuss later in the chapter. Broad recognition of the importance of human dimensions in nature conservation beyond wildlife gave rise to conservation social science, a term adopted recently to encompass the multidisciplinary social science interests in all aspects of nature conservation (Bennett et al. 2017). Herein we focus on social science applied primarily to wildlife management and conservation and describe applications of insights typically captured under the rubric of human dimensions of wildlife management.

Ethics

Wildlife management is made possible by a set of nested decision-making processes aimed at achieving objectives valued by society. Identifying objectives and making decisions about how to achieve them require policy makers and wildlife managers to understand an array of social considerations, among the most important of which are ethical standpoints of various portions of society. Increasingly, ethical considerations are recognized as powerful factors in wildlife management issues and controversies (Jager et al. 2016, Vucetich and Nelson 2017). Human dimensions inquiry can identify ethical considerations that are relevant to different groups of stakeholders, reveal why they are operative in a particular situation, estimate how many people hold various ethical positions, determine how strongly those positions are held, and shed light on the implications for management.

As you progress in your study of real-world wildlife management cases, you will find that social and ecological science alone does not tell a wildlife manager what *should* be done in a given situation. At best, science informs managers of what *could* be done, what stakeholders desire, or what is likely to happen with and without particular interventions. Determining what *should* be done, however, is in the realm of ethics. Ethics are best understood if analyzed rigorously using tools developed in philosophy for examining human behavior, values (i.e., whether certain actions are right or wrong), and morals inherent in motives and actions. Human dimensions inquiry can help wildlife professionals evaluate ethical components by clarifying pertinent values in a particular situation, but it is a different process to determine whose values should prevail in management decisions (e.g., values of all voters, a particular subset of stakeholders or wildlife managers).

Ackoff (2001:345), a highly regarded scholar of management, articulated the following caution for managers attempting to do "the right thing": "Paraphrasing Peter Drucker, there is a big difference between doing things right and doing the right thing. Efficiency is concerned with doing things right; effectiveness with doing the right thing . . . The righter [more efficiently] one does the wrong thing, the wronger one becomes."

Much of a wildlife professional's training is geared toward how to do things right—how to determine number and distribution of a wildlife population, evaluate condition of a habitat, or accurately measure stakeholder attitudes. These are essential skills, and they must be done correctly. Yet, as Ackoff notes, if management concentrates on the wrong things (or less important things), trying harder to do a better job can have the effect of diverting attention away from what should be done. Ethical analysis can be valuable to be certain that focus is placed on the right thing when defining and deliberating about any significant wildlife issue.

While stakeholders with competing value orientations often attempt to pull wildlife managers in different directions, human dimensions inquiry can illuminate the multiple perspectives that exist about what constitutes the right objectives and best outcomes of management in a particular situation. Aiding wildlife management decision makers in their quest to identify the right thing to do and how best to do it in the public interest is a vital role for human dimensions inquiry. This role supports what is known as good governance.

Governance

In the simplest sense, governance refers to mechanisms whereby governments and other organizations direct their activities, including the processes, laws, rules, and policies that collectively guide decisions. Most Western democracies consider good governance to be participatory, transparent, and accountable. Governments, alone or in partnership with nongovernmental organizations (NGOs), conduct most wildlife management. In the United States, elected officials at local, state, and federal levels and government agencies involved in wildlife management rely on human dimensions insights to gauge citizens' expectations of institutions and processes of government with respect to wildlife. Effective governance of wildlife resources in the public interest, as public trust resources, requires insights about human dimensions drawn from research and engagement.

It is important to view wildlife management in the context of public trust resource governance because the legal basis for American wildlife management is the public trust doctrine (The Wildlife Society 2010). Case law and state statutes define public (versus individual) ownership of wildlife and empower governmental agencies to be trustees of wildlife as a public trust resource (Decker et al. 2016). One of the obligations of a trustee for wildlife resources, typically state legislatures and politically appointed commissioners in the United States, is to engage the beneficiaries of the trust (i.e., citizens) when determining goals and objectives for the trust (Decker et al. 2015). In the United States, beneficiaries of the trust are the people of the states in which wildlife resides, including current and future generations. The staff of state wildlife agencies serve as agents of the trustees (i.e., serve as trust managers), mainly playing the critical roles of technical analysts who provide recommendations to decision makers and implementers of management consistent with policy and law (Smith 2011). This arrangement gives rise to a core function of human dimensions inquiry and stakeholder engagement: seeking to understand the human–wildlife experiences and outcomes of management that are desired by stakeholders (Decker et al. 2014). Managers then share their accumulated insight in recommendations offered to decision makers (trustees).

In most instances the specific approach taken to manage wildlife reflects the specific governance procedures that the operative jurisdiction has adopted or has been mandated by law to follow. Wildlife management typically is executed through state and federal wildlife agencies. Increasingly, partnerships among multiple levels of government (including local governments) and NGOs are forged to achieve conservation through a form of participatory governance referred to as collaborative conservation (Lauber et al. 2011). In collaborative conservation, entities other than state and federal government may be deeply involved in wildlife management but typically work cooperatively with or with oversight from those agencies.

The process of wildlife management is sometimes misconstrued because descriptions of it tend to emphasize specific, on-the-ground activities focused on protection or manipulation of wildlife populations, wildlife habitats, and wildlife use. Such activities are necessary to achieve many of the outcomes desired by society, and therefore essential to governance of public wildlife resources, but this depiction leaves out much of what wildlife managers do in practice. Enabling good governance of public wildlife resources includes a greater array of processes (e.g., communication, negotiation, strategic partnerships, decision-making). Most of these critical processes can be improved by insights from human dimensions inquiry and stakeholder engagement.

Impacts Management

Wildlife management enables, regulates, or prohibits various experiences people might have with wildlife. Human experiences with wildlife can be direct or indirect, vary in intensity and type, and occur at many scales. Outcomes of these experiences are perceived effects, the most important of which (i.e., those typically generating strong stakeholder reactions and prompting management attention) are impacts (Riley et al. 2002). Effects take many forms and can be positive or negative (e.g., economic benefits or costs; threats to or enhancement of human health and safety; ecological services that wildlife provide; physical, mental, and social benefits produced by recreational enjoyment of wildlife). Effects arise from many kinds of interactions between humans and wildlife, and from interactions among humans because of wildlife. Influencing these interactions is the function of wildlife management. Understanding these interactions is one of the most important motivators of social science research and stakeholder engagement practices in wildlife management.

An impacts management approach in wildlife management attempts to shape value created by human–wildlife interactions through managing components of social-ecological systems (Riley et al. 2003). The concept of social-ecological systems refers to sociocultural subsystems, ecological subsystems, and their interactions (Berkes and Folke 1998). Essentially, this integrative concept explicitly addresses how linked sociocultural and ecological systems interact to produce diverse outputs. *Social-ecological systems* is a relatively new term in environmental management, but it is not really a new idea in wildlife management. It has been long recognized that

social-ecological systems produce the human–wildlife interactions that create the need for wildlife management (Decker et al. 2006). Identifying the different effects (i.e., sociocultural, biological, ecological) that result from those interactions and understanding how they are perceived by people as desirable or undesirable allows managers to focus management attention. Insights from both natural and social sciences are required to understand human–wildlife interactions and the nature of effects experienced by stakeholders.

Stakeholders have varied reactions to interactions with wildlife, reflecting their perceptions of positive or negative effects, whether the species is common or not (e.g., white-tailed deer [*Odocoileus virginianus*; Decker and Gavin 1987], black bears [*Ursus americanus*; Lischka et al. 2019], cougars [*Puma concolor*; Riley and Decker 2000a], wolves [*Canis lupus*], grizzly bears [*Ursus arctos*; Decker et al. 2006], and tigers [*Panthera tigris*; Carter et al. 2012, Inskip et al. 2016]). For example, the influence of effects was evident in a study about people's acceptance capacity for white-tailed deer in southern Michigan. Rural residents were conflicted in their desire to view deer (78%) as opposed to their concern about the risks of deer–vehicle collisions (i.e., nearly 34% of the respondents reported being involved in a deer–vehicle collision within three years of receiving the survey; Lischka et al. 2008). An effect variable in that study was calculated by summing IMPACT_SCORE (frequency of event x valence [positive or negative] of interactions with deer) for all identified effects, reflecting the overall degree and direction of the mismatch between respondent's desired versus perceived level of effects created by deer. Demographics and the type of stakeholder (i.e., farmer, nonhunting rural landowner, hunter) had minor influence on acceptance capacity when compared to the level and valence of effects.

Attitudes about wildlife are complex, as evidenced in studies where stakeholders (sometimes the same individuals) report both positive and negative effects from their experiences with wildlife such as elk (*Cervus canadensis*), black bears, coyotes (*Canis latrans*), beavers (*Castor canadensis*), and Canada geese (*Branta canadensis*). To be aware of the effects produced by the system, one has to understand the human components. This calls for public input and participation in deliberations about wildlife management (i.e., it requires stakeholder involvement in governance of wildlife resources).

Stakeholder Orientation

As the section on governance indicates, stakeholders are central to why and how wildlife is managed. A stakeholder is any person who significantly affects or is affected by wildlife or wildlife management decisions or actions (Decker et al. 1996). Stakeholders can have various kinds of interests (i.e., stakes) in wildlife, human–wildlife interactions, and management interventions, often framed as recreational, cultural, psychological, social, economic, ecological, or health and safety benefits and costs (Siemer and Decker 2006). Stakeholders may be well organized formal interest groups, individuals joined together in ad hoc, situation-specific grassroots groups, or simply individuals who are unaffiliated and perhaps even unknown to one another but share an interest or stake in a management issue. Although wildlife interest organizations are common, because wildlife are public trust resources, people need not be organized or even aware they have a stake to be stakeholders in public wildlife management.

A wildlife issue might involve a range of stakeholder-identified effects that influence their expectations of management. In beaver management, for example, the range of stakeholders includes forest owners with trees felled by beavers; farmers, homeowners, and highway superintendents who contend with the economic costs and nuisance associated with flooding caused by beavers; public health officials concerned with health threats posed by giardiasis in the water supply; and fur trappers. Sometimes people do not recognize their stakes in wildlife management decisions because they are unaware of the effects they will experience as a consequence of management. This often occurs when effects they never experienced previously arise from management actions (e.g., restoration of beavers to an area where they have long been absent resulting in new crop and timber losses attributable to beaver activity).

Management of wildlife usually occurs in response to stakeholders' expressed need for an agency to influence effects. Understanding what outcomes and actions stakeholders find acceptable can be difficult enough, but there is more to consider. Management actions in response to one stakeholder interest can produce a negative reaction from other stakeholders. This predictably leads to resistance to management or disagreement over management alternatives. For example, calls for management of suburban wildlife that cause negative effects (e.g., deer, Canada geese, coyotes) often are initiated by individuals seeking relief from problems they are experiencing. On the surface, an intervention often seems justifiable, and many suburban residents may agree that a problem exists and something should be done to alleviate it. Nevertheless, management in such situations frequently becomes controversial because some residents find proposed management actions unacceptable (Decker et al. 2004; Siemer et al. 2007). Paradoxically, these secondary stakeholders are in a sense created by the management effort itself. This is common when managing human-wildlife conflicts.

The case of goose management in Puget Sound, Washington (Woodruff et al. 2004), illustrates how management actions to respond to primary stakeholders can create a group of secondary stakeholders (i.e., stakeholders more concerned about management interventions than about negative effects from wildlife). The Washington Department of Fish and Wildlife (WDFW) introduced a nonnative subspecies of Canada geese to the Puget Sound area in the 1960s. By 1998, WDFW estimated the goose population at 20,000 to 25,000. Problems with Canada geese in metropolitan areas such as Seattle increased as growing populations of geese became year-round residents of parks, recreational fields, golf courses, lakes, and shorelines. In response to rising problems with geese, the city of Seattle formed a Metropolitan Waterfowl Committee (MWC) composed of representatives of cities in the Seattle

metro area, King County, the University of Washington, and US Department of Agriculture (USDA) Wildlife Services. The committee commissioned a University of Washington study of the issue, which recommended substantial reduction in the goose population. Between 1989 and 1998, the committee and member cities implemented a range of actions to address negative effects of geese, including goose relocation, oiling eggs to suffocate the embryo inside, addling eggs (shaking eggs so as to damage them to the point of not being viable), landowner education, signs to discourage goose feeding in parks, goose harassment techniques, landscape design changes, repellents, temporary barriers, and lethal removal of individual geese in trouble areas. These measures resulted in only modest or temporary relief from goose-related problems. In 1999 the committee concluded that expanded lethal removal of geese was the only option left to reduce the goose population and goose-related problems. The committee requested that USDA Wildlife Services conduct goose removal on behalf of the cities. Wildlife Services conducted an environmental assessment (EA) to determine whether the goose removal program would have an adverse environmental effect. It was determined that the program would have no adverse environmental effect, and in 2000 the US Fish and Wildlife Service (USFWS) granted Wildlife Services a permit to remove 3,500 resident geese from the Puget Sound area. During the public comment period of the EA process, secondary stakeholders were activated—people with animal welfare and animal rights beliefs who opposed lethal control of geese. A coalition of animal welfare and animal rights groups attempted to stop lethal control actions, first through a legal challenge and then through media campaigns, demonstrations, disruption of goose roundups, and violence against Wildlife Services employees. All of these challenges were eventually overcome, the goose removal was implemented, and steady and sustained reduction in negative effects was achieved.

Even some people who are experiencing the same kinds of negative effects from wildlife may disagree with others on acceptable means of addressing the problems. For example, effects of white-tailed deer in a suburban area (motor vehicle collisions, browse damage to ornamental plants, perceived risk of Lyme disease) may exceed residents' tolerance threshold. The vast majority of residents may even agree that populations need to be reduced markedly. Nevertheless, experience has shown that these consensus sentiments about the existence of a problem and desire for an outcome (fewer deer) do not translate into agreement about acceptable wildlife population control measures—some people may readily support direct control through culling, but others may be vehemently opposed to such action and instead petition for fertility control or trap-and-relocate. Such differences reflect diversity and complexity of ethical positions with respect to humans' relationship with wildlife.

Diversity and complexity of stakeholder perspectives point to another area of human dimensions critical to success of wildlife management: building cooperation and collabora-

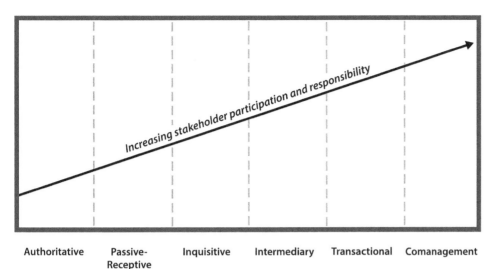

Figure 4.1. Approaches to public participation in wildlife resource governance and relative responsibility of partners and stakeholders versus wildlife management agencies

Authoritative   Passive-Receptive   Inquisitive   Intermediary   Transactional   Comanagement

Stakeholder Participation Approaches

tion among and between stakeholders and wildlife managers. Collaboration is necessary in many situations to assemble required expertise, funding, and staff for effective wildlife management. It is also helpful to overcome jurisdictional impediments and to bridge values gaps between stakeholder groups (Yankelovich 1991a, b; Wondolleck and Yaffee 2000; Beierle and Cayford 2002). Strong, positive relationships among various stakeholders and between them and wildlife agencies are key ingredients of successful collaborative conservation efforts (Lauber et al. 2011). The challenges of creating and managing collaborative ventures can be lessened if managers possess knowledge of the values and motivations of collaborators. This knowledge can aid efforts to forge productive working relationships (Decker et al. 2005). Research conducted in Michigan indicates that people's perceptions of how they are treated during decision processes (i.e., are they treated fairly, do they have a voice, is the decision process transparent) have much greater influence on the levels of trust in the state wildlife agency than perceptions of agency staff's technical competency (Riley et al. 2018).

Stakeholder Engagement: Evolution of Approaches to Input and Involvement

Since the 1990s, wildlife management has sought greater stakeholder involvement in more aspects of management. Wildlife managers are continuously improving their engagement of stakeholders in decision-making (Decker and Chase 1997, Chase et al. 2000). Stakeholder perceptions of transparency and fairness in decision-making, gained through participatory processes and coupled with process accountability (i.e., good governance), increases satisfaction, trust, and support for management (Lauber and Knuth 1997).

Stakeholder engagement practiced by wildlife agencies takes many forms. We describe six general approaches or postures taken with respect to stakeholders: expert authority, passive-receptive, inquisitive, intermediary, transactional, and comanagerial (Fig. 4.1).

EXPERT AUTHORITY APPROACH

Expert authority is a top-down approach in which wildlife managers make decisions and take actions unilaterally. Once the norm, today the expert authority approach is mainly reserved for special circumstances. Most notably, it is practiced in emergency situations, such as when a wildlife disease outbreak (e.g., chronic wasting disease) is discovered and what is referred to as an incident-command structure is triggered to marshal resources to contain the disease (Burgiel 2020).

PASSIVE-RECEPTIVE APPROACH

The passive-receptive approach is where wildlife managers welcome stakeholder input but do not seek it systematically. This approach implicitly assumes that if managers do not hear from stakeholders about their interests and concerns, those stakeholders do not care about the management issue and therefore have little standing in decision-making. This approach advantages stakeholders who can organize themselves to voice their interests and concerns. For example, in the 1970s, few wildlife management agencies systematically collected stakeholder input on white-tailed deer management. They received and considered unsolicited input from organized sportsmen's groups and farmers experiencing deer damage to crops, but they did not receive input from, and thus gave limited consideration to, other kinds of deer management stakeholders (e.g., hunters who were not members of a sportsmen's organization, nonhunters, motorists, gardeners, forest owners).

INQUISITIVE APPROACH

An inquisitive approach recognizes that unsolicited input alone can lead to bias because marginally important stakes can be magnified, and some important stakes can be downplayed or missed. Managers using the inquisitive approach rely on social science research and systematic evaluation to seek broader stakeholder information that can assist with decisions. It has become common practice for wildlife agencies to collect information systematically from wildlife management stakeholders on a range of issues, such as population management, acceptability of management techniques, and species reintroduction.

INTERMEDIARY APPROACH

The intermediary approach encourages two-way communication between stakeholder groups and the wildlife management agency about a wildlife issue but does not emphasize dialogue among stakeholder groups with different concerns. Instead, managers act as intermediaries, assuming much of the responsibility for deciphering, and often communicating, similarities and differences in stakeholder interests and positions. Managers operating in this mode attempt to weigh different, often competing, stakeholder concerns. For example, wildlife managers dealing with a controversial topic such as wolf conservation may choose to work separately with different stakeholder groups (e.g., ranchers, hunters, environmentalists) in an attempt to find common ground prior to or rather than convening representatives of the various interests. In this approach, responsibility for finding compromise falls largely on the shoulders of the manager.

TRANSACTIONAL APPROACH

In the transactional approach, stakeholder representatives describe their stakes to each other and wildlife managers, engaging directly rather than through the manager as intermediary, and they negotiate relative rank or weight of stakes. By learning about various perspectives on (or stakes in) the issue directly, conducting inquiry, discussing viewpoints, debating the issue, and compromising, stakeholder participants frequently reach consensus about appropriate objectives and courses of action (Nelson 1992, Stout et al. 1996). Wildlife managers have convened transactional stakeholder groups to inform a range of programs, from deer and black bear management, to imperiled species management (Siemer and Decker 2006, Haubold 2012, Fleegle et al. 2013, Pomeranz et al. 2014). Transactional approaches are resource intensive and challenging to organize and maintain; as a result, they are used sparingly by wildlife agencies.

COMANAGERIAL APPROACH

In comanagement, wildlife conservation agencies engage other government agencies, NGOs, local communities, and private landowners to share responsibility for management. For example, the Alaska Department of Fish and Game shares authority and comanages some marine mammals with Alaska native organizations and the USFWS. Comanagement is unlike the approaches discussed above, where the role of stakeholders is to provide varying degrees of input for decision-making by the agency. The specifics of collaboration are agreed to on a case-by-case basis, and the responsibility for wildlife conservation is shared in various ways, within legal constraints. Because these collaborations are tailored to individual circumstances, they take several forms (Raik et al. 2005).

Stakeholder engagement is a key part of governance of wildlife resources. Involvement may be as individuals, long-standing NGOs that represent member interests or provisional grassroots groups organically formed around a specific issue. Increasingly, wildlife management is a shared responsibility where good governance of wildlife resources is not simply a government agency exercising authority over, steadfastly retaining control of, or taking sole responsibility for wildlife resources. Good governance is wisely sharing responsibility for wildlife conservation with partners and stakeholders. In any approach taken, insight about stakeholders' beliefs and attitudes, patterns of behavior, and expectations of wildlife management is needed for success. In most complex wildlife issues, social science inquiry and stakeholder engagement processes are used in combination to assist decision-making and management implementation.

Wildlife Management as a Process

By now you likely have a sense that wildlife management is a multifaceted, iterative *process* with many subprocesses and decision points (Fig. 4.2). These include establishing broad goals and policies and involving stakeholders early on to define outcomes desired; setting specific objectives, selecting among alternatives (management actions); implementing those as management interventions; monitoring outcomes; assessing the interventions; and then revisiting objectives, goals, and policies with new insights derived from evaluation (Riley et al. 2003; Decker et al. 2012a:87–96). These facets of management require collecting information (research and monitoring; stakeholder engagement); analysis; planning and decision-making (including articulating fundamental and enabling objectives and selecting actions); implementing various kinds of actions directed at wildlife, habitat, and people; evaluation; and adjustment. Involving partners (e.g., other agencies, NGOs) and other stakeholders in the various facets of the management process adds to the complexity of the process and to the durability of the decisions made throughout the process.

Wildlife management seldom takes a tidy, linear course that unfolds predictably. Management, as depicted in an adaptive impact management process (Riley et al. 2003), is continuous and cyclical over a prescribed period. This is not surprising considering the management process takes place in a management environment (social-ecological system) composed of sociocultural, economic, political, and biotic and abiotic ecological components. An information base, built upon research and experience, includes knowledge about the ecological and human dimensions relevant to the management problem. Because the human dimensions of the management system tend to define management needs and constrain or enable management interventions, they should be well understood and integrated throughout all phases of the management process.

The diversity of expertise necessary to address the complexity of issues in wildlife management is typically greater than

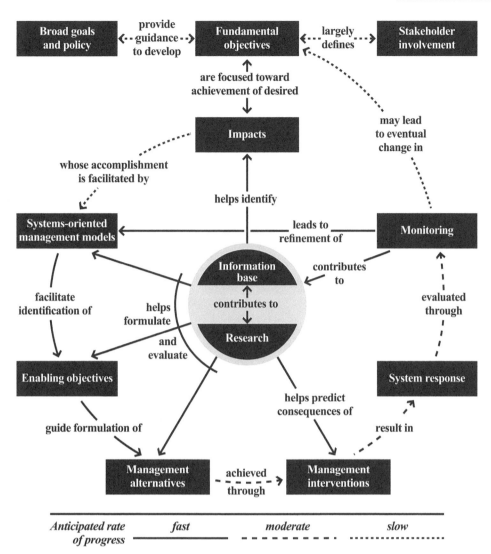

Figure 4.2. A comprehensive model of wildlife management. From Decker et al. (2012*a*)

what can be expected of any single person. Although some wildlife managers operate independently or in groups lacking needed expertise because of constraints imposed by budgets, geography, or other obstacles, much of the work in wildlife management gets done in teams. Preferably, teams of people from disciplines such as wildlife biology and ecology, social psychology, sociology, economics, communication, and law enforcement would be engaged in the multidisciplinary work of wildlife management. Such teams would include partners and other stakeholders. Ideally, this multidisciplinary, multi-party work would start with describing the management system from the diverse perspectives of the management team.

Articulating Management Systems: Thinking Like a Manager

Articulating management systems from a management team's perspective, following some of the tenets of soft-systems methodology (Checkland 1981, Checkland and Scholes 1999), reveals the complex interdependency of the biological and human dimensions of wildlife management. Models of management system descriptions developed by practitioners (i.e.,

concept maps of the systems in which they work; Decker et al. 2006) help identify factors that affect management and the desired outcomes of management interventions (Decker et al. 2012*a*; Fig. 4.3). Taking time to articulate the social-ecological system in which management occurs helps wildlife managers, partners, and stakeholders avoid jumping to familiar or conventional actions without adequately considering efficacy for achieving long-term fundamental objectives and short-term enabling objectives (Riley et al. 2003).

Clear expression of fundamental objectives for wildlife management, including those pertaining to desired human conditions vis-à-vis wildlife (e.g., positive attitudes about wildlife conservation; financial and political support for wildlife management; tolerance of some nuisance associated with human–wildlife coexistence) and wildlife population and habitat conditions, makes possible evaluation of all other considerations in the management process. Fundamental objectives emerge after comparing what is known about actual conditions (derived in part from analysis of the concerns and interests raised by scientists, biologists, managers, and stakeholders) with desired conditions to determine management need (i.e., the dispar-

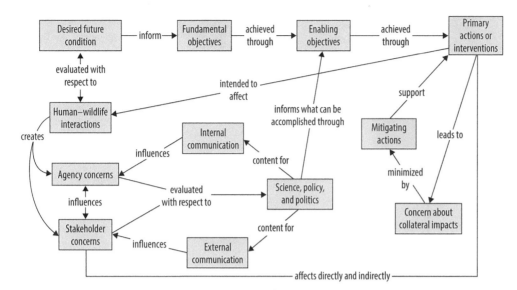

Figure 4.3. General architecture (outline) of a manager's model

ity or gap between actual and desired conditions). If analysis indicates management intervention is necessary, then more specific enabling objectives can be developed and socially acceptable management actions identified.

When seeking socially acceptable and otherwise viable management actions, collateral and subsequent effects of alternative management actions are assessed. Collateral effects occur during implementation of management actions. Subsequent (or secondary) effects occur as a consequence of achieving enabling objectives. Anticipating these effects and calculating their cost beforehand (direct cost to stakeholders and cost to mitigate) such that decisions are better informed is distinctly unlike reacting to effects when they arise. Wildlife management approached comprehensively includes proactive identification of these two types of effects, revealed and addressed in management planning.

Particularly difficult management issues arise when negative collateral or subsequent effects have the potential to offset the benefits of achieving objectives. An example of how collateral effects enter into consideration is the common management issue created by Canada geese. A first approximation of efficient management of overabundant suburban Canada geese can include some form of direct mortality (e.g., rounding up and euthanizing geese), but a study of the local human population may indicate that the collateral effects of taking that course of action would result in a huge controversy, and that alternative methods such as oiling or addling goose eggs would be more acceptable, even if population reduction took longer and with greater cost to the community.

Identifying collateral and subsequent effects often reveals likely additional stakeholders for management—people not necessarily affected by, experiencing problems with, or deriving benefits from wildlife prior to a management intervention. These potential stakeholders are people likely to be affected by management actions as they are implemented or by cascading effects from management achieving enabling objectives. Such stakeholders are created or activated by management inter-

ventions. In the Canada geese scenario, the initial thought of rounding up and killing geese may be acceptable to city park managers and health department officials, yet that approach would likely create and activate a set of vocal stakeholders concerned about the welfare of these waterfowl.

A managers' model lays out these various considerations for review and analysis. Such a conceptual model is not meant to establish final objectives or to select actions without stakeholder involvement; it is intended to help the manager or management team review the broad management situation comprehensively, particularly the many human dimensions to be considered. This professional groundwork allows managers to engage with partners and stakeholders more effectively. Structured decision-making can aid the actual selection of actions during the external part of the management process that includes stakeholder engagement.

Decisions

As emphasized earlier, the core function of management, regardless of the type of management, is making decisions. Wildlife management is no different in that respect. Decisions are needed to set the course of wildlife management (establish goals and objectives), to guide management actions (select biologically and socially feasible courses of action), or to be purposeful in not taking any action. Wildlife managers who are capable of integrating biological and human dimensions' insights into management decisions increase their chances of being successful. Decision analyses integrate scientifically derived knowledge and experience-based insight about human and biological factors to determine which actions are likely to contribute to achieving fundamental objectives (Gregory et al. 2012). Decision analyses that integrate social and biological data have been used in a range of wildlife management decisions (e.g., deer harvest regulations [Robinson et al. 2016]; turkey (*Meleagris gallopavo*) hunting season length and bag limits [Robinson et al. 2017]; duck hunting season dates [Fuller et al. 2021]). Considering only one dimension or the other dimin-

DANIEL J. WITTER (1950–)

Today, the human dimensions of wildlife management are routinely considered in agency decisions and programs, but that was not always the case. As late as the 1970s, wildlife agencies remained wary that considering public opinion would supplant biological considerations and that wildlife management would devolve into a popular vote based on public opinion polling. It would take years of work by trained human dimensions experts to allay those concerns and to demonstrate that integration of human dimensions considerations could be a turning point in wildlife management and conservation. Dan Witter was one of those pioneering experts.

Witter earned an associate's degree at Valley Forge Military College, a bachelor's degree in sociology from Millersville University, a master's degree in resource administration from Penn State, and a doctorate in watershed management from the University of Arizona. His PhD research focused on birdwatchers, hunters, and wildlife professionals' beliefs about the importance and management of wildlife. His dissertation was part of the first wave of research to demonstrate how solid social science could help wildlife managers compare and contrast the social values underlying wildlife management issues.

In 1976 the state of Missouri passed a citizen-initiated referendum that created a sales tax to raise funds for the programs of the Missouri Department of Conservation (MDC). Years before the vote, the MDC unveiled Design for Conservation, a strategic plan explaining how it would use new revenues, beyond traditional angling and hunting income, to serve all Missourians through fisheries, forestry, and wildlife management programs. Witter was well prepared to accept a policy analyst position that the MDC created in 1978 with Design for Conservation funding. He brought social science expertise and a passion for public service that helped his agency understand its many stakeholders and learn how to serve them through the strategic plan. He implemented a comprehensive research program that yielded insights on the behavior and values of the state's diverse publics.

As his career progressed, Dan played a larger role in incorporating such information into agency planning, policies, and regulations. Those efforts, sustained over a 26-year career at MDC, helped the agency fulfill its public service mission and bolster political and financial support for progressive goals. Effective integration of human dimensions contributed directly to sustained growth in the MDC's programs and helped affirm the agency as a national conservation leader. Along the way, Witter has contributed to the profession as an active member of The Wildlife Society and as a mentor to future professionals. Now a consultant with D. J. Case & Associates, he continues to provide guidance to wildlife management agencies that strive to integrate human dimensions considerations into their decisions.

Photo courtesy of D. J. Case & Associates

ishes chances that the management decision will be effective (i.e., successful and durable; Fuller et al. 2020). Combining adaptive resource management (Lancia et al. 1996) and adaptive impact management defined by Riley et al. (2003), where fundamental objectives are expressed in terms of impacts, is a method for integrating insights from biological and human dimensions of wildlife management (Enck et al. 2006). Structured decision-making is discussed further in Chapter 5 of this text; adaptive management is described in Organ et al. (2020).

HUMAN DIMENSIONS INQUIRY AND APPLICATION

Wildlife management generally relies on social science to improve understanding of (1) how and why people value wildlife; (2) benefits stakeholders expect from wildlife management (i.e., desired effects); (3) people's behaviors regarding wildlife (e.g., recreation activities, illegal and criminal activities); (4) social acceptability of management practices; and

(5) how various stakeholders affect or are affected by wildlife and wildlife management decisions. The following subsections describe the main social sciences used to create human dimensions insights and three areas of wildlife management interest: wildlife-related activity participation; human cognition (i.e., values, value orientations, beliefs, attitudes, norms, and motivations); and human dimensions applications (i.e., integrating human dimensions insight into management).

Social Sciences Central to Human Dimensions of Wildlife Management

The entire array of social sciences can assist with wildlife management decisions, but insights about the human dimensions of concern in wildlife management most frequently arise from three disciplines: social psychology, sociology, and economics.

Social Psychology

Improving the likelihood of achieving or sustaining outcomes in terms of impacts desired by stakeholders enhances wildlife managers' effectiveness. This requires knowledge of stakeholders' values, beliefs, attitudes, preferences, and expectations with respect to their interactions with wildlife or their reactions to management actions. Improving understanding of such human attributes is the focus of social psychology. This social science, when applied to wildlife management, offers wildlife managers insight about the bases for stakeholders' perceptions of effects, which are typically expressed in terms of attitudes and preferences. Though not a perfect predictor, understanding people's beliefs and value orientations contributes to anticipating their attitudes and behaviors.

One theoretical perspective in social psychology, the theory of reasoned action or planned behavior, provides a foundation for many human dimensions studies in wildlife management. This theory posits that behavior is a function of people's attitude toward a behavior, assumptions about the likelihood of a behavior happening, and social norms (i.e., established behavioral standards or patterns typical of a social group). This perspective helps us understand variations in stakeholder support for management actions. For example, it explains why people who support lethal control of beavers if the animals are perceived to carry disease do not support it if the reason is to prevent beavers from dropping trees on golf courses.

Social psychology applied to wildlife management seeks answers to practical questions such as: How do people value and evaluate wildlife and human–wildlife interactions? What makes management actions acceptable or not to various stakeholders in different contexts? And why do people participate (or not participate) in various wildlife-related activities? These questions can be approached using a theoretical framework referred to as the cognitive hierarchy (Fulton et al. 1996). Motivation theory also has been used to understand why people engage in wildlife-related activities.

Values, Beliefs, and Value Orientations

Wildlife values are typically regarded as stable, central modes of thought about wildlife (Manfredo et al. 2004). Values are described as fundamental, enduring beliefs that form a basis for evaluation of the desirability of an event or mode of conduct or the outcomes achieved through such action (Brown and Manfredo 1987). Nonetheless, values are not focused on specific objects, interactions, or situations. Rather, values are more abstract; they are concerned with desirable end-states such as equality or respect for life. Values are theorized to be widely shared by all people within a culture or community. But two people who hold the same value of respect for life may orient those values in different ways. One may believe, based on their unique set of life experiences and influences, that all wildlife should have rights similar to humans. Another may believe that wildlife should be used (including killed) humanely for the benefit of humans. In the late 1900s, Fulton et al. (1996) introduced the concept of wildlife value orientations as an explanation for such differences.

Value orientations are the direction and pattern of basic beliefs about wildlife. Basic beliefs are part of a cognitive hierarchy hypothesized to align as follows: values, basic beliefs, attitudes and norms, and behaviors. As mentioned above, people with the same underlying values associated with wildlife could be oriented differently in the expression of those values. Value orientations are better predictors of human behaviors such as participation in wildlife-related activities or how people may evaluate a proposed management actions such as lethal control of problem-causing black bears than are more general values and beliefs. People with a more utilitarian orientation toward wildlife, for example, are more likely to engage in consumptive recreation and support more aggressive forms of wildlife management than those with a protectionist orientation. While values have remained remarkably stable through time, wildlife-associated value orientations that arise from those values have changed to be more positive regarding select species such large carnivores and African (*Loxodonta africana*) and Asian (*Elephas maximus*) elephants, and more negative regarding utilitarian uses of wildlife such as trapping and certain kinds of hunting (George et al. 2016).

Some early human dimensions studies attempted to develop values typologies that described how segments of society relate to wildlife. For example, one characterized people based on a set of ten basic orientations toward animals: naturalistic, ecologistic, humanistic, moralistic, scientistic, aesthetic, utilitarian, dominionistic, negativistic, and neutralistic (Kellert (1980*a*, *b*). Another, the Wildlife Attitudes and Values Scale (Purdy and Decker 1989), sought to describe people's orientations toward wildlife in human–wildlife conflict situations, emphasizing traditional conservation, social benefits, and problem-acceptance perspectives. These attempts to understand how human values vary in the ways they manifest in beliefs, attitudes, and behaviors toward wildlife at a general level were imperfect yet useful in reinforcing the idea that people were not homogeneous in their core perspectives about wildlife and human uses of wildlife.

Attitudes

Values, value orientations, and basic belief patterns are important in shaping a person's attitudes, which can be thought of as favorable or unfavorable evaluations of objects, behaviors, or events (e.g., an encounter with a wild animal in one's yard, a category of wildlife such as predators or charismatic megafauna, a specific management action such as translocation of wildlife). Attitudes are a combination of beliefs about an object or action and negative or positive evaluations of the object or action (Heberlein 2012). Attitudes expressed about specific attitudinal objects (e.g., deer, grizzly bears) or actions (e.g., gray wolf restoration, beaver trapping) are believed to influence their behavior (e.g., participation in an activity, reaction to an encounter with a wild animal, political support, opposition to a policy action).

Public attitudes about wildlife and wildlife management are a major focus of human dimensions research because they provide insight into behaviors. Much of the research on public opinions, stakeholder preferences, and people's perceptions are essentially studies of attitudes. Some of these studies have made important contributions to wildlife managers' understanding of long-standing issues, often yielding surprises.

Managers' concerns about public attitudes toward potentially dangerous wildlife, especially those of people living along the wildland–urban interface, have motivated considerable human dimensions research (Wieczorek Hudenko et al. 2010). Research that illustrates the geographical and species diversity of attitudinal studies includes public attitudes about cougars in Colorado (Manfredo et al. 1998) and Montana (Riley and Decker 2000a); black bears in Colorado (Loker and Decker 1995) and New York (Siemer et al. 2009); wolves in Utah (Bruskotter et al. 2007), Alaska (Decker et al. 2006), and Japan (Sakurai et al. 2020); brown bears in Sweden (Ericsson et al. 2018) and Montana (McCool and Braithwaite 1989); tigers in Nepal (Carter et al. 2012) and Bangladesh (Inskip et al. 2016); rattlesnakes (*Crotalus* spp.) in central Connecticut (Keener-Eck et al. 2020) and Brazil (de Souza et al. 2019); and coyotes in suburban New York (Wieczorek Hudenko et al. 2008) and urban Chicago (Sponarski et al. 2018). Studies indicate that, even despite occurrence of human deaths, the public is divided with respect to how to control human–wildlife interactions. These findings can be partially explained by differences in risk perceptions (Riley and Decker 2000b, Carter et al. 2012), a topic discussed later in this chapter.

Public attitudes also are a critical consideration in decisions about where and how to restore wildlife populations (Riley and Sandström 2016). For instance, research on public attitudes has been used to examine the social feasibility of restoring large herbivores, such as elk and moose (*Alces alces*) that have potential for creating positive and negative effects for humans (Enck et al. 1998; McClafferty and Parkhurst 2001). This line of inquiry often involves predicting respondents' voting intentions on restoration initiatives using attitudinal measures. In the case of moose restoration in the Adirondack Mountain region of northern New York, the concern was less about the presence of moose than the means for enlarging their population (actively capture and relocate animals to aid distribution vs. allow moose to repopulate without management intervention; Lauber and Knuth 1997). In high-profile and controversial efforts to restore large carnivores, such as grizzly bears and wolves in the Greater Yellowstone Ecosystem, knowledge of the basis for polarized public attitudes can be helpful to the wildlife manager.

Norms

Normative beliefs, or social norms, are defined as shared beliefs about the acceptability of a situation, action, or outcome (Shelby et al. 1996). Norms are standards of acceptable behavior that inform people about what they should do or what most people in their social system are doing in a given context (Vaske and Whittaker 2004). Studies of the structural characteristics of norms (e.g., the intensity or strength of the norm and the level of agreement about the norm) in a specific context have practical value to many wildlife management and

wildlife law enforcement contexts (e.g., likely compliance with regulations).

Normative theory has improved managers' grasp of the relationship among tolerance of wildlife problems, acceptance of management actions, and human values with respect to wildlife. The normative model has been used in research to understand interactions among wildlife recreationists (Heberlein et al. 1982, Whittaker 1997), acceptability of wildlife management actions (Wittmann et al. 1998, Zinn et al. 1998, Campbell and Mackay 2003, Dougherty et al. 2003, Jonker et al. 2009), and prediction of people's response to a range of specific human–wildlife interactions or incidents (e.g., seeing wildlife in residential areas, wildlife injuring a person; Zinn et al. 1998).

Tolerance for interactions with wildlife varies across stakeholder groups, situations, species of wildlife, and time (Brenner and Metcalf 2020). This makes it difficult to predict situation-specific stakeholder behavior and expectations of management without inquiry into the particular context of interest or a similar one. Stakeholder acceptance of management actions is as variable as tolerance of human–wildlife interactions. Traditional management methods (e.g., shooting, trapping) can be effective for removing individual problem-causing animals or reducing wildlife populations but may not be socially acceptable to urban residents (Zinn et al. 1998). An understanding of how stakeholders perceive particular management actions can help wildlife agencies minimize controversy when choosing among management alternatives (Loker et al. 1999).

A common question in wildlife management is, How many animals should there be in any given population? For instance, how many wolves should there be in the Great Lakes Ecosystem? How many white-tailed deer should there be in the Cleveland Metroparks? How many, if any, feral hogs should be in Missouri? Wildlife biologists often deliberate about how many animals *can* be or *are* in an area. But how many animals *should* be in an area is another matter because differences in contexts influence norms regarding acceptable population size and management actions.

The wildlife acceptance capacity (WAC) concept illustrates integration of social sciences in wildlife management (Fig. 4.4). The WAC proposes a "maximum wildlife population level in an area that is acceptable to people" (Decker and Purdy 1988:53). Basically, WAC suggests that thresholds of acceptability exist for the effects from human–wildlife interactions produced by a wildlife population of a certain density and distribution in a particular context.

Similar to the concept of range of tolerable conditions in the norm literature (Vaske and Whittaker 2004), the WAC concept suggests that a person's acceptance threshold is situation specific and depends on the severity or desirability of effects produced by human–wildlife interactions associated with the co-occurring populations of humans and wildlife. Accepting this premise, the question for managers becomes, What types of effects should occur and at what frequency? These

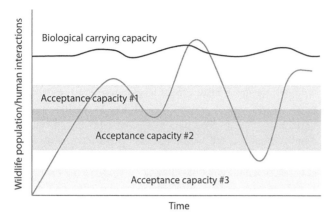

Figure 4.4. A hypothetical model of wildlife stakeholder acceptance capacity for a wildlife species. The gray line depicts a population trend through time of the species in relation to biological carrying capacity of the species and ranges of acceptance capacity expressed by three different types of stakeholders; two overlap in capacities, whereas a third has a separate and lower capacity.

effects, often focusing on negative outcomes (i.e., conflicts, problems), can be arranged along a continuum ranging from nuisance (e.g., bears raiding dumpsters), to economic or aesthetic effects (e.g., elk eating ornamental plants), to health and safety threats (e.g., Lyme disease; Decker 1991).

Frequency and nature of effects influence tolerance and intolerance along a continuum of strength and valence (Riley and Sandström 2016). In one study, the more severe the problem, with severity being reflected by kind of effect (e.g., nuisance versus threat to human safety) or amount (e.g., infrequent versus nightly trash can dumping by raccoons [*Procyon lotor*]), the more likely urban residents were to accept lethal methods for managing wildlife. Residents of suburban communities in New York, for example, were more willing to accept aesthetic or economic wildlife effects (i.e., damage to ornamental plantings) than health risks such as disease (Connelly et al. 1987). Similar findings have emerged in other studies (Decker and Gavin 1987, Stout et al. 1997, Jonker et al. 2009), indicating risks to human health and safety tend to be least acceptable (Wieczorek Hudenko et al. 2010).

The original WAC concept was advanced with a more nuanced view of tolerance, where WAC accounts for multiple types and levels of human–wildlife interactions acceptable to stakeholders in a particular situation (Carpenter et al. 2000, Bruskotter et al. 2015). This refinement recognizes that people perceive negative and positive effects associated with wildlife, and that stakeholders weigh the positives and negatives differentially. Although prior approaches suggest fostering a mix of human–wildlife interactions that minimize negative effects, this second-generation WAC concept focuses on estimating a level of interactions that yields the greatest net benefit while keeping negative effects at socially acceptable levels (often with some type of mitigation involved). Importantly, if causes of the human–wildlife interactions that result in WAC are understood, WAC may be changed in some instances without

having to directly manipulate wildlife populations. For example, research revealed that WAC for black bears in the Catskill Mountains of New York was affected by the extent and frequency of bears gaining access to residential garbage, pet food, and birdfeeders. An educational effort to reduce access to human foods, the New York NeighBEARhood Watch Program, in combination with mass media, influenced risk perceptions and hence WAC (Gore and Knuth 2009). This sort of integrated approach is central to impact management (Riley et al. 2003).

The concept of WAC has been applied also in international settings. Research in Chitwan National Park, Nepal, revealed a distinct spatial pattern of human tolerance for tigers associated with socioeconomic and caste status (Carter et al. 2014). The area with greatest WAC for tigers was correlated with higher caste standing, less dependency on forest resources, and greater influence on park management. Management intensity, aimed at minimizing human–tiger interactions by constructing fencing and initiating more regular ranger patrols in the villages, greatly improved WAC. To increase WAC in those areas with illegal human-induced tiger mortality, a ranger station was proposed to create a visible management presence, and an economic compensation program was initiated for losses due to tigers. All of these actions are meant to increase WAC without having to directly manage (which usually means killing individual animals) the globally threatened tiger population.

Understanding public support or opposition for lethal control of wildlife merits careful inquiry in even the most severe and ostensibly straightforward situations (Zinn et al. 1998, Sponarski et al. 2018). Although the same management tactics can be applied in a variety of situations, the acceptability of the tactic may vary depending on the species of wildlife. Variations in perceptions of the animal's image (weasel [*Mustela* spp.], indiscriminate killer of prey; river otter [*Lontra Canadensis*], playful), abundance, and effect potential may influence norms with respect to acceptability of killing the animal. Coyotes, for example, often are portrayed as scavengers or pests (McIvor and Conover 1994), so it is not surprising that among many people killing a coyote is more acceptable than killing individuals of several other species of wildlife (Wittmann et al. 1998). Nonetheless, even some people who perceive risks to human safety from coyotes may still prefer nonlethal control (Sponarski et al. 2018). These kinds of differences are the ingredients of management controversy and therefore best revealed before management programs are launched.

An animal's potential to create severe effects besides harm to humans also influences social norms about their management. Beavers, coyotes, and cougars can all influence human activities, but in different ways. Beavers alter rural and urban-rural fringe landscapes with dams that cause flooding; coyotes attack pets; and cougars attack livestock, pets, and occasionally humans. Each species' behavior, in conjunction with its reputation and potential for causing problems for humans, influences the acceptability of lethal action as part of a management scheme (Zinn et al. 1998, Larson et al. 2016).

The idea of situation-specific impact dependency as a determinant of management acceptability has been explored in the context of large predator and herbivore management in Alaska (Decker et al. 2006). Research about social norms demonstrated that acceptance of predator control depended on the effects of predation on caribou (*Rangifer tarandus*) and moose, and the perceived secondary effects of this predation for people; that is, the effects of fewer moose and caribou available for human consumption. Alaskans were more likely to support the use of lethal methods to control grizzly bears and wolves in situations where the effect of their predation on moose and caribou had the greatest effect on hunters' chances of harvesting these cervids. Conversely, lethal control of predators was less likely to be supported in situations where the effect of predators on moose and caribou was seen as limited to recreational interests in hunting (hunting for primarily enjoyment vs. reliance on big game for food).

Motivation and Satisfaction

Understanding wildlife recreationists' motivations and satisfactions with respect to participation in an activity is basic knowledge for wildlife program planning. Motivation theory suggests that people are motivated to take actions to achieve particular goals (i.e., they seek certain outcomes or benefits from their experiences). Two lasting approaches to conceptualizing and studying motivations have emerged. One emphasizes a multiple-satisfactions approach to wildlife recreation management (Hendee 1974). This approach suggests that we can identify and manage for categories of wildlife recreationists who differ based on the types of satisfaction they seek and receive. A second line of inquiry emphasizes the importance of understanding the psychological outcomes recreationists desire and derive from participation (Driver et al. 1991). This work also emphasizes the importance of insight gained by examining the context (e.g., setting and activity) associated with those outcomes. Hunters and hunting were a focus of this area of research initially. Planning and managing for different types of hunters were supported by knowledge about the mix of outcomes, activities, and settings sought by participants (Hautaluoma and Brown 1978, Manfredo and Larson 1993).

As indicated, understanding of hunters was deepened by motivation and satisfaction research, from which several key insights emerged. Early research revealed that hunters seek multiple satisfactions from hunting (e.g., companionship, nature appreciation) and can be meaningfully segmented with respect to the satisfactions and benefits sought (Potter et al. 1973, Driver 1976). Hunters also were reported to have a range of primary motivational orientations (i.e., affiliative, achievement, appreciative), often expressed differently by the same hunters in different contexts (Decker et al. 1987). The diversity of hunters' motivations is reflected in their many activity and experience preferences (Hautaluoma and Brown 1978). Knowing the primary motivational orientations of hunters has practical applications; for example, it assists with development of hunting regulations (e.g., the need to issue multiple ant-

lerless deer permits per individual to increase the probability that hunters will harvest at least one antlerless deer; Decker and Connelly 1989). Insights from motivation and satisfaction studies shed light on ways to improve hunter experiences and wildlife population management programs that depend on hunter participation.

While identifying the diversity of experiences desired by participants in wildlife-related activities, researchers uncovered the value of differentiating users into homogeneous and meaningful subgroups (i.e., segmentation). In one conceptualization of subgroups, Bryan (1977:29) defined recreation specialization as a "continuum of behavior from the general to the particular, reflected by equipment and skills used in the sport." Within the continuum, individuals range from the novice to the specialist. User classes differ with respect to motivations, the extent of prior experience with an activity, and commitment to an activity. As people become more specialized, they become more particular in their setting preferences, objectives for various experiences, and equipment. More specialized users are also more likely to have specific managerial requirements and more likely to communicate with managers. Research has applied the concept of specialization to hunting (Miller and Graefe 2000) and wildlife viewing (McFarlane 1996). The work on motivational orientations revealed that an individual can have different motivations for participation in different kinds of hunting and in different social and environmental settings.

Sociology

Values, beliefs, and attitudes described in social psychology are influenced by the societies and social environments in which people live and interact. Similar to other social sciences, sociology is concerned with understanding and predicting human behavior: what people do as members of groups or when interacting with one another, society overall, or the environment (Stedman 2012). Large-scale socioeconomic phenomena such as globalization, urbanization and suburbanization, human population diversity, educational levels, and associated economic affluence all affect the way people interact with and value wildlife (Manfredo et al. 2020a). Sociological inquiry reveals how individuals are influenced by society or their social interactions, and how individuals in turn continually shape societies and institutions, including wildlife management and conservation. It is important to keep in mind that sociological inquiry only points to what can be done, not necessarily what should be done, which is the purview of ethics.

Social differences occur between regions of the United States (e.g., Northeast vs. Southeast) and within them (e.g., urban vs. rural communities) that can significantly influence objectives and approaches for wildlife management. The complexity of wildlife management is caused in large part because the institution of wildlife management operates in a sociological context characterized by diverse and continually changing value orientations with respect to management outcomes desired and methods preferred. Fortunately, sociology can help managers understand why different value orientations emerge in society and through knowledge of social trends predict how value orientations are likely to come into play for wildlife management.

For example, although values remain relatively stable among wildlife stakeholders in the United States, current trends in value orientations, defined and discussed earlier in the social psychology section of this chapter, are changing from predominantly utilitarian orientations (e.g., pro-use of wildlife, interest in benefits from wildlife derived by humans) toward mutualistic orientations (e.g., wildlife living unmolested by humans, wildlife accorded similar if not the same rights as humans, wildlife should exist for their own sake; Manfredo et al. 2009, 2020b). These changes, most notable with respect to select species such as large carnivores and others sometimes thought of as stigmatized or iconic (George et al. 2016), likely will result in more people expressing protectionist attitudes and management preferences. An extension of these findings is that we can expect the United States population as a whole to be less tolerant of lethal management actions or consumptive uses of wildlife. This move toward increased expression of mutualistic orientations is influenced by broad sociocultural factors such as level of education (e.g., increased educational levels correlated with increased mutualistic attitudes), residential setting (e.g., urban being more mutualistic vs. rural being more utilitarian), and level of wealth (e.g., greater annual income is correlated with greater mutualistic attitudes). The change afoot is reinforced by a growing incidence of anthropomorphism, the tendency to assign human characteristics, motives, behaviors, and abilities to nonhuman entities (Manfredo et al. 2020c).

Demography

Demography is the study of population dynamics—in this case human populations—and the relationships between economic, social, cultural, and biological processes that influence a population (Chapter 7 addresses animal demography). It is possible to study the demographic characteristics of humans to better understand and predict human behaviors such as participation in wildlife-related activities or acceptance of various management methods. For instance, although many factors affect participation in hunting, demographic attributes of increasing urbanization, suburbanization, and age structure in hunters are affecting the number of people who purchase hunting licenses (Winkler and Warnke 2013).

Other demographic traits that affect participation in hunting are primarily sociocultural, such as change in use of leisure time with the adoption of technology and organized activities competing for people's time. Not only is leisure time diminishing, but so, too, is regular discretionary time among working parents or potential mentors for young hunters. On the other side of the age spectrum, a growing segment of the human population in the United States is composed of people ≥65 years of age, people who are transitioning away from regular participation in strenuous outdoor-related activities. Furthermore, the fastest-growing ethnic groups are Hispanic and Asian, who are expected to triple in proportion by 2050 (Passel

and Cohn 2008), neither of which have strong cultural traditions of recreational hunting as practiced in the United States. These demographic phenomena affect uses of wildlife and support for conservation.

An example of large-scale sociological transformation is occurring in rural landscapes in the United States. Rural America as a place and a way of life is undergoing major transformation (Fitchen 1991). The changing human population composition of communities is influencing people's desired interactions with wildlife, both in the United States and abroad (Rupprecht 2017, Ericsson et al. 2018). As mentioned previously, urbanization and suburbanization are associated with changes from utilitarian value orientations toward those described as mutualistic or protectionist. In some situations, these changes lead to subcultures of established residents and recent migrants existing side-by-side on the landscape, with potential for creating contentious situations in wildlife management (Adams 2016). Awareness of human demographic trends helps wildlife managers plan for problem-reduction needs and benefits-enhancement opportunities vis-à-vis wildlife resources.

Organizational Behavior

Rapid ecological and social changes happening in the United States and elsewhere exert pressure on many institutions to change so they better respond to societal needs and interests. The wildlife management and conservation institution in the United States is no exception (Jacobson and Decker 2008, Jacobson et al. 2010, Manfredo et al. 2020a). Voices from within and outside the wildlife institution have called for greater program breadth, attention to a wider array of management outcomes, and improved public perceptions of relevancy and value of the institution (particularly state wildlife agencies). Additionally, reform of wildlife resource governance has been advocated, consistent with the ideology of wildlife resources being a public trust. Regardless of reforms desired, they require organizational behavior change at multiple levels and in multiple structural components of the institution, especially state wildlife agencies that are the building blocks of the overall institution (Decker et al. 2016). Change of this extent and nature requires expertise in organizational behavior (Lauricella et al. 2017, Ford et al. 2020a).

Organizational behavior is a field of study that investigates the influence of individuals, groups, and structure on behavior within organizations, for the purpose of applying such knowledge to improve an organization's effectiveness (e.g., meeting stakeholder expectations, achieving conservation goals). The organizational behavior field typically draws on other social sciences such as psychology, sociology, social psychology, anthropology, and political science for theoretical and methodological guidance. As wildlife management has grown in complexity, the importance of applying the sciences, theories, and methods of organizational behavior and organizational change have grown concomitantly. For example, interest in reforming the institution (Jacobson and Decker 2008, Jacobson et al. 2010) and influencing institutional traits that constrain

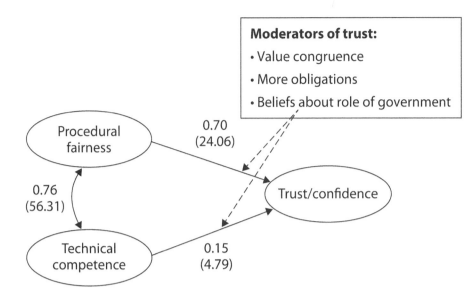

Figure 4.5. A structural equation model of factors that affect trust or confidence in the Michigan Department of Natural Resources' Wildlife Division, Michigan, USA. Data are from a mail-back questionnaire (sample *n* = 2,703) of resident licensed Michigan hunters >18 years old in 2013. Standardized path coefficients associated with direct effects are indicated by arrows, and Z-statistics appear in parentheses.

funding (Jacobson et al. 2007) and public trust in wildlife agencies (Riley et al. 2018) are served by organizational behavior research.

Researchers in Michigan used structural equation modeling (Fig. 4.5) to examine the influence various factors on trust in a state wildlife agency by a prominent category of wildlife stakeholder, licensed hunters ≥18 years old (Riley et al. 2018). The researchers used a randomized survey (*n* = 2,708), stratified by sociocultural regions across Michigan, to test various factors known to affect trust. The model identified the two factors of procedural fairness and competency as the strongest predicators of trust. Stakeholders' perceptions of how they were treated during decision processes (i.e., were they treated fairly, did they believe they had a voice, was the decision process readily understood) had four times more influence on levels of trust in their state wildlife agency than perceptions of the agency's technical competency. Perceived congruency of values between stakeholders and the agency (i.e., how closely aligned stakeholders perceived agency values with their own) was the most influential modifier of the relationship of fairness and competency on trust. The model yielded similar results no matter the geographical or sociocultural location within the diverse state. The complex and dynamic nature of wildlife management assures there always will be factors affecting trust that wildlife professionals can do little to change, yet investments in building effective and inclusive stakeholder engagement can be expected to foster trust and confidence in wildlife managers regardless of geographical or sociocultural location.

Economics

In perhaps the most straightforward type of application, economics helps wildlife managers evaluate costs of management alternatives. For example, cost–benefit analysis was used to compare two types of shotgun hunts for control of white-tailed deer in southwest suburban Carbondale, Illinois, where researchers reported that the cost for a 10-day hunt was $10,575 for one alternative and $7,467 for the other (Hubbard and Nielsen 2011). While such basic analyses are useful in decision-making, the variety of economic tools available to managers can provide much more insight into the economic aspects of impacts management.

Economics is central to an impacts approach to wildlife management because positive and negative impacts from human–wildlife interactions are essentially benefits and costs. Wildlife and management of wildlife have many direct, indirect, and induced economic effects on individuals, communities, states, regions, and even nations. Many human activities can harm wildlife resources. A variety of valuation tools exist to reveal the economic effects of activities that enhance or harm wildlife. These valuation tools can be essential when determining overall costs of proposed land-use projects and economic consequences of regulation or deregulation.

Perhaps the best-known and certainly longest-running effort to monitor expenditures attributable to wildlife-related recreation is the National Survey of Fishing, Hunting, and Wildlife-Associated Recreation. Through a partnership of state wildlife agencies, the USFWS, and national conservation organizations, this national survey estimates expenditures for wildlife-based recreation approximately every five years in the United States (initiated in 1955). The 2016 survey (US Department of the Interior et al. 2018) estimated that 11.5 million Americans ≥16 years hunted, and 86 million were involved in watching wildlife either around the home (81.1 million) or away from home (23.7 million; 18.8 million of these people watched wildlife around the home and away from home). In pursuing their interests in wildlife, people spent more than $102 billion for equipment, travel, licenses, and fees associated with wildlife-related activities ($26.2 billion for hunting and $75.9 billion for wildlife-watching).

These estimates of direct expenditures are useful, but they incompletely describe the overall economic value of wildlife (e.g., nonmarket values). Fortunately, economists have developed ways of valuation that capture attributes of wildlife that cannot be assigned a market value. For example, a comprehensive view of values associated with wildlife includes use values and non-use values (Ready 2012). Use values are both consumptive, such as commercial harvest and recreational harvest, and nonconsumptive. Non-use values include option value (value of retaining options for the future) and existence value (the psychological benefits that accrue from the mere knowledge that a good exists and will continue to exist). Both are important concepts in wildlife conservation (Ready 2012:68–83).

Economics aids wildlife managers in evaluating trade-offs (benefits vs. costs) among competing uses of limited wildlife resources. For example, if management increases the number of osprey (*Pandion haliaetus*) locally, that action will generate benefits (increase the value of osprey) to people who enjoy observing these raptors. Likewise, wild turkey hunters value the opportunity to hunt turkeys, so if more public hunting lands are managed as turkey habitat and produce more turkeys, then the opportunity to hunt turkeys increases, generating benefits to hunters. To determine whether the costs of these viewing and habitat improvements are justified, managers need a way to measure the value of the enhanced opportunities to the users (benefits), even though they do not pay some unit price for the additional ospreys or turkeys. The value of these public goods (goods that people value but that are not sold in markets) can be estimated using methods of nonmarket valuation. These methods are of two types: methods that rely on observed behavior (revealed-preference methods) and methods that rely on survey questionnaires (stated-preference methods). Specific techniques in both of these types quantify nonmarket values of wildlife (e.g., contingency valuation method [willingness to pay], travel cost method). With estimates of value (benefits) in hand, decisions about whether a program is worth the costs can be made objectively (Shwiff et al. 2013).

Another way economics aids wildlife managers is by estimating total economic effects that arise from quantifiable expenditures. Although wildlife may be a public good that is not traded in markets, it often supports associated markets and economic activity. For example, as noted above, the 2016 National Survey of Fishing, Hunting, and Wildlife-Associated Recreation documented that hunters and wildlife viewers spent more than $102 billion on these activities in 2016. These expenditures provide the foundation for economic activity such as employment in manufacturing and sales of equipment used by wildlife enthusiasts and in service industries (e.g., lodging, food service, guides, outfitters, transportation) found in the areas where wildlife-associated recreation occurs. Additionally, economic activity associated with wildlife recreation generates tax revenues for local, state, and federal governments. Methods exist, however, for estimating effects of wildlife recreation expenditures on employment, income, and tax revenues that contribute to local and state economies.

These broad indicators of economic value of wildlife can be valuable data in decisions about projects intended to enhance wildlife or in issuing permits for projects that are expected to have detrimental effects on wildlife.

The economic effect of wildlife management actions in a geographic area is illustrated in a Wyoming case, where an analysis of five management options for bison (*Bison bison*) and elk revealed several potential economic effects (Loomis and Caughlan 2004). The 2004 study found that eliminating elk-feeding (except for emergency purposes during severe winters) and allowing bison hunting on the National Elk Refuge (NER) would result in a decrease in visitation and visitor spending of about 10%. The total (direct, indirect, and induced) decrease in jobs in Teton Counties of Wyoming and Idaho (the local area focused on in the study) was estimated at 1,416, representing a 5.5% decrease in employment in these counties. Expanding feeding from the average of 65 days at the time of the study to 80 days and allowing bison hunting on the NER would result in a decrease in visitation and visitor spending of about 7% to 8%. The projected total decrease in jobs as a consequence of these management actions in the two counties was estimated at 1,078, a 4.2% reduction. The no active management option of never feeding or hunting bison and elk on the NER resulted in a large decrease in visitation and visitor spending of 20%. The loss of jobs in the two counties under a scheme where no active management would be undertaken was estimated at 2,916, an 11.4% decrease. Maintaining the feeding program that existed at the time of the study but allowing bison hunting on the NER resulted in a small positive overall effect on visitor spending of >1%, with a concomitant 1% increase in jobs (204 jobs; Loomis and Caughlan 2004).

Communication in Wildlife Management

Communication is an essential part of wildlife management that relies heavily on human dimensions information. Organizational efficiency and effectiveness depend in part on how well wildlife professionals communicate with one another, with elected and appointed decision makers, and with wildlife management stakeholders. The goals of communication by wildlife professionals and their agencies can be *performative* (i.e., maintaining relationships and image), *informative* (i.e., sharing information, educating), or *persuasive* (i.e., promoting beliefs, attitudes, or behaviors that contribute to wildlife conservation objectives). Applications of communication research in wildlife management are as diverse as the stakeholders served and issues addressed.

The foundation for consistently effective communication is planning, guided by an understanding of how performative, informative, and persuasive communication can help achieve particular management objectives. In today's complex sociocultural and political environment, planning for communication is a necessary, ever-present component of wildlife management program strategy. Communication at minimum seeks to gain intelligence about the management environment through multiway communication with stakeholders and others; keep internal and external stakeholders informed about management processes and activities; and influence knowledge, beliefs, attitudes, and behaviors of partners, cooperators, stakeholders, and others. These communication objectives and associated actions are integral to a comprehensive management program.

Effective communication—meaning effective message development, message delivery, and solicitation of feedback from message recipients—often is supported by context-specific human dimensions inquiry that characterizes the intended audiences and their communication habits or preferences. Issues of health and safety of humans and wildlife illustrate how human dimensions insight has provided a foundation for communication programs delivered by wildlife agencies. First, considerable research effort has been directed to understanding stakeholders' perceptions of wildlife-related health and safety risks (Decker et al. 2010, 2012b; Hanisch-Kirkbride et al. 2013). For example, human dimensions research has helped wildlife agencies understand how laypeople perceive wildlife health (Hanisch-Kirkbride et al. 2014) and how to communicate with target audiences to encourage adoption of behaviors that reduce wildlife disease transmission, including zoonotic diseases (Decker et al. 2011b, Evensen et al. 2012, Lu et al. 2016, Triezenberg et al. 2014). Human dimensions studies also have helped wildlife agencies understand and communicate with specific audiences to reduce interactions that jeopardize the safety of humans and wildlife (e.g., bears [Gore et al. 2006, 2007; Gore and Knuth 2009] and cougars [Gigliotti et al. 2020]).

Attention to risk perception and communication in wildlife management stems from an important human dimensions fact: stakeholders' perceptions are in effect their reality, and that reality informs and conditions their beliefs, attitudes, and behaviors, including their wildlife management preferences. Risk communication—largely informative and persuasive—can serve wildlife management in situations where the public's perceptions of risks are associated with wildlife motivate or affect management. Risk communication theory provides insight about wildlife-related risk information needs of stakeholders and their responses to risk messages. Risk includes a *technical* component that assesses the probability of occurrence and severity of consequence(s) and a *value-based* component that assesses the level of dread (and other emotions) associated with an event. Variations in stakeholders' risk perception can be understood by examining differences in individuals' traits (e.g., sociodemographic characteristics) and sociocultural influences (e.g., social norms, worldview). Public perceptions of risk are as important to risk assessment as determining hazard exposure probabilities. Importantly, public perceptions of wildlife-related risks do not always mirror the perceptions of wildlife managers, indicating the importance of context-specific risk perception research.

Risk assessment efforts link to wildlife decision-making through risk communication, which facilitates incorporating public risk perceptions and expert assessments into the decision-making process. Managers might use a persuasive risk communication approach to inform public beliefs and attitudes about decision alternatives, gain acceptance of a preferred management strategy, or motivate risk-avoidance behaviors (e.g., avoiding the urge to feed ground squirrels [Sciuridae] in western parks and thereby minimize human exposure to the bacterium that causes plague).

Human dimensions researchers have relied on multiple theoretical frameworks in studies designed to improve communication with wildlife management stakeholders. For purposes of illustration, we discuss how two of those theoretical frameworks have guided development of communication about wildlife-related health and safety effects.

SOCIAL AMPLIFICATION OF RISK

Mass media (e.g., print media, television, radio, social media) play a significant role in disclosing risks to society and discussing strategies to reduce risk. This can be a valuable service to public well-being. But exposure to messages conveyed in mass media also may increase public perceptions of risks presented by zoonotic diseases (e.g., rabies, plague, SARS, West Nile Virus, COVID-19) or injurious encounters with wildlife (e.g., bears, coyotes, cougars) through a process known as the social amplification of risk (Kasperson et al. 1988).

Researchers have applied social amplification of risk theory to understand how to communicate with wildlife management stakeholders. For example, Gore and Knuth (2009) applied the theory to their study of a community-based black bear management initiative. They reported that exposure to newspaper articles about black bears was associated with increased perception of bear-related risks in the community, while exposure to more balanced messages from the state bear management program had no measurable effect on risk perceptions. Siemer et al. (2009) also drew upon social amplification of risk theory to study black bear management stakeholders. Using data from a mail survey of New York residents, they explored the extent to which print media use and television viewing could influence people's concern about risks posed by black bears. They reported no evidence that print media coverage was amplifying perceived risks associated with black bears, but television viewing was related to elevated concern.

Hart et al. (2011) applied the social amplification of risk framework to examine the roles that environmental values and media use play in promoting public engagement in wildlife management decision-making. Using structural equation modeling, they analyzed data from a telephone survey with a random sample of residents in Upstate New York. They reported that environmental values influenced people's level of concern, health risk perceptions, and engagement in wildlife management processes. They reported that media use, by amplifying risk perceptions and concern about reducing risks, also increased citizens' interest in participating in wildlife management decision-making.

ELABORATION LIKELIHOOD MODEL

The elaboration likelihood model is a theory of persuasive communication (Petty and Cacioppo 1986). It posits that messages intended to persuade (i.e., to change attitudes and ultimately influence behavior) are processed on a continuum

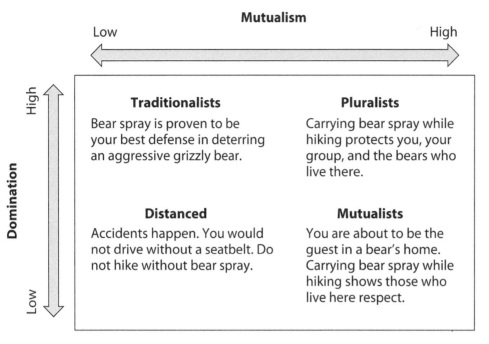

Figure 4.6. Examples of value-framed messages about bear spray designed to increase relevance using wildlife value orientation typology. From Miller et al. (2018)

that ranges from peripheral to central. Elaboration refers to the cognitive process of drawing on one's beliefs and experiences to add additional information to or expand on messages received by the individual. Humans are continually barraged with persuasive communication, yet we have a limited capacity to process information, and we reserve that processing capacity for messages that seem most important or immediately relevant to our lives. Many persuasive messages are processed peripherally—they are given little thought or elaboration (i.e., they do not tap into one's internalized cognitions for enhancement). But a smaller percentage of persuasive messages are given careful consideration; this is referred to as central-route processing (Petty and Cacioppo 1986). Persuasion that results from central-route processing of messages is augmented by elaboration and tends to result in more lasting behavior change than persuasion resulting from peripheral information processing (Petty et al. 1992). People are more motivated to use central-route processing and elaborate on messages when they perceive a message to have personal relevance. Relevance is sensed when messages connect with our values, beliefs, or value orientations. Wildlife-related risks tend to make such connections; thus risk communication focused on wildlife typically is powerful in attracting public attention.

Human dimensions researchers have applied the elaboration likelihood model to help managers understand how and why people may be persuaded to take actions to reduce wildlife-related health and safety risks (Lackey and Ham 2003, 2004; Hockett and Hall 2007; Miller et al. 2018, 2019). Research on hikers in Yellowstone National Park provides a useful example (Miller et al. 2018). Every year many hikers in Yellowstone travel through grizzly bear habitat, creating potential for conflicts that endanger both people and bears. Park managers

conduct ongoing communication campaigns to encourage all hikers to carry bear spray, a nonlethal deterrent to bear attacks. Miller et al. (2018) designed a study to test whether hikers in Yellowstone were more likely to use central-route processing when messages are consistent with a hiker's wildlife value orientation (WVO). In 2016 they systematically sampled 777 groups of hikers during peak hiking season (July–August). With the aid of computer tablets, hikers completed a questionnaire that measured the mutualist WVO and the domination WVO (basically categorized individuals as being either high or low with respect to each WVO). This allowed the researchers to place respondents in one of four cells of a matrix, labeled as follows: traditionalists (high domination, low mutualism), pluralists (high domination, high mutualism), mutualists (high mutualism, low domination), and distanced (low mutualism, low domination) (Fig. 4.6). The questionnaire also asked hikers to report the importance they placed on eight different messages about bears and bear repellent spray (importance was used as a proxy for message relevance). Five of the messages were taken from park communication materials in use at the time, and three others were designed by the researchers. They reported that messages designed to appeal to mutualistic values were rated as more important to mutualist and pluralist hikers than they were to traditionalists. Study results provided strong empirical evidence that messages matching people's WVOs increase message relevancy. The researchers suggested that because hikers tend to hold mutualism WVOs, messages with a mutualistic frame may increase the relevancy of messages for their target audience, which by extension should more likely result in the desired behavior among hikers. For example, communicating that Yellowstone hikers are guests in a bear's home would be useful because that message resonates with the mutualistic WVO of many hikers (Fig. 4.6).

Wildlife-Related Activity Participation

Documenting wildlife-related activity participation is one of the most straightforward kinds of human dimensions studies and provides information useful for several purposes. Studies in this genre of inquiry typically inventory and characterize wildlife users, monitor trends in wildlife recreation participation, and identify use patterns (e.g., location, volume, and kinds of use). Such information is useful for wildlife program planning at various levels.

Following national trends in wildlife recreation in the United States has been facilitated by the National Survey of Fishing, Hunting, and Wildlife-Associated Recreation conducted approximately every five years since 1955 by or for the USFWS (US Department of Interior et al. 2018) and the National Survey on Recreation and the Environment (NSRE), initiated in 1960 (USDA Forest Service). Similar national surveys have been conducted intermittently in Canada.

Data on hunter numbers, types of hunting, hunter days afield, and hunters' expenditures have been collected for more than a half century through the national survey in the United States. Information on nonconsumptive activities (e.g., wildlife observation, photography) has been collected in this survey since 1980. The methodology for these surveys changed at various times, limiting ability to chart precisely the long-term trends in wildlife recreation participation (Chu et al. 1992). Nevertheless, the national surveys conducted since 1955 describe a gradual, multidecadal decline in hunting participation among Americans ≥16 years of age; most recently, 13.7 million hunted in 2011, whereas 11.5 million hunted in 2016. The surveys also reveal that a fairly steady, large portion of the American public of the same age range (80% in 2011; 83% in 2016) enjoys wildlife-viewing, either as a focus of nature-based recreation or as a casual pastime.

Although national- and state-level surveys identified that numbers of hunters were static or declining beginning in the 1980s (Brown et al. 1987, Applegate 1989), the reasons for decline in participation initially were not fully elaborated. Subsequently, human dimensions researchers examined the dynamics of hunting participation in terms of recruitment, retention, continuous versus sporadic participation, and cessation. This line of inquiry indicated that culture strongly influences hunting from initiation through cessation (Decker et al. 1984). Broad sociodemographic trends affecting hunting (e.g., urbanization) suggest decline in participation is likely to continue (Decker et al. 1993, Heberlein and Thomson 1996, Brown et al. 2000). By describing the complexity of hunting as a social phenomenon, this research highlighted the difficulties agencies face when attempting to stimulate recruitment or retention at a level that would effectively curb the decline of hunters. Now adding its influence to the downward trend in hunter participation is an aging population of hunters (a demographic trait) who coincide with the cohort of baby boomers born in the 1950s and early 1960s (Winkler and Warnke 2013).

Nonconsumptive wildlife uses and users (i.e., people engaged in wildlife-associated recreation for other than hunting and trapping purposes) became interests of the wildlife management and research communities beginning in the 1980s (Manfredo 2002). Wildlife observation near people's homes (i.e., viewing wildlife in one's yard or nearby green spaces) accounted for much of the time spent watching wildlife, though trips for the specific purpose of viewing wildlife were also common. While the episodic national survey has since 1980 identified a demand (i.e., number of participants) for wildlife viewing that far exceeded that for hunting, specific studies since the late 1970s have revealed management-relevant insight about potential new beneficiaries for wildlife programs, such as wildlife watchers (Brown et al. 1979, Witter and Shaw 1979) and donors to state and private wildlife program funds (Brown et al. 1979, Witter et al. 1981, Manfredo 1988).

Keeping tabs on wildlife-associated recreation participation rates and demand (usually some of which goes unmet because of wildlife resource inaccessibility) will continue to be an important activity for wildlife managers. As public demand shifts some from one kind of wildlife-associated pursuit to another, managers need to be shifting emphasis to ensure programs align with interests. This is an irrefutable requirement for maintaining the relevancy and value of wildlife and wildlife management in the public's mind.

SUMMARY

Regardless of scale or scope, wildlife management occurs in a social-ecological system having many moving parts and connections. Operating within system constraints (limits and capacities), the wildlife management enterprise aims to produce value from the wildlife resource for society in the form of benefits of various kinds (i.e., positive impacts created, or negative impacts reduced). Making decisions and taking actions that yield such benefits require knowledge about key parts of the social-ecological system (wildlife, habitat, and humans) and how these components are connected. This knowledge typically is used to design interventions for purposes of manipulating the management system. This is an ongoing process because management is cyclical, characterized by planning and assessment, action(s), responses, feedback, and adaptations over time, rather than a discrete, one-shot, linear event. Feedback is key to adaptation and improvement and cannot be left to chance; it comes from continually monitoring the effects of management on humans, wildlife, and habitats. Ultimately, stakeholders evaluate those effects as acceptable or not. Insights from rigorous application of the social sciences and effective stakeholder engagement ensure timely and reliable human dimensions feedback.

One component of the management system is a governance framework. Public wildlife management follows a framework of governance that views wildlife as public trust resources. Stakeholders, as beneficiaries of wildlife resource management, are a fundamental consideration in governance. A stakeholder is any person who is significantly affected by, or can significantly affect, wildlife or wildlife management processes (Decker et al. 1996). Stakeholders interpret through their lens of values and beliefs (values orientations) which potential benefits and costs associated with wildlife merit management consideration (Riley et al. 2002). Durability or sustainability of stakeholder support for wildlife management is linked to achieving desirable impacts and minimizing or avoiding undesirable impacts. Knowing why some interactions with wildlife are perceived as creating certain types of effects can help managers anticipate important issues and integrate biological and human dimensions in wildlife management decision-making.

Human dimensions of wildlife management focus on people—individuals, groups, communities, organizations, and institutions (Decker et al. 2012a). All kinds and scales of human dimensions considerations, from broad social ideals such as good governance and ethical perspectives, to specific traits of people such as value orientations and motivations to participate in wildlife-related activities, are subjects of research to support wildlife management. Often, wildlife managers simply are seeking an answer to the question, What do stakeholders want or expect of management? Insights from human dimensions studies and stakeholder engagement efforts play a major role in answering this question by informing objectives for management and identifying kinds of management actions that are socially acceptable in different contexts. This alignment of management alternatives with stakeholder acceptance is certain to remain important as managers engage a broader, more diverse stakeholder base and consider new technologies for management (e.g., effective fertility control). Wildlife professionals working in a changing sociocultural environment who are continually learning from their experience engaging stakeholders are more likely to create durable and sustained decisions. Encouraging their partners and stakeholders to participate in the management process adds to the sustainability of wildlife management.

Social sciences provide methods of attaining reliable information on which to base wildlife management decisions. Social sciences also aid design of some management actions that are entirely in the domain of human dimensions, such as communication, education, and stakeholder engagement. In a nutshell, social science theory and technique applied under the rubric of human dimensions inquiry routinely serve wildlife management in several broad ways.

1. Influences wildlife policies.
2. Informs program planning.
3. Informs development and improvement of wildlife management practices.
4. Informs design of education and communication efforts.
5. Provides data for evaluation needed for program refinement.
6. Provides guidance in myriad decision processes embedded in wildlife management.

Fundamentally, human dimensions insight improves governance of wildlife resources.

Looking toward the future, the role of community involvement and collaboration with wildlife management agencies is likely to become more important, particularly where overabundant wildlife become a concern. In these and other situations, case-specific research is needed to understand how local communities and agencies can work together and build the necessary trust for effective governance of wildlife to achieve desired impacts.

Trust, usually defined as the willingness of one entity (e.g., beneficiaries of wildlife management) to be vulnerable to another (e.g., wildlife professionals), is a crucial component of effective wildlife governance (Riley et al. 2018). Developing durable conservation programs depends on cultivating good working relationships and trust between key individuals, groups, and government agencies (Lauber and Brown 2006, Ford et al. 2020b). Insights gained from human dimensions research and stakeholder engagement contribute to such collaboration and trust-building. The importance of this contribution cannot be overstated.

Literature Cited

Ackoff, R. L. 2001. OR: after the post-mortem. Systems Dynamics Review 17:341–346.

Adams, C. E. 2016. Urban wildlife management. 3rd edition. CRC Press, Boca Raton, Florida, USA.

Applegate, J. E. 1989. Patterns of early desertion among New Jersey hunters. Wildlife Society Bulletin 17:476–481.

Armitage, D., R. De Loë, and R. Plummer, R. 2012. Environmental governance and its implications for conservation practice. Conservation Letters 5:245–255.

Beierle, T. C., and J. Cayford. 2002. Democracy in practice: public participation in environmental decisions. Resources for the Future, Washington, DC, USA.

Bennett, N. J., R. Roth, S. C. Klain, K. Chan, P. Christie, et al. 2017. Conservation social science: understanding and integrating human dimensions to improve conservation. Biological Conservation 205:93–108.

Berkes, F., and C. Folke, editors. 1998. Linking social and ecological systems: management practices and social mechanisms for building resilience. Cambridge University Press, Cambridge, UK.

Brenner, L. J., and E. C. Metcalf. 2020. Beyond the tolerance/intolerance dichotomy: incorporating attitudes and acceptability into a robust definition of social tolerance of wildlife. Human Dimensions of Wildlife 25:259–267.

Brown, P. J., and M. J. Manfredo, 1987. Social values defined. Pages 12–23 in D. J. Decker and G. R. Goff, editors. Valuing wildlife: economic and social perspectives. Westview Press, Boulder, Colorado, USA.

Brown, T. L., C. P. Dawson, and R. L. Miller. 1979. Interests and attitudes of metropolitan New York residents about wildlife. Transactions of the North American Wildlife and Natural Resources Conference 44:289–297.

Brown, T. L., D. J. Decker, K. G. Purdy, and G. F. Mattfeld. 1987. The future of hunting in New York. Transactions of the North American Wildlife and Natural Resources Conference 52:553–566.

Brown, T. L., D. J. Decker, W. F. Siemer, and J. W. Enck. 2000. Trends in hunting participation and implications for management of game species. Pages 145–154 in W. C. Gartner and D. W. Lime, editors. Trends in outdoor recreation, leisure, and tourism. CABI, New York, New York, USA.

Bruskotter, J. T., R. H. Schmidt, and T. Teel. 2007. Are attitudes toward wolves changing? a case study in Utah. Biological Conservation 139:211–218.

Bruskotter, J. T., A. Singh, D. C. Fulton, and K. Slagle. 2015. Assessing tolerance for wildlife: clarifying relations between concepts and measures. Human Dimensions of Wildlife 20:255–270.

Bryan, H. 1977. Leisure value systems and recreational specialization: the case of trout fishermen. Journal of Leisure Research 9:174–187.

Burgiel, S. W. 2020. The incident command system: a framework for rapid response to biological invasion. Biological Invasions 22:155–165.

Campbell, J. M., and K. J. Mackay. 2003. Attitudinal and normative influences on support for hunting as a wildlife management strategy. Human Dimensions of Wildlife 8:181–197.

Carpenter, L. H., D. J. Decker, and J. F. Lipscomb. 2000. Stakeholder acceptance capacity in wildlife management. Human Dimensions of Wildlife 5:5–19.

Carter, N. H., S. J. Riley, and J. Liu. 2012. Utility of a social-psychological framework for predator conservation. Oryx 46:525–535.

Carter, N. H., S. J. Riley, A. Shortridge, B. K. Shrestha, and J. Liu. 2014. Spatial assessment of attitudes toward tigers in Nepal. Ambio 43:125–137.

Chase, L. C., T. M. Schusler, and D. J. Decker. 2000. Innovations in stakeholder involvement: what's the next step? Wildlife Society Bulletin 28:208–217.

Checkland, P. B. 1981. Systems thinking, systems practice. John Wiley and Sons, New York, New York, USA.

Checkland, P. B., and J. Scholes. 1999. Soft systems methodology in action. John Wiley and Sons, New York, New York, USA.

Chu, A., D. Eisenhower, M. Hay, D. Morganstein, J. Neter, and J. Waksberg. 1992. Measuring the recall error in self-reported fishing and hunting activities. Journal of Official Statistics 8:19–39.

Connelly, N. A., D. J. Decker, and S. Wear. 1987. Public tolerance of deer in a suburban environment: implications for management and control. Eastern Wildlife Damage Control Conference 3:207–218.

Decker, D. J. 1991. Implications of the wildlife acceptance capacity concept for urban wildlife management. Pages 45–53 in E. A. Webb and S. Q. Foster, editors. Perspectives in urban ecology. Denver Museum of Natural History, Denver, Colorado, USA.

Decker, D. J., T. L. Brown, B. L. Driver, and P. J. Brown. 1987. Theoretical developments in assessing social values of wildlife: toward a comprehensive understanding of wildlife recreation involvement. Pages 76–95 in D. J. Decker and G. R. Goff, editors. Valuing wildlife: economic and social perspectives. Westview Press, Boulder, Colorado, USA.

Decker, D. J., T. L. Brown, and W. F. Siemer, editors. 2001. Human dimensions of wildlife management in North America. The Wildlife Society, Bethesda, Maryland, USA.

Decker, D. J., and L. C. Chase. 1997. Human dimensions of living with wildlife—a management challenge for the 21st century. Wildlife Society Bulletin 25:788–795.

Decker, D. J., and N. A. Connelly. 1989. Motivations for deer hunting: implications for antlerless deer harvest as a management tool. Wildlife Society Bulletin 17:455–463.

Decker, D. J., J. W. Enck, and T. L. Brown. 1993. The future of hunting: will we pass on the heritage? Proceedings of the Annual Governor's Symposium on North American Hunting Heritage 2:22–46.

Decker, D. J., D. T. N. Evensen, W. F. Siemer, K. M. Leong, S. J. Riley, M. A. Wild, K. T. Castle, and C. L. Higgins. 2010. Understanding risk perceptions to enhance communication about human–wildlife interactions and the impacts of zoonotic disease. Institute for Laboratory Animal Research 51:255–261.

Decker, D. J., A. B. Forstchen, J. F. Organ, C. A. Smith, S. J. Riley, C. A. Jacobson, G. R. Batcheller, and W. F. Siemer. 2014. Impact management: an approach to fulfilling public trust responsibilities of wildlife agencies. Wildlife Society Bulletin 38:2–8.

Decker, D. J., A. B. Forstchen, E. F. Pomeranz, C. A. Smith, S. J. Riley, C. A. Jacobson, J. F. Organ, and G. R. Batcheller. 2015. Stakeholder engagement in wildlife management: does the public trust doctrine imply limits? Journal of Wildlife Management 79:174–179.

Decker, D. J., and T. Gavin. 1987. Public attitudes toward a suburban deer herd. Wildlife Society Bulletin 15:173–180.

Decker, D. J., C. A. Jacobson, and T. L. Brown. 2006. Situation-specific "impact dependency" as a determinant of management acceptability: insights from wolf and grizzly bear management in Alaska. Wildlife Society Bulletin 34:426–432.

Decker, D. J., C. C. Krueger, R. A. Baer Jr., B. A. Knuth, and M. E. Richmond. 1996. From clients to stakeholders: a philosophical shift for fish and wildlife management. Human Dimensions of Wildlife 1:70–82.

Decker, D. J., R. W. Provencher, and T. L. Brown. 1984. Antecedents to hunting participation: an exploratory study of the social-psychological determinants of initiation, continuation, and desertion in hunting. Outdoor Recreation Research Unit Series No. 84-6. Human Dimensions Research Unit, Cornell University, Ithaca, New York, USA.

Decker, D. J., and K. G. Purdy. 1988. Toward a concept of wildlife acceptance capacity in wildlife management. Wildlife Society Bulletin 1:53–57.

Decker, D. J., D. B. Raik, L. H. Carpenter, J. F. Organ, and T. M. Schusler. 2005. Collaborations for community-based wildlife management. Urban Ecosystems 8:227–236.

Decker, D. J., D. B. Raik, and W. F. Siemer. 2004. Community-based suburban deer management: a practitioner's guide. Northeast Wildlife Damage Management Research and Outreach Cooperative, Ithaca, New York, USA.

Decker, D. J., S. J. Riley, J. F. Organ, W. F. Siemer, and L. H. Carpenter. 2011a. Applying impact management: a practitioner's guide. Human Dimensions Research Unit and Cornell Cooperative Extension, Department of Natural Resources, Cornell University, Ithaca, New York, USA.

Decker, D. J., S. J. Riley, and W. F. Siemer, editors. 2012a. Human dimensions of wildlife management. Johns Hopkins University Press, Baltimore, Maryland, USA.

Decker, D. J., W. F. Siemer, D. T. N. Evensen, R. C. Stedman, K. A. McComas, M. A. Wild, K. T. Castle, and K. M. Leong. 2012b. Public perceptions of wildlife-associated disease: risk communication matters. Human–Wildlife Interactions 6:112–122.

Decker, D. J., W. F. Siemer, K. M. Leong, S. J. Riley, B. A. Rudolph, and L. H. Carpenter. 2008. What is wildlife management? Pages 315–327 in M. J. Manfredo, J. J. Vaske, P. J. Brown, D. J. Decker, and E. A. Duke, editors. Society and wildlife in the 21st century. Island Press, Washington, DC, USA.

Decker, D. J., W. F. Siemer, M. A. Wild, K. T. Castle, D. Wong, K. M. Leong, and D. T. N. Evensen. 2011b. Communicating about zoonotic disease: strategic considerations for wildlife professionals. Wildlife Society Bulletin 35:112–119.

Decker, D. J., C. Smith, A. Forstchen, D. Hare, E. Pomeranz, C. Doyle-Capitman, K. Schuler, and J. Organ. 2016. Governance principles for wildlife conservation in the 21st century. Conservation Letters 9:290–295.

Decker, D. J., M. A. Wild, S. J. Riley, W. F. Siemer, M. M. Miller, K. M. Leong, J. G. Powers, and J. C. Rhyan. 2006. Wildlife disease management: a manager's model. Human Dimensions of Wildlife 11:151–158.

de Souza, P. F., G. Porfirio, and H. M. Herrera. 2019. Perceptions and attitudes of Urucum Settlement residents about local wildlife. Anthrozoös 32:117-127.

Dougherty, E. N., D. C. Fulton, and D. H. Anderson. 2003. The influence of gender on the relationship between wildlife value orientations, beliefs, and the acceptability of lethal deer control in Cuyahoga Valley National Park. Society and Natural Resources 16:603–623.

Driver, B. L. 1976. Toward a better understanding of the social benefits of outdoor recreation participation. Southeastern Forest Experiment Station, USDA Forest Service, North Carolina State University, Ashville, North Carolina, USA.

Driver, B. L., P. J. Brown, and G. L. Peterson, editors. 1991. Benefits of leisure. Venture, State College, Pennsylvania, USA.

Enck, J. W., D. J. Decker, S. J. Riley, J. F. Organ, L. H. Carpenter, and W. F. Siemer. 2006. Integrating ecological and human dimensions in adaptive management of wildlife-related impacts. Wildlife Society Bulletin 34:698–705.

Enck, J. W., W. F. Porter, K. A. Didier, and D. J. Decker. 1998. The feasibility of restoring elk to New York State. College of Agriculture and Life Sciences, Ithaca, New York, and New York State College of Environmental Science and Forestry, Syracuse, New York, USA.

Ericsson, G., C. Sandström, and S. J. Riley. 2018. Rural-urban heterogeneity in stakeholder attitudes towards large carnivores in Sweden, 1976–2014. Pages 190–205 in T. Hovardas, editor. Large carnivore conservation and management in Europe: human dimensions and governance. Routledge Press, London, UK.

Evensen, D. T., D. J. Decker, and K. T. Castle. 2012. Communicating about wildlife-associated disease risks in national parks. George Wright Forum 29:227–235.

Fitchen, J. M. 1991. Endangered spaces, enduring places: change, identity, and survival in rural America. Westview Press, Boulder, Colorado, USA.

Flader, S. L. 1974. Thinking like a mountain. University of Missouri Press, Columbia, USA.

Fleegle, J. T., C. S. Rosenberry, and B. D. Wallingford. 2013. Use of citizen advisory committees to direct deer management in Pennsylvania. Wildlife Society Bulletin 37:129–136.

Ford, J. K., T. Lauricella, J. A. Van Fossen, and S. J. Riley. 2020a. Creating energy for change: the role of perceived leadership support on commitment to an organizational change initiative. Journal of Applied Behavioral Sciences doi:10.1177/0021886320907423.

Ford, J. K., S. J. Riley, T. Lauricella, and J. Van Fossen. 2020b. Factors affecting trust among natural resources stakeholders, partners, and strategic alliance members: a meta-analytic investigation. Frontiers in Communication 5:9. doi.10.3389/fcomm.

Fuller, A. K., D. J. Decker, M. Schiavone, and A. Forstchen. 2020. Ratcheting up rigor in wildlife management decision making. Wildlife Society Bulletin 44:29–41.

Fuller, A. K., J. C. Stiller, W. F. Siemer, and K. A. Perkins. 2021. Engaging hunters in selecting duck season dates using decision science. Chapter 8 in K. L. Pope and L. A. Powell, editors. Harvest of fish and wildlife: new paradigms for sustainable management. CRC Press, Boca Raton, Florida, USA.

Fulton, D. C, M. J. Manfredo, and J. Lipscomb. 1996. Wildlife value orientations: a conceptual and measurement approach. Human Dimensions of Wildlife 1:24–47.

George, K. A., K. M. Slagle, R. S. Wilson, S. J. Moeller, and J. T. Bruskotter. 2016. Changes in attitudes toward animals in the United States from 1978 to 2014. Biological Conservation 201:237–242.

Gigliotti, L., T. Teel, and S. J. Riley. 2020. Human dimensions of cougar management: public attitudes and values. Chapter 8 in J. A. Jenks, editor. Managing cougars in North America. 2nd edition. Western Association of Fish and Wildlife Agencies and Berryman Institute Press, Logan, Utah, USA.

Gore, M. L., and B. A. Knuth. 2009. Mass media effects on the operating environment of a wildlife-related risk-communication campaign. Journal of Wildlife Management 73:1407–1413.

Gore, M. L., B. A. Knuth, P. D. Curtis, and J. E. Shanahan. 2006. Stakeholder perceptions of risk associated with human–black bear conflicts in New York's Adirondack Park campgrounds: implications for theory and practice. Wildlife Society Bulletin 34:36–43.

Gore, M. L., B. A. Knuth, P. D. Curtis, and J. E. Shanahan. 2007. Campground manager and user perceptions of risk associated with human-black bear conflict: implications for communication. Human Dimensions of Wildlife 12:31–43.

Gregory, R., L. Failing, M. Harstone, G. Long, T. McDaniels, and D. Ohlson. 2012. Structured decision making: a practical guide to environmental management choices. Wiley-Blackwell, Oxford, UK.

Hanisch-Kirkbride, S. L., J. P. Burroughs, and S. J. Riley. 2014. What are they thinking? exploring layperson conceptualizations of wildlife health and disease. Human Dimensions of Wildlife 19:253–266.

Hanisch-Kirkbride, S. L., S. J. Riley, and M. L. Gore. 2013. Wildlife disease and risk perception. Journal of Wildlife Diseases 49:841–849.

Hart, P. S., E. C. Nisbet, and J. E. Shanahan. 2011. Environmental values and the social amplification of risk: an examination of how environmental values and media use influence predispositions for public engagement in wildlife management decision making. Society and Natural Resources 24:276–291.

Haubold, E. M. 2012. Using adaptive leadership principles in collaborative conservation with stakeholders to tackle a wicked problem: imperiled species management in Florida. Human Dimensions of Wildlife 17:344–356.

Hautaluoma, J., and P. J. Brown. 1978. Attributes of the deer hunting experience: a cluster-analytic study. Journal of Leisure Research 10:271–287.

Heberlein, T. A. 2012. Navigating environmental attitudes. Oxford University Press, New York, New York, USA.

Heberlein, T. A., and E. Thomson. 1996. Changes in US hunting participation, 1980–90. Human Dimensions of Wildlife 1:85–86.

Heberlein, T. A., J. N. Trent, and R. M. Baumgartner. 1982. The influence of hunter density on firearm deer hunters' satisfaction: a field experiment. Transactions of the North American Wildlife and Natural Resources Conference 47:665–676.

Hendee, J. C. 1974. A multiple satisfaction approach to game management. Wildlife Society Bulletin 1:24–47.

Hockett, K. S., and T. E. Hall. 2007. The effect of moral and fear appeals on park visitors' beliefs about feeding wildlife. Journal of Interpretation Research 12:5–27.

Hubbard, R. D., and C. K. Nielsen. 2011. Cost-benefit analysis of managed shotgun hunts for suburban white-tailed deer. Human–Wildlife Interactions 5:13–22.

Inskip, C., N. H. Carter, S. J. Riley, Z. Fahad, T. M. Roberts, and D. C. Macmillan. 2016. Toward human-carnivore coexistence: understanding tolerance for tigers in Bangladesh. PLoS One 11(1):e0145913.

Jacobson, C. A., and D. J. Decker. 2008. Governance of state wildlife management: reform and revive or resist and retrench? Society and Natural Resources 21:441–448.

Jacobson, C. A., D. J. Decker, and L. H. Carpenter. 2007. Securing alternative funding for wildlife management: insights from agency leaders. Journal of Wildlife Management 71:2106–2113.

Jacobson, C. A., J. F. Organ, D. J. Decker, G. R. Batcheller, and L. Carpenter. 2010. A conservation institution for the 21st century: implications for state wildlife agencies. Journal of Wildlife Management 74:203–209.

Jager, C., M. P. Nelson, L. Goralnik, and M. L. Gore. 2016. Michigan mute swan management: a case study to understand contentious natural resource management issues. Human Dimensions of Wildlife 21:189–202.

Jonker, S. A., J. F. Organ, R. M. Muth, R. R. Zwick, and W. F. Siemer. 2009. Stakeholder norms toward beaver management in Massachusetts. Journal of Wildlife Management 73:1158–1165.

Kasperson, R. E., O. Renn, P. Slovic, H. S. Brown, J. Emel, R. Goble, J. X. Kasperson, and S. Ratick. 1988. The social amplification of risk: a conceptual framework. Risk Analysis 8:177–187.

Keener-Eck, L. S., A. T. Morzillo, and R. A. Christoffel. 2020. A comparison of wildlife value orientations and attitudes toward timber rattlesnakes (*Crotalus horridus*). Human Dimensions of Wildlife 25:47–61.

Kellert, S. R. 1980a. Americans' attitudes and knowledge of animals. Transactions of the North American Wildlife and Natural Resources Conference 45:111–124.

Kellert, S. R. 1980b. Contemporary values of wildlife in American society. Pages 31–60 in W. W. Shaw and E. H. Zube, editors. Wildlife values. Institute Series Report No. 1. Center for Assessment of Non-Commodity Natural Resource Values, University of Arizona, Tucson, USA.

Lackey, B. K., and S. H. Ham. 2003. Contextual analysis of interpretation focused on human–black bear conflicts in Yosemite National Park. Journal of Applied Environmental Education and Communication 2:11–21.

Lackey, B. K., and S. Ham. 2004. Assessment of communication focused on human–black bear conflict at Yosemite National Park. Journal of Interpretation Research 8:25–40.

Lancia, R. A., C. E. Braun, M. W. Callopy, R. D. Dueser, J. G. Kie, C. J. Martinka, J. D. Nichols, T. D. Nudds, W. R. Porath, and N. G. Tilghman. 1996. ARM! for the future: adaptive resource management in the wildlife profession. Wildlife Society Bulletin 24:436–442.

Larson, L. R., C. B. Cooper, and M. E. Hauber. 2016. Emotions as drivers of wildlife stewardship behavior: examining citizen science nest monitors' responses to invasive house sparrows. Human Dimensions of Wildlife 21:18–33.

Lauber, T. B., and T. L. Brown. 2006. Learning by doing: policy learning in community-based deer management. Society and Natural Resources 19:411–428.

Lauber, T. B., and B. A. Knuth. 1997. Fairness in moose management decision-making: the citizen's perspective. Wildlife Society Bulletin 25:776–787.

Lauber, T. B., R. C. Stedman, D. J. Decker, and B. A. Knuth. 2011. Linking knowledge to action in collaborative conservation. Conservation Biology 25:1186–1194.

Lauricella, T. K., J. K. Ford, S. J. Riley, C. L. Powers, and P. E. Lederle. 2017. Employee perceptions regarding an organizational change initiative in a state wildlife agency. Human Dimensions of Wildlife 22:422–437.

Lischka, S. A., S. J. Riley, and B. A. Rudolph. 2008. Effects of impact perception on acceptance capacity for white-tailed deer. Journal of Wildlife Management 72:502–509.

Lischka, S. A., T. L. Teel, H. E. Johnson, and K. R. Crooks. 2019. Understanding and managing human tolerance for a large carnivore in a residential system. Biological Conservation 238:108189.

Loker, C. A., and D. J. Decker. 1995. Colorado black bear hunting referendum: what was behind the vote? Wildlife Society Bulletin 23:370–376.

Loker, C. A., D. J. Decker, and S. J. Schwager. 1999. Social acceptability of wildlife management actions in suburban areas: three cases from New York. Wildlife Society Bulletin 27:152–159.

Loomis, J., and L. Caughlan. 2004. Economic analysis of alternative bison and elk management practices on the national elk refuge and Grand Teton National Park: a comparison of visitor and household responses. Open File Report 2004-1305. Biological Resources Discipline, US Geological Survey, Reston, Virginia, USA.

Lu, H., K. A. McComas, D. E. Buttke, S. Roh, and M. A. Wild. 2016. A One Health message about bats increases intentions to follow public health guidance on bat rabies. PLoS One 11(5):0156205.

Manfredo, M. J. 1988. Second-year analysis of donors to Oregon's nongame tax checkoff. Wildlife Society Bulletin 16:221–224.

Manfredo, M. J. 2002. Planning and managing for wildlife viewing recreation: an introduction. Pages 1–8 in M. J. Manfredo, editor. Wildlife viewing in North America: a management planning handbook. Oregon State University Press, Corvallis, USA.

Manfredo, M. J., and R. A. Larson. 1993. Managing for wildlife viewing recreation experiences: an application in Colorado. Wildlife Society Bulletin 21:226–236.

Manfredo, M. J., L. Sullivan, J. Salerno, and J. Berger. 2020b. Looking forward, not backward in considering the needs for social science in wildlife management. BioScience 70:529–530.

Manfredo, M. J., T. Teel, and A. D. Bright. 2004. Application of the concepts of values and attitudes in human dimensions of natural resources research. Pages 271–282 in M. J. Manfredo, J. J. Vaske, D. R. Field, and P. Brown, editors. Society and natural resources: a summary of knowledge. Modern Litho, Jefferson City, Missouri, USA.

Manfredo, M. J., T. L. Teel, A. W. Don Carlos, L. Sullivan, A. D. Bright, A. M. Dietsch, J. Bruskotter, and D. Fulton. 2020a. The changing sociocultural context of wildlife conservation. Conservation Biology doi:10.1111/cobi.13493.

Manfredo M. J., T. L. Teel, and K. L. Henry. 2009. Value orientations in the western United States. Social Science Quarterly 90:407–427.

Manfredo, M. J., E. G. Urquiza-Haas, A. W. D. Carlos, J. T. Bruskotter, and A. M. Dietsch. 2020c. How anthropomorphism is changing the social context of modern wildlife conservation. Biological Conservation 241:108297.

Manfredo, M. J., H. C. Zinn, L. Sikorowski, and J. Jones. 1998. Public acceptance of cougar management: a case study of Denver, Colorado, and nearby foothills areas. Wildlife Society Bulletin 26:964–970.

Margretta, J. 2002. What management is: how it works and why it's everyone's business. Free Press, New York, New York, USA.

McClafferty, J. A., and J. A. Parkhurst. 2001. Using public surveys and GIS to determine the feasibility of restoring elk to Virginia. Pages 83–98 in D. S. Maehr, R. F. Noss, and J. L. Larkin, editors. Large mammal restoration—ecological and social challenges in the 21st century. Island Press, Washington, DC, USA.

McCool, S. F., and A. M. Braithwaite. 1989. Beliefs and behaviors of backcountry campers in Montana toward grizzly bears. Wildlife Society Bulletin 17:514–519.

McFarlane, B. L. 1996. Socialization influences of specialization among birdwatchers. Human Dimensions of Wildlife 1:35–50.

McIvor, D. E., and M. R. Conover. 1994. Perceptions of farmers and nonfarmers toward management of problem wildlife. Wildlife Society Bulletin 22:212–219.

Miller, C. A., and A. R. Graefe. 2000. Degree and range of specialization across related hunting activities. Leisure Sciences 22:195–204.

Miller, Z. D., W. Freimund, E. Covelli Metcalf, N. Nickerson, and R. B. Powell. 2018. Targeting your audience: wildlife value orientations and the relevance of messages about bear safety. Human Dimensions of Wildlife 23:213–226.

Miller, Z. D., W. Freimund, E. Covelli Metcalf, N. Nickerson, and R. B. Powell. 2019. Merging elaboration and the theory of planned behavior to understand bear spray behavior of day hikers in Yellowstone National Park. Environmental Management 63:366–378.

Nelson, D. 1992. Citizen task forces on deer management: a case study. Northeast Wildlife 49:92–96.

Organ, J. F., D. J. Decker, S. J. Riley, J. E. McDonald, and S. P. Mahoney. 2020. Adaptive management in conservation. Pages 93-106 in N. Silvy, editor. Wildlife techniques manual. 8th edition. Johns Hopkins University Press, Baltimore, Maryland, USA.

Passel, J. S., and V. Cohn. 2008. US Population Projections: 2005-2050. Pew Research Center. https://www.pewresearch.org/hispanic/2008/02/11/us-population-projections-2005-2050/.

Petty, R. E., and J. T. Cacioppo. 1986. The elaboration likelihood model of persuasion. Advances in Experimental Social Psychology 19:123–205.

Petty, R. E., S. McMichael, and L. A. Brannon. 1992. The elaboration likelihood model of persuasion: applications in recreation and tourism. Pages 77–101 in M. J. Manfredo, editor. Influencing human behavior: theory and applications in recreation, tourism, and natural resources management. Sagamore Publishing, Champaign, Illinois, USA.

Pomeranz, E. F., D. J. Decker, W. F. Siemer, A. Kirsch, J. Hurst, and J. Farquhar. 2014. Challenges for multilevel stakeholder engagement in public trust resource governance. Human Dimensions of Wildlife 19:448–457.

Potter, D. R., J. C. Hendee, and R. N. Clark. 1973. Hunting satisfaction: game, guns, or nature? Transactions of the North American Wildlife and Natural Resources Conference 38:220–229.

Purdy, K. G., and D. J. Decker. 1989. Applying wildlife values information in management: the wildlife attitudes and values scale. Wildlife Society Bulletin 17:494–500.

Raik, D. B., W. F. Siemer, and D. J. Decker. 2005. Intervention and capacity considerations in community-based deer management: the stakeholders' perspective. Human Dimensions of Wildlife 10:259–272.

Ready, R. C. 2012. Economic considerations in wildlife management. Pages 68-83 in D. J. Decker, S. J. Riley, and W. F. Siemer, editors. Human dimensions of wildlife management. Johns Hopkins University Press, Baltimore, Maryland, USA.

Riley, S. J., and D. J. Decker. 2000a. Wildlife stakeholder acceptance capacity for cougars in Montana. Wildlife Society Bulletin 28:931–939.

Riley, S. J., and D. J. Decker. 2000b. Risk perception as a factor in wildlife acceptance capacity for cougars in Montana. Human Dimensions of Wildlife 5:50–62.

Riley, S. J., D. J. Decker, L. H. Carpenter, J. F. Organ, W. F. Siemer, G. F. Mattfeld, and G. Parsons. 2002. The essence of wildlife management. Wildlife Society Bulletin 30:585–593.

Riley, S. J., J. K. Ford, H. A. Triezenberg, and P. E. Lederle. 2018. Stakeholder trust in a state wildlife agency. Journal of Wildlife Management 82:1528–1535.

Riley, S. J., and C. Sandström. 2016. Human dimensions insights into reintroduction of wildlife populations. Pages 55–77 in D. S. Jachowski, J. J. Millspaugh, P. L. Angermeir, and R. Slotow, editors. Reintroduction of fish and wildlife populations. University of California Press, Davis, California, USA.

Riley, S. J., W. F. Siemer, D. J. Decker, L. H. Carpenter, J. F. Organ, and L. T. Berchielli. 2003. Adaptive impact management: an integrative approach to wildlife management. Human Dimensions of Wildlife 8:81–95.

Robinson, K. F., A. K. Fuller, J. E. Hurst, B. L. Swift, A. Kirsch, J. Farquhar, D. J. Decker, and W. F. Siemer. 2016. Structured decision making as a framework for large-scale wildlife harvest management decisions. Ecosphere 22:1573–1591.

Robinson, K. F., A. K. Fuller, M. V. Schiavone, B. L. Swift, D. R. Diefenbach, W. F. Siemer, and D. J. Decker. 2017. Addressing wild turkey

population declines using structured decision making. Journal of Wildlife Management 81:393–405.

Rudolph, B. A., M. G. Schechler, and S. J. Riley. 2012. Governance of wildlife resources. Pages 15-25 in D. J. Decker, S. J. Riley, and W. F. Siemer, editors. Human dimensions of wildlife management. Johns Hopkins University Press, Baltimore, Maryland, USA.

Rupprecht, C. D. 2017. Ready for more-than-human? measuring urban residents' willingness to coexist with animals. Fennia-International Journal of Geography 195:142–160.

Sakurai, R., H. Tsunoda, H. Enari, W. F. Siemer, T. Uehara, and R. C. Stedman. 2020. Factors affecting attitudes toward reintroduction of wolves in Japan. Global Ecology and Conservation 22:e01036.

Shelby, B., J. J. Vaske, and M. P. Donnelly. 1996. Norms, standards, and natural resources. Leisure Sciences 18:103–123.

Shwiff, S. A., A. Anderson, R. Cullen, P. C. L. White, and S. S. Shwiff. 2013. Assignment of measurable costs and benefits to wildlife conservation projects. Wildlife Research 40:134–141.

Siemer, W. F., and D. J. Decker. 2006. An assessment of black bear impacts in New York. Human Dimensions Research Unit Series Publication 06-6. Department of Natural Resources, Cornell University, Ithaca, New York, USA.

Siemer, W. F., D. J. Decker, P. Otto, and M. L. Gore. 2007. Working through black bear management issues: a practitioners' guide. Northeast Wildlife Damage Management Research and Outreach Cooperative, Ithaca, New York, USA.

Siemer, W. F., P. S. Hart, D. J. Decker, and J. Shanahan. 2009. Factors that influence concern about human–black bear interactions in residential settings. Human Dimensions of Wildlife 14:185–197.

Smith, C. A. 2011. The role of state wildlife professionals under the public trust doctrine. Journal of Wildlife Management 75:1539–1543.

Sponarski, C. C., C. Miller, and J. J. Vaske. 2018. Perceived risks and coyote management in an urban setting. Journal of Urban Ecology 4:juy025.

Stedman, R. C. 2012. Sociological considerations in wildlife management. Pages 58–67 In D. J. Decker, S. J. Riley, and W. F. Siemer, editors. Human dimensions of wildlife management. Johns Hopkins University Press, Baltimore, Maryland, USA.

Stout, R. J., D. J. Decker, B. A. Knuth, J. C. Proud, and D. H. Nelson. 1996. Comparison of three public-involvement approaches for stakeholder input into deer management decisions: a case study. Wildlife Society Bulletin 24:312–317.

Stout, R. J., B. A. Knuth, and P. D. Curtis. 1997. Preferences of suburban landowners for deer management techniques: a step towards better communication. Wildlife Society Bulletin 25:348–359.

The Wildlife Society. 2010. The public trust doctrine: implications for wildlife management and conservation in the United States and Canada. Technical Review 10-01. The Wildlife Society, Bethesda, Maryland, USA.

Triezenberg, H. A., M. L. Gore, S. J. Riley, and M. K. Lapinski. 2014. Perceived risk from disease and management policies: an expansion of zoonotic disease information seeking and processing model. Human Dimensions of Wildlife 19:123–138.

US Department of the Interior, US Fish and Wildlife Service, US Department of Commerce, and US Census Bureau. 2018. 2016 National Survey of Fishing, Hunting, and Wildlife-Associated Recreation. Report FHW/16-NAT(RV). Washington, DC, USA.

Vaske, J. J., and D. Whittaker. 2004. Normative approaches to natural resources. Pages 283–294 in M. J. Manfredo, J. J. Vaske, D. R. Field, and P. Brown, editors. Society and natural resources: a summary of knowledge. Modern Litho, Jefferson City, Missouri, USA.

Vucetich, J., and M. P. Nelson. 2017. Wolf hunting and the ethics of predator control. Pages 411–429 *in* L. Kalof, editor. The Oxford handbook of animal studies. Oxford University Press, New York, New York, USA.

Whittaker, D. 1997. Capacity norms on bear viewing platforms. Human Dimensions of Wildlife 2:37–49.

Wieczorek Hudenko, H., D. J. Decker, and W. F. Siemer. 2008. Living with coyotes in suburban areas: insights from two New York State counties. Human Dimensions Research Unit Series Publication 08-8. Department of Natural Resources, Cornell University, Ithaca, New York, USA.

Wieczorek Hudenko, H., W. F. Siemer, and D. J. Decker. 2010. Urban carnivore conservation and management: the human dimension. Pages 21–33 *in* S. Gehrt, S. Riley, and B. Cypher, editors. Urban carnivores: ecology, conflict, and conservation. Johns Hopkins University Press, Baltimore, Maryland, USA.

Winkler, R., and K. Warnke. 2013. The future of hunting: an age-period-cohort analysis of deer hunter decline. Population and Environment 34:460–480.

Witter, D. J., and W. W. Shaw. 1979. Beliefs of birders, hunters, and wildlife professionals about wildlife management. Transactions of the North American Wildlife and Natural Resources Conference 44:298–305.

Witter, D. J., D. L. Tylka, and J. E. Werner. 1981. Values of urban wildlife in Missouri. Transactions of the North American Wildlife and Natural Resources Conference 46:424–431.

Wittmann, K., J. J. Vaske, M. J. Manfredo, and H. C. Zinn. 1998. Standards for lethal control of problem wildlife. Human Dimensions of Wildlife 3:29–48.

Wondolleck, J. M., and S. L. Yaffee. 2000. Making collaboration work: lessons from innovation in natural resources management. Island Press, Washington, DC, USA.

Woodruff, R. A., J. Sheler, K. McAllister, D. M. Harris, M. A. Linnell, and K. I. Price. 2004. Resolving urban Canada goose problems in Puget Sound, Washington: a coalition-based approach. Proceedings Vertebrate Pest Control Conference 21:107–112.

Yankelovich, D. 1991*a*. Coming to public judgment. Syracuse University Press, Syracuse, New York, USA.

Yankelovich, D. 1991*b*. The magic of dialogue: transforming conflict into cooperation. Simon and Schuster, New York, New York, USA.

Zinn, H., M. J. Manfredo, J. J. Vaske, and K. Wittmann. 1998. Using normative beliefs to determine the acceptability of wildlife management actions. Society and Natural Resources 11:649–662.

5 STRUCTURED DECISION-MAKING

MICHAEL C. RUNGE, JAMES B. GRAND, AND MICHAEL S. MITCHELL

INTRODUCTION

Wildlife management is an exercise in decision-making. While wildlife science is the pursuit of knowledge about wildlife and its environment, wildlife management is the application of that knowledge in a human social context, application that typically requires a choice of management options. Decisions require the integration of science with values, because in the end any decision is an attempt to achieve some future condition that is desirable to the decision maker (Keeney 1996b). Wildlife management, particularly under the North American Model of Wildlife Conservation (Chapters 2, 9), is often practiced by federal, tribal, state, or private agencies on behalf of the public and thus integrates science, law, and public values (Chapter 4). For example, the development of hunting regulations for white-tailed deer (*Odocoileus virginianus*) in Pennsylvania is a complicated choice among many possible permutations of regulations, a choice designed to balance many desires: hunting opportunity; the long-term conservation of deer; a sense of fair pursuit; fair public access; population levels commensurate with habitat capacity and predator density; wildlife viewing; and tribal, state, and local economic benefits from hunting and tourism. Certainly, there are decades of wildlife science about deer and social science about deer hunters to support the choice of hunting regulations, but they alone cannot identify the best regulations. The decision needs to integrate science- and values-based components (Wagner 1989).

Decision analysis is to wildlife management as the scientific method is to wildlife science, a framework and a theory to guide practice. The field of decision science is broad, with roots in economics stretching back to the 1940s, if not earlier (von Neumann and Morgenstern 1944), and the cross-disciplinary nature of the field became evident in the 1960s, with contributions from cybernetics (computer science), business administration, and mathematics (Raiffa and Schlaifer 1961, Howard 1968). Modern decision science has added expertise in many areas, including psychology, operations research, sociology, risk analysis, and statistics. Decision analysis has been applied in many contexts, including nuclear warfare planning (Dalkey and Helmer 1963), energy planning (Diakoulaki et al. 2005), adoption of health care technologies (Claxton et al. 2002), veterinary disease management (Probert et al. 2016, Webb et al. 2017), epidemiology (Shea et al. 2014, Li et al. 2017), and top-level political decisions in the Finnish parliament (Hämäläinen and Leikola 1996), to name a few. Formal decision analysis techniques are increasingly used in environmental fields (Kiker et al. 2005, Gregory et al. 2012, Runge et al. 2020), particularly fisheries (Bain 1987, Gregory and Long 2009, Runge et al. 2011a), but also in wildlife management (Ralls and Starfield 1995; Johnson et al. 1997; Regan et al. 2005; Lyons et al. 2008; Runge et al. 2009, 2011b; Martin et al. 2010; McDonald-Madden et al. 2010; Moore et al. 2011b, Mitchell et al. 2013, 2018; Sells et al. 2016; Grant et al. 2017; Bernard and Grant 2019; Symstad et al. 2019). But it is perhaps surprising that although wildlife management focuses on integrating values and science to make decisions, formal decision analysis is not applied more often, nor is it a core element in graduate education (van Heezik and Seddon 2005, Johnson et al. 2015).

Is wildlife management an art or a science? There are wildlife managers who will vigorously argue the former, that the decisions they make are the result of years of experience, a deep sense of intuition, and scientific training. This is perhaps a traditional view; the language can be traced to the very beginning of our field. Leopold (1933:3) wrote, "game management is the art of making land produce sustained annual crops of wild game for recreational use." More recently, Bailey (1982:366) similarly described wildlife management: "As an art wildlife management is the application of knowledge to achieve goals . . . In selecting goals, [wildlife managers] compare and judge values." But note that the art that Leopold (1933) and Bailey (1982) describe is the integration of wildlife science with values-based judgments. Leopold's (1933) example embeds three main goals: providing recreational use of wild game, having that use be sustainable, and having that use be consistent (i.e., annual). A deeper question is whether the integration of science and values in making wildlife management decisions can be more than the informal and loosely structured judgment of a decision maker. Are wildlife management decisions transparent and replicable? Does the public know what values were balanced in choosing the decision, and what science was consulted? Would a different decision maker have weighed the evidence and the values in the same way, and would that person come to the same decision? Will the decision maker's successor be able to maintain continuity, or will knowledge be lost every time someone retires? Increasingly, the public is demanding more transparency of natural resource managers, and decision analysis provides the framework for this transparency. This is not to say that the intuitive decision-making of experienced wildlife managers is without merit, only that the modern demands of transparency, accountability, inclusiveness, and efficiency require structured approaches to wildlife management decisions. This chapter introduces the field of decision analysis as a way to transform the art of wildlife management into a science.

A FRAMEWORK FOR DECISION-MAKING

Making decisions is a hallmark of human existence, something we do every day. Decisions are not always difficult to make, but some (e.g., public sector decisions) are sufficiently complex and challenging that the common tools and rules of thumb used by humans in daily decision-making are inadequate for reliably achieving good decisions. Decision analysis, or structured decision-making (SDM; Gregory et al. 2012), is "a formalization of common sense for decision problems which are too complex for informal use of common sense" (Keeney 1982:806). This section describes the elements of decision analysis in the context of wildlife management.

What is a decision? A decision is an "irrevocable allocation of resources . . . not a mental commitment to follow a course of action but rather the actual pursuit of the course of action" (Howard 1966:55). In the United States, the annual federal waterfowl hunting framework and the corresponding state waterfowl hunting seasons are decisions: they irrevocably set in motion harvest of waterfowl. State wildlife action plans (Fontaine 2011) are not themselves decisions, but they give rise to decisions when staff and fiscal resources are dedicated to carrying out actions in the plans. Likewise, recovery plans under the US Endangered Species Act (ESA) are not decisions, but the actions taken under their auspices are.

The PrOACT Framework
There are two hallmarks of structured decision-making: values-focused thinking and problem decomposition. Values-focused thinking emphasizes that all decisions are inherently statements about values, and so discussion of those values should precede other analysis (Keeney 1996a). Problem decomposition breaks a decision into its logical components, allowing identification of impediments to the decision; providing focus when and where needed; and creating an explicit, transparent, and replicable framework for decision-making that improves performance and stands up to scrutiny. The logical components of decision analysis include defining the *Pro*blem, identifying *O*bjectives, defining potential alternative

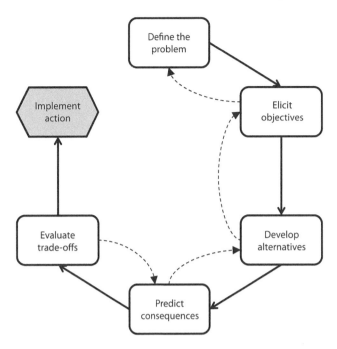

Figure 5.1. The steps of structured decision-making: the PrOACT sequence. From Hammond et al. (1999)

*A*ctions, evaluating *C*onsequences of actions, and assessing *T*rade-offs among alternative actions (Fig. 5.1). These components constitute the PrOACT framework (Hammond et al. 1999). Problem-framing is often an iterative process intended to facilitate insights about a decision throughout development of the analysis. Each step benefits from reevaluation at the completion of subsequent steps (Fig. 5.1).

Defining the problem is the critical first step of SDM that guides the process toward appropriate tools and information, determines appropriate levels of investment, and ensures that the right problem is being solved. Its importance cannot be overstated; time taken to craft a concise yet comprehensive and accurate problem definition pays off (Hammond et al. 1999, Smith 2020a). A good problem statement comprises the actions that need to be taken; legal considerations; who the decision maker is; the scope, frequency, and timing of the decision; goals that need to be met; and the role of uncertainty.

Objectives make explicit what the decision maker cares most about, defining what will constitute successful outcomes in the decision-making process. Along with the problem statement, well-defined objectives are critical to all subsequent steps in structured decision-making, allowing the creation and assessment of alternative actions, identification of pertinent information for making the decision, and explanation of the decision-making process to others.

Actions represent choices available to a decision maker, or alternative approaches to achieving at least a subset of objectives. Good alternative actions address the future (not the past), are unique, encompass a broad range of possible actions, and can be implemented by the decision maker (i.e., are financially, legally, and politically reasonable).

Once alternative actions have been defined, the consequences of taking each action need to be predicted with respect to the objectives. All decisions involve prediction, whether implicit or explicit. One of the strengths of wildlife science is the wealth of tools (e.g., sampling protocols, data analysis methods, and modeling approaches) designed to help managers make predictions.

The final step in the PrOACT sequence is an analysis of trade-offs among alternatives based on their expected performance relative to the objectives, an analysis designed to identify an alternative that best achieves the set of objectives. This analysis can be anywhere from narrative to mathematical, depending on the complexity of the problem. The key role of a decision maker is to integrate the values- and science-based elements of the decision. Done well, this analysis should be transparent, should be comprehensive with respect to all fundamental objectives, should be explicit, should make use of best available information, and should address uncertainty directly.

The PrOACT sequence is simple but surprisingly powerful. In many decision settings, simply framing the problem helps to remove impediments to the decision. But the PrOACT framework also provides direction toward more advanced tools that may be needed in some circumstances.

When Is SDM Appropriate?

Structured decision-making is a broad and flexible set of tools that can be applied in a variety of settings. The PrOACT model provides a useful framework for ordering and deploying these tools, but SDM is not appropriate in all settings. First, SDM assumes that there is a decision to be made, which is not always the case. Strategic planning processes, prioritization schemes, research designs, species status assessments, and compiling of scientific findings are all activities in which a wildlife biologist might participate, and products a wildlife manager might want, but they are not always in service to a specific decision. In those cases, SDM might help guide thinking toward the decisions downstream of those activities, but it might also be frustrating to apply. Second, SDM assumes that there is a single decision maker, a single decision-making body, or multiple decision makers who agree to a spirit of open-mindedness and discovery for the purposes of identifying a common path. In situations where multiple parties to a decision are in substantial conflict, the endeavor might be better served by other facilitation, mediation, joint fact-finding, conflict resolution, or negotiation techniques (Sebenius 2007). In situations where there are multiple decision makers in competition with one another, who have no intention to openly reveal their objectives or search for common ground, another branch of the decision sciences—game theory—provides insights and methods for analysis (Colyvan et al. 2011).

There are numerous other processes meant to support decision-making related to wildlife management, which have overlapping domains of application (Fig. 5.2; Schwartz et al. 2018). Structured decision-making is useful when the objectives are known or can be developed, but conflict resolution

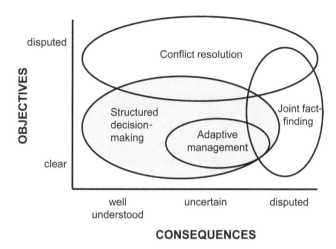

Figure 5.2. When is structured decision-making appropriate?

methods are better when the objectives are deeply disputed. Structured decision-making is broadly applicable whether the scientific aspects of the decision are well known or not; joint fact-finding is sometimes used when the science is disputed, as a way to engage stakeholders and develop common ground (Karl et al. 2007). As discussed later in this chapter, adaptive management is a special case of structured decision-making, valuable for recurrent decisions that are impeded by uncertainty.

Classes of Decisions

One of the values of early attention to problem-framing is the ability to recognize classes of decisions, which can in turn lead to identifying the best analytical tools to support the decision maker. Runge et al. (2020) describe six classes of decision, based on the primary cognitive challenge the decision maker faces (Table 5.1). For many decisions, the primary challenge is *problem-framing*—seeing the structure of the problem; once that is achieved and an evaluation of the alternatives is completed, the preferred alternative may be readily found (Smith 2020a). In *multiple-objective problems*, the challenge is trading off competing objectives; these problems are supported by a broad array of multicriteria decision analysis (MCDA) techniques (Converse 2020). There is a set of decision problems that involve sorting among an almost infinite list of alternatives to find the one that achieves the most benefit within the constraints specified. Such *portfolio problems* include resource allocation questions, prioritization exercises, project selection, and many other decisions (Lyons 2020). For *risk problems*, the challenge to the decision maker is knowing how to act in the face of uncertain outcomes, whether to choose the option with greater risk but higher potential for gain, or the option with less risk and less potential (Gore et al. 2009, Runge and Converse 2020). In *information problems*, the decision maker has the opportunity to acquire more information before committing to a course of action, and the question is whether that information will improve the subsequent decision enough to offset the costs of acquisition (Runge et al. 2011b, Smith 2020b).

Table 5.1. Six classes of decisions, the challenges they present, and some common decision analytic tools associated with them

Class	Cognitive Challenges	Analysis Tools	Wildlife Examples
Framing problems	Understanding the decision context; predicting the consequences	PrOACT,[a] objectives hierarchies, influence diagrams, Bayes nets, single-objective optimization tools	Blomquist et al. (2010), Grant et al. (2017), Bernard and Grant (2019)
Multiple-objective problems	Trading off competing objectives	MCDA[b]	Runge et al. (2011a, 2015), Converse et al. (2013), Robinson et al. (2016)
Portfolio problems	Selecting from among a large set of complex alternatives, often with resource constraints	Linear programming	Converse et al. (2011), Gerber et al. (2018), Lyons et al. (2020)
Risk problems	Understanding risk tolerance	Risk analysis, decision trees, expected utility theory	Mitchell et al. (2013), Sells et al. (2015, 2016, 2020)
Information problems	Evaluating whether new information would improve the decision	EVPI[c]	Runge et al. (2011b, 2017), Rushing et al. (2020)
Dynamic problems	Connecting linked decisions across time and space	Dynamic optimization, ARM[d]	Johnson et al. (1997, 2011), Martin et al. (2010), McDonald-Madden et al. (2010), Moore et al. (2011a), Tyre et al. (2011), Runge (2013)

[a]PrOACT, problem-objectives-alternatives-consequences-tradeoffs.

[b]MCDA, multicriteria decision analysis.

[c]EVPI, expected value of perfect information.

[d]ARM, adaptive resource management.

Finally, in *dynamic problems*, there are decisions that are linked to other decisions, either in a fixed sequence or in a recurrent pattern (Runge 2020). To solve these kinds of problems, we have dynamic optimization methods to address the linkages across time and adaptive methods to account for resolution of uncertainty. Any real decision problem will not fit neatly and solely into one of these classes—often there are multiple challenges a decision maker needs to confront—but the classification scheme helps to focus on the key impediments to a decision. Several of these decision classes are discussed in more detail later in this chapter.

THE VALUES-BASED ASPECTS OF DECISIONS

In the absence of a structured framework for coherently integrating value judgments and scientific judgments, decision makers tend to confound personal preferences and technical predictions (Failing et al. 2007). One of the key benefits of the problem decomposition embodied in PrOACT is the ability to separate the values- and science-based aspects of the decision, which allows those pieces to be analyzed by the right people with the appropriate tools. In the spirit of value-focused thinking (Keeney 1996a), we first discuss the values-based aspects before turning attention to the science-based aspects.

Defining the Problem

How a decision is framed affects how it should be analyzed, and this framing should reflect the values of the decision maker. Framing the decision can be surprisingly difficult and frustrating, but without a full definition of the problem and its context, considerable resources can be invested in solving the wrong problem. Further, a concise framing of the problem can aid clear communication with interested parties. For a simple, widely understood rubric to developing a problem statement, it is useful to refer to the six W's used in journalistic and technical writing. Many of the critical elements of the problem can be identified with explicit statements addressing the who, what, where, when, why, and how of a decision.

One way to begin is to ask, Who needs to make a decision? Sometimes the decision maker is obvious (e.g., where mandated by law or regulations), but other times, identifying the decision maker can be challenging. First, it is useful to distinguish decision makers from those who implement a decision. The decision maker is the authority upon whom responsibility for the decision rests. Second, there may not be a single decision maker. In some collaborative settings, decision-making is the joint responsibility of representatives from multiple agencies or interests; if that is the case, it is important for the decision analyst to understand the governance structure that supports that group. Third, in many public agency settings, the authority for the decision may be delegated. For example, in the United States, the secretary of the interior has statutory responsibility under the ESA, but typically that authority is delegated to the director of the US Fish and Wildlife Service (USFWS), who in turn may further delegate portions of that authority. This can create a challenge, because while the field office supervisor might be the decision maker with the motivation to analyze the decision, it is not clear at the outset how much consultation will be required up the delegated chain.

The question of the decision maker can be broadened considerably by asking, Who is interested in the decision? Stakeholders include anyone with an interest in the outcome of the decision. These include individuals who could be directly or indirectly affected by the actions under consideration. In the case of the private landowner, it may be relatively simple to identify the stakeholders on the basis of familial and business relationships. Many natural resource problems faced by public agencies, however, affect a diverse group of stakeholders, including such consumers as hunters, anglers, hikers, and bird watchers, and groups that are seemingly detached from the natural resources in question but that are intensely interested in their status. For example, few individuals will ever visit arctic Alaska, but interest in the effects of such stressors as mineral extraction and climate change on arctic wildlife has evoked reactions from countless individuals across North America. The field of human dimensions offers methods to identify, understand, and involve stakeholders in decisions (Chapter 4).

The central question that a problem statement needs to address is, What is the decision to be made? To put it differently, What choice does the decision maker face? In wildlife management, decisions can be simple or exceedingly complex. For example, a wildlife manager might be faced with the relatively simple decision of whether to plant wildlife openings with native legumes or to allow old-field succession to take its course. The same manager may be tasked with developing a management plan that involves making decisions about dozens or hundreds of sites that will play out over many years.

Knowing explicitly *where* the affected resources are helps define the geographic and taxonomic scale of the problem. By asking *when* a decision is needed, we define two important aspects of the problem: timing and frequency. The first concerns the urgency for a decision; a short timescale may limit the complexity of the decision analysis. The second concerns whether the decision is made one time or recurrently. In many cases, the decision occurs once, such as the placement of infrastructure—roads, buildings, or dams. In other cases, decisions are recurrent, as in setting annual harvest regulations. In still others, a series of sequential decisions that hinge on the success of previous actions are considered.

The problem statement should address *why* the decision is important. To do so, the consequences of failing to make a decision can be examined. Will it result in strongly negative consequences, such as extinction, loss of hunting opportunities, loss of revenue, or litigation? In some cases, there may be a legal mandate related to agency mission, as in setting harvest regulations, listing species that are candidates for protection as threatened or endangered, or reviewing management alternatives (e.g., an Environmental Impact Statement under the National Environmental Policy Act, or NEPA). In other cases, decisions can be related to meeting an agency's strategic objective, such as providing public hunting or other recreational opportunities. In still other cases, a decision might relate to meeting tactical objectives of an agency, such as minimizing risk to natural resources, maximizing effectiveness of management, or meeting an agreed upon population objective.

The problem statement should also describe *how* to solve the problem. This description should be broad and conceptual; an explicit statement of alternatives and their relative value to solving the problem comes later in the process. A good way to think about this portion of the problem statement is a description of the natural resource management tools that could be implemented in reaching a solution. For example, manipulating harvest regulations at continental scales can maximize harvest of waterfowl. Meeting population objectives for nongame species can be achieved by enhancing habitat quality. These statements may put bounds on the alternatives that will be considered in the analysis, but they might also stimulate discussion and require revision during the development of the decision analysis.

Many insights about the nature of the decision arise out of the analysis, however, so problem definition often evolves. A well-constructed decision process allows the decision maker to revisit the elements of the decision framework repeatedly as the analysis proceeds. Indeed, the notion of rapid, iterated prototyping is often embedded in many structured decision-making processes (Garrard et al. 2017).

Articulating the Objectives

In wildlife management the development of unambiguous, meaningful objectives of the decision makers and the stakeholders is a critical step in the decision-making process. Ambiguous, poorly formed, and hidden objectives often lead to poor decision-making, as does the exclusion of objectives that are important to large or important segments of the community of stakeholders. Clear, concise objectives with measurable attributes are the key to making informed, smart decisions because they define the decision's purpose (Keeney 1996a). When forced to make decisions in natural resources management, however, few individuals may take the time to fully describe the purpose of the actions under consideration. We find it useful to distinguish four steps in the development of objectives: eliciting objectives, classifying objectives, structuring objectives, and developing measurable attributes.

Eliciting Objectives

In developing objectives, it is often useful to start by eliciting the concerns of the decision makers and other stakeholders. Elicitation takes many forms, including workshops, public meetings, and one-on-one interviews. The important concept here is to be inclusive, empowering stakeholders and their representatives to articulate objectives that are important to making an informed decision. A variety of objectives is typical in wildlife management. Traditional concerns relate to the abundance and distribution of wildlife species, the health and quality of individual animals, the resources on which they depend, and their availability for consumptive or nonconsumptive uses (Robinson et al. 2016). During the past several decades, new concerns related to maintaining or increasing biodiversity have made their way into wildlife conservation and management (Gregory et al. 2012). And, increasingly, we recognize that wildlife management takes place in a sociopolitical context, and so a broader set of objectives is important, including economic, cultural, aesthetic, and spiritual concerns (Bengston 2000).

Objectives related to wildlife population abundance usually stem from worries about their viability (e.g., rare species), long-term persistence (e.g., many migratory songbirds), or harvestable surplus (e.g., most game species). Stakeholders often express these types of concerns in terms of declining populations or harvest levels. Concerns over wildlife populations may also stem from overabundance (Wagner and Seal 1992), especially where there are large economic effects— e.g., cormorants (*Phalacrocorax* spp.), white-tailed deer, nutria (*Myocaster coypus*), muskrat (*Ondatra zibethicus*), and raccoons (*Procyon lotor*)—or environmental impacts—e.g., western Canada geese (*Branta canadensis*) and lesser snow goose (*Chen caerulescens*).

Wildlife managers are often concerned about objectives above and beyond wildlife abundance, including distribution and quality of wildlife populations. For example, recovery criteria for listed species usually include a description of the number and distribution of distinct populations—like the red-cockaded woodpecker (*Picoides borealis*; USFWS 2003)—as an indication of viability and as a fundamental desire to see the species restored to its former range. The quality of the individuals in a population is also often a concern, both as an indication of the health of the population and as a fundamental objective. For example, management of wildlife populations for trophy harvest will focus on elements such as age structure, size, and other indicators of individual health (Jenks et al. 2002).

Concerns over biodiversity have increased as the field of wildlife management has been broadened beyond traditional game management. Large-scale programs such as gap analysis (Scott 1993) have increased awareness about the effects of cumulative habitat loss by focusing on land management practices and areas of high biotic diversity. Federal aid programs like state wildlife grants have enabled many state agencies to identify concerns and to develop objectives related to the conservation of biodiversity and populations of concern.

The objectives related to wildlife management, however, transcend concerns about wildlife. Economic concerns, too, are deeply important to stakeholders. The development of the US Forest Service's Northwest Forest Plan needed to consider old-growth habitat for spotted owls (*Strix occidentalis*) in the Pacific Northwest, the viability of the forest products industry, and the livelihood of its employees (Thomas et al. 2006). Reintroduction of wolves (*Canis lupus*) into the northern Rocky Mountains in the United States needed to consider the viability of the wolves and the effects on hunting opportunity for big game, but also economic concerns of cattle and sheep ranchers (Fritts et al. 1997). Social concerns related to the impacts of wildlife management go beyond economic considerations and include spiritual, aesthetic, cultural, and recreational objectives (Bengston 2000; Chapter 4). For example, wildlife and fish management in the Grand Canyon (Arizona, USA) needs to take into consideration spiritual and cultural objectives of native tribes, opportunity for wilderness recreation, and provision of energy and water to the arid Southwest in addition to economic and strictly wildlife-related objectives (Runge et al. 2011*a*, 2015).

Classifying Objectives

Objectives can be classified into four broad categories: strategic, fundamental, means, and process objectives (Keeney 2007). Strategic objectives are the highest-level objectives and are often associated with the mission of the agency or individual. For example, the legal mandates of a state agency associated with the maintenance of imperiled species and productivity of game species would be considered strategic objectives. These objectives are frequently beyond the scope of the management decisions faced by wildlife managers, and as such they often do not help discern among management alternatives. But they do define the context of the fundamental objectives, which are perhaps the most important category. Fundamental objectives are the "ends" of the wildlife management problem and the highest-level objectives incorporated in a decision analysis. Means objectives are the methods by which we achieve the fundamental objectives, but they may not be necessary if there are multiple pathways to achieve the fundamental objectives. Finally, process objectives govern how the decision is made but do not affect discrimination among the alternatives. For example, a decision maker—for legal, strategic, or ethical reasons—may want public meetings and outreach to be included in the decision-making process.

Fundamental objectives are the focus of decision analysis; they alone are used to distinguish among the alternatives. Good fundamental objectives have several key characteristics. First, they are measurable. Attributes can be developed for them that can be measured on an unambiguous scale. Second, good fundamental objectives are controllable; that is, they can be influenced by the management actions under consideration. Third, fundamental objectives are those the decision maker deems essential—there is no acceptable substitute.

It often requires careful thought to distinguish fundamental from means objectives. A useful way to make such distinctions is to ask why each objective is important, which frequently leads to the discovery of new, higher-level objectives that describe the most important, desired outcomes. For example, managers interested in wildlife populations in longleaf pine (*Pinus palustris*) habitats often identify concerns related to the absence or infrequent use of fire in those systems (USFWS 2003). A concise initial objective might be to increase the use of prescribed fire in longleaf pine. When asked why, managers often respond that it improves habitat quality; the restated objective may be to increase foraging habitat for red-cockaded woodpecker and northern bobwhite (*Colinus virginianus*). Asking why again can reveal that there is concern over the productivity or abundance of those populations, suggesting an objective to increase populations of both species. Asking the question yet again may elicit concerns over the viability of the red-cockaded woodpecker population and the size of the harvest of northern bobwhite. Asking why once more may reveal that the agency has a mandate to maintain populations of endangered species (e.g., red-cockaded woodpecker) and to increase harvestable populations of game species (e.g., northern bobwhite). So, classifying objectives identifies two fundamental goals (i.e., to maintain a viable population of red-cockaded woodpecker and to maximize harvest potential of northern bobwhite) from a nested set of means objectives.

Structuring Objectives

A fundamental objectives hierarchy illustrates the relationships among the most important objectives in a decision problem. A generic fundamental objectives hierarchy can be used to stimulate discussion and to identify problem-specific objectives. In many natural resource–related problems, useful generic fundamental objectives include improving or maintaining wildlife populations, minimizing cost, and providing utilitarian and nonutilitarian benefits to stakeholders. A generic fundamental objectives hierarchy (Fig. 5.3) can be modified to develop specific objectives related to a specific problem. Depending on the problem at hand, objectives surrounding the status of wildlife populations may be more specifically defined as one or more of the following: population size, distribution, individual health, genetic diversity, and species diversity. Cost is nearly always a consideration, and given a choice between equally effective solutions, the less expensive option is almost always more desirable. In other situations, where a budget is fixed or cost is viewed as a constraint, the solution that results in the best population status and stakeholder sat-

isfaction for the same cost is the logical choice. A broader set of stakeholder concerns is often a crucial consideration for wildlife populations held in public trust. Notice that the elements of the fundamental objectives hierarchy do not overlap; they express independent elements of concern in the decision problem, so there is no double counting. A fundamental objectives hierarchy must be complete, including all the concerns that bear on the decision.

Developing Measurable Attributes

Attributes are the measurement scales for fundamental objectives. Identifying attributes not only allows measurement of achievement but also forces clarity in the definition of each objective. The purpose of decision analysis is to provide a transparent comparison of the alternatives, and the attributes provide the quantitative measure of the consequences of each alternative for each objective. The capacity to make informed trade-offs is severely compromised if attributes are not clearly described (Keeney 2002). Because fundamental objectives are the focus of decision analysis, measurable attributes should be developed for fundamental objectives. For example, in the case of managing wildlife populations in longleaf pine habitats, achievement of the fundamental objective to increase northern bobwhite harvest might be measured by the annual number of birds taken by hunters, and the achievement of the objective to increase the viability of red-cockaded woodpecker populations might be measured by the probability of persistence over the next 100 years (Table 5.2).

There are three types of measurable attributes: natural attributes, proxy attributes, and constructed attributes. Each of the examples in Table 5.2 is a natural attribute—the scales directly capture the objective of interest, they are easily interpreted by anyone familiar with wildlife management, and there are widely accepted techniques or guidelines for their empirical measurement or estimation. For many objectives, appropriate natural attributes do not exist or are impractical for assessing consequences (e.g., data may not be available). In some cases, an attribute can be constructed on the basis of a relative scale. For example, absent measurement of fitness of individuals in a habitat, no universal scale exists for

---

1. Maximize ecological benefits
   a. Maximize persistence of native species (or communities)
      i. Maximize population size
      ii. Maximize distribution
      iii. Maximize individual quality
      iv. Maintain genetic and species diversity
   b. Minimize non-native and invasive species (or communities)
   c. Maintain ecosystem function
2. Minimize costs
   a. Minimize capital (fixed) costs
   b. Minimize ongoing (variable) costs
3. Maximize public and private benefits (utilitarian benefits)
   a. Maximize consumptive recreational benefit
   b. Maximize nonconsumptive recreational benefit
   c. Maximize public services (e.g., energy generation, water delivery)
   d. Maximize public health and safety
   e. Maximize private economic opportunity
   f. Provide sustainable subsistence use, where appropriate
4. Facilitate cultural values and traditions (nonutilitarian benefits)
   a. Maximize aesthetic and spiritual values
   b. Minimize taking of life
   c. Treat animals in a humane manner

Figure 5.3. Hierarchy of generic fundamental objectives for wildlife management

Table 5.2. Natural attributes for objectives in the longleaf pine example

Objective	Attribute
Increase use of prescribed fire (means)	Return interval or frequency of fires
Increase foraging habitat (means)	Hectares of pine burned in the last four years
Increase northern bobwhite harvest (fundamental)	Number of birds shot by hunters annually
Increase viability of red-cockaded woodpeckers (fundamental)	Probability of persistence over 100 years

Note: Measurable attributes are normally developed only for fundamental objectives, but the attributes for some means objectives are shown, too, for illustrative purposes.

Table 5.3. Example of a constructed scale for habitat quality

Attribute Level	Description
3	Very good: >80% canopy closure, >75% of canopy trees mast-producing oak, hickory, or beech
2	Good: 60% to 80% canopy closure, 26% to 75% of canopy trees mast-producing oak, hickory, or beech
1	Poor: <60% canopy closure, ≤25% of canopy trees mast-producing oak, hickory, or beech
0	No value: no mast-producing trees in forest canopy

measuring the degree to which an area provides habitat for a species because habitat requirements vary among species, and for most species we can only measure what we perceive to be the important requisites for habitat (Hirzel and Le Lay 2008). An attribute for measuring habitat quality might instead be constructed and scored on an ordinal scale. By their very nature, constructed scales are subjective; therefore clear definitions of the levels are required for repeatable, transparent scoring (Table 5.3). By contrast, proxy attributes are usually natural attributes for quantities (sometimes associated with means objectives) that provide an indirect measure of the objective of interest. For example, if our true objective was to increase hunting opportunities on public lands, the number of hectares open to public hunting might be a useful proxy attribute. Although many other factors—weather, access, and habitat condition—influence hunting opportunity, we assume that the area available for public hunting is highly correlated with hunting opportunities on public lands. In general, natural attributes are preferable to proxies or constructed scales. But often this preference needs to be relaxed to achieve a complete description of the decision problem (Keeney 2007).

Generating Alternative Actions

Generating alternatives is both a values-based and scientific exercise. The values-based element recognizes that alternatives are the admissible ways of achieving the objectives. Alternative actions can vary from simple to complex. In some cases, the alternative actions are a small set of discrete options, such as whether to use prescribed fire, mowing, or herbicide to set back succession in a grassland. In other cases, the alternative actions come from a continuous set, such as possible sustained harvest rates for a waterfowl population, which could take any value between zero and the intrinsic growth rate for the population. But often in wildlife management, the alternatives have complex structures. Portfolios are alternative actions that are composed of permutations of like elements. For example, a management agency that allocates resources to invasive species control could consider many potential portfolios of invasive species, each portfolio a list of invasive species targeted by management control. The number of potential portfolios in this case would include all permutations of the set of invasive species in that ecosystem. Strategies (or strategy tables) are alternative actions composed of permutations of unlike

elements. For example, the options considered in an analysis of potential responses to the emergence of white-nose syndrome in bats were strategies composed of such elements as the methods of addressing the fungal agent (*Pseudogymnoascus destructans*), methods of captive propagation, cave access restrictions, and management of disease spread (Szymanski et al. 2009).

Frequently, the need for structured decision-making arises from the desire to compare alternatives that are developed before the problem is well defined, but a thorough analysis of any problem will attempt to consider a wide variety of alternatives. There are various pitfalls that limit our ability to develop creative, potentially valuable alternatives. One of the most common pitfalls is "anchoring." Anchoring is the tendency to conduct business as usual, choosing solutions to recently addressed problems, or grasping at the first suggested alternative (Keeney 1996a). Choices made by anchoring constrain creativity and thoughtful development of alternatives. There are many techniques that can be applied to avoid anchoring and to encourage development of good alternatives (Keeney 1996a, Gregory et al. 2012). One method offers constructive insight: developing creative alternatives may result from broadening the decision context. This usually occurs when the decision maker or analyst determines that additional fundamental objectives exist. For instance, a game manager facing dissatisfied stakeholders (e.g., hunters) may assume that their objective is to harvest trophy animals and may perceive the trigger to be low harvest of trophy animals, which could result in a set of alternatives related to increasing the frequency of trophy characteristics in populations. But if the actual trigger is that hunters are seeing fewer deer, then broadening opportunities to view deer could lead to alternatives that do not result in increased harvest.

In summary, the intent is not to develop an exhaustive set of potential actions, but to develop a set of alternatives for impartial evaluation that represents the spectrum of potential solutions to the problem at hand. The set of alternatives must collectively influence all the fundamental objectives, but it is not necessary to limit alternatives to just those that affect every fundamental objective. It is also possible to find that some important objectives are not controllable within the set of feasible alternatives; in such a case, those objectives may be removed from the analysis (because they do not help discern among the alternatives), or the set of feasible alternatives might be expanded, possibly broadening the context of the problem.

Evaluating the Trade-Offs

The crux of any decision is the set of values placed on the objectives. In a single-objective problem, once the measurable attribute for the objective is established and the values-based aspects of the decision are expressed, the solution is the alternative that best achieves that objective. But a common wildlife management framework that might be cast as a single-objective problem—harvest management—reveals the complexity inherent in objectives. The solution of a maximum sustained yield problem is really a balance between two objectives: maximizing the short-term harvest and sustaining the population in perpetuity. The optimal harvest rate balances these two objectives to produce a maximum annual harvest that can be sustained indefinitely (Runge et al. 2009). But it is possible to ask whether these objectives might be balanced in some other way, or perhaps in deference to even more objectives; such has been the dialogue in the North American waterfowl management community in the 21st century (Runge et al. 2006).

Most wildlife management decisions involve trade-offs among multiple objectives, and meaningful evaluation of those trade-offs is grounded in values preferences among fundamental objectives. It is rare for all the objectives to be achieved under a single alternative; typically, objectives compete, and the challenge for the decision maker is how to choose an alternative that best balances those objectives. The balancing of objectives is a values judgment that should reflect the preferences of the decision maker, which often reflect societal priorities embodied in the organization the decision maker represents (Gregory et al. 2012, Converse 2020).

Several tools from the field of decision analysis are designed to elicit these value judgments from decision makers. A commonly used method is swing weighting (von Winterfeldt and Edwards 1986), which has the desirable property of encouraging decision makers to think about the range of consequences associated with alternatives together with their importance (Keeney 2002). In this method, the decision maker is asked to consider a series of hypothetical scenarios in which the objectives are swung from their worst consequence to their best consequence one at a time; the decision maker ranks these scenarios and then assigns a score that represents how much any scenario is preferred over another. From these scores, weights are derived for each objective, and these weights are used in a multicriteria decision analysis (see below). These weights on the individual objectives explicitly state how much one objective is valued over another and can be used to balance the trade-offs in the analysis.

Another way to examine and value trade-offs is to look at the "efficiency frontier." The efficiency frontier, also called the Pareto frontier, is the set of possible actions for which no gain in one objective can be achieved without a loss in some other objective (Lyons 2020). For two-objective problems, the Pareto frontier is often depicted as a graph of performance on one objective against performance on the other objective. Such a graph makes the trade-off visually evident and can be used to engender discussion about which solution best balances the two objectives.

One important point to emphasize is that the judgment about how to balance competing objectives cannot be answered by science. At its heart, wildlife management is an expression of a rich array of societal objectives that speak to a complex set of economic, recreational, aesthetic, and spiritual values. How these values are expressed in decision analysis is one of the most important things a decision maker needs to be able to judge and communicate.

THE SCIENCE-BASED ASPECTS OF DECISIONS

Although decisions in wildlife management are based on societal objectives and values, wildlife management, as a discipline, is founded in wildlife science; our decisions about how to manage wildlife are, and should be, influenced by our understanding of how natural systems respond to management. The science-based aspects of decisions include three sets of activities: generating alternative actions, predicting the consequences of those actions, and coherently integrating value judgments and technical judgments through reasoned use of decision analysis tools.

Predicting the Consequences

One of the critical roles of science in a decision analysis is the evaluation of the alternatives against the objectives. Often, this involves predicting how the alternative actions will affect the resources in question, and how those effects will influence achievement of the fundamental objectives. These predictions are often made using empirical data, inferring future responses based on past observations, but increasingly we recognize the importance of expert elicitation for predicting consequences. In a full decision analysis, the consequences need to be predicted for all the fundamental objectives. Although predictions about natural resources themselves are the mainstay of traditional wildlife management, predictions about the human responses to wildlife management are also critical.

A central theme in wildlife science is prediction of how individual animals, wildlife populations, and the ecosystems in which they reside respond to management actions (Chapters 7 and 19). The wildlife literature is rich with examples of predictive models based on empirical data, including age-structured population models (Caswell 2001), harvest and take models (Runge et al. 2009), population viability analyses (Beissinger and McCullough 2002), wildlife-habitat models (Morrison et al. 2006), resource selection functions (Boyce and McDonald 1999), and, increasingly, coupled climate-wildlife models (Hunter et al. 2010). There are two steps in the development of these predictive models: development of the model structure and estimation of the parameters. In an applied setting, the model structure is in part determined by the decision context; the alternative actions serve as the inputs to the model, and the measurable attributes of the fundamental objectives are the outputs. The innards of the model structure are an expression of the current understanding of the causal linkages between the actions and the outcomes. Methods for empirical estimation of parameters have flourished in wildlife science since the 1990s (Williams et al. 2002, Royle and Dorazio 2008) and require little comment here.

In a decision-making context, there are often other fundamental objectives besides wildlife resource objectives, and the consequences of the alternatives for these objectives need to be predicted, too. These objectives include economic, recreational, and spiritual objectives, and appropriate methods of prediction need to be found for each. Economic models related to wildlife management are being used more and more

(Pickton and Sikorowski 2004). The nature of human satisfaction with recreational opportunities can be complex, but empirical models are increasingly available (Chapter 4). Models for predicting spiritual and aesthetic outcomes are not common, although some initial attempts have been made (Failing et al. 2007). One of the challenges the human dimensions field faces in incorporating its work into decision-making contexts is moving from descriptive to predictive models (Miller 2017). Many of the current models describe patterns in economic, recreation, and aesthetic outcomes, but they are not yet able to predict those outcomes under alternative management actions.

For all types of outcomes that are important in wildlife management, there are often occasions when there is not enough empirical information to build predictive models, and not enough time to collect new data. In these settings, there is increasing use of methods of expert elicitation (Gore et al. 2009, Kuhnert et al. 2010, Martin et al. 2012). These methods typically rely on the accrued knowledge of a group of experts, rather than on empirical data, to structure a predictive model and to provide parameter estimates. There is a considerable literature on the reliability and fallibility of experts, and from this literature emerges some best practices in expert elicitation (Burgman 2005, Hemming et al. 2018). Briefly, these methods seek to tap into the privileged knowledge of experts while avoiding common cognitive biases to which humans are prone. In the modified Delphi method (MacMillan and Marshall 2006), a group of experts makes individual judgments about a parameter, fact, or relationship; they share their initial responses (often anonymously) with the group; discussion ensues; and then the experts are asked to make a final, private judgment. The feedback step promotes clarity, eliminates linguistic uncertainty, and allows sharing of insights, and the private judgments allow individual insights to be retained, capture uncertainty as expressed by the range of experts, and avoid the effects of damaging group dynamics. Some additional methods of elicitation guard against the overconfidence of experts by asking them to be explicit about their degree of uncertainty (Speirs-Bridge et al. 2010).

There is uncertainty in most predictions, whether they are empirically or expert based, and this uncertainty can affect the identification of a preferred alternative. Several taxonomies of uncertainty have been advanced (Morgan and Henrion 1990, Nichols et al. 1995, Regan et al. 2002); a combination of them is useful here. Broadly, uncertainty can be aleatory or epistemic. Aleatory uncertainty arises from stochastic processes that are outside of the manager's control. For example, environmental stochasticity (e.g., weather patterns), demographic stochasticity (i.e., the chance events that determine which animals survive), and partial controllability (i.e., our inability to completely control the implementation of our actions) give rise to aleatory uncertainty. Epistemic uncertainty arises from our lack of knowledge about the managed system. Structural uncertainty (i.e., uncertainty about how the system works), parametric uncertainty (i.e., imprecision in the model parameters), and partial observability (i.e., the inability to know ex-

actly the condition of the resource) are examples of epistemic uncertainty. The distinction between aleatory and epistemic uncertainty is often important to a decision maker because research or monitoring can theoretically reduce epistemic uncertainty. Incorporation and expression of uncertainty in the consequences are important aspects of prediction; they allow the decision maker to understand—and therefore manage—risk (Runge and Converse 2020).

Generating Alternative Actions

In "The Values-Based Aspects of Decisions" above, we discussed the generation of alternative actions, but there is also a technical side to this step of decision analysis. In some cases, one of the primary impediments to a decision is that none of the available actions can satisfactorily solve the problem, and the decision maker looks to scientists or engineers to craft a novel approach. For example, as noted previously, when white-nose syndrome emerged in cave-dwelling bats in eastern North America, no known method existed for controlling the fungus that causes the disease; one avenue of research was to identify a fungicide that might eradicate it or a vaccine that might prevent its spread (Chaturvedi et al. 2011, Fletcher et al. 2020, Micalizzi and Smith 2020).

The generation of novel alternatives through scientific investigation actually switches the order of analysis implied in PrOACT by putting the consequence analysis before the generation of alternatives. An engineering approach to decision analysis begins with the objectives, works backward through an understanding of how the system works, and then identifies an action that will achieve the objectives. A means-ends network is a useful graphical tool for this approach. The objectives (the ends) are identified, and then means to achieve those ends are drawn based on a current understanding of how the system works. Proceeding backward in this way to more proximate influences leads to actions that might be investigated as potential solutions.

Decision Analysis Tools

In addition to the array of tools it provides to help structure a problem, decision science also provides a diverse set of tools for analysis. These analytical tools offer insight into the nature of decisions and frequently motivate even deeper reflection by decision makers. The complete set of analytical tools is too large to be fully discussed here; what follows is a sampling of some of the most commonly applied techniques: multicriteria decision analysis, decision trees, and value of information. Skilled decision analysts diagnose a decision problem and identify the most appropriate analytical tools to apply.

Multicriteria Decision Analysis

As the field of human dimensions has made evident, wildlife management decisions involve many objectives on the part of many stakeholders. Understanding these objectives, being able to measure these objectives, and predicting the consequences of alternative actions with regard to these objectives are critical steps in evaluation of a multiple-objective problem. Multicriteria decision analysis is a set of techniques to analyze and balance the trade-offs inherent in multiple objectives (Herath and Prato 2006, Converse 2020). A consequence table, which shows the consequences of each alternative action in units of the measurable attribute for each objective, embodies the central expression of the decision problem in MCDA. The analytical question is how to identify the single alternative that best achieves the array of objectives, recognizing that there are trade-offs.

The first step in MCDA is to simplify the problem by examining the structure of the consequence table. A dominated alternative is one for which there is another alternative that is at least as good on all objectives. Given that there should be no reason to choose a dominated alternative, it can be removed from further consideration. An irrelevant objective is one that does not help distinguish the alternatives because they have similar scores on the corresponding measurable attribute. Irrelevant objectives may be important in the absolute sense (the decision maker may care very much about the performance of an action on an irrelevant objective), but they are not important in the relative sense (because they do not help the decision maker choose among the alternatives under consideration); thus they can also be dropped from further analysis. Often, an objective that is initially relevant may become irrelevant as dominated alternatives are identified and removed. When a consequence table no longer has any dominated alternative or irrelevant objectives, the remaining alternatives are said to be "Pareto optimal"; for any alternative, no improvement can be made on one objective without sacrificing another (Converse 2020).

To proceed further with analysis requires grappling with how to trade one objective with another. Some analysts will stop here, and simply ask the decision makers to make an intuitive judgment about the trade-offs and choose a preferred alternative. Quantitative analysis of the trade-offs requires expressing the objectives on a common scale. There are numerous ways to do this—including even swapping, or pricing out, the analytical hierarchy process, and outranking—but perhaps the most common is the weighted additive model embodied in the Simple Multi-Attribute Rating Technique (SMART; Goodwin and Wright 2004). The consequences are first converted to a common scale by normalizing the scores on each objective to a range of zero (i.e., the worst-performing alternative) to one (i.e., the best-performing alternative). Second, the decision maker provides weights, which reflect value judgments about the relative importance of the different objectives, through a process such as swing weighting (von Winterfeldt and Edwards 1986). Third, the weighted sum of the normalized consequences is taken across objectives, using the swing weights, and used to rank the alternatives. Sensitivity analysis can be performed to evaluate the robustness of the preferred alternative to uncertainty in the weights or uncertainty in the consequence values.

Multicriteria decision analysis has been applied extensively in natural resource management (Kiker et al. 2005, Herath and Prato 2006). Specific applications in wildlife management are

increasing (Redpath et al. 2004, Szymanski et al. 2009, Converse et al. 2013, Runge et al. 2015, Robinson et al. 2016). Case study two (see "Wolf Hunting Management in Montana," below) uses an MCDA approach.

Decision Trees

Some decisions must be made in the face of uncertainty, without recourse to resolving the uncertainty first (Gore et al. 2009, Converse 2020). This may be because the uncertainty is aleatory, or because the uncertainty is epistemic, but the decision must be made before the uncertainty can be reduced. In either case, the decision maker must accept the possibility of regret associated with an undesired outcome. Decision trees make clear the risks (and regrets) associated with alternatives, effectively insulating against poor decisions. This setting was the genesis of decision theory in economics (Moscati 2016), but it is just as applicable in wildlife management. For example, imagine a manager of a 400-km² tract of arid land whose primary objective is to provide habitat for pronghorn (*Antilocapra americana*). Without prescribed fire, grasses increase and native forbs, which pronghorns thrive on, decline. Prescribed fire returns nutrients to the soil and encourages growth of forbs, especially in a wet year, but in a dry year, prescribed fire can remove moisture from the system and substantially reduce the total biomass available for forage (Yoakum 1980). The manager has a predictive model for the carrying capacity of the refuge, but whether a particular year will be wet is an uncontrollable uncertainty (Fig. 5.4). The manager can use the decision tree shown in Figure 5.4 to calculate the expected carrying capacity under either decision by taking a weighted average of the carrying capacities in each branch, where the weights describe the likelihood of a wet year. In this particular

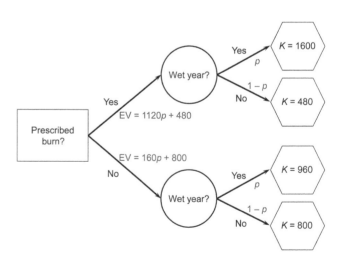

Figure 5.4. Decision tree for pronghorn habitat management. The outcomes are the expected carrying capacities (*K*) of the refuge as a function of whether a prescribed fire was instituted and whether the year was wet. The expected value (EV) of the outcome depends on the probability of a wet year (*p*) and can be used to identify a preferred action. With the values shown here, if *p* > 33%, the expected capacity is higher with a prescribed burn than without.

case, if the probability of a wet year is greater than 33%, the manager should institute the prescribed burn.

There are more advanced methods to make a simple decision tree more realistic. The tree might have additional branches to represent the likelihood of wildfire occurring and restoring some habitat condition. The manager might assign nonlinear values to the outcomes, to reflect a nonneutral attitude toward risk. The tree could be extended to acknowledge that this decision can be made annually, and the manager might care more about the cumulative responses over many years. The value in all these methods is in helping the manager think about how to make decisions in the face of uncontrollable uncertainty.

Expected Value of Information

There may be recourse to resolve uncertainty before having to commit to a decision. As wildlife scientists, we are always interested in reducing uncertainty, but a wildlife manager has a different perspective. The decision maker needs to ask whether the benefits that accrue from acquiring new information are worth the costs of obtaining that information. Decision analysis offers a formal method for answering this question. The expected value of information is the amount by which the outcome can be improved by reducing uncertainty before making the decision (Runge et al. 2011b). A powerful and underutilized method in natural resource management, calculating the value of information can help decide what research is valuable, what monitoring should be instituted, and whether adaptive management is warranted. Applications of the value of information in wildlife management and conservation settings have flourished over the past decade (Runge et al. 2017, Bolam et al. 2019, Rushing et al. 2020).

ADAPTIVE MANAGEMENT

Although use of a broad set of formal decision analysis techniques in wildlife management is only beginning, for decades there has been a widespread call for and use of a special class of decision analysis; namely, adaptive management. Developed in the context of fisheries management in the 1970s (Holling 1978, Walters 1986), adaptive management is now a central tenet of natural resource management in North America, including wildlife management (Lancia et al. 1996, Callicott et al. 1999, Allen et al. 2011, Runge 2020).

A Special Class of Decision

Adaptive management is a special case of structured decision-making for recurrent decisions made under uncertainty (see Fig. 5.2). Many wildlife management decisions have two key features: they are recurrent (a similar decision is made on a regular basis), and they are impeded by uncertainty (the consequences of the alternatives are not fully understood). To address the first feature, a wildlife manager needs to understand and anticipate the dynamics of the system—namely, the immediate costs or rewards from taking an action—and also the future opportunities, costs, and potential rewards attending

subsequent actions that might be taken. System modeling and dynamic optimization are tools that can support recurrent decisions (Runge 2020). To address the second feature, the wildlife manager needs to know how to make decisions in the face of uncertainty, by evaluating and balancing risks. Decision analytical techniques exist for making decisions in the face of uncertainty; in fact, they are the historical basis for the entire discipline (Moscati 2016). When these two key features occur together, when recurrent decisions need to be made in the face of uncertainty, there is an opportunity to learn from actions taken early on to reduce uncertainty so better decisions can be made in the future. The ability to adapt future decisions to information that arises during the course of management is the purpose and foundation of adaptive management.

The PrOACT sequence is central to adaptive management: objectives need to be expressed; alternative actions need to be developed; consequences need to be predicted; and a solution, through optimization or balancing trade-offs, needs to be found. To this sequence adaptive management adds five details: developing dynamic predictive models, articulating and evaluating uncertainty, implementing monitoring to provide feedback, updating the predictive models based on new information, and adapting future decisions based on the updated understanding of how the resource responds to management.

First, the predictive models must be dynamic; that is, they need to predict current rewards and future conditions of the system that could affect subsequent decision-making. Predictive models need to incorporate the temporal linkage among decisions. Predictive models of habitat and population dynamics for wildlife have included such dynamics, even outside of formal decision analysis (Williams et al. 2002).

Second, uncertainty must be articulated and evaluated. What aspects of the predictions are not well known and might impede the decision? Nichols et al. (1995) describe four sources of uncertainty relevant to wildlife management: environmental variation, structural uncertainty, partial observability, and partial controllability. Two of these (i.e., environmental variation and partial controllability) are types of aleatory uncertainty (Helton and Burmaster 1996)—uncertainty that cannot be reduced. The other two (i.e., structural uncertainty and partial observability) are types of epistemic uncertainty—uncertainty owing to our lack of knowledge, which (at least theoretically) can be reduced through investment in monitoring. Formal approaches to decision analysis attempt to express these uncertainties quantitatively, so that the uncertainty in the predictions can be stated clearly. To evaluate the uncertainties, the decision maker wants to know whether reduction of any uncertainty would improve the expected outcome of the decision. In the context of management, relevant uncertainty is uncertainty that affects the *decision*, not simply the *predictions*. The expected value of information measures how much a decision could improve if uncertainty could be reduced, and it is important for identifying the critical uncertainty to address in an adaptive program (Runge et al. 2011b). Key uncertainty is often expressed as a set of plausible models, each of which makes a different prediction about the effects

CARL J. WALTERS (1944–)

Carl J. Walters is a fisheries biologist who pioneered the concept of adaptive management. He received his doctorate from Colorado State University (Fort Collins, Colorado, USA) in 1969 and has been a professor of zoology and fisheries at the University of British Columbia (Vancouver, British Columbia, Canada) ever since. His 1986 book *Adaptive Management of Renewable Resources* offers a full decision-analytical treatment of natural resource management and formally considers how to make optimal recurrent decisions in the face of epistemic uncertainty. His interest in adaptive environmental assessment led to the 2004 publication, with Steven Martell, of *Fisheries Ecology and Management*, a graduate-level textbook on the use of quantitative models in fisheries management. The influence of his work on the practice of wildlife and fisheries management cannot be overstated. Among other honors, Walters received The Wildlife Society's best paper award in 1976, the American Fisheries Society Award of Excellence in 2006, and the Volvo Environment Prize in 2006. He is a fellow of the Royal Society of Canada.

Courtesy of Sandra Buckingham

of management actions on the outcomes that are relevant to the decision maker.

Third, an appropriate monitoring program that provides the necessary feedback to resolve critical uncertainty is central to meeting the promise of adaptive management. The needs of this monitoring program stem from the decision context and serve three fundamental purposes: evaluating performance against the objectives, tracking key variables that are tied to decision thresholds, and reducing key uncertainty (Nichols and Williams 2006). This "targeted" monitoring is important to make efficient use of scarce resources, allocating funds and staff time only to monitoring that is expected to improve management outcomes in the long term. Lyons et al. (2008) provide examples of monitoring design for management on national wildlife refuges that reflect these principles.

Fourth, monitoring data are valuable only if they are analyzed. In an adaptive management setting, analysis consists of confronting the predictive models with the observed data (Hilborn and Mangel 1997). Each of the alternative models makes a prediction about the outcome associated with the action that was last implemented, and the monitoring system provides

information about the actual outcome. The comparison of the observed response to the expected responses allows the predictive models to be updated, often through an application of Bayes's theorem (Williams et al. 2007). The degree of belief increases for those models whose predictions most closely matched the observed response, and decreases for models that performed poorly (Johnson et al. 2002).

Fifth, what makes adaptive management adaptive is the application of learning to subsequent decisions. This adaptation can be anticipated; that is, the decision maker can articulate in advance how future decisions will change as a result of monitoring outcomes. In "active adaptive management," this anticipation goes one step further: in making a decision, the decision maker may choose an action that will accelerate learning, if the long-term gains from that learning are anticipated to offset the short-term costs (Walters and Hilborn 1978, Williams 1996).

Single-, Double-, and Triple-Loop Learning

One of the real challenges of decision analysis is correctly framing the decision. For recurrent decisions, each iteration provides the opportunity to learn and reflect about the framing of the decision, in addition to the predictions of the system models. That is, there is another layer of learning, and hence another layer of adaptive management. This "double-loop" learning (Argyris and Shon 1978) focuses on emerging understanding of the framing of the decision; in particular, the objectives, the set of potential actions, and the relevant uncertainties (Fig. 5.5). In the most challenging natural resource management problems, where the ecological and institutional dynamics are complex, experience managing the system may give rise to insights about the context in which management is occurring. "Triple-loop" learning (Fig. 5.5; Pahl-Wostl 2009) can result in transformative adaptation (e.g., through changes to the institutional relationships, governance structures, regulatory frameworks, or even the social and organizational values that are associated with the managed resources).

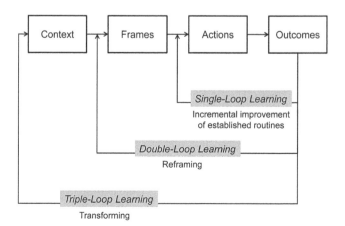

Figure 5.5. Adaptive learning cycles. From Pahl-Wostl (2009)

Schools of Adaptive Management

Adaptive management has seized the imagination of natural resource managers since the phrase was coined, but in the late 20th and early 21st centuries, a number of writers decried its failure to live up to its promise, documenting the challenges and alleged failures of implementation (McLain and Lee 1996, Gregory et al. 2006, Allen and Gunderson 2011). One of the challenges is that there is not a single definition of adaptive management. There are many layers on which learning and adaptive management can occur (Fig. 5.5). So perhaps it is not surprising that there are different schools of thought regarding adaptive management, each focused on a different layer of adaptation. McFadden et al. (2011) provide the beginnings of a long-needed taxonomy of adaptive management, identifying two primary schools of thought: the resilience-experimentalist school, exemplified by Gunderson et al. (1995), and the decision-theoretic school, exemplified by Williams et al. (2007).

The Resilience-Experimentalist School

The resilience-experimentalist (RE) school has arisen from the management challenges in large-scale, complex socioecological systems, where framing the decisions and constructing effective institutional arrangements for collaborative management pose enormous challenges. The ecological dynamics are so complex that the notion of being able to articulate critical uncertainty seems ambitious. So the focus is on double- and triple-loop learning, with an emphasis on collaboration and adaptive governance, with reduction of uncertainty occurring through experimental manipulation. The most-noted examples include management of the Columbia River (Lee and Lawrence 1985), the Everglades (Gunderson and Light 2006), and Grand Canyon (Hughes et al. 2007, Runge et al. 2015). All of these examples reveal the ecosystem focus of the RE school; this broader scope encompasses wildlife management.

The Decision-Theoretic School

The decision-theoretic (DT) school is grounded in the seminal writings of Holling (1978) and Walters (1986), but it was perhaps most profoundly influenced by the adaptive harvest management of waterfowl in North America (Johnson et al. 1997, Nichols et al. 2007). The emphasis is on management of dynamic systems in the face of uncertainty, through explicit use of a decision-theoretic framing of the problem, with reduction of uncertainty not occurring through experimentation but through ongoing monitoring and management. The approach taken by the DT school emphasizes single-loop learning but has the flexibility to accommodate learning and adaptation at all three levels. The DT school is as useful for local decisions with a single decision maker as it is for broad-scale decisions with multiple management partners.

CASE STUDIES

Consider three case studies of the application of structured decision-making that range in scale from local (Skyline Wild-life Management Area, Alabama, USA) to state (wolf harvest management in Montana, USA) to continental (adaptive harvest management of waterfowl in North America). Each of the case studies exemplifies particular elements of the SDM process, but they share the underlying PrOACT structure.

Case Study One: Skyline Wildlife Management Area

The Alabama Department of Conservation and Natural Resources (ADCNR) is charged with management to improve the status of the species of greatest conservation need identified in the state wildlife action plan. In 2005, a team of biologists from ADCNR, researchers from Auburn University, and decision analysts from US Geological Survey was assembled to examine how to balance game and nongame wildlife population objectives for the J. D. Martin Skyline Wildlife Management Area (SWMA), using structured decision-making, as a test case for other state-owned lands. The team undertook this work during several meetings that culminated with a workshop at the National Conservation Training Center in Shepherdstown, West Virginia, in the summer of 2006. The SWMA (170 km²) is located in Jackson County, Alabama, in the Cumberland Plateau region. Most of SWMA was logged at the turn of the 20th century, and only the most inaccessible slopes were spared. Agriculture grew in the region during the 1930s under the auspices of federal programs (Hammer 1967). The majority of the current forest vegetation is the result of natural regeneration, with some planted pine plantations on the plateaus and in the valleys. Even today, forested habitat exists only in narrow valleys, on steep hillsides, and on top of the Cumberland Plateau. Most of the lower, flatter areas in larger valleys have been converted to nonnative pasture or row crops.

Some portions of SWMA are owned and managed by the ADCNR Wildlife and Freshwater Fisheries Division (WFFD). Lands owned by WFFD were purchased with federal aid funds for the purposes of wildlife management and public hunting. Other portions are owned and managed by the ADCNR Lands Division. Lands Division purchases were made with state funds for their potential contribution to parks, nature preserve, wildlife management, and recreation. Other portions of SWMA are under long-term lease from Alabama Power Company, but management decisions are delegated to WFFD in mitigation for the establishment of the R. L. Harris Dam, which flooded 43 km² in central Alabama when it was built in 1983. Each of these entities has a different mandate and approach to wildlife management. The decision in this case was to identify and suggest alternatives for land use and forest management that would benefit greatest conservation need (GCN) species while providing adequate opportunity for hunters.

Objectives

1. Maintain or enhance populations of species of GCN identified in the Alabama Comprehensive Wildlife Conservation Strategy (WFFD 2005).

2. Provide hunting opportunities for large and small game, including white-tailed deer, eastern wild turkey (*Meleagris gallopavo*), eastern cottontail (*Sylvilagus floridanus*), northern bobwhite, and gray squirrel (*Sciurus carolinensis*).

3. Provide nonconsumptive recreational opportunities, including hiking, wildflower viewing, wildlife viewing, horseback riding, and primitive camping.

The measurable attribute for the first objective was the average occupancy of four representative nongame species—cerulean warbler (*Dendroica cerulean*), Kentucky warbler (*Oporornis formosus*), worm-eating warbler (*Helmitheros vermivorum*), and wood thrush (*Hylocichla mustelina*)—equally weighted. The measurable attribute for the second objective was the average occupancy of three representative game species—northern bobwhite, wild turkey, and mourning dove (*Zenaida macroura*)—equally weighted. A measurable attribute for the final objective was not developed for the initial analysis.

Alternatives

The management alternatives considered were combinations of landscape practices that increased the amount of nonforested areas, and treatments to forested and nonforested habitat (Fig. 5.6). The management alternatives considered in forested areas included four options.

1. Status quo, or maintaining the current landscape and forest management practices.
2. Even-aged forest management with large (~60 ha) or small stands (~20 ha) by clear-cutting, seed tree, or shelter wood techniques.
3. Two-aged management system throughout the forest.
4. Uneven-aged forest management using either group or single-tree selection methods.

In nonforested areas the alternatives included:

1. Status quo, or maintaining a mixture of plantings of green fields, row crops, and early successional habitat.

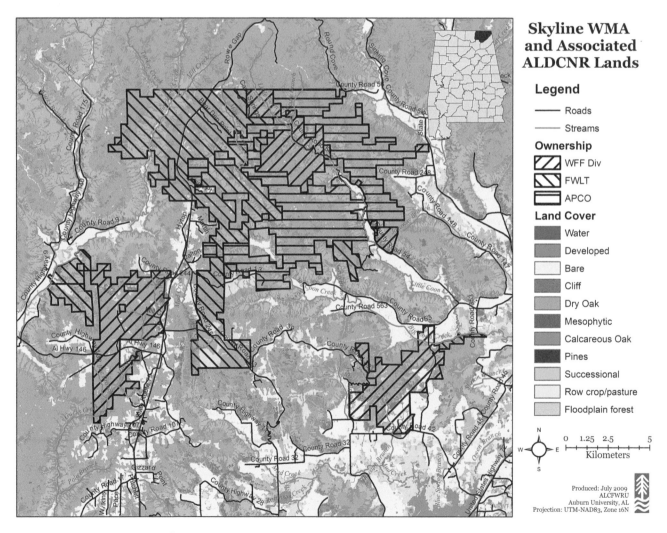

Figure 5.6. J. D. Martin Skyline Wildlife Management Area (WMA) land cover and land ownership boundaries, Jackson County, Alabama, USA. ADCNR, Alabama Department of Conservation and Natural Resources; WFFD, Wildlife and Freshwater Fisheries Division; SLD, Alabama State Lands Division; APCO, Alabama Power Company.

2. Increasing early successional habitats and native warm-season grass meadows.

3. Increasing early successional habitats and native warm-season grass meadows in some areas, and creating oak (*Quercus* spp.) savannah in others.

Modeling Consequences

As a prototype, areas were mapped that met an agreed-upon minimum area requirement for northern bobwhite populations (404 ha) and cerulean warbler (6,000 ha). The consequences of the management actions were predicted and evaluated in terms of the expected population response by the game and nongame populations of interest. Uncertainties included the effect of management practices on the composition and structure of the vegetative cover and the response of the animal populations to the structure, availability, and distribution of suitable areas. Occupancy (i.e., probability of use by each species) was determined to be an acceptable population response for comparing alternatives. For many species of reptiles, amphibians, birds, and small mammals, recent research provided estimates of the relationship between occupancy and many forest characteristics, including composition, structure, and context (Grand et al. 2008). For some game species where occupancy models were not available, expert judgments were elicited to predict wildlife responses. Experts ranked the alternative landscapes with respect to each of the objectives, but it was difficult to predict the effects of forest management on habitat structure and species responses. Therefore a system model employing a Bayesian belief network was developed as a second prototype (Fig. 5.7).

The Bayes net was used because it provided a graphical representation (i.e., influence diagram) of the system, which could be parameterized and converted to a decision model using existing data or expert judgments (Nielsen and Jensen 2009). Each node in the network represents an important characteristic (i.e., state variable) of the system and the linkages among nodes represent relationships between variables. Uncertainty in the relationships between variables, and uncertainty in the estimates of the variables themselves, was incorporated. Decisions were modeled using the Bayes net by adding a decision node, used to manipulate the state of the system under candidate management actions, and a utility node, which was used to assign stakeholder values to the measurable attributes of each top-level objective (game and nongame species). The relative weights on these objectives were not formally elicited from the decision makers; rather, a range of values was explored to understand how the preferred alternative was affected by the weights.

Decision Recommendation and Implementation

Analysis using the second prototype suggested that managing forested areas using a two-aged system would achieve the greatest utility. The model was not sensitive to the size of forest stands, nor did it indicate differences in utility between the alternatives for managing nonforested habitat. As of early 2021, ADCNR is still planning to implement the management alternatives. In addition, the SDM process has been used to develop and evaluate management alternatives for 12 additional wildlife management areas, parks, and nature preserves across the state.

Case Study Two: Wolf Hunting Management in Montana

Gray wolves in the US northern Rocky Mountains (NRM) were first removed from the endangered species list in February 2008, at which point management authority for wolves passed from the USFWS to the states of Montana and Idaho (USFWS 2011). Wolf management in each state included setting harvest quotas and seasons. Lessons learned from the first wolf hunting season in Montana in 2009 suggested that Montana Fish, Wildlife and Parks (MFWP) needed to redefine its wolf management units (WMUs) to better allocate hunter opportunity and harvest and to manage wolf numbers. For the 2009 hunting season, MFWP had defined three WMUs (Fig. 5.8). Because wolves are primarily located in the mountainous portions of western Montana, managers believed that smaller, redistributed WMUs in that portion of the state would be necessary to manage allocation of hunter opportunity and thus the distribution of harvest across the Montana wolf population. Statutory obligations for effective conservation of a game species and often-contentious public attitudes and expectations regarding wolf management in Montana combined to present a challenging context for deciding on new WMUs. The MFWP thus elected to use a structured process to develop the 2010 wolf harvest strategy to ensure explicit consideration of all relevant factors affecting the designation of WMUs, and to provide transparency to the public. A team of specialists, including managers, wolf experts, and decision analysts, was assembled to work through this process.

As part of this structured process, representatives from MFWP—including regional managers, biologists, and wolf specialists—developed the following problem statement: "MFWP must propose a 2010 wolf harvest strategy that maintains a recovered and connected wolf population, minimizes wolf–livestock conflicts, reduces wolf effects on low or declining ungulate populations and ungulate hunting opportunities, and effectively communicates to all parties the relevance and credibility of the harvest while acknowledging the diversity of values among those parties."

Objectives

The group developed a set of fundamental, process, and strategic objectives.

Fundamental objectives

1. Maintain positive and effective working relationships with
    a) livestock producers,
    b) hunters, and
    c) other stakeholders.
2. Reduce wolf effects on big-game populations.
3. Reduce wolf effects on livestock.

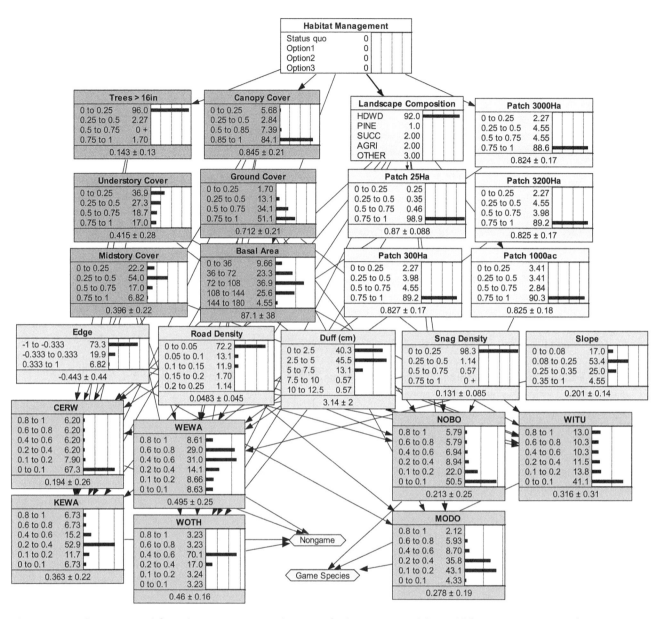

Figure 5.7. Bayes decision network for evaluating management alternatives for the J. D. Martin Skyline Wildlife Management Area, Jackson County, Alabama. The decision network (prototype 2) includes nodes that represent the decision ("Habitat Management"), habitat structure (*top left boxes*), land cover (*top right boxes*), physical characteristics (*central row*), species responses (*bottom boxes*), and utilities (*hexagons*). CERW, cerulean warbler; KEWA, Kentucky warbler; MODO, mourning dove; NOBO, northern bobwhite; WEWA, worm-eating warbler; WITU, wild turkey; WOTH, wood thrush.

4. Maintain hunter opportunity for ungulates.
5. Maintain a viable and connected wolf population in Montana.
6. Maintain hunter opportunity for wolves.

Process objectives

7. Enhance open and effective communication to better inform decisions.
8. Learn and improve as we go.

Strategic objectives

9. Increase broad public acceptance of harvest and hunter opportunity as part of wolf conservation.

10. Gain and maintain authority for the state of Montana to manage wolves.

The group developed five management alternatives to address the set of fundamental objectives. The number and distribution of WMUs affect how finely the state can control the distribution of wolf harvest, which in turn affects wolf density and distribution, and the various effects associated with wolf density. The alternatives focused on the arrangement of WMUs. The first alternative represented the status quo, retaining the same three WMUs used during the 2009 hunting sea-

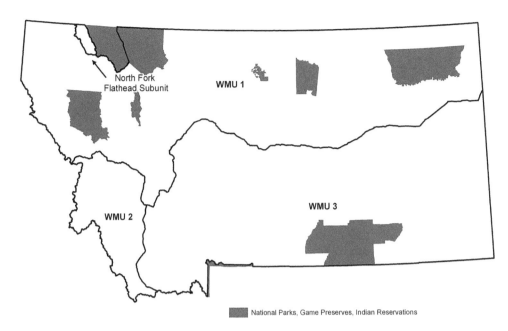

Figure 5.8. Wolf management units in Montana, hunting season 2009

North Fork
Flathead Subunit

WMU 1

WMU 2

WMU 3

National Parks, Game Preserves, Indian Reservations

son. The remaining four options represented alternative ways of dividing Montana into WMUs.

1. Status quo, three WMUs, the same ones used in 2009.
2. Fifteen WMUs, with eastern Montana incorporated into western units.
3. Fourteen WMUs, with eastern Montana incorporated into western units.
4. Thirteen WMUs, with eastern Montana incorporated into western units.
5. Fifteen WMUs, with eastern Montana having its own management unit not incorporated into western units.

The measurable attributes for each fundamental objective were expressed on a constructed scale that ranged from zero (i.e., poor outcome) to one (i.e., ideal outcome). Two fundamental objectives (numbers 5 and 6 above) were not scored, because the group did not believe their consequences varied among the management alternatives and thus did not affect the decision. One of the strategic objectives (number 9 above) was viewed as critical enough to the decision that it was also scored. A panel of experts composed of wildlife managers, biologists, and wolf specialists from MFWP were asked to score individually each alternative against each objective. An average score for each response was taken across experts (Table 5.4).

The status quo (alternative 1) ranked high among alternatives for maintaining relationships with livestock producers, hunters, and other stakeholders, and for public acceptance, but ranked relatively low for reducing effects to big game and livestock while maintaining a sustainable ungulate harvest (Table 5.4). Alternative 2 scored relatively low for maintaining relationships but moderately well for reducing effects of wolves, public acceptance, and maintaining sustainable ungulate harvest. Alternatives 3 and 4 scored comparably across all objectives. Alternative 5 was judged to have strong benefits for reducing effects to big game and maintaining sustainable

ungulate harvests, but it would have the strongest negative effects among alternatives on maintaining relationships with stakeholders and public opinion.

Inspection of the consequence table shows that alternative 3 dominates alternatives 2, 4, and 5 because it scores as well or better than those other alternatives on all objectives. Thus alternatives 2, 4, and 5 can be removed from further consideration, leaving only alternatives 1 and 3 as viable candidates.

At this point, formal multicriteria decision analysis could be used to place weights on the objectives to develop a composite score for each alternative. But the panel chose instead to proceed qualitatively on the basis that identification of dominated alternatives and redundant objectives provided a cognitively accessible trade-off. The group decided alternative 3 (Fig. 5.9) was most likely to satisfy the fundamental objectives for setting WMUs for the 2010 hunting season. This was because the relative benefits of maintaining relationships with hunters, reducing effects, and maintaining sustainable ungulate populations in alternative 3 outweighed the slight advantages in maintaining relationships with livestock producers and stakeholders and public acceptance offered by alternative 1.

The SDM approach allowed decision makers to see the structure of the problem and the major trade-offs among the alternatives; those insights alone were enough to allow the decision to proceed. The decision was presented in July 2010 as a recommendation to the MFWP commission, which adopted it; the SDM process and product were considered clear assets in the public presentation and review of the proposed season structure. The 2010 wolf hunting season was not implemented, however, because wolves in the NRM were returned to the endangered species list by court order in August 2010. With the legislated removal of wolves in the NRM from the endangered species list in May 2011, management of wolves under the 2010 WMUs was implemented in 2011 with minor adjustments.

Table 5.4. Consequence table for case study two, wolf hunting management in Montana, USA

Fundamental Objective	Measurable Attribute	Preferred Direction	Alternative 1	Alternative 2	Alternative 3	Alternative 4	Alternative 5
Maintain Relationships							
Livestock producer	Perception: 0 to 1	Maximize	**0.83**	*0.54*	0.66	0.66	0.63
Stakeholders	Perception: 0 to 1	Maximize	**0.69**	0.60	0.66	0.66	*0.34*
Hunters	Perception: 0 to 1	Maximize	0.80	*0.57*	**0.83**	0.77	0.60
Reduce Impacts							
Big game	Ungulate populations at or near objectives: yes (1) / no (0)	Maximize	*0.60*	**1.00**	**1.00**	0.80	**1.00**
Livestock	Reduction in the number of livestock confirmed injured or killed by wolves: 0 to 1	Maximize	*0.56*	0.72	**0.80**	**0.80**	0.76
Sustainable ungulate harvest	Quota in every wildlife management unit for foreseeable future: yes (1) / no (0)	Maximize	*0.60*	**1.00**	**1.00**	0.80	**1.00**
Public acceptance	Perception: 0 to 1	Maximize	**0.80**	0.72	0.74	0.74	*0.37*

Note: Consequences for each alternative were elicited individually and then averaged over a group of wildlife managers, biologists, and wolf specialists from Montana Fish, Wildlife and Parks. For each objective, alternatives predicted to perform best are indicated by boldface, moderately performing alternatives are indicated by regular type, and alternatives predicted to perform worst are indicated by italics. Alternative 3 dominates all alternatives except alternative 1, which performs relatively well for maintaining relationships but poorest among the alternatives for reducing impacts and maintaining sustainable ungulate harvests.

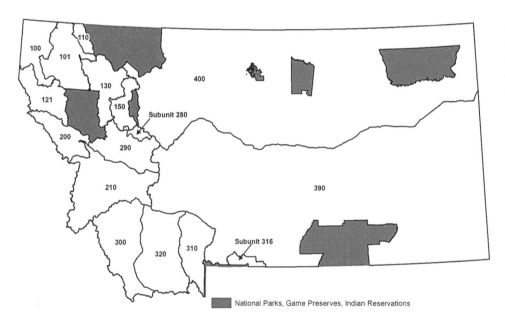

Figure 5.9. Wolf management units adopted for implementation in the 2011 hunting season by the Montana Fish, Wildlife and Parks Commission

Although there was a pressing need to make a recommendation for 2010, this decision can be revisited each year, creating an opportunity to improve the analysis and to reduce uncertainty over time. Future iterations of this process might address three topics: developing better measurable attributes for the objectives, founded on natural scales that are tied to monitoring systems; reinstating the omitted fundamental objectives, which may be more relevant in subsequent years; and analyzing wolf monitoring data over time to evaluate the efficacy of the wolf hunting program.

Case Study Three: Adaptive Harvest Management of Waterfowl in North America

Each year, the USFWS sets harvest regulations for waterfowl based on population and habitat conditions. The USFWS has sole regulatory responsibility for this decision under the Migratory Bird Treaty Act (16 USC 703-712), and a number of NEPA compliance documents govern the regulations-setting process (USFWS 2013). But the USFWS recognizes important management partnerships with the states and flyways, and it has established a formal collaborative structure for garnering input from these partners. In 1995, a prescriptive decision-theoretic approach to setting harvest regulations for midcontinent mallards (*Anas platyrhynchos*) was established (Nichols et al. 1995, Johnson et al. 1997). Referred to as adaptive harvest management (AHM), this process recognizes the dynamic nature of the resource, the recurrent nature of the decisions, and the role that uncertainty plays in impeding decision-making. Because mallards are the most abundant duck species in the midcontinent, AHM also serves as the framework around which regulations for hunting of other duck species is centered.

There are multiple objectives that AHM seeks to achieve: to maximize annual harvest of mallards, to maintain a sustainable level of harvest, to maintain the population size close to or higher than the North American Waterfowl Management Plan (NAWMP) goal, and to prevent closed seasons, except in extreme circumstances (USFWS 2019). These multiple—and competing—objectives have been combined into a single objective function. The objective of AHM is to maximize

$$\sum_{t=0}^{\infty} H_t \min\left(\frac{\hat{N}_{t+1}}{8.5}, 1\right),$$

where H_t is the annual harvest, $\hat{N}_{t+1}$ is the predicted breeding population size in the next year, and 8.5 is the NAWMP goal for midcontinent mallards (in millions). The minimization within the objective function devalues the harvest whenever the population size is predicted to be below the NAWMP goal. Summing the harvest over an infinite time horizon ensures sustainability; the only way to maximize a long-term cumulative harvest is to keep the population extant.

The alternatives are chosen from a small set of regulatory packages: closed, restrictive, moderate, and liberal seasons, which differ in the length of the season and the daily bag limit. The closed season is only permitted when the midcontinent mallard population size falls below 5.5 million. For each of the regulatory packages, an expected harvest rate has been estimated.

The consequences of the different packages are evaluated through predictive models of mallard population dynamics. These models take three input values: two state variables (mallard breeding population size and the number of ponds in prairie Canada) and one decision variable (the regulatory package). They predict two quantities: the expected harvest, H_t, and the breeding population size in the subsequent year, $\hat{N}_{t+1}$ (Runge et al. 2002). One of the motivations for an adaptive management approach was intense disagreement that arose out of uncertainty about the population dynamics. There is uncertainty about the degree of density dependence in recruitment (weak vs. strong density dependence) and uncertainty about the effect of harvest mortality on annual mortality (additive vs. compensatory harvest mortality); in combination, these uncertainties are captured in four alternative population models. This uncertainty matters; the four alternative models lead to very different harvest strategies, and the resolution of

Table 5.5. Optimal regulatory strategy for midcontinent mallards for the 2020 hunting season

B_{pop}	Ponds									
	1.5	2.0	2.5	3.0	3.5	4.0	4.5	5.0	5.5	6.0
≤4.5	C	C	C	C	C	C	C	C	C	C
4.75–6.25	R	R	R	R	R	R	R	R	R	R
6.5	R	R	R	R	R	R	R	R	M	L
6.75	R	R	R	R	R	R	R	L	L	L
7	R	R	R	R	R	L	L	L	L	L
7.25	R	R	R	R	M	L	L	L	L	L
7.5	R	R	R	M	L	L	L	L	L	L
7.75	R	R	M	L	L	L	L	L	L	L
8	R	M	L	L	L	L	L	L	L	L
8.25	M	L	L	L	L	L	L	L	L	L
≥8.5	L	L	**L**	**L**	L	L	L	L	L	L

Source: USFWS (2019).

Note: The two state variables are the breeding population size (B_{pop}, in millions) and the number of ponds in prairie Canada (ponds, in millions). The regulatory packages are closed (C), restrictive (R), moderate (M), and liberal (L). Boldface represents the regulatory prescription for 2020.

the uncertainty has a significant value of information (Johnson et al. 2002).

The optimal strategy is found each year through passive adaptive stochastic dynamic programming (Williams 1996), which produces a state-dependent harvest strategy that stipulates the optimal regulatory package for any combination of breeding population size and number of ponds (Table 5.5). It is a passive adaptive strategy, in that the optimization does not anticipate the effect of learning on future decisions.

The USFWS, Canadian Wildlife Service, US states, and Canadian provinces collaboratively operate an extensive monitoring program for waterfowl, which includes aerial surveys to estimate abundance and habitat conditions, banding and band-recovery programs for survival and related estimates, and harvest surveys for harvest and reproductive estimates. From the standpoint of adaptation, the key annual monitoring data are the breeding population estimates, because these provide the feedback for evaluating the model uncertainty. The weights on the four models have evolved over time as a result of the observed responses to management (Fig. 5.10); the evidence for the weakly density-dependent model has increased significantly, and the evidence for the additive model has increased slightly. These changes in model weights have been accompanied by an evolution in the harvest strategy over time; thus the annual regulations have adapted to the new information.

The AHM program has undergone some technical adjustments and minor policy modifications over the years since its first implementation but has largely remained intact. Currently, the waterfowl management community is engaged in a process of double-loop learning, examining the nature of the objectives, alternatives, and models that underlie the regulations setting process (Anderson et al. 2007).

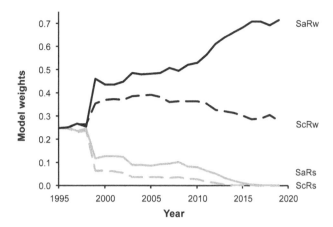

Figure 5.10. Weights on alternative predictive models for midcontinent mallard (*Anas platyrhynchos*) dynamics, 1995–2019 (USFWS 2019). The four models are distinguished by whether the survival model is compensatory (Sc) or additive (Sa), and whether the reproductive model is strongly (Rs) or weakly (Rw) density dependent.

SUMMARY

Wildlife management is a decision-focused discipline. It needs to integrate traditional wildlife science and social science to identify actions that are most likely to achieve the array of desires society has surrounding wildlife populations. Decision science, a vast field with roots in economics, operations research, and psychology, offers a rich set of tools to help wildlife managers frame, decompose, analyze, and synthesize their decisions. The nature of wildlife management as a decision science has been recognized since the inception of the field, but formal methods of decision analysis have been underused. There is tremendous potential for wildlife management to grow further through the use of formal decision analysis. First, the wildlife science and human dimensions of wildlife disciplines can be readily integrated. Second, decisions can become more efficient. Third, decision makers can communicate more clearly with stakeholders and the public. Fourth, intuitive wildlife managers, by explicitly examining how they make decisions, can translate their art into a science that is readily used by the next generation.

Literature Cited

Allen, C. R., J. J. Fontaine, K. L. Pope, and A. S. Garmestani. 2011. Adaptive management for a turbulent future. Journal of Environmental Management 92:1339–1345.

Allen, C. R., and L. H. Gunderson. 2011. Pathology and failure in the design and implementation of adaptive management. Journal of Environmental Management 92:1379–1384.

Anderson, M. G., D. Caswell, J. M. Eadie, J. T. Herbert, M. Huang, et al. 2007. Report from the joint task group for clarifying North American Waterfowl Management Plan population objectives and their use in harvest management. US Fish and Wildlife Service, Department of the Interior, US Geological Survey, Washington, DC, USA.

Argyris, C., and D. Shon. 1978. Organizational learning: a theory of action learning. Addison-Wesley, Reading, Massachusetts, USA.

Bailey, J. A. 1982. Implications of "muddling through" for wildlife management. Wildlife Society Bulletin 10:363–369.

Bain, M. B. 1987. Structured decision making in fisheries management. North American Journal of Fisheries Management 7:475–481.

Beissinger, S. R., and D. R. McCullough, editors. 2002. Population viability analysis. University of Chicago Press, Chicago, Illinois, USA.

Bengston, D. N. 2000. Environmental values related to fish and wildlife lands. Pages 126–132 in D. C. Fulton, K. C. Nelson, D. H. Anderson, and D. W. Lime, editors. Human dimensions of natural resource management: emerging issues and practical applications. Cooperative Park Studies Program, University of Minnesota, Department of Forest Resources, St. Paul, USA.

Bernard, R. F., and E. H. C. Grant. 2019. Identifying common decision problem elements for the management of emerging fungal diseases of wildlife. Society and Natural Resources 32:1040–1055.

Blomquist, S. M., T. D. Johnson, D. R. Smith, G. P. Call, B. N. Miller, W. M. Thurman, J. E. McFadden, M. J. Parkin, and G. S. Boomer. 2010. Structured decision-making and rapid prototyping to plan a management response to an invasive species. Journal of Fish and Wildlife Management 1:19–32.

Bolam, F. C., M. J. Grainger, K. L. Mengersen, G. B. Stewart, W. J. Sutherland, M. C. Runge, and P. J. McGowan. 2019. Using the value of information to improve conservation decision making. Biological Reviews 94:629–647.

Boyce, M. S., and L. L. McDonald. 1999. Relating populations to habitats using resource selection functions. Trends in Ecology and Evolution 14:268–272.

Burgman, M. A. 2005. Risks and decisions for conservation and environmental management. Cambridge University Press, Cambridge, UK.

Callicott, J. B., L. B. Crowder, and K. Mumford. 1999. Current normative concepts in conservation. Conservation Biology 13:22–35.

Caswell, H. 2001. Matrix population models: construction, analysis, and interpretation. Sinauer Associates, Sunderland, Massachusetts, USA.

Chaturvedi, S., S. S. Rajkumar, X. Li, G. J. Hurteau, M. Shtutman, and V. Chaturvedi. 2011. Antifungal testing and high-throughput screening of compound library against *Geomyces destructans*, the etiologic agent of geomycosis (WNS) in bats. PLoS One 6:e17032.

Claxton, K., M. Sculpher, and M. Drummond. 2002. A rational framework for decision making by the National Institute for Clinical Excellence (NICE). Lancet 360:711–715.

Colyvan, M., J. Justus, and H. M. Regan. 2011. The conservation game. Biological Conservation 144:1246–1253.

Converse, S. J. 2020. Introduction to multi-criteria decision analysis. Pages 51–61 in M. C. Runge, S. J. Converse, J. E. Lyons, and D. R. Smith, editors. Structured decision making: case studies in natural resource management. Johns Hopkins University Press, Baltimore, Maryland, USA.

Converse, S. J., C. T. Moore, M. J. Folk, and M. C. Runge. 2013. A matter of tradeoffs: reintroduction as a multiple objective problem. Journal of Wildlife Management 77:1145–1156.

Converse, S. J., K. J. Shelley, S. Morey, J. Chan, A. LaTier, C. Scafidi, D. T. Crouse, and M. C. Runge. 2011. A decision-analytic approach to the optimal allocation of resources for endangered species consultation. Biological Conservation 144:319–329.

Dalkey, N., and O. Helmer. 1963. An experimental application of the Delphi method to the use of experts. Management Science 9:458–467.

Diakoulaki, D., C. H. Antunes, and A. Gomes Martins. 2005. MCDA and energy planning. Pages 859–890 in J. Figueira, S. Greco, and M. Ehrogott, editors. Multiple criteria decision analysis: state of the art surveys. Springer, New York, New York, USA.

Failing, L., R. S. Gregory, and M. Harstone. 2007. Integrating science and local knowledge in environmental risk management: a decision-focused approach. Ecological Economics 64:47–60.

Fletcher, Q. E., Q. M. Webber, and C. K. Willis. 2020. Modelling the potential efficacy of treatments for white-nose syndrome in bats. Journal of Applied Ecology 57:1283–1291.

Fontaine, J. J. 2011. Improving our legacy: incorporation of adaptive management into state wildlife action plans. Journal of Environmental Management 92:1403–1408.

Fritts, S. H., E. E. Bangs, J. A. Fontaine, M. R. Johnson, M. K. Phillips, E. D. Koch, and J. R. Gunson. 1997. Planning and implementing a reintroduction of wolves to Yellowstone National Park and central Idaho. Restoration Ecology 5:7–27.

Garrard, G. E., L. Rumpff, M. C. Runge, and S. J. Converse. 2017. Rapid prototyping for decision structuring: an efficient approach to conservation decision analysis. Pages 46–64 in N. Bunnefeld, E. Nicholson, and E. J. Milner-Gulland, editors. Decision-making in conservation and natural resource management: models for interdisciplinary approaches. Cambridge University Press, Cambridge, UK.

Gerber, L. R., M. C. Runge, R. F. Maloney, C. A. Drew, S. Avery-Gomm, et al. 2018. Endangered species recovery: a resource allocation problem. Science 362:284–286.

Goodwin, P., and G. Wright. 2004. Decision analysis for management judgment. 3rd edition. John Wiley and Sons, West Sussex, UK.

Gore, M. L., R. S. Wilson, W. F. Siemer, H. Wieczorek Hudenko, C. E. Clarke, P. Sol Hart, L. A. Maguire, and B. A. Muter. 2009. Application of risk concepts to wildlife management: special issue introduction. Human Dimensions of Wildlife 14:301–313.

Grand, J. B., Y. Wang, and E. C. Soehren. 2008. Monitoring program for biodiversity of terrestrial vertebrates on conservation lands within the Cumberland Plateau Region of Alabama. US Geological Survey Alabama Cooperative Fish and Wildlife Research Unit, Auburn University, Mobile, Alabama, USA.

Grant, E. H. C., E. Muths, R. A. Katz, S. Canessa, M. J. Adams, et al. 2017. Using decision analysis to support proactive management of emerging infectious wildlife diseases. Frontiers in Ecology and the Environment 15:214–221.

Gregory, R., L. Failing, M. Harstone, G. Long, T. McDaniels, and D. Ohlson. 2012. Structured decision making: a practical guide for environmental management choices. Wiley-Blackwell, West Sussex, UK.

Gregory, R., and G. Long. 2009. Using structured decision making to help implement a precautionary approach to endangered species management. Risk Analysis 29:518–532.

Gregory, R., D. Ohlson, and J. Arvai. 2006. Deconstructing adaptive management: criteria for applications to environmental management. Ecological Applications 16:2411–2425.

Gunderson, L., C. S. Holling, and S. S. Light, editors. 1995. Barriers and bridges to the renewal of ecosystems and institutions. Columbia University Press, New York, New York, USA.

Gunderson, L., and S. S. Light. 2006. Adaptive management and adaptive governance in the Everglades ecosystem. Policy Sciences 39:323–334.

Hämäläinen, R., and O. Leikola. 1996. Spontaneous decision conferencing with top-level politicians. OR Insight 9:24–28.

Hammer, W. 1967. A pictorial walk thru ol' high Jackson: Scottsboro 1868–1968. Limited centennial edition. College Press, Collegedale, Tennessee, USA.

Hammond, J. S., R. L. Keeney, and H. Raiffa. 1999. Smart choices: a practical guide to making better life decisions. Broadway Books, New York, New York, USA.

Helton, J. C., and D. E. Burmaster. 1996. Guest editorial: treatment of aleatory and epistemic uncertainty in performance assessments for complex systems. Reliability Engineering and System Safety 54:91–94.

Hemming, V., M. A. Burgman, A. M. Hanea, M. F. McBride, and B. C. Wintle. 2018. A practical guide to structured expert elicitation using the IDEA protocol. Methods in Ecology and Evolution 9:169–180.

Herath, G., and T. Prato, editors. 2006. Using multi-criteria decision analysis in natural resource management. Ashgate, Hampshire, UK.

Hilborn, R., and M. Mangel. 1997. The ecological detective: confronting models with data. Volume 28. Princeton University Press, Princeton, New Jersey, USA.

Hirzel, A. H., and G. Le Lay. 2008. Habitat suitability modelling and niche theory. Journal of Applied Ecology 45:1372–1381.

Holling, C. S., editor. 1978. Adaptive environmental assessment and management. John Wiley and Sons, London, UK.

Howard, R. A. 1966. Decision analysis: applied decision theory. Pages 55–71 in D. B. Hertz and J. Melese, editors. Proceedings of the fourth international conference on operational research. John Wiley and Sons, New York, New York, USA.

Howard, R. A. 1968. The foundations of decision analysis. IEEE Transactions of Systems Science and Cybernetics 4:211–219.

Hughes, T. P., L. H. Gunderson, C. Folke, A. H. Baird, D. Bellwood, et al. 2007. Adaptive management of the great barrier reef and the Grand Canyon world heritage areas. AMBIO: A Journal of the Human Environment 36:586–592.

Hunter, C. M., H. Caswell, M. C. Runge, E. V. Regehr, S. C. Amstrup, and I. Stirling. 2010. Climate change threatens polar bear populations: a stochastic demographic analysis. Ecology 91:2883–2897.

Jenks, J. A., W. P. Smith, and C. S. DePerno. 2002. Maximum sustained yield harvest versus trophy management. Journal of Wildlife Management 66:528–535.

Johnson, F. A., D. R. Breininger, B. W. Duncan, J. D. Nichols, M. C. Runge, and B. K. Williams. 2011. A Markov decision process for managing habitat for Florida scrub-jays. Journal of Fish and Wildlife Management 2:234–246.

Johnson, F. A., M. J. Eaton, J. H. Williams, G. H. Jensen, and J. Madsen. 2015. Training conservation practitioners to be better decision makers. Sustainability 7:8354–8383.

Johnson, F. A., W. L. Kendall, and J. A. Dubovsky. 2002. Conditions and limitations on learning in the adaptive management of mallard harvests. Wildlife Society Bulletin 30:176–185.

Johnson, F. A., C. T. Moore, W. L. Kendall, J. A. Dubovsky, D. F. Caithamer, J. R. Kelley Jr., and B. K. Williams. 1997. Uncertainty and the management of mallard harvests. Journal of Wildlife Management 61:202–216.

Karl, H. A., L. E. Susskind, and K. H. Wallace. 2007. A dialogue, not a diatribe: effective integration of science and policy through joint fact finding. Environment 49:20–34.

Keeney, R. L. 1982. Decision analysis: an overview. Operations Research 30:803–838.

Keeney, R. L. 1996a. Value-focused thinking: identifying decision opportunities and creating alternatives. European Journal of Operational Research 92:537–549.

Keeney, R. L. 1996b. Value-focused thinking: a path to creative decision-making. Harvard University Press, Cambridge, Massachusetts, USA.

Keeney, R. L. 2002. Common mistakes in making value trade-offs. Operations Research 50:935–945.

Keeney, R. L. 2007. Developing objectives and attributes. Pages 104–128 in W. Edwards, R. F. J. Miles, and D. Von Winterfeldt, editors. Advances in decision analysis: from foundations to applications. Cambridge University Press, Cambridge, UK.

Kiker, G. A., T. S. Bridges, A. Varghese, T. P. Seager, and I. Linkov. 2005. Application of multicriteria decision analysis in environmental decision making. Integrated Environmental Assessment and Management 1:95–108.

Kuhnert, P. M., T. G. Martin, and S. P. Griffiths. 2010. A guide to eliciting and using expert knowledge in Bayesian ecological models. Ecology Letters 13:900–914.

Lancia, R. A., C. E. Braun, M. W. Collopy, R. D. Dueser, J. G. Kie, C. J. Martinka, J. D. Nichols, T. D. Nudds, W. R. Porath, and N. G. Tilghman. 1996. ARM! for the future: adaptive resource management in the wildlife profession. Wildlife Society Bulletin 24:436–442.

Lee, K. N., and J. Lawrence. 1985. Adaptive management: learning from the Columbia River basin fish and wildlife program. Environmental Law 16:431–460.

Leopold, A. 1933. Game management. Charles Scribner's Sons, New York, New York, USA.

Li, S.-L., O. N. Bjørnstad, M. J. Ferrari, R. Mummah, M. C. Runge, C. J. Fonnesbeck, M. J. Tildesley, W. J. Probert, and K. Shea. 2017. Essential information: uncertainty and optimal control of Ebola outbreaks. Proceedings of the National Academy of Sciences 114:5659–5664.

Lyons, J. E. 2020. Introduction to resource allocation. Pages 99–107 in M. C. Runge, S. J. Converse, J. E. Lyons, and D. R. Smith, editors. Structured decision making: case studies in natural resource management. Johns Hopkins University Press, Baltimore, Maryland, USA.

Lyons, J. E., K. S. Kalasz, G. Breese, and C. W. Boal. 2020. Resource allocation for coastal wetland management: confronting uncertainty about sea level rise. Pages 108–123 in M. C. Runge, S. J. Converse, J. E. Lyons, and D. R. Smith, editors. Structured decision making: case studies in natural resource management. Johns Hopkins University Press, Baltimore, Maryland, USA.

Lyons, J. E., M. C. Runge, H. P. Laskowski, and W. L. Kendall. 2008. Monitoring in the context of structured decision-making and adaptive management. Journal of Wildlife Management 72:1683–1692.

MacMillan, D. C., and K. Marshall. 2006. The Delphi process—an expert-based approach to ecological modelling in data-poor environments. Animal Conservation 9:11–19.

Martin, J., A. F. O'Connell, W. L. Kendall, M. C. Runge, T. R. Simons, A. H. Waldstein, S. A. Schulte, S. J. Converse, G. W. Smith, and T. Pinion. 2010. Optimal control of native predators. Biological Conservation 143:1751–1758.

Martin, T. G., M. A. Burgman, F. Fidler, P. M. Kuhnert, S. Low-Choy, M. McBride, and K. Mengersen. 2012. Eliciting expert knowledge in conservation science. Conservation Biology 26:29–38.

McDonald-Madden, E., W. J. M. Probert, C. E. Hauser, M. C. Runge, H. P. Possingham, M. E. Jones, J. L. Moore, T. M. Rout, P. A. Vesk, and B. A. Wintle. 2010. Active adaptive conservation of threatened species in the face of uncertainty. Ecological Applications 20:1476–1489.

McFadden, J. E., T. L. Hiller, and A. J. Tyre. 2011. Evaluating the efficacy of adaptive management approaches: is there a formula for success? Journal of Environmental Management 92:1354–1359.

McLain, R. J., and R. G. Lee. 1996. Adaptive management: promises and pitfalls. Environmental Management 20:437–448.

Micalizzi, E. W., and M. L. Smith. 2020. Volatile organic compounds kill the white-nose syndrome fungus, Pseudogymnoascus destructans, in hibernaculum sediment. Canadian Journal of Microbiology 66:593–599.

Miller, Z. D. 2017. The enduring use of the theory of planned behavior. Human Dimensions of Wildlife 22:583–590.

Mitchell, M. S., H. Cooley, J. A. Gude, J. Kolbe, J. J. Nowak, K. M. Proffitt, S. N. Sells, and M. Thompson. 2018. Distinguishing values from science in decision making: setting harvest quotas for mountain lions in Montana. Wildlife Society Bulletin 42:13–21.

Mitchell, M. S., J. A. Gude, N. J. Anderson, J. M. Ramsey, M. J. Thompson, M. G. Sullivan, V. L. Edwards, C. N. Gower, J. F. Cochrane, and E. R. Irwin. 2013. Using structured decision making to manage disease risk for Montana wildlife. Wildlife Society Bulletin 37:107–114.

Moore, C. T., C. J. Fonnesbeck, K. Shea, K. J. Lah, P. M. McKenzie, L. C. Ball, M. C. Runge, and H. M. Alexander. 2011a. An adaptive decision framework for the conservation of a threatened plant. Journal of Fish and Wildlife Management 2:247–261.

Moore, C. T., E. V. Lonsdorf, M. G. Knutson, H. P. Laskowski, and S. K. Lor. 2011b. Adaptive management in the U.S. National Wildlife Refuge System: science-management partnerships for conservation delivery. Journal of Environmental Management 92:1395–1402.

Morgan, M. G., and M. Henrion. 1990. Uncertainty: a guide to dealing with uncertainty in quantitative risk and policy analysis. Cambridge University Press, Cambridge, UK.

Morrison, M. L., B. G. Marcot, and R. W. Mannan. 2006. Wildlife–habitat relationships: concepts and applications. Island Press, Washington, DC, USA.

Moscati, I. 2016. Retrospectives: how economists came to accept expected utility theory: the case of Samuelson and Savage. Journal of Economic Perspectives 30:219-236.

Nichols, J. D., F. A. Johnson, and B. K. Williams. 1995. Managing North American waterfowl in the face of uncertainty. Annual Review of Ecology and Systematics 26:177–199.

Nichols, J. D., M. C. Runge, F. A. Johnson, and B. K. Williams. 2007. Adaptive harvest management of North American waterfowl populations: a brief history and future prospects. Journal of Ornithology 148:S343–S349.

Nichols, J. D., and B. K. Williams. 2006. Monitoring for conservation. Trends in Ecology and Evolution 21:668–673.

Nielsen, T. D., and F. V. Jensen. 2009. Bayesian networks and decision graphs. Springer, Berlin, Germany.

Pahl-Wostl, C. 2009. A conceptual framework for analysing adaptive capacity and multi-level learning processes in resource governance regimes. Global Environmental Change 19:354–365.

Pickton, T., and L. Sikorowski. 2004. The economic impacts of hunting, fishing and wildlife watching in Colorado. Final report prepared for the Colorado Division of Wildlife. BBC Research and Consulting, Denver, Colorado, USA.

Probert, W. J. M., K. Shea, C. J. Fonnesbeck, M. C. Runge, T. E. Carpenter, et al. 2016. Decision-making for foot-and-mouth disease control: objectives matter. Epidemics 15:10–19.

Raiffa, H., and R. O. Schlaifer. 1961. Applied statistical decision theory. Graduate School of Business Administration, Harvard University, Cambridge, Massachusetts, USA.

Ralls, K., and A. M. Starfield. 1995. Choosing a management strategy: two structured decision-making methods for evaluating the predictions of stochastic simulation models. Conservation Biology 9:175–181.

Redpath, S. M., B. E. Arroyo, F. M. Leckie, P. Bacon, N. Bayfield, R. J. Gutiérrez, and S. J. Thirgood. 2004. Using decision modeling with stakeholders to reduce human–wildlife conflict: a raptor–grouse case study. Conservation Biology 18:350–359.

Regan, H. M., Y. Ben-Haim, B. Langford, W. G. Wilson, P. Lundberg, S. J. Andelman, and M. A. Burgman. 2005. Robust decision making under severe uncertainty for conservation management. Ecological Applications 15:1471–1477.

Regan, H. M., M. Colyvan, and M. A. Burgman. 2002. A taxonomy and treatment of uncertainty for ecology and conservation biology. Ecological Applications 12:618–628.

Robinson, K. F., A. K. Fuller, J. E. Hurst, B. L. Swift, A. Kirsch, J. Farquhar, D. J. Decker, and W. F. Siemer. 2016. Structured decision making as a framework for large-scale wildlife harvest management decisions. Ecosphere 7:e01613.

Royle, J. A., and R. M. Dorazio. 2008. Hierarchical modeling and inference in ecology: the analysis of data from populations, metapopulations and communities. Academic Press, San Diego, California, USA.

Runge, M. C. 2013. Active adaptive management for reintroduction of an animal population. Journal of Wildlife Management 77:1135–1144.

Runge, M. C. 2020. Introduction to linked and dynamic decisions. Pages 227–233 in M. C. Runge, S. J. Converse, J. E. Lyons, and D. R. Smith, editors. Structured decision making: case studies in natural resource management. Johns Hopkins University Press, Baltimore, Maryland, USA.

Runge, M. C., E. Bean, D. R. Smith, and S. Kokos. 2011a. Non-native fish control below Glen Canyon Dam—report from a structured decision making project. Open File Report 2011-1012. US Geological Survey, Reston, Virginia, USA.

Runge, M. C., and S. J. Converse. 2020. Introduction to risk analysis. Pages 149–155 in M. C. Runge, S. J. Converse, J. E. Lyons, and D. R. Smith, editors. Structured decision making: case studies in natural resource management. Johns Hopkins University Press, Baltimore, Maryland, USA.

Runge, M. C., S. J. Converse, and J. E. Lyons. 2011b. Which uncertainty? using expert elicitation and expected value of information to design an adaptive program. Biological Conservation 144:1214–1223.

Runge, M. C., S. J. Converse, J. E. Lyons, and D. R. Smith, editors. 2020. Structured decision making: case studies in natural resource management. Johns Hopkins University Press, Baltimore, Maryland, USA.

Runge, M. C., F. A. Johnson, M. G. Anderson, M. D. Koneff, E. T. Reed, and S. E. Mott. 2006. The need for coherence between waterfowl harvest and habitat management. Wildlife Society Bulletin 34:1231–1237.

Runge, M. C., F. A. Johnson, J. A. Dubovsky, W. L. Kendall, J. Lawrence, and J. Gammonley. 2002. A revised protocol for the adaptive harvest management of mid-continent mallards. Division of Migratory Bird Management, US Fish and Wildlife Service, Laurel, Maryland, USA.

Runge, M. C., K. E. LaGory, K. Russell, J. R. Balsom, R. A. Butler, et al. 2015. Decision analysis to support development of the Glen Canyon Dam Long-Term Experimental and Management Plan. Scientific Investigations Report 2015-5176. US Geological Survey, Reston, Virginia, USA.

Runge, M. C., T. M. Rout, D. A. Spring, and T. Walshe. 2017. Value of information analysis as a decision support tool for biosecurity. Pages 308–333 in A. P. Robinson, T. Walshe, M. A. Burgman, and M. Nunn, editors. Invasive species: risk assessment and management. Cambridge University Press, Cambridge, UK.

Runge, M. C., J. R. Sauer, M. L. Avery, B. F. Blackwell, and M. D. Koneff. 2009. Assessing allowable take of migratory birds. Journal of Wildlife Management 73:556–565.

Rushing, C. S., M. Rubenstein, J. E. Lyons, and M. C. Runge. 2020. Using value of information to prioritize research needs for migratory bird management under climate change: a case study using federal land acquisition in the United States. Biological Reviews 95:1109–1130.

Schwartz, M. W., C. N. Cook, R. L. Pressey, A. S. Pullin, M. C. Runge, N. Salafsky, W. J. Sutherland, and M. A. Williamson. 2018. Decision support frameworks and tools for conservation. Conservation Letters 11:e12385.

Scott, J. M. 1993. Gap analysis: a geographic approach to protection of biological diversity. Wildlife Monographs 123:1–41.

Sebenius, J. K. 2007. Negotiation analysis: between decisions and games. Pages 469–488 in W. Edwards, R. F. J. Miles, and D. Von Winterfeldt, editors. Advances in decision analysis: from foundations to applications. Cambridge University Press, New York, New York, USA.

Sells, S. N., M. S. Mitchell, V. L. Edwards, J. A. Gude, and N. J. Anderson. 2016. Structured decision making for managing pneumonia epizootics in bighorn sheep. Journal of Wildlife Management 80:957–969.

Sells, S. N., M. S. Mitchell, and J. A. Gude. 2020. Addressing disease risk to develop a health program for bighorn sheep in Montana. Pages 156–166 in M. C. Runge, S. J. Converse, J. E. Lyons, and D. R. Smith, editors. Structured decision making: case studies in natural resource management. Johns Hopkins University Press, Baltimore, Maryland, USA.

Sells, S. N., M. S. Mitchell, J. J. Nowak, P. M. Lukacs, N. J. Anderson, J. M. Ramsey, J. A. Gude, and P. R. Krausman. 2015. Modeling risk of pneumonia epizootics in bighorn sheep. Journal of Wildlife Management 79:195–210.

Shea, K., M. J. Tildesley, M. C. Runge, C. J. Fonnesbeck, and M. J. Ferrari. 2014. Adaptive management and the value of information: learning via intervention in epidemiology. PLoS Biology 12:e1001970.

Smith, D. R. 2020a. Introduction to structuring decisions. Pages 15–22 in M. C. Runge, S. J. Converse, J. E. Lyons, and D. R. Smith, editors. Structured decision making: case studies in natural resource management. Johns Hopkins University Press, Baltimore, Maryland, USA.

Smith, D. R. 2020b. Introduction to prediction and the value of information. Pages 189–195 in M. C. Runge, S. J. Converse, J. E. Lyons, and D. R. Smith, editors. Structured decision making: case studies in natural resource management. Johns Hopkins University Press, Baltimore, Maryland, USA.

Speirs-Bridge, A., F. Fidler, M. F. McBride, L. Flander, G. Cumming, and M. A. Burgman. 2010. Reducing overconfidence in the interval judgments of experts. Risk Analysis 30:512–523.

Symstad, A. J., B. W. Miller, T. M. Shenk, N. D. Athearn, and M. C. Runge. 2019. A draft decision framework for the National Park Service Interior Region 5 bison stewardship strategy. Natural Resource Report NPS/NWRO/NRR—2019/2046. National Park Service, Fort Collins, Colorado, USA.

Szymanski, J. A., M. C. Runge, M. J. Parkin, and M. Armstrong. 2009. White-nose syndrome management: report from a structured decision making initiative. US Fish and Wildlife Service, Department of the Interior, Fort Snelling, Minnesota, USA.

Thomas, J. W., J. F. Franklin, J. Gordon, and K. N. Johnson. 2006. The Northwest Forest Plan: origins, components, implementation experience, and suggestions for change. Conservation Biology 20:277–287.

Tyre, A. J., J. T. Peterson, S. J. Converse, T. Bogich, D. Miller, et al. 2011. Adaptive management of bull trout populations in the Lemhi basin. Journal of Fish and Wildlife Management 2:262–281.

USFWS. US Fish and Wildlife Service. 2003. Recovery plan for the red-cockaded woodpecker (Picoides borealis). Second revision. Atlanta, Georgia, USA.

USFWS. US Fish and Wildlife Service. 2011. Endangered and threatened wildlife and plants; reissuance of final rule to identify the Northern Rocky Mountain population of gray wolf as a distinct population segment and to revise the list of endangered and threatened wildlife. Federal Register 76:25590–25592.

USFWS. US Fish and Wildlife Service. 2013. Final supplemental environmental impact statement: issuance of annual regulations permitting the hunting of migratory birds. Department of the Interior, Washington, DC, USA.

USFWS. US Fish and Wildlife Service. 2019. Adaptive harvest management: 2020 hunting season. US Department of the Interior, Washington, DC, USA.

van Heezik, Y., and P. J. Seddon. 2005. Structure and content of graduate wildlife management and conservation biology programs: an international perspective. Conservation Biology 19:7–14.

von Neumann, J., and O. Morgenstern. 1944. Theory of games and economic behavior. Princeton University Press, Princeton, New Jersey, USA.

von Winterfeldt, D., and W. Edwards. 1986. Decision analysis and behavioral research. Cambridge University Press, Cambridge, UK.

Wagner, F. H. 1989. American wildlife management at the crossroads. Wildlife Society Bulletin 17:354–360.

Wagner, F. H., and U. S. Seal. 1992. Values, problems, and methodologies in managing overabundant wildlife populations: an overview. Pages 279–293 *in* D. R. McCullough and R. H. Barrett, editors. Wildlife 2001: populations. Springer, Dordrecht, Netherlands.

Walters, C. J. 1986. Adaptive management of renewable resources. Macmillan, New York, New York, USA.

Walters, C. J., and R. Hilborn. 1978. Ecological optimization and adaptive management. Annual Review of Ecology and Systematics 9:157–188.

Webb, C. T., M. Ferrari, T. Lindström, T. Carpenter, S. Dürr, G. Garner, C. Jewell, M. Stevenson, M. P. Ward, and M. Werkman. 2017. Ensemble modelling and structured decision-making to support emergency disease management. Preventive Veterinary Medicine 138:124–133.

WFFD. Wildlife and Freshwater Fisheries Division, Alabama Department of Conservation and Natural Resources. 2005. Conserving Alabama's wildlife: a comprehensive strategy. Alabama Department of Conservation and Natural Resources, Montgomery, Alabama, USA.

Williams, B. K. 1996. Adaptive optimization and the harvest of biological populations. Mathematical Biosciences 136:1–20.

Williams, B. K., J. D. Nichols, and M. J. Conroy. 2002. Analysis and management of animal populations: modeling, estimation, and decision making. Academic Press, San Diego, California, USA.

Williams, B. K., R. C. Szaro, and C. D. Shapiro. 2007. Adaptive management: the US Department of the Interior technical guide. Adaptive Management Working Group, Department of the Interior, Washington, DC, USA.

Yoakum, J. D. 1980. Habitat management guides for the American pronghorn antelope. Technical Note 347. US Department of Interior, Bureau of Land Management, Denver Service Center, Denver, Colorado, USA.

SCALE IN WILDLIFE MANAGEMENT
The Difficulty with Extrapolation, Replication, and Unappreciated Impediments

JOHN A. BISSONETTE

INTRODUCTION

State and federal land management agencies have legislatively mandated authority for natural resources over broad landscape extents. Given that resource conflicts are manifest worldwide (Bannon and Collier 2003, Humphreys 2005, Clark et al. 2020), agencies will benefit by increasingly relying on research data to guide management options (Holl et al. 2003, MacKenzie 2005, Lebeau et al. 2019). Attempting to understand the relationships between pattern, process, and wildlife response has been an essential component of wildlife ecology and management for almost a century. The problems that managers are usually concerned with are typically large scale (Bissonette 2003), complex (Elliott-Graves 2018), and with nonlinear dynamics (Williams 2013). They may include system-wide changes like invasive species (http://www.invasivespecies info.gov/), changes in species population vital rates caused by anthropogenic land cover loss and fragmentation (Lindenmayer and Fischer 2006), the spread of infectious disease (e.g., chronic wasting disease; Conner et al. 2007), and the longer-term effects of climate change (Weiskopf et al. 2019). Managers set harvest limits for big game that are usually based on hunt units but within a statewide context (Gould et al. 2018, Heffelfinger 2018).

Threefold Problem

It is clear that the problems managers face are complex. Almost every student in any of the environmental sciences reads about the term *scale* (i.e., data resolution and extent) and about its importance in environmental and wildlife studies. The data upon which managers must rely exhibit scale-related problems that are threefold and go to the question of data reliability.

The Extrapolation Problem

First is the extrapolation problem. To understand the relevant dynamics, in the past managers often relied on data that were typically collected at small scales, taken on-site, and over short periods. While still common, this has changed in the past few years (Lindenmayer et al. 2012). Extrapolation from data collected at smaller spatial and temporal resolutions and extents to larger landscapes and longer periods is often problematic. A fundamental mismatch of scale persists. A snapshot in time cannot capture important dynamics.

Number System Mismatch Problem

Second, impediments occur when the problems we study and the methods we use have different structural characteristics (i.e., exist in different number systems; Weinberg 1975; Weinberg and Weinberg 1979). Fundamental differences like these force us to question how closely results from any ecological study conform to the state of nature or measure what we intend. Because we are forced to use small-number approaches to middle-number ecological problems, problems arise (Bissonette 2019).

Big Data Set Problems

Third, the vastly improved technology for monitoring species movements and activities (e.g., lighter transmitters with almost limitless relocations possible and often coupled with auxiliary sensors) lead to big data sets that pose serious problems with reliability, repeatability, and replicability of results (Lewis et al. 2018). Understanding emerging data reliability problems associated with scaling, big data sets, and unappreciated impediments that characterize possible study approaches is integral to evaluating study result reliability (Bissonette 2021).

To provide context for the reader, this chapter will define and briefly address scaling theory and outline the concept of system architecture (i.e., the different conceptual assumptions of hierarchy and network theory). I address how in terrestrial ecology the context in which the scale concept developed is intimately tied to hierarchy theory, but that graph or network theory has emerged as a way to understand the scales within which animals move. I then ask the question whether heterarchies reconcile the different views of system architecture. This chapter then explains the different uses and misuses of the term *scale* in the literature, including its use in measurement and statistical operations, the differences between map and cartographic scale, and a brief explanation of isometric and allometric scaling. I explain the concept of scale dependency (i.e., the effects of changing resolution and extent, with an example of sympatry versus allopatry using data from Big Bend National Park, Texas, and southern Arizona, USA). I then discuss the concepts of space-time diagrams as one of the first comparisons of spatial and temporal scaling. This chapter examines what has been termed the problem of scale and what it means, including how complexity, data measurements, human perceptions, and aggregation error contribute to the scale problem. I then address the idea of finding the right scales (i.e., choosing biological relevant scales for investigation and suggesting that there are disjunctive connections involving observer perceptions). How biologists measure and interpret results is not a trivial exercise and involves imperfect observer perceptions. Using the ideas of ecological neighborhoods and domains of scale provides some relief in simplifying sampling effort. It is possible to choose biologically relevant scale(s) for investigation. I then suggest a circumvention to avoid observer selection of scale by using graph theory and network analysis. Incorporating landscape context in an animal-centered approach is a way forward. de Knegt et al. (2011) provide an example of how spatial context revealed the landscape spatial extent that influenced elephant (*Loxodonta africana*) response. The results were counterintuitive. Finally, I ask what else can go wrong by briefly exploring data reliability problems related to the number system mismatch and those inherent in using big data. I use examples drawn from research studies throughout the chapter to demonstrate these ideas.

SCALING THEORY

Paradigm Shift

The recognition of scale and spatial pattern underwent a paradigm shift in the 1980s (Golley 1989). Early writings by Gould (1979), Allen and Starr (1982), Calder (1983), and Delcourt et al. (1983), and the seminal papers by Wiens (1989) and Levin (1992) clearly established the importance of scale in ecology. Scaling theory is the underpinning for all scale inquiries and refers primarily "to how a system changes when its size changes" (West 2017:15). Log_{10} transformed isometric and allometric scaling plots express these relationships (West et al. 1997), but see Sprugel (1983) for a discussion of bias in allometric equations. Scaling plots show the relationship between some measure of organism size and aspects of its morphology or life history. When the values of the Y_{Log} axis change in equal proportion to the values of the X_{Log} axis, the resulting slope is a straight line. It is isometric. When the values of the Y_{Log} change unequally, the resulting slope is curved and is allometric. Isometric and allometric scaling are discussed below.

System Architecture

What is the architecture (or theoretical structure) of the system? How do wildlife problems fit within the structure? Typically, ecological hierarchical approaches have been assumed or overtly used (Turner et al. 2001). Graph theory (network)

analyses have only recently come into the ecological literature (Urban and Keitt 2001). These two approaches are seemingly at odds with one another, but Cumming (2016) has suggested the concept of heterarchy as a way to reconcile the two different approaches.

Hierarchy Theory

In ecology, the context in which the scale concept has developed is intimately tied (Turner et al. 2001) to hierarchy theory (i.e., system organization). Typically, a hierarchy refers to a top-down or bottom-up approach to causality. Understanding the connections between hierarchical system organization and scaling and the associated terminology is one way to provide clarity when addressing scaling issues. As O'Neill et al. (1986:39) stated, "The task of choosing an appropriate system for investigating a particular phenomenon is inseparable from consideration of underlying organization and complexity." Allen and Starr (1982) argued that hierarchy theory provides the organization framework with which to consider the concept of scale. A level defines a specific organization in a hierarchy (e.g., an organism level or a population level). One way to see the connection is to think about O'Neill and King's (1998:5–6) inescapable conclusion: "If you move far enough across scale, the dominant processes change. It is not just that things get bigger, but the phenomena themselves change. Unstable systems now seem stable . . . Bottom up control turns into top down control . . . Competition becomes less important and climate seems to dominate patterns. These same changes can be observed in aquatic ecosystems . . . terrestrial vegetation . . . and geomorphological dynamics. It is this observation of changing dynamics with scale that formed the basis for the development of hierarchy theory in ecology." Hierarchical systems are typically rate structured (i.e., "the levels are distinguished by differences in the rates, or frequencies, of their characteristic processes"; Turner et al. 2001:35). Triadic approaches (Fig. 6.1) are typically used (Allen and Starr 1982, Delcourt et al. 1983, O'Neill et al. 1986), meaning that whatever the focal level, the next upper level provides constraint, and the next lower level provides the components or mechanisms (i.e., an approximation of causation). Causation can work both ways, however, and the question must always

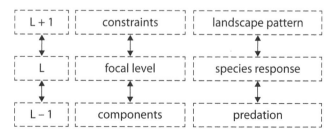

Figure 6.1. A triadic scheme using a hierarchical approach. There is interaction between levels. Even levels are considered rate-structured, and upper levels (e.g., landscape pattern) can influence lower levels (e.g., species response and predation). It is important to develop the focal level of interest and then explore the levels that may influence it.

influence the approach. Hierarchy theory was the basis for much of ecological theory, including the idea of trophic cascades (Pace et al. 1999).

Graph or Network Theory

Another way of addressing wildlife problems involves *network* architecture. If one is interested in animal movement at a very fine-scale resolution, using graph theory (i.e., a network approach) may provide a cleaner answer (Urban and Keitt 2001, Fall et al. 2007, Lookingbill et al. 2010). Mathematically, networks are graphs (Golubski et al. 2016) and are concerned primarily with functional connectivity. Networks represent the third *lattice* (Cressie 2015), where the term refers to the continuous surface of a landscape and the measurement of a variable is assignable to any place on the surface (Urban and Keitt 2001). The two most common representations (lattices) of landscapes are raster and polygon formats. The focus for many wildlife ecologists has been to link change in landscape pattern over time with animal response, and hence the use of raster and polygon formats, which give good representations of landscape pattern. The obvious focus is the landscape. With network approaches, the focus shifts to the animal and typically its movements and use of resources. The landscape is visualized as a network of nodes (locations of interest) and lines (pathways) that represent some process (i.e., gene flow, movement pathways, corridors; Bissonette 2017). Clearly, hierarchy and network theories represent two very different concepts of system architecture. Are they reconcilable?

Do Heterarchies Reconcile Hierarchy and Network Theories?

Cumming (2016:622) suggested that the idea of heterarchies is a new way of "contextualizing and generalizing . . . new methods for analyzing complex structure-function relationships." He recognized that there were different but equally valid observation sets or ways of viewing complex problems, specifically mentioning heterarchies and networks (peer-to-peer interactions), but suggested that system architecture is a continuum and that looking at hierarchies and networks as dichotomous was not valid or productive. Cummins (2016) viewed the continuum as existing on two axes. The vertical axis showed a continuum from individual to networked, while the horizontal axis showed a continuum from flat (i.e., interactions between peer groups to hierarchical structure). Cummins (2016) provides addition details. I suspect that wildlife ecologists will continue to use both hierarchical and network approaches to address pressing species related problems.

DIFFERENT USES OF SCALE IN THE LITERATURE

To a wildlife ecologist, scale is a metric that refers to the spatial or temporal dimensions of an object, pattern, or process (Turner et al. 2001). Schneider (2001:546) described the term as arising from two different etymological roots: a Norse root *skal*, suggesting "measurement by means of pairwise comparisons of objects," and a Latin root *scala*, "suggesting measuring a length by counting steps or subdivisions." There is great variability in the interpretation and meaning of the term scale; its use in the literature is confusing (Montello 2001) because it has several referents. It has variously been referred to in measurement and statistical operations, geographic map scales, isometric and allometric scaling, and in a nonspatial and a spatiotemporal sense of resolution and extent (Schneider 2001). Often terms are used loosely, causing confusion.

Misuse of Terms

Scale and level are often confused. For example, it is easy to find references to population scale and population level, and certainly to landscape scale and level, often in the same publication. It is not uncommon to see phrases like these in the current literature: "at the landscape level"; "to the ecosystem scale"; "we studied habitat selection at the landscape, meso, and plot scales"; "work has been devoted to assessing the relative influence of landscape level on"; "Our goal was to identify brood rearing habitat at the landscape scale." These statements do not distinguish between some important concepts, and they are misleading in their strict interpretation. Although the problem was recognized early (Allen and Hoekstra 1990) and often (King 1997, Allen 1998, O'Neill and King 1998), the problem persists. It is a problem because the terms have different primary meanings. Scale refers to some measurable dimension in space and time, while the term level is more appropriately related to hierarchical structure. When one reads the term *landscape scale* in a journal article, the implication is that the term refers to some landscape extent that is large. Authors seldom define large. The extent of the landscape is seldom given. Furthermore, landscapes exhibit many scales, so the term is imprecise and misleading at best. Because there are two relevant aspects of scale that are most often used (resolution and extent), for clarity it is preferable to use the term *landscape extent* to convey the meaning and give the areal extent of that landscape, or *landscape resolution* to denote the finest level of detail. Early writers often equated resolution with grain. The term *grain* was borrowed from film emulsion when photographic film was in use (https://en.wikipedia.org/wiki/Film_grain). Fine-grain (e.g., a low ISO, for International Standards Organization, rating, along with shutter speed and aperture setting of the camera; https://www.exposureguide.com/iso-sensitivity/) meant higher-resolution film. When digital photography became available, fine grain referred to small pixel size. Satellite imagery is digital. Small pixel size equals higher-resolution images. I suggest that the term resolution replace the older, less precise term grain. Regardless, when referring to scale, an overt reference to which aspect (resolution or extent) is essential for clarity. Additionally, it is important to understand the connection of scale to system organization or structure.

Measurement and Statistical Operations

Scale can refer to nominal, ordinal, interval, or ratio measurement scales, each with permissible operations of addition,

Table 6.1. The scales of measurement with allowable operations

Scale	Basic Operational Determination	Permissible Statistical Operations
Nominal	Equality	Number of cases, mode, contingency, correlation
Ordinal	Greater or less	Median, percentiles
Interval	Equality of intervals or differences	Mean, standard deviation, rank order, and product-moment correlations
Ratio	Equality of ratios	Coefficient of variation

Source: Redrawn from Stevens (1946).

Note: A Spearman rank order correlation is the nonparametric (not involving any assumptions as to the form or parameters of a frequency distribution) version of the Pearson product-moment correlation (makes assumptions about the parameters of a frequency distribution). It measures the strength and direction of association that exists between two variables measured on at least an ordinal scale.

subtraction, division, multiplication (Withers and Meentemeyer 1999) and appropriate statistical tests (Stevens 1946, Siegel 1956). This use of the term scale refers to the theory of measurement (Stevens 1946) where the mathematical or statistical operations allowable on a given set of data are wholly dependent on the level of measurement (Table 6.1). The nominal level of measurement simply implies names or categories for items. Gender, numbers on football jerseys, and handedness (right, left, ambidextrous) are examples.

Classification of range condition (e.g., poor, fair, good, excellent; Dyksterhuis 1949) is an ordinal (ranking) level of measurement because the categories are not only different from each other (nominal), but they also stand in some relationship (better, worse) to each other. Alternatively, one uses a ratio level of measure when measuring the mass of an object; a ratio level of measurement has a definite zero point. Contrast these with interval measurements of temperature in Fahrenheit or centigrade, characterized by a known interval between values, but the zero point and the unit of measurement are arbitrary. For example, the freezing and boiling points in Fahrenheit are 32° and 212° and for centigrade 0° and 100°. Both systems measure temperature, but with different units. To illustrate that the units are actually arbitrary, one can devise an interval measurement with different a zero point and different units. For example, consider a temperature scale called TS with a freezing point of 20° and a boiling point of 180°. If the temperature in Fahrenheit was 70° (~21.1 C°), then the temperature in the new TS is 53.7°. Means, standard deviations, and rank and product-moment correlations are allowable operations (Siegel 1956).

Map versus Cartographic Scale

Scale can refer to map or cartographic scale. In geography the scale concerns are primarily spatial but can also be temporal or thematic (e.g., the grouping of entities such as weather variables; Montello 2001). Lechner and Rhodes (2016) provide an analysis of the effects of spatial and thematic resolution on ecological analysis. Ecologists and geographers use the term scale in very different ways and mean the opposite when they refer to small and large scale. To a geographer, a cartographic scale represents the degree of spatial reduction and is represented as a ratio or representative fraction (Montello 2001) of map distance to Euclidean distance on the surface of the earth; large scale means fine resolution, e.g., 1:24,000. To an ecologist, large scale usually means coarse resolution over large spatial extents, e.g., 1:62,500. The difficulty is in large part semantic. Turner et al. (2001) recommended the use of the terms fine and broad scale to replace small and large scale. Fine scale would then refer to smaller areas, more detail, and greater resolution. Broad scale would refer to larger areas, less detail, and lower resolution (Fig. 6.2). Given the improvement in satellite sensors, it is now possible to view large landscape extents with very fine resolution (Yang et al. 2011).

Isometric and Allometric Scaling

Isometric scaling is characterized by a relationship where the slope of the line representing the relationship between two variables is equal to 1 (i.e., a 45° line). An allometric relationship is disproportionate; one variable increases or decreases more rapidly. In ecology, allometric relationships appear to follow the quarter scaling law (West 2017:93); the "corresponding exponents (of well over 50 scaling laws) are invariably very close to multiples of ¼."

For example, Kleiber (1947) described the allometric scaling relationship between body size and metabolic rate as $Y = Y_o M^b$, where Y = metabolic rate, Y_o is a constant characteristic of the organism, M is body mass, and b is the scaling exponent. Kleiber (1947) reported that metabolic rates scaled as $M^{3/4}$ (see also West et al. 1997). McNab (1963) showed that body mass (W) was related to home range (HR) size (home range size = $k_3 W^{0.75}$, where k_3 is a constant equal to 8.51; hence HR = $8.51 W^{0.75}$. Others have shown scaling relationships important to wildlife biologists. For example, Wolff (1999) and Sutherland et al. (2000) reported a similar relationship between body mass and natal dispersal distance. Recently, Bowman et al. (2002) reported that the slope of the $\log_{10}$ transformed the relationship between maximal dispersal distance, and the home range area was 0.5. "Because dispersal distance is a linear measure and home range area is a square of a linear measure, taking the square root of home range area places both terms on the same linear scale. Thus, maximum dispersal distance is related by a single constant to the linear dimension of the home range," i.e., √HR area (Bowman et al. 2002:2051). Hence the isometric equation is maximum dispersal distance (MxDD) = 40 × linear dimension of the home range, or MxDD = 40 × √HR area. For median dispersal distance (MdDD) the isometric equation is MdDD = 7 × √HR area. Bissonette and Adair (2008) used this fortunate result to determine the optimal spacing of wildlife crossings to restore habitat permeability to roaded landscapes for terrestrial species using scaling concepts. They identified six home range area scale domains; three-quarters of the species clustered in the three smallest domains. They used home range$^{0.5}$ (√HR) to represent a daily movement distance metric; when individual species movements were plotted, >78.1% of 72 species (n = 51) found in North America were included

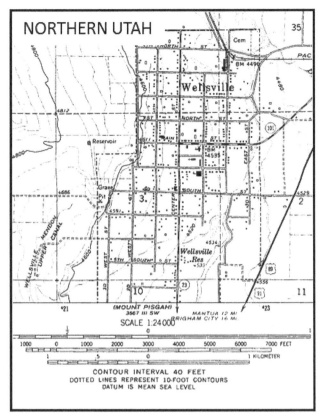

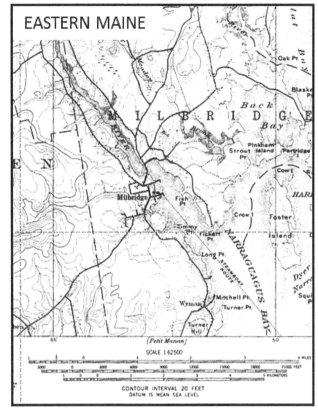

1:24,000 FINE SCALE
small spatial extent
(smaller area represented on map)
greater resolution
more detail

1:62,500 BROAD SCALE
large spatial extent
(larger area represented on map)
lower resolution
less detail

Figure 6.2. Fine and broad scale versus small and large scale

at distances of ≤2 km. Roads in the United States are measured in mile markers; hence 71.2% of species were included at distances of <1 mile. They argued that if the spacing of wildlife crossings in road hotspots of wildlife-vehicle mortality was based on the $HR^{0.5}$ metric, then, along with appropriate auxiliary mitigation, landscape permeability would be greatly improved, thereby facilitating wildlife movement across the roaded landscape and significantly improving road safety by reducing wildlife-vehicle collisions

SCALE DEPENDENCY

Effects of Changing Scale Resolution or Extent

Scale dependency and *scale dependent* are common terms encountered in the literature. Students may be confused as to what these terms mean. Quite simply, the terms refer to the idea that different patterns emerge when the scale resolution (sometimes referred to as the grain) or the scale extent (the area under consideration) of the study is changed. This is especially relevant when the *observation scale* (the scale of mea-

surement and sampling) is changed. There is pattern at every scale, so when the resolution or extent is changed, the patterns change. For example, Turner et al. (1989) reported that rare cover types were lost or homogenized when the resolution of digital maps became coarser, but that landscape pattern influenced the rate of loss. Indices of pattern (e.g., dominance, contagion) also changed as resolution and extent varied. When measurements at finer resolutions and small extents are scaled up or extrapolated to broader landscape extents, transmutation or qualitatively different patterns can emerge (King et al. 1991). This is a manifestation of scale dependency.

An Example from the Real World: Mule Deer

One might think of the broad- and fine-scale relationships between mule deer (*Odocoileus hemionus*) and white-tailed deer (*O. virginianus*). At the spatial extent of species distributions, the pattern is sympatric; the species co-occur in many places. At the resolution of multiple home ranges, the pattern is allopatric, and the distribution of the species do not overlap. Krausman (1976) in Big Bend National Park, Texas, reported that while-tailed and mule deer distributions separated at about

1,494 m elevation, with white-tailed deer locations >1,494 m and mule deer locations <1,494 m. There appeared to be little overlap in their distributions. Anthony and Smith (1977) in southeastern Arizona, however, did not find a similar pattern. Whenever competition between two species results in competitive exclusion, one might expect to find a similar scale relationship. For example, Sherry and Holmes (1988) reported that least flycatchers (*Empidonax minimus*) negatively influenced the distribution of American redstarts (*Setophaga ruticilla*) locally, but the distribution of the two species overlapped in the northeastern United States (Wiens 1989). Clearly, the extent over which the processes are viewed makes a difference.

Resolution and Extent

Nonspatial Measurements
Martinez and Dunne (1998) wrote that ecologists almost exclusively use scale in a spatiotemporal sense in the context of space and time. They suggested that this is restrictive and argued that scale broadly refers to any metric that can quantitatively measure difference between observations. They stated that a variable such as temperature (an interval scale of measurement) can have an extent (range of temps measured) and a resolution (preciseness of the measurement) and that it was possible to apply the metrics of resolution and extent to nonspatial scales (e.g., species richness or primary productivity in the same way). Martinez and Dunne (1998:208) argued that "non-spatio-temporal scales have already advanced food-web research through extent- and resolution-based sensitivity analyses of various food web properties," including species richness and primary productivity. They used the species richness-population stability arguments discussed by Hutchinson (1959) and May (1988), whereby species diversity was related (or not) to stability because of the effects on food web structure. Simply put, Martinez and Dunne (1998) referred to the degree of aggregation of different taxa within a food web as its resolution. The term *extent* referred to the range of species richness values over which specific scaling laws applied. This is a different approach and not discussed as frequently in wildlife research or management discussions as have the more familiar spatial and temporal scales.

Spatiotemporal Frameworks
Perhaps the most common use of scale in the ecological literature and in wildlife science in general has been in a spatiotemporal framework. In this sense, scale has been used in five or more ways (Wu and Li 2006). It has been used to (1) refer to the resolution and extent that the physical problem or process under consideration exists in nature (*characteristic, phenomenon,* or *intrinsic scale*), implying that many, if not most, natural phenomena have their own distinctive spatial extent and event frequencies that characterize their behavior. The term also refers to (2) the scale of measurement and sampling (*observation scale*) as well as (3) the scale that a problem is analyzed or modeled (*analysis* or *analytical scale*), (4) the scale of experiments (*experimental scale*), and (5) the scale of policy-

making (*policy scale*). van Lieshout et al. (2011) give one of the most intriguing discussions about the ramifications of choosing a policy scale. They concluded that scale mismatches about how a policy problem is framed interfered with the decision-making process. In all of these uses, the investigator or manager ultimately selected or determined resolution and extent. In response to the many different components of scale, Dungan et al. (2002) proposed a framework for considering scale terms and approaches. Montello (2001:13501) made essentially the same argument: "Cartographic scale refers to the depicted size of a feature on a map relative to its actual size in the world. Analysis scale refers to the size of the unit at which some problem is analyzed, such as at the county or state level. Phenomenon scale refers to the size at which human or physical earth structures or processes exist, regardless of how they are studied or represented. Although the 3 referents of scale frequently are treated independently, they are in fact interrelated in important ways." Dungan et al. (2002) distinguished between the studied phenomena, the sampling protocol units used to measure the phenomena, and the spatial analysis used to detect pattern. Bissonette (2017) redrew the relationship (Fig. 6.3) to reflect the idea of scaling domains (Wiens 1989:392), which define "a portion of the scale spectrum within which process-pattern relationships are consistent regardless of scale."

The idea of spatiotemporal scaling, although relatively new to ecology, has a long history. Schneider (2001) pointed out that although the term *spatial scaling* first appeared in a paper

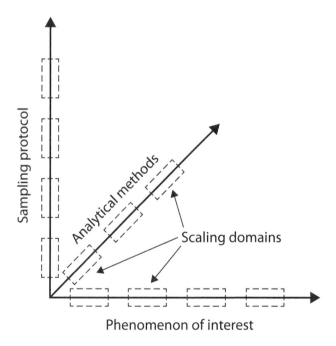

Figure 6.3. A framework in which to consider scale terms and the different components of an ecological question. Ideally, the sampling protocol and analytical methods should accurately reflect the intrinsic scales of the phenomenon of interest. If scale domains are present, then matching sampling and analyses to the problem is theoretically simpler to achieve. Redrawn from Dungan et al. (2002)

on mouse tracking (Martens 1972), explicit treatment of the concept appeared much earlier in works by Collins (1884) and Murphy (1914), who recorded the latitude and longitude for >500 birds collected. Since the early 1970s, use of the terms scale, scaling, and spatial and temporal scale has increased exponentially in the ecology literature. The remainder of this chapter addresses this spatiotemporal use of scale.

SPACE-TIME DIAGRAMS

Perhaps one way to think about the scale is to review the history of its graphical expression: space-time diagrams. A space-time diagram is little more than a comparison of a pattern, phenomenon, or process on space (abscissa) and time (ordinate) axes. Stommel's (1963) schematic diagram of the spectral distribution of sea level (ck) in space and time is considered one of the earliest depictions of a space-time relationship. Delcourt et al. (1983), Wiens (1989), Holling (1992), and Bissonette (1997) give familiar adaptations of space-time diagrams. A space-time diagram (Fig. 6.4) clearly shows the optimal matching of temporal and spatial scales needed to capture the relevant dynamics as closely as possible. In practice, however, it is nearly impossible to optimally match space and timescales. Why? The time frames over which the dynamics operate are usually longer than is practical to measure. Clearly,

HENRY M. STOMMEL (1920–1992)

Henry M. Stommel was a leading theoretician on ocean currents. He graduated from Yale University in 1942 with a bachelor of science degree. After teaching mathematics and astronomy at Yale for two years, he became a research associate at the Woods Hole Oceanographic Institution, leaving in 1960 to become professor of oceanography at Harvard University. In 1963 he joined the faculty of the Massachusetts Institute of Technology, remaining until 1978, when he returned to Woods Hole. Stommel was considered one of the most influential oceanographers of his time. Stommel's (1963) schematic diagram of the spectral distribution of sea level in space and time is considered to be one of the earliest depictions of a space–time relationship. Space–time diagrams clearly show the optimal matching of temporal and spatial scales needed to capture the relevant dynamics as closely as possible and underpin the conceptual theory of scaling.

when the time horizon of the pattern or process becomes very long (e.g., centuries or longer), then it is impossible to closely match sampling scale to phenomenon scale, and a space-for-time substitution is often employed (Pickett 1989). What this means is that rather than studying a phenomenon—e.g., plant succession—over the time periods necessary to capture the relevant dynamics, ecologists simultaneously study different stages in the successional pattern at different sites. For example, the study of how forest-dwelling species are expected to change over time as forests grow and mature is done by studying multiple plots of varying age classes and recording species patterns on each plot (Thompson et al. 1992). The inference of how the community will change over time is based on these shorter-term samples. Damgaard (2019) criticized this practice, especially when used in nonstationary environments. The Missouri Forest Ecosystem Project (http://mdc7.mdc.mo.gov/applications/MOFEP/index.html) is an example of a longer-term study (data collection from 1991 to present) to understand the effects of changing forest dynamics on many components of forest ecology, including small mammals, reptiles and amphibians, forest invertebrates, and interior songbirds. Longer-term studies that match time and space of the dynamics are seldom if ever possible. What is one to do to address scale problems?

THE PROBLEM OF SCALE

Why is scale a problem? The answer is threefold (Schneider 2001) and plays on the system dynamics of complexity, the measurement of data, and scale determination.

Complexity

We live in a complex world (Elliott-Graves 2018). Ecological problems are complex, exist at larger-scale extents, and are characterized by dynamics that have long time horizons. Examples are as diverse as the loss of biodiversity globally (Tilman 2000, Rosa et al. 2020) and the decline of pronghorn (*Antilocapra americana*) in Grand Teton National Park, Wyoming, USA (Berger 2003). Ecology exhibits characteristics of multicausality (Laurance et al. 2007), unexpected thresholds, nonlinear dynamics (Williams 2013), and unpredictable time lags (With 2007, du Toit et al. 2016). The biological and structural characteristics of the landscape mediate animal response (Bissonette 2003, Froehly et al. 2019) and our understanding of the dynamics, and they are "dependent on the temporal and spatial scales of observation" (Bissonette 2019:1019). System history and legacy effects lead to unpredictable events (Ziter et al. 2017). Gutzwiller et al. (2015) suggested that small changes may bring large system changes, systemic complexity, and complex critical thresholds.

Data Measurement

Until recently, ecologists measured most variables directly only in small areas and on-site (Schneider 2001). Point data that are sampled regularly or irregularly in space and are spatially interpolated by a geostatistical technique referred to

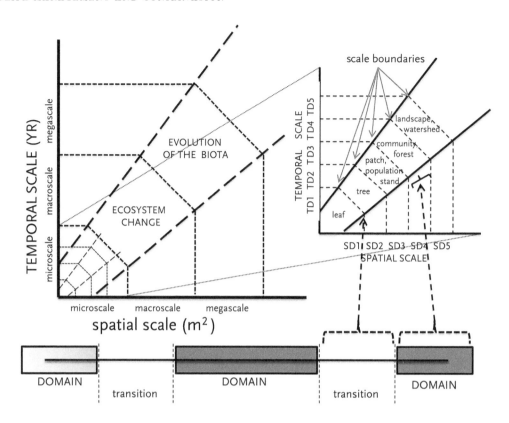

Figure 6.4. Space-time domains for ecological phenomena. The large *X, Y* graph shows scale domains ranging from small to very large spatial and temporal processes. The inset *X, Y* graph illustrates smaller spatial domains (SD) and shorter temporal-scale domains (TD) for processes ranging from gas exchange through stomata on the underside of plant leaves to changes in landscape pattern. The junction between scale domains (i.e., the transitions from one scale to another) represents the locations on the scale gradient where rapid changes or thresholds are expected to occur. The theory suggests that observations and sampling within a scale domain are expected to result in reliable results for processes, phenomena, and patterns operating within that domain. Each process or pattern is characterized by a spatial and temporal domain that indicates the approximate matching of scales required to sample the dynamics. Given the very long timescales involved for larger-scale phenomena, a space for time substitution is often necessary. Adapted from Delcourt et al. (1983), Wiens (1989), and Bissonette (1997)

as kriging (Fortin et al. 1990) are essentially finer-resolution data scaled to large extents (Gustafson 1998). The process produces a trend surface "by computing a weighted average of the known values of the function in the neighborhood of the point" (https://en.wikipedia.org/wiki/Kriging), with the resulting trend surface including both sampled and nonsampled areas. Only a few variables have been measured continuously over large-scale extents. They include the normalized difference vegetation index (NDVI, a remotely sensed indicator of greenness; see https://en.wikipedia.org/wiki/Normalized_difference_vegetation_index); enhanced vegetation index (EVI, a vegetation index enhanced to improve sensitivity; see https://en.wikipedia.org/wiki/Enhanced_vegetation_index); and fraction of absorbed photosynthetically active radiation (FAPAR, the fraction of incoming solar radiation absorbed by plants; see https://en.wikipedia.org/wiki/FAPAR). All are derived indices with often questionable interpretations, in part because of the corrections required to interpret the raw data (http://ivm.cr.usgs.gov/index.php). The relevant point here is to note that ecological problems have their characteristic scale(s) of expression, while our methods for measuring vari-

ables most often do not correspond to their actual spatial and temporal extents.

Scale Determination from a Human Perspective

Additionally, at least in the recent past, we were inclined to study familiar ecological problems at scales that accorded with our perceptions (Wiens 1989) and time constraints and financial resources. Indeed, the nature of research institutions with their characteristic time constraints for MS and PhD research mitigate against longer-term studies (May 1994) and, by extension, studies with larger spatial extents. Tilman (1989) reported that only 13 of 749 studies (1.7%) published in the journal *Ecology* lasted five or more years. Long-term data sets and larger spatial extents increasingly have been used recently, but many are confounded by inconsistency of data collection (Firbank et al. 2017, Katz et al. 2019).

Transmutations and Aggregation Error

Scale is also a problem because it is very difficult to scale up measurements taken at finer-scale resolutions and extents to the larger scale of the real topic of investigation without transmutation or some qualitative change in the resulting pattern (O'Neill 1979, King 1991, King et al. 1991). Transmutation, also called *aggregation error*, was originally defined in terms of hierarchy theory. It occurs most often when a process changes when moved from one hierarchical level to another (O'Neill 1979). For example, assume that the response of some body function in individual animals (e.g., respiration rate, some metabolic function) is influenced by some critical temperature. Now suppose that the response is characterized by a threshold or a step function. If there is variation in the population (i.e., each individual responds to a slightly different critical temperature but in the same way), then by taking an average response across the individuals and projecting it to the population, the shape of the response curve for the population is qualitatively different than for any of the individuals. O'Neill (1979) explained the idea in detail and described transmutations of threshold functions and of more complex discontinuous functions between trophic levels and at the ecosystem level. I argue that the term *transmutation* can also be useful to describe results that are qualitatively different when data collected at smaller spatial extents are extrapolating to larger-scale extents. Miller et al. (2004:310) describe this mismatch as "one of the most formidable challenges confronting environmental scientists." The default assumption appears to be that the areas are either homogeneous, similarly heterogeneous, or the dynamics are linear. For example, when biologists measure species loss at smaller-scale extents and then extrapolate to the species range, problems of data reliability often arise because the number of species often does not scale directly with area (Connor and McCoy 1979). The properties of the system change because scale extent is increased (Rastetter et al. 1992). Additionally, where topography varies within a survey, the effects of area are often difficult to separate from habitat diversity effects (Johnson et al. 2003). The essence of the message is that caution is advisable when scaling essentially small-scale data to larger-scale extents.

These problems are important to consider because scaling is inevitable in research and practice whenever predictions need to be made at a scale extent that is different from the scale at which data are acquired (Wu and Li 2006). Given the above-mentioned problems, Levin's (1992:1959) statement is clear:

> Two fundamental and interconnected themes in ecology are the development and maintenance of spatial and temporal pattern, and the consequences of that pattern for the dynamics of populations and ecosystems. Central to these questions is the issue of how the scale of observation influences the description of pattern; each individual and each species experience the environment on a unique range of scales, and thus responds to variability individualistically . . . Thus, no descriptions of the variability and predictability of the environment make sense without reference to the particular range of scales that are relevant to the organisms or processes being examined.

For example, pronghorns summer in Grand Teton National Park, and most migrate through a narrow corridor to southern Wyoming to spend the winter (Sawyer et al. 2005). They suffer mortality throughout the year. In Grand Teton National Park, however, Berger (2007) and Berger et al. (2008) argued that fawn mortality is mediated through a trophic cascade and meso-predator release involving wolves (*Canis lupus*) and coyotes (*C. latrans*). Areas with higher densities of wolves have lower pronghorn fawn mortalities, apparently because of a negative interaction between wolves and coyotes. Results of fawn mortality of this pronghorn population scaled to the entire Wyoming pronghorn population would likely cause problems because of age-related difference in vulnerability and because wolves are currently either absent or limited in southern Wyoming.

FINDING THE RIGHT SCALE(S)

What, then, is the right scale, and how can we find it? Perhaps the first lesson to note is that there is no single correct scale (Turner et al. 2001). As Wiens (1989:391) stated, "What is an 'appropriate' scale depends in part on the question one asks." Additionally, the specific processes of interest may have their own scales.

Choosing a Biologically Relevant Scale for Investigation

Choosing a biologically relevant scale measures for investigation is imperative. How to do that is less obvious, but a biological rationale seems necessary. One pertinent question involves whether the patterns that emerge from a study of specified resolution and extent appropriately reflect the resolution and extent that the physical problem or process under consideration exists in nature or whether they are simply an artifact of the scale used. It appears that few have explicitly addressed this distinction. Researchers of 79 multiscale wildlife studies from three journals (*Landscape Ecology, Journal of Wildlife Management,* and *Journal of Applied Ecology*) reported that only 29% of the observational scales examined had a biological rationale for their use. In other words, observational scales that were used tended to be "chosen arbitrarily with no biological connection to the system being studied" (Wheatley and Johnson 2009:152).

Disjunctive Connections: Observation, Measurement, and Analysis

There is a disjunct between the scale(s) that characterize animal movements and processes, and their perception and measurement by the observer. Many observers do not distinguish between the *characteristic, phenomenon,* or *intrinsic scale* (the resolution and extent that the physical problems or processes under consideration exist in nature) and the *observation scale* or *analytical scale* (the scale of measurement and sampling or the scale that ecologists analyze and model problems). The disjunct, quite simply, is that problems in nature exist at a characteristic scale extent and resolution, and the way we measure or sample is often at a different scale extent and resolution. What makes this important is that that the scale of our observation (our sampling) fundamentally determines our description and explanation of the natural world (O'Neill and King (1998:7) and should coincide with the scales inherent in the patterns or processes we are measuring. Animals select habitats, forage, migrate, and otherwise fulfill their life requirements within certain domains of scale (Addicott et al. 1987). We observe animals and their processes, and often we measure variables at single or multiple scale resolutions and extents. Consequently, the measurements we make, based on the questions we asked, tend to portray different patterns.

Ecological Neighborhoods

Addicott et al. (1987) argued that there were no well-accepted general procedures or criteria for determining how organisms respond to landscape heterogeneity. They refined the concept of ecological neighborhoods developed earlier by Wright (1943, 1946), Southwood (1977), and Antonovics and Levin (1980). They characterized an ecological neighborhood as defined by three properties: a specific ecological process, a timescale appropriate to that process, and the organism's spatial activity during that period. They suggested that movement (scale extent) defined the neighborhood. For example, foraging activities typically may cover a smaller area than dispersal or alert-alarm movements. The time periods for these activities would also be different. It is not difficult to understand that different processes may need to be measured at different spatial and temporal scales. Understanding foraging dynamics, breeding behavior, or dispersal would seem to require overt thought about the resolution and extent associated with the spatial and temporal dimensions of each process and with the variables or metrics selected for the sampling effort (Martinez and Dunne 1998).

An Organism-Centric Approach

MacNally (1999, 2005) described an organism-centric approach designed to deal with the scale issue that incorporated a species idiosyncratic view of the world based on animal size; for simplicity, he disregarded sensory capabilities. He used maximum body length and expected length of lifetime to develop a characteristic measure that represented the spatial extension of the organism. I interpret this to mean the area over which the animal moves over its existence (i.e., the natural scale that informs how organism uses the landscape). MacNally's approach represents an interesting and heuristic approach to addressing the scale problem.

Multiscale Approaches

To deal with ecological complexity, a multiscale approach is often appropriate. Recent work on chronic wasting disease by Conner et al. (2007) recognized overtly that a multiscale approach was necessary to handle the types of data that were available (Fig. 6.5). First, the investigators developed and adapted methods to handle data appropriate to multijurisdictional or multistate modeling; they termed this scale and extent the regional scale. Then, methods and data appropriate

for within-state areas (wildlife management units or metapopulations) were developed and termed landscape scale. Finally, data and methods adapted for population and individual-based models were developed, and they were termed fine scale. Conner et al. (2007) recognized and reported that course woody debris data have been collected not only over different extents but also at different resolutions. Although the terminology was mixed, their approach was clearly and overtly multiscale.

A Focus on Spatial Context

Åberg et al. (1995), Gascon et al. (1999), Guisan et al. (2006), Wheatley and Johnson (2009), and Driscoll et al. (2013) focused on spatial context, specifically the landscape structure and composition of an area surrounding a site (i.e., the matrix). de Knegt et al. (2011) argued that the landscape surrounding a used resource provided the environmental context for understanding animal response to that resource and proposed the term *range* to represent the ambit (i.e., an area in which something acts or operates or has power or control; http://wordnetweb.princeton.edu/perl/webwn?s=ambit) radii at which context can be considered. It has antecedents with the idea of ecological neighborhoods (Addicott et al. 1987), and the term appears in Haury et al. (1978) and Wiens (1989). Since de Knegt et al. (2011) had no a priori knowledge about the scales that elephants responded to environmental heterogeneity, they suggested that within constraints set by a chosen resolution and extent, measuring environmental or landscape context at different ambit radii centered on used resource sites could possibly reveal the ecologically most relevant scales for studying species–environment relationships. They focused on understanding how elephant responses scaled to food and water resources. They varied the range of environmental context that they measured (i.e., the ambit radii from 0 [no context] to 40 km). Importantly, they reported that the strength of habitat selection was dependent on the size of the ambit radius used to measure the environmental context variables of tree cover, herbaceous biomass, water occurrence, and a proxy measure of vegetation heterogeneity. In other words, elephants appeared to be keying in on resources at a characteristic scale that could be characterized by ambit radius length. They reported that the characteristic scale did not match the scales at which environmental heterogeneity was most dominant (de Knegt et al. 2011). In other words, accounting for the elephants' environmental context revealed relationships that that would not have been exposed had it not been considered. If these results are robust for other species, they suggest that examining a range of environmental contexts, given a specified resolution and extent, will provide a clearer understanding of animal–environment relationships. Franklin and Lindenmayer (2009) argue that many biologists have largely overlooked the pivotal importance of the matrix and the habitat it provides. Brady et al. (2011) reported that matrix development intensity by anthropogenic forces influenced mammal abundance and landscape use, but that the responses tended to be species-specific. Interestingly, native

Chronic Wasting Disease
in North America

**Areas with CWD
infected Cervid populations**

**States/provinces where
CWD has been found in
captive populations**

Figure 6.5. Presence of chronic wasting disease in deer and elk (*Cervus canadensis*) across North America. Conner et al. (2007) developed and adapted methods to handle multiple resolution data appropriate to three different scale extents: multijurisdictional or multistate modeling (the regional scale), within-state areas (i.e., wildlife management units or metapopulations, the landscape scale), and three population and individual-based models (fine scale). Map includes areas with chronic-wasting-disease-infected cervid populations and states and provinces where chronic wasting disease has been found in captive populations. Redrawn from Chronic Wasting Disease Alliance, Map—Chronic Wasting Disease in North America, http://cwd-info .org/map-chronic-wasting -disease-in-north-america

species richness peaked at moderate levels of matrix development, but exotic species richness and feral predators increased with increasing development. Consideration of landscape context is an important influence and can assist with understanding animal responses. de Knegt et al. (2011) have provided a way to incorporate landscape context while at the same time limiting the problems that wildlife ecologists face when trying to conduct scale-sensitive studies.

Domains of Scale

Scaling domains appear to provide some leeway in selecting the appropriate scale(s).

Wiens (1989) pointed to a potentially crippling problem with scale dependency. If there is pattern at every scale, how is one to find generalizations? How can one extrapolate results if the spatial and temporal scale spectrum is a continuum? Understanding patterns would be difficult indeed. Wiens (1989) suggested that if the gradient of spatial and temporal scales is discontinuous, with sections or domains over which patterns do not change or change very little or monotonically with changes in scale, then as ecologists we have a way to deal with extrapolation. As Wiens (1989:393) suggests, "Domains

of scale for particular pattern-process combinations define the boundaries of generalizations." So long as we are within a domain of scale, we can expect relative clean correspondence between the observational scales we use to measure system response and the patterns that emerge (Fig. 6.4). Scale dependence becomes somewhat muted.

Circumventing the Selection of Scale by Using Graph Theory

Bissonette (2017) showed that focusing on the organism and not the landscape allowed the animal to demonstrate the scale(s) it was using. He used developments in movement ecology (Nathan et al. 2008), graph (network) theory and analyses (Urban and Keitt 2001, Blonder et al. 2012, Golubski et al. 2016, Jacoby and Freeman 2016), and important technical advances in biologging (Naito 2004, Rutz and Hays 2009, Kays et al. 2015), coupled with first-time passage (FTP) and area-restricted movement (ARM) methodology and analyses (Fauchald and Tveraa 2003) to show that changes in movements detected by FTP and ARM indicate that relevant scales or animal movement and activity can be calculated and are useful to conservation.

WHAT ELSE CAN GO WRONG?

Unappreciated Impediments

Could it be that the attributes of scale (i.e., resolution and extent) are insufficient by themselves to understand animal response to changing resource availability as reflected in changing landscape patterns? Is there something that we are missing? The manner in how we represent data and the difficulty of obtaining similar results over time influence study results.

Number System Mismatch

Sells et al. (2018) recently raised the argument about reliability in wildlife science and suggested that most scientists accept reductionist and mechanistic approaches as the rigorous way to do science. Bissonette (2019) suggested that there were structural impediments to obtaining rigor that were underappreciated and often impossible to overcome. He referred to the often-misused syllogistic logic, the idea that the assumptions of falsifiability can seldom be met in ecology, the complex nature of causality, and the unavoidable mismatch between the number systems (Weinberg 1975, Weinberg and Weinberg 1979) used to study ecology. Ecology is characterized by being a middle-number system that has an intermediate number of components, with organized interactions and structured relationships among its components. Given the square law of computation, "solving a problem with n variables, with each variable interacting with the other, takes $2n$ units of computation. Five variables would take $25 = 32$, 8 variables $28 = 256$, and 20 variables $220 = 1,048,576$ units of computation" (Bissonette 2019:7). Clearly, although possible, this approach is impractical, so a small-number system is invariably used (i.e., an approach characterized by mathematical analysis of a subset of variables, with each interaction described by a separate equation). Hence an unavoidable structural mismatch occurs with serious implications for data interpretation.

Big Data, Repeatability, and Reliability

It is not my intention to review exhaustively all of the aspects of big data, repeatability, and reliability. A brief explanation of some of the major problems will help set the problems into context. Wüest et al. (2019:1) described big data as "defined by the five Vs: volume, variety (heterogeneity of sources, unstructured data), velocity (speed of data generation and collection), veracity (uncertainty and data quality) and value." The value of big data lies in what we can learn unambiguously from it (Bissonette 2021). Extraction of value from big data, however, is a process fraught with problems that relate to its complexity (see "Complexity," above). Ultimately, issues of spatial and temporal scale of the data collected intrude. Wüest et al. (2019) cite Beck et al. (2012), who identified integration of historical contingencies, explicit consideration of processes, aggregation of large high-quality data sets on a global scale, and the advancement of statistical methods as major challenges for large-scale ecology. Among the major problems in large-scale wildlife ecology studies (and in many other fields

of study) are repeatability (replicability) and reproducibility of studies. In an early paper, Cassey and Blackburn (2006:958) addressed these ideas and argued that misuse, misunderstanding, and careless use of terms were prevalent among scientists. They argued that the search in ecology for general rules and theories was hampered by complexity of natural ecosystems and pointed to "historically based, context specific contingency" as a major problem that required repeatable (or replicable) studies to address the specific theory; a problem that Johnson (2002) recognized early. Cassey and Blackburn (2006) argued that repeatability was fundamentally a different concept than reproducibility. Repeatability does not involve use of the same data set and analyses, while reproducibility does.

SUMMARY

The issue of scale and scaling in wildlife ecology will continue to be a problematic area. First, one easily resolved difficulty is to gain accuracy and consistency in the terminology we use to describe scale concepts (Dungan et al. 2002). Here, context is important. Understanding that scale and scaling are intimately tied to hierarchy theory (i.e., system organization) may account for some of the confusing use of terms. Having a common language with mutually agreed meanings for terms seems a logical first step. Second, understanding that qualitative changes often occur when measurements taken at finer-scale resolutions and extents are scaled larger is the sine qua non for developing scale-sensitive and biologically realistic sampling protocols and analyses. It is important to understand that the field of landscape ecology came into prominence primarily because cross-scale extrapolation was recognized as problematic. Prior to the attention to scale in the early 1980s, ecologists scaled up as if the dynamics they were investigating were linear. But few important relationships in ecology are that simple. Nonlinearity is more often the case, and hence the problem of scaling is much more difficult. It should now be clear that different patterns emerge when the scale resolution and/or the scale extent of the study is changed; this is the concept of scale sensitivity. Perhaps a definitive lesson to note is that there is no single correct scale; the appropriate scale(s) depend on the question one asks and the specific process(es) of interest. Space-time diagrams illustrate the difficulty with achieving a match between the spatial extent over which the dynamics of a process or pattern are expressed and the time it takes for the dynamics to be expressed. Ideally, one attempts to match space and time to understand the dynamics. This is often not possible, so the standard practice appears to involve a space for time substitution, coupled with a sometimes arbitrary selection of the hopefully meaningful scales. If domains of scales exist, then this may be the closest we can come to adjusting our sampling and observational scales to the scales inherent to natural processes and patterns. de Knegt et al. (2011:271) have stated that "no question in spatial ecology can be answered without explicitly referring to these components" (resolution and extent). Wheatley and Johnson (2009), however, make a clear distinction between a strictly scalar study and a spatial study. The distinction deals primarily with holding one or more scale component (resolution or extent) constant while varying the other. When either or both grain or extent or other independent variables are changed across scales as part of a study design, the resulting study is not strictly scalar, and hence cross-scale extrapolation and generalization are usually not valid. It may be that to achieve accurate cross-scale predictions, dealing with adjustment of resolution and extent may be insufficient. Landscape context is an important influence of animal response (Saunders et al. 1991, McIntyre and Barrett 1992, McIntyre et al. 1996, McIntyre and Hobbs 1999, Manning et al. 2004, Fischer and Lindenmayer 2006). If we are interested in cross-scale prediction of animal response, then considering understanding the role environmental context is necessary. Additionally, understanding that ecologists most frequently use small-number system approaches to address complex middle-number systems and the problems attendant with big data leads us to avoid the hubris that study results always provide reliable results. Addressing scale and scaling issues is a necessary part of what wildlife ecologists need to know if they wish to understand the broad-scale management problems facing the profession.

Literature Cited

Åberg, J., G. Jansson, J. E. Swenson, and P. Angelstam. 1995. The effects of matrix on the occurrence of hazel grouse (*Bonasa bonasia*) in isolated habitat fragments. Oecologia 103:235–269.

Addicott, J. F., J. M. Aho, M. F. Antolin, D. K. Padilla, J. S. Richardson, and D. A. Soluk. 1987. Ecological neighborhoods: scaling environmental patterns. Oikos 49:340–346.

Allen, T. F. H. 1998. The landscape "level" is dead: persuading the family to take it off the respirator. Pages 35–54 in D. L. Peterson, and V. T. Parker, editors. Ecological scale: theory and application. Columbia University Press, New York, New York, USA.

Allen, T. F. H., and T. W. Hoekstra. 1990. The confusion between scale-defined levels and conventional levels of organization in ecology. Journal of Vegetation Science 1:5–12.

Allen, T. F. H., and T. B. Starr. 1982. Hierarchy: perspectives for ecological complexity. University of Chicago Press, Chicago, Illinois, USA.

Anthony, R. G., and N. S. Smith. 1977. Ecological relationships between mule deer and white-tailed deer in southeastern Arizona. Ecological Monographs 47:255–277.

Antonovics, J., and D. A. Levin. 1980. The ecological and genetic consequences of density-dependent regulation in plants. Annual Review of Ecology and Systematics 11:411–452.

Bannon, I., and P. Collier, editors. 2003. Natural resources and violent conflict: options and actions. World Bank, Washington, DC, USA.

Beck, J., L. Ballesteros-Mejia, C. M. Buchmann, J. Dengler, S. A. Fritz, et al. 2012. What's on the horizon for macroecology? Ecography 35:673–683.

Berger, J. 2003. Is it acceptable to let a species go extinct in a National Park? Conservation Biology 17:1451.

Berger, K. M. 2007. Conservation implications of food webs involving wolves, coyotes, and pronghorn. PhD dissertation, Utah State University, Logan, USA.

Berger, K. M., E. M. Gese, and J. Berger. 2008. Indirect effects and traditional trophic cascades: a test involving wolves, coyotes, and pronghorn. Ecology 89:818–828.

Bissonette, J. A. 1997. Scale-sensitive properties: historical context, current meaning. Chapter 1, Pages 3–31 in J. A. Bissonette, editor. Wildlife and Landscape ecology: effects of pattern and scale. Springer-Verlag, New York, New York, USA.

Bissonette, J. A. 2003. Linking landscape patterns to biological reality. Pages 15–34 in J. A. Bissonette and I. Storch, editors. Landscape ecology and resource management: linking theory with practice. Island Press, Washington, DC, USA.

Bissonette, J. A. 2017. Avoiding the scale sampling problem: a consilient solution. Journal of Wildlife Management 81:192–205.

Bissonette, J. A. 2019. Additional thoughts on rigor in wildlife science: unappreciated impediments. Journal of Wildlife Management 83:1017–1021.

Bissonette, J. A. 2021. Big data, exploratory data analyses and questionable research practices: suggestion for a foundational principle. Wildlife Society Bulletin 2021:1-5. doi:10.1002/wsb.1201.

Bissonette, J. A., and W. A. Adair. 2008. Restoring habitat permeability to roaded landscapes with isometrically-scaled wildlife crossings. Biological Conservation 141:482–488.

Blonder, B., T. W. Wey, A. Dornhaus, R. James, and A. Sih. 2012. Temporal dynamics and network analysis. Methods in Ecology and Evolution 3:958–972.

Bowman, J., J. A. G. Jaeger, and L. Fahrig. 2002. Dispersal distance of mammals is proportional to home range size. Ecology 83:2049–2055.

Brady, H. J., C. A. McAlpine, C. J. Miller, H. P. Possingham, and G. S. Baxter. 2011. Mammal responses to matrix development intensity. Austral Ecology 36:35–45.

Calder, W. A., III. 1983. Ecological scaling: mammals and birds. Annual Review of Ecology and Systematics 4:213–230.

Cassey, P., and T. M. Blackburn. 2006. Reproducibility and repeatability in ecology. Bioscience 56:958–959.

Clark, P. E., J. Chigbrow, D. E. Johnson, L. L. Larson, R. M. Nielson, M. Louhaichi, T. Roland, and J. Williams. 2020. Predicting spatial risk of wolf-cattle encounters and depredation. Rangeland Ecology and Management 73:30–52.

Collins, J. W. 1884. Notes on the habits and methods of capture of various species of sea birds that occur on the fishing banks off the eastern coast of North America, and which are used as bait for catching codfish by New England fisherman. Report of the Commissioner of Fish and Fisheries for 1882 13:311–335.

Conner, M. M., J. E. Gross, P. C. Cross, M. R. Ebinger, R. R. Gillies, M. D. Samuel, and M. W. Miller. 2007. Scale-dependent approaches to modeling spatial epidemiology of chronic wasting disease. Special Report 2007. Utah Division of Wildlife Resources, Salt Lake City, USA.

Connor, E. G., and E. D. McCoy. 1979. The statistics and biology of the species-area relationship. American Naturalist 113:791–833.

Cressie, N. A. 2015. Statistics for spatial data. Revised edition. John Wiley and Sons, New York, New York, USA.

Cumming, G. S. 2016. Heterarchies: reconciling networks and hierarchies. Trends in Ecology and Evolution 31:622–632.

Damgaard, C. 2019. A critique of the space-for-time substitution practice in community ecology. Trends in Ecology and Evolution 34:416–421.

de Knegt, H. J., F. van Langevelde, A. K. Skidmore, A. Delsink, R. Slotow, et al. 2011. The spatial scaling of habitat selection of African elephants. Journal Animal Ecology 80:270–281.

Delcourt, H. R., P. A. Delcourt, and T. Webb. 1983. Dynamic plant ecology: the spectrum of vegetational change in space and time. Quaternary Science Review 1:153–175.

Driscoll, D. A., S. C. Banks, P. S. Barton, D. B. Lindenmayer, and A. L. Smith. 2013. Conceptual domain of the matrix in fragmented landscapes. Trends in Ecology and Evolution 28:605–613.

Dungan, J. L., J. N. Perry, M. R. T. Dale, P. Legendre, S. Citron-Pousty, M.-J. Fortin, A. Jakomulska, M. Miriti, and M. S. Rosenberg. 2002. A balanced view of scale in spatial statistical analysis. Ecography 25:626–640.

du Toit, M. J., D. J. Kotze, and S. S. Cilliers. 2016. Landscape history, time lags and drivers of change: urban natural grassland remnants in Potchefstroom, South Africa. Landscape Ecology 31:2133–2150.

Dyksterhuis, E. J. 1949. Condition and management of range land based on quantitative ecology. Journal of Range Management 2:104–115.

Elliott-Graves, A. 2018. Generality and causal interdependence in ecology. Philosophy of Science 85:1102–1114.

Fall, A. M., M.-J. Fortin, M. Manseau, and D. O'Brien. 2007. Spatial graphs: principles and applications for habitat connectivity. Ecosystems 10:448–461.

Fauchald, P., and T. Tveraa. 2003. Using first-passage time in the analysis of area-restricted search and habitat selection. Ecology 84:282–288.

Firbank, L. G., C. Bertora, D. Blankman, G. Delle Vedove, M. Frenzel, et al. 2017. Towards the co-ordination of terrestrial ecosystem protocols across European research infrastructures. Ecology and Evolution 7:3967–3975.

Fischer, J., and D. B. Lindenmayer. 2006. Beyond fragmentation: the continuum model for fauna research and conservation in human-modified landscapes. Oikos 112:473–480.

Fortin, M. J., P. Drapeau, and P. Legendre. 1990. Spatial autocorrelation and sampling design in plant ecology. Pages 209–222 in G. Grabherr, L. Mucina, M. B. Dale, C. J. F Ter Braak, editors. Progress in theoretical vegetation science. Advances in Vegetation Science 11. Springer, Dordrecht, The Netherlands.

Franklin, J. F., and D. B. Lindenmayer. 2009. Importance of matrix habitats in maintaining biological diversity. Proceedings of the National Academy of Sciences 106:349–350.

Froehly, J. L., A. K. Tegeler, C. M. B. Jachowski, and D. S. Jachowski. 2019. Effects of scale and land cover on loggerhead shrike occupancy. Journal of Wildlife Management 83:426–434.

Gascon, C., T. E. Lovejoy, R. O. Bierregaard, J. R. Malcom, P. C. Stouffer, H. L. Vasconcelos, W. F. Laurance, B. Zimmerman, M. Tocher, and S. Borges. 1999. Matrix habitat and species richness in tropical forest remnants. Biological Conservation 91:223–229.

Golubski, A. J., E. E. Westlund, J. Vandermeer, and M. Pascual. 2016. Ecological networks over the edge: hypergraph trait-mediated indirect interaction (TMII) structure. Trends in Ecology and Evolution 31:344–354.

Golley, F. B. 1989. Paradigm shift. Landscape Ecology 3:65–66.

Gould, M. J., J. W. Cain, G. W. Roemer, W. R. Gould, and S. G. Liley. 2018. Density of American black bears in New Mexico. Journal of Wildlife Management 82:775–788.

Gould, S. J. 1979. An allometric interpretation of species-area curves: the meaning of the coefficient. American Naturalist 114:335–343.

Guisan, A., A. Lehmann, S. Ferrier, M. Austin, J. MC. C. Overton, R. Aspinall, and T. Hastie. 2006. Making better biogeographical predictions of species' distributions. Journal of Applied Ecology 43:386–392.

Gustafson, E. J. 1998. Quantifying landscape spatial pattern: what is the state of the art? Ecosystems 1:143–156.

Gutzwiller, K. J., S. K. Riffell, and C. H. Flather. 2015. Avian abundance thresholds, human-altered landscapes, and the challenge of assemblage-level conservation. Landscape Ecology 30:2095–2110.

Haury, L. R., J. A. McGowan, and P. H. Wiebe. 1978. Patterns and processes in the time-space scales of plankton distribution. Pages 227–327 in J. H. Steele, editor. Spatial pattern in plankton communities. Plenum, New York, New York, USA.

Heffelfinger, J. R. 2018. Inefficiency of evolutionarily relevant selection in ungulate trophy hunting. Journal of Wildlife Management 82:57–66.

Holl, K. D., E. E. Crone, and C. B. Schultz. 2003. Landscape restoration: moving from generalities to methodologies. Bioscience 53:491–502.

Holling, C. S. 1992. Cross-scale morphology, geometry, and dynamics of ecosystems. Ecological Monographs 62:447–502.

Humphreys, M. 2005. Natural resources, conflict, and conflict resolution: uncovering the mechanisms. Journal of Conflict Resolution 49:508–537.

Hutchinson, G. E. 1959. Homage to Santa Rosalia. American Naturalist 93:145–159.

Jacoby, D. M. P., and R. Freeman. 2016. Emerging network-based tools in movement ecology Trends in Ecology and Evolution 31:301–314.

Johnson, D. H. 2002. The importance of replication in wildlife research. Journal of Wildlife Management 66:919–932.

Johnson, M. P., N. J. Frost, M. W. J. Mosley, M. F. Roberts, and S. J. Hawkins. 2003. The area-independent effects of habitat complexity on biodiversity vary between regions. Ecology Letters 6:126–132.

Katz, S. L., K. A. Barnas, M. Diaz, and S. E. Hampton. 2019. Data system design alters meaning in ecological data: salmon habitat restoration across the U.S. Pacific Northwest. Ecosphere 10:e02920.

Kays, R., M. C. Crofoot, W. Jetz, and M. Wikelski. 2015. Terrestrial animal tracking as an eye on life and planet. Science 348:6240.

King, A. W. 1991. Translating models across scales in the landscape. Pages 479–517 in M. G. Turner and R. H. Gardner, editors. Quantitative methods in landscape ecology. Springer-Verlag, New York, New York, USA.

King, A. W. 1997. Hierarchy theory: a guide to system structure for wildlife biologists. Pages 185–212 in J. A. Bissonette, editor. Wildlife and landscape ecology: effects of pattern and scale. Springer, New York, New York, USA.

King, A. W., A. R. Johnson, and R. V. O'Neill. 1991. Transmutation and functional representation of heterogeneous landscapes. Landscape Ecology 5:239–353.

Kleiber, M. 1947. Body size and metabolic rate. Physiological Reviews 27:511–541.

Krausman, P. R. 1976. Ecology of the Carmen Mountains white-tailed deer. PhD dissertation, University of Idaho, Moscow, USA.

Laurance, W. F., H. E. M. Nascimento, S. G. Laurance, A. Andrade, R. M. Ewers, K. E. Harms, R. C. C Luizão, and J. E. Ribeiro. 2007. Habitat fragmentation, variable edge effects, and the landscape-divergence hypothesis. PLoS ONE 2:e1017.

Lebeau, C. W., K. T. Smith, M. J. Holloran, M. E. Kauffman, G. D. Johnson. 2019. Greater sage-grouse habitat function relative to 230-kV transmission lines. Journal of Wildlife Management 83:1773–1786.

Lechner, A. M., and J. R. Rhodes. 2016. Recent progress on spatial and thematic resolution in landscape ecology. Current Landscape Ecology Reports 1:98–105.

Levin, S. A. 1992. The problem of pattern and scale in ecology: the Robert H. MacArthur Award Lecture. Ecology 73:1943–1967.

Lewis, K. P., E. Vander Wal, and D. A. Fifield. 2018. Wildlife biology, big data, and reproducible results. Wildlife Society Bulletin 42:172–179.

Lindenmayer, D. B., and J. Fischer. 2006. Habitat fragmentation and landscape change. Island Press, Washington, DC, USA.

Lindenmayer, D. B., G. E. Likens, A. Andersen, D. Bowman, C. M. Bull, et al. 2012. Value of long-term ecological studies. Austral Ecology 37:745–757.

Lookingbill, T. R., R. H. Gardner, J. R. Ferrari, and C. E. Keller. 2010. Combining a dispersal model with network theory to assess habitat connectivity. Ecological Applications 20:427–441

MacKenzie, D. I. 2005. What are the issues with presence–absence data for wildlife managers? Journal of Wildlife Management 69:849–860.

MacNally, R. 1999. Dealing with scale in ecology. Pages 10–17 *in* J. A. Wiens and M. R. Moss, editors. Issues in landscape ecology. International Association for Landscape Ecology, Pioneer Press, Greeley, Colorado, USA.

MacNally, R. 2005. Scale and an organism-centric focus for studying interspecific interactions in landscapes. Pages 52–69 *in* J. A. Wiens and M. R. Moss, editors. Issues and perspectives in landscape ecology. Cambridge University Press, Cambridge, UK.

Manning, A. D., D. B. Lindenmayer, and H. A. Nix. 2004. Continua and Umwelt: novel perspectives on viewing landscapes. Oikos 104:621–628.

Martens, G. G. 1972. Censusing mouse populations by means of tracking. Ecology 53:859–867.

Martinez, N. D., and J. A. Dunne. 1998. Time, space, and beyond: scale issues in food-web research. Pages 207–226 *in* D. L. Peterson and V. T. Parker, editors. Ecological scale: theory and applications. Columbia University Press, New York, New York, USA.

May, R. M. 1988. How many species are there on earth? Science 241:1441–1449.

May, R. M. 1994. The effects of spatial scale on ecological questions and answers. Pages 1–17 *in* P. J. Edwards, R. M. May, and N. R. Webb, editors. Large-scale ecology and conservation biology. Blackwell Scientific, London, UK.

McIntyre, S., and G. W. Barrett. 1992. Habitat variegation, an alternative to fragmentation. Conservation Biology 6:146–147.

McIntyre, S., G. W. Barrett, and H. A. Ford. 1996. Communities and ecosystems. Pages 154–170 *in* I. F. Spellerberg, editor. Conservation biology. Longman Group, Essex, UK.

McIntyre, S., and R. J. Hobbs. 1999. A framework for conceptualizing human effects on landscapes and its relevance to management and research models. Conservation Biology 13:1282–1292.

McNab, B. K. 1963. Bioenergetics and the determination of home range size. American Naturalist 97:133–140.

Miller, J. R., M. G. Turner, E. A. H. Smithwick, C. L. Dent, and E. H. Stanley. 2004. Spatial extrapolation: the science of predicting ecological patterns and processes. Bioscience 54:310–320.

Montello, D. R. 2001. Scale in geography. Pages 13,501–13,504 *in* N. J. Smelser and P. B. Baltes, editors. International encyclopedia of the social and behavioral sciences. Pergamon Press, Oxford, UK.

Murphy, R. C. 1914. Observations on birds of the south Atlantic. Auk 31:439–457.

Naito, Y. 2004. New steps in bio-logging science. Memoirs of the National Institute of Polar Research 58:50–57.

Nathan, R., W. M. Getz, E. Revilla, M. Holyoak, R. Kadmon, D. Saltz, and P. E. Smouse. 2008. A movement ecology paradigm for unifying organismal movement research. Proceedings National Academy of Sciences 105:19052–19059.

O'Neill, R. V. 1979. Transmutations across hierarchical levels. Pages 59–78 *in* G. S. Innis and R. V. O'Neill, editors. Systems analysis of ecosystems. International Cooperative Publishing House, Fairland, Maryland, USA.

O'Neill, R. V., D. L. De Angelis, J. B. Waide, and T. F. H. Allen. 1986. A hierarchical concept of ecosystems. Monographs in Population Ecology 23. Princeton University Press, Princeton, New Jersey, USA.

O'Neill, R. V., and A. W. King. 1998. Homage to St. Michael; or, why are there so many books on scale? Pages 3–15 *in* D. L. Peterson and V. T. Parker, editors. Ecological scale: theory and application. Columbia University Press, New York, New York, USA.

Pace, M. L., J. J. Cole, S. R. Carpenter, and J. F. Kitchell. 1999. Trophic cascades revealed in diverse ecosystems. Trends in Ecology and Evolution 14:483–488.

Pickett, S. T. A. 1989. Space-for-time substitution as an alternative to long term studies. Pages 110–135 in G. E. Likens, editor. Long-term studies in ecology: approaches and alternatives. Springer-Verlag, New York, New York, USA.

Rastetter, E. B., A. W. King., F. J. Cosby, G. M. Hornberger, R. B. O'Neill, and J. E. Hobbie. 1992. Aggregating fine-scale ecological knowledge to model coarser-scale attributes of ecosystems. Ecological Applications 2:55–70.

Rosa, I. M. D., A. Purvis, R. Alkemade, R. Chaplin Kramer, S. Ferrier, et al. 2020. Challenges in producing policy-relevant global scenarios of biodiversity and ecosystem services. Global Ecology and Conservation 22:1–11.

Rutz, C., and G. C. Hays. 2009. New frontiers in biologging science. Biology Letters 5:289–292.

Saunders, D. A., R. J. Hobbs, and C. R. Margules. 1991. Biological consequences of ecosystem fragmentation: a review. Conservation Biology 5:18–32.

Sawyer, H., F. Lindzey, and D. McWhirter. 2005. Mule deer and pronghorn migration in western Wyoming. Wildlife Society Bulletin 33:1266–1273.

Schneider, D. C. 2001. The rise of the concept of scale in ecology. Bioscience 51:545–553.

Sells, S. N., S. B. Bassing, K. J. Barker, S. C. Forshee, A. C. Keever, J. W. Goerz, and M. S. Mitchell. 2018. Increased scientific rigor will improve reliability of research and effectiveness of management. Journal of Wildlife Management 82:485–494.

Sherry, T. W., and R. T. Holmes. 1988. Habitat selection by breeding American Redstarts in response to a dominant competitor, the Least Flycatcher. Auk 105:350–364.

Siegel, S. 1956. Nonparametric statistics for the behavioral sciences. McGraw-Hill, New York, New York, USA.

Southwood, T. R. E. 1977. Habitat, the templet for ecological strategies? Journal of Animal Ecology 46:337–365.

Sprugel, D. G. 1983. Correcting for bias in log-transformed allometric equations. Ecology 64:209–210.

Stevens, S. S. 1946. On the theory of scales of measurement. Science 103:677–680.

Stommel, H. 1963. The varieties of oceanographic experience. Science 139:572–557.

Sutherland, G. D., A. S. Harestad, K. Price, and K. P. Lertzman. 2000. Scaling of natal dispersal distance in terrestrial birds and mammals. Conservation Ecology 4:1–16.

Thompson, F. R., III, W. D. Dijak, T. G. Kulowiec, and D. A. Hamilton. 1992. Breeding bird populations in Missouri Ozark forests with and without clearcutting. Journal of Wildlife Management 56:23–30.

Tilman, D. 1989. Ecological experimentation: strengths and conceptional problems. Pages 136–157 *in* G. E. Likens, editor. Long-term studies in ecology. Springer, New York, New York, USA.

Tilman, D. 2000. Causes, consequences and the ethics of biodiversity. Nature 405:208–211.

Turner, M. G., R. H. Gardner, and R. V. O'Neill. 2001. Landscape ecology in theory and practice: pattern and process. Springer, New York, New York, USA.

Turner, M. G., R. V. O'Neill, R. H. Gardner, and B. T. Milne. 1989. Effects of changing spatial scale on the analysis of landscape pattern. Landscape Ecology 3:153–162.

Urban, D., and T. Keitt. 2001. Landscape connectivity: a graph-theoretic perspective. Ecology 82:1205–1218.

van Lieshout, M., A. Dewulf, N. Aarts, and C. Termeer. 2011. Do scale frames matter? scale frame mismatches in the decision making process about a "mega farm" in a small Dutch village. Ecology and Society 16:38.

Weinberg, G, M. 1975. An introduction to general systems thinking. John Wiley and Sons, New York, New York, USA.

Weinberg, G. M., and D. Weinberg. 1979. On the design of stable systems. John Wiley and Sons, New York, New York, USA.

Weiskopf, S. R., O. E. Ledee, and L. M. Thompson. 2019. Climate change effects on deer and moose in the Midwest. Journal of Wildlife Management 83:769–781.

West, G. 2017. Scale: the universal laws of growth, innovation, sustainability, and the pace of life in organisms, cities, economics, and companies. Penguin, New York, New York, USA.

West, G. B., J. H. Brown, and B. J. Enquist. 1997. A general model for the origin of allometric scaling laws in biology. Science 276:122–126.

Wheatley, M., and C. Johnson. 2009. Factors limiting our understanding of ecological scale. Ecological Complexity 6:150–159.

Wiens, J. A. 1989. Spatial scaling in ecology. Functional Ecology 3:385–397.

Williams, C. K. 2013. Accounting for wildlife life-history strategies when modeling stochastic density-dependent populations: a review. Journal of Wildlife Management 77:4–11.

With, K. A. 2007. Invoking the ghosts of landscapes past to understand the landscape ecology of the present. Pages 43–58 in J. A. Bissonette and I. Storch, editors. Temporal dimensions of landscape ecology: wildlife responses to variable resources. Springer, New York, New York, USA.

Withers, M. A., and V. Meentemeyer. 1999. Concepts of scale in landscape ecology. Pages 205–252 in J. M. Klopatek and R. H. Gardner, editors. Landscape ecological analysis: issues and applications. Springer, New York, New York, USA.

Wolff, J. O. 1999. Behavioral model systems. Pages 11–40 in G. W. Barrett, and J. D. Peles, editors. Landscape ecology of small mammals. Springer, New York, New York, USA.

Wright, S. 1943. Isolation by distance. Genetics 28:114–138.

Wright, S. 1946. Isolation by distance under diverse systems of mating. Genetics 31:39–59.

Wu, J., and H. Li. 2006. Concepts of scale and scaling. Pages 3–13 in J. Wu, K. B. Jones, H. Li, and O. L. Loucks, editors. Scaling and uncertainty analysis in ecology: methods and applications. Springer, New York, New York, USA.

Wüest, R. O., N. E. Zimmermann, D. Zurell, J. M. Alexander, S. A. Frits, et al. 2019. Macroecology in the age of big data—where do we go from here? Journal Biogeography 47:1–12.

Yang, J., H. S. He, S. R. Shifley, F. R. Thompson, Y. Zhang. 2011. An innovative computer design for modeling forest landscape change in very large spatial extents with fine resolutions. Ecological Modelling 222:2623–2630.

Ziter, C., R. A. Graves, and M. G. Turner. 2017. How do land–use legacies affect ecosystem services in United States cultural landscapes? Landscape Ecology 32:2205–2218.

WILDLIFE POPULATION DYNAMICS

L. SCOTT MILLS AND HEATHER E. JOHNSON

INTRODUCTION

Are the recovery and recolonization of gray wolves (*Canis lupus*) reducing the number of elk (*Cervus canadensis*) available to hunters? Does oil and gas development influence sage grouse (*Centrocercus urophasianus*) populations, and if so, how great is the effect? How fast are raccoons (*Procyon lotor*) proliferating and spreading diseases that affect humans? These are just a few cases where wildlife management and conservation depend critically on understanding population dynamics. Without this knowledge, wildlifers do not know even the most basic things about populations: whether a population is increasing or decreasing in size, how a particular stressor may affect the persistence of a population, or the quality of the habitat that supports a population.

In short, wildlife cannot be conserved or managed effectively without understanding population dynamics. But to understand population dynamics, biologists must master many fields. They must understand basic ecology, the conceptual foundation upon which our current knowledge has been built. They must know some math, because mathematic models reveal processes underlying population dynamics that cannot be seen with even a keen eye or sharp intuition. They must understand the natural history of different species, to help apply insights from one species to another and to avoid wandering into irrelevance when applying models. And they must grasp the fundamentals of disparate fields, including population genetics, quantitative biology, animal behavior, animal physiology, plant ecology, and human dimensions, because the mechanisms influencing population dynamics are best understood by deciphering the interactions animals have with each other and their environment.

Readers will approach this chapter from different backgrounds and have varying levels of training or knowledge in the different pieces needed to understand wildlife population dynamics. Therefore we assume relatively little background, building up from the basics and providing some references to more advanced readings once you master the fundamentals. Expanded coverage of most of these topics is presented by Mills (2007, 2013).

In this chapter we provide an overview of some core concepts, describe exponential growth as the basic foundation for understanding population dynamics, and discuss some of the factors that can affect wildlife population dynamics. We then show how management insights can be gained from analyzing the dynamics of individual age or stage classes, examine dynamics of multiple populations across a landscape, consider key aspects of monitoring wildlife population dynamics, and close with a case study that applies many of the topics in the chapter. Throughout we stress a few key themes: (1) variation is as important as the mean in understanding population dynamics (embrace uncertainty!); (2) some of the most powerful insights into outcomes of wildlife management actions are nonintuitive, revealed by applying data to models; and (3) because different management actions influence population dynamics in different ways, we must understand population processes to identify the most effective actions to meet population objectives.

KEY DEFINITIONS AND CONCEPTS

A discussion of wildlife population dynamics requires precise language, often stated in mathematical terms for clarity. We begin by introducing some of the key definitions and concepts to provide a solid foundation to build on throughout the chapter.

First, what is a population? We prefer a broad and practical definition, and refer to a population as being a collection of individuals of a species occupying a defined area, for which it is meaningful to refer to birth and death rates, sex ratios, abundances, and age structures (Cole 1957). For certain applications, more restrictive definitions of populations are needed (e.g., limited gene flow with other populations or long-term occupancy), but for our purposes this general definition allows us to focus on the wide range of population dynamics that form the basis of wildlife management and conservation.

Fundamentally, all population dynamics can be determined through births, immigration, deaths, and emigration (BIDE). The abundance (N) of a population at time $t + 1$ equals the abundance in the previous time step (t), plus the number of animals that have been added to the population through births (B) and immigration (I), and minus those animals that have died (D) or emigrated (E):

$$N_{t+1} = N_t + B + I - D - E.$$

Almost everything in this chapter will revolve around using these components to interpret the status of wildlife populations and how management actions are most likely to affect their dynamics.

The pieces of BIDE make up some of the vital rates that influence population dynamics. Births are quantified as litter size or clutch size or, as time passes and some of the newborns die, as the recruitment of juveniles into a population (i.e., the net number of new animals added to a population). Deaths are often described as a mortality rate, and the flip side of mortality is survival (survival rate = [1 − mortality rate]). Immigration refers to individuals permanently moving into a population, and emigration refers to individuals permanently leaving a population. Both of these interpopulation vital rates are mediated by dispersal, defined as the permanent movement of an individual from one population to another; dispersal can be different from gene flow, where animals not only move into a new population, but also reproduce successfully.

Other vital rates that influence the dynamics of a population include age structure and sex ratio. Age structure refers to the proportion of individuals of each age in a population; often, age structure is generalized to stage structure, grouping categories not by age but by size or developmental stage. For example, a male deer might live for 10 years, but the stage classes most relevant for management agencies to collect data and evaluate population dynamics might be fawns, yearlings, and adults. These "stages" of deer have different vital rates (e.g., the average annual probability of fawn survival may be 50%, while adult male survival may be 75%) and are easily distinguished during field surveys. Another key characteristic of populations is the sex ratio, or the proportion of the population that is males versus females. The sex ratio can differ among different stage classes of animals, from birth to weaning to adulthood, particularly in populations with a sex-biased harvest.

Collectively, these vital rates determine the abundance or density of a population and allow us to track demographic trends through time. Abundance is the number of individuals in a population; easy to say but remarkably hard to estimate because we almost never count every animal present (we will return to this idea in "Monitoring Population Dynamics," below). Density is the abundance scaled to unit area, allowing comparisons on the same scale (e.g., 3 mice/m² in a hay barn compared to 0.1 mice/m² in a forest).

The population growth rate describes the trend in abundance (or density) over time. While this is arguably the most critical parameter when determining the dynamics of a population, the term is a bit misleading, because population growth can refer to an increasing population or to a decreasing or stationary population (i.e., neither increasing nor decreasing). We will come back to this concept throughout the chapter,

but it is so crucial that we introduce two key descriptors of population growth (or population change) here.

The first is the geometric growth rate, or discrete growth rate, referred to as lambda (λ), which describes the proportional change in abundance from one year (or other appropriate time step) to the next. In a simple equation, with abundance this year (N_t) and abundance next year (N_{t+1}):

$$\lambda = N_{t+1}/N_t. \tag{1}$$

If $\lambda = 1$, the population is stationary in size; $\lambda < 1$ indicates a declining population, and $\lambda > 1$ indicates an increasing population. Lambda is easy to work with because it easily converts to percentage change per year: percentage change = ($\lambda - 1$) × 100. For example, if $\lambda = 1.25$, the population will increase by 25% next year; if $\lambda = 0.75$, it will decrease by 25%.

Although λ is intuitive and easy to understand as a proportional change in population size, the discrete growth represented by λ has some awkward mathematical properties. As a result, a solid understanding of population growth requires a second descriptor, the calculus-based continuous time analog of λ, defined by r and called the exponential growth rate or the instantaneous per capita growth rate. The two measures, λ and r, are interchangeable after a simple conversion:

$$\lambda = e^r \text{ or } r = \ln \lambda \tag{2A or 2B}$$

Here, ln is the natural logarithm, with the base e (which is about 2.718). A population with $r > 0$ is increasing, and a population with $r < 0$ is decreasing.

When should we use λ versus r? Typically, biologists use λ when describing population growth to managers or the public because it is intuitive and easy to interpret. Conversely, r is often used when doing mathematical calculations of population growth, as it is independent of a specific time interval, can be easily compared among taxa, and values can be added across time intervals or averaged among them.

Other core concepts related to population dynamics are variation and uncertainty (Mills 2013). Population dynamics are inherently variable over space and time, and biologists must understand and embrace that variation. No understanding of wildlife population dynamics can emerge without understanding how variation affects dynamics, where it comes from, and how managers might use it to their advantage. We provide examples throughout the chapter but start with a lively anecdote, first quoted by Ankney (1996:41), to describe how a clever, if imagined, waterfowl biologist might respond to setting a duck population objective based only on the average abundance, ignoring variation in abundance over time:

> Did someone say that the "average" numerical standing of the North American mallard population over some period of years would be a good standard for management to try to maintain? Don't let the Old Forecaster hear such talk. Not long ago, he got involved in certain philosophical deliberations regarding the "average" condition of dynamite. Which commodity, in its quiescent state, was a small cylinder having a volume of a few cubic inches, yet at its peak of explosion occupied hundreds of

cubic yards. Seemingly, the "average" condition of dynamite could be determined by adding together measurements taken at various levels between these two extremes, and dividing their sum by the number of measurements. The forecaster took one look at the results of all this arithmetic, turned slightly purple, and then decided ruefully that dynamite in its "average condition" must be one helluva thing to crate, ship, and otherwise handle. He has assiduously avoided "averages" ever since that unfortunate experience.

Indeed, variation around an average (or "mean") can be much more important than the average itself. In the broadest sense, there are two main sources of variation in population dynamics: process variation and sampling variation.

Process variance is the real variation in population processes that comes from nature or management. Often we describe process variance as arising from stochastic, or random, events in nature such as those imposed by weather. For example, in barn owls (*Tyto alba*), harsh winter weather is associated with declines in juvenile and adult survival, which has resulted in population crashes in some years (Altwegg et al. 2006). Process variance can also arise from more predictable factors, called deterministic factors, which can occur from nature (e.g., the influence of predators on a population) or wildlife management activities that attempt to increase or decrease a population (e.g., a change in harvest or a habitat enhancement project).

In addition to real process variance, population dynamics may appear variable due to difficulties in accurately estimating population parameters. Sample variation, or observation error, is the inevitable uncertainty that arises from estimates based on incomplete sampling of animals. Animals avoid detection by moving around, making it hard to know who has already been sampled, and by spreading widely across landscapes so that only a portion of the population can be sampled. Sample variance or observation error is a constant challenge in wildlife studies but can often be accounted for with sound field methods and appropriate statistical tools. It is important to embrace uncertainty from sampling variation

as an honest representation of the amount of confidence you have in a population parameter estimate.

Conclusions about wildlife population dynamics must always be couched in terms of uncertainty about the ecological system (process variation), given limitations in our abilities to estimate parameters in the field (sample variation). In sum, in population dynamics, as with dynamite, the conditions influencing temporal or spatial variation can be at least as important as the average.

PROJECTING POPULATION CHANGE: THE BASICS

The simplest dynamics a population can exhibit is a constant change in abundance through time. For just one time step, and building off equations (1) and (2), this geometric or exponential change would be:

$$N_{t+1} = N_t \lambda \quad \text{or equivalently} \quad N_{t+1} = N_t e^r. \quad \text{(3A or 3B)}$$

Next, extend this constant growth rate (λ or r) for T time steps (say, years) into the future, starting from the initial abundance at time (N_0):

$$N_T = N_0 \lambda^T \text{ or equivalently } N_T = N_0 e^{rT}. \quad \text{(4A or 4B)}$$

As an example, if a population of 20 black-footed ferrets (*Mustela nigripes*) experienced a constant change of $\lambda = 1.6$ or $r = 0.47$ (Grenier et al. 2007), how many ferrets would be expected in five years?

$$N_T = 20 \times 1.6^5 = 210 \text{ ferrets}$$

or equivalently

$$N_T = 20 \times e^{0.47 \times 5} = 210 \text{ ferrets}$$

This example illustrates the power exponential growth can have in leading to rapid increases in population size (Fig. 7.1A). At the same time, however, do not make the common mistake of confusing exponential growth with really big growth. For example, after five years, our initial 20 ferrets would only be

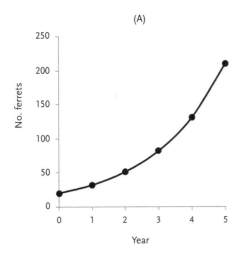

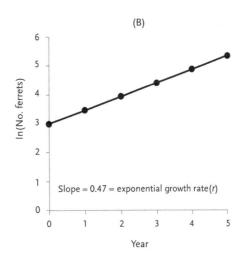

Figure 7.1. Exponential population growth of ferrets for five years with a starting size of 20 and constant $\lambda = 1.6$. (A) Abundance over time. (B) Logarithm of abundance over time, producing a straight line with the slope equal to the exponential growth rate.

21 if $\lambda = 1.006$, and only 12 ferrets if $\lambda = 0.9$ (try the projection yourself with equation [4A]).

Because population growth is a geometric or exponential process (where the growth rate is multiplied by, not added to, N; Fig. 7.1A), a straight line based on the abundances plotted as logarithms can replace a curved line (Fig. 7.1B). In this case, the slope of the line is equal to r (our old friend the exponential growth rate from equation [2]). To make clear how r is the slope of the log-transformed abundance over time, let's rewrite equation (4B):

$$\ln(N_T) = \ln(N_0) + rT.$$

This is an equation for a straight line of log abundance over time, with slope r.

Because of process variance, population growth is not constant over time for any wild population, and this has two strong implications for population dynamics. First, greater variation in λ over time makes it more likely that a population will stumble toward extinction, regardless of trend (Boyce 1977, Vucetich et al. 2000). Second, in addition to the direct effect of increasing extinction probability, process variance has the subtle, yet important, effect of decreasing the future expected population size below that expected from the arithmetic mean λ (Morris and Doak 2002, Mills 2013). This interesting phenomenon follows the same laws that cause financial planners to advise investors to choose less volatile stocks, even if they have a lower average net return, because the most likely growth of any multiplicative process over time (like money in an interest-bearing account, or numbers in a wildlife population) is better described by the geometric mean growth rate than by the arithmetic mean. This is because variability in the growth rate causes the probability of a population being any particular size in the future to be more skewed over time, with most populations likely to be relatively small (they cannot go below zero) but a few getting incredibly large. Those few large populations can inflate the average arithmetic mean population size, making it larger than the most likely future population size. Without variation, the most likely population size is the same as the average, but with variation, the most likely population size will be less than the average. The geometric mean reflects this reality, becoming smaller than the arithmetic mean when variation is present (Lande et al. 2003, Doak et al. 2005). The geometric mean λ, sometimes called the stochastic growth rate (λ_s), is calculated by multiplying λ values over T time steps and taking the T root of the product:

$$\lambda_s = \sqrt[T]{(\lambda_1 \times \lambda_2 \times \lambda_3 \times \ldots \lambda_T)}. \quad (5)$$

The stochastic growth rate based on the geometric mean contrasts with the more familiar arithmetic mean (λ_A):

$$\lambda_A = \frac{(\lambda_1 + \lambda_2 + \lambda_3 + \ldots \lambda_T)}{T}. \quad (6)$$

For an example of how λ_s becomes smaller than λ_A when λ is variable, consider a population changing across three time

intervals: in the first interval, $\lambda = 1.1$; in the second, $\lambda = 0.3$; and in the third, $\lambda = 1.9$. Across all time periods, the arithmetic mean population growth rate is $\lambda_A = 1.1$ (implying our population would *increase* by an average of 10%/yr), but the geometric mean stochastic growth rate is $\lambda_s = 0.86$ (correctly implying an expected average *decline* of 14%/yr). The implications of this phenomenon are profound because it means that ignoring variability may lead to an expectation of population increase, when in fact real-world variability could make the population most likely to decline. The effect of stochasticity on population growth rates gets complicated, but this basic and nonintuitive phenomenon is fundamental to understanding population dynamics and must be accounted for when projecting wildlife population trajectories for management and conservation purposes. Just remember, the most likely population trajectory in a variable environment (as experienced by any wildlife population) is governed by λ_s, which will nearly always be smaller than λ_A.

FACTORS THAT AFFECT POPULATION DYNAMICS

Many factors affect the vital rates of a population, and therefore its growth rate, over time. Some of these are considered stochastic, or random, while others are deterministic, having a predictable outcome. We first discuss the main sources of stochastic variation in populations (e.g., demographic, genetic, and environmental stochasticity) and then discuss several sources of deterministic variation in populations (e.g., habitat quality, predation, disease, interspecific interactions, anthropogenic factors, and density dependence). Although we will talk about them individually, realize that these all can affect population dynamics in simultaneous and interactive ways.

Stochastic Factors
When factors have random or unpredictable outcomes on population dynamics, they are considered stochastic.

Demographic Stochasticity
Demographic stochasticity represents random deviations from mean birth and death rates that arise from sampling whole animals that experience probabilistic demographic rates (e.g., an animal with a 70% probability of survival cannot 70% survive; it either lives or dies). For small populations, demographic stochasticity causes variation in population growth even in a constant environment, with no change in mean birth or death rates. One of the easiest ways to understand demographic stochasticity and its special effects on small populations is by example. If you toss a fair coin, the expected probability of heads is a constant 50%. But if you toss the coin only 3 times, you cannot possibly get 50% heads: you can only get 0, 33%, 67%, or 100% heads, by chance. Even if you tossed it 4 times, you would not be too surprised to get something other than 50:50. If you tossed the coin 100 times, however, you would expect the percentage of heads to be much closer

to 50%, and if you tossed it 1,000 times, you would be highly confident of converging on 50%, the expected probability of heads. Likewise, you can see how, for example, demographic stochasticity could cause the sex ratio in a small population to bounce randomly around from year to year, even if the mean or expected sex ratio remains a constant 50:50. For a biological example of demographic stochasticity, in 2009 there were only five mountain caribou (*Rangifer tarandus*) left in Banff National Park. All five died in a single avalanche, extirpating the herd (Hebblewhite et al. 2010), an outcome that would be highly unlikely in a larger population.

Genetic Stochasticity and Inbreeding Depression

In small populations, the frequencies of alternative forms of genes (alleles) change randomly, causing one allele or the other to be lost, by chance. Over time, this stochastic process, also called genetic drift, can overwhelm natural selection and lead to a loss of genetic variation (heterozygosity and allelic diversity). Genetic drift is a form of inbreeding that arises even when mating is random. In many (but not all) instances, loss of heterozygosity due to inbreeding can lead to inbreeding depression by causing reduced survival, fecundity, and population growth rates (Allendorf et al. 2013, Mills 2013). For example, the Arctic fox (*Vulpes lagopus*) population in Sweden was severely reduced in the early 20th century due to hunting associated with the fur trade. Despite protections, their numbers remained low in part because of inbreeding depression that had reduced survival and reproductive rates, and inhibited population recovery (Norén et al. 2016). In such cases, inbreeding depression can be a powerful influence of change in the dynamics of small populations, potentially requiring management intervention.

Environmental Stochasticity

While the effects of demographic and genetic stochasticity are limited to relatively small populations, environmental stochasticity can cause variation in the vital rates of a population of any size. Environmental stochasticity refers to unpredictable changes that alter mean vital rates from year to year. Environmental stochasticity is often caused by variation in weather (e.g., wet years versus dry, heavy snow versus light snow), either through its direct effects on animals or by influencing food resources, but it can also be caused by unexpected events such as disease outbreaks or changes in predator abundance. For example, late-spring frosts and drought can cause annual mast (e.g., berries, nuts) shortages for black bears (*Ursus americanus*) in Colorado. During these years, bears increase their use of human development as they forage on anthropogenic foods (i.e., garbage, bird seed, fruit trees). This behavior has been associated with reduced survival and population declines, as bears within developed landscapes are more susceptible to vehicle collisions, lethal removal, and other mortality factors (Laufenberg et al. 2018, Johnson et al. 2020). For all kinds of species, environmental variability is being exacerbated by climate change in ways that strongly influence the dynamics of populations (see Chapter 17).

Deterministic Factors

When factors are well enough understood to have relatively predictable outcomes on population dynamics, they are considered deterministic.

Habitat Quality

All wildlife species need certain food items, cover types, climatic conditions, and other characteristics (e.g., sites for nesting, hibernating, lekking) to survive and reproduce successfully. As a result, habitat quality is highly species-specific, varying across the landscape for each species in accordance with the availability of different resources. This inherent variation in habitat quality can account for differences in the vital rates of populations. For example, Ozgul et al. (2006) reported that juvenile survival in yellow-bellied marmots (*Marmota flaviventris*) was highly variable based on habitat conditions, with survival decreasing at higher elevations and on northeast-facing slopes. Such differences in habitat had large effects on the growth rates of different populations, as λ ranged from 0.96 to 1.09 depending on the conditions of the site. Human-induced changes to habitat, such as loss, degradation, or fragmentation, can also have major effects on wildlife populations. These changes pose some of the greatest threats to wildlife (Wilcove et al. 1998, Crooks et al. 2017) and have been associated with declining population trends (Wittmer et al. 2007, Harper et al. 2008, Rushing et al. 2016).

Predation

Predators can affect prey population dynamics both by directly killing prey and through indirect behavioral effects. The direct population-level effects of predators killing prey depend on three key factors: the rate of predation, which individuals are killed, and whether predation is additive or compensatory (Mills 2013). The first piece, the predation rate, is simply expressed as:

$$\frac{\text{number of prey killed}}{\text{prey abundance}} \times 100, \qquad (7)$$

where the number of prey killed is a function of both the number of predators present at a particular prey density (i.e., the numerical response) and the number of prey killed per predator (i.e., the functional response, or kill rate; Holling 1959).

The second piece that determines how predation may affect prey numbers reflects the fact that individual prey differ in their value to the growth rate of a population and in their susceptibility to predation. For example, a reproductive adult female Kemp's Ridley sea turtle (*Lepidochelys kempii*) that can lay more than 100 eggs/yr is worth much more to the growth rate of a population than a newborn hatchling with a low survival rate (Heppell et al. 1996; see "Accounting for Effects of Ages and Stages on Population Dynamics," below).

The third factor that determines the direct effect of predation on a prey population is whether those animals that are killed were likely to die anyway, a phenomenon known as compensation. Compensatory mortality arises when factors such as disease or the availability of territories, cover, or food inherently limit the number of animals that could survive in a population, so that deaths due to predation have little effect on the prey survival rate (Errington 1956). At the other extreme, if predation acts independent of other forms of mortality, then it is considered "additive" and leads to depressed survival rates.

In addition to the direct effects of predators on prey population dynamics, predators can exert indirect effects on prey behavior and distributions. For example, prey might avoid high-quality habitat, increase their vigilance, or alter their foraging patterns to avoid predators (Fortin et al. 2005, Hamel and Côté 2007, Bourbeau-Lemineux et al. 2011). These behaviors have the potential to shift prey populations spatially, and even to decrease survival and reproductive rates (Preisser et al. 2005).

Disease

Disease can have a major influence on wildlife population dynamics, with outbreaks exacerbated by factors such as population density, environmental conditions, and pathogen exposure and transmission dynamics. As with predation, certain stage classes or vital rates may be more or less susceptible to a particular disease. For example, outbreaks of respiratory disease in bighorn sheep (*Ovis canadensis*) typically have the greatest effect on juvenile animals, potentially causing a reduction in juvenile survival of ~60%, while adults might only suffer a reduction of ~5% (Festa-Bianchet 1988, Singer et al. 2000, George et al. 2008). The duration and the magnitude of disease events can also be highly variable, as respiratory disease may chronically affect bighorn sheep populations at low levels (Cassirer and Sinclair 2007) or cause catastrophic die-offs (Martin et al. 1996). Diseases introduced from domesticated animals and invasive species have traditionally posed some of the greatest threats to wildlife populations (Pedersen et al. 2007, Smith et al. 2009), with the influence of a disease event being particularly pronounced when a population is exposed to a pathogen that the species did not evolve with. Animals can also transmit zoonotic (animal-borne) diseases to humans (Han et al. 2016, Allen et al. 2017), as recently experienced with the dramatic global spread of the novel coronavirus COVID-19 (Ali et al. 2020). While humans can also transmit pathogens to wildlife, there is relatively little research on the topic despite increasing reports of occurrence (Messenger et al. 2014).

Other Interspecific Interactions

Predation and disease represent interactions where one or more species benefit at the expense of others. Interspecific competition, arising from different species using a finite resource, can also cause negative interactions among one or more interacting species. The ways that species can compete vary as much as life itself and include consuming a shared resource, preempting space, and physically excluding other species by fights, perhaps to the death (e.g., wolves excluding coyotes [*Canis latrans*]; Berger and Gese 2007).

But interspecific interactions are not always negative. Other types of species interactions that can affect population dynamics include mutualisms, where both species benefit each other, and facilitation or commensalisms, where one species benefits and the other is unaffected.

A classic example of a mutualism occurs between white-bark pine (*Pinus albicaulis*) and Clark's nutcracker (*Nucifraga columbiana*; Hutchins and Lanner 1982, Lorenz et al. 2011). The pine has evolved mechanisms that allow only Clark's nutcrackers to disperse the seeds successfully away from the tree, and the nutcracker has evolved spectacular adaptations to harvest pine seeds. The nutcrackers carry up to 50 seeds in a cheek pouch especially suited for the purpose; in a single year, one bird will cache up to 98,000 seeds at distances up to 22 km from the tree of origin! The birds benefit by eating the seeds, although more than 66% of the seeds are never eaten and so remain available for germination, thus benefiting the pines.

An example of facilitation or commensalism can be found in the life cycle of hermit crabs (Paguroidea family). Most hermit crab species are dependent upon finding and using snail shells to protect themselves from predators. While the crabs do not influence snail populations, the abundance and distribution of different types of snails (and thus their shells) can dramatically influence the dynamics of local hermit crabs, affecting their reproduction, growth, and survival (Williams and McDermott 2004).

Anthropogenic Effects

Of the deterministic factors that influence wildlife populations, human effects have the greatest impact. Human population growth, development, illegal poaching, and per capita consumption (see box on page 116) result in habitat loss, degradation and fragmentation, pollution, the introduction of non-native species, overexploitation, and, ultimately, a change in global climate and biogeochemical cycles. Even species that have historically experienced minimal direct effects from human development are being affected through human-induced climate change, arguably the greatest threat to global biodiversity (Running and Mills 2009, Bellard et al. 2012, Field et al. 2014; see Chapter 18). For example, polar bears (*Ursus maritimus*) in the Arctic are experiencing a reduction in the available sea ice, a key habitat component for successful foraging, caused by warming temperatures. As a result, researchers suspect that the continual loss of sea ice will be associated with reduced polar bear body condition, reproductive success, and survival, ultimately leading to population declines (Stirling and Derocher 2012). The effects of climate change are expected not only to have singular effects on wildlife populations, but also to interact synergistically with many of the other factors that influence populations, like disease, overexploitation, and habitat loss (Brook et al. 2008, Smith et al. 2009, Northrup et al. 2019).

Unlike the unintentional consequences of anthropogenic factors on wildlife species, people also intentionally influence wildlife populations through management, often working to decrease or reverse the negative effects of humans. Managers may work to increase, decrease, or stabilize populations using a variety of approaches, including habitat enhancement, public harvest, translocations, captive breeding, contraception, and the enforcement of various regulatory mechanisms. For some species termed "conservation reliant" (Scott et al. 2005, 2020; Horne et al. 2020), continuing active management is necessary for persistence. As we stress throughout the chapter, management actions will be most effective when applied with a sound understanding of their influence on the dynamics of a population. Ultimately, however, whether anthropogenic effects on wildlife populations are unintentional (e.g., increase in road density associated with urban development) or intentional (e.g., change in harvest strategy), it is critical to recognize that the human footprint is pervasive across the globe (Venter et al. 2016, Waters et al. 2016, Ceballos et al. 2020). Managers need to acknowledge and incorporate human factors into wildlife management strategies to be successful at meeting population objectives.

Density Dependence: Positive and Negative

Density dependence occurs when a population's density or abundance affects the vital rates (e.g., reproduction, survival) of individuals in the population, which in turn can affect the population growth rate (Fig. 7.2). Density dependence can be positive (vital rates increase with density), negative (vital rates decrease with density), or both (Fig. 7.2).

Positive density dependence, where vital rates (and population growth) are positively related to density, is often referred to as an Allee effect (Allee 1931). Although sometimes referred to as "inverse density dependence" or "depensation," we prefer positive density dependence because it reflects the fact that density and vital rates are positively associated; when

density goes up or down, vital rates follow. Positive density dependence is often caused by mechanisms such as cooperative defense, foraging efficiency, or group rearing of offspring (Table 7.1; Kramer et al. 2009, Gregory et al. 2010) because an increase in the number of animals that can participate in these activities leads to an increase in individual survival and reproduction that ultimately boosts population growth. Positive density dependence can occur in populations of any size. For example, in house finches (*Carpodacus mexicanus*), positive density dependence has been reported to lead to rapid popu-

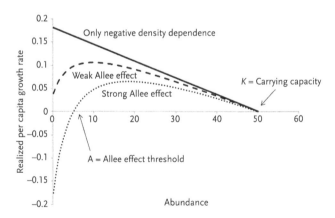

Figure 7.2. The influence of density dependence on a population with an exponential growth rate of *r* = 0.18 and carrying capacity of 50. Changes in realized per capita population growth rate [ln(*Nt* + 1 / *Nt*)] are plotted. The solid line shows pure logistic (negative density–dependent) dynamics. The dashed and dotted lines show positive and negative density dependence operating simultaneously. Although both of these exhibit decreases in per capita growth rate at small abundances expected under positive density dependence (Allee effects), only the dotted line shows strong Allee effects that create a threshold abundance below which the population would actually decline toward extinction. From Mills (2013)

Table 7.1. Mechanisms that can lead to positive density dependence (Allee effects) in wild populations

Mechanism	Example
Minimizing Predation	
Increased predator detection and defense	Survival rates in dwarf mongoose (*Helogale parvula*) increase with group size because guard mongoose decrease predator attacks (Rasa 1989).
Greater confusion for predator	Largemouth bass (*Micropterus salmoides*) take longer to capture silvery minnows as minnow school size increases (Landeau and Terborgh 1986).
Predator satiation	Black brant (*Branta bernicla nigricans*) per capita nest mortality declines as colony size increases and predators become satiated (Raveling 1989).
Foraging Advantages	
Access to foods	Small colonies of blind mole rats (*Cryptomys damarensis*) are more likely to fail because they are unable to rapidly extend burrow systems to obtain food during the brief period when the soil is moist and easily worked (Jarvis et al. 1998).
Increased resource detection	After being reduced in numbers by hunters and habitat alteration, a cause of further decline in passenger pigeons (*Ectopistes migratorius*) may have been that small flocks were compromised in their ability to find their patchy and sporadic food sources (e.g., acorns and nuts; Reed 1999).
Cooperative resource defense	Larger groups of coyotes (*Canis latrans*) are better able to defend carcasses against intruder coyotes (Bekoff and Wells 1986).
Mating and Caring for Young	
Finding mates	Glanville fritillary butterflies (*Melitaea cinxia*) are less likely to locate mates and successfully reproduce in small populations (Kuussaari et al. 1998).
Caring for young	The number of young fledged per nest increases with group size in white-fronted bee eaters (*Merops bullockoides*) because helpers provision food and assist in nest excavation, nest defense, and egg incubation (Emlen 1990).
Conditioning of Environment	
Temperature tolerance	Bobwhite quail (*Colinus virginianus*) in large coveys standing in a circle with their tails toward the center of the circle are better able to survive extreme low temperatures (Allee et al. 1949).

Source: Modified from Mills (2013).

lation growth and faster spread of large populations (Veit and Lewis 1996), showing how density dependence can exacerbate rapid increases of invasive species.

Although positive density dependence can occur at any population size, some of the most important instances for wildlife management occur at very small numbers (Fig. 7.2). In these cases, vital rates decrease as densities decrease (still a positive relationship between density and vital rates), which can destabilize a population, sending it spiraling toward extinction. This pattern has been observed in several wildlife species (Kramer et al. 2009), including caribou, where smaller populations have declined faster than larger populations (Wittmer et al. 2005). Critical management actions in these instances would focus on increasing abundance, potentially through translocations or captive breeding, to increase vital rates and reverse the extinction threat.

Negative density dependence arises when increases in density lead to decreases in a population's vital rates and growth rate, and vice versa (Fig. 7.2). Factors that elicit negative den-

sity dependence include intraspecific competition (i.e., competition among individuals within a species for resources or mates) and heightened susceptibility to predation, parasites, and disease. Because resources are finite, any population, including humans, must eventually exhibit negative density dependence (see box on page 117). Intraspecific competition sometimes occurs as direct interference or contests among individuals, as in fights for food, mates, or territories; winners obtain sufficient resources to survive and reproduce, while losers may not. In other instances, the competitive interaction is exploitative, or based on a scramble for resources, with simultaneous use of a common resource (often food) lowering the amount of that resource available to each individual.

Whether influenced by intraspecific competition or other processes, negative density dependence tends to regulate abundance around an equilibrium size range, called the carrying capacity, or K (Fig. 7.3). Theoretically, the carrying capacity is the population size at which reproduction and mortality are equal. That said, it is best to think of the carrying capacity as a

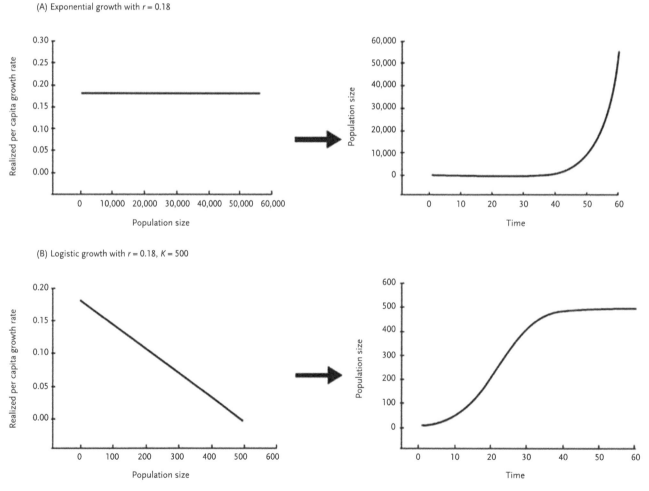

Figure 7.3. The contrast between exponential and logistic growth. (*A*) Exponential growth. The realized per capita population growth rate measured from a time series ($\ln[Nt + 1 \, / \, Nt]$) is equal to the intrinsic growth rate (r), no matter the population size. This lack of density dependence leads to exponential growth of the population over time. (*B*) Logistic growth. With a linear decline in the realized per capita population growth rate as the population size increases, the population increases exponentially at first, then slows its growth as it approaches carrying capacity. From Mills (2013)

range of values dependent on conditions, not as a single number. The K for any population is determined both by factors that become more limiting to population growth as density increases (i.e., food, nesting sites, territories) and by factors independent of density (i.e., weather). For example, negative density dependence in the endangered San Joaquin kit fox (*Vulpes macrotis mutica*) arises both from a restricted number of available territories and from rainfall, which influences the abundance of small mammals that foxes eat (White and Garrott 1999, Dennis and Otten 2000).

The simplest way to represent negative density dependence in a predictive model of population dynamics is to assume a linear decline in the per capita growth rate as density increases (see Fig. 7.2). In this case, the per capita growth rate is positive and exponential at very small densities, zero at the carrying capacity, and negative at densities above K (Fig. 7.3). This specific form of negative density dependence leads to an S-shaped logistic curve of abundance plotted against time, with rapid growth at small population sizes (r_0, the exponential growth at small abundance) and dampened growth as the population size approaches K (Fig. 7.3B). Skipping the math that derives and details this relationship, here is a discrete time form of the logistic equation, called the Ricker equation, which projects the population one time step forward:

$$N_{t+1} = N_t e^{r_0[1-(N_t/K)]}. \qquad (8)$$

Compare equation (8) to equation (3B), and notice that the only difference is that the exponential r now becomes penalized in a simple linear way as density increases. When density is zero, there is no penalty ($1 - 0 / K = 1$), so r_0 is unaffected by density, and growth is exponential. When abundance reaches the carrying capacity, however, the growth term becomes zero ($[1 - N/K] = [1 - 1] = 0$, so $r \times 0 = 0$). In short, a linear decline in per capita growth rate occurs as density goes from zero to K (Fig. 7.3B).

Notice also in equation (8) that a one-year time lag exists between the operation of density dependence and abundance the following year. Biologically, it makes sense that birth and death rates would respond to past population density. Returning to the San Joaquin kit foxes, rainfall affects small-mammal abundance with a one-year time lag, which in turn affects kit fox abundance with an additional one-year lag (i.e., rainfall affects foxes with a two-year lag; Dennis and Otten 2000). It turns out that a population following logistic growth with a time lag can exhibit some crazy dynamics that include cycles and even unpredictable chaotic dynamics (Mills 2013). Wildlife cycles are interesting population dynamics phenomena in their own right, and the inherent dynamics from negative density dependence are one factor contributing to population cycles (see box on page 117).

Realize that the logistic curve is just one of a nearly infinite number of ways that density dependence might manifest in wildlife populations (Mills 2013). The logistic model of population growth assumes that only negative density dependence happens, and that its effects become gradually worse from 0

to K in a linear way. No law dictates this to be so; in fact, most species probably exhibit positive density dependence at small numbers, and some form of negative density dependence at high numbers (see Fig. 7.2; Sibly et al. 2005, Peacock and Garshelis 2006, Cayuela et al. 2019).

Although the logistic curve should not be overinterpreted as a law of population growth, so long as the assumptions are recognized, the logistic equation can be useful for generating general predictions of how negative density dependence may affect populations and management actions. One such prediction of direct relevance to management is that a population following logistic growth will recruit the most individuals into the population when abundance is at intermediate values of K (Fig. 7.4). Why? Because at small densities, few young can be added to the population because there are fewer females present. At high densities, near K, the per capita growth rate is severely penalized by negative density dependence so that, again, few young are added to the population. But at intermediate densities, recruitment is maximized because density effects are not yet severe, and the number of females is reasonably high (Fig. 7.4). This phenomenon was used for many years in harvest management, especially in fisheries, to determine the maximum sustainable yield. The idea was that if the population was harvested down to 50% K, then the population would add the most new recruits, which meant that humans could harvest the maximum number of animals. The general idea is useful for informing us that under pure negative density dependence, the population will provide the most recruits at some intermediate density. But an appropriate level of humility about the assumptions of the logistic equation (and uncertainties in estimating densities and K) should caution us to not treat 50% K as any sort of magic number of maximum recruitment. To do so is to risk overharvest toward extinction, an all-too-common occurrence especially in ocean fisheries. For these reasons, the concept of sustainable yield has empha-

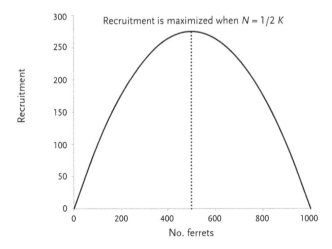

Figure 7.4. Under ideal logistic growth with no variability, recruitment of new individuals into the population is maximized at 50% K. Here the logistic curve, with $r = 1.1$, is used to show recruitment relative to population size.

PAUL EHRLICH (1932–) AND HUMAN POPULATION DYNAMICS

By 1800, aided by the industrial revolution, the development of global agriculture, and improved nutrition and hygiene, the human population numbered 1 billion and began a period of strong population growth. By the mid-1960s, human numbers increased to 3.5 billion and reached a peak annual growth rate of just over 2%, a rate that would double the population in about 34 years.

In 1968 Dr. Paul Ehrlich published *The Population Bomb*. Although only one of more than forty books written over an astonishingly productive career—including works on coevolution, population dynamics, genetics of checkerspot butterflies (*Euphydryas editha bayensis*), natural history of birds and insects, human ecology, and evolutionary biology—*The Population Bomb* has been Ehrlich's highest-impact publication. It introduced the public to the fundamental ecological premise that no species, including humans, can sustain exponential growth indefinitely, and that exceeding the earth's carrying capacity will negatively affect humans and other organisms. The book sold more than two million copies and has been the source of political controversy from all sides. As Ehrlich and Ehrlich (2009:64) state, "none of those constituencies seemed to understand that the fundamental issue was whether an overpopulated society, capitalist or socialist, sexually repressed or soaked, egalitarian or racist/sexist, religious or atheist, could avoid collapse . . . Perhaps the biggest barrier to acceptance of the central arguments of *The Bomb* was—and still is—an unwillingness of the vast majority of people to do simple math and take seriously the problems of exponential growth."

Ehrlich also popularized another critical point fundamental to population dynamics: the overall effect of any population is a function of both population size and per capita consumption (Ehrlich 1968, Ehrlich and Holdren 1971). Eventually, this concept—that effect depends on both population size and resource use per person—became the famous IPAT equation, where *impact* = *population* × *affluence* × *technology*.

What about now? Ehrlich was right that the rapid exponential growth of the mid-1960s could not and would not be sustained. Our growth rate has fallen steadily since that peak, to about 1%, with our numbers approaching 8 billion in late 2021. Collectively, the ever-increasing product of IPAT continues to manifest as accelerated human-caused extinction rates of species. We do not know what our carrying capacity will be or how it will be realized, but there are predictions that our numbers will stabilize around 10–12 billion by the end of the 21st century (United Nations 2019). Clearly, the subjects Ehrlich raised—human numbers and resource use confronting a finite global carrying capacity—are still relevant and pressing.

Courtesy of Paul Ehrlich

---

sized the need to avoid harvesting at the theoretical maximum sustainable yield of 50% *K* and instead consider multiple scenarios of density dependence and other complexities (Mills 2013; Chapter 5).

ACCOUNTING FOR EFFECTS OF AGES AND STAGES ON POPULATION DYNAMICS

So far, we have seen how various outside factors can act on the mean and variance of population growth rates. But for three main reasons, wildlife managers are often as interested in the dynamics of specific age or stage classes as they are in the dynamics of the population as a whole. First, different stages may have different values to the public. Consider big-game hunting regulations, where populations are managed based on the harvest of certain sexes or stages (e.g., branched-antlered, spike, or non-antlered animals). Second, stressors and management actions affect different stages of animals in unique ways. In woodpeckers, for example, drilling nesting cavities in trees might increase fecundity, while removing invasive pred-

ators might improve adult survival. This leads to the third, and most important, reason why wildlifers must consider dynamics of different ages or stages: all stage classes and stage-specific vital rates are not equal in their effects on population growth, which means that management actions that target different stage classes will not be equal in their effectiveness.

A great story to convey these points is based on the management of threatened loggerhead sea turtles (*Caretta caretta*) in the Atlantic. Prompted by long-term declines in population numbers, management efforts focused on reducing the stressors that seemed intuitively to be causing the decline: mortality of the eggs and newborn hatchlings that were killed by predators, crushed by vehicles, and disoriented by lights as they tried to make their way from the nest to the ocean. In one of the best examples of population models informing management, an analysis by Crouse et al. (1987) demonstrated that even large increases in egg or hatchling survival would do little to reverse the population decline because hatchling survival had such a minimal effect on the overall population growth rate. A small increase in the survival of young adults, however, was

CYCLES IN WILDLIFE POPULATIONS

Many species show dramatic population fluctuations over time. For many species—including forest insects, lemmings, voles, hares, and grouse—the fluctuations are cyclic, with a repeatable periodicity (often 3–4 or 8–10 years). In North America, the most prominent cycles involve snowshoe hares (*Lepus americanus*) and their specialized predators, Canada lynx (*Lynx canadensis*; Elton 1924, Krebs et al. 1995). In the northern part of their range, including Alaska and Canada, snowshoe hares undergo cyclic density changes of 15- to 100-fold over a 10-year period (Keith 1990, Royama 1992). The hare cycles are caused by complex interactions from the top down (primarily lynx, but also other predators and disease), from the bottom up (vegetation defense compounds and dynamics), and from negative density dependence or self-regulation (Krebs et al. 2001, Post et al. 2002; Chapter 14). For many cyclic species including snowshoe hares, cycles observed in northern populations are dampened or eliminated in the southern parts of the range (Kumar 2020). The mechanisms dampening snowshoe hare cycles are described in more detail in "Sources and Sinks," below.

The population matrix is conventionally denoted by M (Fig. 7.5), a square of k columns and k rows, where k is the total number of stage classes. Each element (or cell) of the matrix contains vital rate values that are used to project the individuals in each stage class forward one time step (for more details on the structure and projection of wildlife population matrices, see Mills 2013). Fig. 7.5 provides a simple example of a matrix and population projection for bighorn sheep. Notice that the matrix has three stage classes based on physical characteristics that can be identified by managers in the field: lambs, yearlings, and adults.

If conditions are relatively constant through time, the matrix provides several useful population descriptors known as asymptotic properties. First, a population defined by a given matrix will converge on a stable stage distribution (SSD, analogous to a stable age distribution) that describes the expected constant proportion of animals in each stage class. Given the matrix of bighorn sheep vital rates (Fig. 7.5), the SSD would maintain 18% of the population as lambs, 12% as yearlings, and 70% as adults. Second, when the population reaches SSD, it will grow at a constant growth rate that is characteristic for that set of vital rates; this asymptotic growth rate at SSD is symbolized by λ_{SSD}. For the vital rates in the bighorn sheep

From this stage...

$$
\begin{array}{c}
\text{Lambs} \quad \text{Yearlings} \quad \text{Adults}
\end{array}
$$

$$
\begin{array}{c}
\text{Lambs} \\
\text{Yearlings} \\
\text{Adults}
\end{array}
\begin{bmatrix}
0 & 0 & 0.27 \\
0.74 & 0 & 0 \\
0 & 0.92 & 0.92
\end{bmatrix}
\times
\begin{bmatrix}
N_{L(t)} \\
N_{Y(t)} \\
N_{A(t)}
\end{bmatrix}
=
\begin{bmatrix}
N_{L(t+1)} \\
N_{Y(t+1)} \\
N_{A(t+1)}
\end{bmatrix}
$$

To this stage...

Figure 7.5. Anatomy of a female-based projection matrix, using bighorn sheep as the species of interest (see case study "Recovering Endangered Sierra Nevada Bighorn Sheep"). This species has three stages that can be identified by wildlife managers in the field and that were used to develop the matrix: lambs (first year of life), yearlings (the second year of life), and adults (individuals over two years of age). The first row of the matrix represents reproduction from each stage class that contributes to the number of lambs in the population the following year. In this example, we assume that bighorn sheep do not consistently reproduce until they are adults, so $a1,1$ (the element in row 1, column 1) and $a1,2$ are both 0, as neither lambs nor yearlings contribute to reproduction. Each adult female, however, does contribute an average of 0.27 lambs to the population, as seen in $a1,3$. The diagonal ($a1,1 = 0$, $a2,2 = 0$, and $a3,3 = 0.92$) represents the proportion of individuals in a stage class that will survive and will still be in the same stage next year (all lambs and yearlings transition to the next stage, while adults can remain in that stage class for several years), while the sub-diagonal ($a2,1 = 0.74$ and $a3,2 = 0.92$) represents the proportion of the stage class that survives and advances to the next stage next year. The matrix is then multiplied by a vector of abundances of each stage class in year t ($N_{L(t)}$, $N_{Y(t)}$, $N_{A(t)}$) to obtain a vector of abundances of each stage class in year $t + 1$ ($N_{L(t+1)}$, $N_{Y(t+1)}$, $N_{A(t+1)}$).

found to lead to substantial increases in population growth. This scientific finding was translated into a management action to reduce young adult mortality via "turtle excluder devices" on shrimp nets to reduce lethal entanglement. Though loggerhead turtles still face many challenges, the identification of vital rates that most affect population growth led to more efficient management actions to facilitate recovery.

Because the effectiveness of different management actions depends upon their influence on the vital rates of specific age or stage classes and are not necessarily intuitive, we need to use population models that account for these effects. In the next section, we give an overview of population projection models, which are commonly used to help connect management actions to their outcomes on wildlife population dynamics (Mills 2013).

Fundamentals of Population Projection Matrices

Because vital rates differ among animals of different ages or stages, we need to keep track of them in a specific way. Historically, ecologists used life tables to keep track of age- or stage-specific survival and birth rates, but current approaches often use population projection models. A population matrix provides a convenient accounting system to track stage-specific vital rates and abundances, and allows us to project populations through time to determine how different vital rates and stages affect dynamics.

matrix (Fig. 7.5), $\lambda_{SSD} = 1.08$, indicating an expected 8% increase per year once the population achieves SSD. Finally, under asymptotic (constant) conditions, the matrix can reveal reproductive values for each stage (or age) class, which can be interpreted as the relative contribution individuals in each stage will make to future population growth, values that add to 1 across all stages. From this matrix, the reproductive value of the lamb stage class is 0.24, for yearlings it is 0.35, and for adults it is 0.41. In other words, these values indicate that each adult is nearly twice as valuable as each lamb as a seed for future population growth.

Of course, conditions often are not constant through time, and when vital rates do change due to the imposition of the "Factors Affecting Population Dynamics," described above, the non-asymptotic dynamics that occur can be profound. For example, the proportion of individuals in different stage classes can get knocked out of SSD, perhaps following a translocation, disease event, poaching, or predators killing certain stages disproportionately. The new distribution of individuals across stage classes can substantially affect population dynamics, a phenomenon known as "population inertia" (Ezard et al. 2010). The effects of population inertia can be readily seen in Figure 7.5. If we conducted a translocation with 10 bighorn sheep, how would the composition of those introduced sheep affect expected abundance 20 years later? Assuming a constant growth rate, if the initial 10 sheep were approximately at SSD (18% lambs, 12% yearlings, and 70% adults rounds to 2 lambs, 1 yearling, and 7 adults), the population would be expected to grow by the asymptotic growth rate, $\lambda_{SSD} = 1.08$, and after 20 years we would expect 21 bighorn sheep. By contrast, if the initial 10 translocated sheep were all lambs (the stage class with the lowest reproductive value), then after 20 years the expectation would be only 14 bighorn sheep in the population, while starting with all adult sheep (the stage class with the highest reproductive value), we would expect 23 bighorn sheep after 20 years. In all cases, with constant vital rates the population would converge on asymptotic SSD and λ_{SSD}, but the different reproductive values of translocated individuals cause different initial population changes. These insights can be used to maximize the effect of management to reach population objectives.

Although we have described stage-structured matrix models in their simplest form, they are quite flexible and can incorporate many other processes, such as density dependence (described above) and multiple population dynamics (discussed below). Also, for species with ample data, individual-based models (IBMs) of population dynamics can be used (e.g., van de Kerk et al. 2019).

Evaluating Vital Rate Importance and Management Actions through Sensitivity Analysis

In a broad sense, population sensitivity analysis refers to how changes in vital rates, due to either a natural change or through management, would change population growth or persistence (Mills 2013, Manlik et al. 2018). This is one of the most useful contributions of population ecology to wildlife management because it quantifies how particular management actions, which change certain vital rates, would be expected to affect population growth. As a result, it allows us to assess the relative effects of different management actions for meeting our population goals. We mention several approaches for conducting sensitivity analysis and return to examples at the end of the chapter.

Early approaches to sensitivity analysis focused on matrix-based analytical calculations of "elasticity" and "sensitivity" based on assumed asymptotic properties of the matrix and equal absolute or proportional changes in vital rates. Although these metrics were useful in showing that all vital rates are not equal in how they affect population dynamics, their assumptions were limiting in the real world. First, sensitivities and elasticities do not account for how much different vital rates can be changed through management. For example, for a woodpecker, it may be that fire management increases nest success by 8%, removal of invasive predators improves adult survival by 23%, and so on (see Hartway and Mills 2012 for examples with managing bird populations). Second, the reliance on asymptotic properties of the matrix is limiting because variability in vital rates often prevents populations from remaining at stable stage distribution. To account for these limitations, several alternative methods of conducting sensitivity analysis avoid analytical sensitivities and elasticities, directly quantifying how changes in vital rates affect population growth. For example, one could simply perturb vital rates in a matrix in a sensible way and ask how the future growth rate or abundance is affected. Drawing on the woodpecker example above, we could see the relative effects of the management options on the population growth rate, either increasing nest success by 8%, or increasing adult survival by 23%. This manual perturbation method is a flexible and powerful approach to sensitivity analysis, quantifying the outcome from any number of "what if" scenarios. Different vital rates can be projected to change by any amount either naturally (i.e., environmental stochasticity, disease) or under management. Additionally, you can start with any age structure, and no assumptions of SSD are necessary.

Another simulation-based sensitivity analysis method is called life-stage simulation analysis (LSA; Wisdom et al. 2000, Mills 2013, Manlik et al. 2018). LSA is in some ways a formalized version of the manual perturbation method. Many plausible matrices are simulated (perhaps 1,000 or more), embracing the full range of variation that is possible for that population by building each matrix with randomly chosen vital rates drawn from distributions with the means and variances based on field estimates. The population growth rate is then calculated for each matrix, so one can determine what percentage of replicates have positive versus negative population growth. Baseline scenarios can then be compared to those where realistic management alternatives are simulated given changes in different vital rates, and the outcomes of those scenarios can be evaluated. Additionally, λ values from each of the 1,000 matrices can be regressed against each vital rate for that particular matrix. The coefficient of determination (R^2)

between λ's and their associated vital rate values represents the proportion of the variation in the population growth rate explained by variation in each vital rate, with all other vital rates varying simultaneously as they would be expected to do in nature. When all main effects and interactions are included, the R^2 values sum to 1. Most applications of LSA have been based on asymptotic growth, but nothing in the method restricts it to that (see the case study at end of the chapter).

MULTIPLE POPULATION DYNAMICS

Again, recall the BIDE equation: births, immigration, deaths, and emigration. So far, we have focused on births and deaths within populations, but a fundamental truth in wildlife population dynamics is that animals move a lot, resulting in immigration and emigration. In fact, it could be said that the more we study dispersal, through improved tools ranging from GPS radio collars to sophisticated mark-recapture analyses to genetic sampling (Mills et al. 2003, Crooks and Sanjayan 2006, Lowe and Allendorf 2010), the more we see that long-distance movements among populations are surprisingly common. For example, juvenile Columbia spotted frogs (*Rana luteiventris*) in Montana commonly move among ponds each year, traveling up to 5 km and gaining 750 m in elevation (Funk et al. 2005); a young male wolverine (*Gulo gulo*) radio-collared in northern Wyoming traveled more than 500 km to Colorado, and then more than 800 km to North Dakota (Packila et al. 2017); a female burrowing owl (*Athene cunicularia*) traveled 1,860 km from Arizona to Saskatchewan (Holroyd et al. 2011); and a mountain lion (*Puma concolor*) moved almost 3,200 km from South Dakota to Connecticut (Drajem 2011). In this section we describe how these movements could affect wildlife population dynamics.

Metapopulations

Populations across a landscape can have distinct characteristics, just as individuals making up a population do. At one extreme, populations can be isolated from each other. This might occur in oceanic islands, or for invasive species colonizing a new area, or in habitat fragments or protected areas surrounded by an impenetrable, hostile landscape. If populations are entirely isolated and small, they may be vulnerable to loss of genetic variation and inbreeding depression, as mentioned above. Such populations are also highly susceptible to other catastrophic events, such as a severe weather episode, a disease outbreak, or the introduction and expansion of an exotic predator or competitor. Persistence of these populations may also be hindered by the lack of recolonization, as isolated populations can easily go extinct.

In contrast to isolates, multiple populations can exist with some level of connectivity, or dispersal and gene flow, among them. The term *metapopulation* was coined by Levins (1970) to refer to a population of populations, where the dynamics of multiple populations depend on vital rates within—and also among—populations (immigration and emigration). Connec-

tivity between populations can therefore influence local population dynamics and overall metapopulation persistence. The concept highlights the importance of connectivity as a dominant force in population ecology, and captures the nonintuitive fact that sometimes management on areas where a species does not even live (but travels through) may be as important to persistence as occupied habitat. For example, for small and scattered populations of US federally threatened Canada lynx, high gene flow into the US populations from Canada may be as important for recovering this species as management to improve the local habitat conditions of the US populations (Schwartz et al. 2002). Other key concepts in wildlife population ecology, such as sources and sinks and ecological traps, spin off of the metapopulation concept.

Sources and Sinks

A form of metapopulation dynamics of special interest to wildlife biologists is captured by the metaphor of sources and sinks (Lidicker 1975, Pulliam 1988). If some populations are strong contributors to the metapopulation (sources that contribute to positive overall population growth) while others are drains on the system (sinks whose dynamics can decrease overall population growth), then identifying these areas becomes a priority for managers. Sinks cannot be identified based only on abundance or density because abundance can vary in different seasons, and poor habitats may have high numbers of animals for a variety of reasons (e.g., adults may exclude juveniles from high-quality habitat areas; Van Horne 1983).

To identify from field data the sources contributing to metapopulation growth and the sinks that drain the metapopulation, we must account for the two ways that a population can contribute to the greater metapopulation (Runge et al. 2006, Mills 2013). First, a population can contribute to its own growth (call this self-recruitment of population x, or R^x) via births. Second, a local population x can contribute by providing successful emigrants to other subpopulations (E^x). Jumping over a fair bit of details, a source and sink can be defined by a metapopulation contribution metric that is analogous to λ. A population where $R^x + E^x > 1$ is a source, while a population where $R^x + E^x < 1$ is a sink.

Snowshoe hares (*Lepus americanus*) provide an example of how this concept can be extended to heterogeneous landscapes. The hares exhibit remarkable population cycles in northern latitudes but dampened cycles in the southern portion of their range (see box on page 120), a pattern attributed to the role of different vegetation patch types in driving source–sink dynamics. For northern populations the landscape primarily consists of relatively continuous boreal forest, while the landscape becomes increasingly patchy south of the US-Canadian border. In Montana, source–sink dynamics were quantified for hares in dense versus open forest stands (Griffin and Mills 2009). Dense stands operated as sources that had positive contributions to the metapopulation through self-recruitment and emigration, while open stands were sinks where immigrating hares were more susceptible to predation. The maintenance of the overall hare metapopulation across the landscape will

QUANTIFYING SOURCE–SINK DYNAMICS FOR HARVEST MANAGEMENT

In the western United States, many wildlife management agencies have promoted a source–sink metapopulation approach in the harvest management of mountain lions (Sweanor et al. 2000, Laundre and Clark 2003, Beausoleil et al. 2013). The approach assumes that areas with little to no harvest will serve as sources for the overall metapopulation, supplying emigrants to sink areas that are more heavily harvested. Hunting has been found to induce source–sink dynamics in other carnivore species (Boyd and Pletscher 1999, Loveridge et al. 2007, Adams et al. 2008), and in mountain lions it seemed likely that immigration might sustain populations that were heavily harvested (Robinson et al. 2008). To quantify the effects of harvest on connectivity among populations, and to quantify how source–sink dynamics might be created, researchers used vital rate data across multiple years from hundreds of radio-collared mountain lions in hunted and unhunted areas in Wyoming and Montana (Robinson and DeSimone 2011, Newby et al. 2013). Mountain lions were captured and treed with hounds, immobilized, fitted with telemetry collars, and monitored for survival and movements. Field crews then tracked the dispersal of subadults to quantify emigration rates, dispersal distance, and disperser success.

Hunter harvest was by far the leading cause of mountain lion mortality, affecting not only within-population dynamics but also the dynamics of the greater metapopulation (see the figure below). Hunting led to decreased emigration, dispersal distance, and disperser success, which in turn affected whether hunted populations acted as sources or sinks in the landscape (Newby et al. 2013). Before harvest closures (1997–2000), the heavily hunted Garnet population was strongly declining with low survival and little emigration, resulting in a population sink ($Rx + Ex$ = contribution metric < 1). After hunting closures were enacted (2000–2006), however, the population exhibited positive population growth and increased emigration, making it a source that positively contributed to overall metapopulation growth (see also Robinson and DeSimone 2011).

The influence of connectivity was also clear in the region around Yellowstone National Park studied during two different time periods. During 1987–1993 (phase I), the mountain lion population was a source, because the sum of its self-recruitment and emigration to other areas ($Rx + Ex$) exceeded 1. Interestingly, the population would decline if it did not also receive immigrants from other areas; for example, the self-recruitment (Rx) in phase I was 0.98, and the positive population growth ($\lambda = 1.11$) only occurred because the population was subsidized by a high average per capita annual immigration rate. Thus connectivity both supported the Yellowstone mountain lion population and allowed it to support other surrounding populations (Newby et al. 2013), underscoring the wisdom of conserving multiple, mutually supportive source areas across a landscape.

Overall, these results show how harvest can affect not only local population dynamics but also dispersal and dynamics across populations. To manage wildlife populations, we need to understand and account for these between-population dynamics (immigration and emigration).

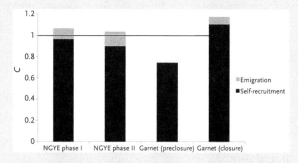

The estimated contribution metric (C) based on self-recruitment into a population and on successful emigration (dispersal) from that population to others. The solid horizontal line indicates $C = 1$; a population with $C > 1$ is a source, and one with $C < 1$ is a sink. The contribution is estimated from field data on mountain lions in the Northern Greater Yellowstone Ecosystem (NGYE) in 1987–1993 (phase I) and 1998–2005 (phase II), and in the Garnet, Montana, region before and after closure of a 915-km^2 area to hunting. From Newby et al. (2013)

require an appropriate threshold of high-producing source stands (Wirsing et al. 2002, Griffin and Mills 2003).

A key point to recognize when identifying population sources and sinks is that one cannot use λ alone to make the determination because it does not account for how much of a population's growth rate comes from immigrants, or how much a local population contributes to the metapopulation via emigration (see box on page 120). For example, headwater salamanders (*Gyrinophilus porphyriticus*) in upper reaches of first-order streams in New Hampshire have positive population growth rates only because downstream salamanders have high reproduction and preferentially move upstream (Lowe 2003). To truly discriminate between a population source and sink, we must account for all four of the components of population growth—births, immigration, deaths, and emigration.

Ecological Traps

Sometimes a sink habitat is preferred over better habitats because formerly reliable cues for habitat selection become mismatched with current fitness consequences in a modified landscape. This phenomenon creates a metapopulation dynamic known as an ecological trap, often associated with human-modified habitats (Robertson and Hutto 2006, Hale and Swearer 2016). The drain created when "good animals love bad habitats" (Battin 2004) can overwhelm even strong sources and cause entire metapopulations to decline toward extinction. As a result, the identification of ecological traps can lead to critical management actions for recovering species of concern (Mills 2013). For example, if a grassland bird preferentially nests in agricultural fields with abundant fences or poles as perches (attractive habitat cues) but mowing of the fields destroys nests, management solutions might include adjusting mowing schedules to allow chicks to fledge or eliminating perches and scaring birds to reduce attraction to the trap (Battin 2004). Indigo buntings (*Passerina cyanea*) provide another example of a species that prefers to nest along man-made forest edges. These edges are often ecological traps because they support a large number of subsidized and introduced predators (Weldon and Haddad 2005). The attractiveness of such a trap could be reduced by making clear-cut edges straight, minimizing the edge-to-area ratio, and thus reducing the area affected by the traps.

How Much Connectivity Is Needed among Multiple Populations?

Connectivity among populations, whether through a landscape porous to movement or via particular movement corridors, plays multiple critical roles in population dynamics across a species' range (Haddad et al. 2011, Keeley et al. 2019). These roles include recolonizing extinct patches, colonizing new patches as the environment changes, minimizing inbreeding depression, and maintaining small populations via demographic rescue. Conversely, issues that could arise from too much connectivity (typically less of a problem than too little connectivity) may include swamping of local adaptation and transmission of disease.

A common question in wildlife management is how connected populations need to be (Mills et al. 2003). One answer for identifying appropriate levels of connectivity might be to mimic predisturbance, or historical levels. Historical levels of connectivity might be inferred by estimating movement rates using radiotelemetry or genetic tools (Lowe and Allendorf 2010) across an undisturbed or unfragmented natural area. An example of this approach is the use of genetic connectivity indices in areas with and without major highway development to infer how highway corridors may fracture grizzly bear (*Ursus arctos*) movements (Proctor et al. 2012). Another way to infer historical connectivity takes advantage of the ability of genetic sampling to compare current levels of connectivity to those in the past based on genetic signatures from museum or other historic specimens. For example, a break in gene flow due to habitat fragmentation and loss was quantified for greater prairie chickens (*Tympanuchus cupido pinnatus*) and Attwater's prairie chickens (*T. c. attwateri*) by comparing samples from 12 contemporary populations to samples collected between 1936 and 1970 (Bouzat et al. 2009). While these approaches have been useful, one must recognize that intact habitat areas are increasingly hard to find for many species and that natural fragmentation may still be present because of fire, topography, or other factors.

In cases where there is no knowledge of background or historical levels of connectivity, a genetic rule of thumb does exist. One migrant per generation, where a migrant in the genetic sense is a breeding disperser, allows local adaptation to proceed in populations while minimizing the negative effects of inbreeding depression (Mills and Allendorf 1996). One migrant per generation may translate to 10 or more actual dispersers per generation, depending on the survival and reproductive success of individual animals. While this guideline has been useful, the ecological and demographic needs of different populations will often mandate higher or lower levels of connectivity (Vucetich and Waite 2000).

MONITORING POPULATION DYNAMICS

We have seen some of the complications in understanding wildlife population dynamics and the considerable body of science that can be harnessed to deepen our insights. The full power of wildlife population ecology can be realized when monitoring wildlife population dynamics over time. Wildlife monitoring can increase our understanding of population processes, allow us to assess the factors influencing those processes, and determine how populations will change in response to management actions. Some of the management-based motivations for wildlife monitoring include evaluating effects of a stressor on one or more species, developing regulations for the sustainable harvest of a game species, assessing the recovery of a threatened species, evaluating the outcome of a reintroduction program, determining the status of an invasive or pest species, or quantifying changes in biodiversity over time. Often, monitoring is a government mandate or policy. For example, many governmental agencies must conduct monitoring as part of the legal requirements outlined by legislation, such as the US Endangered Species Act or the National Environmental Policy Act (Schultz 2010).

If approached casually, wildlife monitoring can suffer from poorly defined objectives and deficient statistical design and analysis, which in turn lead to weak inferences that are impossible to defend in court or in the eyes of the public. Happily, monitoring has been the subject of intense attention in mainstream wildlife population ecology, resulting in a rigorous scientific basis for its application. Here we describe some keys to a successful and efficient program to monitor wildlife population dynamics and the influence of management actions. In the process, we draw on some of the fundamental concepts of wildlife population dynamics already discussed in this chapter.

To place wildlife monitoring on the rigorous scientific footing it requires and deserves, there must be thorough consideration of why the monitoring is occurring, what will be monitored, and how it will be done. To help develop these goals, we consider a couple of different forms of monitoring and describe some commonly used field and analytical techniques.

Why Monitor? Targeted versus Surveillance Monitoring

The first step in monitoring a wildlife population is to define the purpose for a project. By identifying a priori (before the start) hypotheses related to the mechanisms influencing population dynamics and the effects of management on a species of interest, and including (when possible) natural or induced manipulations, targeted monitoring programs can efficiently lead to a deep understanding of biological and management questions (Nichols and Williams 2006). A classic example of this approach can be found in the monitoring of midcontinent mallard ducks (Anas platyrhynchos), a species for which managers want to maximize hunting opportunities while maintaining a sustainable population. To meet these objectives, numerous federal, state, and provincial agencies in the United States and Canada annually survey the duck population and use that survey data and harvest information to evaluate a suite of hypotheses that describe how harvest influences mallards (i.e., whether harvests have additive or compensatory effects on mallard survival). By applying monitoring data to a specific set of ecological and management-related questions, biologists have gained valuable information about the dynamics of the system and about sustainable harvest regulations (Nichols et al. 1995, Nichols and Williams 2006).

In many cases, however, biologists use surveillance monitoring of wildlife, where there are no a priori hypotheses about population dynamics or management actions. This is often used when little is known about the species, the relevant spatial scale is unknown, management effects are particularly complex, or the goals of monitoring are poorly defined. In many cases, numerous attributes of a system are measured over time, usually without manipulations directed at testing specific mechanistic hypotheses, resulting in "omnibus" data

collection that could be used for multiple purposes. Different management factors (e.g., harvest levels, patch size, road density, fire suppression, human access, invasive species), occurring cumulatively and simultaneously across several spatial scales and species, may be monitored concurrently to track numerous processes. In some cases, this approach can be useful for discovering previously unknown phenomena (Wintle et al. 2010). Additionally, when research capacity or public support is limited, surveillance monitoring can help raise public interest in biology, empower local people in natural resource decision-making, and may be the only practical form of monitoring if knowledge of population ecology study design is limited (Danielsen et al. 2009).

While surveillance monitoring is routinely conducted, biologists should strive to initiate targeted monitoring as much as possible to answer mechanistic, hypothesis-influenced questions about the factors (natural and management induced) that affect population dynamics (Yoccoz et al. 2001, Nichols and Williams 2006). Even though it is harder to do, a study designed around causal hypotheses will be better situated to separate multiple population influences or stressors and to determine the effects of management actions. Targeted monitoring is a central component of adaptive management (Lindenmayer and Likens 2009), a synthetic framework to understand population dynamics and effects of management actions (Chapter 5 covers this topic in detail).

Field and Analytical Techniques for Monitoring

After identifying the objectives for wildlife population monitoring, one must decide how to collect and analyze field data. These issues of study design and statistical analysis of carefully specified variables are too often ignored in wildlife monitoring studies (Marsh and Trenham 2008). Although details of study design and analysis are beyond the scope of this chapter, some key points warrant mention. First, the study should be at an appropriate scale and pay attention to the basic principles of statistical sampling (Yoccoz et al. 2001, Garton et al. 2020). Typically, inferences for wildlife populations are desired across some large area (e.g., a national park, a state, a province, a country), but it is possible to sample only a small portion of the area. There are many ways to link limited sampling frames to broader areas of inference, but all methods are rooted in the basic principle of random sampling, the most powerful approach for strong statistical inference. A pitfall to avoid is the temptation to initiate convenience or subjective sampling. A monitoring program where field sites are chosen for convenience (e.g., easy road access) or because they are subjectively felt to be representative of the larger area will always compromise inferences from that study to a larger area.

A second point to emphasize is the need to consider, before the study begins, whether sample sizes will be sufficient to accurately estimate a population parameter of interest (e.g., abundance, survival) or to detect whatever change in a population you are interested in (e.g., a 20% change in density or distribution). Without this a priori assessment, a wildlife monitoring program will be doomed from the start, unable to estimate population parameters with desired precision or to detect meaningful trends in the dynamics of populations (Taylor and Gerrodette 1993, Gibbs et al. 1998).

What are the solutions? One is to initiate discussions between scientists and managers about the desired precision in a population estimate and the costs and benefits of detecting, or failing to detect, a change in population parameters through monitoring. This can be formalized in a decision theory framework. For example, koalas (*Phascolarctos cinereus*) were declining in much of eastern Australia despite being a major tourist attraction. Field et al. (2004) conducted power analyses that incorporated the financial effect of lost tourism if an undetected decline of koalas occurred. The researchers reported that classical approaches to monitor trends would always fail to detect economically important declines of koalas owing to low statistical power. In fact, $5 million would be saved in this case if monitoring were abandoned in favor of direct management action to prevent a decline from occurring. The decision theory framework was critical to point out in this case where statistical power was inadequate given the economic consequences of koala decline.

Another way to increase statistical power and to improve inferences from a monitoring program is to carefully choose the monitoring field methodology. Although traditional monitoring techniques to capture and mark animals continue to be useful (e.g., trapping or sighting transects for mammals, cover boards for amphibians, mist netting for birds and bats), many powerful new approaches have emerged in noninvasive sampling (no capture or handling of animals; Long et al. 2008; Kelly et al. 2012). Noninvasive sampling can include photos from remote cameras, or genetic sampling of animals (e.g., scats, eggshells) or their habitats (i.e., environmental DNA; Adams et al. 2019). Especially for species that are rare, elusive, and hard to monitor with conventional approaches, noninvasive sampling can provide data on the species present, individual identity, gender, relatedness, and genetic distinctiveness (Oyler-McCance et al. 2020).

The final decisions to make before implementing monitoring include choosing the specific measures that summarize the status of a population of interest (state variables) and other covariates of interest to understanding the population's dynamics (auxiliary information). State variables include those that address abundance, the presence, distribution, growth rate, or viability of a population(s). At the same time, auxiliary information is collected to facilitate tests of how the system works or the influence of management. For example, harvest management of North American mallards measures population size as the state variable but also includes mallard harvest and number of wetlands on key breeding areas to help inform the models that connect the monitoring data to an understanding of harvest and mallard population dynamics (Yoccoz et al. 2001). In the next few sections, we describe the most common state variables: abundance, trends in population growth, presence/absence of the target species, and risk of decline.

Monitoring Abundance or Density

One of the most common targets of wildlife studies in general, and monitoring in particular, is an estimate of abundance or density (abundance per unit area). Although abundance may seem to be an obvious and simple population attribute to measure, it is actually quite challenging because of the simple fact that animals move, hide, and generally avoid being detected. This means that a simple count of animals in an area will be less than the number of animals actually present. Said more formally, detection probabilities are typically less than 1. In fact, the case where detection probability equals 1, so that we detect all individuals in an area, is so unusual that we use a special word to describe it: census.

In addition to accounting for incomplete detection, animal abundance estimates must also account for incomplete sampling. Typically, neither the time nor resources are available to sample the entire population of interest (e.g., Big Bend National Park, state of Wisconsin). If the study area represents only a portion of the population, then the proportion of area sampled (a) is less than 1. Both a and the detection probability ($\hat{p}$; the hat over the p is a standard convention indicating an estimate) are incorporated in this generalized equation for estimating abundance ($\hat{N}$):

$$\text{estimate of abundance} = \hat{N} = \frac{\text{count of animals}}{\hat{p} \times a}. \quad (9)$$

For example, suppose 40 rabbits on an island are counted on a spotlight transect that includes the entire island ($a = 1.0$), but detection probability is 0.5; the estimate of abundance ($\hat{N}$) would be 80 (40/0.5). If the transect included only 25% of the area occupied by the rabbit population, the abundance estimate would be:

$$\hat{N} = \frac{40}{0.5 \times 0.25} = 320.$$

Counting animals can be achieved through a variety of methods, including transect sampling (e.g., sighting animals by plane, foot, vehicle, horseback), live trapping, and noninvasive genetic and camera sampling. In all cases, detection probability must be estimated, requiring thoughtful methodologies (e.g., mark-recapture, double sampling) and oftentimes sophisticated statistical analyses. Although the rich analytical details of estimating detection probability (and abundance) with different methods are beyond the scope of this chapter (see overviews in Williams et al. 2002, Mills 2013), an example using the simplest possible approach appears in the box on page 124.

In some cases, a statistically rigorous estimate of detection probability (and abundance) is not possible. For example, perhaps only raw counts of animals are available, or observations of animal signs such as numbers of pellets, tracks, calls, nests, scrapes, or burrow entrances. These are index counts, which in general should be avoided in monitoring studies because they can give misleading signals of changes in abundance. The problem is that an index is a function of not only the abun-

ESTIMATING ABUNDANCE WITH THE LINCOLN-PETERSEN METHOD

The Lincoln-Petersen (LP) estimator is based on two sampling occasions where animals are captured and marked on the first occasion, and then captured again on the second occasion. The population is assumed to be closed to births, deaths, immigration, and emigration, and all individuals are assumed to be equally detected (captured) and unaffected by the capture or marking procedure (these assumptions can be relaxed when using more complex abundance estimators). The capture and marking could be physical live-capturing or alternatives such as mark-resight or noninvasive sampling with cameras or genetic material.

Suppose you are estimating abundance from two sampling occasions that occur on consecutive days. Denote the number of animals captured, marked, and released on day one as n_1. During day two, a total of n_2 animals are captured, of which m_2 have marks from the previous day. The capture probability, , is simply the proportion of the animals marked the first day that are captured on the second day (m_2/n_1). Taking the total captures on the second day (n_2) as the count, the estimate of abundance is:

$$\hat{N} = \frac{n_2}{\hat{p}} = \frac{n_2}{m_2 / n_1} = \frac{n_1 n_2}{m_2}.$$

Applying this equation to an example, consider an estimate of abundance of flying squirrels (*Glaucomys sabrinus*) in a park. In one night of trapping, 32 squirrels were captured, ear-tagged, and released. One week later, a total of 44 squirrels were captured, 20 of which had been ear-tagged during the first trapping occasion.

$$\hat{N} = \frac{n_1 n_2}{m_2} = \frac{32 \times 44}{20} = 70.4.$$

Given these numbers, the abundance of flying squirrels would be estimated to be 70. It turns out that this simple and intuitive equation needs to be corrected for statistical reasons, so the actual LP estimator and its variance for actual application are

$$\hat{N} = \left[\frac{(n_1 + 1)(n_2 + 1)}{(m_2 + 1)} \right] - 1$$

and

$$\operatorname{var}(\hat{N}) = \frac{(n_1 + 1)(n_2 + 1)(n_1 - m_2)(n_2 - m_2)}{(m_2 + 1)(m_2 + 2)}.$$

For our flying squirrel example, the actual estimate of abundance would be 69.7 squirrels with a variance of 44.1.

dance of animals, but also the relationship between the index and true abundance. Unless a biologist can be sure that the relationship between the index and abundance stays constant across space and time, any change in the index could be caused by a change in that relationship *or* by a change in abundance. For example, suppose in one year you record an average of 60 pheasants (*Phasianus colchicus*) at calling stations. The next year you record 80. You cannot assume the population increased, because the change in the index from 60 to 80 could have been caused by an increase in call rate or call detection (maybe each bird was calling more, or observers were better trained to hear them, or weather conditions were more favorable for sound to travel). The bottom line is that indices are best avoided for monitoring studies, but if they must be used, one should do everything possible to ensure that the relationship between the index and abundance stays constant.

Monitoring Trends in Abundance (Population Growth over Time)

If the monitoring objective includes evaluating population growth over time, two primary approaches can be used. One way uses vital rates and population projection models, following the methods described above in the section on stage-structured population dynamics. The second and more common way to estimate trend uses abundance estimates over time from a monitoring program. It is this abundance-based trend estimator approach that we cover briefly below (for more details, see Humbert et al. 2009, Mills 2013). Although trend estimators can incorporate real-world complications such as density dependence, observer effects, and other covariates, for simplicity we assume exponential growth without additional effects.

The most commonly used method of estimating trend from a time series of abundance values is a simple linear regression of the natural log (ln) of abundances against time (the natural log accounts for the fact that birth and death processes cause wildlife populations to change geometrically, not arithmetically). The slope of the regression represents the estimated average rate of population change $\hat{\bar{r}}$. The simplicity of the method explains its popularity, but the method has a major limitation: it assumes that all variation in the trend arises only from the uncertainty in estimating N (i.e., pure observation error or sample variance; Humbert et al. 2009). That is, this method assumes that population growth is constant over time, unaffected by process variance arising from weather, predators, or other environmental conditions. A suite of other widely used methods make the opposite assumption, that no observation error exists and that all variation in the trend arises from process variance or process noise (e.g., the diffusion approximation; Dennis et al. 1991).

More recent developments permit exponential trend estimation with both process and observation error occurring simultaneously. These methods include state-space statistical models that contain a component to account for the stochastic fluctuations due to process noise and a component to accommodate observation error in abundance estimates (Dennis et al. 2006, Humbert et al. 2009, Mills 2013, Hostetler and Chandler 2015). Investigators also estimate wildlife population trends using approaches such as Bayesian hierarchical modeling (Link and Sauer 2002, Royle and Dorazio 2008), population reconstruction (Skalski et al. 2005), and by combining data on abundance with data on other demographic parameters to develop integrated population models (Schaub and Abadi 2011, Zipkin and Saunders 2018).

How long should a time series be to reliably estimate population trends? The answer depends on many factors, of course, but for most wildlife species, the state-space model should use a bare minimum of 10 years, with at least five samples of abundance during that time (Humbert et al. 2009). Of course, unusual events that affect process variance (e.g., 20-year floods or 15-year fire events) will only be picked up with longer sampling.

Monitoring Species Distribution or Community Composition

Another set of state variables in wildlife population monitoring is the distribution of a single species, or the diversity of a suite of species. Instead of estimating abundance or vital rates for assessing population growth, these approaches estimate changes in occupancy, or presence versus absence, of target species in a sampled area.

Although the distribution or occupancy of a species in an area would seem to be straightforward, it is (like abundance) complicated by incomplete detection probability. Just as a raw count of animals does not describe abundance when individuals are undetected, estimates of occupancy are also biased low if species presence is detected imperfectly. If a species is detected, then of course it is present. But if it is not detected, it is not necessarily absent. Therefore a single presence/absence survey should be called a present/not detected survey. The challenge, then, is to adjust counts of presences of the species to account for the probability that they were present but not detected.

The scientific framework for such estimates is called occupancy modeling. Space prevents us from detailing the methods of occupancy modeling (MacKenzie and Royle 2005, MacKenzie et al. 2018). In brief, however, data collection for occupancy surveys involves searching for evidence of presence in sampling units, with detection methods ranging from visual observations or captures to photographs and indirect evidence of presence such as hair, feces, or tracks. Detections and non-detections at multiple sample units form the detection history used to estimate the detection probability of the species, which is used to estimate occupancy. Covariates that may influence detection and/or occupancy, such as vegetation type or weather, can also be incorporated into models.

Monitoring Extinction Risk and Population Viability Management

Monitoring abundance, trend, or distribution can be important for many reasons in applied wildlife population studies. But in other cases, the ultimate goal centers on risk assessment, or estimating the probability that a population will decline or go extinct. Applied population ecology has a variety of tools for assessing the viability of wildlife populations, including quantitative approaches of population viability analysis (PVA). PVA involves the application of data and models to estimate likelihoods of a population crossing specified abundance or persistence thresholds within various time spans, and gives insights into factors that constitute the biggest threats (Mills 2013). Models used in PVA typically simulate the dynamics of a population to quantify the probability that it goes extinct or declines below some specified value (e.g., 100 individuals) within a period of management interest (e.g., 20, 50, or 100 years). The two main ways to conduct PVAs with monitoring data are by analyzing count data from time series and using population projection models based on estimated vital rates and interacting factors (Horne et al. 2020).

An example of a comprehensive approach to PVA can be found in efforts to recover the endangered island fox (*Urocyon littoralis*) on the Channel Islands in California (Bakker and Doak 2009, Bakker et al. 2009). The first step in the PVA process was for managers and researchers to specify recovery criteria in terms of acceptable risk, accounting for the inevitably complex sociopolitical and biological considerations. Next, readily monitored population attributes were selected (e.g., adult population size and adult mortality) and modeled relative to extinction thresholds (where extinction could be a number greater than zero; e.g., it may be the number of animals that would trigger a captive breeding program to be initiated). Finally, management actions thought to affect population attributes were simulated; for example, how the removal of golden eagles (*Aquila chrysaetos*) would increase adult fox survival and subsequently reduce extinction risk. This approach provided a platform for adaptive decision-making, whereby the influence of different management strategies could be simulated and compared, actions could then be implemented, and the results could feed back into demographic models of future extinction risk. Their approach embraced various types of parameter and process uncertainty, while presenting managers with straightforward results that directly related to recovery criteria. Intensive management intervention combined with rigorous population monitoring and modeling yielded a success story for the island fox, as their populations quickly recovered (Coonan et al. 2014), and in 2016 foxes on three islands were delisted (US Fish and Wildlife Service 2016).

CASE STUDY: RECOVERING ENDANGERED SIERRA NEVADA BIGHORN SHEEP

We close with a case study to demonstrate how a mechanistic understanding of population dynamics (as described in this chapter) can inform on-the-ground management decisions. It highlights how monitoring data can be used to develop stage-structured models that in turn help assess population responses to different management scenarios, and how those responses can be used to prioritize conservation actions. This example also illustrates the complexity of deciphering the underlying factors driving populations, a constant challenge for wildlife managers.

Figure 7.6. Federally endangered female Sierra Nevada bighorn sheep. The ear tag is used for estimating population size using mark-resight field surveys. Courtesy of Art Lawrence

Sierra Nevada bighorn sheep (*Ovis canadensis sierrae*) are the rarest subspecies of mountain sheep in North America (Fig. 7.6), endemic to the Sierra Nevada mountain range that spans eastern California (Fig. 7.7). The subspecies was listed as endangered under the US Endangered Species Act in 1999, when biologists could locate only about 100 adults in the wild (US Fish and Wildlife Service 2007). Since then, government agencies, nonprofit organizations, and university scientists have been collecting detailed monitoring data on the population dynamics of this endangered subspecies. To identify and prioritize management actions that would be most effective at increasing the sizes of Sierra Nevada bighorn sheep populations, researchers outlined three objectives:

1. Identify which stage-specific vital rates were most important for influencing the growth rates of bighorn sheep populations.
2. Determine the stochastic and deterministic factors responsible for spatial and temporal variation in those key vital rates influencing population growth rates.
3. Use information from objectives 1 and 2 to develop effective management strategies to increase population sizes and reach recovery goals.

To meet these objectives, researchers used a variety of field methods to obtain demographic data on Sierra Nevada bighorn sheep populations. Since 1980, they conducted systematic ground counts each year, with attempts to annually census the number of bighorn sheep in each small population. Start-

ing in 2001, they captured bighorn sheep using net-guns fired from helicopters and fit them with telemetry collars to monitor their annual survival, reproduction, and movements. Additionally, starting in 2006, biologists conducted annual mark-resight surveys (analogous to mark-recapture, except animals are resighted as opposed to recaptured) to obtain estimates of abundance with detection probabilities and associated measurements of error. Using information from these different data types, biologists estimated annual stage-specific vital rates for different populations of bighorn sheep, which could then be used in population models to assess their growth rates (Johnson et al. 2010a, b). In this case study, we will focus specifically on data from the Mono Basin, Wheeler, and Langley populations of Sierra Nevada bighorn sheep (Fig. 7.8).

Identifying Vital Rates Most Important for Influencing Population Growth

To develop effective management strategies for the recovery of small or declining populations, it is necessary to identify those vital rates responsible for poor population performance and those that can be increased to most efficiently change a population's trajectory. To do that for Sierra Nevada bighorn sheep, biologists monitored three vital rates that captured their basic life cycle and population dynamics and were relevant to stage classes that could be reliably distinguished in the field: adult female survival (S_A, the number of adults that survived to year t given the number of adults and yearlings in year $t - 1$); yearling female survival (S_Y, the number of lambs in year $t - 1$ that survived to be yearlings in t); and adult female fecundity (F_A, the number of lambs born per adult female in year t; only adult females produced offspring). These vital rates were estimated from collared individuals and grounds counts of lambs, yearlings, and adults.

The field data revealed significant differences in the means and variances of these vital rates in different bighorn sheep populations, and even within the same population for different phases of population growth (Johnson et al. 2010a). For example, the Mono Basin population experienced two very different phases of growth. The herd dramatically increased after it was initially reintroduced in 1986, but it then declined precipitously around 1994 (Fig. 7.8). Below are two matrices based on the mean vital rates measured during the increasing phase (from 1986 to 1993) and the decreasing phase (from 1994 to 2007). Although we will not delve into the details of the population matrix, we point out that fecundity (matrix element $a_{1,3}$) is multiplied by adult survival because field surveys occurred just after lambs were born, so that the number of lambs, in this case, is also a function of the number of adult females that survived to reproduce (see Mills 2013 for details about post-birth pulse matrices).

Vital Rate Symbols			Increasing Phase			Decreasing Phase		
0	0	$F_A * S_A$	0	0	0.27	0	0	0.32
S_Y	0	0	0.74	0	0	0.51	0	0
0	S_A	S_A	0	0.92	0.92	0	0.84	0.84

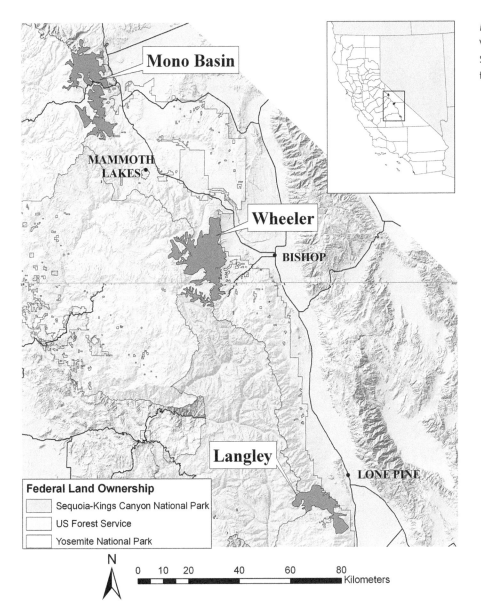

Figure 7.7. Locations of the Mono Basin, Wheeler, and Langley populations of Sierra Nevada bighorn sheep in California. From Johnson et al. (2010*a*)

During the increasing phase in Mono Basin, the asymptotic growth rate was 1.08 (the population grew at 8%/yr), while during the decreasing phase it was 0.98 (declining by 2%/yr). Notice that the fecundity rate was higher during the decreasing phase, but yearling survival (element $a_{2,1}$) and adult survival (elements $a_{3,2}$ and $a_{3,3}$) were much lower. Which vital rate was most responsible for these different population trends, and, especially, which rate was associated with the period of population decline? Which vital rate should managers try hardest to improve in order to boost the size of this population?

To answer these questions, researchers turned to sensitivity analyses. First, they calculated the analytical elasticities of each vital rate and found that the vital rate with the highest elasticity, across all populations and phases of growth, was adult female survival (Fig. 7.9), which corresponds to general patterns from long-lived species (Sæther and Bakke 2000, Crone 2001). Indeed, given that adult females have the highest reproductive value and comprise a large proportion of the

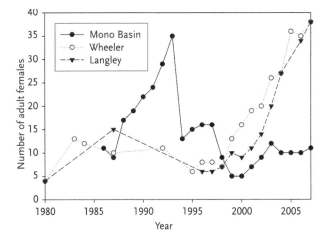

Figure 7.8. Number of adult female Sierra Nevada bighorn sheep in the Mono Basin, Wheeler, and Langley populations, 1980–2007. From Johnson et al. (2010*a*)

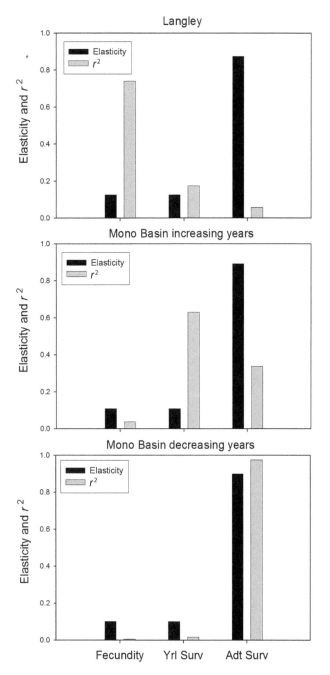

Figure 7.9. Elasticity and r^2 values for fecundity, yearling survival, and adult survival of different Sierra Nevada bighorn sheep populations and different phases of growth. From Johnson et al. (2010*a*)

population (>65% of the females were adults), it is not surprising that a small change in adult female survival would yield a proportionately greater change in the population growth rate than a small change in either yearling survival or fecundity.

While this pattern has been well recognized in the scientific literature, it has limitations for wildlife managers. For example, the temptation to conclude that adult female survival should be the vital rate targeted for efficient population increase must be tempered by the fact that elasticity does not account for how much that rate can realistically be manipulated

through management. During the increasing phase in Mono Basin, adult survival was 92%. This is close to the expected biological maximum for adult survival for bighorn sheep, with little room for improvement. If managers wanted to further increase population growth but could not easily increase adult survival, what other vital rate should be the focus?

To incorporate the realistic range of variation that is observed in different vital rates, researchers turned to LSA. This analysis determines how observed variation in different vital rates is associated with observed variation in population growth rates. In other words, which vital rates are responsible for variation in population growth? During the increasing phase, researchers reported that variation in yearling survival at Mono Basin explained most of the variation in λ (63%), even though adult female survival had the highest elasticity (Fig. 7.9). If managers wanted to increase the population growth rate during this phase, they would have been most successful by working to boost yearling survival. During the decreasing phase, however, variation in adult survival explained 98% of the variation in λ (Fig. 7.9), as the decrease in adult survival from 92% to 84% was primarily responsible for the drop in population growth from 1.08 to 0.98. During this kind of declining phase, managers should work to increase adult female survival to increase overall growth rates.

Across the study herds, researchers reported that LSA patterns were highly population-specific, and each of the three vital rates was found to be the primary population influence in different herds and for different phases of growth (Fig. 7.9; Johnson et al. 2010*a*). Wildlife managers often focus on improving the same vital rate in different populations of the same species, or in populations of similar species. This analysis of the dynamics of different endangered populations, however, demonstrates that appropriate management targets may often be idiosyncratic in space and time.

Determining the Factors Responsible for Variation in the Vital Rates That Influence Population Growth

Once biologists identified which vital rates were most important in driving the dynamics of Sierra Nevada bighorn sheep populations, they wanted to identify the factors that influence variation in those vital rates. Some of the factors hypothesized to influence bighorn sheep population dynamics were density dependence, predation, inbreeding depression, and weather (i.e., winter severity). A suite of analyses revealed that all of these factors were significant, but their degree of influence varied among the different populations and for different vital rates (Johnson et al. 2010*b*, 2011, 2013).

For example, adult female survival had the greatest influence on population growth in Wheeler and during the decline in Mono Basin, but the factors that affected this vital rate were dramatically different between the herds. In Mono Basin, adult female survival was positively associated with population density (positive density dependence) and summer rainfall (an index of summer forage quality), and negatively associated with winter snow depth (an index of winter severity). Meanwhile, adult survival in Wheeler was negatively associated with

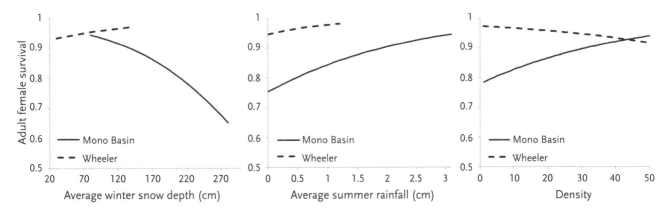

Figure 7.10. Predicted effects of winter snow depth, summer rainfall, and density on adult female survival for the Mono Basin and Wheeler bighorn sheep populations. Predictions for weather covariates were only modeled for the observed range of variation in each population.

population density (negative density dependence) and mountain lion predation (Johnson et al. 2013), and positively associated with winter snow depth and summer rainfall (Fig. 7.10).

Why would adult female survival of bighorn sheep in two populations be affected so differently by the same set of factors? The explanation is that habitat conditions were highly variable between the herds, particularly during winter. The Mono Basin population persisted at high elevations (~3,400 m) during winter, relying on high plateaus that were blown free of snow. Meanwhile, most bighorn sheep in the Wheeler population spent the winter at lower elevations (~2,550 m), on south-facing slopes below the snowline. This difference explains the negative effect of winter snow depth on survival at Mono Basin compared to Wheeler. That same distinction in winter ranges also accounted for the observed disparity in mountain lion predation. The winter range of bighorn sheep in the Wheeler population overlapped with the winter range of thousands of mule deer that supported a healthy mountain lion population, the primary predator of bighorn sheep in this system. Meanwhile, the Mono Basin population spent winters at high elevations, far away from mule deer and mountain lions (Johnson et al. 2013). As for density dependence, bighorn sheep are a species that probably experiences both positive and negative density dependence at different population sizes. As a gregarious species, foraging efficiency and predator detection are likely enhanced above a threshold population size (Mooring et al. 2004), and the small size of the Mono Basin population (~10 adult females in recent years) may have induced Allee effects. For larger populations like Wheeler, limitations in forage resources may have induced negative density dependence.

As with populations of Sierra Nevada bighorn sheep, biologists often find that different factors have variable effects on vital rates in different populations. The key is to maintain focus on those factors that influence the vital rates most consequential to population growth. Factors that influence vital rates that are largely inconsequential to λ should not be a priority for managers, as they will not be able to significantly influence the trajectory of a population. For example, researchers found evidence of inbreeding depression in fecundity rates of

Sierra Nevada bighorn sheep. Using population models, however, they determined that genetic management (i.e., increasing gene flow among populations to bolster genetic diversity) would have limited utility in the near term because fecundity was not an important influence of growth rates for most populations (Johnson et al. 2011). Instead, they recommended that managers focus on actions that would yield greater short-term benefits to populations while planning for increased gene flow among herds in the future.

Developing Effective Management Strategies to Increase Population Sizes and Reach Recovery Goals

Vital rates can change either naturally (i.e., the influence of winter severity on adult survival in Mono Basin) or as a result of management. Managers cannot change the weather (although knowing the importance of winter severity may be critical for identifying future reintroduction sites), but they often have a suite of potential actions that can be used to manipulate the trajectories of populations. The key, however, is to identify which of those potential actions will be most effective at influencing the dynamics of populations, and thus reaching management objectives. Often, well-intentioned management actions have wasted critical resources because they had little to no influence on population dynamics (see the sea turtle example above).

For Sierra Nevada bighorn sheep, the list of potential management options included mountain lion removal, augmentation (i.e., increasing the size of an existing population), reintroduction (i.e., starting a new population), genetic management, disease prevention, and prescribed fire to enhance forage quality. To determine which of those activities would be most effective for increasing population growth, researchers simulated their effects using stage-structured population models. To simulate these "what if" scenarios, managers must have information about how different activities influence key vital rates and the magnitude of their effects. For Sierra Nevada bighorn sheep, managers used models to compare the effects of two specific actions for which they had reliable data, mountain lion removal and population augmentation. Both

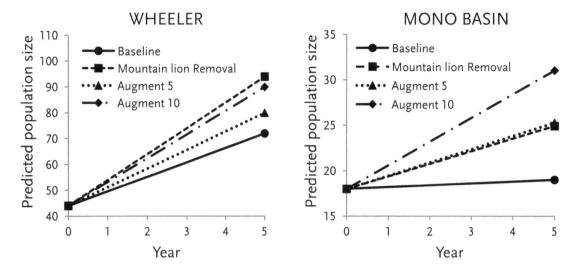

Figure 7.11. Predicted increase in population size resulting from mountain lion removal and augmentation (adding five and ten adult females to the population) for the Wheeler and Mono Basin populations of Sierra Nevada bighorn sheep over a five-year period.

actions could be readily implemented in the field and had clear and quantifiable influences on vital rates or abundance.

Researchers evaluated the relative influence of these two actions over a five-year period on Wheeler and Mono Basin, given their unique dynamics (Johnson et al. 2010a). In this analysis, different vital rates were manipulated in a population model to reflect expected management outcomes, and their anticipated effects on population growth were compared. Biologists simulated mountain lion removal by conservatively modeling a 5% increase in each vital rate, and an augmentation by adding either five or ten adult females to each population, a realistic number given limited source stock for translocations. These kinds of demographic modifications can be simulated from field data, or when data are not available, from values published in the scientific literature. For Sierra Nevada bighorn sheep, over the modeled period, investigators reported that predator control would be most effective for boosting population growth in the Wheeler herd, while an augmentation would be more beneficial in Mono Basin (Fig. 7.11). Effective management actions were unique to each herd based on their individual dynamics.

Even once the best management action has been identified and implemented, the job of a biologist is not finished. They must carefully monitor the demographic effects of those actions on populations over time. This information can then be used within the iterative process of monitoring and adaptive management to improve modeling efforts, so that better predictions of the responses of populations to management can be used for sound decision-making in the future (Williams 2001, Lyons et al. 2008; Chapter 4). In recent years, managers of Sierra Nevada bighorn sheep have conducted mountain lion removals in Wheeler and other populations facing high predation rates, augmented the Mono Basin population, and reintroduced bighorn sheep into areas with low-elevation winter range that had limited overlap with mule deer and mountain lions. These and other management activities are successfully enabling the subspecies to recover, such that there were estimated to be ~550 Sierra Nevada bighorn sheep in 2018 (Greene et al. 2019).

SUMMARY

We have covered a lot of ground, from the basics of population growth and dynamics, to the factors that influence populations, to stage-structured models, to considerations for monitoring population dynamics. The field of wildlife population dynamics is broad and complex. While we could only briefly introduce many of the key concepts here, we aim to instill an appreciation for the importance of population dynamics in understanding and managing wildlife populations. Before taking any management action intended to change the size or distribution of a population, one should ask, What is our specific population objective? Which vital rate(s) should be altered to achieve that objective? And which management action(s) can most effectively change the vital rate(s) driving population growth? In those cases where we do not have a clear understanding of how a management action will influence the dynamics of a population, biologists can use targeted monitoring and adaptive management as powerful tools for elucidating such information.

The application of carefully collected field data to population models can yield critical and nonintuitive insights about the mechanisms driving populations. As human development, climate change, the expansion of nonnative species, and other factors continue to affect wildlife populations, understanding the processes that govern population dynamics is increasingly critical for preserving biodiversity. Given the available suite of powerful field, statistical, and modeling tools, biologists can apply information about the dynamics of populations to maximize their management success and achieve population objectives.

Literature Cited

Adams, C. I. M., M. Knapp, N. J. Gemmell, G.-J. Jeunen, M. Bunce, M. D. Lamare, and H. R. Taylor. 2019. Beyond biodiversity: can environmental DNA (eDNA) cut it as a population genetics tool? Genees 10:192.

Adams, L. G., R. O. Stephenson, B. W. Dale, R. T. Ahgook, and D. J. Demma. 2008. Population dynamics and harvest characteristics of wolves in the Central Brooks Range, Alaska. Wildlife Monographs 170:1–25.

Ali, S. A., N. Ahmed, A. A. Ali, and A. Iqbal. 2020. The outbreak of coronavirus disease 2019 (COVID-19)—an emerging global health threat. Journal of Infection and Public Health 13:644–646.

Allee, W. C. 1931. Animal aggregations: a study in general sociology. University of Chicago Press, Chicago, Illinois, USA.

Allee, W. C., O. Park, A. E. Emerson, T. Park, and K. P. Schmidt. 1949. Principles of animal ecology. W. B. Saunders, Philadelphia, Pennsylvania, USA.

Allen, T., K. A. Murray, C. Zambrana-Torrelio, S. S. Morse, C. Rondinini, M. Di Marco, N. Breit, K. J. Olival, and P. Daszak. 2017. Global hotspots and correlates of emerging zoonotic diseases. Nature Communications 8:1124.

Allendorf, F. W., G. Luikart, and S. N. Aitken. 2013. Conservation and the genetics of populations. 2nd edition. Wiley-Blackwell, Cambridge, Massachusetts, USA.

Altwegg, R., A. Roulin, M. Kestenholz, and L. Jenni. 2006. Demographic effects of extreme winter weather in the barn owl. Oecologia 149:44–51.

Ankney, C. D. 1996. Why did the ducks come back in 1994 and 1995: was Johnny Lynch right? Proceedings of the Seventh International Waterfowl Symposium 7:40–44.

Bakker, V. J., and D. F. Doak. 2009. Population viability management: ecological standards to guide adaptive management for rare species. Frontiers in Ecology and Environment 7:158–165.

Bakker, V. J., D. F. Doak, G. W. Roemer, D. K. Garcelon, T. J. Coonan, S. A. Morrison, C. Lynch, K. Ralls, and R. Shaw. 2009. Estimating and incorporating ecological drivers and parameter uncertainty into a demographic population viability analysis for the Island Fox (*Urocyon littoralis*). Ecological Monographs 79:77–108.

Battin, J. 2004. When good animals love bad habitats: ecological traps and the conservation of animal populations. Conservation Biology 18:1482–1491.

Beausoleil, R. A., G. M. Koehler, B. T. Maletzke, B. N. Kertson, and R. B. Weilgus. 2013. Research to regulation: cougar social behavior as a guide for management. Wildlife Society Bulletin 37:680–688.

Bekoff, M., and M. C. Wells. 1986. Social ecology and behavior of coyotes. Advances in the Study of Behavior 16:251–338.

Bellard, C., C. Bertelsmeier, P. Leadley, W. Thuiller, and F. Courchamp. 2012. Impacts of climate change on the future of biodiversity. Ecology Letters 15:365–377.

Berger, K. M., and E. M. Gese. 2007. Does interference competition with wolves limit the distribution and abundance of coyotes? Journal of Animal Ecology 76:1075–1085.

Bourbeau-Lemieux, A., M. Festa-Bianchet, J.-M. Gaillard, and F. Pelletier. 2011. Predator-driven component Allee effects in a wild ungulate. Ecology Letters 14:358–363.

Bouzat, J. L., J. A. Johnson, J. E. Toepfer, S. A. Simpson, T. L. Esker, and R. L. Westemeier. 2009. Beyond the beneficial effects of translocations as an effective tool for the genetic restoration of isolated populations. Conservation Genetics 10:191–201.

Boyce, M. S. 1977. Population growth with stochastic fluctuations in the life table. Theoretical Population Biology 12:366–373.

Boyd, D. K., and D. H. Pletscher. 1999. Characteristics of dispersal in a colonizing wolf population in the central Rocky Mountains. Journal of Wildlife Management 63:1094–1108.

Brook, B. W., N. S. Sodhi, and C. J. A. Bradshaw. 2008. Synergies among extinction drivers under global change. Trends in Ecology and Evolution 23:453–460.

Cassirer, E. F., and A. R. E. Sinclair. 2007. Dynamics of pneumonia in a bighorn sheep population. Journal of Wildlife Management 71:1080–1088.

Cayuela, H., B. R. Schmidt, A. Weinbach, A. Besnard, and P. Joly. 2019. Multiple density-dependent processes shape the dynamics of a spatially structured amphibian population. Journal of Animal Ecology 88:164–177.

Ceballos, G., P. R. Ehrlich, and P. H. Raven. 2020. Vertebrates on the brink as indicators of biological annihilation and the sixth mass extinction. Proceedings of the National Academy of Sciences 117:13596–13602.

Cole, L. C. 1957. Sketches of general and comparative demography. Cold Spring Harbor Symposia on Quantitative Biology 22:1–15.

Coonan, T. J., V. Bakker, B. Hudgens, C. L. Boser, D. K. Garcelon, and S. A. Morrison. 2014. On the fast track to recovery: island foxes on the northern Channel Islands. Monographs of the Western North American Naturalist 7:373–381.

Crooks, K. R., C. L. Burdett, D. M. Theobald, S. R. B. King, M. Di Marco, C. Rondinini, and L. Boitani. 2017. Quantification of habitat fragmentation reveals extinction risk in terrestrial mammals. Proceedings of the National Academy of Sciences 114:7635–7640.

Crooks, K. R., and M. Sanjayan, editors. 2006. Connectivity conservation. Cambridge University Press, Cambridge, UK.

Crone, E. 2001. Is survivorship a better fitness surrogate than fecundity? Evolution 55:2611–2614.

Crouse, D. T., L. B. Crowder, and H. Caswell. 1987. A stage-based population model for loggerhead sea turtles and implications for conservation. Ecology 68:1412–1423.

Danielsen, F., N. D. Burgess, A. Balmford, P. F. Donald, M. Funder, et al. 2009. Local participation in natural resource monitoring: a characterization of approaches. Conservation Biology 23:31–42.

Dennis, B., P. L. Munholland, and J. M. Scott. 1991. Estimation of growth and extinction parameters for endangered species. Ecological Monographs 61:115–143.

Dennis, B., and M. R. M. Otten. 2000. Joint effects of density dependence and rainfall on abundance of San Joaquin kit fox. Journal of Wildlife Management 64:388–400.

Dennis, B., J. M. Ponciano, S. R. Lele, M. L. Taper, and D. F. Staples. 2006. Estimating density dependence, process noise, and observation error. Ecological Monographs 76:323–341.

Doak, D. F., W. F. Morris, C. Pfister, B. E. Kendall, and E. M. Bruna. 2005. Correctly estimating how environmental stochasticity influences fitness and population growth. American Naturalist 166:E14–E21.

Drajem, B. 2011. A cougar in Connecticut. Science News. Web Edition 2:August.

Ehrlich, P. R. 1968. The population bomb. Ballantine Books, New York, New York, USA.

Ehrlich, P. R., and H. H. Ehrlich. 2009. The population bomb revisited. Electronic Journal of Sustainable Development 1:63–71.

Ehrlich, P. R., and J. P. Holdren. 1971. Impact of population growth. Science 171:1212–1217.

Elton, C. 1924. Periodic fluctuations in the numbers of animals: their causes and effects. British Journal of Experimental Biology 2:119–163.

Emlen, S. T. 1990. White-fronted bee-eaters: helping in a colonially nesting species. Pages 489–526 in R. B. Stacey and W. D. Koenig, editors. Cooperative breeding in birds: long-term studies of ecology and behavior. Cambridge University Press, Cambridge, UK.

Errington, P. L. 1956. Factors limiting higher vertebrate populations. Science 124:304–307.

Ezard, T. H. G., J. M. Bullock, H. J. Dalgleish, A. Millon, F. Pelletier, A. Ozgul, and D. N. Koons. 2010. Matrix models for a changeable world: the importance of transient dynamics in population management. Journal of Applied Ecology 47:515–523.

Festa-Bianchet, M. 1988. A pneumonia epizootic in bighorn sheep, with comments on preventative management. Proceedings of Biennial Symposium of the Northern Wild Sheep and Goat Council 6:66–76.

Field, C. B., V. R. Barros, K. J. Mach, M. D. Mastrandea, M. K. van Aalst, et al. 2014. Technical summary. Climate change 2014: impacts, adaptation, and vulnerability. Contribution of Working Group II to the Fifth Assessment Report of the Intergovernmental Panel on Climate Change. Intergovernmental Panel on Climate Change, Geneva, Switzerland.

Field, S. A., A. J. Tyre, N. Jonzén, J. R. Rhodes, and H. P. Possingham. 2004. Minimizing the cost of environmental management decisions by optimizing statistical threshold. Ecology Letters 7:669–675.

Fortin, D., H. L. Beyer, M. S. Boyce, D. W. Smith, T. Duchesne, and J. S. Mao. 2005. Wolves influence elk movements: behavior shapes a trophic cascade in Yellowstone National Park. Ecology 86:1320–1330.

Funk, W. C., A. E. Greene, P. S. Corn, and F. W. Allendorf. 2005. High dispersal in a frog species suggests that it is vulnerable to habitat fragmentation. Biology Letters 1:13–16.

Garton, E. O., J. L. Aycrigg, C. Conway, and J. S. Horne. 2020. Research and experimental design. Pages 1–40 in N. J. Silvy, editor. The wildlife techniques manual. 8th edition. Volume 1. Johns Hopkins University Press, Baltimore, Maryland, USA.

George, J. L., D. J. Martin, P. M. Lukacs, and M. W. Miller. 2008. Epidemic pasteurellosis in a bighorn sheep population coinciding with the appearance of a domestic sheep. Journal of Wildlife Diseases 44:388–403.

Gibbs, J. P., S. Droege, and P. Eagle. 1998. Monitoring populations of plants and animals. Bioscience 48:935–940.

Greene, L. E., C. P. Massing, D. W. German, D. Gammons, K. Anderson, E. A. Siemion, and T. R. Stephenson. 2019. 2017–18 annual report of the Sierra Nevada bighorn sheep recovery program. California Department of Fish and Wildlife, Sacramento, California, USA.

Gregory, S. D., C. J. A. Bradshaw, B. W. Brook, and F. Courchamp. 2010. Limited evidence for the demographic Allee effect from numerous species across taxa. Ecology 91:2151–2161.

Grenier, M. B., D. B. McDonald, and S. W. Buskirk. 2007. Rapid population growth of a critically endangered carnivore. Science 317:779.

Griffin, P. C., and L. S. Mills. 2003. Snowshoe hares in a dynamic managed landscape. Pages 438–449 in H. R. Akcakaya, M. A. Burgman, O. Kindvall, et al., editors. Species conservation and management: case studies. Oxford University Press, Oxford, UK.

Griffin, P. C., and L. S. Mills. 2009. Sinks without borders: snowshoe hare dynamics in a complex landscape. Oikos 118:1487–1498.

Haddad, N. M., B. Hudgens, E. I. Damschen, D. J. Levey, J. L. Orrock, J. J. Tewksbury, and A. J. Weldon. 2011. Assessing positive and negative ecological effects of corridors. Pages 475–503 in J. Liu, V. Hull, A. Morzillo, and J. Wiens, editors. Sources, sinks, and sustainability across landscapes. Cambridge University Press, Cambridge, UK.

Hale, R., and S. E. Swearer. 2016. Ecological traps: current evidence and future directions. Proceedings of the Royal Society Series B 283:20152647.

Hamel, S., and S. D. Côté. 2007. Habitat use patterns in relation to escape terrain: are alpine ungulate females trading off better foraging sites for safety? Canadian Journal of Zoology 85:933–943.

Han, B. A., A. M. Kramer, and J. M. 2016. Global patterns of zoonotic disease in mammals. Trends in Parasitology 32:565–577.

Harper, E. B., T. A. G. Rittenhouse, and R. D. Semlitsch. 2008. Demographic consequences to terrestrial habitat loss for pool-breeding amphibians: predicting extinction risks associated with inadequate size of buffer zones. Conservation Biology 22:1205–1215.

Hartway, C., and L. S. Mills. 2012. A meta-analysis assessing the effects of common management actions on the nest success of North American birds. Conservation Biology 26:657–666.

Hebblewhite, M., C. White, and M. Musiani. 2010. Revisiting extinction in national parks: mountain caribou in Banff. Conservation Biology 24:341–344.

Heppell, S. S., L. B. Crowder, and D. T. Crouse. 1996. Models to evaluate headstarting as a management tool for long-lived turtles. Ecological Applications 6:556–565.

Holling, C. S. 1959. The components of predation as revealed by a study of small-mammal predation of the European pie sawfly. Canadian Entomologist 91:293–320.

Holroyd, G. L., C. J. Conway, and H. E. Trefry. 2011. Breeding dispersal of a burrowing owl from Arizona to Saskatchewan. Wilson Journal of Ornithology 123:378–381.

Horne, J. S., L. S. Mills, J. M. Scott, K. M. Strickler, and S. A. Temple. 2020. Ecology and management of small populations. Pages 270–292 in N. J. Silvy, editor. The wildlife techniques manual. 8th edition. Volume 2. Johns Hopkins University Press, Baltimore, Maryland, USA.

Hostetler, J. A., and R. B. Chandler. 2015. Improved state-space models for inference about spatial and temporal variation in abundance from count data. Ecology 96:1713–1723.

Humbert, J.-Y., L. S. Mills, J. S. Horne, and B. Dennis. 2009. A better way to estimate population trend. Oikos 118:1487–1498.

Hutchins, H. E., and R. M. Lanner. 1982. The central role of Clark's nutcracker in the dispersal and establishment of whitebark pine. Oecologia 55:192–201.

Jarvis, J. U. M., N. C. Bennett, and A. Spinks. 1998. Food availability and foraging by wild colonies of Damaraland mole-rats (Cryptomys amarensis): implications for sociality. Oecologia 113:290–298.

Johnson, H. E., M. Hebblewhite, T. R. Stephenson, D. W. German, B. M. Pierce, and V. C. Bleich. 2013. Evaluating apparent competition in limiting the recovery of an endangered ungulate. Oecologia 171:295–307.

Johnson, H. E., D. L. Lewis, and S. W. Breck. 2020. Individual and population fitness consequence associated with large carnivore use of residential development. Ecosphere 11:e03098.

Johnson, H. E., L. S. Mills, J. Wehausen, and T. R. Stephenson. 2010a. Population-specific vital rate contributions influence management of an endangered ungulate. Ecological Applications 20:1753–1765.

Johnson, H. E., L. S. Mills, J. Wehausen, and T. R. Stephenson. 2010b. Combining ground count, telemetry, and mark-resight data to infer population dynamics in an endangered species. Journal of Applied Ecology 47:1083–1093.

Johnson, H. E., L. S. Mills, J. Wehausen, T. R. Stephenson, and G. Luikart. 2011. Translating inbreeding depression on component vital rates to overall population growth in endangered bighorn sheep. Conservation Biology 25:1240–1249.

Keeley, A. T. H., P. Beier, T. Creech, K. Jones, R. H. G. Jongman, G. Stonecipher, and G. M. Tabor. 2019. Thirty years of connectivity conservation planning: an assessment of factors influencing plan implementation. Environmental Research Letters 14:103001.

Keith, L. B. 1990. Dynamics of snowshoe hare populations. Pages 119–195 in H. H. Genoways, editor. Current mammalogy. Plenum, New York, New York, USA.

Kelly, M. A., J. Betsch, C. Wultsch, B. Mesa, and L. S. Mills. 2012. Noninvasive sampling for carnivores. Pages 47–69 in L. Boitani and R. Powell, editors. Carnivore ecology and conservation. Oxford University Press, Oxford, UK.

Kramer, A. M., B. Dennis, A. M. Liebhold, and J. M. Drake. 2009. The evidence for Allee effects. Society of Population Ecology 51:341–354.

Krebs, C. J., R. Boonstra, S. Boutin, and A. R. E. Sinclair. 2001. What drives the 10-year cycle of snowshoe hares? BioScience 51:25–35.

Krebs, C. J., S. Boutin, R. Boonstra, A. R. E. Sinclair, J. N. M. Smith, M. R. T. Dale, K. Martin, and R. Turkington. 1995. Impact of food and predation on the snowshoe hare cycle. Science 269:1112–1115.

Kumar, A. V. 2020. Biotic and abiotic drivers of acyclic snowshoe hare population dynamics in a spatiotemporally complex system. PhD dissertation, University of Montana, Missoula, USA.

Kuussaari, M., I. Saccheri, M. Camara, and I. Hanski. 1998. Allee effects and population dynamics in the Glanville fritillary butterfly. Oikos 82:384–392.

Lande, R., S. Engen, and B.-E. Sæther. 2003. Stochastic population dynamics in ecology and conservation. Oxford University Press, Oxford, UK.

Landeau, L., and J. Terborgh. 1986. Oddity and the "confusion effect" in predation. Animal Behaviour 34:1372–1380.

Laufenberg, J. S., H. E. Johnson, P. F. Doherty Jr., and S. W. Breck. 2018. Compounding effects of human development and a natural food shortage on a black bear population along a human development-wildland interface. Biological Conservation 224:118–198.

Laundre, J., and T. W. Clark. 2003. Managing puma hunting in the western United States: through a metapopulation approach. Animal Conservation 6:159–170.

Levins, R. 1970. Extinction. Lectures on Mathematics in the Life Sciences 2:75–107.

Lidicker, W. Z., Jr. 1975. The role of dispersal in the demography of small mammals. Pages 103–128 in B. Golley, K. Petrusewicz, and L. Ryszkowski, editors. Small mammals: their productivity and population dynamics. Cambridge University Press, Cambridge, UK.

Lindenmayer, D. B., and G. E. Likens. 2009. Adaptive monitoring: a new paradigm for long-term research and monitoring. Trends in Ecology and Evolution 24:482–486.

Link, W. A., and J. R. Sauer. 2002. A heirarchical analysis of populaton change with application to cerulean warblers. Ecology 83:2832–2840.

Long, R. A., P. MacKay, W. J. Zielinski, and J. C. Ray. 2008. Noninvasive survey methods for carnivores. Island Press, Washington DC, USA.

Lorenz, T. J., K. A. Sullivan, A. V. Bakian, and C. A. Aubry. 2011. Cache-site selection in Clark's nutcracker (*Nucifraga columbiana*). Auk 128:237–247.

Loveridge, A. J., A. W. Searle, F. Murindagomo, and D. W. Macdonald. 2007. The impact of sport-hunting on the population dynamics of an African lion population in a protected area. Biological Conservation 134:548–558.

Lowe, W. H. 2003. Linking dispersal to local population dynamics: a case study using a headwater salamander system. Ecology 84:2145–2154.

Lowe, W. H., and F. W. Allendorf. 2010. What can genetics tell us about population connectivity? Molecular Ecology 19:3038–3051.

Lyons, J. E., M. C. Runge, H. P. Laskowski, and W. L. Kendall. 2008. Monitoring in the context of structured decision-making and adaptive management. Journal of Wildlife Management 72:1683–1692.

MacKenzie, D. I., J. D. Nichols, J. A. Royle, K. H. Pollock, L. L. Bailey, and J. E. Hines. 2018. Occupancy estimation and modeling: inferring patterns and dynamics of species occurrence. 2nd edition. Elsevier, London, UK.

MacKenzie, D. I., and J. A. Royle. 2005. Designing occupancy studies: general advice and allocating survey efforts. Journal of Applied Ecology 42:1105–1114.

Manlik, O., R. C. Lacy, and W. B. Sherwin. 2018. Applicability and limitations of sensitivity analyses for wildlife management. Journal of Applied Ecology 55:1430–1440.

Marsh, D. M., and P. C. Trenham. 2008. Current trends in plant and animal population monitoring. Conservation Biology 22:647–655.

Martin, K. D., T. Schommer, and V. L. Coggins. 1996. Literature review regarding the compatibility between bighorn and domestic sheep. Biennal Symposium of the Northern Wild Sheep and Goat Council 10:72–77.

Messenger, A. M., A. N. Barnes, and G. C. Gray. 2014. Reverse zoonotic disease transmission (zooanthroponosis): a systematic review of seldom-documented human biological threats to animals. PLoS One 9:e89055.

Mills, L. S. 2007. Conservation of wildlife populations: demography, genetics, and management. Wiley-Blackwell, Malden, Massachusetts, USA.

Mills, L. S. 2013. Conservation of wildlife populations: demography, genetics, and management. 2nd edition. Wiley-Blackwell, Malden, Massachusetts, USA.

Mills, L. S., and F. W. Allendorf. 1996. The one-migrant-per-generation rule in conservation and management. Conservation Biology 10:1509–1518.

Mills, L. S., M. K. Schwartz, D. A. Tallmon, and K. P. Lair. 2003. Measuring and interpreting connectivity for mammals in coniferous forests. Pages 587–613 in C. J. Zabel and R. G. Anthony, editors. Mammal community dynamics: management and conservation in the coniferous rorests of western North America. Cambridge University Press, Cambridge, UK.

Mooring, M. S., T. A. Fitzpatrick, T. T. Nishihira, and D. D. Reisig. 2004. Vigilance, predation risk, and the Allee effect in desert bighorn sheep. Journal of Wildlife Management 68:519–532.

Morris, W. F., and D. F. Doak. 2002. Quantitative conservation biology: theory and practice of population viability analysis. Sinauer Associates, Sunderland, Massachusetts, USA.

Newby, J. R., L. S. Mills, T. K. Ruth, D. H. Pletscher, M. S. Mitchell, H. B. Quigly, K. M. Murphy, and R. DeSimone. 2013. Human-caused mortality influences spatial population dynamics: pumas in landscapes with varying mortality risks. Biological Conservation 159:230–239.

Nichols, J. D., F. A. Johnson, and B. K. Williams. 1995. Managing North American waterfowl in the face of uncertainty. Annual Review of Ecology and Systematics 26:177–199.

Nichols, J. D., and B. K. Williams. 2006. Monitoring for conservation. Trends in Ecology and Evolution 21:668–673.

Norén, K., E. Godoy, L. Dalén, T. Meijer, and A. Angerbjörn. 2016. Inbreeding depression in a critically endangered carnivore. Molecular Ecology 25:3309–3318.

Northrup, J. M., J. W. Rivers, Z. Yang, and M. G. Betts. 2019. Synergistic effects of climate and land-use change influence broad-scale avian population declines. Global Change Biology 25:1561–1575.

Oyler-McCance, S. J., E. K. Latch, and P. L. Leberg. 2020. Conservation genetics and molecular ecology in wildlife management. Pages 526–546 in N. J. Silvy, editor. The wildlife techniques manual. 8th edition. Volume 1. Johns Hopkins University Press, Baltimore, Maryland, USA.

Ozgul, A., K. B. Armitage, D. T. Blumstein, and M. K. Oli. 2006. Spatiotemporal variation in survival rates: implications for population dynamics of yellow-bellied marmots. Ecology 87:1027–1037.

Packila, M. L., M. D. Riley, R. S. Spence, and R. M. Inman. 2017. Long-distance wolverine dispersal from Wyoming to historic range in Colorado. Northwest Science 91:399–407.

Peacock, E., and D. L. Garshelis. 2006. Comment on "On the regulation of populations of mammals, birds, fish, and insects IV." Science 313:45.

Pedersen, A. B., K. E. Jones, C. L. Nunn, and S. A. Altizer. 2007. Infectious disease and mammalian extinction risk. Conservation Biology 21:1269–1279.

Post, E., N. C. Stenseth, R. O. Peterson, J. A. Vucetich, and A. M. Ellis. 2002. Phase dependence and population cycles in a large-mammal predator-prey system. Ecology 83:2997–3002.

Preisser, E. L., D. I. Bolnick, and M. F. Benard. 2005. Scared to death? the effects of intimidation and consumption in predator-prey interactions. Ecology 86:501–509.

Proctor, M. F., D. Paetkau, B. N. McLellan, G. B. Stenhouse, K. C. Kendall, et al. 2012. Population fragmentation and inter-ecosystem movements of grizzly bears in western Canada and the northern United States. Wildlife Monographs 180:1–46.

Pulliam, H. R. 1988. Sources, sinks, and population regulation. American Naturalist 132:652–661.

Rasa, O. A. E. 1989. The costs and effectiveness of vigilance behavior in the dwarf mongoose: implications for fitness and optimal group size. Ethology Ecology and Evolution 1:265–282.

Raveling, D. G. 1989. Nest-predation rates in relation to colony size of black brant. Journal of Wildlife Management 53:87–90.

Reed, J. M. 1999. The role of behavior in recent avian extinctions and endangerments. Conservation Biology 13:232–241.

Robertson, B. A., and R. L. Hutto. 2006. A framework for understanding ecological traps and an evaluation of existing evidence. Ecology 87:1075–1085.

Robinson, H. S., and R. M. DeSimone. 2011. The Garnet Range Mountain Lion Study: characteristics of a hunted population in west-central Montana. Final report. Montana Department of Fish, Wildlife and Parks, Helena, USA.

Robinson, H. S., R. B. Wielgus, H. S. Cooley, and S. W. Cooley. 2008. Sink populations in carnivore management: cougar demography and immigration in a hunted population. Ecological Applications 18:1028–1037.

Royama, T. 1992. Analytical population dynamics. Chapman and Hall, London, UK.

Royle, J. A., and R. M. Dorazio. 2008. Hierarchical modeling and inference in ecology: the analysis of data from populations, metapopulations and communities. Elsevier, London, UK.

Runge, J. P., M. C. Runge, and J. D. Nichols. 2006. The role of local populations within a landscape context: defining and classifying sources and sinks. American Naturalist 167:925–938.

Running, S., and L. S. Mills. 2009. Terrestrial ecosystem adaptation. Resources for the Future, Washington, DC, USA. http://www.rff.org/News/Features/Pages/09-07-08-Managing-for-Resilience.aspx.

Rushing, C.S., T. B. Ryder, and P. P. Marra. 2016. Quantifying drivers of population dynamics for a migratory bird throughout the annual cycle. Proceedings of the Royal Society Series B 283:20152846.

Sæther, B.-E., and Ø. Bakke. 2000. Avian life history variation and contribution of demographic traits to the population growth rate. Ecology 81:642–653.

Schaub, M., and F. Abadi. 2011. Integrated population models: a novel analysis framework for deeper insights into population dynamics. Journal of Ornithology 152 (Suppl. 1):S227–S237.

Schultz, C. 2010. Challenges in connecting cumulative effects analysis to effective wildlife conservation planning. BioScience 60:545–551.

Schwartz, M. K., L. S. Mills, K. S. McKelvey, L. F. Ruggiero, and F. W. Allendorf. 2002. DNA reveals high dispersal synchronizing the population dynamics of Canada lynx. Nature 415:520–522.

Scott, J. M., D. D. Goble, J. A. Wiens, D. S. Wilcove, M. Bean, and T. Male. 2005. Recovery of imperiled species under the Endangered Species Act: the need for a new approach. Frontiers in Ecology and the Environment 3:383–389.

Scott, J. M., J. A. Wiens, B. Van Horne, and D. D. Goble. 2020. Shepherding nature: the challenge of conservation reliance. Cambridge University Press, Cambridge, UK.

Sibly, R. M., D. Barker, M. C. Denham, J. Hone, and M. Pagel. 2005. On the regulation of populations of mammals, birds, fish and insects. Science 309:607–610.

Singer, F. J., E. Williams, M. W. Miller, and L. C. Zeigenfuss. 2000. Population growth, fecundity, and survivorship in recovering populations of bighorn sheep. Restoration Ecology 8:75–84.

Skalski, J. R., K. E. Ryding, and J. J. Millspaugh. 2005. Wildlife demography: an analysis of sex, age, and count data. Elsevier Academic Press, Burlington, Massachusetts, USA.

Smith, K. F., K. Acevedo-Whitehouse, and A. B. Pedersen. 2009. The role of infectious diseases in biological conservation. Animal Conservation 12:1–12.

Stirling, I., and A. E. Derocher. 2012. Effects of climate warming on polar bears: a review of the evidence. Global Change Biology 18:2694–2706.

Sweanor, L. L., K. A. Logan, and M. G. Hornocker. 2000. Cougar dispersal patterns, metapopulation dynamics, and conservation. Conservation Biology 14:798–808.

Taylor, B. L., and T. Gerrodette. 1993. The uses of statistical power in conservation biology: the vaquita and northern spotted owl. Conservation Biology 7:489–500.

United Nations. 2019. World population prospects 2019: highlights. Department of Economic and Social Affairs, Population Division. https://population.un.org/wpp/Publications/Files/WPP2019_Highlights.pdf.

US Fish and Wildlife Service. 2007. Recovery plan for the Sierra Nevada bighorn sheep. Sacramento, California, USA.

US Fish and Wildlife Service. 2016. Endangered and threatened wildlife and plants: removing the San Miguel island fox, Santa Rosa island fox, and Santa Cruz island fox from the federal list of endangered and threatened wildlife, and reclassifying the Santa Catalina island fox from endangered to threatened. Federal Register 81:53315.

van de Kerk, M., D. P. Onorato, J. A. Hostetler, B. M. Bolker, and M. K. Oli. 2019. Dynamics, persistence, and genetic management of the endangered Florida panther population. Wildlife Monographs 203:3–35.

Van Horne, B. 1983. Density as a misleading indicator of habitat quality. Journal of Wildlife Management 47:893–901.

Veit, R. R., and M. A. Lewis. 1996. Dispersal, population growth, and the Allee effect: dynamics of the house finch invasion of eastern North America. American Naturalist 148:255–274.

Venter, O., E. W. Sanderson, A. Magrach, J. R. Allan, J. Beher, et al. 2016. Sixteen years of change in the global terrestrial human footprint and implications for biodiversity conservation. Nature Communications 7:12558.

Vucetich, J. A., and T. A. Waite. 2000. Is one migrant per generation sufficient for the genetic management of fluctuating populations? Animal Conservation 3:261–266.

Vucetich, J. A., T. A. Waite, L. Qvarnemark, and S. Ibargüen. 2000. Population variability and extinction risk. Conservation Biology 14:1704–1714.

Waters, C. N., J. Zalasiewicz, C. Summerhayes, A. D. Barnosky, C. Poirier, et al. 2016. The Anthropocene is functionally and stratigraphically distinct from the Holocene. Science 351:aad2622.

Weldon, A. J., and N. M. Haddad. 2005. The effects of patch shape on indigo buntings: evidence for an ecological trap. Ecology 86:1422–1431.

White, P. J., and R. A. Garrott. 1999. Population dynamics of kit foxes. Canadian Journal of Zoology 77:486–493.

Wilcove, D. S., D. Rothstein, J. Dubow, A. Phillips, and E. Losos. 1998. Quantifying threats to imperiled species in the United States. Bioscience 48:607–615.

Williams, B. K. 2001. Uncertainty, learning, and optimization in wildlife management. Environmental and Ecological Statistics 8:269–288.

Williams, B. K., J. D. Nichols, and M. J. Conroy. 2002. Analysis and management of animal populations. Academic Press, San Diego, California, USA.

Williams, J. D., and J. J. McDermott. 2004. Hermit crab biocoenoses: a worldwide review of the diversity and natural history of hermit crab associates. Journal of Experimental Marine Biology and Ecology 305:1–128.

Wintle, B. A., M. C. Runge, and S. A. Bekessy. 2010. Allocating monitoring effort in the face of unknown unknowns. Ecology Letters 13:1325–1337.

Wirsing, A. J., T. D. Steury, and D. L. Murray. 2002. A demographic analysis of a southern snowshoe hare population in a fragmented habitat: evaluating the refugium model. Canadian Journal of Zoology 80:169–177.

Wisdom, M. J., L. S. Mills, and D. F. Doak. 2000. Life-stage simulation analysis: estimating vital rate effects on population growth for conservation. Ecology 81:628–641.

Wittmer, H. U., B. N. McLellan, R. Serrouya, and C. D. Apps. 2007. Changes in landscape composition influence the decline of a threatened woodland caribou population. Journal of Animal Ecology 76:568–579.

Wittmer, H. U., A. R. E. Sinclair, and B. N. McLellan. 2005. The role of predation in the decline and extirpation of woodland caribou. Oecologia 144:257–262.

Yoccoz, N. G., J. D. Nichols, and T. Boulinier. 2001. Monitoring of biological diversity in space and time. Trends in Ecology and Evolution 16:446–453.

Zipkin, E. F., and S. P. Saunders. 2018. Synthesizing multiple data types for biological conservation using integrated population models. Biological Conservation 217:240–250.

WILDLIFE HEALTH AND DISEASES

DAVID A. JESSUP

INTRODUCTION

"The role of disease in wildlife conservation has probably been radically underestimated" (Leopold 1933:325). Leopold's words certainly seem prescient today. With the SARS-CoV-2 pandemic, our world is forever changed, and this certainly applies to the traditional wildlife management perspective that disease is just a form of compensatory loss and of no significance to wildlife populations. Diseases can reduce wildlife populations and harvestable surplus, at times profoundly enough to affect ecosystem health and sustainability. And, through effects on domestic animal and human health, disease can radically alter conservation priorities. This is also what the One World–One Health concept tells us—the health of animals, people, and environments are inexorably bound together (Centers for Disease Control and Prevention 2021).

New diseases emerge, existing diseases get moved around, and others lie in wait for the right conditions to reemerge. The conditions for disease emergence include host, agent, and environmental factors, and to have serious effects, the disease process must be able to propagate in susceptible host populations through time and space. Many disease-causing organisms, and most parasites, have coevolved with their hosts to some degree, limiting their pathogenicity. But what balance may have existed can be upset by disease agents entering new hosts and new ecosystems, by hosts entering new ecosystems or coming in contact with existing disease they have no experience with, by changes in host immunity or agent pathogenicity, or because of changes in ecosystems that alter established host and agent relationships. Diseases may be a consequence of, or work in concert with, climate change and other forms of anthropogenic wildlife exploitation, including but not limited to poaching, bushmeat and illegal trade in live and dead wildlife, habitat degradation for monoculture or industrial farming, and many forms of pollution. Optimizing the health of wildlife and that of the ecosystems on which they (and we) depend (One Health) is one of the great challenges facing wildlife management in the 21st century.

In the past decade, biologists have seen a number of instances where wildlife health and disease have transcended the perceived borders between biology and the social, legal, financial, and political realms. As little as 5–10 years ago, few would have believed this could become the new normal. Even the epidemiologists and microbe hunters who warned us of the medical consequences of species jumping pathogens probably did not expect that a disease of apparent wildlife origin would change the world's economy and political order in 2020. This chapter will first examine some of the biological, social, legal, and financial consequences of selected wildlife diseases. Following this examination are seven case studies of the biomedical aspects and biological effects of wildlife health-disease problems, and then a discussion of methods for wildlife disease investigation and management.

THE BIG PICTURE

The Wildlife Disease That Changed Everything: SARS-CoV-2

Late in 2019, a coronavirus that appears to have evolved in bats, and perhaps transited other species, began to change the world as we had known it. It did so not by its effect on bats, bat populations, or ecological processes dependent on bats, but by its effect on human health, world and regional travel and economics, laws and policies, social order and politics. For reasons not yet well understood, SARS-CoV-2 began replicating in humans and passing rapidly from person to person. Although less lethal than the first severe acute respiratory syndrome coronavirus (SARS-CoV-1) in 2003, whose source appears to have been civet cats (Viverridae), SARS-CoV-2 was able to infect some people without causing overt disease and thus spread undetected. Epidemiologist would describe SARS-CoV-2 as having a high infectivity rate, like cold and seasonal flu viruses. It is perhaps ironic that far more lethal viruses like Ebola, hanta, and Marburg, which can kill approximately 50% or more of infected people, may be of less danger to society in general and the world economy, than less lethal (1% to 2%) viruses that have higher infectivity rates. Although avian influenza provided some warning, this has surprised traditional wildlife management thinking about wildlife diseases, and we are all now witness to the power and importance of zoonotic pandemics of wildlife origin, a potential previously relegated to the realm of science fiction.

At the time of the writing of this chapter more than 216 million people worldwide have been diagnosed with COVID-19 (the disease caused by SARS-CoV-2), almost 4.5 million have died, and the numbers continue to rise. It is approaching the morbidity and mortality levels of the 1918–1920 great influenza pandemic, despite huge advances in communications, medical knowledge, and infrastructure. It has shut down or vastly reduced travel within and between nations, has altered world trade, and has cost more than $16 trillion. The pandemic has caused major social and some political changes in many nations, including the United States. On the conservation front, it is hoped that COVID-19, or the threat of something like it, will change people's values and behaviors toward wildlife and natural ecosystems. Whether it will reduce the demand for live and dead wild animal tissues (poaching, bushmeat, wet markets) is an unanswered question.

It is worth remembering that zoonotic pandemics are not just a foreign or wildlife phenomenon. The great influenza of 1918–1920 appears to have originated in pigs from farms near Fort Riley, Kansas, USA. Commonly accepted practices like concentration of wildlife and exotic animals under stress at auction yards and at breeding farms and warehouses; or holding animal waste in sewage lagoons from industrial-scale farming, and human waste management that allows bacteria, viruses, and parasites to sit and percolate with antibiotics, antivirals, and other drugs have the potential to serve as sources for epidemics too. So, there are many reasons why conservationists should be interested in the health and diseases of animals, particularly wildlife.

Population- and Ecosystem-Level Effects: Rinderpest

Perhaps the best historic example of a disease profoundly affecting many wildlife species, the ecosystems they live in, and the interconnected human populations on a continental scale is that of rinderpest in Africa, particularly East Africa (https://journals.plos.org/plosbiology/article?id=10.1371/journal.pbio.1000210). Rinderpest virus came to Africa from Asia in the 1890s during a period of intense colonization and military activity by the imperial nations via cattle brought along as food on the hoof. It quickly spread to a wide variety of Africa ungulates, including Cape buffalo (*Syncerus caffer*), giraffe (*Giraffe giraffe*), warthog (*Phacochoerus africanus*), and all Tragelaphinae (spiral-horned antelope, ~70 species), which were particularly susceptible. Wildebeest (*Connochaetes taurinus*) were affected in the first wave of infection and subsequently suffered a more subtle yearling disease in the Serengeti up until rinderpest elimination in the 1960s. Gazelles (*Gazella* spp.) and small antelope are generally resistant yet show greater susceptibility to the similar small ruminant virus that causes goat plague, which is somewhat difficult to distinguish

from rinderpest. Rinderpest evolved with domestication of cattle in Eurasia, so native African ungulates had no innate or acquired immunity, which in part explains their susceptibility. In a few years, the disease killed more than 90% of cattle and cloven-hooved wild animal species from the Horn of Africa down to the Cape of Good Hope. The initial wave through eastern Africa was swift, and then the miombo (i.e., woodlands) slowed down the epidemic until it broke through to Rhodesia, possibly via wildlife carrying the virus, and a second wave ensued. Rinderpest vastly altered ecological processes in ecosystems that evolved with grazing, with trickle-down effects on wildlife abundance and distribution, including carnivores, and the lives of hunter-gatherers, herders, and agriculturalists who depended on cattle (for which wildlife remained a possible source of infection). In Ethiopia alone, a third of its human population was lost, and rinderpest impoverished hundreds of millions of people for more than 100 years (Food and Agriculture Organization 2010).

Like other morbilliviruses (e.g., measles, canine distemper, phocine and dolphin morbilliviruses), rinderpest does not survive long outside a living host and relies on rapid spread between animals to maintain itself in an environment. Thus, reduced contact and immunization were tools that could be used to combat it. Early efforts to control or eliminate rinderpest in the 1980s showed promise, but the vaccine required refrigeration to remain potent, remote regions and cattle populations were neglected, and the program failed. Beginning in 1993, efforts to use a new heat-stabile rinderpest vaccine in cattle in Africa and Asia began. The vaccine was also tested on wildlife species. The Food and Agriculture Organization, supported mostly by the European Union, established the Pan African Rinderpest Campaign, which succeeded in restricting the disease to a few hot spots (e.g., Sudan). The Somali region of eastern Africa was the last stronghold where a mixed population of wildlife and pastoral livestock created the conditions for the virus to circulate. The final push, which focused on Kenya, Tanzania, Ethiopia, and Somalia, under the Program for the Pan African Control of Epizootics, was able to build veterinary surveillance capacity in wildlife. It and the Community-Based Animal Health and Participatory Epidemiology Project for development of veterinary service, vaccination, and disease surveillance in remote and conflict-prone areas resulted in near elimination of rinderpest in livestock. In the late 1990s recurrent outbreaks occurred in wildlife, mainly Cape buffalo, suggesting asymptomatic cattle herds could be reinfecting wildlife populations, and asymptomatic wildlife could be reinfecting cattle. Decades of effort costing hundreds of millions of Euros hung in the balance. Work by the Kenya Wildlife Service showed the patterns of infection among wildlife and how populations of Cape buffalo were biologically competent to maintain the virus while showing only mild clinical signs of disease. Once these situations were dealt with, the last roadblock to elimination of rinderpest fell. By 2000 the last known case of rinderpest in Asia was detected in Pakistan. In 2001 the last known case of rinderpest in Africa was confirmed in Cape buffalo in Meru National Park, Kenya. Intensive vaccination of this zone and eastward into Somalia ensued. The successful elimination of live circulating rinderpest virus in wildlife and livestock was confirmed by 2003–2004. Intensive surveillance continued until 2010 without evidence of any viral circulation, and the world now appears free of the rinderpest virus, only the second disease scourge (after smallpox) ever eradicated completely.

Africa is rapidly losing wildlife habitat to development, mostly for crop agriculture, human population growth, and infrastructure. Livestock herding persists in many pastoral communities, in part because of the success of rinderpest eradication. Pastoral herding is not a major threat to conservation, and it reduces the likelihood of further development of these critical rangelands on the continent. Removal of rinderpest has increased the resilience of African wildlife populations, which tolerate most indigenous diseases such as foot-and-mouth disease and trypanosomiasis (lethal to cattle), and wild ungulate populations have shown some recovery. Although eliminating rinderpest, an introduced pathogen, may be a mixed blessing, removal of indigenous microorganisms may not serve conservation well because they may be essential to a balanced ecosystem and biodiversity. One side effect of the rinderpest elimination efforts was significant improvement of wildlife health and domestic animal health infrastructure in some African countries, a better understanding of the role of wildlife in disease epidemiology and as a sentinel for livestock diseases. It also increased regional cooperation on diseases, the involvement of wildlife health professionals, and the consideration of ecological processes in disease control.

Other Examples

In the latter half of the 20th century, type C botulism and avian cholera waterfowl mortalities were of such a magnitude in some North American flyways that harvest of some species of waterfowl was reduced or curtailed. Avian influenza (AI) has the potential to cause similar large-scale waterfowl die-offs as occurred in barred geese (*Anser indicus*) and other species at Lake Quinghai, China, in 2005. Like chronic wasting disease (CWD), AI has the potential to negatively affect wildlife management programs as the result of the public perception (or reality) that contact with wild ducks and geese may be a serious human health risk. It is not possible to differentiate the gross lesions of avian influenza from those of avian cholera or duck virus enteritis in the field (Fig. 8.1).

For a few years in the 1950s and 1960s, bluetongue (actually two viral diseases) in white-tailed deer (*Odocoileus virginianus*) was believed to be a serious threat to deer population recovery in the southeastern United States. Even today, major outbreaks of bluetongue or epizootic hemorrhagic disease can result in altered hunting regulations.

The most striking recent examples of the negative effects of wildlife diseases at the population level do not involve harvested species. The global and prolonged mass mortality—and in some cases extinction of multiple amphibian species because of chythrid (*Batrachochytrium dendrobatidis*) fungus infections (Berger et al. 1998)—and massive bat mortality in North America caused by another fungal disease, white-nose syndrome (Frick et al. 2010), are cases in point. Both diseases are so lethal and persistent that they can greatly reduce the ecosystem services provided by bats (e.g., insectivory, pollination) and amphibians. Ironically and somewhat mysteriously, fungal organisms are seldom highly infectious or highly pathogenic to their hosts, so the emergence and rapid spread of both of these diseases could be related to larger phenomena (e.g., immune function effects of climate change or contaminants) that compromise disease resistance.

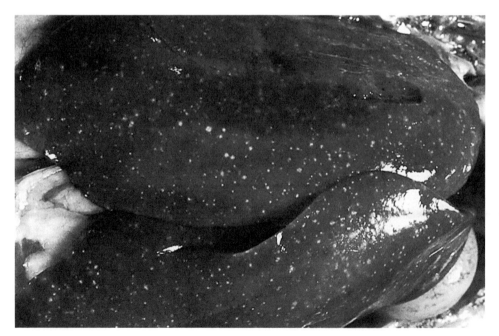

Figure 8.1. The lesions of fowl cholera are relatively obvious on gross postmortem examination, but similar lesions can be caused by avian influenza and duck virus enteritis. Affected ducks, geese, and swans usually die peracutely, in good body condition, have small (petechial to ecchymotic) hemorrhages on the heart and great vessels, and have multiple small white spots (septic infarcts) on the liver (*shown*). A blood smear stained with Wright's or Geimsa reveals many bipolar staining rod-shaped bacteria. Courtesy of Cornell University

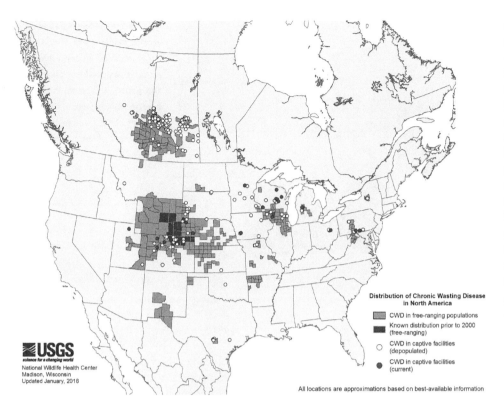

Figure 8.2. This map of the distribution of chronic wasting disease (CWD) in wild and captive cervids in North America shows the extent to which this disease has spread in approximately 20 years and the apparent association of infected captive facilities with infection in free-ranging cervids. The discontinuous nature of the distribution suggests movement of infected animals across state and national borders. Reprinted with permission of the US Geological Survey–National Wildlife Health Center; map also available at https://www.usgs.gov/media/images/distribution-chronic-wasting-disease-north-america-0

Perception Is Reality: Chronic Wasting Disease

Chronic wasting disease has geographically advanced so inexorably and has such great potential to negatively affect wildlife management it has garnered attention of all management agencies and the public (Fig. 8.2). This is only in small part because of the morbidity and premature mortality it causes in cervids. There has never been a case of encephalopathy caused by CWD diagnosed in humans, but some believe that deer (*Odocoileus* spp.), elk (*Cervus canadensis*), moose (*Alces alces*), and other cervids might harbor an untreatable prion organism similar to the cause of mad cow disease and early-onset Creutzfeldt-Jakob disease in humans that could profoundly affect big-game hunting across North America. Although CWD is not a known human health risk (like SARS-CoV-2 or rabies) or threat to livestock health (like rinderpest or brucellosis), both modeling and field observation have shown it is capable of reducing deer population sizes, and thus the huntable surplus. Harvest of wildlife is big business, in part because people view wild meat as a healthy and a desirable luxury. If that perception is significantly altered, and harvested wildlife are viewed as unhealthy, the financial consequences could be dire for recreation-based communities, hunting and fishing equipment manufacturers and suppliers, and government wildlife management agencies whose budgets are supported by hunting licenses and fees. With disease—as with other areas of conservation—controversy, public perception, and social values can sometimes override biological realities.

Chronic wasting disease has also revealed weakness in the North American model of wildlife conservation (the Model). The commercialization of wildlife (anathema to the Model) in the form of cervid ranching or farming has contributed to the movement of CWD in North America, with CWD-infected captive cervids serving as a source (nidus) of infection for free-ranging cervids. The major geographic spread of CWD began in the mid- to late 1980s with increasing legal and illegal transport of deer and elk across state lines. For more than three decades, there was no nonlethal test available to help identify infected animals and premises, making surveillance difficult and time-consuming. The reclassification of captive deer and elk as alternative livestock in many states, to remove them from jurisdiction of wildlife management agencies, has contributed to the spread of disease and problems cleaning up infected premises. Ironically, CWD has done as much or more than legal or political efforts to limit commercial deer farming for venison and other products by making it economically unviable. But it has done less to limit the raising of trophy deer for shooting within fenced facilities. More on CWD is found in "Case Studies," below.

Fear and Social Expectations: Rabies

At the interface between public health and wildlife disease, public fears and expectations are important. Rabies in human beings is a terrible and almost uniformly fatal viral disease, but it is now exceedingly rare in North America. And public expects it to stay that way. Vaccination greatly diminished the role of domestic animals in rabies transmission in the 1960s and 1970s, and wildlife inherited the weight of social and political commitment to stamp it out. The result has been decades of studies on rabies ecology and development of vaccine and bait technologies, laboratory and field trials, and field ap-

Figure 8.3. Modified live rabies vaccine bait for carnivores. This approach to rabies control has proven considerably more effective and socially acceptable than trapping and killing were in the 1950s through 1980s. Courtesy of the US Department of Agriculture Wildlife Services

plications that yielded positive results in controlling rabies in North American meso-carnivores (fox, skunk, and racoon). Experimental work with recombinant rabies vaccines shows promise for potential preventive vaccination of bats, including but not limited to vampire bat (Stading et al. 2017). Because rabies is a public health issue (Fig. 8.3), traditional wildlife management agencies have had limited involvement. From a traditional wildlife management perspective, rabies is not of significance to wildlife population numbers, but wildlife management agencies are clearly expected to be responsive to public fears and expectations.

Disruption of Commerce: African Swine Fever

African swine fever (ASF), caused by an Asfivirus, is lethal (90% to 95%) to all swine, infectious, and persistent (months to years). It is a disease the US Department of Agriculture (USDA) has attempted to exclude from North America for decades, and until 2007 it was thought to be limited to Africa, largely in a sylvatic transmission cycle involving warthogs and soft ticks (*Ornithodoros* spp.). A genotype II ASF virus emerged in domestic swine in the Caucasus in 2007 and then spread to adjacent eastern European countries, where it developed into a wild boar–habitat cycle (Guberti et al. 2019). By 2014 it was in Belgium, Hungary, Poland, and Romania; by 2018 had spread to China; and by 2019 to Cambodia, Mongolia, and Vietnam (Guberti et al. 2019). The efforts to control it with massive depopulation and slaughter programs cost many billions of dollars (Brown and Bevins 2018). African swine fever has most recently (2021) been detected in Haiti.

The ASF virus is highly infectious and lasts for months to years in urine, feces, blood, carcasses, soils, prepared and frozen meats, and on knives, boots, tires, and other surfaces (Guberti et al. 2019). Flesh flies may also spread the virus for several days. When the virus is introduced into an ASF-free wild boar (*Sus scrofa*) population, an epidemic is likely, and the more effective the spread of the virus, the sooner it leads to a relatively rapid decline of the wild boar population (Guberti et al. 2019). In hunted populations the reduction of wild boar numbers becomes evident even more quickly (Guberti et al. 2019). As a result of decreasing populations, the number of interspecific contacts also declines, and the epidemic moves into an endemic phase (Guberti et al. 2019). A fade-out of the disease may be apparent, but its reappearance within months is a common occurrence, likely determined by wild boars moving within the infected area and contacting the dormant virus in the infectious wild boar carcasses (Guberti et al. 2019). While the virus tends to remain endemic in previously infected areas (mainly because of infected carcasses), it also spreads by direct contact into the yet unaffected, neighboring wild boar groups (Guberti et al. 2019). Most experts believe it is not a question of whether ASF will break out in North America, but when. This is particularly troubling as over the past three decades feral swine have become widespread in the United States, and we have native species of soft ticks whose vector competence is unknown (Brown and Bevins 2018). The susceptibility of javelina (*Tayassu tajacu*) populations to ASF is unknown.

Beginning in the 1980s, concurrent with the commercialization of cervids and growth in commercial hunting operations in many states, wild pig and wild boar hybrids began showing up in many areas of the United States and Canada where they had not been found before (Fig. 8.4), many likely because of illegal transport. Wild pigs with European boar characteristics survive well in harsh climates and reproduce prolifically, making control by hunting nearly impossible. Wild pigs, which are non-native omnivores, create serious ecological problems. Should (or when) ASF gets a foothold in North America, it will be extremely difficult to control or eliminate, with losses to domestic pork producers and cost of control potentially enormous. Management of ASF will result in tremendous financial, political, and legal pressure on wildlife agencies

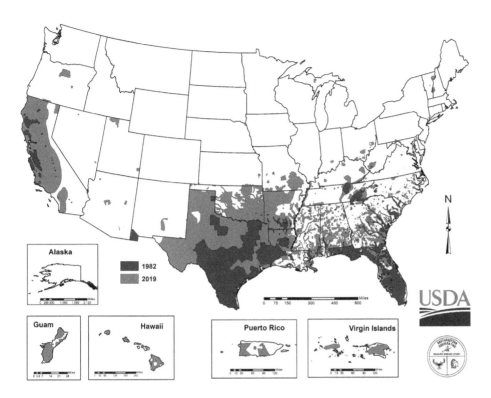

Figure 8.4. The distribution of wild pigs in North America has changed dramatically over the past approximately 40 years, much of it a result of anthropogenic activities (e.g., illegal relocation). This change has implications for management of many diseases that exist in, or may be introduced to, North America. Courtesy of Mark Lutman, US Department of Agriculture

to reduce and control wild pig populations. The recent development of a safe and effective vaccine for domestic swine may provide some hope.

Wild pig populations in some areas of the United States are chronically infected with brucellosis (*Brucella suis*) or with pseudorabies virus (Clark et al. 1983), which may also be transmitted to domestic pigs, and the former to people. African swine fever, like foot-and-mouth disease, rinderpest, and to perhaps a lesser extent brucellosis and avian influenza, is a disease at the wildlife–livestock interface, where wildlife are potential reservoirs of persistent infection, and any response or control effort must deal with them effectively.

CASE STUDIES

Seven case studies of wildlife diseases are provided to illustrate agent–host–environment relationships; emergence or recognition of disease problems; how they are diagnosed and managed; and comments on biological, social, financial, or political consequences.

White-Nose Syndrome

White-nose syndrome (WNS) is a fungal disease that has killed many millions of bats in the United States and Canada between 2006 and 2020. First identified in New York in 2006, WNS spread rapidly in eastern North America. Some bat species declined by 90% or more within five years of the disease reaching communal roosts (Langwig et al. 2012). By 2020, WNS was present in 33 states and 7 Canadian provinces.

The disease's name derives from the distinctive fungal

growth on the muzzles and wings of hibernating bats (Fig. 8.5). It is caused by the fungus *Pseudogymnoascus destructans* (PD), which colonizes the bat's skin. As of 2019, thirteen North American bat species, including two endangered and one threatened species, are known to be affected. Some individuals within a population may be resistant to infection, and several treatments, currently more applicable to individuals in captivity than populations in caves, have been developed. The US Geological Survey (USGS) National Wildlife Health Center is working on a vaccine that may be applicable to wild bats.

In 2011 researchers reported that 100% of healthy North American bats infected with the fungus that causes WNS exhibit lesions consistent with the disease (Lorch et al. 2011). These experiments suggest that physical contact is required for bat-to-bat transmission; the fungus is not transmitted by aerosol (Lorch et al. 2011) or contact with an infected substrate. When hibernating bats are disturbed, it raises their body temperature, depleting fat reserves, so even a single, seemingly quiet visit can kill bats that would otherwise survive the winter.

Pseudogymnoascus destructans grows best in the 4°C to 15°C range and not above 20°C. Initially identified as being in the genus *Geomyces* (Blehert et al. 2009), later genetic analysis resulted in it being reclassified within *Pseudogymnoascus* (Minnis and Linder 2013). *Pseudogymnoascus destructans* has been found on bats in Europe and Asia (Young 2011), but no mortality in those bats has been reported (Wibbelt et al. 2013). Genetic research suggests the fungus was brought to North America from Europe, probably via human activities (Drees et al. 2017), as bats do not normally migrate between Europe and North

Figure 8.5. Brown bats with white-nose syndrome, a highly fatal fungal infection that has spread rapidly across the United States. The ecological consequences of losing large numbers of insectivorous and plant-pollinating bats may be serious. Progress is being made on treatment and vaccination, but no population-level control or treatment program yet exists.

America. Translocations of bats by plane and ship have been documented (Constantine 2003). Researchers reported that the fungus can persist on human clothing and caving equipment. The US Fish and Wildlife Service (USFWS) called for decontamination of clothing or equipment. Submersing clothes and gear in hot 60°C water for 20 minutes effectively kills the fungus (Geboy 2008). In some cases, access to caves by humans is closed (Munger 2008).

Symptoms of WNS include unusual winter behavior like frequent or prolonged arousal from torpor, flying, loss of body fat, damage and pitting of the wing membranes, and excessive mortality. As early as 2011, it was hypothesized that premature expenditure of fat reserves needed for winter survival may be a cause for death (USFWS 2011), but electrolyte imbalance also contributes to mortality. Increased energy expenditure and elevated blood pCO^2 and bicarbonate cause chronic respiratory acidosis. Hyperkalemia (elevated blood potassium) ensues owing to leakage of potassium from dying cells. Water and electrolyte loss from open lesions results in further dehydration and a positive feedback loop, leading to death (Verant et al. 2014). Treating individual bats with electrolyte and fluid replacement can be effective but obviously is not applicable to populations.

White-nose-syndrome-infected bats have been treated with a probiotic bacterium (*Pseudomonas fluorescens*) previously used in chytridiomycosis treatments. Treated bats were five times more likely to survive than untreated bats (Hoyt et al. 2019). Early-stage research on vaccines that would prevent WNS or increase bats' resistance to infection and that can be externally applied to populations is under way (Rocke et al. 2019).

Beyond its effects on bat populations, WNS has broader ecological implications. In 2008 the US Forest Service estimated that reductions in bat numbers due to WNS would

result in at least 1.1 million kg of insects not eaten, with resultant damage to crops, losses to farmers, and possible increased use of pesticides. Bats save farmers in the United States $3 billion a year in pest control, with no estimate available for the value of pollination and seed dispersal services (Washington Department of Fish and Wildlife 2019).

Conservation of bats, particularly prevention and treatment of WNS, has garnered great public support, which has translated into funding for research and monitoring. But beyond sympathy for small flying mammals, the financial and ecological implications of massive and sustained loss of bats have resulted in support from agricultural sectors. Given this, WNS is an example of a disease that can unite agriculture and conservationists that might otherwise not recognize common ground. Comparisons have been made to colony collapse disorder, a poorly understood cause of mortality in western honeybee (*Apis mellifera*) colonies (Mann 2008) and with chytridiomycosis, a fungal skin disease linked with worldwide declines in amphibian populations (Blehert et al. 2009). The reasons for the emergence of these three diseases at this time in history, and the circumstances that promoted them, are not well understood. And, as noted previously, fungi have not been considered aggressive pathogens. All three diseases suggest fundamental disturbances of ecological processes and environmental sustainability.

Brucellosis

In the 19th and early 20th centuries, cattle infections with brucellosis bacteria (*Brucella abortus*) were relatively common, as was the resulting human disease, undulant fever. As the bacteria's Latin name suggests, the most common sign of infection in cattle was abortion. Brucellosis was considered an occupational hazard for stockmen and for those who consumed unpasteurized milk. Efforts to eradicate brucellosis included test and quarantine, test and slaughter, farm sanitation, national meat inspection, and milk pasteurization (National Academies of Sciences, Engineering, and Medicine 2017). Because the focus of the disease was livestock and public health, the agencies in charge of brucellosis programs were within the USDA and the various state agriculture and livestock agencies. Sometime in the late 19th century, elk and bison (*Bison bison*) became infected in a number of locations. Since that time, brucellosis has persisted in wildlife in the Greater Yellowstone Area (National Academies of Sciences, Engineering, and Medicine 2017). As eradication efforts in livestock progressed, the remaining infections in wildlife became the focus of research and control efforts. Disease transmission occurs when animals (elk, cattle, bison) come into contact with the fetal membranes from an aborted calf or from an infected normal birthing event, which are heavily contaminated with the brucellosis bacteria (Williams and Barker 2001). The bacteria can survive outside the body for hours to days, depending on environmental conditions. Transmission from wildlife to cattle is relatively rare but does occur in winter, particularly when elk enter ranches to feed off haystacks and may abort, or late spring when infected elk in their second trimester are migrating through foothill and mountain meadows where cattle have been turned out to graze.

In the 1970s and 1980s, the Wyoming Game and Fish Department began a series of landmark experiments at its Sybille Wildlife Research Center. These experiments revealed the infectivity of aborted tissues, how long the bacteria could last under environmental conditions, and the relative susceptibility of other species (Williams and Barker 2001). Vaccination trials were conducted using the cattle product known as Strain 19 vaccine on captive elk, and in biobullet form for nearly a dozen years on free-ranging elk on winter feeding grounds. When Strain 19 proved ineffective at protecting elk from infection, another vaccine (RB 51) was tried experimentally, but it was neither protective nor effective (National Academies of Sciences, Engineering, and Medicine 2017).

For about two decades, the National Park Service, USDA, US Forest Service, USFWS, and the governors of Idaho, Montana, and Wyoming cooperated in the formation of an interagency working group to coordinate the management of brucellosis in the Greater Yellowstone Area with tribal and advocacy groups as well as nongovernmental organizations. Brucellosis management efforts in the 1980s to 2000s involved a suite of actions that together were thought to be capable of accomplishing what no single action had (i.e., reduce the prevalence of infection in wildlife and reduce or eliminate the likelihood of elk or bison transmitting brucellosis to cattle). Bison were assumed to be the core source of infections to cattle, and a facility was set up to capture and test bison as they attempted to leave the northern end of Yellowstone National Park near Gardner, Montana, in winter, and to vaccinate and cull. Most bison that continued to move north into Montana were shot. As noted, it was soon recognized that vaccination of bison was ineffective. Better management of elk winter-

ing on feeding grounds (a substitute for winter pastures that had become ranches) was thought to have potential to reduce infection (Fig. 8.6). This included spreading feed lines and remote vaccination using Strain 19 biobullets. But subsequent research found that the immune system of elk differs from that of cattle, and no currently available approved or experimental vaccine was effective. More recently it has been reported that essentially all infections in cattle over the past 20 years (most in southern Montana, USA) have come from elk, not bison; almost entirely from elk in herds that do not use Wyoming feeding grounds; and about half of contacts and infections appear to have occurred on public lands (most of it managed for multiple use by the US Forest Service), not privately owned ranches (National Academies of Sciences, Engineering, and Medicine 2017). Further, many ranchers using public lands do not choose to vaccinate or double vaccinate cattle for brucellosis, and many are resistant to delaying cattle turnout dates to reduce potential overlap with migrating elk that may be infected or abort (National Academies of Sciences, Engineering, and Medicine 2017).

One approach suggested for reducing brucellosis risk to cattle has been reducing elk populations by liberal hunting of female elk. The vast land areas in the Greater Yellowstone Area, multiple jurisdictions and constituencies, and the migratory nature of the elk make this approach problematic at best. Further, large private land inholdings now serve as refuges during hunting seasons, protecting elk herds from efforts to decrease their numbers. A trial effort to test whether culling could reduce brucellosis prevalence in elk was conducted at the Muddy Creek feeding ground, Wyoming, over a five-year period (2006–2010). In all, 646 of 1,321 (46%) of the elk were trapped, and of these 107 were killed because they were seropositive (National Academies of Sciences, Engineering, and Medicine 2017). This reduced seroprevalence from 37% to 5% over five years, but within four years of ceasing the culling, seroprevalence rates rebounded to baseline levels (National Academies of Sciences, Engineering, and Medicine 2017). If culling of known positive female elk over a five-year period provides only temporary reduction in infection rate, increased killing of females of unknown infections status (which may vary from 15% to 40%) seems very unlikely to succeed. Thus heavy-handed hunting offtake and professional culling are considered by many to be ineffective and costly methods, and are unpopular with hunters and many other stakeholders.

Much has been written on problems of wildlife feeding, but the dense aggregation of elk on government-run winter feeding grounds seriously exacerbates disease transmission problems. Feeding grounds were initially made necessary by loss of natural winter ranges to ranching, which has not improved in the ~90 years since. But many ranchers feel they help keep elk from mixing with wintering cattle and their feed. The translocation of wolves (*Canis lupus*) into Yellowstone National Park and the recovery of grizzly bear (*Ursus arctos*) has reduced elk numbers in the core of Yellowstone and dispersed them, potentially reducing transmission. This could theoretically help reduce elk numbers elsewhere, but there is substantial public

Figure 8.6. Elk on winter feeding grounds in Wyoming. Elk infected with brucellosis may abort in late winter and early spring, and concentration around feed increases likelihood of disease transmission. If abortion occurs when elk are visiting haystacks or on open range with cattle, it may result in cattle becoming infected. Courtesy of the Wyoming Game and Fish Department

resistance to the translocation or reestablishment of predators in Idaho, Montana, and Wyoming.

Another potential strategy could be development of a more efficacious brucellosis vaccine for elk and bison, or a vaccine for cattle that better reduces potential for disease transmission at the livestock–wildlife interface (National Academies of Sciences, Engineering, and Medicine 2017). The removal of brucellosis from the list of potential biological warfare agents by the Department of Homeland Security is a positive step. Although USDA shut down its own elk vaccine and brucellosis management research station in 2017, more recently (2021) it has advertised availability of funding for wildlife vaccine development.

Brucellosis has a political component. Those with agriculture interests had the impression that National Park Service (and conservationists) were to some degree responsible for brucellosis transmission to livestock, and serious polarization and legal challenges resulted. Although experimental conditions suggested that bison could transmit brucellosis to livestock, that was never proven definitively to occur under free-ranging conditions. Even so, state agriculture authorities in Montana and Idaho were adamant that bison wandering out of Yellowstone must be shot. Now it appears all the focus on bison and National Park Service management of them was not appropriate or effective, as elk-to-cattle transmission well away from park boundaries is the real problem. To some degree, the real underlying battle is over who controls and benefits from public grazing lands in the Greater Yellowstone Area (i.e., livestock interests or conservation interests), and brucellosis serves to focus and magnify these larger conflicts.

The disease control programs mandated by state and federal agriculture government agencies have historically made eradication of brucellosis the goal, at times placing extreme financial pressure on stockmen in states where the disease oc-

curs, thus making them risk averse. Wildlife interest groups point out that brucellosis is not a threat to the viability of wildlife populations, only rarely gets transmitted to livestock, and that wildlife culling and other proposed draconian measures are unlikely to be successful or sustainable. Although more than $3.8 billion has been spent on efforts to eradicate brucellosis since 1934, it is becoming clear that eradication is unlikely, and that even management of such a disease in large, free-ranging ungulate populations, over vast and rugged geographic areas, and with varying land ownership patterns is a difficult task. Unfortunately, a number of the basic assumptions on which the state and federal cooperative efforts at brucellosis eradication were based proved to be false: elk, not bison, are the primary source of cattle exposure and infections over the past 15–20 years; elk living outside Yellowstone National Park are heavily infected, expanding the geographic range and frequency of cattle outbreaks; vaccination of elk on feeding grounds proved to be an ineffective disease management tool; and limiting bison to the confines of the park has had relatively little positive effect while attracting strong public criticism (National Academies of Sciences, Engineering, and Medicine 2017). This situation remains unresolved, but emphasis is now being placed on risk management and contact reduction, rather than disease eradication.

Rabbit Hemorrhagic Disease

Rabbit hemorrhagic disease (RHD) is caused by a calicivirus, resulting in systemic infections in rabbits characterized by liver necrosis, disseminated intravascular coagulation, and rapid death. It is highly lethal, and because dying rabbits may hemorrhage from the mouth (Fig. 8.7), nose, and other orifices, it is sometimes called bunny Ebola, although it has no relationship to the Ebola virus. At about the same time (early

Figure 8.7. Hemorrhagic disease of rabbits is a highly fatal, environmentally resistant disease that, as of 2020, is rapidly spreading across North America. The virus causes bleeding from various orifices and is nearly uniformly fatal. The potential effects on rabbits, particularly endangered and sensitive species of cottontail, will cause population declines. Courtesy of the US Department of Agriculture

in 2020) that SARS-CoV-2 began spreading from China across the world, a new rabbit hemorrhagic disease virus (RHDV) emerged in wild rabbits in North America, and it may have profound implications for wildlife and ecosystems.

Classic RHD caused by RHDV1 (or RHDVa) affects only adult European rabbits (*Oryctolagus cuniculus*). It was first reported in China in 1984 (Liu et al. 1984) and spread throughout much of Asia, Europe, Australia, and elsewhere. Prior to 2020 only a few isolated outbreaks of RHDV had occurred in the United States and Mexico, and they were in domestic rabbits, remained localized, and were eradicated.

A new virus (RHDV2) emerged in France in 2010. Importantly, it killed rabbits previously vaccinated for RHDV1 and affected young European rabbits and hares (*Lepus* spp.; Bárcena et al. 2015). Rabbit hemorrhagic disease virus 2 has since spread to the majority of Europe and to Australia. The first report of RHDV2 virus in North America was on a farm in Québec, Canada, in 2016. In 2018, a larger outbreak occurred in feral European rabbits on Delta and Vancouver Islands, Canada (USDA 2020). In early 2020, outbreaks of RHD occurred in Mexico, and within four months cases were reported in Arizona, California, Colorado, Nevada, New Mexico, Texas, and

Utah. But in these outbreaks, wild rabbits including mountain cottontail rabbits (*Sylvilagus nutalli*), desert cottontail rabbits (*S. audubonii*), antelope jackrabbits (*L. alleni*), and black-tailed jackrabbits (*L. californicus*) were affected (Office International des Epizooties 2020). At the time of this writing, cases have been detected in wild rabbits in Colorado and Wyoming, but easterly spread appears stalled at the Mississippi River. The rapid spread and high lethality of this virus and environmental persistence suggest that native rabbit populations in North America may plummet and stay depressed for years, with major consequences for predators and ecosystems that are shaped by rabbits.

Both strains of RHDV replicate in the liver and cause hepatic necrosis, liver failure, and jaundice. Clotting factors and platelets may be depleted, resulting in disseminated intravascular coagulation and marked bleeding from various locations. A presumptive diagnosis of RHD can often be made on the basis of clinical presentation and postmortem lesions. Definitive diagnosis requires detection of the virus. Because most caliciviruses cannot be grown in cell culture, molecular and serologic methods of viral detection are often used (Gleeson and Petritz 2020). The classic postmortem lesion seen in rabbits with RHD is extensive hepatic necrosis and jaundice. Also, multifocal hemorrhages, splenomegaly, bronchopneumonia, pulmonary hemorrhage or edema, and myocardial necrosis may be observed (Gleeson and Petritz 2020).

The molecular confirmatory test for RHD viruses is reverse transcriptase–polymerase chain reaction, but other tests include enzyme-linked immunosorbent assay, electron microscopy, immunostaining, western blot, and in situ hybridization (Gleeson and Petritz 2020). The tissue of choice for molecular testing is fresh or frozen liver because it usually contains the largest numbers of virus, but if this is not available, spleen and serum can also be used.

Both RHD viruses are extremely contagious and spread as a result of contact with infected animals, carcasses, bodily fluids (urine, feces, respiratory secretions), and hair. Surviving rabbits may be contagious for up to two months (Gleeson and Petritz 2020). Clothing, food, cages, bedding, feeders, and water also spread the virus. Flies, fleas, and mosquitoes can carry the virus between rabbits (Kerr and Donnelly 2013). Predators and scavengers can also spread the virus by shedding it in their feces (Kerr and Donnelly 2013). Caliciviruses are highly resistant in the environment and can survive freezing for prolonged periods. Importation of rabbit meat may be a major contributor in the spread of the virus to new geographic regions (Gleeson and Petritz 2020). Virus can persist in infected meat for months and for prolonged periods in decomposing carcasses. Some rabbits infected with RHDV1 or RHDV2 do not develop signs of disease (Kerr and Donnelly 2013). Surviving rabbits develop a strong immunity to the viral variant with which they were infected (Gleeson and Petritz 2020), but asymptomatic infected rabbits can shed virus for months and infect others.

Caliciviruses are difficult to inactivate. Common household disinfectants like bleach do not work against these viruses. A list of disinfectants that are effective against calicivirus can be obtained from the Environmental Protection Agency. Because of the highly infectious nature of the disease, strict quarantine is necessary when outbreaks occur. Deceased rabbits must be removed immediately and discarded in a safe manner. Surviving rabbits should be quarantined or humanely euthanized.

Vaccines for RHDV1 and RHDV2 have been developed, but their use is restricted by countries in which RHD is not endemic. Importation of RHDV vaccines into the United States is only by approval of the USDA and the appropriate state veterinarian (US Animal Health Association 2020). Currently an RHDV vaccine is being used experimentally in California to try and protect several small core populations of listed subspecies of cottontail rabbits.

Across much of North America, rabbits of various species make up a major portion of the prey base for a variety of carnivores. Their high fecundity and geological engineering make them a keystone species, particularly in the arid ecosystems in the western United States. There are numerous threatened or endangered cottontail rabbit subspecies that could be rendered extinct or extinct in the wild even if biosecurity and captive breeding manage to save them. At the writing of this chapter, RHD has spread from the southwestern United States north to Washington and east to Wyoming. The effects of RHD on rabbit populations and their ecosystems in North America are potentially immense and likely to be prolonged.

Distemper and Plague

Canine distemper has been common in Europe for almost 200 years (Williams and Barker 2001). Caused by a morbillivirus that infects domestic and wild canids, and all mustelids and procyonids, canine distemper is capable of infecting other species, including wild felids (Williams and Barker 2001). The incidence of distemper declined considerably in dogs in North America with the implementation of widespread and effective vaccination in the 1950s and 1960s. Transmitted by aerosol or contact with fluids of infected and dying animals, the virus cannot survive outside a living host for long, so its survival depends on repeated transmission to and between animals in susceptible populations. Periodic outbreaks in suburban and rural red (*Vulpes vulpes*) and gray fox (*Urocyon cinereoargenteus*) and raccoon (*Procyon lotor*) occur in various parts of North America, often in summer and when populations are relatively high (Davidson 2006). The absence of domestic dog outbreaks suggests it has become endemic in wildlife.

Once inhaled or ingested, the virus invades local mucous membranes and is picked up by white blood cells (i.e., lymphocytes). It multiplies in lymphoid organs and spreads to the lymph nodes throughout the body, eventually lodging in the central nervous system (Williams and Barker 2001). Although some of the signs and lesions of distemper (oculonasal discharge, diarrhea, hard pad) are fairly characteristic (Davidson 2006), the most common signs are not exclusive to distemper. Depression, seizures, and partial paralysis are also potential signs of rabies. Coinfections of distemper and rabies can occur. Professional postmortem examination is the only way to differentiate between the two diseases with confidence.

When black-tailed prairie dog (*Cynomys ludovicianus*) ecosystems covered 20% of the western rangelands, the black-footed ferret (*Mustela nigripes*) was common in 12 western states and 2 Canadian provinces. Ferrets are exquisitely sensitive to distemper virus, and infections are fatal in 95% to 100% of cases. On shortgrass prairies, distemper outbreaks naturally occur every three to seven years, affecting coyote (*Canis latrans*), swift fox (*Vulpes velox*), badger (*Taxidea taxus*), skunk (*Mephitis* spp.), and ferrets (Williams and Barker 2001). Black-footed ferrets were placed on the endangered species list in 1967 but were thought to have become extinct in the 1970s when the last known captive individuals died after receiving an inadequately attenuated modified live canine distemper vaccine. But in 1981, a remnant wild population of black-footed ferrets was discovered near Meeteetse, Wyoming.

Narrowly avoiding a distemper outbreak, 18 wild ferrets were taken into captivity between 1985 and 1987 by the USFWS and Wyoming Game and Fish Department. Subsequently, the National Black-footed Ferret Conservation Center was established, and additional breeding programs in zoos (e.g., Cheyenne Mountain, Colorado; Front Royal, Virginia; Louisville, Kentucky; Phoenix, Arizona; Toronto, Canada) produced hundreds of kits. Eventually, an effective ferret-safe distemper vaccine was developed that made captive management easier and release into infected environments safer (Williams and Thorne 1999). These successes made possible several successful translocations of black-footed ferrets to historical habitat in the northern Great Plains. Currently, there are more than 500 adult black-footed ferrets in self-sustaining wild populations in Montana and Wyoming. A successful translocation occurred to the Grasslands National Park, Canada, and the success of another in northern Mexico is still being assessed. The management goal is to establish 3,000 adults scattered across historical ferret habitat.

Highly infectious and highly lethal in many species, canine distemper has been blamed for catastrophic die-offs of African wild dogs (*Lycaon pictus*) and African lions (*Panthera leo*). It is one of the diseases with the potential to destroy whole populations of susceptible species, and because asymptomatic animals can be carriers, it is difficult to predict or limit in wildlife. When canine distemper destroys or seriously limits large carnivore populations, it can alter predator–prey relationships and cause serious ecological ripple effects.

The survival of wild ferret populations is intimately connected to the presence of prairie dog colonies. Another disease, plague (caused by the bacteria *Yersinia pestis*), threatens the survival of prairie dogs and ferrets. It can be lethal to ferrets and frequently eliminates entire prairie dog colonies, heavily affecting species that are dependent on them. Plague was introduced into North America in the late 1800s (Friend 2006). In the past several decades, an effective injectable plague vaccine has been used to protect black-footed ferrets (Rocke et al.

ELIZABETH S. WILLIAMS (1951–2004)

Dr. Elizabeth S. Williams graduated from Purdue University with a DVM in 1977 and completed a residency and PhD in veterinary pathology from Colorado State University in 1981. While working on her PhD on *Mycobacterium paratuberculosis* in wildlife, Beth astutely recognized wasting disease of captive mule deer and elk as a spongiform encephalopathy. This event led to her research interests in this field and culminated in her recognition as the foremost expert on chronic wasting disease in deer and elk in the United States. Beth was active and skilled in research, diagnostic veterinary pathology, and teaching. She was a professor in veterinary sciences and an adjunct professor in zoology and physiology at the University of Wyoming and an adjunct professor in veterinary pathology at Colorado State University. Beth was recognized as an outstanding mentor of students. She was a diplomate of the American College of Veterinary Pathologists, coedited the latest edition of *Infectious Diseases of Wild Mammals*, and was editor of the *Journal of Wildlife Diseases* at the time of her death.

E. TOM THORNE (1941–2004)

Dr. Tom Thorne graduated with a DVM from Oklahoma State University in 1967 and started his career as a wildlife veterinarian by supervising research projects and providing veterinary care for the Wyoming Game and Fish Department. Tom was a prominent researcher of brucellosis, tuberculosis, and chronic wasting disease in wild ungulates; was involved in management of bighorn sheep; and initiated the successful captive breeding program for endangered black-footed ferrets. He progressed to services division chief and was acting director of the department prior to his retirement. He authored and coauthored many publications. Over the years he was also vice president of the Wildlife Disease Association, president and chairman of the Advisory Council for the American Association of Wildlife Veterinarians, and chairman of the US Animal Health Association's Wildlife Diseases Committee. In 2003, Tom retired from the Wyoming Game and Fish Department after 35 years of service.

    Tom Thorne and Beth Williams were renowned for their collaborative work in infectious diseases and management of wildlife, as well as for important contributions to conservation of the black-footed ferret, Wyoming toad, and other sensitive species. They died together in a car accident in late 2004 and are remembered as two thoughtful, generous, and productive scientists who loved wildlife and the people who work with them, and were loved by them in turn.

2008), and an oral vaccine and delivery system for prairie dogs has been developed (Rocke et al. 2017). Although vaccination has significant limitations as a general wildlife treatment, the successful recovery of black-footed ferrets is dependent on vaccination of these animals for plague, distemper, and other intensive wildlife management strategies. Oral immunization of prairie dogs through distribution of vaccine-laden baits may significantly enhance ferret recovery by protecting their prey base (Rocke et al. 2017). Field studies aimed at getting approval for widespread use of this vaccine were underway at the time of the writing of this chapter.

    The presence of an endangered species like black-footed ferrets provides justification for protection of public lands on which they live. Protecting larger areas to ensure their prey base is available contributes to the amount of land set aside for conservation. As with brucellosis and pneumonias of bighorn sheep (*Ovis canadensis*), this allocation of public lands for wildlife has political, social, and financial implications, and show diseases are an important component of land use.

Chronic Wasting Disease

Chronic wasting is a transmissible spongiform encephalopathy (TSE) that affects deer. This family of diseases is thought to be caused by misfolded proteins called prions. Similar diseases include Kuru and Creutzfeldt-Jakob disease in humans, bovine spongiform encephalopathy (BSE, also known as mad cow disease), and scrapie of sheep (Williams and Young 1980). Chronic wasting disease was first recognized as a wasting syndrome in mule deer (*Odocoileus hemionus*) at a university research facility in Colorado in 1967. Williams and Young (1980) recognized that the brain lesions were scrapie-like and consistent with other TSEs. Chronic wasting disease affects mule deer, white-tailed deer, elk, moose, red deer (*Cervus elaphus*), and reindeer (*Rangifer tarandus*; Klein 2013). Fallow (*Dama dama*), roe deer (*Capreolis capreolis*), and sika deer (*Cervus nippon*) have been experimentally infected with fallow deer, showing notable resistance to CWD.

    There was little known spread of CWD through the mid-1990s, but it is now in free-ranging and captive cervids in 23 states and 3 Canadian provinces (USGS 2019), much of it likely the result of moving infected animals or tissues. In addition, CWD has been found in South Korea in elk imported from Canada (Centers for Disease Control and Prevention 2019). A wild reindeer herd in Norway as well as wild moose hundreds of kilometers away were reported to be infected in 2016. The CWD variant in the reindeer closely matches isolates from North America, where isolates from moose are unique. Single cases of CWD were reported in moose in Finland in 2018 and Sweden in 2019.

No causal relationship between CWD and other TSE diseases of animals or people is known. Although transmission to humans has not been documented, as a precaution, the Centers for Disease Control and Prevention along with other public health organizations warn hunters to avoid consuming tissues from deer or elk known to harbor the CWD prion (particularly brain, spinal cord, eyes, spleen, tonsils, lymph nodes; Belay et al. 2004).

The cause of CWD (like other TSEs) is a prion, a misfolded form of a normal protein that is most commonly found in the central and peripheral nervous systems. These misfolded prion proteins are capable of converting normally folded proteins into the abnormal form in a self-perpetuating process (mimicking an infectious agent like a virus). The end result is widespread neurodegeneration. In elk there is an allele that codes for leucine that provides some resistance to infection (Klein 2013). In white-tailed deer there is an allele that delays disease progression (Johnson et al. 2011). Studies suggest that some populations of certain deer and reindeer may be more resistant to infection with CWD.

Chronic wasting disease signs are usually seen in adult animals, with the earliest recognized case at 15 months (Haley and Hoover 2015). The most consistent presentation is weight loss over time, easily confused with starvation and inanition in free-living wild cervids. Neurologic and behavioral changes result in agitation, confusion, depression, lowering of the head, tremors, and stereotypic walking patterns. Excessive salivation and grinding of the teeth also occur, and aspiration pneumonia, likely from damage to nerves that control swallowing, is a common finding at postmortem. Most deer show increased drinking and urination, and prions are shed in both saliva and urine.

Although signs and gross lesions may be suggestive of CWD, definitive diagnosis is based on complete postmortem examination and testing. Microscopic examination shows spongiform changes in various central nervous system tissues. Immunohistochemistry can be used to test brain, lymph nodes, and neuroendocrine tissues for the presence of CWD prion proteins, with positive findings in the obex considered the gold standard (USGS 2019). A tonsillar biopsy technique has been a reliable method of testing for CWD in mule and white-tailed deer, but it is not reliable in elk (Wolfe et al. 2002). Biopsies of the rectal mucosa have also been effective at detecting CWD in live mule, white-tailed deer, and elk, though reliability may be influenced by numerous factors (Spraker et al. 2009, Monello et al. 2013).

Chronic wasting disease may be directly transmitted by contact with infected animals or their bodily tissues or fluids (Saunders et al. 2012), regardless of whether they are symptomatic (Selariu et al. 2015). Saliva from infected deer can spread the CWD prions (Mathiason et al. 2006), and sharing of feed or water results in transmission from diseased deer (Ernest et al. 2010). Horizontal transmission from female elk to calves occurs under experimental conditions (Selariu et al. 2015). Early observations that deer and elk from various sources became infected with CWD when housed in pens previously used by CWD animals led to experiments that verified that leaving pens vacant for as long as four years was not sufficient to interrupt the transmission cycle. Bones and cast antlers were subsequently reported to be infectious. It is now known that CWD prions adhere to soil particles, and this may be a major route of transmission as cervids may graze on or close to the ground, and prions may be taken up by grasses (USGS 2019).

Once in the environment, CWD prions may remain infectious for many years. While the longevity of CWD prion is unknown, the scrapie prion has been measured to endure for 16 years (Prusiner 2001, Klein 2013). Thus decomposition of diseased carcasses; infected gut piles from hunters who field dress their cervid harvests; and urine, saliva, feces, and antler velvet of infected individuals that are deposited in the environment all have the potential to become sources of CWD (USGS 2019). The use of deer and elk urine for scent baits has come under question as these are collected from captive animals, are not tested for presence of prions, or come from certified CWD-free stock. Indeed, import and use of scent bait is a plausible explanation for how the North American strain of CWD got to Norway. But the CWD variants found in moose in Norway are unique and likely the result of local spontaneous mutation.

Chronic wasting disease prions are nearly insoluble in solvents, not killed by formalin, resistant to incineration, and highly resistant to digestion by proteases (Klein 2013). At least one scavenger species, the American crow (*Corvus brachyrhynchos*), has been able to pass viable CWD prions through its digestive tract (VerCauteren et al. 2012). Considering the environmental resistance of CWD to degradation; continued movement of cervids and their tissues; and the extensive geographic range of crows, their scavenging habits, communal roosting, and mobility, continued spread of CWD in North America now seems inevitable. A real test case will be the efforts in Norway to eliminate CWD from vast areas of harsh terrain where a herd of CWD-infected reindeer (more than 3,000 animals) was slaughtered and carcasses removed.

By 2012, a voluntary system of control was published by US Department of Agriculture–Animal Plant Health Inspection Service, with captive cervid herd certification to try to avoid interstate CWD movement. Federal funding for CWD research and management has been inconsistent and seldom available to the state agencies to monitor interstate travel for infectious CWD materials. State wildlife agencies attempting to inspect deer and elk carcasses at border check stations have had to fund these efforts out of sources meant for other management programs. In 2019 Minnesota started providing a no-cost deer carcass incineration program to help stem the spread of CWD in one region. Hence one can see why wildlife agencies fear CWD will degrade the value of hunting; public appreciation for deer, elk, and moose; and the threat that this poses to traditional funding for wildlife management.

Multicausality of Pneumonia in Bighorn Sheep

Bighorn sheep populations across the western United States and Canada have declined by more than 90% from historical population numbers. There were likely many causes, but the most prominent appear to have been market hunting, habitat loss and degradation, and diseases contracted from domestic sheep (Jessup 2011). The most severe, persistent, and deadly of these diseases are pneumonias (Williams and Barker 2001). Historical observations of massive die-offs of bighorn sheep of all ages following contact with domestic sheep date back over a century, but causes were poorly documented. Beginning in the early 1980s, die-off investigations, and experimental exposures by contact and inoculation, affirmed that bighorn frequently died of bacterial pneumonia within days to weeks following contact with domestic sheep (Foreyt and Jessup 1982) or their oral flora. The association between domestic sheep contact as a cause of pneumonia in bighorn has been confirmed in 13 captive commingling experiments (Cassirer et al. 2018). In free-ranging bighorn sheep the result is often massive mortality that afflicts all age classes, with recent data showing median mortality of 47%, but fatality rates vary from 15% to 100% (Cassirer et al. 2018).

Several different *Pasteurella* species could be isolated from sick and dying animals, and investigators noted its close parallel to shipping fever in cattle (Williams and Barker 2001, Jessup 2011). Although wildlife agencies sought to develop agreements to allow better geographic separation for protection of wild sheep, domestic sheep advocates responded first with denial, then with alternative explanations exculpating domestic sheep, and then with legal challenges and demands for absolute proof, beyond even that attainable for shipping fever in cattle after more than 90 years of research and observation (Jessup 2011).

Several bacteria of the *Pasteurellacea* family (*Pasteurella multocida*, *Pasteurella trehalose*, and *Mannheimia*—formerly *Pasteurella*—*haemolytica*) can cause pneumonia in domestic sheep, goats, cattle, and bighorn sheep (Williams and Barker 2001). Bacterial pneumonia in bighorn may be secondary to lungworm infestation, but it may also be a primary cause of pneumonia epidemics, including all-age die-offs. Researchers have shown that white blood cells (neutrophils) of bighorn sheep are less capable of killing these bacteria than those of domestic sheep (Jessup 2011). When apparently healthy bighorn sheep are experimentally placed in contact with domestic sheep, bighorn contract pneumonia and die, while the domestic sheep remain healthy (Foreyt and Jessup 1982, Wehausen et al. 2011, Cassirer et al. 2018). This does not occur when bighorn sheep are placed in contact with cattle, deer, elk, horses,

or llamas (*Lama glama*). Contact with domestic goats is less often fatal but can result in serious diseases, including pneumonia, and bacterial keratoconjunctivitis (e.g., eye infections; Jansen et al. 2006).

A *Mannheimia haemolytica* originally isolated from healthy domestic sheep was genetically marked and labeled with a fluorescent dye and inoculated back into the oropharynx of domestic sheep. They remained healthy. Nose-to-nose contact between these domestic sheep and bighorn sheep across a fence line allowed the bacteria to spread and cause pneumonia that was uniformly fatal for bighorn (Lawrence et al. 2010). The dyed and genetically marked bacteria were subsequently isolated from the bighorn that died of pneumonia (Lawrence et al. 2010), thus fulfilling Koch's postulates (a century-old set of conditions used to prove a specific organism causes a specific disease). Infection did not occur when separation barriers of 3 m or more were maintained.

In the past decade it has become clear that the bighorn sheep pneumonia complex may involve several organisms. Winter penning of Dall sheep (*Ovis dalli*) with domestic sheep in a zoo resulted in all the Dall sheep dying of pneumonia caused by *Mycoplasma ovis ovipneumoniae* (Black et al. 1988). This primitive bacterial organism, which is capable of facilitating or exacerbating pneumonia in bighorn sheep, is host-specific to Caprinae and commonly carried by healthy domestic sheep and goats. Although experimental mycoplasma infection alone does not consistently cause fatal respiratory disease in exposure trials, it appears to be the primary agent of most bighorn pneumonia outbreaks, likely even more fundamental than *Pasteurellacea* bacteria. *M. ovipneumoniae* infection can persist in bighorn sheep populations for decades. Carrier females transmit it to their susceptible lambs, triggering fatal pneumonia outbreaks in nursery groups, which limits recruitment and slows or prevents population recovery for years (Cassirer et al. 2018). The result is that demographic costs of pathogen persistence can outweigh the effects of the initial infection and die-off (Cassirer et al. 2018). Mycoplasma has recently been found in mountain goats (*Oreamnos americanus*), a species that may occupy the same habitat as some bighorn sheep (Wolff et al. 2019).

Diagnosis of pneumonia is reasonably straightforward; affected sheep often cough and have nasal discharge. Acute cases can result in death within days with few outward signs. But respiratory disease signs in bighorn sheep do not reveal the cause or the likely outcome for the individual or the populations. At postmortem examination, classic fibrinopurulent bronchopneumonia is commonly found when mycoplasma and bacteria are involved. Professional postmortem examination can help reveal the cause or causes, but as the disease progresses, the initiating organisms may become harder to identify. Identification of the bacterial and mycoplasmal causes of pneumonia in bighorn sheep requires excellent laboratory support for isolation and typing of the organisms and to determine potential origins of infection.

Many efforts to develop vaccines that can elicit sufficient immune response to *Pasteurella* spp. and *Mycoplasma* spp. to protect domestic livestock and bighorn sheep have been made. None have been successful. Nor is there any effective antibiotic treatment for mycoplasma infections. Further, vaccination of bighorn places the responsibility of preventing disease transmission on public trust agencies rather than on those private parties who would use public land for their financial gain. No wildlife management actions to date have been successful in reducing morbidity, mortality, or disease spread once a mycoplasma and/or bacterial pneumonia outbreak is in progress (Cassirer et al. 2018). At present, spatial and temporal separation of domestic sheep and goats from bighorn sheep, as a means to reduce risk of contact and pneumonia transmission, appears to be the most effective management tool. A series of court decisions over the past three decades have consistently supported government agency actions to alter, restrict, or abolish public land grazing permits to protect bighorn sheep from pneumonia. Like brucellosis, the conflict between woolgrowers and conservationists over protecting bighorn sheep from pneumonia appears to involve larger issues about use of public rangelands in western states, the legal precedents, and the financial and political consequences.

It is not easy to detect mycoplasma, and apparently healthy bighorn sheep may harbor it and (rarely) virulent forms of *Pasteurella*. In light of this, the common wildlife management practice of relocating surplus bighorn sheep into remnant herds or populations should be seriously questioned. The ready provision of substantial funds to support bighorn sheep relocations by advocacy groups provides a perverse incentive for wildlife agencies to seriously weigh potential benefit against inherent disease risks. Relocated animals will carry with them their flora of bacteria, viruses, mycoplasma, and parasites, and the stress of capture and relocation may reduce immune function. Respiratory disease organisms expressed by translocated animals may then kill off the remnant populations, with the loss of their genetic heritage. Conversely, a remnant population may be small because of previous health problems and may harbor organisms that will cause mortality in translocated animals. The end result of these supplementation efforts may not be increased genetic diversity, but rather disease-related morbidity and mortality, and occasionally loss of founder stock (and genetic diversity). The theoretical value of mixing bighorn populations should be weighed carefully as to the amount of genetic diversity desired, whether that can be measured to determine if it already exists in the herd, and whether the genes for which diversity might be beneficial are known and present in the animals to be translocated. If you are not able to measure or monitor genetic diversity, it is hard to justify taking serious risks of disease transmission by admixing bighorn of varying origins. Testing and establishing pathogen and genetic profiles for potential source and recipient bighorn populations over time can be used to better ensure safety of these practices.

Although it is now clear that domestic sheep can carry respiratory bacteria that are frequently fatal to bighorn sheep, a number of western state wildlife management agencies have been slow to fully investigate pneumonia in bighorn popula-

tions and take responsibility for their own contribution to it. If sheep grazers on public land are to be expected to do all they can to reduce potential for their animals to cause bighorn die-offs, wildlife management agencies need to do all they can to prevent them as well. This means giving more thought to where bighorn populations are started and the frequent but seemingly poorly considered admixing of bighorn populations.

Geographic separation of bighorn sheep from domestic sheep seems to hold the greatest promise for protecting bighorn in the short run, but the propensity of bighorn to wander, and for stray domestic sheep and goats to seek out the company of bighorn, makes separation inherently difficult. For land-use planning purposes, 40–50 km is generally seen as an optimal separation distance. But a number of domestic sheep management changes, like trucking them from the high-elevation summer grazing ranges instead of trailing them, can also help reduce risk. The controversy over optimal management of bighorn sheep and the pneumonia that plagues them has yet to be resolved.

Pneumonia in bighorn sheep is an example of multiple organisms causing one recognized disease process. The next case study involves multiple diseases and parasites, toxins, and anthropogenic and natural factors that negatively affect the health and recovery of a depleted marine mammal population.

Sea Otters, Diseases, and Habitat Degradation

Failure of a wildlife population to recover when complex wildlife health problems and diseases are present can involve complex interactions between the host, its nutrition, and environment, disease organisms, toxins, and predators. Despite decades of legal protection, the recovery of California's southern sea otter (*Enhydra lutris nereis*) population has been hindered by high mortality in prime-aged adult animals. Up to 50% of sea otter mortality in some years has been attributed to protozoal and bacterial infections (Thomas and Cole 1996, Kreuder et al. 2003). The connections between diseases and various sources and types of pollution have been documented (Jessup et al. 2004, 2007; Miller et al. 2010a). Some southern sea otter mortality caused by ingestion of marine biotoxins appears to be influenced by nutrient loading that comes from coastal freshwater and estuarian sources, and directly from biotoxins originating from freshwater sources (Kudela et al. 2008, Miller et al. 2010c). Most persistent organic pollutants and contaminants of ecological concern are found in the blood of live (and in liver samples of dead) southern sea otters at levels 20–50 times higher than those of sea otters from more pristine areas of Alaska (Kannan et al. 2006, 2007, 2008; Jessup et al. 2010). Some sources of persistent organic pollutants and contaminants of ecological concern are known (point sources), but most come from nonpoint sources on land. These disease and health problems all appear to be more prevalent along urbanized coastlines and near river mouths, suggesting that land–sea flow of pathogens, toxins, and nutrients are important components of southern sea otter morbidity and mortality (Miller et al. 2010b, Jessup et al. 2004).

Some pathogens appear to be evolutionarily new. One cause of fatal systemic and neurologic protozoal infections in sea otters is *Sarcocystis neurona* (Miller et al. 2010*b*), which was inadvertently introduced to California via its nonnative invasive species definitive host, the Virginia opossum (*Didelphis virginiana*; Dubey et al. 2001), about 100 years ago. A second, similar protozoal parasite is *Toxoplasma gondii* (Fig. 8.8). Although native felid species, bobcat (*Lynx rufus*) and mountain lion (*Puma concolor*), coevolved alongside sea otters in coastal California and are competent hosts of *T. gondii*, the introduction of millions of domestic cats into California occurred in the past 150 years, and at a time when sea otters were recovering from the brink of extinction. Most municipal sewage generated along coastal California is released into the ocean after treatment, but primary and secondary treatment does not kill many pathogens, notably *Toxoplasma* and *Sarcocystis* (Conrad et al. 2005).

Bacterial infections, some of them caused by organisms commonly associated with feces, are another cause of sea otter illness and death (Thomas and Cole 1996, Kreuder et al. 2003, Jessup et al. 2007, Miller et al. 2010*a*). Old or inadequate sewage infrastructure, seasonally heavy precipitation, and accidents can result in uncontrolled releases of sewage. Many residences in rural portions of California's coastal counties where sea otters live are connected to old septic systems, and some boats discharge untreated sewage directly into the ocean. Feces from pets and wildlife are periodically flushed into the ocean with storm runoff. This fecal matter, plus human and livestock feces, all may be sources of bacterial and protozoal pathogens (Sercu et al. 2009).

Some of the most productive and intensively farmed land in the United States is adjacent to southern sea otter habitats. The heavy use of nitrogen fertilizers, phosphates, and other nutrients; light and porous soils; and seasonally heavy rainfall result in significant nutrient pulses in embayment areas. Higher levels of urea promote blooms of, and domoic acid toxin production by, the marine diatom (*Pseudonitzschia aus-*

tralis) that causes amnesic shellfish poisoning in humans and major mortality events in various marine birds and mammals (Kudela et al. 2008), including sea otters. Cyanotoxins produced by *Microcystis* spp. in nutrient- (phosphate-)rich freshwater lakes and watercourses has reached the ocean, killing sea otters (Miller et al. 2010*a*). Cyanotoxins are 1,000 times more potent than domoic acid and kill fish, dogs, people, and many other living organisms. More than 30 sea otter deaths have been traced back to cyanotoxin blooms draining into the ocean.

A growing body of evidence suggests that the cumulative effects of all these forms of pollution (i.e., protozoa, bacteria, biotoxins, persistent organic pollutants, contaminants of ecological concern) and habitat degradation contribute to southern sea otter morbidity and mortality, and may slow or limit population recovery. Southern sea otters also show evidence of being food limited. They respond to limited abundance of preferred prey like abalone (*Haliotis* spp.) by diversifying their diets. Unfortunately, many of the less preferred prey items are filter feeders or pseudo–filter feeders and are capable of harboring or concentrating protozoa, pathogenic bacteria, biotoxins, and organic pollutants (Johnson et al. 2009). So, pathogen and pollution issues are inseparable from prey abundance and diversity.

Several syndromes or complex disease processes influenced by multiple factors have been identified in southern sea otters, including cardiomyopathy syndrome that either or both domoic acid intoxication and protozoal infection may contribute to (Miller et al. 2020). End lactation syndrome in adult females results from combinations of food limitation, the energetic cost of nursing pups, mating trauma, bacterial infections, and comorbidity from some of the other toxins and diseases discussed above (Miller et al. 2020). Predation on sea otters by great white sharks (*Carcharodon carcharias*), which apparently kill them by mistake as they do not consume them, began to rise in the early 2000s and has now eclipsed many other causes of death (Miller et al. 2020). The cause for this is unknown. The important point to recognize is that the health of individual wild animals and small groups and populations is subject to change and influenced by many factors; any single study or set of findings may only represent a snapshot in what is a dynamic process.

Southern sea otter problems are a classic example of One Health—the concept that human, animal, and environmental health is inexorably linked (Jessup et al. 2007). Most of the organisms and toxins that kill sea otters also sicken and kill other marine mammals, domestic species, and humans. Sea otters are a keystone species for kelp forests, protecting them from kelp stipe grazers like sea urchins (*Stronglyocentrotus* spp.). Kelp forests and the biological diversity they support are healthier with sea otters than without them (Estes 2005). Kelp forests reduce storm damage and beach erosion. In the eastern Pacific Ocean, kelp forests are one of the major marine carbon-sequestering macroorganisms. Thus diseases and intoxications that reduce sea otter health and abundance may contribute to larger environmental and social problems.

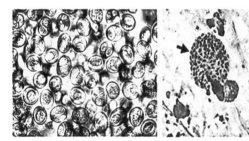

Figure 8.8. Oocyst (*left*) and tissue cyst (*right*) phases of *Toxoplasma gondii*. The oocysts are shed only in the feces of cat species and have been shown to bioaccumulate in filter-feeding shellfish, which appears to be the route by which sea otters and other marine mammals become infected. The tissue cysts in brain can lay dormant for years before other stressors allow them to break out and cause fatal neurologic and systemic disease in sea otters. Courtesy of the California Department of Fish and Wildlife

The efforts to understand the causes of southern sea otter mortality, and potential management options to foster recovery, have taken more than 30 years to develop and have required the resources of the USGS; USFWS; California Department of Fish and Wildlife; Monterey Bay Aquarium; Marine Mammal Center; University of California (UC), Davis, Wildlife Health Center, UC Santa Cruz Institute for Marine Science; and many other organizations. More than 40 years of consistent population surveys, carcass pickup and examination, and studies on particular pathogens and toxins have provided an understanding of sea otter health and population recovery challenges. This work has required the merging of ecological and epidemiological data sets with those developed by pathologists and clinical veterinarians. It has benefited from data collected on live, dead and stranded, captive, and free-ranging sea otters. The overall effort is a classic example of what widely collaborative and cooperative wildlife health programs can accomplish. And it is quintessential One Health thinking in action.

METHODS OF INVESTIGATION

Any wildlife disease investigation must ask three primary questions. Is a disease or health problem present? What is causing it? What effect is it having on the population? Wildlife population surveys and demographic data that show whether numbers and habitat use are increasing or decreasing can be indirect measures of health. They can also be used to investigate whether a disease might be present and a measure of effects at the population level. Effective wildlife management requires conducting population surveys, determining age class distribution, and measuring changes in those data over time using different methods. Individual animals and age classes can be determined by direct count of small populations at a point in time repeated yearly to establish a minimum number (e.g., southern sea otters). Alternatively, surveys that involve standardized methods and routes as well as a multiplication or correction factor can be used to estimate larger populations (e.g., waterfowl). And mark-recapture methods (e.g., Lincoln-Pettersen Index) are often used when a known number of marked animals are in a population (e.g., collared bighorn sheep, deer, elk; Wobeser 1994). When populations are very large and remote, and when disease processes are subtle, a decline in population numbers or changes in age distribution may be one of the best, or only, early indicators of population health problems.

Field observation of die-offs and on-the-ground investigations often provide information that is difficult to gather in other ways. These include observation of behaviors and health status of live individuals, numbers and freshness of carcasses (approximate time of death), age and population structure of affected individuals, involvement of more than one species, proximity to water, water quality, presence of vectors, and many other vital observations (Wobeser 1994). Investigators familiar with the terrain, ecology, and general management can make huge contributions to such investigations. A team

made up of biologists and a veterinarian or disease specialist is recommended. Local ranchers and farmers can be helpful if properly motivated.

Postmortem examination (i.e., necropsy) is the primary method of establishing a cause of mortality. Fresh carcasses representative of the population and event are needed, particularly if several diseases may be present or if several factors compromise health at the same time. Each carcass must be examined methodically, opening body cavities and examining all organs. Frequently, tissue samples are taken and preserved in formalin or placed in culture media or frozen for further analysis. Veterinary pathologists specialize in this type of work. Under most circumstances, submitting wildlife carcasses to a veterinary diagnostic laboratory and submitting samples taken from field necropsies conducted under professional supervision are the most effective ways to determine the cause of illness or death.

In some cases, gross lesions, or those apparent by visual observation, and basic on-site testing may be enough to establish a diagnosis. Clinical pathology including the rapid analysis of blood counts, chemistries, and stains may be a useful diagnostic adjunct. More often, definitive diagnosis requires laboratory culture for bacterial, fungal, viral, or protozoal organisms; microscopic examination of sectioned and stained tissues; and possibly immunohistochemistry, toxicological, or genetic sequence analysis. To better understand population and management implications, the genus and species of parasites may need to be identified. For cases or die-offs that involve complicated or multiple disease processes, identification can take days to weeks or months to complete.

Epidemiology, the study of disease and health in populations, often takes over after the diagnostic process has established a cause or causes. Spatial and temporal patterns of disease or mortality over time are useful in establishing answers to the primary questions of presence, cause, and population effects. Incidence (i.e., the number of new cases per unit of time) and prevalence rates (i.e., number of cases at any one point in time), and their change over time, are key measures of the progress of a disease epidemic (i.e., major outbreak). Epidemiology may involve a field component (shoe leather epidemiology) and a modeling (mathematical and statistical) component.

There are a variety of models that can be used to predict the course of a given disease in a population through time, and these can be used to develop intervention and treatment strategies (e.g., when and how many susceptible individuals would need to be vaccinated to stop the spread of the disease). The susceptible, infectious, recovered compartmental model is one of the more fundamental ones. It relies on the fact that individuals (animals or people) move from one stage of disease to the next through time at a predictable rate. Measles is a classic example that shows how the number of susceptible individuals in a naive population drops quickly and predictably as the number of infected begins to peak, followed by a rise and peak of recovered individuals and the end of the epidemic. The rate at which individuals transition from susceptible to infected is called the force of infection (F), but for many infectious diseases, to calculate F, it is more realistic to consider the fraction of individuals in the population that are susceptible rather than the absolute number. A more sophisticated version of susceptible, infectious, and recovered takes into account that, with many diseases, susceptible individuals are exposed and might or might not become infected. Another model factors in the effects of maternal-derived immunity on susceptibility. Modeling is a powerful tool that can shape and scale disease intervention strategies, and indicate when and which interventions might or might not be successful. The COVID-19 pandemic has brought epidemiology, disease modeling, and preventive medicine—otherwise relatively ignored sectors of medicine—to the forefront of the world's attention.

At some point along the continuum of disciplines that are used in wildlife health investigations, veterinary epidemiology begins to blend with disease ecology. A thorough understanding of various aspects of the science of ecology is vital to understand wildlife diseases and the ecosystems in which they occur. Disease ecology is in turn a special area of ecology that has grown and greatly matured in the last two decades.

MANAGEMENT OF HEALTH AND DISEASE

Theoretically, wildlife populations in pristine environments at or below carrying capacity should have few significant disease problems beyond perhaps a low level of well-adapted parasites. This idyllic set of circumstances is rare to nonexistent, in no small part because there are few pristine wildlife populations and ecosystems left in the world. The general principles of wildlife management—providing sufficient food, water, cover, and space—are generally good methods for reducing health problems at the population level. More than any other basic factor, crowding and concentrating wildlife contribute to the occurrence, incidence, and severity of many wildlife diseases (e.g., waterfowl diseases, brucellosis, CWD, distemper, parasitism in general, pneumonias).

Agent and Vector Reduction

One of the most important contributions of wildlife disease research to wildlife conservation is that, by knowing the intimate details of the biology of disease agents or the vectors that transmit them, cycles can be broken and the effects of disease lessened or managed. One classic example is the extensive effort to manage cattle trypanosomiasis and human sleeping sickness in southern and eastern Africa, which are transmitted by the tsetse fly (*Glossina* spp.). Tsetse flies can be attracted to large, dark-colored cloth targets that pivot and move in the breeze and smell like cow's breath, mimicking the host, such as cow, Cape buffalo, or other large ruminant. With insecticide permeating these targets, the local tsetse fly population and transmission of trypanosomes (vectors and disease agent) can be reduced.

Another example in the United States is the switch to steel shot for waterfowl hunting nationwide, and to copper rifle bullets for hunting in California, primarily to protect the Cali-

fornia condor (*Gymnogyps californianus*). Both manage disease (i.e., lead poisoning) by removal of the agent from the environment. Banning the use of several persistent organic pollutants (e.g., dichlorodiphenyltrichloroethane, or DDT, and chlordane) is another example. Interestingly, the loss of DDT has made it harder to control fleas that transmit wildlife plague in the United States. The eventual development of insecticide resistance is an expected consequence of repeated and continual use of pesticides, which can have unintended side effects or be unpredictable.

Some states also try to reduce wildlife disease by informing the public and enlisting their support. California state law requires a warning on cat litter that cat feces should not be flushed down the toilet or placed where it can be washed into the ocean in an attempt to reduce the levels of *Toxoplasma* oocysts ingested by shellfish and sea otters. Californians who hunt elk and deer in Rocky Mountain states are informed about CWD in various ways and may be stopped at border check stations to be sure that have not retained potentially infectious tissues.

Host Management

In some cases, the behavior, spatial or temporal abundance, or other critical features of affected wildlife can be managed. Historically, efforts to manage outbreaks of avian cholera and botulism have relied on dispersal of birds to reduce contact and transmission (Friend 2006). Pyro techniques, carbide cannons, airboats, and aerial hazing used to disperse waterfowl are examples of host management. Dispersal can reduce exposure and mortality, but animals often adapt to repetitive disturbance and threat, and it is difficult to permanently haze off birds or many other animals (Wobeser 1994).

Permanent separation by 3 m double fencing between domestic sheep and bighorn can prevent transmission of pneumonia causing organisms. Potential for contact with domestic sheep on unfenced ranges can be significantly reduced if they are not allowed to graze within 40 km of desert bighorn populations. Bighorn sheep that have suffered pneumonia outbreaks should not be used to augment other bighorn populations, as they may carry lethal respiratory disease agents. These are further examples of host management to prevent disease.

Many wildlife disease management efforts have relied on population reduction as a form of host management. These have included focal population reduction, zonal reduction (i.e., creation of barriers), and general population reduction (Wobeser 1994). Once popular, these types of programs are used less commonly today.

General population reduction is a blunt tool for disease management. An example of general population reduction was the widespread and extensive effort to kill skunk (*Mephitis* spp.), fox, raccoon, and coyote by shooting, poisoning, and trapping as a means of combating rabies. It was carried out for decades until it was finally determined that killing mesopredators on a massive scale had little or no effect on rabies prevalence. Recent efforts to reduce the prevalence and spread of

CWD in Wisconsin by encouraging and focusing high levels of hunter harvest have also not proven effective. An example of a more successful program might be the widespread culling of brushtail possum (*Trichosurus vulpecula*) that harbor tuberculosis in New Zealand. Because these possums are not native to the habitats from which they are being culled, this program is more popular than culling of native species. The problem with host reduction as a disease management strategy is that, to be effective, the target animals must be relatively easy to identify and remove, or the disease must be relatively host specific, not very environmentally resistant, or not very infectious. Where this is the case, diseases often die out spontaneously without much culling.

Immunization

The use of vaccines to create or boost an animals' immune response to a disease is a specific case or type of host management. Treatment of individual animals has not been a traditional tool of wildlife management, but treatments that can be applied to populations have significant promise. Rabies vaccines for dogs were developed before the virus itself was even identified, and modern killed or modified live rabies vaccines are effective at stimulating an immune response. When years of large-scale population reduction efforts failed to reduce the incidence of red fox rabies in Europe, vaccination was tried. The first vaccines were delivered by putting them in chicken heads and feeding them to foxes. Better vaccines and more stable artificial baits were subsequently developed for fox, skunk, and raccoon. Splicing portions of the rabies gene into other viruses has improved both safety and efficacy of vaccination. Wildlife rabies control and outbreak response is a major program of the USDA, Wildlife Services, and a major success of wildlife management from a public perspective (Slate et al. 2009). Because vampire bats are not attracted to bait used for carnivores, large-scale killing has continued to be the preferred method to control rabies in this species. To reduce the need for such indiscriminate practices, Constantine (2003) coupled a high-pitched sound-emitting device, which specifically attracted rabid bats to a harp trap to capture and kill them. A bat-specific rabies vaccine that can be dusted on bats and that will be spread through a roost by normal bat activities is being developed at the National Wildlife Health Center.

As discussed above, the restoration of the black-footed ferret to its native ranges would probably have failed if ferret-safe canine distemper vaccines had not been developed. By the 1980s, wildlife managers recognized that modified live canine vaccines were lethal to ferrets, gray foxes, and some other species, and that killed vaccines were not antigenic enough to offer much protection. Extensive efforts led to development of ferret-safe gene-spliced vaccines in use today. Vaccination, along with captive breeding programs, appears to have secured the future of black-footed ferrets.

The emergence of plague as a threat to remnant prairie dog communities and the species that depend on them is a further and perhaps even more profound example of how vaccination

can contribute to wildlife conservation. It is also worth noting that by controlling plague, conservation is served, and human and domestic animal health are protected. This is another example of the One Health approach (Rocke et al. 2008).

Some disease organisms do not stimulate sufficient host immune response for vaccination to reliably provide immunity. No effective or widely usable vaccine for protection of human or animal populations against tuberculosis has been developed, but not for lack of effort. Some host species are less responsive to vaccines than others, as noted regarding brucellosis vaccination of elk and bison with Strain 19 and RB 51 cattle vaccines.

Rinderpest virus infection in wildlife and livestock has now been eliminated from Africa through large-scale and well-coordinated vaccination campaigns. Although it took over a decade, the vaccination campaign provided sufficient numbers of immune animals that the virus could not continue to move between naive populations. Every susceptible animal, wild and domestic, did not need to be vaccinated, just enough so that the virus could not survive. Wildlife vaccination programs are inherently expensive and risky, but when properly designed and carried out, they can be effective where other ecologically damaging efforts like large-scale, long-lasting depopulation or cordon fencing (Table 8.1) failed.

Environmental Management

Diseases occur or become epidemic when conditions of the host animal, agent, and environment allow it. Manipulation of the environment is a tool used frequently to help control or manage disease outbreaks.

One example is type C botulism poisoning of waterfowl. In the western United States and Canada, conditions often favor the conversion of the nearly ubiquitous spore form of *Clostridium botulinum* type C into the vegetative state (Jessup 1986, Wobeser 1994). A warm, moist alkaline environment and a source of decomposing protein are key ingredients. Under these conditions the spores germinate and the vegetative form replicates rapidly, and if activated by a common bacteriophage, it produces an extremely potent neurotoxin (Jessup 1986, Friend 2006). Under natural conditions this process occurs when alkaline lake beds become flooded and small mammals, reptiles, or insects die, providing a decomposing protein source. Flooding of agricultural fields creates the same conditions.

Fly maggots feed on the toxic carcasses and uptake high levels of botulinum toxin. In feeding trials, as few as two maggots contain enough toxin to paralyze and kill a duck in a few hours (Jessup 1986). Flies lay eggs on the dead duck carcasses, which are steeping in the warm alkaline waters of 'botulism soup', and the resulting maggots are eaten by more ducks. In a few days, mortality can increase from a few ducks to thousands. The toxin will also paralyze and kill geese, swans, and wading and shore birds (Friend 2006). Hawks are somewhat resistant; vultures are extremely resistant and nearly impossible to kill with type C toxin. Humans and other mammals, including dogs used for retrieving, are also resistant to ingested type C toxin.

Table 8.1. Examples of the use of depopulation on wild animals as a method of disease management

Species	Disease	Method	Location	Year Reported
(a) Depopulation of Focal Area(s)				
Striped skunk	Rabies	T	Wisconsin	1966
		P	New Mexico	1970
		T, P, S	Alberta	1986
Vampire bat (*Desmodus rotundus*)	Rabies	P	Nicaragua	1976
Ground squirrel	Plague	S	Colorado	1982
American coot (*Fulica americana*)	Avian cholera	W	Virginia	1976
European badger (*Meles meles*)	Tuberculosis	T, G	England	1974
				1982
				1984
Blackbirds	Histoplasmosis	W	Tennessee	1985
(b) Creation of a Barrier or Control Zone				
Striped skunk	Rabies	P, G, S, T	Alberta	1978
Red fox		T	New York	1960
		G, S	Europe	1974
Vampire bat		G	Argentina	1974
		P	Latin America	1980
Cape buffalo	Rinderpest	S	Uganda	1953
(c) General Depopulation over a Large Area				
Coyote, red fox, wolf, bear	Rabies	P, T, S	Alberta	1954
Red fox		T	Virginia	1966
		G, S	Germany	1981
Mongoose		T	Puerto Rico	1966
Vampire bat		S, P	Trinidad	1960
Ungulates	Rinderpest	S	Africa	1982
Deer	FMD	S	California	1921

Source: Modified from Wobeser (1994).

Abbreviations are as follows: FMD, foot-and-mouth disease; G, gassing of dens or caves; P, poisoning; S, shooting; T, trapping; W, wetting with surfactants.

In what were once alkaline semidesert areas of much of the western United States, irrigation has brought vast wealth via farming (e.g., California). But practices like flood irrigating in summer or early fall create the perfect conditions for type C botulism die-offs. Sprinkler irrigation, periodic ditch irrigation, and flood irrigating when temperatures are cooler all help reduce the conditions favoring botulism die-offs. When flood irrigation is necessary for crops like rice, the steeper the slope of the check or earth wall, the less area for dead animals to collect that start die-offs. Irrigation practices are not the only human activity that can exacerbate potential for botulism die-offs. Anything that kills ducks and leaves their dead carcasses in warm, alkaline water can cause a die-off that can expand if conditions are right. Power lines, windmills, or towers can be the initial source of bird mortality (Friend 2006). Adjusting farming practices and design of infrastructure can help keep waterfowl die-offs from occurring.

Perhaps the most infamous effort to control or contain disease by manipulation of the environment has been the attempt to keep foot-and-mouth disease out of Botswana and other southern African countries by constructing huge cordon fences across much of the country. This expensive effort helped protect a highly subsidized cattle industry that uses arid land largely unsuited to cattle grazing. The practice has continued at the cost of millions of dead wildlife, severe disturbance of normal ecological processes, and the ending of some nomadic people's traditional lifestyles. It has been fairly effective in protecting those who depend on the cattle export economy, but at a high environmental price.

In California, native tule elk (*Cervus canadensis nannodes*) were repatriated to the swamps of the Sacramento–San Joaquin River Delta when a population was put back onto the Grizzly Island Wildlife Area between Sacramento and San Francisco. The herd grew and expanded from a few dozens to several hundred. Their grazing effect was noticeable, and plans for hunts and translocations to reduce and manage the population were being developed when a die-off killed about 20% of the herd, primarily younger animals. Investigation revealed that lack of fire or tilling to remove the dead grass thatch and a long, wet, foggy winter had not allowed the emergence of green grasses or forbs (Jessup et al. 1986). The primary green feed available in many locations was poison hemlock (*Conium maculatum*), which contains the neurotoxin conine (Jessup et al. 1986). This case is an example of the failure to foresee the need for—and subsequent use of—environmental management. The emergency response included provision of alfalfa hay, discing and spraying of hemlock patches, and hazing of elk from hemlock-rich areas. These initial steps stopped the die-off, and this was followed with population reduction by relocation and either-sex hunts to control numbers. These measures have kept large-scale hemlock poisoning from reoccurring, even though the plant is still common on the wildlife area.

As noted above, one of the more dangerous environmental management practices, unless one takes the perspective of the disease organism, is concentrating wild animals. Concentrating wild animals results in greatly increased animal-to-animal contact, stress that lowers immune function, ingestion of unnatural feeds, and unnatural feeding and watering behaviors that may exacerbate disease transmission. Tuberculosis and brucellosis have already been mentioned. Both organisms can survive outside the host, particularly in feces or fluids. It should be no surprise that these diseases, along with paratuberculosis (Johne's disease) and CWD, which are also both environmentally persistent, are a serious concern in captive settings such as in zoos and on game farms. Even under the most stringent conditions of observation, treatment, and sanitation, these diseases are difficult to manage or eliminate in captive or crowded animal populations.

Another dramatic example is *Mycoplasma gallisepticum* infections in finches and seed-eating migratory birds. This mycoplasma is highly infectious and causes conjunctivitis; the bacterial discharge from the eyes, mouth, and nares is the source of infection (Friend 2006). Although bird feeding is extremely popular, the concentration of birds around feeders—and repeated contact of the eye and nares of infected birds with contaminated feeder surfaces—is partially responsible for the toll it takes on finches and other species (Davidson 2006). Often, dozens to hundreds of birds sicken and die in less than a week around a few adjacent feeders. Only removal and sanitation of the feeders can help control the infection. The inherent characteristics of mycoplasmas make development of effective treatment or vaccines extremely unlikely.

Like dispersion of feed and natural feed sources, dispersion of water and clean water sources can reduce prevalence of wildlife diseases and parasites. Irrigated pastures exacerbate infestations of lungworm (*Dictyocaulus* spp.) in deer and the giant liver fluke (*Fascioloides magna*; Wobeser 1994). Providing multiple small water sources like spring boxes or drinkers instead of one large muddy stock pond appears to reduce the prevalence and severity of hemorrhagic diseases of deer, particularly where water is scarce and shared with livestock.

Fire, properly managed, can improve feed conditions for grazing wildlife, disperse them, and in some habitats reduce tick infestations and the diseases they transmit. When parasites and chronic diseases recur on lands managed for wildlife, it is sometimes valuable to review how food, water, cover, and habitat are being managed, including carrying capacity, and predator–prey balance.

Some wildlife management activities may strongly influence the manifestation, emergence, or severity of disease. A case in point is wildlife translocation. The translocated host animal may harbor disease agents and allow their introduction to new environments (Jessup 1993). Conversely, translocated animals may encounter new agents or vectors in a new environment. There may be health implications for the translocated animals, conspecifics for other species in the new location, or conspecifics for domestic species or humans (Jessup 1993). When tule elk were relocated to historic ranges at Point Reyes National Seashore, California, the soils of the peninsula where they were released had a high molybdenum content that led to fatal and debilitating copper deficiency (Gogan et al. 1989). The presence of Johne's disease (paratuberculosis) in the adjacent dairy cattle resulted in infection of these elk. Temporary diet supplementation of copper, separating elk and cattle, and extensive disease screening and culling of infected elk herds resulted in effective control but not eradication of these two diseases (Gogan et al. 1989, Jessup and Williams 1999).

Many other examples of transplantation and disease could be cited: the emergence of raccoon rabies in the Eastern Seaboard states because of the relocation of raccoons for hunting (Davidson 2006); potential transplantation of *Mycoplasma gallisepticum* and *M. meleagridis* along with wild turkeys in western states (Jessup et al. 1983); *Brucella suis*, *Leptospira* spp., and pseudorabies virus spread with movement and stocking of feral hogs (Clark et al. 1983); the emergence of neurologic disease in horses and sea otters caused by *Sarcocystis neurona* shed by Virginia opossum transplanted to California; and the withering syndrome rickettsia to abalone off California via release of infected ship ballast water or possibly aquaculture practices. Disease analysis and health evaluation should be part of any major stocking or translocation effort. Significant wildlife disease outbreaks and problems have convinced many wildlife management agencies to incorporate disease investigation and control into their core programs.

Translocation is only one example of many human activities that can upset the host–agent–environment relationship with regard to wildlife disease. More pervasive and profound are common agricultural or resource extraction practices that transform environments and bring people, domestic animals, and disease agents into contact with wildlife in novel ways (Friend 2006). The emergence of Nipah virus and fatal epidemics in pigs and people largely resulted from the development of swine feeding facilities in fruit bat (*Epomophorus* spp.) habitats where the disease was endemic and where bat feces and carcasses could get into pig sheds. Other examples include trapping and consumption of civets (*Paradoxurus* spp.) and the outbreak of severe acute respiratory syndrome in China and its rapid spread to North America via airline travel; the recent worldwide COVID-19 pandemic and the wet market trade of China and Southeast Asia; the trade in African bushmeat and potential for outbreaks of Ebola and Marburg virus in major North American cities; and live bird importations into North America and avian influenza. At least 30 viruses appear to have been introduced to North America by the exotic bird trade.

Integrated Approach

Much like integrated pest management has proven effective in dealing with more intractable agricultural disease problems, integrated approaches hold promise for managing wildlife diseases, where efforts are made to manage various host, agent, and environmental factors at the same time. Rinderpest vaccination was only effective after infected populations of wildlife and livestock were identified, reduced, or eliminated—or limited in movement by traditional regulatory, testing, and population reduction methods. The comprehensive brucellosis control efforts in the Greater Yellowstone Area provide an example of coordinated and integrated disease control efforts, even if they have so far been unsuccessful. Management of distemper and plague in prairie dog and black-footed ferret habitats also includes host, agent, and environmental components (Williams and Barker 2001). The examples given of tule elk management to mitigate hemlock poisoning (Jessup et al. 1986), molybdenum-mediated copper deficiency (Gogan et al. 1989), and Johne's disease (Jessup and Williams 1999) were more primitive integrated efforts.

One Health

Although what is now called the One Health concept has been around under various names and in various forms for about 50 years, it seems to have recently gained serious traction and acceptance in veterinary medicine, and at least the public health sector of human medicine. The concept is that we need to seek and implement optimal solutions to health problems that maximize benefits to human health, animal health, environmental health, and sustainability. Political, social, and financial realities cause wildlife managers to seek One Health solutions to wildlife health problems. The mission statement of the Wildlife Disease Associations is "to acquire, disseminate, and apply knowledge of the health and diseases of wild animals in relation to their biology, conservation, and ecology, including interaction with humans and domestic animals." Although written more than 30 years ago, this statement clearly embodies One Health (Jessup et al. 2007).

Considering and accommodating wildlife and the environment when human and animal health programs are developed and implemented exemplify the One Health concept. From an environmental perspective this represents a significant advance over the days when population reduction was the primary tool used to enhance wildlife health, or when DDT and other persistent pesticides were widely used, without consideration of ecological and vector resistance consequences. Wildlife health professionals often need to consider whether actions to reduce or control disease organisms are warranted. In general, introduced pathogens, like invasive species, are the most dangerous and worthy of energetically responding to. Wildlife health problems resulting from or significantly contributed to by human activities, particularly those that alter normal ecosystem processes, are an increasing problem that we can, and in some cases legally must, respond to. Most toxins, chemicals, and pollutants have pernicious effects on health of animals, including

people and wildlife. Wildlife diseases that affect keystone species, sensitive or threatened species, or jeopardize wildlife recovery programs are also ones most often needing treatment and other management.

Treating or eliminating all parasites and diseases, particularly if they have coevolved with their host, may not be desirable or improve wildlife population health in the long run. Outbreaks of human and domestic animal disease could have roots in disturbed ecosystems or disrupted environments and may result in health programs that are more ecologically sound and sustainable. One Health should be kept in mind should outbreaks of highly pathogenic avian influenza in waterfowl or foot-and-mouth disease in wild ungulates occur in North America. Wildlife management and conservation participation in One Health is just beginning. One Health represents potential positive change and should be incorporated into wildlife health, management, and conservation programs.

SUMMARY

The more rapidly we move people, animals, diseases, and disease vectors around the world, and the deeper we penetrate once-pristine ecosystems, the more likely exotic and zoonotic wildlife diseases are to emerge. Seventy percent of emerging human diseases in the past five decades have been zoonotic (i.e., of animal origin), and the majority of these are associated with wildlife, not domestic, species (Friend 2006). This seems unlikely to change in the coming decades.

Some of the larger centers for wildlife disease diagnosis and research (Table 8.2) and other laboratories provide excellent wildlife health diagnostic work, research, and conservation. The discovery of the emerging West Nile virus epidemic 20 years ago resulted from the simultaneous workup of diseased exotic birds and crows at a major zoo and a state wildlife health investigation facility (McLean et al. 2002).

The most appropriate group to assist with investigation of a wildlife health problem will largely be determined by the species involved and the location. In general, all native wildlife is managed under the authority of that state's wildlife or natural resource department. Twenty states in the southeastern and midwestern United States use the Southeastern Wildlife Cooperative Disease Study at the University of Georgia for some of their wildlife health investigations, even some that have their own wildlife veterinarian or health professional. Seven states in the Northeast use the services of the Northeastern Wildlife Disease Cooperative. Health of endangered species, migratory birds, and on lands under the US Department of the Interior (e.g., national wildlife refuges) is the responsibility of the USFWS. Much of the USFWS and other Department of the Interior work is done at the USGS National Wildlife Health Center (NWHC). National parks have their own wildlife disease programs but often cooperate with USGS/NWHC. Marine mammals and marine fish are under the National Oceanic and Atmospheric Administration's National Marine Fisheries Service. Canadian wildlife falls under

Table 8.2. Selected large centers of wildlife health / disease research and diagnosis in North America

Group	Location(s)	Services
California Department of Fish and Game	Rancho Cordova, CA	State management (terrestrial)
	Santa Cruz, CA	Diagnosis, state management (marine)
Canadian Cooperative Wildlife Health Centres	Guelph, Quebec, Saskatoon, Calgary, Prince Edward Island	Diagnosis, epidemiology, research management advice, provides all services for Canada
Colorado State University Veterinary School and Labs	Fort Collins, CO	Diagnosis, epidemiology, and state management of Colorado and cooperating states
Cornell University	Cornell, NY	Diagnosis, epidemiology, research
Michigan Department of Natural Resources and Michigan State University Diagnostic Lab	East Lansing, MI	Diagnosis, epidemiology, research, state management
National Animal Disease Center (USDA-APHIS, ARS, VS)	Ames, IA	Diagnosis, epidemiology, research, foreign animal and program diseases
National Wildlife Research Center (USDA-APHIS and WS)	Fort Collins, CO	Research, disease and damage management, epidemiology
National Wildlife Health Center (USDI-USGS/BRD), University of Wisconsin	Madison, WI	Diagnosis, epidemiology, research, information technology services, USDI agency management advice
National Park Service, Veterinary Services	Fort Collins, CO	All national parks
NOAA, National Marine Fisheries Service	Beltsville, MD	Marine animal diseases, diagnosis, epidemiology, management
Washington State University WADDL, Global Programs	Pullman, WA	Diagnosis, epidemiology, research
Southeastern Cooperative Wildlife Disease Study (SCWDS)	Athens, GA	Diagnosis, epidemiology, research, management advice for 20 southeastern and midwestern states
University of Florida Florida Game and Fish	Gainesville, FL	Diagnosis, epidemiology, research
University of Georgia	Athens, GA	Diagnosis and research
University of California, Davis, Veterinary School, Wildlife Health Center, Animal Health and Food Safety (CAHFS) and other labs	Davis, CA	Diagnosis, epidemiology, research, OWCN, Sea Doc, mountain gorilla management advice
Tufts Cummings School of Veterinary Medicine	North Grafton, MA	Diagnosis, epidemiology, research, Center for Conservation Medicine, wildlife clinic, international programs
Wyoming Department of Game and Fish with Wyoming State Veterinary Lab, Sybille Research Unit	Wheatland, WY Laramie, WY	Diagnosis, epidemiology, research
Colorado Division of Wildlife	Fort Collins, CO	Diagnosis, epidemiology, research

the purview of Fish and Wildlife Canada, but the cooperative formed by all the veterinary schools and provincial agencies in Canada (Canadian Wildlife Health Cooperative) do their disease investigations.

Increasingly rapid, worldwide movement of people and animals has resulted in widespread and accelerating introduction of potential diseases, and brought organisms into wildlife habitats and exposed people to once-rare disease organisms. Wildlife pathogens are twice as likely to become emerging diseases of humans as pathogens without wildlife hosts. Since the 1980s, there has been increasing recognition that the ecology of wildlife diseases is a complex and significant field of study within the field of ecology. During this same period, key wildlife management programs—including big-game hunting, wildlife translocations, sensitive species recovery, upland game bird stocking, waterfowl hunting, and many endangered species programs—have been shaped by economic, legal, financial, political, and biological aspects of wildlife diseases. Wildlife diseases can have profound ecological and wildlife management implications. Optimizing the health of wildlife populations requires the cooperation of many parts of society if the goals of One Health and the protection of public and

ecosystem health—and healthy and sustainable wildlife and domestic animal populations—are to be met.

Literature Cited

Bárcena, J., B. Guerra, I. Angulo, J. González, F. Valcárcel, et al. 2015. Comparative analysis of rabbit hemorrhagic disease virus (RHDV) and new RHDV2 virus antigenicity, using specific virus-like particles. Veterinary Research 46:1–6.

Belay, E. D., R. A. Maddox, E. S. Williams, M. W. Miller, P. Gambetti, and L. B. Schonberger. 2004. Chronic wasting disease and potential transmission to humans. Emerging Infectious Diseases 10:977–984.

Berger, L. R., R. Spear, P. Daszak, D. E. Green, A. A. Cunningham, et al. 1998. Chytridiomycosis causes amphibian mortality associated with population declines in rain forests of Australia and Central America. Proceedings of the National Academy of Science 95:9031–9036.

Black, S.R., I. K. Barker, K. G. Mehren, G. J. Crawshaw, S. Rosendal, et al. 1988. An epizootic of *Mycoplasma ovipneumoniae* infection in captive Dall's sheep (*Ovis dalli dalli*). Journal of Wildlife Diseases 24:627–635.

Blehert, D. S., A. C. Hicks, M. Behr, C. U. Meteyer, B. M Berlowski-Zier, et al. 2009. Bat white-nose syndrome: an emerging fungal pathogen? Science 323:227.

Brown, V., and S. N. Bevins. 2018. A review of African swine fever and the potential for introduction into the United States and possibility of subsequent establishment in feral swine and native ticks. Frontiers in Veterinary Science. doi:10-3389/fvets.2018.00011.

Cassirer, E. F., R. K. Manlove, E. S. Almberg, P. Kamath, M. Cox, et al. 2018. Pneumonia in bighorn sheep: risk and resilience. Journal of Wildlife Management 82:32–45.

Centers for Disease Control and Prevention. 2019. Chronic wasting disease: occurrence. https://www.cdc.gov/prions/cwd/occurrence.html.

Centers for Disease Control and Prevention. 2021. One Health basics. https://www.cdc.gov/onehealth/basics/index.html.

Clark, R. K., D. A. Jessup, D. W. Hird, R. Ruppanner, and M. E. Meyer. 1983. Serologic survey of California wild hogs for antibodies against selected zoonotic disease agents. Journal of the American Veterinary Medical Association. 183:1248–1251.

Conrad, P. A., M. A. Miller, C. Kreuder, E. R. James, J. Mazet, H. Dabritz, D. A. Jessup, F. Gulland, and M. E. Grigg. 2005. Transmission of *Toxoplasma*: clues from the study of sea otters as sentinels of *Toxoplasma gondii* flow into the marine environment. International Journal of Parasitology 35:1125–1168.

Constantine, D. G. 2003. Geographic translocation of bats: known and potential problems. Emerging Infectious Diseases 9:17–21.

Davidson, W. R. 2006. Field manual of wildlife diseases in the southeastern United States. 3rd edition. Southeastern Cooperative Wildlife Disease Study, Athens, Georgia, USA.

Drees, K. P., J. M. Lorch, S. J. Puechmaille, K. L. Parise, G. Wibbelt, et al. 2017. Phylogenetics of a fungal invasion: origins and widespread dispersal of white-nose syndrome. Molecular Biology 8:e01941-17.

Dubey, J. P., D. S. Lindsay, W. J. Saville, S. M. Reed, D. E. Granstrom, and C. A. Speer. 2001. A review of *Sarcocystis neurona* and equine protozoal myeloencephalitis (EPM). Veterinary Parasitology 95:89–131.

Ernest, H. B., B. Hoar, J. A. Well, and K. I. O'Rourke. 2010. Molecular genealogy tools for white-tailed deer with chronic wasting disease. Canadian Journal of Veterinary Research 74:153–156.

Estes, J. A. 2005. Carnivory and trophic connectivity in kelp forests. Pages 61–81 *in* J. C. Ray, K. H. Redford, R. S. Steneck, and J. Berger, editors. Large carnivores and the conservation of biodiversity. Island Press, Washington, DC, USA.

Food and Agriculture Organization. 2010. Ridding the world of rinderpest. https://www.rgs.org/CMSPages/GetFile.aspx?nodeguid=41e8c1b4-754d-4fc0-a36e-617275825d99&lang=enGB#:~:text=Africa%20remained%20free%20of%20the,third%20of%20its%20human%20population.

Foreyt, W. J., and D. A. Jessup. 1982. Fatal pneumonia of bighorn sheep following association with domestic sheep. Journal of Wildlife Diseases 18:163–168.

Frick, W. F., J. F. Pollock, A. C. Hicks, K. E. Langwig, D. S. Reynolds, G. R. Turner, C. M. Butchkoski, and T. H. Kunz. 2010. An emerging disease causes regional population collapse of a common North American bat species. Science 329:679–682.

Friend, M. 2006. Disease emergence and resurgence: the wildlife human connection. Circular 1285. US Geological Survey, Reston, Virginia, USA. http://www.nwhc.usgs.gov/publications/disease_emergence/index.jsp.

Geboy, R. 2008. National decontamination protocol update. Midwest Regional WNS Coordinator, US Fish and Wildlife Service, Washington, DC, USA.

Gleeson, M., and O. A. Petritz. 2020. Emerging infectious diseases of rabbits. Veterinary Clinics of North America: Exotic Animal Practice 23:249–261.

Gogan, P. J. P., D. A. Jessup, and M. Akeson. 1989. Copper deficiency in tule elk at Point Reyes, California. Journal of Range Management 42:233–238.

Guberti, V., S. Khomenko, M. Masiulis, and S. Kerba. 2019. African swine fever in wild boar ecology and biosecurity. FAO Animal Production and Health Manual No. 22. Food and Agriculture Organization, International Office of Epizootics, European Community, Rome, Italy.

Haley, N. J., and E. A. Hoover. 2015. Chronic wasting disease of cervids: current knowledge and future perspectives. Annual Review of Animal Biosciences 3:305–325.

Hoyt, J. R., K. E. Langwig, J. P. White, H. M. Kaarakka, J. A. Redell, K. L. Parise, W. F. Frick, J. T. Foster, and A. M. Kilpatrick. 2019. Field trial of a probiotic bacteria to protect bats from white-nose syndrome. Scientific Reports 9.

Jansen, B. D., J. R. Hefelfinger, T. H. Noon, P. R. Krausman, and J. C. deVos. 2006. Infectious keratoconjunctivitis in bighorn sheep, Silver Bell Mountains, Arizona, USA. Journal of Wildlife Diseases 42:407–411.

Jessup, D. A. 1986. Anseriformes: avian cholera, waterfowl botulism, duck virus enteritis. Pages 342–353 *in* M. E. Fowler, editor. Zoo and wildlife medicine. 2nd edition. Iowa State University Press, Ames, USA.

Jessup, D. A. 1993. Translocation of wildlife. Pages 493–499 in M. E. Fowler, editor. Zoo and wild animal medicine. 3rd edition. W. B. Saunders. Philadelphia, Pennsylvania, USA.

Jessup, D. A. 2011. Wild and domestic sheep disease information. American Association of Wildlife Veterinarians. http://aawv.net/bighorn/.

Jessup, D. A., J. H. Boermans, and N. D. Kock. 1986. Toxicosis in tule elk caused by ingestion of poison hemlock. Journal of the American Veterinary Medical Association 189:1173–1175.

Jessup, D. A., A. J. DaMassa, R. Lewis, and K. R. Jones. 1983. Mycoplasma gallisepticum infection in wild-type turkeys living in close contact with domestic fowl. Journal of the American Veterinary Medical Association 183:1245–1247.

Jessup, D. A., C. K. Johnson, J. Estes, D. Carlson-Bremer, W. Jarman, S. Reese, E. Dodd, M. T. Tinker, and M. H. Ziccardi. 2010. Persistent organic pollutants and other contaminants of concern in the blood of free ranging sea otters (*Enhydra lutris*) in Alaska and California. Journal of Wildlife Diseases 46:1214–1233.

Jessup, D. A., M. Miller, J. Ames, M. Harris, P. Conrad, C. Kreuder, and J. A. K. Mazet. 2004. The southern sea otter (*Enhydra lutris nereis*) as a sentinel of marine ecosystem health. Ecohealth 1:239–245.

Jessup, D. A., M. A. Miller, C. Kreuder-Johnson, P. Conrad, T. Tinker, J. Estes, and J. Mazet. 2007. Sea otters in a dirty ocean. Journal of the American Veterinary Medical Association 231:1648–1652.

Jessup, D. A., and E. S. Williams. 1999. Paratuberculosis in free-ranging wildlife in North America. Pages 616–620 *in* M. E. Fowler and R. E. Miller, editors. Zoo and wildlife medicine. 4th edition. W. B. Saunders, Philadelphia, Pennsylvania, USA.

Johnson, C. J., A. Herbst, C. Duque-Velasquez, J. P. Vanderloo, P. Bochsler, R. Chappell, and D. McKenzie. 2011. Prion protein polymorphisms affect chronic wasting disease progression. PLoS One 6(3):e17450.

Johnson, C. K., M. T. Tinker, J. A. Estes, P. A. Conrad, M. Staedler, M. A. Miller, D. A. Jessup, and J. K. Mazet. 2009. Prey choice and habitat use drive sea otter pathogen exposure in a resource-limited coastal system. Proceedings of the National Academy of Science 106:2242–2247.

Kannan, K., H. B. Moon, S. H. Yun, T. Agusa, N. J. Thomas, and S. Tanabe. 2008. Chlorinated, brominated, and perfluorinated compounds, polycyclic aromatic hydrocarbons and trace elements in livers of sea otters from California, Washington, and Alaska (USA), and Kamchatka (Russia). Journal of Environmental Monitoring 10:552–558.

Kannan, K., E. Perrotta, and N. J. Thomas. 2006. Association between perfluorinated compounds and pathological conditions in southern sea otters. Environmental Science and Technology 40:4943–4948.

Kannan, K., E. Perrotta, N. Thomas, and K. Aldous. 2007. A comparative analysis of polybrominated diphenyl ethers and polychlorinated biphenyls in southern sea otters that died of infectious diseases and noninfectious causes. Archives of Environmental Contamination and Toxicology 53:293–302.

Kerr, P. J., and T. M. Donnelly. 2013. Viral infections of rabbits. Veterinary Clinics of North America: Exotic Animal Practice 16:437–468.

Klein, P. M. 2013. Chronic wasting disease—APHIS proposed rule to align BSE import regulations to OIE. CWD Program Manager, US Department of Agriculture / Animal and Plant Health Inspection Service, Washington, DC.

Kreuder, C., M. A. Miller, D. A. Jessup, L. J. Lowenstein, M. D. Harris, J. A. Ames, T. E. Carpenter, P. A. Conrad, and J. A. K. Mazet. 2003. Patterns of mortality in southern sea otters (Enhydra lutris nereis) from 1998–2001. Journal of Wildlife Diseases 39:495–509.

Kudela, R. M., J. Q. Lane, and W. P. Cochlan. 2008. The potential role of anthropogenically derived nitrogen in the growth of harmful algae in California, USA. Harmful Algae 8:103–110.

Langwig, K. E., W. F. Frick, J. T. Bried, A. C. Hicks, T. H. Kunz, and A. M. Kilpatrick. 2012. Sociality, density-dependence and microclimates determine the persistence of populations suffering from a novel fungal disease, white-nose syndrome. Ecology Letters 15:1050–1057.

Lawrence, P. K., S. Shanthalingham, R. Dassanyake, R. Subramaniam, C. N. Herndon, et al. 2010. Transmission of Mannheimia haemolytica from domestic sheep (Ovis aries) to bighorn sheep (Ovis canadensis): unequivocal demonstration with green fluorescent protein tagged organisms. Journal of Wildlife Diseases 46:706–717.

Leopold, A. 1933. Game management. Charles Scribner's Sons, New York, New York, USA.

Liu, S. J., H. P. Xue, B. Q. Pu, and N. H. Qian. 1984. A new viral disease in rabbits. Animal Husbandry and Veterinary Medicine 16:253–255.

Lorch, J. M., C. U. Meteyer, M. J. Behr, J. G. Boyles, P. M. Cryan, et al. 2011. Experimental infection of bats with Geomyces destructans causes white-nose syndrome. Nature 480:376–378.

Mann, B. 2008. Northeast bat die-off mirrors honeybee collapse. All Things Considered. National Public Radio.

Mathiason, C. K., J. G. Powers, S. J. Dahmes, D. A. Osborn, K. V. Miller, et al. 2006. Infectious prions in the saliva and blood of deer with chronic wasting disease. Science 314:133–136.

McLean, R. G., S. R. Ubico, D. Bourne, and N. Komar. 2002. West Nile virus in livestock and wildlife. Current Topics in Microbiology and Immunology 267:271–308.

Miller, M. A., B. A. Byrne, S. S. Jang, E. M. Dodd, E. Dorfmeier, et al. 2010a. Enteric bacterial pathogen detection in southern sea otters (Enhydra lutris nereis) is associated with coastal urbanization and freshwater runoff. Veterinary Research 41:01. doi:10.1051/vetres/2009049.

Miller, M. A., P. A. Conrad, M. Harris, B. Hatfield, G. Langlois, et al. 2010b. Localized epizootic of meningoencephalitis in southern sea otters (Enhydra lutris nereis) caused by Sarcocystis neurona. Veterinary Parasitology 172:183–194.

Miller, M. A., R. M. Kudela, A. Mekebri, D. Crane, S. C. Oates, et al. 2010c. Evidence for a novel marine harmful algal bloom: cyanotoxin (Microcystin) transfer from land to sea otters. PLoS One 5:e12576. doi:10.1371/journal.pone.0012576.

Miller, M. A., M. E. Moriarty, L. Henkel, M. T. Tinker, T. L. Burgess, et al. 2020. Predators, disease, and environmental change in the nearshore ecosystem: mortality in southern sea otters (Enhydra lutris nereis) from

1998–2012. Frontiers in Marine Science. https://doi.org/10.3389/fmars.2020.00582.

Minnis A. M., and D. L. Lindner. 2013. Phylogenetic evaluation of Geomyces and allies reveals no close relatives of Pseudogymnoascus destructans in bat hibernacula of eastern North America. Fungal Biology 117:638–649.

Monello, R. J., J. G. Powers, N. T. Hobbs, T. R. Spraker, K. I. O'Rourke, and M. A. Wild. 2013. Efficacy of antemortem rectal biopsies to diagnose and estimate prevalence of chronic wasting disease in free-ranging cow elk (Cervus elaphus nelsoni). Journal of Wildlife Diseases 49:270–278.

Munger, E. 2008. Group asking cavers to keep out. Daily Gazette. https://dailygazette.com/2008/02/14/0214_caves/.

National Academies of Sciences, Engineering, and Medicine. 2017. Revisiting brucellosis in the Greater Yellowstone Area. National Academies Press, Washington, DC, USA. https://doi.org/10.17226/24750.

Office International des Epizooties. 2020. Rabbit haemorrhagic disease, United States of America. https://www.oie.int.

Prusiner, S. B. 2001. Neurodegenerative diseases and prions. New England Journal of Medicine 344:1516–1526.

Rocke T. E., B. Kingstad-Bakke, M. Wüthrich, B. Stading, R. C. Abbott, et al. 2019. Virally-vectored vaccine candidates against white-nose syndrome induce anti-fungal immune response in little brown bats (Myotis lucifugus). Science Reports 6788. doi:10.1038/s41598-019-43210-w.

Rocke, T. E., S. Smith, P. Marinari, J. Kreeger, J. T. Enama, and B. S. Powell. 2008. Vaccination with the F1-V fusion protein protects black-footed ferrets (Mustela nigripes) against plague upon oral challenge. Journal of Wildlife Diseases 44:1–7.

Rocke, T. E., D. W. Tripp, R. E. Russel, R. C. Abbott, K. L. D. Richgels, et al. 2017. Sylvatic plague vaccine partially protects prairie dogs (Cynomys spp.) in field trials. EcoHealth 14:438–450.

Saunders, S. E., S. L. Bartelt-Hunt, and J. C. Bartz, 2012. Occurrence, transmission, and zoonotic potential of chronic wasting disease. Emerging Infectious Diseases 18:369–376.

Selariu, A., J. G. Powers, A. Nalls, M. Brandhuber, A. Mayfield, and C. K. Mathiason. 2015. In utero transmission and tissue distribution of chronic wasting disease-associated prions in free-ranging Rocky Mountain elk. Journal of General Virology 96:3444–3455.

Sercu, B., L. C. Van De Werfhorst, J. Murray, and D. Holden. 2009. Storm drains are sources of human fecal pollution during dry weather in three urban Southern California watersheds. Environmental Science and Technology 43:293–298.

Slate, D., T. P. Aldeo, K. M. Nelson, R. B. Chipman, D. Donavan, J. D. Blanton, M. Niezgoda, and C. C. Rupprecht. 2009. Oral rabies vaccination in North America: opportunities, complexities and challenges. PLoS Neglected Tropical Diseases 3:e549. doi:10.1371/journal.pntd.0000549.

Spraker, T. R., K. C. VerCauteren, T. Gidlewski, D. A. Schneider, R. Munger, A. Balachandran, and K. I. O'Rourke. 2009. Antemortem detection of PrP CWD in preclinical, ranch-raised Rocky Mountain elk (Cevvus Elaphus nelsoni) by biopsy of the rectal mucosa. Journal of Veterinary Diagnostic Investigation 21:15–41.

Stading, B., J. A. Ellison, W. C. Carson, P. S. Satheshkumar, T. E. Rocke, and J. E. Osorio. 2017. Protection of bats (Eptesicus fuscus) against rabies following topical or oronasal exposure to a recombinant raccoon poxvirus vaccine. PLoS Neglected Tropical Diseases Oct. 4;11:e0005958. doi:10.1371/journal.pntd.0005958.

Thomas, N. J., and R. A. Cole. 1996. The risk of disease and threats to the wild population. Endangered Species Update 13:23–27.

US Animal Health Association. 2020. State animal health officials. https://www.usaha.org/federal-and-state-animal-health.

USDA. US Department of Agriculture. 2020. Rabbit hemorrhagic disease in British Columbia, Canada. https://www.aphis.usda.gov/animal_health/downloads/Rabbit-Hemorrhagic-Disease_062018.pdf.

USFWS. US Fish and Wildlife Service. 2011. A national plan for assisting states, federal agencies, and tribes in managing white-nose syndrome in bats. US Fish and Wildlife Service. https://pubs.er.usgs.gov/publication/70039214.

USGS. US Geological Survey. 2013. Frequently asked questions concerning chronic wasting disease (CWD). USGS National Wildlife Health Center. Accessed 23 Apr 2017.

USGS. US Geological Survey. 2019. Distribution of chronic wasting disease in North America. https://www.usgs.gov/media/images/distribution-chronic-wasting-disease-north-america-0.

Verant, M. L., C. U. Meteyer, J. R. Speakman, P. M. Cryan, J. M. Lorch, and D. S. Blehert. 2014. White-nose syndrome initiates a cascade of physiologic disturbances in the hibernating bat host. BMC Physiology 14:10.

VerCauteren, K. C., J. L. Pilon, P. B. Nash, G. E. Phillips, and J. W. Fischer. 2012. Prion remains infectious after passage through digestive system of American crows (*Corvus brachyrhynchos*). PLoS One 7(10): e45774.

Washington Department of Fish and Wildlife. 2019. White nose syndrome of bats fact sheet. wdfw.wa.gov.

Wehausen, J. D., S. T. Kelly, and R. R. Ramey. 2011. Domestic sheep, bighorn sheep and respiratory disease: a review of experimental evidence. California Fish and Game Quarterly 97:7–24.

Wibbelt, G., S. J. Puechmaille, B. Ohlendorf, K. Mühldorfer, T. Bosch, T. Görföl, K. Passior, A. Kurth, D. Lacremans, and F. Forget. 2013. Skin lesions in European hibernating bats associated with *Geomyces destructans*, the etiologic agent of white-nose syndrome. PLoS One 8:e74105.

Williams, E. S., and I. K. Barker. 2001. Infectious diseases of wild mammals. 3rd edition. Iowa State University Press, Ames, USA.

Williams, E. S., and E. T. Thorne. 1999. Veterinary contributions to the black-footed ferret conservation program. Pages 460–463 *in* M. E. Fowler and R. E. Miller, editors. Zoo and wildlife medicine. 4th edition. W. B. Saunders, Philadelphia, Pennsylvania, USA.

Williams, E. S., and S. Young. 1980. Chronic wasting disease of captive mule deer: a spongiform encephalopathy. Journal of Wildlife Diseases 16:89–98.

Wobeser, G. A., 1994. Investigation and management of diseases in wild animals. Plenum Press, New York, New York, USA.

Wolfe, L. L., M. M. Conner, T. H. Baker, V. J. Dreitz, K. P. Burnham, E. S. Williams, N. T. Hobbs, and M. W. Miller. 2002. Evaluation of antemortem sampling to estimate chronic wasting disease prevalence in free-ranging mule deer. Journal of Wildlife Management 66:564–573.

Wolff, P. L., J. A. Blanchong, D. D. Nelson, P. J. Plummer, C. McAdoo, M. Cox, T. E. Besser, J. Muñoz-Gutiérrez, and C. A. Anderson. 2019. Detection of *Mycoplasma ovipneumoniae* in pneumonic mountain goat (*Oreamnos americanus*) kids. Journal of Wildlife Diseases 55:206–212.

Young, S. 2011. Culprit behind bat scourge confirmed. Nature doi: 10.1038/news.613.

HUNTING AND TRAPPING

JAMES R. HEFFELFINGER

INTRODUCTION

Hunting and trapping in all their various forms provide the very foundation of a system of wildlife conservation developed in North America to stem the loss of native wildlife species caused by overexploitation and habitat alteration. As restrictive regulations halted wildlife population declines, the field of wildlife management shifted from a protective paradigm to a focus on consumptive use of wildlife to manage population abundance and demography within desired goals. Beyond actual population management, hunting provides the financial and social support for continued sustainable conservation of habitat and nonhunted species that benefits all members of the public.

The Human Hunter

Hunting and trapping have been integral to human existence since early hominids emerged from the forest and began carrying weapons to facilitate intake of an increasingly protein-rich diet (Ardrey 1976). The reduction in the size of canine teeth throughout human evolution (Haile-Selassie et al. 2004) corresponds to an increasing use of tools to harvest and process animal protein. The incredible development of the human brain is thought to be the result of an increased need to communicate and coordinate abstract plans associated with hunting animals (Watson 1971). The high-protein diet that hunting provided would have also allowed for rapid gains in physical size, strength, and intelligence (Leonard and Robertson 1994). Through time, cultures differentiated and specialized in hunting and trapping the wildlife species with which they coexisted.

As technology developed more effective weapons and trapping techniques, humans were able to exploit a wider variety of quarry. Several sites in southeastern Arizona contain juvenile mammoth fossils with Clovis spear points and butchering tools in association with the bones (Haury et al. 1959). Some researchers believe these cases explain the megafaunal extinctions at the close of the Pleistocene as being caused by humans (Martin and Klein 1984). Although no single cause easily explains these mass extinctions, it is unlikely that Pleistocene

humans armed only with spears were able to single-handedly cause that many species of large mammals to become extinct (Grayson and Meltzer 2002, 2003).

Technological advances were not limited to hunting and gathering because humans eventually learned they could grow some of their own plants and even domesticate animals for food. With the genesis of primitive farming and animal husbandry, humans, edible plants, and animals inevitably began to domesticate one another. Domestication began the quiet and nearly imperceptible shift of hunting and trapping being a survival necessity to an activity that, for most cultures, merely supplemented food obtained more easily elsewhere.

From Sustenance to Conservation

At the early stages of Western civilization in North America, hunting and trapping were still important survival skills that supplemented supplies, but the harvest of wildlife was also one of the leading contributors to the economy of the young, growing nations. With a complete lack of limits to harvest, wildlife populations began to diminish rapidly. Unlike their spear-wielding predecessors, Europeans armed with rifles easily overharvested some species of wildlife that had the misfortune to be edible, wrapped in fur and leather, adorned by ornate feathers, or otherwise desirable. Out of these early years of unmitigated slaughter for market and subsistence use came the realization by the young nations that something had to change. Concerned citizenry, led by some of the most prominent hunters and other conservationists of the day, developed a collective sense of stewardship and conservation (Roosevelt et al. 1902, Grinnell 1913, Leopold 1933, Reiger 1986, Bradley 1995, Thomas 2010). The desire to maintain wildlife populations evolved into an incredibly successful system whereby consumptive use of wildlife in the form of hunting and trapping became the cornerstone of what is now known as the North American Model of Wildlife Conservation.

The North American Model was not an a priori strategic plan to direct conservation efforts, but rather a description of a set of laws and practices that made conservation of native wildlife so successful in the United States and Canada. Seven principles collectively provide the conceptual and philosophical basis for the North American Model. Although the exact wording sometimes varies, the important and relevant elements are consistently represented as: (1) wildlife is held in public trust for all, (2) deleterious commerce in wildlife is prohibited, (3) appropriate use of wildlife is regulated by law, (4) wildlife can only be killed for legitimate purposes, (5) wildlife is managed as an international resource, (6) science is the appropriate basis for wildlife policy, and (7) opportunity to harvest wildlife is fairly allocated (Geist et al. 2001, Mahoney et al. 2019).

This conservation paradigm built around scientifically regulated harvest has benefited nearly all species of native wildlife because as wildlifers conserved, restored, and managed habitat for wildlife species, they also conserved a landscape that provided habitat for nonhunted species. The North American Model of Wildlife Conservation has been so successful that recognition is growing worldwide that hunting and trapping are proven and valuable tools for maintaining the sustainable use of wildlife in perpetuity. Management in other countries differs from the system that evolved in North America (Putnam et al. 2011, Cooney 2019). At worldwide symposia in London and Namibia, participants reaffirmed the importance of hunting to wildlife conservation and urged other countries to model their programs after the North American model (Geist 2006, Patterson 2009, Duda et al. 2010b).

The support for wildlife by hunters and trappers arises partly out of self-interest because of the importance of harvest to them personally (Geist 2006). Participants accrue many benefits—psychological, physical, sociological, and nutritional—from hunting and trapping, but the real benefit in North America is collective stewardship of wildlife and the habitats on which they rely (Leopold 1943). Consumptive use of wildlife provides several important functions beyond the personal benefits to participants (Duda et al. 1998). Hunters and trappers have voluntarily and willingly contributed billions of dollars to support conservation for all wildlife species, not just those that are hunted (Southwick and Allen 2010, USFWS 2020). They have been the central pillars of this conservation paradigm and thus are responsible for supporting a wide variety of conservation activities that the public values.

CLINTON HART MERRIAM (1855–1942)

Clinton Hart Merriam was born the son of a US congressman in 1855 and began studying animals as a small boy. By the time he was 17 years old, he took part in the Hayden Survey exploring the Yellowstone area. He entered medical school at Yale and Columbia, graduating with his MD in 1879. Merriam practiced medicine for a few years, but by the age of 30 he had turned to the study of wildlife and was leading the Division of Economic Ornithology and Mammalogy, which later became the US Fish and Wildlife Service. He was a founding member of the National Geographic Society and was president of many scientific organizations of the time.

For most of his career, Merriam focused on collecting and describing mammals, including approximately 660 he named as previously unknown and new to science. He was known as a taxonomic "splitter," and many of these "new" species were invalidated through further study. Still, Merriam's lasting contribution to science was in perfecting study methods that relied on large specimen samples with accurate geographic data and a combination of field and laboratory studies. His published writings number nearly 500, but he may be best known for developing the concept of life zones, showing that altitudinal changes in local vegetation associations corresponded to latitudinal zones from the equator to the poles. He retired from government service in 1910 and was given an annual pension of $12,000 to study anything he desired. By then his interests were shifting from mammalogy to ethnology, and he studied and published on Native tribes in California well into his 80s. Merriam died in 1942 at the age of 87.

CONSERVATION THROUGH HUNTING AND TRAPPING

Conservation by Consumption

Both the need and success of the North American Model of Wildlife Conservation have revolved around sustainability of harvest (Mahoney 2009). Depletion of wildlife resources in the late 1800s spurred the invention of a unique system of conservation in which wildlife could be used in a sustainable manner that was closely regulated by law and based on the best available science. In its infancy, wildlife management began as a system of harvest limits to stop the rapid decline of wildlife

populations during the era of overexploitation (Baughman and King 2008). As the laws gained in effectiveness, management programs evolved slowly to restore and manage hunted species as they responded positively to early hunting restrictions. It is a common misconception that hunting caused the extinction or extirpation of some species. On the contrary, when unregulated killing was controlled in concert with growing restoration efforts, these populations rebounded vigorously only after regulated hunting was used to influence a bold, experimental system of comprehensive wildlife conservation (Fig. 9.1).

Figure 9.1. Hunters contribute to conservation through funding, management of wildlife populations, and advocacy for wild things in wild places. Courtesy of Lakeisha Woodard

This system of conservation has been so undeniably successful it has been recognized as a model to emulate worldwide. The World Conservation Strategy of the International Union for Conservation of Nature (1980:18) defines conservation as "the management of human use of the biosphere so that it may yield the greatest sustainable benefit to present generations while maintaining its potential to meet the needs and aspirations of future generations." A symposium in Namibia, South Africa, on the ecological and economic benefits of hunting reaffirmed the superiority of managed conservation over strictly protection. Participants recognized that "sound scientific information demonstrates the importance of hunting to the future of wildlife" (Rowe 2009:392).

Legal Basis

There is sometimes considerable confusion about the function and purpose of the various natural resource agencies as it relates to conservation. The National Park Service, US Forest Service, and Bureau of Land Management are all federal agencies. They are referred to as land management agencies because they are responsible for managing land-based resources (e.g., timber, vegetation, recreation) rather than wildlife populations. The US Fish and Wildlife Service (USFWS) has land management (national wildlife refuges) and wildlife management (endangered species and migratory birds) responsibilities. The Canadian Wildlife Service has wildlife and land management responsibilities almost identical to the USFWS. Having a federal agency in charge of migratory birds makes sense because they can be managed in large flyways that span a large portion of the continent. States, provinces, and territories have agencies responsible for managing resident wildlife in their jurisdictions (excluding tribal or First Nation lands). The authority of the state and provincial agencies to manage native wildlife evolved because wildlife belongs to the public and is held in public trust (Decker et al. 1996; Smith 2011, 2019). Additionally, because of the complex mosaic of land ownership patterns, it would make no sense for this authority to change as wildlife moved through parcels of land owned by different agencies and entities.

State and provincial wildlife agencies use research, population monitoring, adaptive management, emerging technologies, and experience to develop management guidelines and protocols to determine the appropriate level of harvest for each population. The allowable level of harvest might simply be that which is sustainable with no ill effects to the population, or it might be a prescription to reach specific management goals of animal abundance and demography. Societal pressures play a role in some cases, when such issues as vehicle collisions, agricultural crop damage, or simply hunter preferences affect management goals. Management goals are normally set through an open and transparent process where the public has ample opportunity to provide input. In addition, agencies use surveys to obtain the opinions and desires of hunters, anglers, and the public (Decker et al. 2001). In most wildlife management scenarios, there is a wide range of management goals that are appropriate ecologically and biologically. Because wildlife are held in public trust and managed on behalf of all citizens, it is wholly appropriate that the public have a voice in how wildlife are managed within this wide range of possible goals (Decker et al. 1996, 2019; Smith 2011).

Once goals are determined, proper management involves monitoring wildlife population abundance and demography and harvest-related parameters. Both survey (e.g., abundance, distribution, sex and age ratios) and harvest (e.g., number harvested, age structure, harvest/unit effort) information represent vital data that help biologists manage wildlife populations. No single piece of information alone is sufficient; managers must use all available information in concert to make informed choices about how to achieve management goals. Managers look at the current conditions and, more importantly, the recent trends in population parameters. Trend data from male:female ratios, fawn:female ratios, age structure, population estimates or density, and abundance should all be monitored to measure current performance against the intended goals of the agency. Using these points of information, the manager can learn from and predict how certain management actions will affect the population.

The management goals and guidelines, and the actual data collected, vary greatly by species throughout North America because of a diversity of financial, logistical, and social constraints. However, all state and provincial agencies monitor populations of hunted species and apply monitoring data to management goals. Each jurisdiction has a commission or wildlife board made up of citizens or political appointees. These commissions receive recommendations from agency biologists and consider research and input from the public to make wildlife management decisions, to direct agencies, and to set the annual hunting and trapping rules and regulations.

Managing Animal Abundance

Today, hunters and trappers are the cornerstone of North American wildlife conservation because of the funds and advocacy they bring to the table, and because they remain the most effective logistical agents of actual population management. The early days of North American wildlife management were spent stopping declines of species that humans deemed worth saving and encouraging population growth with limited seasons, male-only hunting for some species, daily bag limits, and other restrictions. As successful law enforcement, habitat restoration and enhancement, and wildlife management programs grew, so did most wildlife populations. After the initial protections resulted in overpopulations of some species, biologists, hunters, and the public realized that their efforts were too successful in some cases, and that local populations had been allowed to exceed the social and biological carrying capacity of the habitat (Meine 1991).

When some large mammal populations were too abundant for the amount of habitat available, reproduction decreased and mortality increased because of intraspecific competition for resources available (McCullough 1979, Dusek et al. 1989). Reducing densities lessened competition and increased the population growth rate by improving reproduction and survival. Early biologists saw this compensatory effect of harvest as evidence that wildlife populations could be managed as a renewable natural resource where the population could replace the portion removed by hunters or trappers.

It is often necessary to maintain wildlife populations below environmental and social carrying capacity to reduce die-offs, to provide for more productive populations, to protect habitat, to reduce the spread of disease, or to reduce conflicts with humans (Conover 2001). In cases where population reduction is the management goal, managers must implement harvest beyond the level at which the population can replace itself in the short term (Carpenter 2000). Population reductions or main-

tenance at appropriate levels are examples of hunters acting as partners in wildlife management. Conflicts with humans in the form of vehicle collisions, nuisance wildlife, livestock depredation, conflicts with agricultural production, and human and domestic animal health and safety may result in a goal to manage at a social carrying capacity lower than the biological limit of the habitat (Conover 2002, Duda et al. 2010b). For example, the number of deer–vehicle collisions is estimated to exceed 1.5 million every year on US roadways (Conover et al. 1995). This tremendous loss of life and property illustrates the importance to all society of effectively managing wildlife abundance to appropriate levels.

In recent decades, there are many examples of hunting alone being ineffective in controlling a few species, most obviously white-tailed deer (*Odocoileus virginianus*; Brown et al. 2000) and white geese (*Chen* spp.). White-tailed deer are a highly adaptable and prolific species that benefit from habitat disturbance (including agriculture) and urban refugia. Declining access to hunting opportunities, reluctance to harvest females, and a counterproductive protectionist attitude toward wildlife exacerbate the problems related to managing their abundance. This is not an indictment of the failure of hunting as a wildlife management tool in general, but an example of how socially and biologically complex wildlife management can be.

Overabundant wildlife populations can dramatically alter the habitat to the detriment of many other sympatric species (Horsley et al. 2003, Rawinski 2008), underlining the importance of hunting in controlling wildlife populations. In some areas, deer overpopulation is exacerbated by trends in landowners charging access fees or otherwise restricting hunting access and thereby greatly reducing the number of hunters on the landscape (Carpenter 2000). Hunters paying for access expect lower hunter densities and high deer densities, which makes it more difficult for agencies to control populations without full cooperation of the private landowner.

Like hunting, trapping is scientifically regulated to ensure it does not affect populations, unless that is the management goal (Todd 1981). Furbearer species are renewable natural resources, as are other harvested species, and trapping contributes to supporting the North American Model. When furbearer or predator population reduction is needed, trapping provides a different tool to manage some species of wildlife that are difficult to harvest or capture any other way (Payne 1980). Participation in trapping has historically varied with changes in fur prices, but more recent societal changes have resulted in far fewer trappers in recent decades across the continent (Novak 1987, Armstrong and Rossi 2000).

For many years, hunting and trapping were partially justified as necessary management actions to save animals from a lingering death by starvation. That is certainly true in many cases, but also not true in many more. If prey species have to be hunted because predator populations were reduced, why are predators still being hunted? The truth is more complicated than the simplistic idea of wildlife overpopulation. In reality, the importance of hunting to conservation in the broad sense is not tied simply to population control. One has to understand that a simple deer season or duck season might seem like an isolated activity, but it is merely a component—a critical one—of a much larger wildlife conservation model. Properly managed, wildlife populations are renewable resources that literally pay the bills for a far-reaching, comprehensive system of sustainable wildlife conservation that has proven itself superior to any other widely implemented model.

Demographic Effects of Consumptive Use

Disproportionate harvesting of certain sex or age classes can affect population demographics. Heavily hunted populations might have age structures and sex ratios that are very different from unexploited populations. Males of many species naturally have a higher mortality rate, resulting in more females in the population even when not hunted, which becomes exaggerated in populations with a predominately male harvest. This is much less of an issue with trapping because trapping is not as sex biased.

Many ungulate populations are managed for high hunter opportunity, which often results in more females than males in the population. Also, heavy exploitation of the male segment will lower the average age of that part of the population. Low male:female ratios affect reproductive behavior but do not significantly affect productivity (Desimone et al. 1993, Noyes et al. 1996, Bender and Miller 1999, Freeman et al. 2014). In heavily hunted populations of white-tailed and mule deer (*Odocoileus hemionus*), there is no indication that a low number of males negatively affects reproductive rates or overall population robustness (Ozoga and Verme 1985, White 2001).

Changes to population demographics may alter social structure and breeding behavior. Researchers have suggested that white-tailed deer populations with a young male age structure and low male:female ratios experience a longer, later, and less intensive breeding season in the southeastern United States, where photoperiod changes less than northern latitudes (Guynn et al. 1988; Jacobson 1992). When hunting occurs prior to the breeding season, this effect may be more pronounced. Concerns have been raised that delaying breeding dates in northern climates may produce younger, smaller offspring entering the harsh winter, resulting in lower winter survival rates. Although more research is needed, there is currently little empirical evidence to support this concern (Bender 2002). In fact, some of the populations with the heaviest male exploitation also have the highest reproductive rates (McCaffery et al. 1998). Low male:female ratios in ungulates appear to be less of a biological concern and much more of a social concern in terms of hunter perception and satisfaction.

Hunting older ungulate males appears to be increasing in popularity, encouraged by a segment of society that places heavy emphasis on large horns and antlers (Knox 2011). Individuals who prefer hunting large, mature animals sometimes pressure wildlife agencies to manage the entire state or province more conservatively to provide for a mature male age structure and higher hunt success. Such requests by a minority of hunters present a dilemma for agencies because conservative management would dramatically reduce hunter opportu-

nity, which in turn negatively affects the financial support and advocacy for wildlife and their habitat. For example, states like Arizona issue hunting permits through a lottery-style drawing where 123,296 applicants competed for only 51,503 deer permits in 2019. Increasing the amount of conservative hunting over large areas runs contrary to the foundational success of the North American Model, whereby everyone has access to hunting opportunity.

Genetic Effects of Hunting

Harvest of animals with specific traits (e.g., large horns or antlers) more so than those without has led some to argue that such selection can cause evolutionary change that may be detrimental to the species, especially if those traits are related positively to individual fitness (Coltman et al. 2003; Festa-Bianchet 2003, 2008; Darimont et al. 2009; Mysterud and Bischof 2010; Mysterud 2011). Using information from a broad range of aquatic and terrestrial systems exposed to myriad potential and operational selective pressures, several authors have made expansive generalizations about selective harvest and its applicability to ungulates (Festa-Bianchet 2003, 2008; Darimont et al. 2009, 2015; Allendorf and Hard 2009; Minard 2009). Harvest-based selection can be intensive enough to be relevant in an evolutionary sense, but phenotypic changes consistent with hunter selection often are confounded with multiple environmental influences (Postma 2006, Coltman 2008, Heffelfinger 2018, LaSharr et al. 2019).

Inferences derived from long-term research in a single, small, isolated bighorn sheep population in Alberta (Coltman et al. 2003, Darimont et al. 2009, Douhard et al. 2016, Pigeon et al. 2016) have created a misconception that trophy hunters are ubiquitous, selecting genetically superior specimens in most populations and imposing a detrimental footprint on the evolutionary trajectory of horned and antlered ungulates. Human dimensions surveys do not support these broad extrapolations (Duda et al. 2010b). Most hunters are not selective in the type of animal they take, satisfied instead to harvest any legal animal (Heffelfinger 2018). In a few exceptions, however, regulations may force the harvest of animals of a minimum size or age regardless of the hunters' personal choice. Factors such as age, genetic contribution of females, nutrition, maternal effects, epigenetics, patterns of mating success, gene linkage, movements, refugia, date of birth, and other selective pressures interact with harvest to impede unidirectional evolution of a trait (Heffelfinger 2018).

The intensity of selection determines potential for evolutionary change in a meaningful temporal framework (Festa-Bianchet 2016). Indeed, only under severe intensity and strict selection on a trait could harvest prompt evolutionary changes in that trait in the face of other more influential factors. Broad generalizations across populations or ecological systems can yield erroneous extrapolations and inappropriate assumptions. Removal of males expressing a variety of horn or antler sizes, including some very large males, does not inevitably represent artificial selection unless the selective pressures are intensive enough to cause a unidirectional shift in phenotype or allele frequencies that may act on some relevant life-history trait or process (Heffelfinger 2018).

Although many speculative papers have been written (Festa-Bianchet 2003, 2008; Darimont et al. 2009, 2015; Mysterud and Bischof 2010; Mysterud 2011), researchers that demonstrate an effect of selective harvest on the size of horns or antlers are rare and mostly limited to wild sheep in one Canadian province (Coltman et al. 2003, Pigeon et al. 2016). There is no empirical evidence of selective harvest causing evolutionary change in cervids (Festa-Bianchet 2016). Despite all the speculative discussion papers, no other researchers have shown significant phenotypic changes in ungulates that could be attributed to selective hunting, while several show no effect or contradictory trends (Brown et al. 2010, Loehr et al. 2010, Rughetti and Festa-Bianchet 2010). Singer and Zeigenfuss (2002:695) offer an alternative view regarding genetic diversity and removal of old males: "trophy hunting permits more subdominant and smaller-horned rams to obtain copulations, and thus may increase the ratio of effective population size to census population size (Ne:N) and thus increase total genetic diversity."

FUNDING THE NORTH AMERICAN MODEL

Nearly everyone enjoys wildlife, but many people are unaware of the financial contributions made by the hunters, trappers, anglers, and recreational shooters to support sustainable conservation. Neither do they realize the fundamental role hunters and trappers play in preserving the wildlife and wild places they enjoy (Duda et al. 1998, Brennan et al. 2019). Although many consumptive users know that their financial contributions from licenses and some equipment helps pay for wildlife management, they do not always fully appreciate their key role in wildlife conservation on the broader scale, far beyond the species they pursue.

During 2020, $601 million was distributed to state wildlife agencies in the United States from the excise tax collected on hunting, fishing, and shooting purchases through the Federal Aid in Wildlife Restoration Funds, or Pittman-Robertson Act (USFWS 2020). The sale of hunting and trapping licenses and private donations by hunters for conservation efforts also contributed, bringing the total of direct conservation dollars to $1.2 billion annually (Southwick and Allen 2010). Some of these contributions are voluntary, but most are a requirement of participation. Hunters who saw the value of a conservation model that protects nature through collective public stewardship built these requirements into the system (Geist 1994).

There are about 11.5 million hunters in the United States alone (USFWS 2016), and their annual expenditures provide significant support to rural communities in Canada, Mexico, and the United States. Overall, hunting and trapping voluntarily redistributes wealth from urban centers to smaller rural communities, where it is multiplied through the local economy (Duda et al. 2010b). Economic multipliers are commonly used to estimate this compounding ripple effect. In 2016 it was estimated that $27.1 billion spent in the United States by 11.5 million hunters supported 525,000 jobs (Southwick Associates 2018). If similar figures were available for Canada, the total contribution would be staggering.

In addition to institutionalized programs, nongovernmental organizations raise and contribute additional money for specific research projects, habitat acquisition and enhancement, and population monitoring. For example, the Wild Sheep Foundation (WSF) has raised and contributed more than $115 million in the past 40 years to activities that benefit wild sheep and other wildlife (Wild Sheep Foundation 2021). Similarly, the Rocky Mountain Elk Foundation has funded 12,700 individual projects that protect or enhance more than 3.6 million ha of wildlife habitat (Rocky Mountain Elk Foundation 2021). Since 1973, the National Wild Turkey Federation (NWTF), in cooperation with state and federal partners, has spent more than $488 million to restore wild turkeys (*Meleagris gallopavo*) and to conserve more than 7.7 million ha of habitat (National Wild Turkey Federation 2021).

Agency Infrastructure

In recent decades, some state and provincial wildlife agencies became creative in their ability to garner additional funding sources to supplement the long-standing contribution of hunters, anglers, and shooting enthusiasts. Lottery sources, state income tax, special stamps, and similar funds are sometimes channeled to wildlife agencies and earmarked for programs that have not received adequate financial support in the past (e.g., nongame or habitat acquisition and management). These recent supplemental funds are an important addition to the budgets of wildlife agencies, but they are vulnerable to legislative meddling and do not replace or negate the importance of the base funds from consumptive activities. Supplemental funds are only effective because there is an agency infrastructure in place that can take additional money and apply it directly to a specific program area. Funds from consumptive wildlife support law enforcement, personnel resources, and all other day-to-day agency operations, allowing supplemental dollars to be effective. The agencies most effective in conserving all resources, whether hunted or not, are those with a solid financial foundation provided by the well-regulated consumption of a few wildlife species.

Law Enforcement

One of the important contributions to wildlife conservation that hunters, trappers, and anglers have made is the creation and maintenance of law enforcement officers to uphold the massive amount of legal restrictions to wildlife harvest (Paz and Heffelfinger 2018). Regulated hunting is only regulated if the laws are obeyed. Surveys consistently report that about 50% of hunters and anglers have had recent contact with these enforcement personnel in the field and hold them in high regard (Duda et al. 1998).

Currently, thousands of wildlife conservation law enforcement officers are actively working in the United States and Canada, and most are paid with income from the sale of hunting, trapping, and fishing licenses. This ubiquitous wildlife law enforcement presence is almost always independent of other law enforcement agencies and allows for the majority of their time and energy to be devoted to protecting natural resources. Besides policing hunters and anglers, they also perform duties related to water quality, habitat protection, public safety, search and rescue, littering, vandalism, trade in threatened and endangered species, and providing backup to other local law enforcement agencies (Paz and Heffelfinger 2018). Opponents of hunting rarely offer alternatives for funding trained officers to protect wildlife against exploitation. If hunting were made illegal in North America, we would immediately lose this massive protection force and likely degrade into the unregulated destruction that was common before hunting was institutionalized as the basis (rather than the bane) of conservation.

Population Restoration

The restoration of wildlife populations across North America is the greatest wildlife success story in the history of conservation anywhere. Conservation has restored nearly all of the badly overexploited populations before the development and implementation of the North American model. Species like Canada geese (*Branta canadensis*), wood ducks (*Aix sponsa*), white-tailed deer, pronghorns (*Antilocapra americana*), bighorn sheep, and wild turkeys all represent important species whose restoration was made possible by funding and advocacy generated by hunting.

The state of Arizona began restoring desert bighorn sheep in 1955 with translocations to historical habitat. Since then, more than 100 translocations of at least 1,800 bighorn sheep have restored this iconic species in all previously occupied habitat and many areas of suitable habitat (O'Dell 2007). Across North America, wild sheep populations have been restored with more than 1,460 translocations involving about 21,600 wild sheep since 1922 (Wild Sheep Foundation 2021), exemplifying the type of restoration activity that has occurred for decades throughout all states and provinces in North America with the funds generated from the regulated consumption of a few species.

North America has a nearly full complement of native wildlife living in habitat that has changed remarkably little in the past 300 years, compared to other continents. Restoration of large mammal populations continues today, with elk (*Cervus canadensis*) being successfully translocated into eastern historical ranges for the enjoyment of all residents. Work also continues for other species, such as bison (*Bison bison*) and large predators whose restoration comes with significant societal controversy. Individual hunters or even some organizations might not be supportive of the restoration of some large predators, but they support the system of collective stewardship that works to restore native species. With the restoration success of hunted species, focus has shifted to restoring nonhunted species, with threatened and endangered animals receiving the highest priority.

Monitoring and Management of Wildlife Populations

Monitoring wildlife populations and accumulating baseline trend data are the bases of well-informed, science-based decisions that are foundational to the North American Model (Baughman and King 2008). Hunted species are not the only ones monitored, but they generally do receive the most attention. State, provincial, and federal agencies have a history of monitoring wildlife populations, beginning at the very genesis of wildlife conservation in North America. Many agencies have examples of monitoring programs that have remained relatively consistent for decades and provide valuable trend data. The Canadian Wildlife Service and USFWS have conducted continent-wide aerial waterfowl surveys since 1955. This cooperative survey effort involves flying more than 128,000 km of survey each year throughout waterfowl areas from southern Mexico to northern Canada.

It is usually not necessary to monitor intensively the abundance and demographics of populations that are not being annually harvested, unless such species are causing significant damage, are negatively affecting habitats or other species, or are threatened or endangered. Under the current system, wildlife species are monitored and managed intensively by agencies financially solvent enough to use resources for any other species as the need arises.

Habitat Acquisition, Protection, Restoration, and Enhancement

Land management agencies manage wildlife habitat on millions of hectares of federal and Crown land. Many states and provinces have also purchased wildlife habitat with the proceeds from hunting licenses and taxes on some specific hunting, fishing, and shooting equipment. Funding sources for habitat vary among Canadian provinces. For example, in British Columbia, surcharges collected on hunting, angling, trapping, and guide-outfitting licenses go into a trust fund managed by an independent organization, the Habitat Conservation Trust Foundation (HCTF). Since its inception in 1981, the HCTF has invested $189 million in more than 2,980 habitat acquisition, restoration, and enhancement and other priority conservation projects throughout the province (Habitat Conservation Trust Foundation 2021).

During a five-year period (2005–2009) in the United States, $58.5 million from Federal Aid in Wildlife Restoration funds was apportioned to the states for the acquisition of more than 12.2 million ha of wildlife habitat (USFWS 2010a). In addition, such wildlife conservation organizations as Canadian Wildlife Federation, Ducks Unlimited, Mule Deer Foundation, National Wild Turkey Federation, The Nature Conservancy, Pheasants Forever, the Rocky Mountain Elk Foundation, Wildlife Habitat Canada, the Wild Sheep Foundation, and myriad state and provincial wildlife organizations used private donations to purchase land or conservation easements on large tracts of wildlife habitat. Most of these areas are purchased with hunted species in mind, but wetlands acquired for waterfowl, forests purchased for deer or turkeys, mountainous areas protected for wild sheep, and grasslands restored for quail and pronghorn have benefited countless nongame and endangered species that rely on those habitat associations. Recent estimates indicate about 70% of users in these areas do not hunt, and in some properties the percentage may be as high as 95% (USFWS 2010b). Ironically, there are sometimes conflicts when nonconsumptive users express concern about seeing hunters on these properties during the few days or weeks each year the hunting seasons are open.

Research

One of the foundations of the North American Model is that management decisions are based in science. In the United States, about $57 million was apportioned in 2009 to state wildlife agencies from the Federal Aid in Wildlife Restoration Program for conducting more than 10,000 wildlife research projects (USFWS 2010a). In the early years of the wildlife management profession, money was spent exclusively on learning more about species that were at low levels because of the lack of a comprehensive system of wildlife conservation. As more was learned about managing those species back to abundance, focus shifted to all species and their habitats.

Canadian wildlife researchers obtain funding from a wide variety of sources. All revenue from hunting and fishing in Canada is placed into the general revenue and then distributed to each fish and wildlife agency. Most provinces have an association or trust fund that funds research and land acquisition. Wildlife conservation organizations (e.g., Rocky Mountain Elk Foundation, Alberta Conservation Association, WSF) and extractive industries (e.g., oil companies) also contribute money for wildlife research.

To facilitate meaningful research in the United States, a series of cooperative research units were established in 1935 at universities to provide an opportunity for the federal government, state wildlife agencies, universities, and nongovernmental organizations to work together. These units receive federal funds to employ two to five scientists, but most of the annual baseline operating budget comes from hunters and anglers through the state wildlife agencies. This system began with seven co-op units and has grown to a network of units at 40 campuses distributed across 38 states (Organ et al. 2016). Canada has also established a Cooperative Fish and Wildlife Research unit in New Brunswick modeled after this US system. Over the past 75 years, funding from hunters has provided information that influences intelligent management decisions to better understand and conserve hunted and nonhunted species.

Hunter and Trapper Education Programs

Hunter education programs are important to wildlife conservation because illegal acts by a few hunters can cast all hunters in a bad light and erode public support for this successful conservation paradigm. North America has a network of hunter education programs that deliver information and coursework on wildlife management, hunter ethics, firearms safety, and hunting and trapping techniques. Each year a vol-

unteer hunter education instructor force of more than 70,000 trains about 650,000 hunters (W. East, International Hunter Education Association, personal communication). An annual apportionment of Federal Aid in Wildlife Restoration provides funding for such training in the United States, funds that exceeded $160 million in 2018 (USFWS 2018). Each state receives between $2 million and $20 million, depending on their size and need (USFWS 2010a). In Canada, these programs are paid for by users or funded through the provincial wildlife agencies. The successful completion of a hunter education course is mandatory for certain age classes and certain kinds of hunting in all 50 states and 10 Canadian provinces, and has resulted in more than 35 million students being trained since the beginning of the program (International Hunter Education Association 2010a, b).

Nongame Including Threatened or Endangered Species

A preponderance of hunter-generated money is still expended on the management and protection of hunted species. This is appropriate because populations of species that are being annually hunted generally require a greater intensity of monitoring, law enforcement, research, and management.

Various sources—including income tax check-offs, special stamps, independent grants, donations, lottery or gambling revenue, some sales tax, and hunters' dollars from the Federal Aid in Wildlife Restoration Program—fund nongame activities. In Canada, funding comes from provincial and federal sources with the management responsibility remaining with provinces (except for migratory birds). A portion of Wildlife Restoration funds in the United States is available to the states for conservation of birds and mammals that are not hunted. This funding mechanism provides millions of dollars annually for the conservation of nonhunted species. This is a small percentage of what is needed and does not begin to address fully the needs of all other taxonomic groups such as native fish, songbirds, amphibians, and reptiles. As a consequence, wildlife agencies must be creative to find and maintain additional funds for management of species that were not obviously exploited historically and therefore are not the immediate focus of the Wildlife Restoration Program.

Through the current authorization of the Wildlife Restoration Program, hunters' dollars contribute to the restoration of many threatened and endangered species such as the California condor (*Gymnogyps californianus*), Mexican gray wolf (*Canis lupus baileyi*), and the black-footed ferret (*Mustela nigripes*). Under this system, conservation actions have the potential to protect other wildlife species before they become threatened or endangered (Baughman and King 2008). The future of conservation in North America will have to include the existing model of using hunters' dollars to conserve all wildlife for all people.

Information and Public Relations

Communicating with the public and considering human dimensions in wildlife management have become vital to the

OLOF CHARLES WALLMO (1919–1982)

O. C. "Charlie" Wallmo was born in Iowa in 1919 and studied forestry and wildlife at the Universities of Wisconsin and Montana before completing his bachelor's degree at Utah State University in 1947. He returned to the University of Wisconsin for his master's degree and then attended Texas A&M University, where he earned a PhD. Through his work in Alaska, Arizona, the Rocky Mountains, and Texas, Wallmo pioneered research that resulted in many of the fundamental and foundational concepts in wildlife management.

He conducted the first comprehensive study of scaled quail ecology early in his career. He was also one of the first to use free-ranging tame deer as research tools to elucidate diet, behavior, and metabolism of mule deer. Wallmo was sought-after for his knowledge of mule deer nutrition and the effects of habitat manipulations on deer population dynamics. His work in the central Rockies showed the benefits of small forest clear-cuts to deer nutrition, and early work on deer survey methodology formed the basis for improved management of deer populations. His efforts in southeast Alaska demonstrated the value of overstory cover for black-tailed deer during winter.

Wallmo published more than 50 significant publications, and his edited tome *Mule and Black-Tailed Deer of North America* served as the primary source of basic information about that species for four decades. Even though he was known for his dedication to science and the scientific process, his lasting legacy is not volumes of esoteric scientific publications or reams of data analysis, but important contributions to the body of knowledge that wildlife managers used for decades as the foundation for improved management. In addition, many of his former graduate students have become known for their work with cervids across North America. Finally, every two years, North America's leading black-tailed and mule deer biologist is honored with the O. C. Wallmo Award.

Courtesy of Joe B. Wallmo

effectiveness of management agencies (Decker et al. 2019). All wildlife agencies have some public information officers on staff to disseminate wildlife information and to update stakeholders on agency activities through press releases, websites, social networking media, radio, television, and a multitude of publications for diverse audiences. Some wildlife agencies use Federal Aid in Wildlife Restoration funds, but most simply use money garnered from the sale of hunting and fishing licenses. In this way, the entire public benefits from the baseline funding provided by regulated hunting. In Canada, much of the funding for these kinds of activities comes from taxes, which poses a problem for wildlife agencies as funding cuts and downsizing continually erodes their ability to be effective.

ADVOCACY FOR WILDLIFE AND WILDLANDS

Management guidelines control hunting and trapping efforts and law enforcement officers enforce regulations, but much of North America's conservation success arises from incentives based on self-interest and personal ethics (Leopold 1933). Many outdoorspeople in North America go far beyond what the law requires. Historically, hunting has been the greatest force in ensuring wildlife a place on the landscape. Consumptive use fosters attention, and wildlife thrives with attention and withers from neglect (Geist 2006). The strong desire to hunt wildlife appears to be deeply primordial. Most commonly, the passion to hunt expresses itself as a deep, lifelong interest

in and devotion to wildlife, often accompanied by considerable work, even sacrifice, by the hunter on behalf of wildlife. Witness the many organizations dedicated to the conservation of wildlife in North America. Arizona's first desert bighorn sheep hunt occurred in 1953, and it was probably no coincidence that two years later the state wildlife agency began its aggressive translocation program that has now restored all of the state's historical sheep populations. There are endless examples of hunter advocacy for being instrumental in implementing wildlife conservation on a broad scale. It comes as no surprise that Aldo Leopold, considered the founder of the wildlife management profession, was an avid, lifelong hunter (Bradley 1995, Peyton 2000).

No collective group is composed entirely of active leaders, and hunting is no different. A small percentage of the hunting community rises to the position of spokespersons or leaders of conservation organizations. As in any group, the majority is happy to follow leaders and follow the rules. For this reason, it may be unreasonable to expect every hunter to act as a steward of broad-scale biodiversity at the individual level (Holsman 2000). The hunting community, acting as collective stewards for the greater good of wildlife and their habitat, has influenced this successful conservation paradigm.

Political Support

Early groups of organized hunters were instrumental in providing the political support necessary to implement many of the laws that coalesced into the system of conservation we have today (Chapter 2). For example, Theodore Roosevelt organized the Boone & Crockett Club by assembling most of the powerful and influential conservation-minded people of the day, many of them hunters. The Boone & Crockett Club successfully lobbied for the establishment of Yellowstone National Park, the preservation of the bison, cessation of market hunting, and much more in the 20th century.

It is difficult to maintain separation between the sometimes-detrimental world of politics and wildlife management. When political influence threatens proper wildlife conservation efforts, sportsmen at the local and national level have shown themselves willing and able to come together in support of wildlife. There are many examples of wildlife agency funds or commission structure coming under attack by politicians, only to have the organized wildlife groups step up to its defense. A survey of outdoor user groups in the southwestern United States asked respondents if they would be willing to write a letter if wildlife conservation funding was threatened with diversion to other uses. Seventy-four percent of hunters responded they would be likely or very likely to write a letter protesting such action (Responsive Management 1995).

Nonhunters are often involved, but it is the organizational infrastructure of hunting organizations that is frequently the vehicle that influences such coordinated defensive activities. This infrastructure is also used for mounting campaigns in defense of crucial wildlife habitat threatened by conflicting interests. With declines in the proportion of the population that hunts, wildlife may have a less organized, less effective voice on their behalf. As wildlife and their habitat are subjected to increasing pressures, hunters and nonhunters will need to focus on common goals and combine their collective resources for the good of the wildlife they both enjoy.

Biological Samples and Information

Hunters and trappers are an important source of biological information for wildlife managers. Harvest data such as total number harvested, sex and age ratios, body weight or condition, and harvest location have been collected at check stations since the early years of wildlife management. Other hunt-related information is routinely collected at check stations, in the field, or with posthunt questionnaires by phone or mail. Whether the hunter was successful, the number of hunter-days expended, area hunted, and other information can be used to track trends in population parameters or abundance. In some cases, these check stations and questionnaires are mandatory, but in many instances, hunters go to great lengths to provide information that might help managers.

Biological samples can help determine prior disease exposure, parasite loads, nutritional status, genetic relationships or diversity, and approximate age. Hunters themselves sometimes collect these samples, requiring a certain amount of cooperation and commitment. Along with other members of the public, hunters routinely provide information on species distribution and sources of unusual mortality. Such input is valuable in tracking changes in wildlife distributions in the face of habitat loss, natural disasters, climate change, and emerging disease issues.

Volunteerism

Hunters and trappers individually, and the organizations to which they belong, have always been active in volunteering for habitat improvement projects, constructing nesting structures or boxes, altering fences to be wildlife friendly, teaching hunting and trapping education courses, conducting wildlife surveys, working check stations, performing routine facility maintenance, cleaning up trash, and many other beneficial activities (Bleich 1990). These volunteer efforts benefit wildlife directly and allow wildlife management agencies to stretch their conservation dollars further to accomplish additional goals.

Most of these projects benefit more than hunted species. For example, big-game hunters throughout the West have installed water collection and retention devices, but uncounted numbers of bird, mammal, insect, reptile, and amphibian species use them, too. Designs of water catchments for large mammals have been altered through the years to accommodate the needs of bats and smaller terrestrial animals (Chapter 13).

Residents in the state of Maryland were asked if they would be interested in volunteering their time to help the state wildlife agency (Responsive Management 1993). Results revealed that 22% of hunters were very likely to volunteer, compared to 7% of the nonhunters. This does not imply nonhunters do not care about wildlife, but it does illustrate the level of commitment to collective stewardship of natural resources that is inherent in the hunting community.

THE CONSISTENCY OF CHANGE

One constant throughout the history of wildlife management is the societal change to which managers must adapt. The contribution of hunters and trappers to wildlife and habitat conservation is undeniable. Consumptive users' contributions have been steady and consistent through time, even as society at large has changed. These sociological trends are not likely to abate or reverse themselves, so resource managers must accept these sociological shifts as the new reality (Brennan et al. 2019, Decker et al. 2019).

Foundational Changes and Trends

The number of hunters has decreased for decades, while the overall population continues to grow (Duda et al. 2010a). In 2016 the number (11.5 million) and percentage (4%) of Americans who hunted was at the lowest level in at least the past 25 years (USFWS 2016). This decline had many causes, mostly related to changes in society and not directly to the act of hunting. Popularity of wildlife watching around the home increased between 2011 and 2016, but the percentage of people who engaged in wildlife watching away from their home remained static at 9% (USFWS 2016).

Trapping as practiced avocationally has been on the decline for decades. Trapping conducted for the control of nuisance wildlife, however, is on the rise because of increasing conflicts with urban dwellers (Armstrong and Rossi 2000). Trapping is also an important research and management technique used by wildlife professionals for specific objectives that would be difficult to accomplish any other way. It was estimated there were about 145,000 avocational trappers in the United States in 1999, with only 18 states having more than 1,500 trappers (Armstrong and Rossi 2000). Although a survey by Responsive Management (2015) reported more trappers in the United States in 2014 (176,573) compared with 2004 (142,287), this is far fewer than the 485,285 trappers reported in the United States and 95,129 in Canada during the 1983–1984 season (Novak 1987:fig. 1). Declines in avocational trapping are the result of many things, including depressed fur prices, less access to lands, and societal changes that have led to trapping bans for many species in Arizona (1994), California (1998), Colorado (1996), Massachusetts (1996; Duda et al. 2010a:197), and Washington (2000). Even when regulated and sustainable, trapping has long been a source of controversy because of perceptions of cruelty and because of incidences of bycatch. The evolution of trapping management and trap technology has progressed in response to societal pressure, and a comprehensive set of best management practices now guide trapping rules and regulations. Trapping is now a heavily regulated activity coupled with analysis of harvest data and population monitoring (White et al. 2021).

The public's perceptions of, and interactions with, nature have changed tremendously in only a few decades (White et al. 2021). After more than 4 million years of interacting firsthand with animals and nature, humans became detached from the natural world in a geologic blink of an eye. More people moved to urban settings where they no longer hunted for food or butchered their own livestock. An increasing number of people obtain their information about nature from digital media in the form of television and the Internet. Television presentations of nature focus on individual animals and not on the realities of managing populations and executing successful conservation programs. Many television programs depict the rescue of individual animals, sometimes members of extremely abundant species that may be overpopulated at the time. These programs, coupled with incorrect information available online, create a public with distorted views of traditional wildlife management. This severing of firsthand ties to nature has had a profound effect on how the latest generations perceive the natural world as a whole and specific wildlife management actions in particular.

Television programs in the 1950s depicted hunters as rugged and self-sufficient, skilled to thrive in the outdoors. They were portrayed as admirable individuals who cared about nature. Today, hunters and trappers in television and movie roles are more often portrayed as criminals or unsavory characters. This incremental decline in the status of hunters in popular culture is partially because of the societal changes above, but also because wildlife educators, agencies, and professionals have not done an effective job articulating and illustrating how important hunting and trapping are to the successful North American system of conservation.

As society changed, the profession of wildlife management was changing, too. Wildlife professionals in universities and state, provincial, and federal agencies grew more diversified to meet the challenges of all wildlife, and an increasing percentage of them did not come from farming, ranching, or hunting backgrounds. Individuals who have had no exposure to consumptive use of wildlife or even those who come from an animal protectionist background are ascending into the ranks of tenured university professors, researchers, scientific consultants, and agency administrators. With them comes an unfamiliarity and sometimes bias against some or all forms of consumptive wildlife use. Unfortunately, these biases make their way into inferences made in research and the recommendations made to wildlife management practitioners. Increasingly, we see a false dichotomy expressed by such people between hunters and conservationists, implying that hunters are the opposite of conservationists.

Some are unwilling or reluctant to appropriately credit hunters and trappers for the success this continent has enjoyed in the conservation of wildlife and the habitat upon which they depend (Nelson et al. 2011). Unfortunately, criticisms are often rooted in personal bias rather than scholarly critiques or credible faults in the North American Model. For example, Nelson et al. (2011) begin their historical interpretation in the 1960s, decades after the North American Model was constructed around the principle of controlled consumptive use of wildlife. They further obfuscate the issue by comparing the historical role of hunters in conservation to child labor and by asking esoteric questions about social justice, human liberty, and ethical reasoning. Peer-reviewed publications with poorly supported criticisms of the scientific underpinnings of wildlife conservation (Artelle et al. 2018) and of some forms of hunting (Darimont and Child 2014) based merely on Internet searches do not meet the standard of science upon which such discussions should be based (Mawdsley et al. 2018). Other criticisms will follow, and there is certainly room for improvement in our current implementation of conservation (Brennan et al. 2019), but conversations must be fact based, transparent, and honest.

Future Role of Hunters and Trappers in Conservation

Because hunters had a lead role in the development of the most successful system of wildlife conservation does not mean they own the future of this paradigm. Societal changes are already challenging the foundation of conservation through consumption. Those truly interested in perpetuating a realistic and proven conservation model will need to work to preserve it. Future efforts to conserve wildlife and wild places will not succeed without a broad base of public support (Mahoney 2007, Brennan et al. 2019, Decker et al. 2019). History has illustrated the failure of conservation prescribed by the elite and fashioned after the desires of the minority. Stewardship of wildlife has always taken a lot of hard work and that will not change, nor will the need for advocacy backed by science-based research, education, and management programs.

Internationally, the superiority of the North American Model has not gone unnoticed. Africa, Asia, Russia, and many other parts of the world are making attempts to implement what has worked so well in North America (Mahoney 2007). Many symposia have addressed the success and application of hunting-based conservation, recognizing the success of this conservation model (Mahoney 2009, Patterson 2009). Decades of attempting to build conservation programs based on a preservation paradigm without local grassroots support or funding have largely failed.

Regulated hunting currently enjoys a broad base of public support in North America. Several surveys have consistently reported that 75% to 81% of respondents support hunting and agree it should continue (Duda et al. 2010a:32). Trend data from sequential surveys indicate there may be an increasing proportion of Americans who approve of legal fair-chase hunting when the harvest is utilized (Duda et al. 2010a:44). The reemergence of positive media portrayals of hunters and the political attention paid to hunters by presidential candidates reflect this trend in subtle ways. Wildlife management agencies and conservation organizations have been discussing and celebrating the continued success of the North American model for the past two decades, and perhaps the general public has noticed.

The continued success of hunters, trappers, and anglers supporting conservation will depend on a public that understands how the system works. Everyone enjoys seeing wildlife and beautiful wild places when spending time outdoors. A camping, hiking, or hunting trip is much more enjoyable when wildlife are seen. And yet most people do not understand what influences and funds the programs that provide the foundation for the nature-based opportunities they enjoy. The founders of the North American model worked hard to turn the tide of public opinion in support of a widespread conservation ethic (Reiger 1986, Mahoney 2007). Those interested in wildlife must continue that hard work by increasing the awareness of this uniquely effective system among an increasingly disengaged public.

The future success of the North American model will require hunters and trappers to remain relevant to conservation (Mahoney 2007, 2019; Decker et al. 2019). Hunters and trappers must be recognized for their past, present, and potential future contributions (Geist et al. 2001). Remaining relevant also means they are not seen as degrading or obstructing wildlife conservation efforts. Without an alternative paradigm, no serious wildlife advocate is calling for fundamental changes in the North American Model, but there is ample room for improvement in several areas. Consumptive wildlife users will have to do a better job of policing themselves so as not to represent hunters in a bad light, thereby eroding public confidence in a hunter-based system of conservation. Hunters and trappers must continue to demonstrate and articulate that they are truly stewards of all wildlife (Posewitz 2002). How the hunter communicates and positions himself in the minds of the nonhunting public will decide whether hunting will be supported far into the future.

Producing deer, ducks, and turkeys in abundance is not what wildlife conservation has been about in the past, nor will it be in the future. Wildlife managers and their constituent public need to support conservation of all native wildlife, not just those that are hunted. Currently, carnivores are testing the public's support for a full complement of native wildlife (predators and prey) and our ability to manage predators and prey without being crippled by laws, lawsuits, and regulations that can grind proper management efforts to a halt. Completing the recovery of all native wildlife in North America continues, but it will not be possible to restore all species to all former habitat, because humans are now well distributed across the landscape. We should support restoration of large carnivores, but they must be subject to the same scientific management that has been so successful for all other restored species of large mammals.

Human dimensions research to assess the desires of the public has become an important part of modern resource management (Chapter 4). Gauging public sentiment, helping the public understand the trade-offs, and then determining how the public would like their wildlife resources managed will become even more critical in the future. Professionals must continue to engage human dimensions experts and focus on the values of the public, especially when they conflict with traditional uses. The past litmus test of whether something affects the population is no longer valid in the face of a rising level of empathy for individual animals in society today. Most wildlife enthusiasts have a common goal; everyone has an investment in sustainable wildlife resources for the future. Resource managers, consumptive users, nonconsumptive users, and all other stakeholders and beneficiaries will have to work together to guide conservation actions that result in abundant and properly managed populations of native wildlife that persist in minimally disturbed habitat for generations to come.

SUMMARY

Hunting and trapping have been a vital part of human existence long before they acquired conservation or recreational relevance. In recent centuries, technology evolved rapidly to the point where humans could have an effect on the abundance of animals around them. On most continents, unsustainable exploitation of wildlife resources was not recognized and halted in time to retain most native species. In North America, however, individuals concerned about natural resources recognized that something had to change in order to protect major species of native birds and mammals. This concern, expressed by some of the most prominent hunters and other conservationists of the day, developed into a collective sense of stewardship and conservation ethic. The desire to maintain wildlife populations into perpetuity led to North America's unique and incredibly successful system, whereby consumptive use of wildlife in the form of hunting and trapping became the cornerstone of what is now known as the North American Model of Wildlife Conservation. Seven principles define the North American Model: (1) wildlife is held in public trust for all, (2) deleterious commerce in wildlife is prohibited, (3) appropriate use of wildlife is regulated by law, (4) wildlife can only be killed for legitimate purposes, (5) wildlife is managed as an international resource, (6) science is the appropriate basis for wildlife policy, and (7) the opportunity to harvest wildlife is allocated fairly.

This conservation paradigm, built around scientifically regulated harvest, has benefited nearly all species of native wildlife because advocates conserved, restored, and managed habitat for hunted species. They also conserved a landscape that provided habitat for nonhunted species. This system of conservation has been so successful it has been recognized as a model to emulate worldwide. State and provincial wildlife agencies use research, population monitoring, adaptive management, emerging technologies, and experience to develop management guidelines and protocols to determine the appropriate level of harvest for each population. The allowable level of harvest might simply be that which is sustainable to the population, or it might be a prescription to reach specific management goals of animal abundance and demography. All of this is determined through a public process that sets management goals and guidelines. The types of management data collected vary greatly by species throughout North America because of a diversity of financial, logistical, and social constraints; however, all state and provincial agencies monitor populations of hunted species and apply monitoring data to management goals.

Today, hunters and trappers are the bedrock of North American wildlife conservation because of the funds and advocacy they contribute, and because they remain the most effective logistical agents of actual population management. The early days of North American wildlife management were spent stopping declines of those species that humans deemed important and worth saving. Early protections resulted in recovery of populations and even overpopulations of some species, and so management agencies switched from recovery actions to management programs. Continual population management through hunting is mindful of genetic effects and strives to maintain demographic and abundance characteristics defined by management goals.

Most people are not aware that the wildlife they enjoy viewing depend on the financial contributions made by hunters to support sustainable conservation. Hunters, trappers, anglers, and recreational shooters play a critical role in preserving wild things in wild places. The agencies most effective in conserving all resources, whether hunted or not, are those with a solid financial foundation provided by the well-regulated consumption of a few wildlife species. Hundreds of millions of dollars are generated for wildlife conservation as part of a program that levies an excise tax on hunting, fishing, and shooting equipment and distributes these funds to wildlife management agencies. In addition, the sales of hunting and fishing licenses and private contributions raise more than $1.2 billion each year. This money supports law enforcement officers and agency infrastructure, helps monitor trends in wildlife populations, restores species to historical habitat, and finances research. It also goes toward conserving nonhunted species, both threatened and endangered; informing the public; and acquiring, protecting, and enhancing wildlife habitat.

Hunters and trappers and the organizations to which they belong are strong advocates for wildlife and their habitat. They provide volunteer labor for improving habitat, educating hunters and trappers, and surveying wildlife, among other activities. Consumptive users of wildlife also consistently provide biological samples and information that biologists use to manage wildlife populations. Through these activities emerges a community-based infrastructure advocating for wildlife conservation.

Regulated hunting currently enjoys a broad base of public support in North America. But societal change brings with it new problems, and wildlife managers must be prepared to respond. The contribution of hunters and trappers to wildlife and habitat conservation is undeniable, but sociological trends will likely pose challenges in the future. Human dimensions research to assess the desires of stakeholders and beneficiaries plays an increasingly important role in resource management. Most wildlife enthusiasts have a common goal; everyone has an investment in sustainable wildlife resources for the future. As wildlife and their habitat are subjected to increasing pressures, hunters and nonhunters will need to focus on common goals and combine their collective resources for the good of the wildlife they all enjoy.

Literature Cited

Allendorf, F. W., and J. J. Hard. 2009. Human-induced evolution caused by unnatural selection through harvest of wild animals. Proceedings of the National Academy of Sciences 106:9987–9994.

Ardrey, R. 1976. The hunting hypothesis. McClelland and Stewart, Toronto, Ontario, Canada.

Armstrong, J. B., and A. N. Rossi. 2000. Status of avocational trapping based on the perspectives of state furbearer biologists. Wildlife Society Bulletin 28:825–832.

Artelle, K. A., J. D. Reynolds, A. Treves, J. C. Walsh, P. C. Paquet, and C. T. Darimont. 2018. Hallmarks of science missing from North American wildlife management. Science Advances 4:eaao0167.

Baughman, J., and M. King. 2008. Funding the North American Model of Wildlife Conservation in the United States. Pages 57–64 in J. Nobile and M. D. Duda, editors. Strengthening America's hunting heritage and wildlife conservation in the 21st century: challenges and opportunities. Responsive Management, Harrisonburg, Virginia, USA.

Bender, L. C. 2002. Effects of bull elk demographics on age categories of harem bulls. Wildlife Society Bulletin 30:193–199.

Bender, L. C., and P. J. Miller. 1999. Effects of elk harvest strategy on bull demographics and herd composition. Wildlife Society Bulletin 27:1032–1037.

Bleich, V. C. 1990. Affiliations of volunteers participating in California wildlife water development projects. Pages 187–192 in G. K. Tsukamoto and S. J. Stiver, editors. Wildlife water development. Nevada Department of Wildlife, Reno, USA.

Bradley, N. L. 1995. How hunting affected Aldo Leopold's thinking and his commitment to a land ethic. Pages 10–13 in Proceedings of the fourth annual governor's symposium on North American hunting. North American Hunting Club, Minnetonka, Minnesota, USA.

Brennan, L. A., D. G. Hewitt, and S. P. Mahoney. 2019. Social, economic, and ecological challenges to the North American model of wildlife conservation. Pages 130–147 *in* S. P. Mahoney and V. Geist, editors. The North American Model of Wildlife Conservation. Johns Hopkins University Press, Baltimore, Maryland, USA.

Brown, D. E., W. C. Keebler, and C. D. Mitchell. 2010. Hunting and trophy horn size in male pronghorn. Proceedings of the Pronghorn Workshop 24:30–45.

Brown, T. L., D. J. Decker, S. J. Riley, J. W. Enck, T. B. Lauber, P. D. Curtis, and G. F. Mattfeld. 2000. The future of hunting as a mechanism to control white-tailed deer populations. Wildlife Society Bulletin 28:797–807.

Carpenter, L. H. 2000. Harvest management goals. Pages 192–213 *in* S. Demarais and P. R. Krausman, editors. Ecology and management of large mammals in North America. Prentice-Hall, Upper Saddle River, New Jersey, USA.

Coltman, D. W. 2008. Molecular ecological approaches to studying the evolutionary impacts of selective harvesting in wildlife. Molecular Ecology 16:221–235.

Coltman, D. W., P. O'Donoghue, J. T. Jorgenson, J. T. Hogg, C. Strobeck, and M. Festa-Bianchet. 2003. Undesirable evolutionary consequences of trophy hunting. Nature 426:655–658.

Conover, M. R. 2001. Effect of hunting and trapping on wildlife damage. Wildlife Society Bulletin 29:521–532.

Conover, M. R. 2002. Resolving human–wildlife conflicts: the science of wildlife damage management. CRC Press, Boca Raton, Florida, USA.

Conover, M. R., W. C. Pitt, K. K. Kessler, T. J. DuBow, and W. A. Sanborn. 1995. Review of human injuries, illnesses, and economic losses caused by wildlife in the United States. Wildlife Society Bulletin 23:407–414.

Cooney, R. 2019. A comparison of the North American Model to other conservation approaches. Pages 148–155 *in* S. P. Mahoney and V. Geist, editors. The North American Model of Wildlife Conservation. Johns Hopkins University Press, Baltimore, Maryland, USA.

Darimont, C. T., S. M. Carlson, M. T. Kinnison, P. C. Paquet, T. E. Reimchen, and C. C. Wilmers. 2009. Human predators outpace other agents of trait change in the wild. Proceedings of the National Academy of Sciences 106:952–954.

Darimont, C. T., and K. R. Child. 2014. What enables size-selective trophy hunting of wildlife? PLoS One 9(8):e103487. doi:10.1371/journal.pone.0103487.

Darimont, C. T., C. H. Fox, H. M. Bryan, and T. E. Reimchen. 2015. The unique ecology of human predators. Science 349:858.

Decker, D. J., T. L. Brown, and W. F. Siemer. 2001. Human dimensions of wildlife management in North America. The Wildlife Society, Bethesda, Maryland, USA.

Decker D. J., A. B. Forstchen, W. F. Siemer. C. A. Smith, R. K. Frohlich, M. V. Schiavone, P. E. Lederle, and E. F. Pomeranz. 1996. From clients to stakeholders: a philosophical shift for fish and wildlife management. Human Dimensions of Wildlife 1:70–82.

Decker, D. J., A. B. Forstchen, W. F. Siemer, C. A. Smith, R. K. Frohlich, M. V. Schiavone, P. E. Lederle, and E. F. Pomeranz. 2019. Moving the paradigm from stakeholders to beneficiaries in wildlife management. Journal of Wildlife Management 83:513–518.

Desimone, R., J. Vore, and T. Carlson. 1993. Older bulls—who needs them? Pages 29–35 *in* J. D. Cada, J. G. Peterson, and T. N. Lonner, editors. Proceedings of the Western States and Provinces Elk Workshop. Montana Wildlife, Fisheries and Parks, Helena, USA.

Douhard, M., M. Festa-Bianchet, F. Pelletier, J.-M. Gaillard, and C. Bonenfant. 2016. Changes in horn size of Stone's sheep over four decades correlate with trophy hunting pressure. Ecological Applications 26:309–321.

Duda, M. D., S. J. Bissell, and K. C. Young. 1998. Wildlife and the American mind: public opinions on and attitudes toward fish and wildlife management. Responsive Management, Harrisonburg, Virginia, USA.

Duda, M. D., M. F. Jones, and A. Criscione. 2010*a*. The sportsman's voice. Venture, State College, Pennsylvania, USA.

Duda, M. D., M. Jones, A. Criscione, and A. Ritchie. 2010*b*. The importance of hunting and the shooting sports on state, national and global economies. Pages 276–293 *in* World symposium: ecologic and economic benefits of hunting. World Forum on the Future of Sport Shooting Activities, Windhoek, Namibia.

Dusek, G. L., R. J. Mackie, J. D. Herringes Jr., and B. B. Compton. 1989. Population ecology of white-tailed deer along the lower Yellowstone River. Wildlife Monographs 104:1–68.

Festa-Bianchet, M. 2003. Exploitative wildlife management as a selective pressure for the life history evolution of large mammals. Pages 191–207 *in* M. Festa-Bianchet and M. Apollonio, editors. Animal behavior and wildlife conservation. Island Press, Washington, DC, USA.

Festa-Bianchet, M. 2008. Ecology, evolution, economics, and ungulate management. Pages 183–202 *in* T. E. Fulbright and D. G. Hewitt, editors. Wildlife science: linking theory and management applications. CRC Press, Boca Raton, Florida, USA.

Festa-Bianchet, M. 2016. When does selective hunting select, how can we tell, and what should we do about it? Mammal Review 47:76–81. https://doi.org/10.1111/mam.12078.

Freeman, E. D., R. T. Larsen, M. E. Peterson, C. R. Anderson Jr., K. R. Hersey, and B. R. McMillan. 2014. Effects of male-biased harvest on mule deer: implications for rates of pregnancy, synchrony, and timing of parturition. Wildlife Society Bulletin 38:806–811. https://doi.org/10.1002/wsb.450.

Geist, V. 1994. Wildlife conservation as wealth. Nature 368:491–492.

Geist, V. 2006. The North American Model of Wildlife Conservation: a means of creating wealth and protecting public health while generating biodiversity. Pages 285–293 *in* D. M. Lavigne, editor. Gaining ground: in pursuit of ecological sustainability. International Fund for Animal Welfare, University of Limerick, Limerick, Ireland.

Geist, V., S. P. Mahoney, and J. F. Organ, 2001. Why hunting has defined the North American model of wildlife conservation. Transactions of the North American Wildlife and Natural Resources Conference 66:175–185.

Grayson, D. K., and D. J. Meltzer. 2002. Clovis hunting and large mammal extinction: a critical review of the evidence. Journal of World Prehistory 16:313–359.

Grayson, D. K., and D. J. Meltzer. 2003. A requiem for North American overkill. Journal of Archaeological Science 30:585–593.

Grinnell, G. B. 1913. The game preservation committee. Pages 421–432 *in* G. B. Grinnell, editor. Hunting at high altitudes. Harper and Brothers, New York, New York, USA.

Guynn, D. C., Jr., J. R. Sweeney, R. J. Hamilton, and R. L. Marchington. 1988. A case study in quality deer management. South Carolina White-Tailed Deer Management Workshop 2:72–79.

Habitat Conservation Trust Foundation. 2021. Grants. https://hctf.ca/grants/.

Haile-Selassie, Y., G. Suwa, and T. D. White. 2004. Late Miocene teeth from Middle Awash, Ethiopia, and early hominid dental evolution. Science 303:1503–1505.

Haury, E. W., E. B. Sayles, and W. W. Wasley. 1959. The Lehner mammoth site, southeastern Arizona. American Antiquity 25:2–32.

Heffelfinger, J. R. 2018. Inefficiency of evolutionarily relevant selection in trophy ungulate hunting. Journal of Wildlife Management 82:57–66.

Holsman, R. H. 2000. Goodwill hunting? exploring the role of hunters as ecosystem stewards. Wildlife Society Bulletin 28:808–816.

Horsley, S. B., S. L. Stout, and D. S. deCalesta, 2003. White-tailed deer impact on the vegetation dynamics of a northern hardwood forest. Ecological Applications 13:98–118.

International Hunter Education Association. 2010a. Who we are. http://ihea-usa.org/about-ihea/usa/who-we-are.

International Hunter Education Association. 2010b. Hunter education requirements. http://ihea-usa.org/hunting-and-shooting/hunter-education.

International Union for Conservation of Nature. 1980. World conservation strategy. Gland, Switzerland.

Jacobson, H. A. 1992. Deer condition response to changing harvest strategy, Davis Island, Mississippi. Pages 48–55 in R. D. Brown, editor. The biology of deer. Springer-Verlag, New York, New York, USA.

Knox, M. W. 2011. The antler religion. Wildlife Society Bulletin 35:45–48.

LaSharr, T. N., R. A. Long, J. R. Heffelfinger, V. C. Bleich, P. R. Krausman, et al. 2019. Hunting and mountain sheep: do current harvest practices affect horn growth? Evolutionary Applications 12:1823–1836.

Leonard, W. R., and M. L. Robertson. 1994. Evolutionary perspectives on human nutrition: the influence of brain and body size on diet and metabolism. American Journal of Human Biology 6:77–88.

Leopold, A. 1933. Game management. Charles Scribner's Sons, New York, New York, USA.

Leopold, A. 1943. Wildlife in American culture. Journal of Wildlife Management 7:1–6.

Loehr, J., J. Carey, R. B. O'Hara, and D. S. Hik. 2010. The role of phenotypic plasticity in responses of hunted thinhorn sheep ram horn growth to changing climate conditions. Journal of Evolutionary Biology 23:783–790.

Mahoney, S. P. 2007. The importance of how society views hunting. Pages 62–67 in Proceedings of the Western Association of Fish and Wildlife Agencies. Arizona Game and Fish Department, Flagstaff, Arizona, USA.

Mahoney, S. P. 2009. Recreational hunting and sustainable wildlife use in North America. Pages 266–281 in B. Dickson, J. Hutton, and W. M. Adams, editors. Recreational hunting, conservation and rural livelihoods: science and practice. Blackwell, Chichester, West Sussex, UK.

Mahoney, S. P. 2019. The model in transition: from proactive leadership to reactive conservation. Pages 156–160 in S. P. Mahoney and V. Geist, editors. The North American Model of Wildlife Conservation. Johns Hopkins University Press, Baltimore, Maryland, USA.

Mahoney, S. P., V. Geist, and P. R. Krausman. 2019. The North American Model of Wildlife Conservation: setting the stage for evaluation. Pages 1–8 in S. P. Mahoney and V. Geist, editors. The North American Model of Wildlife Conservation. Johns Hopkins University Press, Baltimore, Maryland, USA.

Martin, P. S., and R. G. Klein. 1984. Quaternary extinctions. University of Arizona Press, Tucson, Arizona, USA.

Mawdsley, J. R., J. F. Organ, D. J. Decker, A. B. Forstchen, R. J. Regan, S. J. Riley, M. S. Boyce, J. E. McDonald Jr., C. Dwyer, S. P. Mahoney. 2018. Artelle et al. (2018) miss the science underlying North American wildlife management. Science Advances 4:eaat8281.

McCaffery, K. R., J. E. Ashbrenner, and R. E. Rolley. 1998. Deer reproduction in Wisconsin. Transactions of the Wisconsin Academy of Sciences, Arts and Letters 86:249–261.

McCullough, D. R. 1979. The George Reserve deer herd: population ecology of a k-selected species. University of Michigan Press, Ann Arbor, USA.

Meine, C. D. 1991. Aldo Leopold: his life and work. University of Wisconsin Press, Madison, USA.

Minard, A. 2009. Hunters speeding up evolution of trophy prey? National Geographic News. http://news.nationalgeographic.com/new/2009/01/090112-trophy-hunting.html.

Mysterud, A. 2011. Selective harvesting of large mammals: how often does it result in directional selection? Journal of Applied Ecology 48:827–834.

Mysterud, A., and R. Bischof. 2010. Can compensatory culling offset undesirable evolutionary consequences of trophy hunting? Journal of Animal Ecology 79:148–160.

National Wild Turkey Federation. 2021. Our history. https://www.nwtf.org/about/know-us/our-history.

Nelson, M. P., J. A. Vucetich, P. C. Paquet, and J. K. Bump. 2011. An inadequate construct? the wildlife professional. The Wildlife Society, Bethesda, Maryland, USA.

Novak, M. 1987. The future of trapping. Pages 89–97 in M. Novak, J. A. Baker, M. E. Obbard, and B. Malloch, editors. Wild furbearer management and conservation in North America. Ontario Ministry of Natural Resources and Ontario Trappers Association, Toronto, Canada.

Noyes, J. H., B. K. Johnson, L. D. Bryant, S. L. Findholt, and J. W. Thomas. 1996. Effects of bull age on conception dates and pregnancy rates of cow elk. Journal of Wildlife Management 60:508–517.

O'Dell, J. 2007. 50 years and a lot of sheep later. Arizona Wildlife Views November–December. Arizona Game and Fish Department, Phoenix, USA.

Organ, J., S. A. Williams, J. R. Mawdsley, E. M. Hallerman. D. J. Austen, B. K. Williams, P. Souza, and A. Kinsinger. 2016. The future of cooperative fish and wildlife research. Transactions of the 80th North American Wildlife and Natural Resources Conference 80:133–141.

Ozoga, J. J., and L. J. Verme. 1985. Comparative breeding behavior and performance of yearling vs. prime-age white-tailed bucks. Journal of Wildlife Management 49:364–372.

Patterson, R. 2009. Executive summary. Pages 391–392 in World symposium: ecologic and economic benefits of hunting. World Forum on the Future of Sport Shooting Activities, Windhoek, Namibia.

Payne, N. F. 1980. Furbearer management and trapping. Wildlife Society Bulletin 8:345–348.

Paz, G., and J. R. Heffelfinger. 2018. Enforcement of laws and policies. Pages 90–101 in B. D. Leopold, W. B. Kessler, and J. L. Cummins, editors. North American wildlife policy and law. Boone & Crockett Club, Missoula, Montana, USA.

Peyton, R. B. 2000. Wildlife management: cropping to manage or managing to crop? Wildlife Society Bulletin 28:774–779.

Pigeon, G., M. Festa-Bianchet, D. W. Coltman, and F. Pelletier. 2016. Intense selective hunting leads to artificial evolution in horn size. Evolutionary Applications 9:521–530. https://doi.org/10.1111/eva.12358

Posewitz, J. 2002. Beyond fair chase: the ethics and tradition of hunting. Falcon, Kingwood, Texas, USA.

Postma, E. 2006. Implications of the difference between true and predicted breeding values for the study of natural selection and microevolution. Journal of Evolutionary Biology 19:309–320.

Putnam, R., M. Apollonio, and R. Andersen. 2011. Ungulate management in Europe: problems and practices. Cambridge University Press, Cambridge, UK.

Rawinski, T. J. 2008. Impacts of white-tailed deer overabundance in forested ecosystem: an overview. Northeastern Area State and Private

Forestry, US Forest Service, US Department of Agriculture. http://www.na.fs.fed.us/fhp/special_interests/white_tailed_deer.pdf.

Reiger, J. F. 1986. American sportsmen and the origins of conservation. Revised edition. University of Oklahoma Press, Norman, USA.

Responsive Management. 1993. Wildlife viewing in Maryland: participation, opinions and attitudes of adult Maryland residents towards a watchable wildlife program. Report Prepared for the Maryland Wildlife Division. Harrisonburg, Virginia, USA.

Responsive Management. 1995. Federal aid outreach survey, region II: Arizona anglers, boaters, and hunters; New Mexico anglers, boaters and hunters; Oklahoma anglers, boaters and hunters; Texas anglers, boaters, hunters and passport holders. Report Prepared for US Fish and Wildlife Service. Harrisonburg, Virginia, USA.

Responsive Management. 2015. Trap use, furbearers trapped, and trapper characteristics in the United States in 2015. Association of Fish and Wildlife Agencies, Washington, DC, USA. https://www.dec.ny.gov/docs/wildlife_pdf/afwatrapuserpt15.pdf.

Rocky Mountain Elk Foundation. 2021. RMEF history. https://www.rmef.org/rmef-history/.

Roosevelt, T., T. S. Van Dyke, D. G. Eliot, and A. J. Stone. 1902. The deer family. MacMillan, New York, New York, USA.

Rowe, T. 2009. WFSA President Ted Rowe's closing remarks. Page 392 in World symposium: ecologic and economic benefits of hunting. World Forum on the Future of Sport Shooting Activities, Windhoek, Namibia.

Rughetti, M., and M. Festa-Bianchet. 2010. Compensatory growth limits opportunities for artificial selection in alpine chamois. Journal of Wildlife Management 74:1024–1029.

Singer, F. J., and L. C. Zeigenfuss. 2002. Influence of trophy hunting and horn size on mating behavior and survivorship of mountain sheep. Journal of Mammalogy 83:682–698.

Smith, C. A. 2011. The role of state wildlife professionals under the public trust doctrine. Journal of Wildlife Management 75:1539–1543.

Southwick, R., and T. Allen. 2010. Expenditures, economic impacts and conservation contributions of hunters in the United States. Pages 308–313 in World symposium: ecologic and economic benefits of hunting. World Forum on the Future of Sport Shooting Activities, Windhoek, Namibia.

Southwick Associates 2018. Hunting in America: an economic force. Fernandina Beach, Florida, USA.

Thomas, E. D., Jr. 2010. How sportsmen saved the world: the unsung conservation effort of hunters and anglers. Lyons Press, Guilford, Connecticut, USA.

Todd, A. W. 1981. Ecological arguments for fur-trapping in boreal wilderness regions. Wildlife Society Bulletin 9:116–124.

USFWS. US Fish and Wildlife Service. 2010a. National summary of accomplishments, 2005–2009. Washington, DC, USA.

USFWS. US Fish and Wildlife Service. 2010b. Federal Aid Division—The Pittman-Robertson Federal Aid in Wildlife Restoration Act. http://www.fws.gov/southeast/federalaid/pittmanrobertson.html.

USFWS. US Fish and Wildlife Service. 2016. 2016 National Survey of Fishing, Hunting, and Wildlife-Associated Recreation. Washington, DC, USA.

USFWS. US Fish and Wildlife Service. 2018. Hunter education (WR Program)—funding. https://www.fws.gov/wsfrprograms/Subpages/GrantPrograms/HunterEd/HE_Funding.html.

USFWS. US Fish and Wildlife Service. 2020. Final apportionment of Pittman-Robertson wildlife restoration Funds for fiscal year 2020. Federal Aid Division—The Pittman-Robertson Federal Aid in Wildlife Restoration Act. https://www.fws.gov/wsfrprograms/Subpages/GrantPrograms/WR/WRFinalApportionment2020.pdf.

Watson, L. 1971. The omnivorous ape. Coward, McCann and Geoghegan, New York, New York, USA.

White, G. C. 2001. Effect of adult sex ratio on mule deer and elk productivity in Colorado. Journal of Wildlife Management 65:543–551.

White, H. B., G. R. Batcheller, E. K. Boggess, C. L. Brown, J. W. Butfiloski, et al. 2021. Best management practices for trapping furbearers in the United States. Wildlife Monographs 207:3–59. https://doi.org/10.1002/wmon.1057.

Wild Sheep Foundation. 2021. About WSF. https://www.wildsheepfoundation.org/about.

10 EFFECTS OF WEATHER AND ACCIDENTS ON WILDLIFE

MICHAEL R. CONOVER, JONATHAN B. DINKINS, AND
MICHAEL J. HANEY

INTRODUCTION

Weather is variation in ambient temperature, precipitation, and wind across time and space. A weather event is a specific weather phenomenon that lasts no more than a few days (e.g., thunderstorms, hurricanes, windstorms, blizzards). A weather pattern is a weather phenomenon that lasts weeks or months, and climate is the prevailing weather at a point on the earth's surface over the course of many years or decades.

HOW WEATHER EVENTS AFFECT WILDLIFE

Unusually cold or hot temperatures can kill animals, especially those living along the edge of the geographic range for their species. During 2010, a cold snap in Florida killed 431 manatees (*Trichechus manatus latirostris*)—13% of the population (Segelson 2010)—and stunned more than 3,500 sea turtles (McNulty 2010). High temperatures also kill wildlife. Hyperthermia caused the death of four Sonoran pronghorn (*Antilocapra americana sonoriensis*) fawns that died during the three hottest days of 2005 (with highs of 44.1°C, 44.1°C, and 43.6°C; Wilson and Krausman 2008).

In painted turtles (*Chrysemys picta*), nests exposed to extreme heat produced more hatchings with abnormal shell morphologies than normal, regardless of whether the mother had any abnormalities. A continual increase in heat waves and higher temperatures worldwide may hinder the development in the next generation of the painted turtle (Telemeco et al. 2012).

Hail and lightning pose risks to animals (Glasrud 1976, Roth 1976). One hailstorm killed >35 tundra swans (*Cygnus columbianus*; Hochbaum 1955), while another killed >100 snow geese (Krause 1959). In July 1953, two hailstorms in Alberta covered 2,500 km² and killed many birds in the area, including more than 60,000 waterfowl (Smith and Webster 1955). More than 40,000 migrants representing 45 species were killed during a tornado and storm on 8 April 1993 off Louisiana (Wiedenfeld and Wiedenfeld 1995). Sandstorms caused by high winds can be a problem for birds moving through arid regions. Sandstorms have killed barn swallows (*Hirundo rustica*), north-

ern wheatears (*Oenanthe oenanthe*), and white storks (*Ciconia ciconia*; Moreau 1928, Schüz et al. 1971).

In a study performed in Wyoming on sage thrashers (*Oreoscoptes montanus*), Brewer's sparrows (*Spizella breweri*), and vesper sparrows (*Pooecetes gramineus*), 17% of nests failed because of hailstorms. In the area where the hailstorm was the strongest, 45% of the nests failed. Nests with stronger structure or those under shrub canopies were better protected from the storm and were more likely to survive (Hightower et al. 2018). Several large hailstorms passed through India during 2014, resulting in high mortality of wildlife, especially where the size of the hail was 2.5–5 cm in diameter. In total, 62,250 birds from 35 different species and 1,076 mammals from 9 different species were killed by the hailstorms. Two species of bats made up 79% of mammalian deaths (Narwade et al. 2014).

Many weather events kill animals but have little effect on the population dynamics of a species unless all individuals of that species are restricted to a small area. One example of a species with a restricted range is the whooping crane (*Grus americana*). All free-ranging whooping cranes winter in a single area along the Texas coast. Even a localized weather event might cause extinction of the species, and the US Fish and Wildlife Service has been trying to establish populations of whooping cranes in other areas to reduce this risk.

Effect of Weather Events on Foraging Efficiency

Waterfowl and shorebirds die when cold temperatures cause their aquatic foraging areas to freeze, reducing their ability to forage (Smith 1964, Barry 1968, Fredrickson 1969). In spring 1964, an estimated 100,000 king eiders (*Somateria spectabilis*)—13% of the population—died when sea ice in the Beaufort Sea refroze, preventing the birds from feeding (Barry 1968).

Weather events also can put an animal or species at a competitive disadvantage by changing its foraging behavior and efficiency. During subfreezing temperatures and high levels of snow, eastern bluebirds (*Sialia sialis*) increase their foraging group size and shift their focus from foraging on insects to foraging for fruit (Weinkam et al. 2017). Lesser kestrels (*Falco naumanni*) switch among foraging tactics based on weather, especially wind velocity and solar radiation (Cecere et al. 2020). As another example, rattlesnakes need to keep their body temperatures within a narrow range to hunt small mammals effectively. When it is too cold for them, mammalian predators that hunt the same species have a competitive advantage over rattlesnakes because mammals can better regulate their body temperature. But when ambient temperatures are optimal, rattlesnakes have a competitive advantage over mammalian predators because snakes have a lower metabolic rate and do not need to catch as much prey as do mammalian predators. Birds and mammals must also balance the cost of thermoregulation when temperatures are either very hot or cold with the need to find food. Among southern yellow-billed hornbills (*Tockus leucomelas*), body condition and foraging success decrease during hot days because of the energetic costs of panting and the need to select cooler foraging sites (Van de Ven et al. 2019). The burrowing owl (*Athene cunicularia*) is an endangered species in Canada; increased precipitation in the Canadian breeding grounds increases vegetation height and makes prey less accessible, increasing starvation and death in burrowing owls (Wellicome et al. 2014).

Sunny days warm the ground and create updrafts, which in turn provide optimal hunting conditions for ferruginous hawks (*Buteo regalis*), a species that hunts for prey while soaring. Their hunting efficiency declines when cloudy skies and the lack of wind prevent the creation of updrafts. But the lack of updrafts does not have the same adverse effect on other hawk species that hunt from perches, such as red-tailed hawks (*Buteo jamaicensis*).

Effect of Weather Events on Risk of Depredation

Weather events can alter an animal's risk of being killed by a predator. Rain and wind increase ambient noise levels and can muffle the sound of an approaching predator. For this reason, many herbivores remain in secure cover and refrain from foraging under such conditions. Changing ambient light levels can also increase an animal's risk of depredation. Leopards (*Panthera pardus*) have better night vision than their prey. This advantage is particularly acute when it is completely dark, such as when clouds block the moonlight, because herbivores can no longer see a stalking leopard, but the leopard can still see them.

Other predators, such as raccoons (*Procyon lotor*), striped skunks (*Mephitis mephitis*), and feral hogs (*Sus scrofa*), use olfaction to locate prey when weather conditions favor its use. Ideal atmospheric conditions for the use of olfaction include high humidity, a light breeze, and a temperature inversion close to the ground (Conover 2007). This may explain why turkey nests are more likely to be depredated during periods of wet weather (Palmer et al. 1993; Roberts et al. 1995; Roberts and Porter 1998a, b).

Weather can influence depredation rates in unexpected ways. Loss of eider (*Somateria* spp.) ducklings to predators doubles during rainy and windy days because the ducklings do not stay as close to their mother as during calm, dry days (Mendenhall and Milne 1985). Stormy weather also can increase depredation rates on ducklings by forcing the broods into calm bays, where they are depredated by gulls (Bergman 1982 as cited by Johnson et al. 1992).

Effect of Weather Events on Reproduction

Weather events have a pronounced effect if they occur when animals are particularly vulnerable, such as when they are only a few days old (i.e., neonates) or when they are migrating. Eggs are especially vulnerable to weather events because of their immobility. A severe snowstorm in the Northwest Territories of Canada that occurred when goose nests were hatching killed 100% of snow goose (*Chen caerulescens*) goslings, 75% of white-fronted goose (*Anser albifrons*), and 50% of brant (*Branta bernicla*; Barry 1967, Sargeant and Raveling 1992). Nests built along shorelines, on islands, or in low-lying areas can become submerged during periods of heavy rain, floods, high tides, or storm surges. These calamities may destroy nearly all low-lying waterfowl nests (Hansen 1961, Sargeant and Raveling 1992). Also vulnerable to flooding are birds that build nests over water. In Manitoba, 6% of canvasback duck (*Aythya valisineria*) nests are lost to flooding (Stoudt 1982). Flooding is a problem for ground-nesting birds in upland areas; each year, 2% of dabbling duck nests in the prairies and parklands of North America are lost to flooding (Sargeant and Raveling 1992). Birds nesting in tidal marshes face a constant threat of flooding, but this threat can be reduced by starting to nest when spring tides peak. Snowy plovers (*Charadrius nivosus*) do this, but even so, 6% of their nests are lost to tidal flooding (Plaschke et. 2019).

While nesting in trees provides protection from floods, high winds can destroy nests. Mourning doves (*Zenaida macroura*) that build structurally sound nests in secure locations (e.g., in a tree fork) are more likely to reproduce successfully than doves that build poor nests or locate them on thin branches where they are vulnerable to being blown out by the wind (Coon et al. 1981).

Neonates have only a limited ability to thermoregulate; rain or cold temperatures can cause mass mortality of young if these conditions occur during periods when the young are vulnerable. Exposure to adverse weather is believed to kill more waterfowl broods than starvation, diseases, or parasites (Johnson et al. 1992). In house sparrows, hatching success was reduced during periods of cold weather ($<16°C$) during incubation or hatching (Pipoly et al. 2013). Cold weather can reduce the body condition of neonates, while hot weather can have the same effect on older chicks; in both cases, the young birds must allocate more energy to thermoregulation (Pipoly et al. 2013).

Effect of Weather Events on Migrating Animals

Common weather events rarely cause mortality of adult wildlife located within their own home range. Migration, however, is a dangerous time for many animals. When geese nest in the Arctic or Subarctic, a short nesting season and long distance to the wintering grounds combine to make a fledgling's first fall migration dangerous; it is a major cause of mortality among fledglings. Heavier fledglings are more likely to survive fall migration than lighter ones in greater snow geese (*Chen caerulescens atlantica*; Menu et al. 2005), lesser snow geese (*C. c. caerulescens*; Francis et al. 1992), Ross's geese (*C. rossii*; Slattery 2002), emperor geese (*C. canagicus*; Schmutz 1993), and barnacle geese (*Branta leucopsis*; Owen and Black 1989, Loonen et al. 1999).

Headwinds can force migrating birds to fly longer, depleting their energy reserves, while heavy rains can saturate plumage and cause excessive wing loading (Cottam 1929, Williams 1950, Kennedy 1970). These stresses can cause birds to fly lower than they otherwise would, increasing their risk of flying into obstacles (e.g., communication towers) or forcing migrants to land in unsuitable areas. Animals that must migrate across inhospitable areas in adverse weather are particularly at risk. Many of the recorded mortalities of migrating birds occurred when terrestrial birds encountered storms while crossing large water bodies (Dick and Pienkowski 1979, Morse 1980, Evans and Pienkowski 1984, Spendelow 1985).

At least 106 avian species have washed up on beaches following storms over the Gulf of Mexico (James 1956, Webster 1974, King 1976, Wiedenfeld and Wiedenfeld 1995). Small birds are especially vulnerable to storms when migrating across water, yet large birds can succumb to oceanic storms, too. For instance, 4,600 dead rooks (*Corvus frugilegus*) were found in southern Sweden following a storm (Alestam 1988). In April 1980 more than 1,300 birds of prey (including eagles) died off Israel when strong winds blew them over the sea from their usual landward route (Kerlinger 1989).

Likewise, migrating waterbirds may encounter problems when they need to migrate over land. In western North America, eared grebes (*Podiceps nigricollis*) migrate over hundreds of kilometers of desert with few places to land in an emergency. Snowstorms have occasionally brought down thousands of these birds during a single night. On 13 December 2011, thousands of eared grebes crash-landed in southern Utah during a snowstorm. Once down, most of the eared grebes were unable to take flight again and died (Roberts et al. 2014). Migratory behavior persists in many species despite the heavy losses of migrating birds because the fitness costs (mortality) are more than offset by the fitness benefits in terms of improved overall survival and breeding success that accrue from being able to move between areas (Newton 2007).

HOW WEATHER PATTERNS AFFECT WILDLIFE

Spring weather patterns often cause annual variation in reproductive rates of wildlife, with reduced success during periods of wet and cold weather. For example, snow and ice conditions in June adversely affect reproduction of several species of geese that nest in the Arctic (Reeves et al. 1976). Annual nest success rates of wild turkey (*Meleagris gallopavo*) are correlated with the rainfall amounts during spring (Roberts et al. 1995;

Roberts and Porter 1998*a*, *b*). Precipitation and wind account for more than 90% of annual variation in survival of eider ducklings (*Somateria mollissima*; Mendenhall and Milne 1985). In contrast, wet years benefit some species. Annual production of mallards (*Anas platyrhynchos*) in the Mississippi Flyway correlates with the number of ponds containing water in the Northern Prairies, and more ponds fill with water during wet years (Bellrose et al. 1961). In the Mississippi Flyway, winter survival of mallards is highest during years when mild temperatures are combined with heavy winter rains; the heavy rains create more areas where the ducks can forage (Reinecke et al. 1987).

Winter weather patterns account for annual variation in survival of many wildlife species, including caribou (*Rangifer tarandus caribou*; McLoughlin et al. 2003), Dall sheep (*Ovis dalli*; Burles and Hoefs 1984), and alpine ibex (*Capra ibex*; Jacobson et al. 2004). Ground-feeding birds and mammals can starve when deep snows cover the ground for an extended period (Roseberry 1962, Vepsäläinen 1968, Bull and Dawson 1969). In Sweden the two worst winters recorded in 68 years (>60 days with >5 cm of snow depth) caused a population of barn owls (*Tyto alba*) to decline by more than 70% when their main prey (i.e., rodents) became unavailable under a thick layer of snow (Altwegg et al. 2006).

In arid regions the amount and timing of precipitation are critical to reproduction and survival of many wildlife species. In New Mexico, Bender et al. (2011) reported that reproduction and survival of mule deer (*Odocoileus hemionus*) were reduced during droughts. To mitigate the effects of dry periods, some birds and mammals in arid regions delay nesting and reproduction until the onset of rain (Johnson et al. 1992).

CLIMATE CHANGE AND WILDLIFE

Climate change has occurred at a slow rate throughout earth's history. Humans are changing the climate much faster than under natural conditions (Root and Schneider 2002). Burning fossil fuels, deforestation, and other land-use practices (e.g., conversion of forested areas to urban or agricultural areas) can affect climate, but burning fossil fuels has the greatest effect by increasing atmospheric levels of carbon dioxide. This gas heats the earth by capturing a large percentage of long-wavelength heat energy near the earth's surface, producing a greenhouse effect (Root and Schneider 2002). The global average temperature has increased 0.6°C to 0.7°C over the past century (Green et al. 2001, Root et al. 2005); 1°C to >2°C have been projected as possible increases by 2050 (Intergovernmental Panel on Climate Change 2014). This is much greater than during most of the Holocene period (the past 10,000 years), when the average global temperature changed by <0.01°C per year.

With a warming climate, the range of plant species is expected to shift northward in the Northern Hemisphere, southward in the Southern Hemisphere, and move higher in elevation on mountains. Wildlife populations are expected to follow (Root et al. 2005). Parmesan and Yohe (2003) verified movement of animals toward the poles when they conducted a meta-analysis involving 99 avian species, 16 butterfly species, and 17 alpine plant species and reported movement of 6.1 km (±2.4 km) per decade of range limits toward the poles during the 20th century. Arctic and antarctic wildlife species, such as polar bears (*Ursus maritimus*) and Adélie penguins (*Pygoscelis adeliae*), have distributions that are at least seasonally tied or restricted to areas with sea ice (Green et al. 2001). As sea ice melts with a warming climate, there will be less suitable habitat left for these species. In contrast, some tropical species will likely benefit from climate change by having a larger area of suitable habitat that they can occupy.

One danger is that unprecedented rates of climate change may outpace the natural adaptive capacities of some wildlife species. Not all vegetation and wildlife will be able to move to new areas in response to climate change; some species will be trapped, and especially vulnerable will be species that occupy small ranges. For example, many species of amphibians, reptiles, and small mammals that occupy isolated mountains (ecological islands) will be forced to shift their range to higher elevations as the climate warms, but this will result in a reduction of the species' range and genetic diversity (Root and Schneider 2002, Moritz et al. 2008, Rowe 2009, Rowe et al. 2011, Morelli et al. 2012).

Climate change will produce trophic mismatches that affect many wildlife species. A trophic mismatch occurs when the availability of food and habitat resources in an area is no longer in synchrony with when wildlife needs those resources (Parmesan 2006). One group of birds that are vulnerable to trophic mismatches are those that have to feed during their migration between winter areas and breeding areas, including many neotropical migrants. For example, red knots (*Calidris canutus*) winter in Argentina and Brazil but breed in the Arctic. They lack the energy to make this trip nonstop, so they stop in Delaware to gorge themselves on the eggs of horseshoe crabs that are laying eggs on the beaches at this exact time. The concern is that if global warming causes horseshoe crabs to change their breeding schedule, the red knots may arrive in Delaware and find no horseshoe crab eggs. Other species at risk for trophic mismatches are obligate hibernators and species that rely on ephemeral resources (Ozgul et al. 2010).

Warmer temperatures from climate change are expected to increase rainfall worldwide, but the effects of climate changes will vary by location. Some areas will become drier or colder than current conditions, depending on changes in wind patterns. Chapter 18 addresses climate change and its influence on wildlife in more detail.

EFFECTS OF ACCIDENTAL DEATHS ON WILDLIFE

Accidents are a fact of life, and accidental deaths befall every species, but unless animals are tracked using radio collars, it is difficult to determine how frequently accidents kill wildlife. What little is known about accidental deaths comes from species that from a human perspective live in dangerous areas, such as bighorn sheep (*Ovis canadensis*). Jokinen et al. (2008) radio-collared 46 female bighorn sheep in Alberta and tracked them from 2003 until 2005. Eleven females died during the study; three were killed in accidents (one fell to its death, one was killed in an avalanche, and one broke its leg), one died from an unknown cause, and seven were killed by predators. In Hells Canyon, 22% of bighorn sheep died from falls or injuries (Cassirer and Sinclair 2007). In the Sierra Nevada Mountains, the cause of death was determined for 105 bighorn sheep—19 were killed in avalanches or accidents, five died of exposure, one was struck by a vehicle, and 80 were killed by predators (US Fish and Wildlife Service 2003). In Arizona, Kamler et al. (2003) reported the cause of death for 46 bighorn sheep; nine of them (20%) died from climbing accidents. Among adults, accidents often peak during the mating season, when animals are distracted or engage in territorial or dominance fights and mating. Among muskoxen (*Ovibos moschatus*) and bighorn sheep, 5% to 10% of adult males are killed annually from injuries sustained during the rut (Geist 1971, Wilkinson and Shank 1976).

Young animals are particularly vulnerable to accidents because neonates have limited ability to escape the hazards in their environment. In a suburban population of Canada geese (*Branta canadensis*), accidents were the second leading cause of death among young goslings; accidents resulted from being hit by vehicles, falling into holes, being struck by golf balls, and becoming separated from their parents (M. R. Conover, Utah State University, unpublished data). Accidents can even befall eggs inside nests. Parents normally roll the eggs several times a day, because if eggs remain stationary for too long, the membranes surrounding the developing embryo will adhere to the shell, killing the embryo. But rolling the eggs increases the chance that an egg might tumble out of the nest. When it does, Canada geese will use their bill to roll an egg back into the nest, a difficult task when the eggs roll downhill, slip over ledges, or roll into the water.

Arboreal animals, such as chimpanzees (*Pan troglodytes*), run the risk of accidently falling out of trees, but it does not happen often. Although chimpanzees travel both on the ground and in the trees, mothers are more cautious on the ground owing to the higher risk of depredation, and infants also do not travel as far away from their mothers on the ground as they do in the trees. Increased freedom and travel away from their mothers in the trees by young chimps increase the risk of infants accidently falling and dying. Arboreal accidents also kill healthy chimpanzees or result in skeletal injuries. Although these accidents are not witnessed and even less often reported, such deaths may be more common than realized (Nakamura and Ramadhani 2014).

ACCIDENTAL DEATHS CAUSED BY HUMAN ENDEAVORS

One of the main goals of wildlife management is to lessen the effects of human activities on wildlife. For this reason, accidental deaths that result from human endeavors are a greater source of concern than deaths from the types of accidents that have befallen wildlife for eons. Likewise, most of the research on accidental deaths of wildlife has investigated the effect of human-induced accidents. For instance, dozens of papers have been written about how many birds die from flying into windows or power lines, but we are not aware of any reports about how many birds are killed from flying into cliffs. For the rest of this chapter we will focus on accidental deaths from anthropogenic activities and structures.

Oil Ponds

When gas and oil are pumped from the ground, water often comes up with the petroleum. Oil pits, ponds, or tanks are used to separate the oil from the water and then store the contaminated water. Waste-oil ponds also result from oil spills, oil drips, or from the flaring of hydrogen sulfide. In this chapter we refer to these collectively as oil ponds. Birds and mammals are attracted to oil ponds by thirst, struggling insects, or other animals that are caught in the oil. Migratory waterfowl can become trapped by landing in the ponds before they realize that the ponds are covered by oil. Death results from becoming trapped in the oil and drowning, getting oil on their feathers or fur and ingesting the toxic oil when animals try to clean themselves, or swallowing too much oil when drinking oily water. More than half (62%) of the birds recovered from oil ponds were members of the order Passeriformes; 10% were Anseriformes, 6% Columbiformes, 5% Strigiformes, 4% Ciconiiformes, 3% Charadriiformes, 2% Falconiformes, and 2% Cuculiformes (Trail 2006).

The US Fish and Wildlife Service estimated that oil ponds killed 2 million migratory birds during 1997 (Ramirez 1999). Since then, oil companies have taken steps to address the problem by replacing many oil ponds with closed tanks or by using nets to keep birds out of open oil ponds. Trail (2006) reported that these measures have decreased annual mortality rates from 1 million birds to 500,000.

Communication Towers

There are more than 50,000 communication towers in the United States that stand higher than 66 m above ground level; they are used mostly by the radio, television, and wireless telephone industries. During inclement weather, migratory birds collide with these towers; most victims (97%) are passerines (Longcore et al. 2013). Nocturnal migrants, including thrushes, vireos, tanagers, cuckoos, and sparrows, are particularly vulnerable to colliding with communication towers (Avery et al. 1976). Longcore et al. (2012) estimated that 6.8 million birds are killed annually in the United States and Canada from this type of mishap; 13 avian species of conservation concern suffer a mortality of 1% to 9% of their total population annually by communication towers (Longcore et al. 2013).

Power Lines

Power lines include transmission lines and distribution lines. The former conduct high-voltage electricity (115–500 kV) over long distances and are typically held aloft by tall wooden or metal towers. Distribution lines deliver electricity from substations to individual buildings and carry up to 70 kV (Manville

2005, Edison Electric Institute 2009). There are more than 1,217,000 km of transmission lines and 9,612,000 km of distribution lines in the United States (Edison Electric Institute 2009).

A bird must complete the circuit of an electrified power line to be electrocuted, which often happens when a large bird perches on a power line and its wings touch another energized or grounded line (Bevanger 1995, 1998; Harness and Wilson 2001; Lehman 2001; Manville 2005). In rural western United States, eagles, hawks, and owls account for 96% of all electrocutions (Harness and Wilson 2001). In the United States, electrocution is the second-greatest cause of mortality for golden eagles (*Aquila chrysaetos*) and the fourth-greatest for bald eagles (*Haliaeetus leucocephalus*; LaRoe et al. 1995, Lehman 2001). In southwestern Spain, Ferrer et al. (1991) estimates that 400 raptors and vultures are electrocuted each year on a 1-km stretch of power line and 1,200 birds of prey are electrocuted annually in Doñana National Park. Between 1974 and 1982, 52% of banded raptors and 69% of eagles found dead in the Doñana area had been electrocuted (Ferrer et al. 1991). Rubolini et al. (2005) estimates that 0.15 birds died per power pole in Italy, 0.21 per pole in Spain, and 0.15–5.2 per pole in the United States.

Birds also are killed by flying into power lines, and the list of birds that are susceptible to collision with power lines is much more diverse than birds susceptible to electrocution. Most vulnerable to collisions are less maneuverable fliers: birds with small wings relative to their mass, such as ducks and grebes, and birds with long wings (e.g., cranes). In prairies, waterfowl are more likely to collide with power lines than other avian groups (Faanes 1987). Collisions with power lines are a considerable problem for many threatened or endangered birds. For example, the two largest sources of mortality for California condors (*Gymnogyps californianus*) were collisions with power lines and lead poisoning (Meretsky et al. 2000).

Mortality rates per kilometer of power line have been calculated at 124 birds/km in North Dakota prairie wetlands (Faanes 1987), 5.3 willow ptarmigan (*Lagopus lagopus*) in subalpine Norway (Bevanger and Brøseth 2004), 2.3–5.8 common cranes (*Grus grus*) in Spain (Janss and Ferrer 2000), and 1.6–4.0 great bustards (*Otis tarda*) in Spain (Janss and Ferrer 1998, 2000). The most thorough studies on avian mortality rates caused by power lines were conducted in the Netherlands. Annual estimates of birds killed in the Netherlands were 113 birds/km of transmission line in grasslands, 58/km in agriculture lands, and 489 at river crossings (Koops 1987). Koops (1987) calculated that between 750,000 and 1 million birds were killed annually by flying into transmission lines in the Netherlands by combining these fatality rates with the country's 4,600 km of transmission lines. Erickson et al. (2001) hypothesized that the fatality rate per kilometer of transmission line in the United States is similar to the rate in the Netherlands. Considering that there are 800,000 km of transmission lines in the United States, Erickson et al. (2001) and Manville (2005) estimated that 130 to 175 million birds are killed annually by these lines. This mortality rate is higher than we would have expected. It excludes, however, mortalities from

RICHARD A. DOLBEER (1945–)

Dr. Richard A. Dolbeer served at the forefront of research, management, and policy development to reduce wildlife hazards to aviation. He was one of first individuals to recognize that the cost in property and lives due to wildlife–aircraft collisions was avoidable. In the mid-1980s, Dolbeer helped create the Aviation Project of the US Department of Agriculture–Wildlife Services (WS) National Wildlife Research Center and served as its leader until 2002. He then served as the WS national airports coordinator for the Aviation Safety and Assistance Program. His work was international in scope, including collaborative efforts with five foreign governments, and it has produced advances in how airport habitats should be managed to reduce use by wildlife, considerations for the design of turbine-powered engines and airframes to withstand bird strikes, and advances in how wildlife strike data are reported and interpreted.

Dolbeer's work with management of airport habitats to reduce use by wildlife has produced a dramatic reduction in aircraft collisions with birds at John F. Kennedy International Airport (JFKIA). Based on his research findings, JFKIA began clearing large stands of woody vegetation to eliminate food sources, roosting sites, and perching sites for birds that pose bird strike risks and adopted recommendations for maintaining vegetation 6 to 10 inches high to reduce bird and small mammal use of grasslands.

In collaboration with the Federal Aviation Administration, Dolbeer and scientists under his guidance produced a series of publications that culminated in the registration by the US Environmental Protection Agency of two chemical foraging repellents and one investigational new permit by the US Food and Drug Administration for the wildlife-capture drug alpha-chloralose. In addition, Dolbeer and collaborators developed a laser product for bird dispersal and were awarded several patents. Two wildlife foraging repellent applications received patents, as did a collaborative project with Precise Flight, in Bend, Oregon, to develop an aircraft-mounted lighting system. Another patent was issued for a hazard avoidance system to increase the visibility of aircraft to birds and to provide hazard data to pilots.

Dolbeer coauthored with Edward Cleary *Wildlife Hazard Management at Airports: A Manual for Airport Personnel*. This manual provided for the first time a detailed course of action for understanding and managing wildlife hazards at airports. In 2001 the manual was translated into Spanish for distribution to airports throughout Mexico and Latin America, and into French for use in France, Quebec, and West Africa.

Since 1995, Dolbeer supervised the editing and entry of strike reports into the national Wildlife Strike Database. These data were critical to understanding the circumstances of wildlife collisions with aircraft. Wildlife biologists who work at airports are now encouraged to increase efforts to detect, remove, and disperse large species of birds from the airport environment. This transition from research findings to applied management is a hallmark of Dolbeer's research career.

Adapted from Blackwell and DeVault (2009)

Courtesy of the Federal Aviation Administration

---

all of the more than 1 million km of distribution lines in the United States.

Wind Turbines

Thousands of wind turbines have been built in North America to generate electric power. Almost 5,000 are in the Altamont Pass Wind Resource Area (APWRA) in California. Smallwood and Thelander (2008) reported that APWRA wind turbines collectively kill over 2,700 birds annually, including 67 golden eagles, 188 red-tailed hawks, 348 American kestrels (*Falco sparverius*), and 440 burrowing owls. Loss et al. (2013) determined that monopole wind turbines kill approximately 234,000 birds each year. Smallwood (2013) estimated that all wind turbines in the United States (both monopole and lattice) kill 888,000 bats and 573,000 birds (including 83,000 rap-

tors) and 888,000 bats (Table 10.1). Voigt et al. (2015) reported that 250,000 bats in Germany are killed annually from wind turbines. Migratory bats made up 70% of that number as a major migratory route of bats crosses the country (Voigt et al. 2015). On a related note, utility-scale solar power plants in the United States are estimated to kill between 37,800 and 138,600 birds annually. Mortality results from bird collisions with infrastructure at power plants and from burning or singeing from exposure to concentrated sunlight (Walston et al. 2016).

Windows

Klem et al. (2004) and Klem (2006) asserted that window collisions are the second-largest cause of avian mortality related to human endeavors (after habitat loss). Loss et al. (2015) also listed windows in second place, but they listed feral and do-

Table 10.1. Estimates of the number of birds and mammals killed annually in the United States by human activities and collisions with man-made objects

Objects	Birds	Mammals
Oil ponds	500,000	Unknown
Communication towers	6.8 million	Unknown
Power lines	130–175 million	Unknown
Wind turbines	450,000	888,000
Solar energy facilities	37,800–138,600	Unknown
Windows	0.1–1 billion	Unknown
Automobiles and trucks	100 million	200 million
Fences	Unknown	110,000
Aircraft	75,000	4,000
Hunting	56 million	7 million
Households	Unknown	78 million
Total	0.4–1.3 billion	286 million

Note: Data are based on the most recent year for which data are available; see text for specific dates.

mestic cats as the greatest killer. Passerines strike windows more than any other avian group, probably because large numbers of them live near human dwellings. Mortalities caused by striking windows are a greater concern when rare or endangered birds are involved. More than 1.5% of all endangered swift parrots (*Lathamus discolor*; only 1,000 breeding pairs remain) are killed annually by collisions with windows (Klem 2006). Six species on the national Birds of Conservation Concern have been identified as vulnerable to collisions with windows including the golden-winged warbler (*Vermivora chrysoptera*), painted bunting (*Passerina ciris*), Canada warbler (*Cardellina canadensis*), wood thrush (*Hylocichla mustelina*), Kentucky warbler (*Geothlypis formosa*), and worm-eating warbler (*Helmitheros vermivorum*; Loss et al. 2014).

Klem (1990, 2006) estimated that each building in the United States kills an average of one to ten birds annually from window collisions. He based his estimate on counts he made of birds killed by striking windows located on residential and commercial buildings in southern Illinois and New York from 1974 to 1986. Dunn (1993) independently verified Klem's estimated fatality rate during Project Feeder Watch. During the winter of 1989–1990, Project Feeder Watch documented 995 fatal bird–window strikes based on 1,165 reports from participants across the United States and Canada. Dunn (1993) refined her estimate to 0.7–7.7 birds strike windows in an average house each year by taking into account possible underestimates caused by scavengers removing bird carcasses before they are counted, and overestimates of bird–window strike frequency owing to observer bias. Based on the number of buildings in North America, Klem (1990), Dunn (1993), and Loss et al. (2014) estimate that 100 million to 1 billion birds are killed annually in North America by striking windows.

Factors that affect the occurrence and frequency of avian collisions with windows include weather, type of glass, orientation of glass, time of day, time of year, and proximity of bird

feeders to windows (Klem 1989). Typically, birds collide with windows when attempting to fly through them, or to reach habitat that they can see through the glass or habitat reflected by the glass (Klem 2006). Territorial birds also may collide with windows because they mistake their reflected image for an intruding bird. Birds that attack their reflection usually collide with the glass at lower velocities than birds that try to fly through the glass and are less likely to be injured (Klem 2006). Migrating birds collide into windows on tall, commercial buildings at a greater frequency than resident birds (O'Connell 2001; Gelb and Delacrétaz 2006), whereas resident birds collide into windows on residential houses with greater frequency than migrants (Klem 1990; Dunn 1993). Methods to mitigate bird deaths include turning off lights during migrations, limiting deep alcoves in buildings, reducing parts of buildings where birds can see through windows to the opposite side of the building, and placing objects in front of or on the windows (Klem and Saenger 2013, Loss et al. 2015).

Cars and Trucks

State Farm Insurance Company (2018) reported that 1,332,322 claims were filed annually with insurance companies for accidents involving deer–vehicle collisions (DVCs) during from June 2017 to July 2018. Only about half of all DVCs are reported to insurance companies or to the police (Decker et al. 1990, Marcoux and Riley 2010); hence the actual number of DVCs occurring annually in the United States is closer to 2.6 million annually. Being hit by a vehicle is fatal to deer 92% of the time (Allen and McCullough 1976), which means that more than 2.3 million deer die each year on US highways. Each year, there are about 1,500 collisions between vehicles and moose (*Alces alces*) in Alaska, Maine, and Newfoundland (Rattey and Turner 1991, Pelletier 2006).

So many Florida Key deer are killed annually from DVCs that the US Fish and Wildlife Service classifies the species as threatened. This deer occurs on a small number of lower Florida keys, such as Big Pine Key, that have substantial human densities. These keys are crossed by US Highway 1, which provides access between the city of Key West and the Florida mainland. Because of the traffic on this highway, DVCs account for over 25% of all Florida Key deer deaths. To reduce the frequency of DVCs, a continuous 2.6-km fence was erected along Highway 1 on Big Pine Key, and two underpasses were built that allowed the deer to move from one side of the road to the other without having to set foot on the road itself. The construction effort was successful. During 2009 more than 1,300 deer were photographed using the underpasses. Deer–vehicle collisions decreased 90% along the highway after the fences and underpasses were constructed (Parker et al. 2011).

There are over a half million reported DVCs in Europe (Linnell et al. 2020). These involve more than 360,000 collisions with roe deer (*Capreolus capreolus*), 73,000 feral hogs (*Sus scrofa*), 13,000 moose, 13,000 red deer (*Cervus canadensis*), and 7,000 fallow deer (*Dama dama*). In some European countries, collisions occur regularly with chamois (*Rupicarpra rupicarpra*) and reindeer (*Rangifer tarandus*; Groot Bruinderink and

Hazelbrook 1996). In Saudi Arabia, 600 free-ranging camels (*Camelus dromedarius*) die annually in vehicle collisions out of a population of over 500,000 camels (Al-Ghamdi and AlGadhi 2003, Al Shimemeri and Arabi 2012).

It is much harder to estimate the number of small vertebrates that are killed in road accidents because such collisions often escape the driver's attention and are rarely reported. The number of small mammals killed greatly exceeds the number for larger animals because they are not excluded by highway fences, they take longer to cross the road, and drivers are less likely to see them. Data on how frequently small animals are struck by vehicles usually come from investigators counting carcasses on the same stretch of road at regular intervals and are expressed as the number killed per kilometer per day. Vertebrate death rates on a road adjacent to the Big Creek National Wildlife Area in Ontario were estimated at 13/km/day from April through October (Ashley and Robinson 1996). Vertebrate death rate was 5/km/day on a highway through Payne's Prairie State Park, Florida (Smith and Dodd 2003), and 1/km/day on roads in Saguaro National Park in Arizona (Gerow et al. 2010). Most of these reports, however, come from parks or other areas with high wildlife densities and might not be applicable to all highways in the United States.

Banks (1979) estimated that the annual avian mortality rate was 9/km of road in the United States, the same rate Hodson and Snow (1965) estimated for Britain. Since Bank's study, the network of US roads has increased at a rate of 0.2% annually, to 6.7 million km by 2006; vehicle kilometers traveled in the United States have increased at a much more rapid rate of 1.9% annually, to 5.0 trillion in 2006 (US Federal Highway Administration 2008). If we assume that there is a 1:1 relationship between the rate at which birds are struck by vehicles and the number of vehicle kilometers traveled, then during 2006 the avian mortality rate was 14.9/km of road; this means that 100 million birds are killed annually on US roads. This is between the 89 to 340 million birds that Loss et al. (2015) estimated die from vehicle collisions in the United States. An estimated 9.4 million birds are killed by vehicles in Germany (Fuellhaas et al. 1989 as cited by Vidal-Vallés et al. 2018), 7 million in Bulgaria (Nankinov and Todorov 1983), and 14 million in Canada (Bishop and Brogan 2013).

In Saguaro National Park, Arizona, five times as many mammals as birds were killed by vehicle collisions (Gerow et al. 2010). We believe a conservative estimate is that twice as many mammals are killed on roads as birds (i.e., 200 million mammals annually).

Reptiles and amphibians that try to cross a road are more vulnerable to being struck by a vehicle than birds and mammals because their slower speeds require them to take more time to cross roads. Birds and mammals also have a greater ability to evade a vehicle that is heading toward them. In Arizona, Gerow et al. (2010) reported 12,264 dead reptiles and 12,208 dead amphibians along the same stretch of road that contained just 759 dead birds and 4,146 dead mammals. During the amphibian breeding season, Hels and Buchwald (2001) estimated that 4–16 amphibians were killed per kilometer per

day along a section of road in Denmark. Rosen and Lowe (1994) estimated that 22.5 snakes were killed annually per kilometer on a stretch of highway running mainly within Organ Pipe Cactus National Monument in Arizona. In Alabama, Dodd et al. (1989) drove 19,000 km over 135 days in Alabama and reported 239 dead reptiles and 64 live reptiles on the road, or 0.016 reptiles/km of road. Victims were turtles (0.008/km), snakes (0.008/km), and lizards (0.001/km). Three studies of snakes in California reported 0.025 dead snakes/km, 0.080 dead snakes/km, and 0.175 dead snakes/km; other studies reported 0.007 snake carcasses/km in New Mexico, 0.136/km in Kansas, 0.010/km in Louisiana, 0.008/km in Mississippi (0.008), and 0.007 in Alabama (Dodd et al. 1989). The average of all of these snake reports is 0.046 snakes/km. We doubt that a snake carcass is recognizable after two weeks from being run over multiple times or being eaten by scavengers. If so, the annual mortality rate would be 1.2 snakes/km/year. According to the US Federal Highway Administration, there were 7.4 million km of highways in the United States during 2018; multiplying this value by the annual rate (1.2/km/year) yields an estimated 9 million snake fatalities annually. We suspect that the number of other reptiles (mainly lizards and turtles) and amphibians (mainly frogs and toads) killed on US highways is similar to the number of snake fatalities (i.e., 9 million snakes, 9 million other reptiles, 9 million amphibians).

Fences

One of the major man-made changes to the North American landscape is the construction of fences. Wire fences confine livestock on farms and ranches, mark property boundaries, and line many roads and railroads. These fences pose a barrier to ungulates as they move across their home range, causing a threat to them if they fail to jump successfully over the fence. In Colorado the annual fatality rate for was 0.11/km of fence for pronghorns (*Antilocapra americana*), 0.08/km for mule deer, and 0.06/km for elk (*Cervus canadensis*; Harrington and Conover 2006). Juveniles were eight times more likely to be killed by fences than adults. Woven-wire fences topped by a single strand of barbed wire were more lethal than woven-wire fences topped with two- or four-strand barbed-wire fences. This happens because when a leg is caught between two stands of barbed wire, there is enough play between the two wires that the ungulate can often pull free, but woven wire is much more rigid, making it harder to extract a leg when it is cinched between a stand of barbed wire and woven wire. Considering the millions of kilometers of fences erected in North America, the number of ungulates killed in fences is significant. Most US highways have fences on each side of the road. With 6.7 million km of roads in the United States, this works out to 13 million km of fencing. Given data from Harrington and Conover (2006), we estimate that road fences kill 5,000 elk, 5,000 pronghorns, and 100,000 white-tailed and mule deer each year in the United States from fence entanglements. Fences are also used to mark property boundaries and to enclose pastures, but the total length of boundary and pasture fences in the United States is unknown, as is the number of wildlife killed by them. Fences also cause injuries, which may not be immediately fatal but may decrease the animal's long-term probability of survival. Even hair loss from crawling through fences can make thermoregulation more energetically expensive (Jones 2014).

Unfortunately, ungulates are not the only animals caught in fences. In Australia, a fence built for the exclusion of pests killed more than 1,000 reptiles in a period of 16 months; most of the victims were eastern long-necked turtles (*Chelodina longicollis*) followed by lizards and snakes. The fence was estimated to kill 3.3% of the long-necked turtle population and disrupt the movements of another 20%. Many turtles were caught in fences near wetlands and wildlife preserves, which may effectively isolate turtle populations from each other (Ferronato et al. 2014).

Grouse and other low-flying birds are also at risk of colliding with a barbed-wire livestock fence (Robinson et al. 2016). Lesser prairie-chickens (*Tympanuchus pallidicinctus*) have been decreasing in numbers for several decades. Wolfe et al. (2007) determined the cause of death for 260 lesser prairie chickens in Oklahoma and New Mexico; 33% of mortalities resulted from fence collisions, but other studies have reported that only 1% to 4% of mortalities resulted from fences.

Aircraft

During 2018 the federal government received 15,799 reports of wildlife being killed civil aircraft in the United States; 95% were birds, 3% bats, and 2% terrestrial mammals (Dolbeer et al. 2019). Terrestrial mammals were struck when they wandered onto a runway at the wrong time. During a 29-year period (1990–2018), reported fatalities were most common among mourning doves (10,187), killdeer (*Charadrius vociferous*; 6,357), American kestrels (6,155), barn owls (6,036), horned larks (*Eremophila alpestris*; 5,149), and European starlings (*Sturnus* vulgaris; 4,947). Most bird strikes (80%) at passenger-certificated airports and 95% of strikes at general aviation airports are never reported (Dolbeer and Wright 2009), suggesting that approximately 75,000 birds and 4,000 mammals are killed by civil aircraft annually. The number killed by military aircraft is unknown.

Trains

Large mammalian herbivores are also common victims of train accidents; moose are hit in Alaska, Canada, Norway, and Sweden, causing damage to both the train and the moose population. In Alaska, almost 25% of moose deaths were caused by trains. In winters with greater snowfall, moose will often walk along the train tracks where snow is not as deep. Elk, deer, and bighorn sheep are common victims in Canada, while moose, roe deer, reindeer and muskox are struck by Norwegian trains. Railroads and trains may also kill a younger age class of ungulates than is normal for the species, as many ungulates die of old age, sickness, or depredation (Santos et al. 2017). In India, elephants (*Elephas maximus*), tigers (*Panthera tigris*), leopards, rhinoceros (*Rhinoceros unicornis*), and gaur (*Bos gaurus*) are killed every year by trains. From 1987 to 2013 in Odisha, India, 150 elephants were killed by trains. To reduce the risk of elephant–train collisions, Palei et al. (2013) recommended clearing vegetation from the railroad right-of-way, constructing overpasses and underpasses, and preventing passengers from throwing of edible waste from tourism cars because it attracts elephants to the tracks.

In North America and Europe, bears are frequently killed by trains because they are attracted to the railways by grain spills from trains. In Europe, many other animals are often hit by trains, although they are less often documented, including rabbits, badgers, red fox (*Vulpes vulpes*), stoats, hedgehogs, moles, squirrels, muskrats, rats, martens, bats, and wild boar (Kušta et al. 2011).

Birds are also frequently reported to be killed by trains. In the Netherlands, trains commonly kill swans, coots, gulls, hawks, and owls. In Spain, buzzards, black kites (*Milvus migrans*), and griffon vultures (*Gyps fulvus*) are vulnerable to trains, perhaps because they scavenge other dead animals along the rails. Owls are at a higher risk from being hit, perhaps because they become easily disoriented by train's lights (Santos et al. 2017).

Reptiles and amphibians are also hit by trains, although their carcasses are often never found. In the United States, leopard frogs (*Lithobates* spp.) and American toads (*Anaxyrus americanus*) are common victims, especially after it rains. In Poland, most amphibian carcasses found on roads are common toads, common frogs (*Bufo bufo*), and green frogs (*Pelophylax* spp.) because they are all abundant. Researchers report that eastern box turtles (*Terrapene carolina*) and small amphibians can become trapped between the rails (Santos et al. 2017).

Wildlife Control by Households

There are more than 60 million households in the 100 largest metropolitan centers in the United States; each year 42% of them (25 million) take action to try to solve a pest problem involving wildlife on their property (Conover 1997, 2002). Mice are the most common culprits, followed by squirrels, raccoons, moles, pigeons, starlings, and skunks. There are also 34 million households in smaller cities, towns, or rural areas (Conover 2002). Because wildlife populations are higher in towns and in rural areas, we assume that at least 42% of rural households (14 million) take action annually to solve a wildlife pest problem (39 million counting both metropolitan and rural areas). If we assume that each year 39 million households in metropolitan or rural areas kill two mice or other mammals, the total mortality equals 78 million mammals annually.

SUMMARY

Every year, weather patterns and specific weather events kill millions of birds and mammals; millions more die from accidents. Most of these deaths are unseen by humans and go unreported. The result is that we know little about how often animals die from weather events or accidents.

Table 10.2. Number of birds harvested by hunters in the United States and Canada during 2019–2020

Country	Species	Number Killed Annually
USA	Duck	9,720,000
	Geese	2,691,000
	Sea duck	95,000
	Brant	20,000
	Mourning dove	9,980,000
	White-winged dove	1,748,000
	Band-tailed pigeon	10,000
	Woodcock	171,000
	Snipe	93,000
	Coot	243,000
	Gallinule	20,000
	Rail	30,000
	Pheasants	20,000,000
	Total	56,230,000
Canada	Ducks	989,000
	Geese	927,000
	Total	1,916,000

Source: Migratory bird data are from Raftovich et al. (2020).

Each year, 400 million to 1.3 billion birds and 286 million mammals are killed by human endeavors (Table 10.1). Banks (1979) estimated that there are 10 billion birds in the United States during the start of the breeding season and probably 20 billion in the fall, with the addition of young that hatch each spring. This implies that less than 10% of all birds in the United States are killed accidently by human endeavors. By way of comparison, US hunters annually harvest more than 56 million birds (Table 10.2) and 15 million mammals, including 7 million ungulates (Adams and Hamilton 2011) and 6 million eastern gray squirrels (*Sciurus carolinensis*), 125,000 eastern diamond rattlesnakes (*Crotalus adamanteus*; Means 2010), and 39,000 free-ranging alligators (*Alligator mississippiensis*; Louisiana Wildlife and Fisheries 2008, Florida Fish and Wildlife Conservation Commission 2010).

In contrast to the major sources of mortality listed above, few birds (975,000) or mammals (4,000) die annually from colliding with aircraft because of a concerted effort by the US Federal Aviation Agency, US Department of Agriculture Wildlife Services, and the US Armed Forces to prevent bird–aircraft collisions at airports. Similar efforts should be undertaken to reduce wildlife mortality rates from other human endeavors. For comparative purposes, free-ranging domestic cats kill between 1.3 and 4.0 billion birds and 6.3 and 22.3 billion mammals (Loss et al. 2013) and an additional 100–350 million in Canada (Blancher 2013).

Literature Cited

Adams, K., and J. Hamilton. 2011. Management history. Pages 355–377 *in* D. G. Hewitt, editor. Biology and management of white-tailed deer. CRC Press, Boca Raton, Florida, USA.

Alestam, T. 1988. Findings of dead birds drifted ashore reveal catastrophic mortality among early spring migrants, especially rooks *Corvus frugilegus*, over southern Baltic Sea. Anser 27:181–218.

Al Shimemeri, A., and Y. Arabi. 2012. A review of large animal vehicle accidents with special attention to Arabian camels. Journal of Emergency Trauma and Acute Care 2012:21.

Al-Ghamdi, A. S., and S. A. AlGadhi. 2003. Warning signs as countermeasures to camel-vehicle collisions in Saudi Arabia. Accident Analysis and Prevention 36:749–760.

Allen, R. E., and D. R. McCullough. 1976. Deer-car accidents in southern Michigan. Journal of Wildlife Management 40:317–325.

Altwegg, R., A. Roulin, M. Kestenholz, and L. Jenni. 2006. Demographic effects of extreme winter weather in the barn owl. Oecologia 149:44–51.

Ashley, E. P., and J. T. Robinson. 1996. Road mortality of amphibians, reptiles and other wildlife on the Long Point Causeway, Lake Erie, Ontario. Canadian Field Naturalist 110:403–412.

Avery, M., P. F. Springer, and J. F. Chassel. 1976. The effect of a tall tower on nocturnal bird migration—a portable ceilometer study. Auk 93:281–291.

Banks, R. C. 1979. Human related mortality of birds in the United States. Special Scientific Report—Wildlife No. 215. US Fish and Wildlife Service, Department of the Interior, Washington, DC, USA.

Barry, T. W. 1967. The geese of the Anderson River Delta, Northwest Territories. Dissertation, University of Alberta, Edmonton, Canada.

Barry, T. W. 1968. Observations on natural mortality and native use of eider ducks along the Beaufort Sea coast. Canadian Field Naturalist 82:140–144.

Bellrose, F. C., T. G. Scott, A. S. Hawkins, and J. B. Low. 1961. Sex ratios and age ratios in North American ducks. Illinois Natural History Survey Bulletin 27:391–474.

Bender, L. C., J. C. Boren, H. Halbritter, and S. Cox. 2011. Condition, survival, and productivity of mule deer in semiarid grassland-woodland in east-central New Mexico. Human–Wildlife Interactions 5:276–286.

Bergman, G. 1982. Inter-relationships between ducks and gulls. Pages 241–247 *in* D. A. Scott, editor. Managing wetlands and their birds: a manual of wetland and waterfowl management. International Waterfowl Research Bureau, Slimbridge, UK.

Bevanger, K. 1995. Estimates and population consequences of tetraonid mortality caused by collisions with high tension power lines in Norway. Journal of Applied Ecology 32:745–753.

Bevanger. K. 1998. Biological and conservation aspects of bird mortality caused by electricity power lines: a review. Biological Conservation 86:67–76.

Bevanger, K., and H. Brøseth. 2004. Impact of power lines on bird mortality in a subalpine area. Animal Biodiversity and Conservation 27:67–77.

Bishop, C. A., and Brogan, J. M., 2013. Estimates of avian mortality attributed to vehicle collisions in Canada. Avian Conservation and Ecology 8(2):2.

Blackwell, B. F., and T. L. DeVault. 2009. A tribute to Richard A. Dolbeer. Human–Wildlife Interactions 3:296–297.

Blancher, P., 2013. Estimated number of birds killed by house cats (*Felis catus*) in Canada. Avian Conservation and Ecology 8(2).

Bull, P. C., and P. G. Dawson. 1969. Mortality and survival of birds during an unseasonable snow storm in South Canterbury, November 1967. Notornis 14:172–179.

Burles, D. W., and M. Hoefs. 1984. Winter mortality of Dall's sheep, *Ovis dalli*, in Kluane National Park, Yukon. Canadian Field Naturalist 98:479–484.

Cassirer, E. F., and A. R. E. Sinclair. 2007. Dynamics of pneumonia in a bighorn sheep metapopulation. Journal of Wildlife Management 71:1080–1088.

Cecere, J. G., F. De Pascalis, S. Imperio, D. Ménard, C. Catoni, M. Griggio, and D. Rubolini. 2020. Inter-individual differences in foraging tactics of a colonial raptor: consistency, weather effects, and fitness correlates. Movement ecology 8:1–13.

Conover, M. R. 1997. Wildlife management by metropolitan residents in the United States: practices, perceptions, costs, and values. Wildlife Society Bulletin 25:306–311.

Conover, M. R. 2002. Resolving human–wildlife conflicts: the science of wildlife damage management. Lewis Brothers, Boca Raton, Florida, USA.

Conover, M. R. 2007. Predator–prey dynamics: the role of olfaction. CRC Press, Boca Raton, Florida, USA.

Coon, R. A., J. D. Nichols, and H. F. Percival. 1981. Importance of structural stability to success in mourning dove nests. Auk 98:389–391.

Cottam, C. 1929. A shower of grebe. Condor 31:80–81.

Decker, D. J., K. M. Loconti-Lee, and N. A. Connelly. 1990. Incidence and costs of deer-related vehicular accidents in Tompkins County, New York. Human Dimensions Research Group 89-7. Cornell University, Ithaca, New York, USA.

Dick, W. J. A., and M. W. Pienkowski. 1979. Autumn and early winter weights of waders in north-west Africa. Ornis Scandinavica 10:117–123.

Dodd, C. K., Jr., K. M. Enge, and J. N. Stuart. 1989. Reptiles on highways in north-central Alabama, USA. Journal of Herpetology 23:197–200.

Dolbeer, R. A., M. J. Begier, P. R. Miller, J. R. Weller, and A. L. Anderson. 2019. Wildlife strikes to civil aircraft in the United States 1990–2018. National Wildlife Strike Database, Serial Report Number 25. US Department of Transportation, Federal Aviation Administration, Washington, DC, USA.

Dolbeer, R. A., and S. E. Wright. 2009. Safety management systems: how useful will the FAA National Wildlife Strike database be? Human–Wildlife Interactions 3:167–178.

Dunn, E. H. 1993. Bird mortality from striking residential windows in winter. Journal of Field Ornithology 64:302–309.

Edison Electric Institute. 2009. Out of sight, out of mind revisited: an updated study on the undergrounding of overhead power lines. Washington, DC, USA.

Erickson, W. P., G. D. Johnson, M. D. Strickland, K. J. Sernka, and R. E. Good. 2001. Avian collisions with wind turbines: a summary of existing studies and comparisons to other sources of avian collision mortality in the United States. National Wind Coordinating Committee Resource Document. Western EcoSystems Technology, Cheyenne, Wyoming, USA.

Evans, P. R., and M. W. Pienkowski. 1984. Population dynamics of shorebirds. Behavior of Marine Animals 5:83–123.

Faanes, C. A. 1987. Bird behavior and mortality in relation to power lines in prairie habitats. General Technical Report 7. US Fish and Wildlife Service, Department of the Interior, Washington, DC, USA.

Ferrer, M., M. de la Riva, and J. Castroviejo. 1991. Electrocution of raptors on power lines in southwestern Spain. Journal of Field Ornithology 62:181–190.

Ferronato, B. O., J. H. Roe, and A. Georges. 2014. Reptile bycatch in a pest-exclusion fence established for wildlife reintroductions. Journal for Nature Conservation 22:577–585.

Fuellhaas, U., C. Klemp, A. Kordes, H. Ottersber, M. Pirmann, A. Thiessen, C. Tshoetschel, and H. Zucchi. 1989. Untersuchungen zum Strassentod von Vögeln, Säugetieren, Amphibien und Reptilien. Beiträge Naturkunde Niedersachsens 42:129–147.

Florida Fish and Wildlife Conservation Commission. 2010. Alligator harvest summary for 2009. Gainesville, Florida, USA.

Francis, C. M., M. H. Richards, F. Cooke, and R. F. Rockwell. 1992. Long-term changes in survival rates of lesser snow geese. Ecology 73:1346–1362.

Fredrickson, L. H. 1969. Mortality of coots during severe spring weather. Wilson Bulletin 81:450–453.

Geist, V. 1971. Mountain sheep: a study in behavior and evolution. University of Chicago, Chicago, Illinois, USA.

Gelb, Y., and N. Delacrétaz. 2006. Avian window strike mortality at an urban office building. Kingbird 56:190–198.

Gerow, K., N. C. Kline, D. E. Swann, and M. Pokorny. 2010. Estimating annual vertebrate mortality on roads at Saguaro National Park, Arizona. Human–Wildlife Interactions 4:283–292.

Glasrud, R. D. 1976. Canada geese killed during lightning storm. Canadian Field Naturalist 90:503.

Green, R. E., M. Harley, M. Spalding, and C. Zöckler. 2001. Impacts of climate change on wildlife. United Nations Environment Programme, World Conservation Monitoring Centre and the Royal Society for the Protection of Birds, Sandy, UK.

Groot Bruinderink, G. W. T. A., and E. Hazelbrook. 1996. Ungulate traffic collisions in Europe. Conservation Biology 10:1059–1067.

Hansen, H. A. 1961. Loss of waterfowl production to tide floods. Journal of Wildlife Management 25:242–248.

Harness, R. E., and K. R. Wilson. 2001. Electric-utility structures associated with raptor electrocutions in rural areas. Wildlife Society Bulletin 29:612–623.

Harrington, J. L., and M. R. Conover. 2006. Characteristics of ungulate behavior and mortality associated with wire fences. Wildlife Society Bulletin 34:1295–1305.

Hels, T., and E. Buchwald. 2001. The effect of road kills on amphibian populations. Biological Conservation 99:331–340.

Hightower, J. N., J. D. Carlisle, and A. D. Chalfoun. 2018. Nest mortality of sagebrush songbirds due to a severe hailstorm. Wilson Journal of Ornithology 130:561–567.

Hochbaum, H. A. 1955. Travels and traditions of waterfowl. University Minnesota Press, Minneapolis, Minnesota, USA.

Hodson, N. L., and D. W. Snow. 1965. The road deaths enquiry, 1960–1961. Bird Study 12:90–99.

Intergovernmental Panel on Climate Change. 2014. Climate change 2014: synthesis report. R. K. Pachauri and L. A. Meyer, editors. Geneva, Switzerland.

Jacobson, A. R., A. Provenzale, A. von Hardenberg, B. Bassano, and M. Festa-Bianchet. 2004. Climate forcing and density dependence in a mountain ungulate population. Ecology 85:1598–1610.

James, P. 1956. Destruction of warblers on Parade Island, Texas, in May 1951. Wilson Bulletin 68:224–227.

Janss, G. F. E., and M. Ferrer. 1998. Rate of bird collision with power lines: effects of conductor-marking and static wire-marking. Journal of Field Ornithology 69:8–17.

Janss, G. F. E., and M. Ferrer. 2000. Common crane and great bustard collision with power lines: collision rate and risk exposure. Wildlife Society Bulletin 28:675–680.

Johnson, D. H., J. D. Nichols, and M. D. Schwartz. 1992. Population dynamics of breeding waterfowl. Pages 446–485 in B. D. J. Batt, A. D. Afton, M. G. Anderson, et al., editors. Ecology and management of breeding waterfowl. University of Minnesota Press, Minneapolis, Minnesota, USA.

Jokinen, M. E., P. F. Jones, and D. Dorge. 2008. Evaluating survival and demography of a bighorn sheep (Ovis canadensis) population. Biennial Symposium of the Northern Wild Sheep and Goat Council 16:138–159.

Jones, P. F. 2014. Scarred for life: the other side of the fence debate. Human–Wildlife Interactions 8:150–154.

Kamler, J. F., R. M. Lee, J. C. deVos Jr., W. B. Ballard, and H. A. Whitlaw. 2003. Mortalities from climbing accidents of translocated bighorn sheep in Arizona. Southwestern Naturalist 48:145–147.

Kennedy, R. J. 1970. Direct effects of rain on birds: a review. British Birds 63:401–414.

Kerlinger, P. 1989. Flight strategies of migrating hawks. Chicago University Press, Chicago, Illinois, USA.

King, K. A. 1976. Bird mortality, Galveston Island, Texas. Southwest Naturalist 21:414.

Klem, D., Jr. 1989. Bird–window collisions. Wilson Bulletin 101:606–620.

Klem, D., Jr. 1990. Collisions between birds and windows: mortality and prevention. Journal of Field Ornithology 61:120–128.

Klem, D., Jr. 2006. Glass: a deadly conservation issue for birds. Bird Observer 34:73–81.

Klem, D., Jr., and P. G. Saenger. 2013. Evaluating the effectiveness of select visual signals to prevent bird-window collisions. Wilson Journal of Ornithology 125:406–411.

Klem, D., Jr., D. C. Keck, K. L. Marty, A. J. Miller Ball, E. E. Niciu, and C. T. Platt. 2004. Effects of window angling, feeder placement, and scavengers on avian mortality at plate glass. Wilson Bulletin 116:69–73.

Koops, F. B. J. 1987. Collision victims of high-tension lines in the Netherlands and effects of marking. Report 01282-mob 86-3048. KEMA, Arnheim, Netherlands.

Krause, H. 1959. Northern Great Plains region. Audubon Field Notes 13:380–381.

Kušta, T., M. Ježek, and Z. Keken. 2011. Mortality of large mammals on railway tracks. Scientia Agriculturae Bohemica 42:12–18.

LaRoe, E. T., G. S. Farris, C. E. Puckett, P. D. Doran, and M. J. Mac. 1995. Our living resources: a report to the nation on the distribution, abundance and health of U.S. plants, animals, and ecosystems. National Biological Service, US Department of the Interior, Washington, DC, USA.

Lehman, R. N. 2001. Electrocutions on power lines: current issues and outlook. Wildlife Society Bulletin 29:804–813.

Linnell, J. D., B. Cretois, E. B. Nilsen, C. M. Rolandsen, E. J. Solberg, et al. 2020. The challenges and opportunities of coexisting with wild ungulates in the human-dominated landscapes of Europe's Anthropocene. Biological Conservation 244:108500.

Longcore, T., C. Rich, P. Mineau, B. MacDonald, D. G. Bert, et al. 2012. An estimate of avian mortality at communication towers in the United States and Canada. PLoS One 7(4):e34025.

Longcore, T., C. Rich, P. Mineau, B. MacDonald, D. G. Bert, et al. 2013. Avian mortality at communication tower in the United States and Canada: which species, how many, and where? Biological Conservation 158:410–419.

Loonen, M. J. J. E., L. W. Bruinzeel, J. M. Black, and R. H. Drent. 1999. The benefits of large broods in barnacle geese: a study using natural and experimental manipulations. Journal of Animal Ecology 68:753–768.

Loss, S. R., T. Will, S. S. Loss, and P. P. Marra. 2014. Bird-building collisions in the United States: estimates of annual mortality and species vulnerability. Condor 116:8–23.

Loss, S. R., T. Will, and P. P. Marra. 2013. Estimates of bird collision mortality at wind facilities in the contiguous United States. Biological Conservation 168:201–209.

Loss, S. R., T. Will, and P. P. Marra. 2015. Direct mortality of birds from anthropogenic causes. Annual Review of Ecology, Evolution and Systematics 46:99–120.

Louisiana Wildlife and Fisheries. 2008. Louisiana's alligator management program: 2007–2008 annual report. Office of Wildlife, Coastal and Nongame Resources Division, Louisiana Department of Wildlife and Fisheries, Baton Rouge, Louisiana, USA.

Manville, A. M., II. 2005. Bird strike and electrocutions at power lines, communication towers, and wind turbines: state of the art and state of the science-next steps toward mitigation. General Technical Report PSW-GTR-191:1051–1064. US Department of Agriculture Forest Service, Washington, DC, USA.

Marcoux, A., and S. J. Riley. 2010. Driver knowledge, beliefs, and attitudes about deer-vehicle collisions in southern Michigan. Human–Wildlife Interactions 4:47–55.

McLoughlin, P. D., E. Dzus, B. Wynes, and S. Boutin. 2003. Declines in populations of woodland caribou. Journal of Wildlife Management 67:755–761.

McNulty, S. 2010. Florida sea turtle cold stunning. Wildlife Data Integration Network, Madison, Wisconsin, USA.

Means, D. B. 2010. Time to end rattlesnake roundups. Wildlife Professional 4:64–67.

Mendenhall, V. M., and H. Milne. 1985. Factors affecting duckling survival of eiders Somateria mollissima in northeast Scotland. Ibis 127:148–158.

Menu, S., G. Gauthier, and A. Reed. 2005. Survival of young greater snow geese (Chen caerulescens atlantica) during fall migration. Auk 122:479–496.

Meretsky, V. J., N. F. R. Snyder, S. R. Beissinger, D. A. Clendenen, and J. W. Wiley. 2000. Demography of the California condor: implications for reestablishment. Conservation Biology 14:957–967.

Moreau, R. E. 1928. Some further notes from the Egyptian deserts. Ibis 4:453–475.

Morelli, T. L., A. B. Smith, and C. R. Kastely. 2012. Anthropogenic refugia ameliorate the severe ameliorate the severe climate related decline of a mountain mammal along its trailing edge. Proceedings of the Royal Society B 279(1745). doi:10.1098/rspb.2012.1301.

Moritz, C., J. L. Patton, and C. J. Conroy. 2008. Impact of a century of climate change on small mammal communities in Yosemite National Park, USA. Science 322:261–264.

Morse, D. H. 1980. Population limitation: breeding or wintering grounds? Pages 505–516 in A. Keast and E. S. Morton, editors. Migrant birds in the Neotropics: ecology, behavior and distribution. Smithsonian Institution Press, Washington, DC, USA.

Nakamura, M., and A. Ramadhani. 2014. Hidden risk of arboreality? an arboreal death of an infant chimpanzee at Mahale. Pan African News 21:17–19.

Nankinov, D. N., and N. M. Todorov. 1983. Bird casualties on highways. Soviet Journal of Ecology 14:388

Narwade, S., M. C. Gaikwad, K. Fartade, S. Pawar, M. Sawdeker, and P. Ingale. 2014. Mass mortality of wildlife due to hailstorms in Maharastra, India. Bird Populations 13:28–35.

Newton, I. 2007. Weather-related mass-mortality events in migrants. Ibis 149:453–467.

O'Connell, T. 2001. Avian window strike mortality at a suburban office park. Raven 72:141–149.

Owen, M., and J. M. Black. 1989. Factors affecting survival of barnacle geese on migration from the breeding grounds. Journal of Animal Ecology 58:603–617.

Ozgul, A., D. Z. Childs, and M. K. Oli. 2010. Coupled dynamics of body mass and population growth in response to environmental change. Nature 466:482–485.

Palei, N. C., B. P. Rath, and C. S. Kar. 2013. Death of elephants due to railway accidents in Odisha, India. Gajah 38:39–41.

Palmer, W. E., S. R. Priest, R. S. Seiss, P. S. Phalen, and G. A. Hurst. 1993. Reproductive effort and success in a declining wild turkey population. Proceedings of the Annual Conference of the Southeastern Association of Fish and Wildlife Agencies 47:138–147.

Parker, I. D., R. R. Lopez, N. J. Silvy, D. S. Davis, and C. B. Owen. 2011. Long-term effectiveness of US 1 crossing project in reducing Florida key deer mortality. Wildlife Society Bulletin 35:296–302.

Parmesan, C. 2006. Ecological and evolutionary responses to recent climate change. Annual Review of Ecology, Evolution, and Systematics 37:637–669.

Parmesan, C., and G. Yohe. 2003. A globally coherent fingerprint of climate change impacts across natural systems. Nature 421:37–42.

Pelletier, A. 2006. Injuries from motor-vehicle collisions with moose—Maine, 2000–2004. Morbidity and Mortality Weekly Report 55:1272–1274.

Pipoly, I., V. Bókony, G. Seress, K. Szabó, and A. Liker. 2013. Effects of extreme weather on reproductive success in a temperate-breeding songbird. PLoS One 8(11).

Plaschke, S., M. Bulla, M. Cruz-López, S. G. del Ángel, and C. Küpper. 2019. Nest initiation and flooding in response to season and semilunar spring tides in a ground-nesting shorebird. Frontiers in Zoology 16(1):15.

Raftovich, R. V., K. A. Wilkins, K. D. Richkus, S. S. Williams, and H. L. Spriggs. 2020. Migratory bird hunting activity and harvest during the 2018–19 and 2019–20 hunting seasons. US Fish and Wildlife Service, Laurel, Maryland, USA.

Ramirez, P., Jr. 1999. Fatal attraction: oil field waste pits. Endangered Species Bulletin 24:10–11.

Rattey, T. E., and N. E. Turner. 1991. Vehicle–moose accidents in Newfoundland. Journal of Bone and Joint Surgery 73:1487–1491.

Reeves, H. M., F. G. Cooch, and R. E. Munro. 1976. Monitoring Arctic habitat and goose production by satellite imagery. Journal of Wildlife Management 40:532–541.

Reinecke, K. J., C. W. Shaiffer, and D. Delnicki. 1987. Winter survival of female mallards in the lower Mississippi Valley. Transactions of the North American Wildlife and Natural Resources Conference 52:258–263.

Roberts, A. J., M. R. Conover, and J. L. Fusaro. 2014. Factors influencing mortality of eared grebes (Podiceps nigricollis) during a mass downing. Wilson Journal of Ornithology 126:584–591.

Roberts, S. D., J. M. Coffey, and W. F. Porter. 1995. Survival and reproduction of female wild turkeys in New York. Journal of Wildlife Management 59:437–447.

Roberts, S. D., and W. F. Porter. 1998a. Influence of temperature and precipitation on survival of wild turkey poults. Journal of Wildlife Management 62:1499–1505.

Roberts, S. D., and W. F. Porter. 1998b. Relationship between weather and survival of wild turkey nests. Journal of Wildlife Management 62:1492–1498.

Robinson, S. G., D. A. Haukos, R. T. Plumb, C. A. Hagen, J. C. Pitman, J. M. Lautenbach, D. S. Sullins, J. D. Kraft, and J. D. Lautenbach. 2016. Lesser prairie-chicken fence collision risk across its northern distribution. Journal of Wildlife Management 80:906–915.

Root, T. L., J. T. Price, K. R. Hall, S. H. Schneider, C. Rosenzweig, and J. A. Pounds. 2005. The impact of climatic change on wild animals and plants: a meta-analysis. General Technical Report PSW-GTR-191:1115–1118. US Department of Agriculture Forest Service, Washington, DC, USA.

Root, T. L., and S. H. Schneider. 2002. Climate change: overview and implications for wildlife. Pages 1–56 in S. H. Schneider and T. L. Root, editors. Wildlife responses to climate change: North American case studies. Island Press, Washington DC, USA.

Roseberry, J. L. 1962. Avian mortality in southern Illinois resulting from severe weather conditions. Ecology 43:739–740.

Rosen, P. C., and C. H. Lowe. 1994. Highway mortality of snakes in the Sonoran Desert of southern Arizona. Biological Conservation 68:143–148.

Roth, R. R. 1976. Effects of a severe thunderstorm on airborne ducks. Wilson Bulletin 88:654–656.

Rowe, R. J. 2009. Environmental and geometric drivers of small mammal diversity along elevational gradients in Utah. Ecography 32:411–422.

Rowe, R. J., R. C. Terry, and E. A. Rickart. 2011. Environmental change and declining resource availability for small-mammal communities in the Great Basin. Ecology 92:1366–1375.

Rubolini, D., M. Gustin, G. Bogliani, and R. Garavaglia. 2005. Birds and power lines in Italy: an assessment. Bird Conservation International 15:131–145.

Santos, S. M., F. Carvalho, and A. Mira 2017. Current knowledge on wildlife ecology in railways. Pages 11–22 in L. Borda-de-Água, R. Barrientos, P. Beja, and H. M. Pereira, editors. Railway ecology. Springer Open, Cham, Switzerland.

Sargeant, A. B., and D. G. Raveling. 1992. Mortality during the breeding season. Pages 396–422 in B. D. J. Batt, A. D. Afton, M. G. Anderson, et al., editors. Ecology and management of breeding waterfowl. University of Minnesota Press, Minneapolis, USA.

Schmutz, J. A. 1993. Survival and pre-fledging body mass in juvenile emperor geese. Condor 95:222–225.

Schüz, E., P. Berthold, E. Gwinner, and H. Oelke. 1971. Grundirss der Voelzugskunde. Paul Parey, Berlin, Germany.

Segelson, C. 2010. Record cold leads to record numbers of manatee deaths. Florida Fish and Wildlife Conservation Commission, Tallahassee, Florida, USA.

Slattery, S. M. 2000. Factors affecting first-year survival in Ross's Goose. Dissertation, University of Saskatchewan, Saskatoon, Canada.

Smallwood, K. S. 2013. Comparing bird and bat fatality-rate estimates among North American wind-energy projects. Wildlife Society Bulletin 37:19–33.

Smallwood, K. S., and C. Thelander. 2008. Bird mortality in the Altamont Pass Wind Resource Area, California. Journal of Wildlife Management 72:215–223.

Smith, A. G., and H. R. Webster. 1955. Effects of hail storms on waterfowl population in Alberta, Canada—1953. Journal of Wildlife Management 19:368–374.

Smith, L. L., and C. K. Dodd Jr. 2003. Wildlife mortality on U.S. Highway 441 across Payne's Prairie, Alachua County, Florida. Florida Scientist 66:128–140.

Smith, M. A. 1964. Cohoe-Alaska. Audubon Field Notes 18:478–479.

Spendelow, P. 1985. Starvation of a flock of chimney swifts on a very small Caribbean island. Auk 102:387–388.

State Farm Insurance Company. 2018. Deer crashes down: annual report from State Farm shows reduction in deer-related crashes. Bloomberg, Indiana, USA. https://newsroom.statefarm.com/2018-deer-crashes-down/.

Stoudt, J. H. 1982. Habitat use and productivity of canvasbacks in southwestern Manitoba, 1961–72. Special Scientific Report—Wildlife 248. US Fish and Wildlife Service, Department of the Interior, Washington, DC, USA.

Telemeco, R. S., D. A. Warner, M. K. Reida, and F. J. Janzen. 2012. Extreme developmental temperatures result in morphological abnormalities in painted turtles (Chrysemys picta): a climate change perspective. Integrative Zoology 8:197–208.

Trail, P. 2006. Avian mortality at oil pits in the United States: a review of the problem and efforts for its solution. Environmental Management 38:532–544.

US Federal Highway Administration. 2008. Status of the nation's highways, bridges, and transit: conditions and performance. Report to Congress. US Department of Transportation, Washington, DC, USA.

US Fish and Wildlife Service. 2003. Draft recovery plan for the Sierra Nevada bighorn sheep. Portland, Oregon, USA.

Van de Ven, T. M. F. N., A. E. McKechnie, and S. J. Cunningham. 2019. The costs of keeping cool: behavioural trade-offs between foraging and thermoregulation are associated with significant mass losses in an arid-zone bird. Oecologia 191:205–215.

Vepsäläinen, K. 1968. The effect of the cold spring 1966 upon the lapwing *Vanellus vanellus* in Finland. Ornis Fennica 45:33–47.

Vidal-Vallés, D., E. Pérez-Collazos, and A. Rodríguez. 2018. Bird roadkill occurrences in Aragon, Spain. Animal Biodiversity and Conservation 41:379–388.

Voigt, C. C., L. S. Lehnert, G. Petersons, F. Adorf, and L. Bach. 2015. Wildlife and renewable energy: German politics cross migratory bats. European Journal of Wildlife Research 61:213–219.

Walston Jr, L. J., K. E. Rollins, K. E. LaGory, K. P. Smith, and S. A. Meyers. 2016. A preliminary assessment of avian mortality at utility-scale solar energy facilities in the United States. Renewable Energy 92:405–414.

Webster, F. S. 1974. The spring migration, April 1–May 31, 1974, South Texas region. American Birds 28:822–825.

Weinkam, T. J., G. A. Janos, and D. R. Brown. 2017. Habitat use and foraging behavior of Easter bluebirds (*Sialia sialis*) in relation to winter weather. Northeastern Naturalist 24:sp7.

Wellicome, T. I., R. J. Risher, R. G. Poulin, L. D. Todd, E. M. Bayne, D. T. T. Flockhart, J. K. Schmutz, K. De Smet, and P. C. James. 2014. Apparent survival of adult burrowing owls that breed in Canada is influenced by weather during migration and on their wintering grounds. Condor 116:446–458.

Wiedenfeld, D. A., and M. G., Wiedenfeld. 1995. Large kill of Neotropical migrants by tornado and storm in Louisiana, April 1993. Journal of Field Ornithology 66:70–80.

Wilkinson, P. F., and C. C. Shank. 1976. Rutting-fight mortality among musk oxen on Banks Island, Northwest Territories, Canada. Animal Behaviour 24:756–758.

Williams, G. C. 1950. Weather and spring migration. Auk 67:52–65.

Wilson, R. R., and P. R. Krausman. 2008. Possibility of heat related mortality in desert ungulates. Journal of the Arizona-Nevada Academy of Science 40:12–15.

Wolfe, D. H., M. A. Patten, E. Shochat, C. L. Pruett, and S. K. Sherrod. 2007. Causes and patterns of mortality in lesser prairie-chickens *Tympanuchus pallidicinctus* and implications for management. Wildlife Biology 13:95–104.

NUTRITIONAL ECOLOGY

KATHERINE L. PARKER

INTRODUCTION

Nutritional ecology is the science of relating an animal to its environment through nutritional interactions. This field of study is important to wildlife ecologists because nutrition affects animal survival and reproduction. The life histories of animals (e.g., their activities, feeding patterns, movements, and reproductive strategies) are shaped by nutritional requirements and constrained by the foods available to meet them. The goal of this chapter is to provide readers with an overview of basic concepts in nutritional ecology, including examples that help explain why free-ranging animals use their environments as they do.

ENERGY REQUIREMENTS

Energy is the universal currency for survival. It is needed to support basic body functions and activities, to stay warm, to reproduce, and to grow and produce tissues. Animals transform chemical energy from food, with the addition of oxygen, into products that help meet energy demands and into heat that helps maintain an acceptable body temperature. Energy costs (i.e., the energy needed per unit time, or metabolic rate) have been measured indirectly as heat produced (kilojoules, kJ, or kilocalories, kcal; 1 kcal = 4.184 kJ) or directly as oxygen consumed. These measurements for specific behaviors and ambient conditions are made on animals in sealed metabolic chambers or tanks, or on animals wearing respiratory masks. The animals may be on treadmills for locomotion studies or in wind tunnels for flight research.

Basal Metabolism

Small species and young individuals have relatively high energy demands (per unit mass) compared to older or larger ones, in part because the energy needed to support basal functions such as respiration, blood circulation, and muscle tone is higher in small animals. Basal metabolic rate (BMR), as the energy used when animals are lying, calm, fasted, and not thermally stressed, has been measured for numerous wildlife species in captivity. From these measurements, the interspecific

relationship commonly used to approximate basal metabolic rate from body mass in placental mammals is BMR = 293 × kg$^{0.75}$ kJ/day (Kleiber 1947). Birds, especially small passerines, tend to have higher basal metabolism than this relationship predicts; most marsupials have slightly lower rates (Barboza et al. 2009:220). Deviations from the interspecific relationship can be adaptive. Compared to arctic canids, for example, desert canids have lower metabolism, which would reduce water loss and risk of dehydration (Careau et al. 2007). The high metabolic rates in arctic ungulates enable rapid growth in environments with a pulsed forage supply in a short growing season (Lawler and White 2003).

Activity

Energy costs for daily existence incorporate numerous activities throughout the day and therefore are higher than BMR alone. Energy costs of standing by game birds and mammals are approximately 17% to 22% higher than costs for sitting or lying. Costs of feeding by birds and ungulates range from 10% to 60% above perching or standing, depending on the type, composition, and availability of food (Fancy and White 1985, Robbins 1993:143). For all species, energy costs of travel increase at higher speeds. This increase is generally linear for species that use terrestrial locomotion (Fig. 11.1A). More energy is required to move uphill, and less is needed going downslope. Data obtained from animals walking on treadmills have helped quantify energetic costs as diverse as waddling by penguins (*Aptenodytes forsteri*; Griffin and Kram 2000) and hunting strategies used by cougars (*Puma concolor*; Williams et al. 2014). In species such as kangaroos (*Macropus* spp.), which change from bipedal locomotion to hopping at higher speeds, energy costs actually decline as energy is recovered on the hop (Dawson and Taylor 1973). For swimming species, energy costs increase exponentially at higher speeds because of increasing drag in the water, which results in greater resistance than when in air (Fig. 11.1B). Leaping out of the water or wave-riding by dolphins (Delphinidae) may allow them to expend less energy because of decreased drag and a

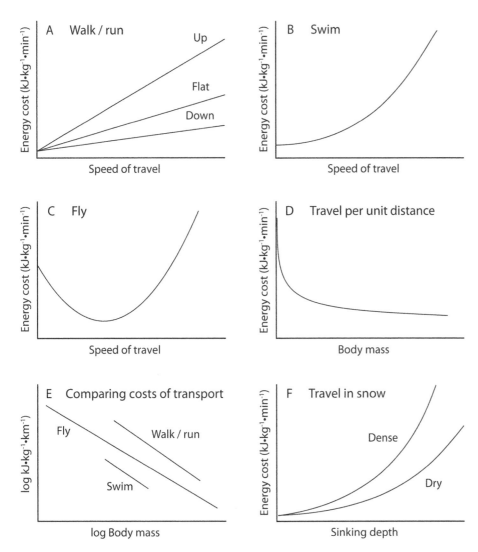

Figure 11.1. Energy costs of travel in relation to speed for (*A*) walking and running (uphill, horizontal, downhill), (*B*) swimming, and (*C*) flying; in relation to (*D*) body mass when moving a kilometer for (*E*) three forms of locomotion; and in relation to (*F*) sinking depths in dense and dry snow.

lower frequency of swimming strokes (Williams et al. 1992). For birds and bats, energy costs of flight are represented as a U-shaped function relative to speed (Fig. 11.1*C*). When first taking off, costs are elevated because an individual must support its weight against gravity and overcome aerodynamic drag; energy costs then increase steadily with increasing wing-flapping and flying speed. During energy-saving migrations with minimal stopovers or foraging opportunities, the optimal speed of flight for long-distance migrants is assumed to be the speed at which energy costs are lowest (Hedenström and Alerstam 1995). For each of these forms of locomotion—running, swimming, and flying—relative energy costs of traveling a known distance are highest for small species and individuals (Fig. 11.1*D*). It takes more energy per kilogram for a neonate to travel a kilometer than it does for its mother. When comparing the three forms of locomotion, net energy costs of moving a kilometer are least to swim and most to run or walk (Fig. 11.1*E*).

Substrate may exacerbate the energy needed for terrestrial movement. Travel by caribou (*Rangifer tarandus*) in bogs can cost 30% to 40% more energy than when moving on solid footing (Fancy and White 1985). In snow, energy costs of travel increase exponentially as the sinking depth of the animal increases (Fig. 11.1*F*). When snow depths reach front knee height (~60% of leg length), energy costs of travel may double for cervids and bovids (Parker et al. 1984; Dailey and Hobbs 1989). When snow and cooling temperatures trigger autumn migrations of mule deer (*Odocoileus hemionus hemionus*) to areas with less snow, higher energy costs are avoided (Monteith et al. 2011). If animals sink to brisket height (leg length), energy costs can be three to seven times higher than when walking on bare ground, depending on snow density (Robbins 1993:135). Juveniles with shorter leg lengths are compensated to some extent by having a lower foot loading (i.e., body mass per unit foot area), therefore reducing their sinking depths in snow (Fancy and White 1985). Nonetheless, snowy winters compounded by lower food availability take their greatest toll on young animals that have relatively high metabolism. Ungulates are disadvantaged compared to predators such as wolves (*Canis lupus*) because of their generally heavier foot loading (Telfer and Kelsall 1984) and consequently greater sinking depths. Delayed snowmelt in spring further increases population mortality rates if energy costs remain high, body stores have been depleted, and spring food supplies are delayed.

Thermoregulation

Very cold, very hot, and wet, rainy environments increase energy demands for wildlife. These supplemental energy costs for thermoregulation are typically small compared to activity costs but nonetheless contribute to energetic drains. Most birds and mammals exhibit a characteristic thermal response in which energy costs remain relatively constant over a range of ambient temperatures (i.e., the thermoneutral zone; Fig. 11.2). In the cold, animals piloerect their feathers or hair to increase thermal depth and insulation value, in addition to behavioral and postural adjustments that minimize heat loss. Below a lower critical temperature, animals shiver to increase metabolic rate and heat production physiologically, and to maintain an acceptable body temperature. Lower critical temperatures, below which animals are thermally stressed, are lower in winter when animals have more thermal insulation from winter pelage and fat depots than in summer. Among northern ungulates, lower critical temperatures in winter range from −10°C for bighorn sheep (*Ovis canadensis*) and Columbian black-tailed deer (*Odocoileus hemionus columbianus*) to below −40°C for moose (*Alces alces*) and reindeer (*Rangifer tarandus*; Mautz et al. 1985, Parker and Robbins 1985, Renecker and Hudson 1986). Fed animals can withstand lower temperatures than animals without food because of the heat produced during food digestion, which can be beneficial to survival during cold winters. At temperatures above an upper critical temperature, animals usually pant or sweat to rid the body of excess heat. A potential advantage of sweating over panting in non-arid environments where water is not limited is that animals can continue to forage efficiently at high temperatures.

Some wildlife species avoid the increases in energy demands needed to maintain a relatively constant body temperature at cold ambient temperatures by going into torpor. They allow

body temperature to decline to a regulated lower temperature and are capable of arousing back to normal body temperature. Bats (e.g., big brown bat, *Eptesicus fuscus*) and hummingbirds (e.g., *Selasphorus* spp.) often use torpor at night; chickadees (*Poecile* spp.) use torpor during extreme weather conditions, and bears (*Ursus* spp.) and ground squirrels (*Spermophilus* spp.) employ extended periods of torpor throughout winter (e.g., hibernation). At high ambient temperatures, some desert ungulates such as gazelles (*Gazella* spp.) allow body temperature to rise above normal to minimize the added energy costs (and inherent water loss) of trying to maintain a stable body temperature.

The ambient thermal environment experienced by free-ranging animals includes more than air temperature. Rather, animals are exposed to the combined effects of air temperature, wind, and solar radiation. High winds decrease the effective temperature that animals experience; sunshine increases it. Mule deer, which shiver violently at −30°C in early morning in winter, cease shivering with incoming direct sunshine (Parker and Robbins 1984). A useful index to quantify the thermal environment that animals experience is standard operative temperature (Bakken et al. 1985), which is similar to a wind chill index but includes the added influence of solar radiation (it does not accommodate rainfall). This index can be calculated for any combination of air temperature, wind, and radiation to compare thermal values of different habitats. Species- and habitat-specific models then can be used to predict when animals are metabolically stressed outside the thermoneutral zone and therefore when populations could be adversely affected if food resources are limited (Parker and Gillingham 1990, Beaver et al. 1996). For resource managers, modeling exercises are useful to determine the frequency of elevated thermal costs and whether suitable cover is available. Supplementary energy costs of thermoregulation are even higher for animals under rainy, wet conditions, depending on temperature and season (Parker 1988), particularly for neonates and nestlings.

Thermal cover is considered to be a manageable entity for furbearers and ungulates. Coarse, woody debris on the forest floor and cavities in snags provide added insulation for American martens (*Martes americana*), which have relatively high lower critical temperatures (Buskirk et al. 1989). In the deeryards of eastern North America, forest cover is managed to ameliorate the effects of cold temperatures and wind, and to reduce snow depths. Vegetative thermal cover also provides shelter at high operative temperatures, such as for moose, which are heat-stressed above freezing temperatures in winter when winter pelage and body fat provide much greater insulation (Renecker and Hudson 1986). Managing for thermal cover, however, is difficult because animals are not always found in the thermoneutral zone. Animals in good body condition may remain in thermally stressful environments if the energy gain from the food exceeds the short-term thermal costs (Long et al. 2014). Attempts to validate the influence of forest thermal cover on the body condition of elk (*Cervus canadensis*) were confounded by variability in small, infrequent

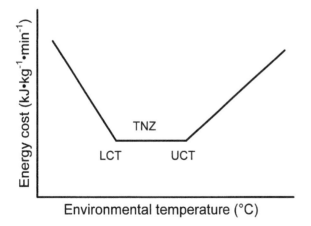

Figure 11.2. Common energetic thermoregulatory response of birds and mammals to environmental temperature. LCT, lower critical temperature; TNZ, thermoneutral zone; UCT, upper critical temperature.

weather effects, but access to incoming solar radiation in open areas decreased mortality in cold winters (Cook et al. 1998).

Reproduction and Production

Reproduction increases energy requirements. For birds and mammals, energy costs for reproductive females are relatively higher (per kilogram) for small species than larger ones because the eggs or offspring usually are larger in proportion to body mass. In birds, energy costs depend on the number and composition of the eggs. Eggs with large yolks are more costly to produce. Precocial species such as waterfowl have large yolks to support the development of functionally advanced chicks capable of searching for their own food upon hatching (Vleck and Bucher 1998). Depending on the number of eggs laid, energy requirements for egg production by precocial bird species could be more than five times higher than those of altricial species (Robbins 1993:192). During incubation and chick rearing, energy costs vary with the development of the chicks and with uniparental versus biparental care (Tulp et al. 2009). The amount of time spent foraging versus incubating can affect males and females energetically. One or both sexes also may invest additional energy into the production of plumage color that signals mate quality.

In mammals, energy costs for reproductive females increase exponentially during fetal development. More than 90% of the energy requirements of gestation occurs in the last trimester. Small mammals have shorter gestation periods but relatively larger young compared to large mammals. Energy costs during lactation are higher than for gestation because they support larger mass gains by offspring than those of the fetus. Lactation is considered to be the most energetically demanding period of life for adult reproducing females. Daily energy expenditures can rise to four to seven times BMR depending on milk composition and stage of lactation (Robbins 1993:203). Because of the relatively long period of gestation overlapping the environmental constraints of winter, and the subsequent period of lactation, highest total daily energy costs for reproducing female ungulates usually occur from late winter to midsummer. In contrast, highest daily energy costs for male ungulates typically occur when they are most active during the breeding period in autumn, often at the expense of foraging. Males also may expend energy and nutrients in secondary characteristics to attract mates (e.g., antlers in cervids).

Free-ranging animals require energy for growth and molt, although these two biological processes are largely protein demands. Requirements for energy and nutrients by juveniles are highest when the proportional changes in body mass are highest—when cells are rapidly dividing or expanding (Barboza et al. 2009:249). The energy content of mass gain varies depending on the deposition of fat and protein. Juveniles that are still growing deposit less fat and a higher proportion of protein toward body mass than do adults. The highest energy content of gain (associated with fat) occurs in marine mammals and species that fatten before hibernation. Mammals and birds also may need supplemental energy for seasonal pelage and plumage replacement, and for corresponding thermoregulatory costs that increase during that time.

Energy Budgets of Free-Ranging Animals

Wildlife biologists can multiply the time spent by free-ranging wildlife in different behaviors (i.e., the activity budget) by published activity-specific energy costs to calculate an energy budget. These types of calculations provide insights into the effects of disturbance and changing environments. For example, biologists can quantify the effect of winter recreationists on ungulates by knowing the distances that animals flee and the energy costs of travel in snow. Energy costs in response to snow machine travel are often lower than in response to less predictable cross-country skiers (Reimers et al. 2003). Plowed roadways, which decrease energy costs in winter, may have facilitated the northward expansion of coyotes (Canis latrans) in eastern North America (Crête and Larivière 2003). The energetic effects associated with aircraft or fishermen on behavior vary among waterfowl species (Conomy et al. 1998).

When researchers cannot monitor daily activities but are able to handle animals for sampling, doubly labeled (isotopic) water can be used to determine energy expenditures (Nagy et al. 1999). This method involves injecting an animal with two labels (i.e., hydrogen—^{2}H or radioactive ^{3}H—and oxygen—^{18}O) and measuring the rate of loss of the two isotopes in one or more subsequent water samples (usually blood or urine). Because hydrogen is lost from the body as water (H_2O, via excretion and evaporation), and oxygen (O_2) is lost as both H_2O and carbon dioxide (CO_2), the difference in the rates of these losses is a measure of CO_2 production. The measure of CO_2 produced is used to derive O_2 consumed, a measure of energy cost. Doubly labeled water can accurately quantify energy costs of free-ranging animals, but there are cautions when using the technique. In addition to knowing the time needed for the labels to equilibrate in the animal's body water before sampling, small sampling errors can easily influence the slopes that describe rates of isotopic decline and therefore the calculations of energy costs. In spring and summer, if hydrogen is incorporated into milk fat production or body fat depots (and therefore not lost from the body water pool only as water), the rate of isotopic decline can be affected enough to cause impossible estimates of energy expenditure (Parker et al. 1999). Heart-rate telemetry also has been used to index daily energy costs because heart rates typically increase with energy expenditures (Butler et al. 2004). This relationship varies widely among individuals and by season, however, making extrapolation to the population difficult.

When activity budgets and measures of doubly labeled water are not obtainable, average daily energy expenditures are approximated to be between 2 and 3 × BMR for mature birds and mammals without reproductive requirements or highly elevated thermal and locomotion costs (Robbins 1993:159).

PROTEIN REQUIREMENTS

Protein is important to wildlife because it is a major constituent of the animal body. Proteins are in pelage and plumage, hooves and claws, the antibodies for disease resistance, blood clotting factors, and all hormones. Proteins enable movement of muscle fibers and biochemical reactions with enzymes. Composed of sequences of amino acids, all proteins contain nitrogen.

For an animal to maintain body protein during routine, nonstressed activity, protein requirements can be approximated from protein losses. Losses caused by the normal breakdown of body proteins are excreted in the urine (i.e., endogenous urinary nitrogen, or EUN) and are related to body mass and metabolism. Losses also occur from the gastrointestinal tract (as metabolic fecal nitrogen, or MFN) because of the loss of digestive enzymes, bacterial byproducts, and sloughing of cells along the tract. Protein lost as MFN is of nonfood origin, but it is related to food quality because highly fibrous diets place more wear and tear on the digestive lining. Protein lost as MFN by browsers is therefore generally higher than by grazers (Robbins 1993:177). To estimate protein requirements of free-ranging wildlife, biologists use species-specific equations for EUN and MFN (with some adjustment for dietary nitrogen that is not protein) determined from animals in captivity based on body mass and daily intakes (Spalinger et al. 2010). In birds, nitrogen losses are excreted together.

For all species, relative protein requirements (per unit mass) are typically highest during growth. Game bird chicks require 30% dietary protein compared to the 12% needed for maintenance of adults (Robbins 1993:175). Diets that are more than 75% insects help meet this demand (Johnson and Boyce 1990). Reproduction also increases protein requirements. Many avian species, including hummingbirds, increase the high-protein invertebrate content of their diets during egg laying. Hence the timing of egg laying and the growth of nestlings coincide with and rely on high-protein diets in spring and summer. For mammals, protein requirements needed during lactation are second only to juvenile growth. The protein incorporated in milk to support offspring is more than the protein demands of producing a fetus during gestation. If requirements for fetal growth and milk production cannot be met by food intake, species such as caribou must rely on maternal body protein stores (Barboza and Parker 2008). Furthermore, when animals are food stressed in winter, they may mobilize body protein to meet energy requirements if body fat stores have been significantly depleted.

MINERALS AND VITAMINS

All wildlife species require minerals and vitamins. Deficiencies and toxicities are not that common in wild animals in natural habitats, but they do occur. Concentrations of minerals in particular vary in different areas, and therefore concentrations needed in the diet depend on intake. Requirements vary among species.

Water-soluble vitamins (C, B vitamins) rarely pose deficiencies in free-ranging wildlife and are not usually toxic because any excesses are excreted in the urine. Fat-soluble vitamins (A, D, E, K) are absorbed in the intestinal tract with fat and can be stored in large concentrations. They also do not usually pose nutritional problems for free-ranging species. Vitamin A was assumed to synchronize reproduction in quail (*Colinus* spp.) populations, but more recent findings suggest that plant estrogens instead are more likely to be a regulating factor (Robbins 1993:84). Vitamin D is synthesized in the presence of ultraviolet light; young mammals that are born in burrows without access to sunlight obtain vitamin D through the mother's milk. Deficiencies of vitamin E, which is important in maintaining cell membranes, can precipitate tissue damage during the stress of wildlife captures. Vitamin K is important in blood clotting. Warfarin, which is used as a rodent poison, inhibits vitamin K and thus clotting, and poisoned animals bleed to death. Barboza et al. (2009:192) and Robbins (1993:83, 96) give more detailed descriptions of the functions of vitamins and signs of deficiency and toxicity.

Twenty-six of the 90 elements in the chemical periodic table are essential to life. These minerals are associated with many essential functions (Barboza et al. 2009:172; Robbins 1993:55). Important macroelements (i.e., elements usually re-

CHARLES T. ROBBINS (1946–)

Charles T. Robbins is a world-renowned nutritional ecologist who has worked in the fields of wildlife bioenergetics, digestive physiology, foraging ecology, and nutritional balance. After completing BS (Colorado State University) and MS (Syracuse University) degrees, he worked on the biological basis of carrying capacity for his doctoral research (Cornell University). Since 1974, Robbins has been at Washington State University.

Robbins developed an internationally recognized research program in wildlife nutritional ecology, which spans specific wildlife requirements and adaptations as well as applied implications and management. His initial emphasis on a variety of wild ruminants has extended to in-depth studies on grizzly, black, and polar bears. In addition to more than 150 publications, Robbins has emphasized the relevance of nutritional ecology and brought this science to the forefront for many wildlife biologists with his book *Wildlife Feeding and Nutrition*.

Courtesy of Darin Watkins

quired as ≥1 mg/g of food) include calcium (Ca), phosphorus (P), magnesium (Mg), and sodium (Na). Ratios of 1:1 to 2:1 for Ca:P are required for normal skeletal formation and commonly increase to 4:1 in laying birds. Calcium appears to be a major determinant of reproductive success in some avian species. Providing supplemental Ca to tree swallows (*Tachycineta bicolor*) in nest boxes, for example, resulted in earlier laying dates, larger eggs, larger clutches, and higher growth rates of larger chicks (Bidwell and Dawson 2005). Actively growing plants are usually high in P, and therefore deficiencies are most likely to occur in herbivores in winter. Porcupines (*Erethizon dorsatum*) and bighorn sheep may chew on antlers and bones to balance P requirements. Magnesium is abundant in plants and animals, but very high potassium levels in herbaceous forage can induce a Mg deficiency known as grass tetany, which results in limited muscle contraction, uncoordinated movements, convulsions, and often death. In contrast, most plants cannot accumulate Na, and when coupled with the leaching of minerals from alpine and mountain soils may be one reason why ungulates travel to licks to obtain nutrients that cannot be found in forage alone. Besides Na, lick soils often contain elevated levels of Ca, Mg, carbonates, and clay, all of which can help ameliorate gastrointestinal disturbances associated with the transition from winter browse to lush green forage (Ayotte et al. 2006). In addition to movements to licks by northern cervids and bovids and the congregations of tropical parrots at licks, minerals may partially influence the long migrations of wildebeests (*Connochaetes taurinus*; McNaughton 1990). Soil consumption at licks also appears to enable snowshoe hares (*Lepus americanus*) to consume higher amounts of willow (*Salix* spp.) leaves that have high anti-herbivore defense compounds (Worker et al. 2015). Aquatic plants containing high levels of Na (50–500 times higher than most terrestrial plants) reportedly influence the movements and abundance of moose in some areas (Belovsky and Jordan 1981).

Trace elements are required in smaller amounts (<1 mg/g of food) than macroelements. Important links to free-ranging wildlife have been made for iron (Fe), copper (Cu), fluoride (F), and selenium (Se). Iron is obtained by geophagia (i.e., consumption of soil) and is adequate in most forages. Copper deficiencies, causing defective keratinization of hair and hooves, occur in moose and deer in areas with Cu-deficient soils and can be precipitated in mine-spoil areas where levels of molybdenum are high. Around hot springs where F accumulates naturally, dental problems associated with tooth enamel have been reported in elk and bison (*Bison bison*). Selenium interacts with vitamin E to maintain the integrity of tissues and cell membranes. When Se is deficient (e.g., in high-rainfall areas), there is increased susceptibility to muscle damage during capture for species such as mountain goats (*Oreamnos americanus*), and white muscle disease may result. In contrast, during periods of poor forage conditions, ingesting some toxic species of milkvetch (*Astragalus* spp.), which contain thousands of times the Se levels needed for maintenance, can cause either chronic (alkali disease) or acute (blind staggers) Se toxicity.

ESSENTIAL FATTY ACIDS

Fatty acids are important to wildlife because they are components of body fat stores and are found in all cell membranes. Although difficult to define structurally, fatty acids are grouped as saturated (single bonds between all carbon atoms) and unsaturated (having one or more double bond). Because of their structure, unsaturated fatty acids do not stack in an orderly manner and remain more fluid especially at low temperatures. The legs of arctic caribou have high levels of mono-unsaturated fatty acids, which help to facilitate movement in very cold temperatures (Blix 2005). Alpine marmots (*Marmota marmota*) store high levels of polyunsaturated fatty acids and can mobilize those fat depots more easily, thereby tolerating longer bouts of hibernation (Geiser et al. 1994). The primary essential fatty acids, which must be obtained in the diet because animals lack the enzymes needed to create specific double bonds, are linoleic, linolenic, arachidonic, eicosapentaenoic, and docosahexaenoic acids (Barboza et al. 2009:121). Because plants synthesize fatty acids, however, deficiencies of essential fatty acids are uncommon in free-ranging herbivores and their predators.

WATER

Water is the most essential nutrient for all wildlife and influences the ecological life histories of many desert species. Small and young animals generally require more water per unit mass than larger species because they have higher metabolic rates. Although water can be obtained from freestanding sources, many wild animals rarely drink free water. Instead, they acquire preformed water in the tissues of food and metabolic water from the breakdown of carbohydrates, proteins, and fats. Small mammals in arid environments are commonly seedeaters because high carbohydrate content produces the greatest net amount of water for the animal. Nocturnal or fossorial desert species take advantage of lower temperatures and higher humidity to conserve water. Ungulates that time parturition to avoid the driest periods of the year have access to higher food abundance and quality. As examples, fawning is delayed by three to four weeks for mule deer in the southwestern United States compared to northern populations (Bowyer 1991), and numbers of white-tailed deer (*Odocoileus virginianus*) track the amount of precipitation in the southern United States (Teer et al. 1965). Huge congregations of bison remained on the North American Great Plains, where the timing of lactation coincided with the highest precipitation levels (supporting abundant high-quality forage), instead of in the drier northwestern river valleys (Mack and Thompson 1982). Chapter 13 discusses water management for wildlife.

FOOD INTAKE TO MEET NUTRITIONAL REQUIREMENTS

Nutritional interactions between animals and their environments are complex because food supplies, weather, and metabolic states and rates are all variable. For birds and mammals subjected to contrasting seasonal environments, there also are underlying endogenous programs that govern annual cycles of appetite, such that animals have an adaptive response to seasonal changes in food supply (Tyler et al. 2020). In summer, increased appetite and a greater volume of the digestive system coincide with seasonal abundance of food, all of which facilitate growth and mass gain. In winter when food supplies are typically lower, a reduced appetite to feed and voluntary mass loss result in lower levels of activity and lower energy expenditures. Hence animals perform (grow, survive, and reproduce) by adjusting to changing environmental conditions through internal regulatory mechanisms and behavioral and physiological responses.

Food resources and body stores are used to meet nutritional requirements. Whether food alone meets animal requirements depends on its availability and the quantity and quality of the food consumed. Food intake influences what is possible in terms of body mass and condition of the animal. The animal's body stores help meet nutrient demands when food supplies are limited.

Diet

Most animals feed selectively. Herbivores select different plant species and different plant parts. Large carnivores often concentrate their feeding on portions of their prey, as with bears selecting the energy-rich tissues—brains and eggs—of spawning salmon (*Oncorhynchus* spp.; Gende et al. 2001). Defining specifically what free-ranging animals eat can be done through assessments of feeding sites, visual observations of feeding, and by using cameras mounted on animals or their collars (Table 11.1). As an example, video cameras have documented harbor seals (*Phoca vitulina*) consuming five different types of fish and the swim speeds they used to pursue them (Bowen et al. 2002). More often, reporting on diets involves the use of hair, bone, skull, and exoskeleton keys for carnivores and insectivores, and requires knowledge of plant epidermal characteristics for herbivores. Samples are obtained from regurgitated pellets in birds, esophageal contents, stomach and rumen contents, and cheek pouches and crops. Diets determined by microhistology of fecal (or scat) samples are limited to plant fragments that are least digested by herbivores and to identifiable remains for carnivores. Estimating the proportions of items in diets requires corrections for digestibility because highly digestible items will be underrepresented in feces. For carnivores, biases associated with using just frequency of prey occurrence in scat samples can be reduced using calculations adjusted for prey biomass. Additional details of techniques to assess diet composition, their application, and analyses are given in Garnick et al. (2018) and Shipley et al. (2020).

Stable isotopes are increasingly used to reconstruct diets, particularly for carnivores, for which residues in fecal material are limited if meat is entirely digested. Further, scat samples give a relatively short-term dietary assessment of recent ingestion compared to longer-term estimates from stable isotopes

that index the assimilation of prey material into consumer tissues. Stable isotopes of a chemical element contain the same number of protons but a different number of neutrons. Wildlife ecologists typically use nitrogen ($^{15}N/^{14}N$) and carbon ($^{13}C/^{12}C$) to define diets. Isotopic ratios are expressed in delta notation (δ), comparing the ratio of the heavy:light isotope in a sample with the ratio of a nonvariable standard. Delta values are positive if the ratio is higher than the standard, and negative if the ratio is lower than the standard. Samples collected for isotopic analyses depend on the time frame of interest. Signatures in plasma generally reflect what has been nutritionally absorbed during the previous seven to ten days; red blood cells and muscle are indicative of the last three to four months; bone represents assimilation over a lifetime; and hair, feathers, and nails reflect the period when they were grown. Segmenting hairs that began growing in the spring and finished growing several months later can indicate seasonal changes in diets of grizzly bears (*U. arctos*; Milakovic and Parker 2013) and wolves (Darimont et al. 2008). Analyzing eggs can help quantify avian diets during egg laying (Hobson 1995).

Nitrogen isotopes ($^{15}N/^{14}N$) are particularly important because they help determine trophic position. Animals preferentially excrete the lighter nitrogen isotope (^{14}N) in nitrogenous wastes during normal enzymatic breakdown. Relatively more of the heavier isotope (^{15}N) is incorporated into tissues, and therefore there is an increase in $\delta^{15}N$ with each trophic level. Carbon isotopes ($^{13}C/^{12}C$) distinguish broadly between plants with different photosynthetic pathways. The $\delta^{13}C$ of C_3 plants (e.g., many herbaceous plants, shrubs, cool-season grasses) usually differs from that of C_4 plants (e.g., sedges, warm-season grasses). In tropical areas where grasses are primarily C_4 plants and other plants use C_3 pathways for photosynthesis, the $\delta^{13}C$ of grazers differs from browsers, as do the signatures incorporated into their predators. This distinction among herbivores is less clear in colder environments where there are few C_4 plants. Nonetheless, sections of antler and hoof tissue have shown switches between shrub-based, graminoid, and lichen-based diets of reindeer, and from woody browse to leafy and aquatic diets of moose (Kielland and Finstad 2000, Kielland 2001). The contribution to diets by some aquatic plants and cacti, which use a third photosynthetic pathway, can be quantified using $\delta^{13}C$ when combined with hydrogen isotopes (δ^2H, $^2H/^1H$; Karasov and Martinez del Rio 2007:449).

As natural markers, stable isotopes provide important information on geographic, temporal, and age-specific variation in diets. Hydrogen signatures in bird feathers are a means of tracking origins of migratory species because of a North American continent-wide gradient in δ^2H in precipitation (Karasov and Martinez del Rio 2007:462). High-sulfur signatures ($\delta^{34}S$, $^{34}S/^{32}S$) in whitebark pine (*Pinus albicaulis*) nuts helped quantify the importance of that food resource to grizzly bears in Yellowstone National Park (Felicetti et al. 2003). Strontium ($^{87}S/^{86}S$) isotopes, as geochemical signatures, were used to track the diets and locations of African elephants (*Loxodonta africana*; Koch et al. 1995). Nitrogen, carbon, oxygen, hydrogen, sulfur, and oxygen ($^{18}O/^{16}O$) isotopes are

Table 11.1. Some commonly used methods to quantify diets, food intake, and food quality for free-ranging wildlife

	Index	Premise	Cautions for Biologists	Source
Diet and intake	Feeding site methods	What has been removed by herbivores or what remains after predation can be quantified	Biased toward foods in which removal is easiest to detect; problems with missing items and regrowth	Holechek et al. (1982)
	Contents of stomach, rumen, crop, cheek pouches	Foods that have recently been consumed will be in the digestive tract	Biased toward foods that are least digestible and easiest to identify	Holechek et al. (1982)
	Microhistology of fecal analyses	Foods can be identified by indigestible residues voided in the feces	Biased toward foods least digestible and most likely to fragment; detailed microhistological keys needed	Holechek et al. (1982), Klare et al. (2011), Tollit et al. (2015)
	Remote monitoring of feeding behavior	Time spent feeding can be determined using telemetry transmitters and cameras	Variation around the relation between signal pattern and behavior can be high; difficulties positioning cameras	Hassall et al. (2001)
	Direct observations of feeding by free-ranging animals	Bite counts or time spent feeding reflect food intake	Biased toward plants that are easiest to see at a distance; difficult to stay close to wild animals; hand-picked bites are assumed to simulate amounts (and quality) of forages consumed	Holechek et al. (1982)
	Direct observations of feeding by tractable animals in the wild	Bite-by-bite estimates of intake are determined by visual estimates of plant unit size or hand-picked estimates of bite size	Training needed to recognize sizes of plant units or bite sizes consumed; animals must not be naive to native foods and must allow close observers	Collins et al. (1978), Parker et al. (1993a)
	Stable isotopes	Isotopic signatures in consumer tissues reflect assimilation of foods	All dietary sources must be identified and have isotopic signatures that differ from each other; discrimination between animal and diet is needed; variation around food estimates depends on the mixing model (e.g., IsoSource, SIAR, SISUS, MixSIR)[a]	Karasov and Martinez del Rio (2007:433)
	Fatty acids	Fatty acids in consumer tissues reflect fatty acids of the prey consumed	Species with foregut microbial digestion may break down dietary fatty acids so that they are not absorbed intact in the consumer's tissues, biasing results of mixing models (QFASA)[b]	Iverson et al. (2004), Tollit et al. (2006)
	n-alkanes	Alkane profiles in the feces reflect alkane profiles in the diet	Dietary components are limited by the number of n-alkane markers; dosing with an even-chained alkane helps correct for digestibility	Bugalho et al. (2005), Dove and Mayes (2005)
	Genetic analyses of fecal material	DNA in feces gives high specificity and taxonomic resolution of food items.	Amounts of DNA can vary by tissue and species, affecting dietary estimates determined by different assays (e.g., barcoding, qPCR, HTS).[c]	Valentini et al. (2008), Murray et al. (2011)
Food quality	Energy content by bomb calorimetry	Gross energy is released by complete oxidation of food	Not all gross energy is available to the animal depending on what is digested and metabolized	Robbins (1993:9)
	Protein content	Kjeldahl procedure or total elemental nitrogen analysis provides a crude estimate of dietary protein	Non-protein nitrogen is included in varying amounts; not all protein is available to the animal	Barboza et al. (2009:138)
	Total dietary fiber	Higher fiber decreases digestibility	Some fiber may be digested by microbial fermentation	Prosky et al. (1984)
	Digestibility by in vivo digestion trials	Digestion trials with live animals document specific digestibilities of different foods	Requires relatively tame animals and sufficient time to equilibrate to foods before trials	Servello et al. (2005), Shipley et al. (2020)
	Digestibility by internal markers	Marker concentrations in the food and feces can be used to calculate indigestibility	Food intake and internal marker concentration must be recorded over an extended period	Hupp et al. (1996), Mayes and Dove (2000)
	Digestibility by in vitro digestion	Digestion in a test tube simulates digestion in an animal	Apparent digestibility depends on an adapted donor inoculum (microbes must be habituated to foods)	Tilley and Terry (1963)
	Proximate analysis	Food fiber and non-fiber (water, ash, crude protein, crude fat, nitrogen free extract) components can be quantified	Components may not be separated completely; outdated procedure	Robbins (1993:257)
	Digestibility by sequential detergent analyses	Nutritive value of plants depends on cell contents and cell wall constituents	Plant secondary metabolites may confound estimates of the digestible fraction	Goering and Van Soest (1970), Mould and Robbins (1981)
	Tannin content based on bovine serum albumin (BSA) precipitate	Protein-precipitating capacity indexes the reduction in protein availability	Quantifiable effects of tannins on digestion are species-specific	Hanley et al. (1992), Spalinger et al. (2010)
	Fecal protein	Higher fecal nitrogen reflects higher dietary nitrogen	Limited use for species consuming mostly tannin-containing forages because tannins bind to protein, elevating fecal protein	Robbins et al. (1987), Leslie et al. (2008).
	Fecal 2,6-diaminopimelic acid (DAPA)	DAPA (a component of non-digested rumen microflora) tracks changes in dietary energy	Should not be used to track dietary nitrogen	Osborn and Ginnett (2001)
	Near-infrared reflectance spectroscopy (NIRS)	Spectral properties index diet components and quality (digestible energy and protein) from fecal samples	Calibration equations for the relationships between constituents in the sample and NIRS spectral information must be developed; these vary with botanical composition	Kamler et al. (2003)

[a]IsoSource, stable isotope mixing model; SIAR, stable isotope analysis in R; SISUS, stable isotope sourcing using sampling; MixSIR, mixing stable isotope analysis in R.

[b]QFASA, quantitative fatty acid signature analysis.

[c]qPCR, quantitative polymerase chain reaction; HTS, high-throughput sequencing.

all enriched in marine systems. Consequently, they have been used to assess the degree of terrestrial foraging by polar bears (*U. maritimus*) as the sea ice declines (Ramsay and Hobson 1991) and the contribution of marine foods when terrestrial sources are limited for species such as Arctic foxes (*Vulpes lagopus*; Angerbjörn et al. 1994).

Fatty acids also are used to reconstruct shifts in diets. They are most useful in simple-stomached carnivores for which fatty acids are deposited in the tissues with minimal modification from the diet. Fatty acids in the blubber of marine mammals and bears reflect the fatty-acid signatures of their major food items (Iverson et al. 2001, 2004). Vibrissae, which continually grow throughout an animal's lifetime, can be a time series of seasonal shifts in fatty acids related to diet composition (Newsome et al. 2010). In non-simple-stomached animals, such as ruminants, fatty acids might not always remain intact because of microbial metabolism, and therefore distinguishing among different types of forages eaten might not be possible.

Besides stable isotopes and fatty acids, another molecular-based approach to investigate diet is the use of genetic sequence data from fecal samples. DNA barcoding and DNA analyses provide semi-quantitative diet composition data. Such DNA technology has most commonly been used to distinguish among species eaten by carnivorous animals, although not exclusively (Valentini et al. 2008). A study on sea lions (*Eumetopias jubatus*) compared estimates of diet composition based on three of these technologies: fatty acid signatures in blubber, DNA extractions from scats; and prey remnants in scats (Tollit et al. 2006).

Available Food and Quantity Consumed

Food availability, which depends on how accessible the food is, typically changes with season. For carnivores, when one prey species declines, predators shift to other prey species, move to other areas, or also decline. Decreases in lemming (*Lemmus* spp.) and ptarmigan (*Lagopus* spp.) populations trigger snowy owls (*Bubo scandiacus*), and other raptors to move southward from arctic tundra (Gessaman 1972). Canada lynx (*Lynx canadensis*) populations cycle in abundance relative to snowshoe hare populations (Stenseth et al. 1997). For herbivores, food availability is constrained by the subset of forage species that they ingest from the total plant biomass, which varies with season. Caribou, for example, avoid more than 50% of understory species in boreal forests in summer and select most frequently for deciduous shrubs (Denryter et al. 2017); in winter, caribou also are selective, particularly for specific lichen species within large mats of ground-dwelling lichens (Johnson et al. 2000). Biologists tasked with quantifying food availability for either carnivores or herbivores are furthered challenged by patchy distributions of food and changing availability associated with environmental factors such as snow.

As the abundance of accepted foods increases, food intake by animals per unit time typically increases. For most species, intake rates level off at a maximum rate (Fig. 11.3A; Wickstrom et al. 1984), although for some species no asymptote has been observed under natural foraging conditions (Trudell and White 1981). This functional response and the variability around it depend on an herbivore's requirements and characteristics of the food (i.e., quality, structure, and spatial arrangement). Per capita kill rates by predators change similarly with prey density. At very low food biomass, animals may choose not to feed because the total value of the food obtained over time is less than the costs of obtaining it (Barboza et al. 2009:34). Some studies define thresholds of food abundance (biomass per unit area) below which the environment is not suitable to meet intake requirements.

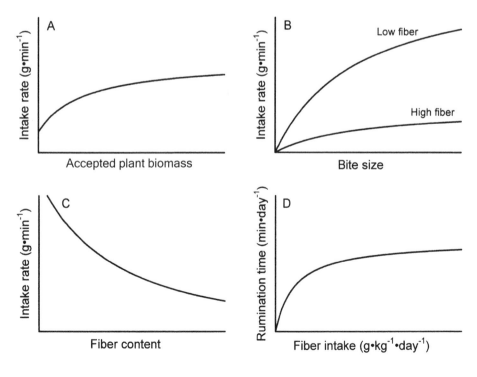

Figure 11.3. Food intake rate in relation to (*A*) accepted plant biomass and (*B*) bite size, and the influence of plant fiber on (*C*) intake rate and (*D*) rumination time.

The quantity of food an animal eats is also regulated by physical and physiological factors. For species from snowshoe hares to moose, intake rates increase asymptotically with bigger bite sizes (Fig. 11.3B; Shipley and Spalinger 1992). Intake rates by Sitka black-tailed deer (*Odocoileus hemionus sitkensis*) on large skunk cabbage (*Lysichiton americanum*) or devil's club (*Oplopanax horridus*) leaves can be five times higher than when picking individual forb leaves (Gillingham et al. 1997). Fibrous foods lower intake rates because of the processing constraints of chewing and digestion (Fig. 11.3B, C, D). Ruminants in particular must break down food into small particles before it can move through the ruminant stomach and the remainder of the gastrointestinal tract. As such, black-tailed deer eating low-quality fibrous food in winter spend up to 10 hours each day chewing their cud (Parker et al. 1999), precluding intake of forage during this time. The number, size, and nutritional value of bites taken ultimately influence the duration of feeding bouts and time spent feeding each day (Shipley 2007). Some carnivores such as badgers (*Taxidea taxus*) can increase digestive efficiency by slowing the passage of prey through the gut, thereby increasing nutrient gain during periods of prey scarcity (Harlow 1981).

Highest voluntary food intakes are typically associated with highest nutrient demands. For mammals, the quantity of food consumed per kilogram by lactating females can be 50% higher than what it is for males (Parker et al. 1999). For birds, intake rates to meet daily energy expenditures during the incubation phase in species that incubate without assistance from their mates can be twice as high as during the nestling phase (Tinbergen and Williams 2002). Young animals and small species, with relatively higher metabolism, must eat more per unit mass than adults or large species to meet their requirements. Daily intakes when corrected to dry matter generally range from 2% of body mass per day for large mature animals without reproductive demands to 10% of body mass for smaller or younger animals, although this is dependent on food quality (Robbins 1993:333). Intakes are even higher when animals put on body stores prior to migration, hibernation, or winter.

Food intakes for free-ranging animals can be approximated from the mass of bites consumed while watching individuals feed (Table 11.1). These observations have most frequently been accomplished with habituated or tractable animals, notably ungulates. Usually such estimates are for only short periods (minutes or hours) because of logistical constraints; extrapolations are then made to total daily intakes based on daily times spent foraging. The latter are typically estimated using activity sensors attached to the head or neck. Similarly, the intakes of different insects by human-imprinted gamebirds have been quantified (Doxon and Carroll 2010). Less commonly used to quantify food intakes by wildlife species are indigestible markers or the plant wax compounds known as *n*-alkanes. Most of the *n*-alkanes in forage have naturally present odd-numbered carbon chains with 25–35 carbons. Because there are differences in the alkane profiles of plants, the proportion of intake from different dietary components can be discriminated by multiple plant wax markers.

Food Quality

Nutritive value varies with plant and animal parts and changes with season. Importantly, food quality influences the nutrients that can be extracted from the food. Even when food quantity is sufficient, food quality can limit the growth of individuals and populations. Quality is commonly expressed in terms of energy content per gram and percent protein or nitrogen (N) content on a dry matter basis (Table 11.1). Forages generally range between 18.8 and 20.5 kJ gross energy per gram dry weight (slightly higher for foods containing resins, waxes, or oils), but the digestibility of that forage usually declines from spring to autumn as structural and chemical characteristics of the forage change. Protein content is highest in the new growing buds and shoots (20% to 30%) of grasses, forbs, and shrubs in spring, and declines with plant maturity and senescence (3% to 4%). Because proteins average 16% N, percentage crude protein is calculated as grams N per gram feed × 6.25 (from 100 ÷ 16). Seeds are relatively high in both energy and protein. Animal tissue, which is largely protein, usually increases in energy content with age of the animal because of increasing fat content. For carnivores and simple-stomached species that do not consume highly fibrous foods, available energy in prey can be roughly approximated from average fuel values: 16.7 kJ/g carbohydrate, 16.7 kJ/g protein, and 37.7 kJ/g fat based on dry mass of the prey.

The higher the fiber content of the food, typically the lower is its digestibility. For bears consuming fish, meat, berries, pine nuts, and plants, fiber content varied fivefold and digestibilities ranged from 90% to 40% (Pritchard and Robbins 1990). Fiber content of forage generally increases from new growth in spring to senescence in winter; stems usually have more fiber than leaves. Consequently, the digestibilities of common forages for herbivores in winter may be only 30% to 40% of summer forages (Cook 2002). And therefore food value for grazers and browsers, most commonly indexed by the energy and protein content that is digestible, often varies markedly between seasons.

Nutritive value of forages may be further reduced by plant defensive compounds, also called plant secondary metabolites (PSMs) or secondary plant compounds because initially they were not known to have primary metabolic functions in plants and are presumably produced to deter herbivory. These compounds include toxic alkaloids (found in lupine [*Lupinus* spp.], larkspur [*Delphinium* spp.], nicotine, morphine); flavonoids that alter reproductive patterns; terpenoids (in volatile oils of sagebrush [*Artemisia* spp.], eucalyptus, conifers, and citrus); and tannins. Terpenoids generally have toxic effects on microbes that ferment cellulose. Perhaps uniquely, though, the microbial populations habituated to terpenoids in pronghorns (*Antilocapra americana*), pygmy rabbits (*Brachylagus idahoensis*), and sage grouse (*Centrocercus urophasianus*) allow those species to consume diets of mostly sagebrush in winter. Much attention has been given to understanding the influence of tannins on forage quality because they are present in the leaves of most deciduous woody species and some forbs. Their levels vary depending on whether they are produced under shade versus sun conditions (Happe et al. 1990). Tannin levels are minimal in grasses and relatively low in winter browse. Tannins in other forage, however, bind to forage protein (making it unavailable to the animal) and decrease digestibility. For example, the tannin content of forages consumed by moose increases over the summer, resulting in a decrease in protein availability (McArt et al. 2009).

Some animals avoid plants with PSMs; others that frequently consume them have counteradaptations to modify plant defenses (Dearing et al. 2005, Marsh et al. 2006). Moose and deer, for example, produce salivary proteins that bind tannins to reduce their absorption and effects on digestion. Some other species (usually specialists) absorb PSMs and must detoxify them for excretion. The rate of detoxification can determine the rate of feeding. So why do animals eat toxic compounds that can influence food quality, digestion, and potentially reproduction? A uniform threshold for ingestion of toxic PSMs does not appear to exist. Rather, the value of the nutrient content presumably is greater than the cost of detoxification (McArthur et al. 1993). There are important implications of the current increasing levels of CO_2 on the planet to landscape dynamics because some PSM concentrations increase at higher levels of CO_2, and consumers may attempt to avoid elevated defensive compounds. Chickadees, for example, prefer eating caterpillars that feed on leaves with low PSM content. Avoidance by avian predators of gypsy moth (*Lymantria dispar*) caterpillars consuming highly defended leaves may lead to even larger areas of defoliation (Müller et al. 2006).

In the past, resource management professionals managed those food species that were most limited during critical times. Enhancing willow and aspen (*Populus* spp.) stands generally increased or maintained populations of moose and ruffed grouse (*Bonasa umbellus*), respectively. Now there is more emphasis placed on understanding the nutritional components that influence feeding patterns. Both energy and protein are influencers of food selection, with confounding influences of fiber and PSMs. Ruffed grouse feed in spring on the large male aspen (*P. tremuloides* and *P. grandidentata*) catkins that have the highest energy content (Svoboda and Gullion 1972). Spruce grouse (*Canachites canadensis*) select needles of jack pine (*Pinus banksiana*), which have significantly higher protein and mineral content than other conifers (Gurchinoff and Robinson 1972). Pronghorns are adapted to consume big sagebrush (*A. tridentata*) despite its terpenoid content and may benefit from the higher protein levels that are found in sagebrush compared to other winter browse species (Bailey 1984:96). Mule deer and black-tailed deer eat conifer seedlings and branches blown from treetops but tend to avoid the medium-aged, low branches of conifers that have higher terpenoid content (Parker et al. 1999). Koalas (*Phascolarctos cinereus*), even as highly specialized folivores, visit eucalyptus (*Eucalyptus* spp.) trees that are least defended (Moore and Foley 2005). In the Subarctic, ptarmigan, snowshoe hares, and beavers select foods to avoid plant defensive compounds (Bryant and Kuropat 1980).

In all likelihood, food selection is further confounded by the need to obtain multiple nutrients such that animals may not maximize consumption of just energy or just protein, but instead balance food constituents using intake targets that vary with physiological state (Felton et al. 2018). As such, selection to obtain nutrients varies seasonally. The forage maturation hypothesis suggests that herbivores should balance forage quality and quantity by selecting earlier phenological stages at intermediate biomass, thereby maximizing consumption of digestible nutrients (Fryxell 1991). Temperate ungulates that migrate, such as mule deer, elk, moose, bison, and bighorn sheep, follow the seasonal wave of new vascular plants in spring (Merkle et al. 2016). This "green-wave surfing" usually occurs on landscapes where green-up of new vegetation is short-lived locally but progresses sequentially across larger areas. In contrast, red deer (*Cervus elaphus*) appear to "jump the green wave" and then track phenological spring as it arrives at the end of their migration route (Bischof et al. 2012). On landscapes with less wave-like green-up, resident individuals may derive foraging benefits similar to those of migrators because the green-up of new high-quality forage lasts longer (Aikens et al. 2020). Other species such as Stone's sheep (*Ovis dalli stonei*) females track forage quality by moving up in elevation as the growing season progresses (Walker et al. 2006). Pregnant barren-ground caribou (*R. t. granti*) move across the Alaskan coastal plain to calve in areas where rates of spring green-up and forage quality will be highest (Griffith et al. 2002), enabling them to meet the energy demands of lactation and regain lost body mass (protein); later in summer they select more for forage quantity to accrue body fat and increase the likelihood of breeding in autumn (Johnson et al. 2021).

Understanding the value of different foods to free-ranging wildlife requires some knowledge of how the animal processes and retains foods. Feeding trials with captive carnivores and herbivores are used to quantify food value specifically from measures of all food consumed minus any digestive (fecal) and metabolic (urine and digestive gas) losses. In birds, digestive and metabolic losses are excreted together. The apparent energy available to a free-ranging animal (metabolizable energy) is calculated by multiplying the gross energy content (kJ per dry gram) of the prey or forage by the percentage that is digestible (excluding fecal losses) and metabolizable (excluding urinary and gaseous losses). Digestive fecal losses from food not absorbed by the animal tend to be substantially higher and more variable than metabolic urinary losses (i.e., food digested but not metabolized) and are more easily quantified with laboratory measures. Hence food value is often presented simply as digestible energy (and digestible protein), recognizing that animals retain slightly less than these estimates. Such calculations enable comparison of the values of different food types. One snowshoe hare, for example, is worth the energetic equivalent of 49 small voles (*Microtus* and *Clethrionomys* spp.) to a bobcat (*Lynx rufus*; Powers et al. 1989).

Digestion trials that use live animals to determine digestibility by measuring total food intake and fecal output or the concentration of an internal marker in diet and feces are not always possible, and many foods have not been fed during experimental trials. Instead, researchers have simulated digestion in test tubes, quantified fiber and nonfiber components, and chemically digested the plant materials (Table 11.1; Shipley et al. 2020). The contents of plant cells (i.e., the cell solubles) are highly digestible, but plant cell walls composed of a cellulose matrix provide nutritional value only to wildlife species that can partially digest them using microbial fermentation. These species include all ruminants and species with microbes in pouches of the hindgut—horses, porcupines and beavers (*Castor canadensis*), giraffes (*Giraffa camelopardalis*), elephants, kangaroos—or microbes in caeca, such as in rabbits and hares (Leporidae), and grouse (Tetraoninae). Sequential detergent analysis on forage is the most common method for estimating what herbivores with microbes can potentially digest. In this approach, a neutral detergent digests the cell contents, leaving neutral detergent fiber (NDF, or the plant cell walls) as the residue. A sequential acid detergent added to the NDF digests a portion of the cell walls that microbes can digest, leaving an acid detergent fiber (ADF). Additional protocols are used to quantify other constituents of the fiber (e.g., cellulose, lignin, cutin). Equations that predict digestibility of different forages, validated with digestion trials, commonly incorporate the components of sequential detergent analysis. Lab procedures using bovine serum albumin (BSA) mimic how much protein is bound to tannins; BSA then can be incorporated as a correction factor in equations predicting protein availability and digestibility (Table 11.1). Wildlife ecologists frequently estimate food value from chemical analyses of foods for species such as voles, ruffed grouse, bears, and ungulates (Servello et al. 2005).

Indexing the quality of a diet composed of multiple foods is one of the greatest challenges for wildlife ecologists (Table 11.1). Composite diet quality can be estimated based on the proportions of each food and its nutrient value. The use of fecal samples for this purpose has met with mixed success. Fecal nitrogen can serve as a general index to dietary nitrogen (and therefore protein), depending on animal and forage species. Studies that assume diet quality varies with fecal protein, however, and can overestimate digestibility and protein value for wildlife species that consume mostly tanniferous browse leaves in summer. Fecal 2,6-diaminopimelic acid (DAPA) generally tracks dietary energy but not dietary nitrogen. Near-infrared reflectance spectroscopy (NIRS), with extensive calibration, might index various aspects of dietary quality. Fecal chlorophyll has been used to document consumption of green biomass in relation to timing of spring green-up (Christianson and Creel 2009). Values of fecal protein also appear to correspond with changes in large-scale indices of green biomass derived from satellite imagery (e.g., normalized difference vegetation index, or NDVI; Pettorelli et al. 2005, Hamel et al. 2009). Not all green plant biomass is food, however, and changes in NDVI provide only a general measure of food quality (Johnson et al. 2018, Shipley et al. 2020).

Food Intake to Meet Requirements and Estimates of Nutritional Carrying Capacity

Wildlife ecologists estimate the amount of food that an individual needs to consume on the basis of nutritional requirements and food value. For example, energy requirements can be determined after calculating an energy budget derived from observations of all activities, each multiplied by energy-specific costs (or as determined from doubly labeled water or a general estimate of $2–3 \times BMR$). Food energy value is its energy content multiplied by its digestibility (and metabolizability, if known). From these calculations the amount of food required per unit time can be determined for carnivores or herbivores, recognizing that requirements vary with sex and age, and typically with season. Food needed to meet protein requirements can be assessed from equations in a similar manner. Based on such process models of nutrient requirements and food values, long-term changes in benthic communities associated with climate shifts are expected to affect arctic wintering sea ducks (spectacled eiders, *Somateria fischeri*) that rely on bivalve prey (Lovvorn et al. 2009). In northern temperate areas, digestible energy is reported to be the greatest nutritionally limiting factor for populations of black-tailed deer and elk (Parker et al. 1999, Cook et al. 2004). In contrast, inadequate forage protein appears to constrain northern moose and arctic caribou populations (McArt et al. 2009, Barboza et al. 2018).

The ability to predict nutritional value helps define the importance of seasonal and changing habitats, but biologists are challenged by how to describe such foodscapes. At landscape scales, indices extracted from remotely sensed data such as NDVI correlate with plant biomass in different communities and with relative changes in plant phenology (Pettorelli et al. 2011). Spring, summer, and autumn metrics of NDVI have been related to demographic parameters in several species of birds and mammals (Hurley et al. 2014). In contrast, behavioral indicators can be used to measure the value of foodscapes (Searle et al. 2007). Very fine-scale data from close observations of foraging animals document actual species eaten, and these data, when coupled with the nutritional value of foods, can be used to approximate foodscapes in terms of food quality (energy or protein content of diets) or energy and protein intakes (Cook et al. 2016). Whether there is *enough* energy and protein depends on whether the landscape meets daily requirements. Mapping thresholds of food value as nutrient intakes is useful in showing where animals can and cannot meet their requirements. It is important to recognize, however, that behavioral and foraging decisions could allow individuals to exploit the spatiotemporal variation in food abundance and quality at finer scales than those depicted with maps. Nonetheless, nutritional-landscape models have related limited availability of high-quality summer forage to declines in mule deer populations (Merems et al. 2020) and to trends in moose populations (Schrempp et al. 2019). Implications of foodscapes to animal performance are explicitly examined in the energy-protein model originally developed to simulate growth and reproduction of Porcupine barren-ground caribou in Alaska, with potential applications related to environmental change, disturbance, and displacement (White et al. 2014). Extrapolation to a nutritional carrying capacity (i.e., the number of individuals supported per unit area) can be made if there are reasonable estimates of individual animal requirements and of food abundance and its quality, adjusted for changing food availability within and across seasons (Hobbs and Swift 1985, Hanley and Rogers 1989, Guthery 1999, Hanley et al. 2012). Estimates of the food resources needed to support an individual or group of individuals would be lower if animals are able to mobilize body stores, particularly during seasons when food availability and quality are low.

BODY CONDITION

Body condition is a general term for an animal's level of energy stores and results from the integration of nutrient requirements and food intake. Body fat, body protein, and body mass have direct consequences for reproduction and population dynamics. In ungulates, food resources in summer strongly influence body size and condition in autumn, affecting the timing and probability of conception by females and then carrying a fetus to term (Cook et al. 2004). Juvenile growth in summer directly affects subsequent adult body mass and age of first reproduction. In winter, food resources and environmental conditions have direct consequences to survival and body condition in spring, and to timing of parturition and birth mass of young (Parker et al. 2009). Likewise for waterfowl, habitats needed to acquire body stores prior to breeding are critical for breeding success and survival of young.

The body capital–income continuum helps explain timing of reproduction and the importance of different habitats toward population recruitment (Barboza et al. 2009:246). Energy and nutrient supplies that birds and mammals use to support reproduction can be obtained from body stores accumulated before breeding (i.e., capital breeders), food supplies available at the time of reproduction (i.e., income breeders), or

both. Migrating waterfowl and shorebirds that arrive early on northern nesting areas often use body stores (capital) acquired at southern wintering areas to support egg production. Income breeders that arrive later rely on food supplies (income) at staging areas and nesting areas. Similarly, reindeer that calve before spring green-up produce calves and the milk to support them by mobilizing maternal tissues deposited the previous autumn. Caribou with calves born later can meet more of those nutritional demands with spring forage (Barboza and Parker 2008).

Animals rely on body stores to help meet nutrient demands when food supplies are limited. Body fat is the primary energy store of the body. In Mautz's classic 1978 paper, the annual cycle of changes in body fat in seasonal deer was likened to "sliding on a bushy hillside." Ungulates in northern environments that acquire more fat over summer (the uphill climb) have fat stores that last longer, and these reserves help slow the downhill slide in body condition over winter. When winter habitats are poor or winter severity is high, body stores are depleted faster (Fig. 11.4A), threatening adult survival and fetal development. Animals regain less condition when summer habitats are deficient (Fig. 11.4B). As a consequence, breeding in autumn may not occur or it may be delayed, subsequently delaying parturition. Lactating animals that support a neonate are often in poorer condition than non-lactating adults during summer and into autumn (Fig. 11.4C). Our understanding of seasonal changes in body condition is further complicated by carryover or temporally lagged effects of previous nutrition from multiple seasons and reproductive state that affect the annual cycle in condition and extend to subsequent years (Parker et al. 2009, Monteith et al. 2013). It is commonly assumed that northern ungulates are in their "best" body condition in autumn following the food abundance of summer. In areas where food is sufficient to continue gaining condition in autumn (or hold high condition from autumn into winter), however, individuals could be in as good or better condition in midwinter than autumn, as observed in some caribou populations (Russell et al. 1993, Kelly 2020).

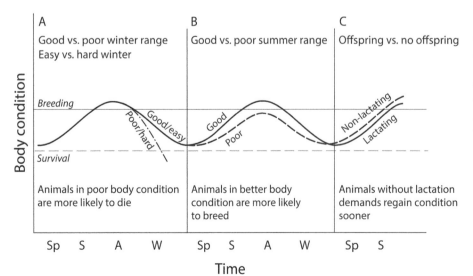

Figure 11.4. Seasonal changes in body condition of ungulates as influenced by (A) winter range and winter severity, (B) summer nutrition, and (C) the energetic demands of lactation, relative to thresholds for breeding and survival. Sp, spring; S, summer; A, autumn; W, winter.

When fat is mobilized for energy over the winter period, animals tend to draw down subcutaneous fat depots first (such as rump fat in ungulates), followed by abdominal and visceral fat, and lastly bone marrow fat, although there is overlap in these processes. Traditionally it was assumed that fat stores were depleted before animals reverted to the use of body protein. The use of body protein and body fat occurs simultaneously, albeit protein stores are catabolized at much lower rates and to a greater extent when fat is depleted (Parker et al. 1993b). The use of both fuels allows animals to "buy time" as

a buffer against winter (Torbit et al. 1985). In lean-bodied animals with sparse fat depots, such as American martens, labile proteins are used to sustain the animals during fasting periods between acquiring prey items (Buskirk and Harlow 1989).

Wildlife biologists use a variety of morphological and physiological indices to estimate relative body condition (Table 11.2; Parker 2003, Servello et al. 2005, Barboza et al. 2009, Shipley et al. 2020). Body mass and skeletal measurements provide indirect insights to condition. Linear and circumferential measures (e.g., chest girth, body length, hind

Table 11.2. Premise and limitations of common indices used to assess body condition

	Index	Premise	Cautions for Biologists	Source
Morphological measurements	Body mass	Comparative changes in mass or carcass mass among years provide a relative index of changing conditions	Varies with age and physiology; does not index body fat	Servello et al. (2005), Shipley et al. (2020)
	Chest girth, hindfoot length, femur length, wing length, body length	Structural measures index long-term nutrition	Considerable variability among individuals	Servello et al. (2005), Shipley et al. (2020)
	Body length:mass ratio	Size adjusted for mass provides a sound index of condition	Difficult to interpret biologically because ratio is confounded by effects of body mass	Servello et al. (2005), Shipley et al. (2020)
	Antler beam width and diameter	Antler size reflects spring and summer range conditions	Most affected by nutrition just prior to antler growth; confounded by genetics	Scribner et al. (1989)
Fat indices	Visual condition scoring or manual palpation	Visual scoring of avian fat depots and profiles indexes subcutaneous fat; ungulate body condition scores index body fat	Requires extensive training for accuracy	Krementz and Pendelton (1990), Gerhardt et al. (1996)
	Kidney:fat ratio	General indicator of total body fat	Confounded by seasonal variation in kidney mass; not useful in young animals without fat reserves or very fat animals	Van Vuren and Coblentz (1985)
	Bone marrow:fat ratio	Body condition based on color and consistency of bone marrow	Limited use in young animals because bone marrow is active in red blood cell formation, and for animals in very good condition	Harder and Kirkpatrick (1994)
	Total body fat by whole body grinding and lipid extraction	Absolute chemical measure of percent fat	Time-consuming, specialized facilities required	Shipley et al. (2020)
	Total body fat by water dilution using hydrogen isotopes	Total fat can be predicted from an estimate of body water after injecting H isotope	Estimate of the time needed for the isotope to equilibrate in the animal is required	Torbit et al. (1985), Hildebrand et al. (1998)
	Total body fat by total body electrical conductivity (TOBEC) or bioelectrical impedance assay (BIA)	Electrical conductivity through lipids is much lower than other body components	Most useful on animals that distribute fat relatively evenly; requires training to consistently position animals for measurements	Scott et al. (2001)
	Ultrasonography	Rump fat thickness coupled with body condition scoring in ungulates is related to total body fat	Limited use when rump fat is depleted; need for species-specific equations	Cook et al. (2010), Stephenson et al. (2020)
	Dual emission x-ray absorptiometry (DEXA)	X-rays are transmitted differentially depending on tissue type	Equipment generally not field portable	Nagy (2001)
	Quantitative magnetic resonance (QMR)	Magnetic resonance scan distinguishes between fat and wet lean mass.	Limited accuracy for total body water content of very small species	McGuire and Guglielmo (2010)
Physiological indices	Urinary urea nitrogen:creatinine (UN:C) ratio	In winter, when forage protein remains relatively constant, increases in UN:C ratios occur when body protein is broken down	May be confounded if protein intake is variable; assumes creatinine excretion is stable	DelGiudice (1995)
	Urinary cortisol:creatinine ratio	Cortisol excretion increases during nutritional stress and prior to death	May be confounded by non-nutritional stressors	Saltz and White (1991)
	Urinary allantoin:creatinine ratios	Excretion of allantoin, a product of microbial fermentation, increases with higher intakes of digestible nutrients	Levels of allantoin in some species are too low to be used as an index	Vagnoni et al. (1996)
	Urinary glucuronic acid:creatinine ratios	Because glucuronic acid is a by-product of digesting plant secondary metabolites, higher levels indicate reduced dietary quality	Confounded by level of food intake; not produced from consumption of all plant defensive compounds	Guglielmo et al. (1996), Servello and Schneider (2000)
	Urinary $\delta^{15}N$ concentration	Higher concentrations of $\delta^{15}N$ are excreted when body protein is being catabolized	Potentially high variance around model parameters	Gustine et al. (2011)

foot length) are often taken on ungulates, lagomorphs, and bears. The most commonly used indices that describe body condition are associated with body fat depots, however, and animals with more fat are assumed to be in better condition. Total body fat can be determined from dead animals on the basis of fat extraction of homogenized tissues (i.e., obtained using a grinder). Kidney fat is a general index of body fat for adult ungulates and lagomorphs, but it does not reflect body condition at extremely low body fat levels. Gizzard fat provides a similar index in game birds. Mass of xiphoid fat at the base of the sternum indexes total body fat in canids. Bone marrow fat based on color and consistency is sensitive for animals only in very poor condition. On live animals, total body fat can be predicted from total body water content after injecting animals with hydrogen isotopes (^{2}H or ^{3}H; Speakman et al. 2001). Total body fat also can be indexed reliably in species that deposit fat relatively evenly throughout the body—such as bears, seals (*Phoca* spp.), wombats (*Lasiorhinus latifrons*), and porcupines (*Erethizon dorsatum*)—using measures of electrical conductivity through the body, because the conductivity of lipids is only 4% to 5% that of other body tissues (Robbins 1993:236). In ungulates, ultrasound measures of rump fat and body condition scores based on palpation are used to predict total body fat, often in conjunction with other morphological measures. Visual scoring of avian fat stores and profiles is commonly used for waterfowl and songbirds. Servello et al. (2005) and Shipley et al. (2020) describe many of these techniques.

Physiological indicators when calibrated and validated can be used to infer nutritional deprivation and, indirectly, body condition (Table 11.2). As animals resort to breaking down more body protein to meet energy demands when fat stores are mostly depleted, higher levels of urinary nitrogen are excreted, as are metabolites of muscle breakdown. The use of nitrogen isotopes can further distinguish urinary nitrogen that results from breakdown of body protein versus nitrogen from food intake because of the preferential excretion of the lighter nitrogen isotope (^{14}N) in nitrogenous wastes, as mentioned previously. Hence in winter, when the isotopic content of winter forage remains relatively constant, a progressive increase in the excretion of the heavier isotope (^{15}N) would indicate greater mobilization of body protein (Parker et al. 2005). Glucocorticoid (cortisol or corticosterone) levels also may increase with increasing nutritional stress, particularly when population densities are high and food becomes more limiting. The excretion of allantoin (a urinary metabolite resulting from microbial fermentation and digestion of nucleic acids in forage) declines as digestible energy intakes decrease. For bison and elk, consecutive measures of allantoin have been used to index changes in available forage resources over time or among ranges. Higher levels of glucuronic acid are excreted when the consumption of plant secondary metabolites is high, which presumably reflects lower forage quality for such species as deer, snowshoe hares, and grouse. Each of these physiological indicators—urea δ^{15}N, glucocorticoids, allantoin, glucaronic acid—can be obtained noninvasively from urine samples in snow and is commonly presented as a ratio of creatinine. Creatinine is a normal excretory product proportional to muscle mass, and it serves as a background to avoid differences in dilution or hydration state. Urea nitrogen and glucocorticoid indicators also can be obtained using blood samples, but animal handling is required, and the indicators in blood typically vary more in response to short-term changes. Cook (2002) and Servello et al. (2005) report on other blood indices. Whenever possible, wildlife ecologists should use multiple indices of nutritional condition and available food resources to best assess nutritional status and constraints of study animals and study areas.

NUTRITIONAL ECOLOGY IN THE BIG PICTURE

For free-ranging species, life is a trade-off between costs and benefits. Over different spatial and temporal scales, animals choose what and where to eat, depending on quality and quantity of available food, energy spent to obtain it, and risks associated with weather and predation (Felton et al. 2018). Animals generally use seasonal strategies that minimize the greatest detriments to fitness at different times of the year (Parker et al. 2009). In northern winters, ungulates minimize energy expenditures by decreasing movement rates, particularly in deep snow areas and by avoiding the most thermally stressful environments where and when possible. At a time when food supplies are usually reduced, minimizing energy costs and the loss of protein stores increases the likelihood of survival and fetal development. Following successful birthing in spring, adult females must regain their body mass and replenish body fat stores for subsequent breeding in autumn. Males also must regain condition for successful breeding, and juveniles must amass sufficient size to ensure survival though their first winter. In spring through autumn, animals usually maximize intake of high-quality forage. In species that migrate from summer to winter ranges, strategies to lessen the primary deterrents to fitness depend on body condition. Animals in good nutritional condition may delay autumn migration to benefit from higher food quality at the risk of early winter snows. Animals in poor condition migrate earlier, trading off the potential of higher energy costs for the predictable albeit lower-quality food supply (Monteith et al. 2011). Under extreme conditions of hard winters or poor summers, some females may conserve maternal condition at the expense of producing offspring, ultimately enabling higher future reproduction (Clutton-Brock et al. 1989, Festa-Bianchet and Côté 2008). Pauses in breeding or embryonic mortality in early pregnancy represent trade-offs between investment in current versus future offspring for females with too few body stores to support both (Cameron 1994, Russell et al. 1998). In avian species, food quantity and quality also affect body stores for migration, and the ability to support income or restore capital during the reproductive period (Anteau and Afton 2004).

For many species, predation risks often are highest when young are produced, and reducing this risk coincides with the need to obtain nutrients to support the young (e.g., lactation). The northern mountain ecotype of woodland caribou (*R. t.*

caribou) chooses calving areas that have relatively low wolf risk and selects the highest-quality vegetation from what is available. Within two months of parturition, solitary female caribou and their calves move to form large groups in areas with higher food abundance (Gustine et al. 2006), minimizing risk while maximizing intake. Similarly, African ungulates birth when forage quality is high and then select for abundant plant biomass. Predation risk is lower during the highly synchronous birthing of large-herd species that swamps predator response and when parturition by small-herd species is spatially and temporally variable (Sinclair et al. 2000). But predation does not negate the importance of nutrition. In some cases, predation may even mask the nutritional value provided (or not provided) by different vegetation communities. In areas where nutritional value of forage is marginal for population recruitment, good body condition of mothers that have been relieved of the high energetic costs of lactation following neonatal loss as by predation might suggest that nutritional value is higher for the reproductive segment of the population than it is. In other cases, antipredator behavior may increase the nutritional deficits of winter, most affecting animals with high reproductive demands (Christianson and Creel 2010).

Small differences in food value and distribution compound the effects on animal performance (White 1983). Nutrient selection influences body mass and in turn conception rate, age at first breeding, breeding pauses and parturition, subsequent growth and survival of young, and survival of adults. As such, nutrition affects individuals, populations, and ecosystem dynamics. Not surprisingly, food intakes and habitat segregation may differ among females, males, and nonreproductive animals because nutritional requirements differ between sexes and with reproductive status (Barboza and Bowyer 2001). Feeding pressures can modify communities enough to change the complement of herbivores, their predators, and associated species on the landscape (Berger et al. 2001, Allombert et al. 2005).

Ever-increasing rates of environmental change pose challenges for wildlife. The ability to accommodate these changes depends on adaptation through nutritional interactions (i.e., meeting requirements when food resources change in timing, abundance, and quality). Climatic changes in winter (e.g., deep snows, freezing rains, icing events) can increase energy requirements and decrease access to food. Body stores acquired in summer help buffer these nutritional deficits. Earlier plant green-up induced by globally warmer temperatures in spring, however, may cause trophic mismatches between nutritional supplies and demands if animals are unable to alter the timing of reproduction (and potentially the timing of migration) so that it corresponds with peak food values. In these cases, life histories could be altered because of subsequent declines in body condition. Anthropogenic habitat changes that alter plant succession and plant community composition also inevitably influence the nutritional ecology of free-ranging wildlife. Wildlife species with high behavioral and physiological plasticity to change will be those that are most likely to survive and reproduce on changing landscapes. The concepts and tools referred to in this chapter help wildlife ecologists

assess nutritional constraints on habitat value by quantifying shifts in food abundance and quality as well as body condition that accompany environmental change.

SUMMARY

Nutritional requirements shape the activities, feeding patterns, and movements of wildlife. Because of their higher metabolism, small and young animals generally require more energy, more protein, and more water per unit mass than larger species. Reproductive demands also are relatively higher for small species than larger ones because eggs or new offspring typically are larger in proportion to body size. Wildlife species use food resources and body stores to meet these energetic and protein demands, which vary with life history and seasonally changing environments. Even when food quantity is sufficient, food quality may limit the growth of individuals and populations as it influences the nutrients that can be extracted from food. Food quality varies with plant and animal parts and also changes with season. Both the quantity and quality of food consumed influence what is possible in terms of body mass and condition of the animal, which then have direct consequences for survival and reproduction. As such, there are cascading effects of nutrition that link body condition, probability of pregnancy, over-winter survival, timing of parturition, neonatal birth mass and survival, juvenile growth rates, and adult mass gain. The field of nutritional ecology relates wildlife to their environments through such nutritional interactions.

The ability of biologists to predict nutritional value helps define the importance of seasonal habitats. Even small differences in nutrient supplies can have multiplier effects on animal performance, population size, and ecosystem dynamics. Nutrition remains important even in systems with predator influences because foodscapes affect vital rates for survival and reproduction. Understanding the trade-offs that wildlife species make to minimize deterrents to fitness enables biologists to better gauge the effects of changing environments. Foundational scientists of nutritional ecology, such as C. T. Robbins, R. G. White, and R. J. Hudson in the Northern Hemisphere and I. D. Hume and W. J. Foley in the Southern Hemisphere, have documented a wide range of adaptations that wild species use to survive and reproduce on different landscapes.

Literature Cited

Aikens, E. O., A Mysterud, J. A. Merkle, F. Cagnacci, I. M. Rivrud, et al. 2020. Wave-like patterns of plant phenology determine ungulate movement tactics. Current Biology 30:3444–3449.e4. https://doi.org/10.1016/j.cub.2020.06.032.

Allombert, S., A. J. Gaston, and J.-L. Martin. 2005. A natural experiment on the impact of overabundant deer on songbird populations. Biological Conservation 126:1–13.

Angerbjörn, A., P. Hersteinsson, K. Lidén, and E. Nelson. 1994. Dietary variation in Arctic foxes (*Alopex lagopus*)—an analysis of stable carbon isotopes. Oecologia 99:226–232.

Anteau, M. J., and A. D. Afton. 2004. Nutrient reserves of lesser scaup (*Aythya affinis*) during spring migration in the Mississippi Flyway: a test of the spring condition hypothesis. Auk 121:917–929.

Ayotte, J. B., K. L. Parker, J. M. Arocena, and M. P. Gillingham. 2006. Chemical composition of lick soils: functions of soil ingestion by four ungulate species. Journal of Mammalogy 87:878–888.

Bailey, J. A. 1984. Principles of wildlife management. John Wiley and Sons, New York, New York, USA.

Bakken, G. S., W. R. Santee, and D. J. Erskine. 1985. Operative and standard operative temperature: tools for thermal energetic studies. American Zoologist 25:933–943.

Barboza, P. S., and R. T. Bowyer. 2001. Seasonality of sexual segregation in dimorphic deer: extending the gastrocentric model. Alces 37:275–292.

Barboza, P. S., and K. L. Parker. 2008. Allocating protein to reproduction in arctic reindeer and caribou. Physiological and Biochemical Zoology 79:628–644.

Barboza, P. S., K. L. Parker, and I. D. Hume. 2009. Integrative wildlife nutrition. Springer-Verlag, Heidelberg, Germany.

Barboza, P. S., L. L. Van Someren, D. D. Gustine, and M. S. Bret-Harte. 2018. The nitrogen window for arctic herbivores: plant phenology and protein gain of migratory caribou (*Rangifer tarandus*). Ecosphere 9(1):e02073. doi:10.1002/ecs2.2073.

Beaver, J. M., B. E. Olson, and J. M. Wraith. 1996. A simple index of standard operative temperature for mule deer and cattle in winter. Journal of Thermal Biology 21:345–352.

Belovsky, G. E., and P. A. Jordan. 1981. Sodium dynamics and adaptations of a moose population. Journal of Mammalogy 62:613–621.

Berger, J., P. B. Stacey, L. Bellis, and M. P. Johnson. 2001. A mammalian predator–prey imbalance: grizzly bear and wolf extinction affect avian neotropical migrants. Ecological Applications 11:947–960.

Bidwell, M. T., and R. D. Dawson. 2005. Calcium availability limits reproductive output of tree swallows (*Tachycineta bicolor*) in a nonacidified landscape. Auk 122:246–254.

Bischof, R., L. E. Loe, E. L. Meisingset, B. Zimmerman, B. Van Moorter, and A. Mysterud. 2012. A migratory northern ungulate in the pursuit of spring: jumping or surfing the green wave? American Naturalist 180:407–424.

Blix, A. S. 2005. Arctic animals and their adaptations to life on the edge. Tapir Academic Press, Trondheim, Norway.

Bowen, W. D., D. Tully, D. J. Boness, D. Bulheier, and G. Marshal. 2002. Prey dependent foraging tactics and prey profitability in a marine mammal. Marine Ecology Progress Series 244:235–245.

Bowyer, R. T. 1991. Timing of parturition and lactation in southern mule deer. Journal of Mammalogy 72:138–145.

Bryant, J. P., and P. J. Kuropat. 1980. Selection of winter forage by subarctic browsing vertebrates: the role of plant chemistry. Annual Review of Ecology and Systematics 11:261–285.

Bugalho, M. N., J. A. Milne, R. A. Mayes, and F. C. Rego. 2005. Plant-wax alkanes as seasonal markers of red deer dietary components. Canadian Journal of Zoology 83:465–473.

Buskirk, S. W., S. C. Forrest, M. G. Raphael, and H. J. Harlow. 1989. Winter resting site ecology of marten in the central Rocky Mountains. Journal of Wildlife Management 53:191–196.

Buskirk, S. W., and H. H. Harlow. 1989. Body fat dynamics of the American marten (*Martes americana*) in winter. Journal of Mammalogy 70:191–193.

Butler, P. J., J. A. Green, I. L. Boyd, and J. R. Speakman. 2004. Measuring metabolic rate in the field: the pros and cons of the doubly labeled water and heart rate methods. Functional Ecology 18:168–183.

Cameron, R. D. 1994. Reproductive pauses by female caribou. Journal of Mammalogy 75:10–13.

Careau, V., J. Morand-Ferron, and D. Thomas. 2007. Basal metabolic rate of canidae from hot deserts to cold Arctic climates. Journal of Mammalogy 88:394–400.

Christianson, D., and S. Creel. 2009. Fecal chlorophyll describes the link between primary production and consumption in a terrestrial herbivore. Ecological Applications 19:1323–1335.

Christianson, D., and S. Creel. 2010. A nutritionally mediated risk effect of wolves on elk. Ecology 91:1184–1191.

Clutton-Brock, T. H., S. D. Albon, and F. E. Guinness. 1989. Fitness costs of gestation and lactation in wild mammals. Nature 337:260–262.

Collins, W. B., P. J. Urness, and D. D. Austin. 1978. Elk diets and activities on different lodgepole pine habitat segments. Journal of Wildlife Management 42:799–810.

Conomy, J. T., J. A. Collazo, J. A. Dubovsky, and W. J. Fleming. 1998. Dabbling duck behavior and aircraft activity in coastal North Carolina. Journal of Wildlife Management 62:1127–1134.

Cook, J. G. 2002. Nutrition and food. Pages 259–349 *in* D. E. Toweill and J. W. Thomas, editors. North American elk: ecology and management. Smithsonian Institution Press, Washington, DC, USA.

Cook, J. G., R. C. Cook, R. W. Davis, and L. L. Irwin. 2016. Nutritional ecology of elk during summer and autumn in the Pacific Northwest. Wildlife Monographs 195:1–81.

Cook, J. G., L. L. Irwin, L. D. Bryant, R. A. Riggs, and J. W. Thomas. 1998. Relations of forest cover and condition of elk: a test of the thermal cover hypothesis in summer and winter. Wildlife Monographs 141:1–61.

Cook, J. G., B. K. Johnson, R. C. Cook, R. A. Riggs, T. Delcurto, L. D. Bryant, and L. L. Irwin. 2004. Effects of summer-autumn nutrition and parturition date on reproduction and survival of elk. Wildlife Monographs 155:1–61.

Cook, R. C., J. G. Cook, T. R. Stephenson, W. L. Myers, S. M. McCorquodale, et al. 2010. Revisions of rump fat and body scoring indices for deer, elk, and moose. Journal of Wildlife Management 74:880–896.

Crête, M., and S. Larivière. 2003. Estimating the costs of locomotion in snow for coyotes. Canadian Journal of Zoology 81:1808–1814.

Dailey, T. V., and N. T. Hobbs. 1989. Travel in alpine terrain: energy expenditures for locomotion by mountain goats and bighorn sheep. Canadian Journal of Zoology 67:2368–2375.

Darimont, C. T., P. C. Paquet, and T. E. Reimchen. 2008. Spawning salmon disrupt trophic coupling between wolves and ungulate prey in coastal British Columbia. BioMed Central Ecology 8:14.

Dawson, T. J., and C. R. Taylor. 1973. Energetic cost of locomotion in kangaroos. Nature 246:313–314.

Dearing, M. D., W. J. Foley, and S. McLean. 2005. The influence of plant secondary metabolites on the nutritional ecology of herbivorous terrestrial vertebrates. Annual Review of Ecology, Evolution, and Systematics 36:169–189.

DelGiudice, G. D. 1995. Assessing winter nutritional restriction of northern deer with urine in snow: considerations, potential and limitations. Wildlife Society Bulletin 23:687–693.

Denryter, K. L., R. C. Cook, J. G. Cook, and K. L. Parker. 2017. Straight from the caribou's (*Rangifer tarandus*) mouth: detailed observations of tame caribou reveal new insights into summer-autumn diets. Canadian Journal of Zoology 95:81–94.

Dove, H., and R. W. Mayes. 2005. Using *n*-alkanes and other plant wax components to estimate intake, digestibility and diet composition of grazing/browsing sheep and goats. Small Ruminant Research 59:123–139.

Doxon, E. D., and J. P. Carroll. 2010. Feeding ecology of ring-necked pheasant and northern bobwhite chicks in conservation reserve program fields. Journal of Wildlife Management 74:249–256.

Fancy, S. G., and R. G. White. 1985. Incremental cost of activity. Pages 143–159 in R. J. Hudson, and R. G. White, editors. Bioenergetics of wild herbivores. CRC Press, Boca Raton, Florida, USA.

Felicetti, L. A., C. C. Schwartz, R. O. Rye, M. A. Haroldson, K. A. Gunther, D. L. Phillips, and C. T. Robbins. 2003. Use of sulfur and nitrogen stable isotopes to determine the importance of whitebark pine nuts to Yellowstone grizzly bears. Canadian Journal of Zoology 81:763–770.

Felton, A. M., H. K. Wam, C. Stolter, K. M. Mathisen, and M. Wallgren. 2018. The complexity of interacting nutritional drivers behind food selection, a review of northern cervids. Ecosphere 9(5):e02230. doi:10.1002/ecs2.2230.

Festa-Bianchet, M., and S. D. Côté. 2008. Mountain goats: ecology, behavior, and conservation of an alpine ungulate. Island Press, Washington, DC, USA.

Fryxell, J. M. 1991. Forage quality and aggregation by large herbivores. American Naturalist 138:478–498.

Garnick, S., P. S. Barboza, and J. W. Walker. 2018. Assessment of animal-based methods used for estimating and monitoring rangeland herbivore diet composition. Rangeland Ecology and Management 71:449–457.

Geiser, F., B. M. McAllan, and G. J. Kenagy. 1994. The degree of dietary fatty acid unsaturation affects torpor patterns and lipid composition of a hibernator. Journal of Comparative Physiology B 164:299–305.

Gende, S. M., T. P. Quinn, and M. F. Willson. 2001. Consumption choice by bears feeding on salmon. Oecologia 127:372–382.

Gerhardt, K. L., R. G. White, R. D. Cameron, and D. E. Russell. 1996. Estimating fat content of caribou from body condition scores. Journal of Wildlife Management 60:713–718.

Gessaman, J. A. 1972. Bioenergetics of the snowy owl (Nyctea scandiaca). Arctic and Alpine Research 4:223–238.

Gillingham, M. P., K. L. Parker, and T. A. Hanley. 1997. Forage intake by large herbivores in a natural environment: bout dynamics. Canadian Journal of Zoology 75:1118–1128.

Goering, H. K., and P. J. Van Soest. 1970. Forage analyses (apparatus, reagents, procedures, and some applications). Agricultural Handbook 379. US Department of Agriculture, Washington, DC, USA.

Griffin, T. M., and R. Kram. 2000. Penguin waddling is not wasteful. Nature 408:929.

Griffith, B., D. C. Douglas, N. E. Walsh, D. D. Young, T. R. McCabe, D. E. Russell, R. G. White, R. D. Cameron, and K. R. Whitten. 2002. The porcupine caribou herd. Pages 8–37 in Arctic refuge coastal plain terrestrial wildlife research summaries. Biological Science Report 2002-0001. Biological Resources Division, US Geological Survey, Anchorage, Alaska, USA.

Guglielmo, C. G., W. H. Karasov, and W. J. Jakubas. 1996. Nutritional costs of a plant secondary metabolite explain selective foraging by ruffed grouse. Ecology 77:1103–1115.

Gurchinoff, S., and W. L. Robinson. 1972. Chemical characteristics of jackpine needles selected by feeding spruce grouse. Journal of Wildlife Management 36:80–87.

Gustine, D. D., P. S. Barboza, L. G. Adams, R. G. Farnell, and K. L. Parker. 2011. An isotopic approach to measuring nitrogen balance in caribou. Journal of Wildlife Management 75:178–188.

Gustine, D. D., K. L. Parker, R. J. Lay, M. P. Gillingham, and D. C. Heard. 2006. Calf survival of woodland caribou in a multi-predator ecosystem. Wildlife Monographs 165:1–32.

Guthery, F. S. 1999. Energy-based carrying capacity for quails. Journal of Wildlife Management 63:664–674.

Hamel, S., M. Garel, M. Festa-Bianchet, J.-M. Gaillard, and S. D. Côté. 2009. Spring normalized difference vegetation index (NDVI) predicts annual variation in timing of peak faecal crude protein in mountain ungulates. Journal of Applied Ecology 46:582–589.

Hanley, T. A., C. T. Robbins, A. E. Hagerman, and C. McArthur. 1992. Predicting digestible protein and digestible dry matter in tannin-containing forages consumed by ruminants. Ecology 73:537–541.

Hanley, T. A., and J. J. Rogers. 1989. Estimating carrying capacity with simultaneous nutritional constraints. Research Note PNW-RN-485. Pacific Northwest Research Station, US Forest Service, Portland, Oregon, USA.

Hanley, T. A., D. E. Spalinger, K. J. Mock, O. L. Weaver, and G. M. Harris. 2012. Forage resource evaluation system for habitat–deer: an interactive deer–habitat model. General Technical Report PNW-GTR-858. Pacific Northwest Research Station, US Forest Service, Portland, Oregon, USA. http://treesearch.fs.fed.us/pubs/40300.

Happe, P. J., K. J. Jenkins, E. E. Starkey, and S. H. Sharrow. 1990. Nutritional quality and tannin astringency of browse in clear-cuts and old-growth forests. Journal of Wildlife Management 54:557–566.

Harder, J. D., and R. L. Kirkpatrick. 1994. Physiological indices in wildlife research. Pages 275–306 in T. A. Bookhout, editor. Research and management techniques for wildlife and habitats. 5th edition. The Wildlife Society, Bethesda, Maryland, USA.

Harlow, H. 1981. Effect of fasting on rate of food passage and assimilation efficiency in badgers. Journal of Mammalogy 62:173–177.

Hassall, M. A., S. J. Lane, M. Stock, S. M. Percival, and B. Pohl. 2001. Monitoring feeding behaviour of Brent geese Branta bernicla using position-sensitive radio transmitters. Wildlife Biology 7:77–86.

Hedenström, A., and T. Alerstam. 1995. Optimal flight speed of birds. Philosophical Transactions of the Royal Society of London B 348:471–487.

Hildebrand, G. V., S. D. Farley, and C. T. Robbins. 1998. Predicting body condition of bears via two field methods. Journal of Wildlife Management 62:406–409.

Hobbs, N. T., and D. M. Swift. 1985. Estimates of habitat carrying capacity incorporating explicit nutritional constraints. Journal of Wildlife Management 49:814–822.

Hobson, K. A. 1995. Reconstructing avian diets using stable-carbon and nitrogen isotope analysis of egg components: patterns of isotopic fractionation and turnover. Condor 97:752–762.

Holechek, J. L., M. Vavra, and R. D. Pieper. 1982. Botanical composition determination of range herbivore diets: a review. Journal of Range Management 35:309–315.

Hupp, J. W., R. G. White, J. S. Sedinger, and D. G. Robertson. 1996. Forage digestibility and intake by lesser snow geese: effects of dominance and resource heterogeneity. Oecologia 108:232–240.

Hurley, M. A., M. Hebblewhite, J.-M. Gaillard, S. Dray, K. A. Taylor, W. K. Smith, P. Zager, and C. Bonenfant. 2014. Functional analysis of normalized difference vegetation index curves reveals overwinter mule deer survival is driven by both spring and autumn phenology. Philosophical Transactions of the Royal Society B 369:20130196.

Iverson, S. J., C. Field, W. D. Bowen, and W. Blanchard. 2004. Quantitative fatty acid signature analysis: a new method of estimating predator diets. Ecological Monographs 74:211–235.

Iverson, S. J., D. E. McDonald, and L. H. Smith. 2001. Changes in the diet of free-ranging black bears in years of contrasting food availability revealed through milk fatty acids. Canadian Journal of Zoology 79:2268–2279.

Johnson, C. J., K. L. Parker, and D. C. Heard. 2000. Feeding site selection by woodland caribou in north-central British Columbia. Rangifer, Special Issue 12:159–172.

Johnson, G. D., and M. S. Boyce. 1990. Feeding trials with insects in the diet of sage grouse chicks. Journal of Wildlife Management 54:89–91.

Johnson, H. E., T. S. Golden, L. G. Adams, D. D. Gustine, E. A. Lenart, and P. S. Barboza. 2021. Dynamic selection for forage quality and quantity in response to phenology and insects in an arctic ungulate. Ecology and Evolution 11 11664–11688. doi:https://doi.org/10.1002/ece3.7852.

Johnson, H. E., D. D. Gustine, T. S. Golden, L. G. Adams, L. S. Parrett, E. A. Lenart, and P. S. Barboza. 2018. NDVI exhibits mixed success in prediction spatiotemporal variation in caribou summer forage quality and quantity. Ecosphere 9(10):e02461. doi:10.1002/ecs2.2461.

Kamler, J., M. Homolka, and D. Čižmár. 2003. Suitability of NIRS analysis for estimating diet quality of free-living red deer *Cervus elaphus* and roe deer *Capreolus capreolus*. Wildlife Biology 10:235–240.

Karasov, W. H., and C. Martinez del Rio. 2007. Physiological ecology: how animal process energy, nutrients, and toxins. Princeton University Press, Princeton, New Jersey, USA.

Kelly, A. 2020. Seasonal patterns of mortality for boreal caribou (*Rangifer tarandus caribou*) in an intact environment. MSc thesis. University of Alberta, Edmonton, Canada.

Kielland, K. 2001. Stable isotope signatures of moose in relation to seasonal forage composition: a hypothesis. Alces 37:329–337.

Kielland, K., and G. Finstad. 2000. Differences in 15N natural abundance reveal seasonal shifts in diet choice of reindeer and caribou. Rangifer 12:145.

Klare, U., J. F. Kamler, and D. W. McDonald. 2011. A comparison and critique of different scat-analysis methods for determining carnivore diet. Mammal Review 41:294–312.

Kleiber, M. 1947. Body size and metabolic rate. Physiological Reviews 27:511–541.

Koch, P. L., J. Heisinger, C. Moss, R. W. Carlson, M. L. Fogel, and A. K. Behrensmeyer. 1995. Isotope tracking of change in diet and habitat use in African elephants. Science 267:1340–1343.

Krementz, D. G., and G. W. Pendelton. 1990. Fat scoring: sources of variability. Condor 92:500–507.

Lawler, J. P., and R. G. White. 2003. Temporal responses in energy expenditure and respiratory quotient following feeding in muskox: influence of season on energy costs of eating and standing and an endogenous heat increment. Canadian Journal of Zoology 81:1524–1538.

Leslie, D. M., R. T. Bowyer, and J. A. Jenks. 2008. Facts from feces: nitrogen still measures up as a nutritional index for mammalian herbivores. Journal of Wildlife Management 72:1420–1433.

Long, R. A., R. T. Bowyer, W. P. Porter, P. Mathewson, K. L. Monteith, and J. G. Kie. 2014. Behavior and nutritional condition buffer a large-bodied endotherm against direct and indirect effects of climate. Ecological Monographs 84:513–532.

Lovvorn, J. R., J. M. Grebmeier, L. W. Cooper, J. K. Bump, and S. E. Richman. 2009. Modeling marine protected areas for threatened eiders in a climatically changing Bering Sea. Ecological Applications 19:1596–1613.

Mack, R. N., and J. N. Thompson. 1982. Evolution in steppe with few large, hooved mammals. American Naturalist 119:757–773.

Marsh, K. J., I. R. Wallis, R. L. Andrew, and W. J. Foley. 2006. The detoxification limitation hypothesis: where did it come from and where is it going? Journal of Chemical Ecology 32:1247–1266.

Mautz, W. W. 1978. Sledding on a bushy hillside: the fat cycle in deer. Wildlife Society Bulletin 6:88–90.

Mautz, W. W., P. J. Pekins, and J. A. Warren. 1985. Cold temperature effects on metabolic rate of white-tailed deer, mule deer and black-tailed deer in winter. Pages 453–457 in P. F. Fennessy and K. R. Drew, editors. Biology of deer production. Royal Society of New Zealand Bulletin 22:453–457.

Mayes, R. W., and H. Dove. 2000. Measurement of dietary nutrient intake in free-ranging mammalian herbivores. Nutrition Research Reviews 13:107–138.

McArt, S. H., D. E. Spalinger, W. B. Collins, E. R. Schoen, T. Stevenson, and M. Bucho. 2009. Summer dietary nitrogen availability as a potential bottom-up constraint on moose in south-central Alaska. Ecology 90:1400–1411.

McArthur, C., C. T. Robbins, A. E. Hagerman, and T. A. Hanley. 1993. Diet selection by a ruminant generalist browser in relation to plant chemistry. Canadian Journal of Zoology 71:2236–2243.

McGuire, L. P., and C. G. Guglielmo. 2010. Quantitative magnetic resonance: a rapid, noninvasive body composition analysis technique for live and salvaged bats. Journal of Mammalogy 91:1375–1380.

McNaughton, S. J. 1990. Mineral nutrition and seasonal movements of African migratory ungulates. Nature 345:613–615.

Merems, J. L. L. A. Shipley, T. Levi, J. Ruprecht, D. A. Clark, M. J. Wisdom, N. J. Jackson, K. M. Stewart, and R. A. Long. 2020. Nutritional-landscape models link habitat use to condition of mule deer (*Odocoileus hemionus*). Frontiers in Ecology and Evolution 8:98. https://doi.org/10.3389/fevo.2020.00098.

Merkle, J., K. L. Monteith, E. O. Aikens, M. M. Hayes, K. R. Hersey, A. Middleton, B. A. Oates, H. Sawyer, B. Scurlock, and M. Kauffman. 2016. Large herbivores surf waves of green-up during spring. Proceedings of the Royal Society B 283:20160456.

Milakovic, B., and K. L. Parker. 2013. Quantifying carnivory by grizzly bears in a multi-ungulate system. Journal of Wildlife Management 77:39–47.

Monteith, K. L., V. C. Bleich, T. R. Stephenson, B. M. Pierce, M. M. Conner, R. W. Klaver, and R. T. Bowyer. 2011. Timing of seasonal migration in mule deer: effects of climate, plant phenology, and life-history characteristics. Ecosphere 2:1–34.

Monteith, K. L., T. R. Stephenson, V. C. Bleich, M. M. Conner, B. M. Pierce, and R. T. Bowyer. 2013. Risk-sensitive allocation in seasonal dynamics of fat and protein reserves in a long-lived mammal. Journal of Animal Ecology 82:377–388.

Moore, B. D., and W. J. Foley. 2005. Tree use by koalas in a chemically complex landscape. Nature 435:488–490.

Mould, E. D., and C. T. Robbins. 1981. Evaluation of detergent analysis in estimating nutritional value of browse. Journal of Wildlife Management 45:323–334.

Müller, M. S., S. R. McWilliams, D. W. Podlesak, J. R. Donaldson, H. M. Bothwell, and R. L. Lindroth. 2006. Tri-trophic effects of plant defenses: chickadees consume caterpillars based on host leaf chemistry. Oikos 114:507–517.

Murray, D. C., M. Bunce, B. L. Cannell, R. Oliver, J. Houston, N. E. White, R. A. Barrero, M. I. Bellgard, and J. Haile. 2011. DNA-based faecal dietary analysis: a comparison of qPCR and high throughput sequencing approaches. PLoS One 6(10):e25776.

Nagy, K. A., I. A. Girard, and T. K. Brown. 1999. Energetics of free-ranging mammals, reptiles, and birds. Annual Review of Nutrition 19:247–277.

Nagy, T. R. 2001. The use of dual-energy X-ray absorptiometry for the measurement of body composition. Pages 211–229 in J. R. Speakman, editor. Body composition analysis of animals: a handbook of non-destructive methods. Cambridge University Press, Cambridge, UK.

Newsome, S. D., M. T. Clementz, and P. L. Koch. 2010. Using stable isotope biogeochemistry to study marine mammal ecology. Marine Mammal Science 26:509–572.

Osborn, R. G., and T. F. Ginnett. 2001. Fecal nitrogen and 2,6-diaminopimelic acid as indices to dietary nitrogen in white-tailed deer. Wildlife Society Bulletin 29:1131–1139.

Parker, K. L. 1988. Effects of heat, cold, and rain on black-tailed deer. Canadian Journal of Zoology 66:2475–2483.

Parker, K. L. 2003. Advances in the nutritional ecology of cervids at different scales. Ecoscience 10:395–411.

Parker, K. L., P. S. Barboza, and M. P. Gillingham. 2009. Nutrition integrates environmental responses of ungulates. Functional Ecology 23:57–69.

Parker, K. L., P. S. Barboza, and T. R. Stephenson. 2005. Protein conservation in female caribou (Rangifer tarandus): effects of decreasing diet quality during winter. Journal of Mammalogy 86:610–622.

Parker, K. L., and M. P. Gillingham. 1990. Estimates of critical thermal environments for mule deer. Journal of Range Management 43:73–81.

Parker, K. L., M. P. Gillingham, and T. A. Hanley. 1993a. An accurate technique for estimating forage intake of tractable animals. Canadian Journal of Zoology 71:1462–1465.

Parker, K. L., M. P. Gillingham, T. A. Hanley, and C. T. Robbins. 1993b. Seasonal patterns in body weight, body condition, and water transfer rates of free-ranging and captive black-tailed deer (Odocoileus hemionus sitkensis) in Alaska. Canadian Journal of Zoology 71:1397–1404.

Parker, K. L., M. P. Gillingham, T. A. Hanley, and C. T. Robbins. 1999. Energy and protein balance of free-ranging black-tailed deer in a natural forest environment. Wildlife Monographs 143:1–48.

Parker, K. L., and C. T. Robbins. 1984. Thermoregulation in mule deer and elk. Canadian Journal of Zoology 62:1409–1422.

Parker, K. L., and C. T. Robbins. 1985. Thermoregulation in ungulates. Pages 161–182 in R. J. Hudson and R. G. White, editors. Bioenergetics of wild herbivores. CRC Press, Boca Raton, Florida, USA.

Parker, K. L., C. T. Robbins, and T. A. Hanley. 1984. Energy expenditures for locomotion by mule deer and elk. Journal of Wildlife Management 48:474–488.

Pettorelli, N., S. Ryan, T. Mueller, N. Bunnefeld, B. A. Jedrzejewska, M. Lima, and K. Kausrud. 2011. The normalized difference vegetation index (NDVI): unforeseen successes in animal ecology. Climate Research 46:15–27.

Pettorelli, N., J. O. Vik, A. Mysterud, J.-M. Gaillard, C. J. Tucker, and N. C. Stenseth. 2005. Using the satellite-derived NDVI to assess ecological responses to environmental change. Trends in Ecology and Evolution 20:503–510.

Powers, J. G., W. W. Mautz, and P. J. Pekins. 1989. Nutrient and energy assimilation of prey by bobcats. Journal of Wildlife Management 53:1004–1008.

Pritchard, G. T., and C. T. Robbins. 1990. Digestive and metabolic efficiencies of grizzly and black bears. Canadian Journal of Zoology 68:1645–1651.

Prosky, L., N. Asp, I. Furda, J. W. Devries, T. F. Schweizer, and B. F. Harland. 1984. Determination of total dietary fiber in foods, food products, and total diets: interlaboratory study. Journal of the Association of Official Agricultural Chemists 67:1044–1052.

Ramsay, M. A., and K. A. Hobson. 1991. Polar bears make little use of terrestrial food webs: evidence from stable-carbon isotope analysis. Oecologia 86:598–600.

Reimers, E., S. Eftestøl, and J. E. Colman. 2003. Behavior responses of wild reindeer to direct provocation by a snowmobile or skier. Journal of Wildlife Management 67:747–754.

Renecker, L. A., and R. J. Hudson. 1986. Seasonal energy expenditures and thermoregulatory responses of moose. Canadian Journal of Zoology 64:322–327.

Robbins, C. T. 1993. Wildlife feeding and nutrition. 2nd edition. Academic Press, San Diego, California, USA.

Robbins, C. T., T. A. Hanley, A. E. Hagerman, O. Hjeljord, D. L. Baker, C. C. Schwartz, and W. W. Mautz. 1987. Role of tannins in defending plants against ruminants: reduction in protein availability. Ecology 68:98–107.

Russell, D. E., K. L. Gerhart, R. G. White, and D. Van de Wetering. 1998. Detection of early pregnancy in caribou: evidence for embryonic mortality. Journal of Wildlife Management 62:1066–1076.

Russell, D. E., A. M. Martel, and W. A. C. Nixon. 1993. Range ecology of the porcupine caribou herd in Canada. Rangifer, Special Issue 8:1–168.

Saltz, D., and G. C. White. 1991. Urinary cortisol and urea nitrogen responses in irreversibly undernourished mule deer fawns. Journal of Wildlife Diseases 27:41–46.

Schrempp, T. V., J. L. Rachlow, T. R. Johnson, L. A. Shipley, R. A. Long, J. L. Aycrigg, and M. A. Hurley. 2019. Linking forest management to moose population trends: the role of the nutritional landscape. PLoS One 14(7):e0219128. https://doi.org/10.1371/journal.pone.0219128.

Scott, I., C. Selman, P. I. Mitchell, and P. R. Evans. 2001. The use of total body electrical conductivity (TOBEC) to determine body condition in vertebrates. Pages 127–160 in J. R. Speakman, editor. Body composition analysis of animals: a handbook of non-destructive methods. Cambridge University Press, Cambridge, UK.

Scribner, K. T., M. H. Smith, and P. E. Jones. 1989. Environmental and genetic components of antler growth in white-tailed deer. Journal of Mammalogy 70:284–291.

Searle, K. R., N. T. Hobbs, and I. J. Gordon. 2007. It's the "foodscape," not the landscape: using foraging behavior to make functional assessments of landscape condition. Israel Journal of Ecology and Evolution 53:297–316.

Servello, F. A., E. C. Hellgren, and S. R. McWilliams. 2005. Techniques for wildlife nutritional ecology. Pages 554–590 in C. E. Braun, editor. Techniques for wildlife investigations and management. 6th edition. The Wildlife Society, Bethesda, Maryland, USA.

Servello, F. A., and J. W. Schneider. 2000. Evaluation of urinary indices of nutritional status for white-tailed deer: test with winter browse diets. Journal of Wildlife Management 64:137–145.

Shipley, L. A. 2007. The influence of bite size on foraging at larger spatial and temporal scales by mammalian herbivores. Oikos 116:1964–1974.

Shipley, L. A., R. C. Cook, and D. G. Hewitt. 2020. Techniques for wildlife nutritional ecology. Volume 1: Research. Pages 439–482 in N. J. Silvy, editor. The wildlife techniques manual. 8th edition. The Wildlife Society and Johns Hopkins University Press, Baltimore, Maryland, USA.

Shipley, L. A., and D. E. Spalinger. 1992. Mechanics of browsing in dense food patches: effects of plant and animal morphology on intake rate. Canadian Journal of Zoology 70:1743–1752.

Sinclair, A. R. E., S. A. R. Mduma, and P. Arcese. 2000. What determines phenology and synchrony of ungulate breeding in Serengeti? Ecology 81:2100–2111.

Spalinger, D. E., W. B. Collins, T. A. Hanley, N. E. Cassara, and A. M. Carnahan. 2010. The impact of tannins on protein, dry matter, and energy digestion in moose (Alces alces). Canadian Journal of Zoology 88:977–987.

Speakman, J. R., G. H. Visser, S. Ward, and E. Krol. 2001. The isotope dilution method for the evaluation of body composition. Pages 56–98 in J. R. Speakman, editor. Body composition analysis of animals: a handbook of non-destructive methods. Cambridge University Press, Cambridge, UK.

Stenseth, C. C., W. Falck, O. N. Bjørnstad, and C. J. Krebs. 1997. Population regulation in snowshoe hare and Canadian lynx: asymmetric food web configurations between hare and lynx. Proceedings of the National Academy of Sciences 94:5147–5152.

Stephenson, T. R., D. W. German, E. F. Cassirer, D. P. Walsh, M. E. Blum, M. Cox, K. M. Stewart, and K. L. Monteith. 2020. Linking population performance to nutritional condition in an alpine ungulate. Journal of Mammalogy 101:1244–1256.

Svoboda, F. J., and G. W. Gullion. 1972. Preferential use of aspen by ruffed grouse in northern Minnesota. Journal of Wildlife Management 36:1166–1180.

Teer, J. G., J. W. Thomas, and E. A. Walker. 1965. Ecology and management of white-tailed deer in the Llano Basin of Texas. Wildlife Monographs 15:1–62.

Telfer, E. S., and J. P. Kelsall. 1984. Adaptation of some large North American mammals for survival in snow. Ecology 65:1828–1834.

Tilley, J. M. A., and R. A. Terry. 1963. A two-stage technique for *in vitro* digestion of forage crops. Journal of the British Grassland Society 18:104–111.

Tinbergen, J. M., and J. B. Williams. 2002. Energetics of incubation. Pages 299–313 in D. C. Deeming, editor. Avian incubation: behaviour, environment and evolution. Oxford University Press, Oxford, UK.

Tollit, D. J., S. G. Heaslip, B. Deagle, S. J. Iverson, R. Joy, D. A. S. Rosen, and A. W. Trites. 2006. Estimating diet composition in sea lions: which technique to choose? Pages 293–308 in A. W. Trites, S. K. Atkinson, D. P. DeMaster, et al., editors. Sea lions of the world. Alaska Sea Grant College Program, University of Alaska Fairbanks, Anchorage, USA.

Tollit, D. J., M. A. Wong, and A. W. Trities. 2015. Diet composition of Steller sea lions (*Eumetopias jubatus*) in Frederick Sound, southeast Alaska: a comparison of quantification methods using scats to describe temporal and spatial variabilities. Canadian Journal of Zoology 93:361–376.

Torbit, S. C., L. H. Carpenter, A. W. Alldredge, and D. M. Swift. 1985. Mule deer body composition—a comparison of methods. Journal of Wildlife Management 49:86–91.

Trudell, J., and R. G. White. 1981. The effect of forage structure and availability on food intake, biting rate, bite size and daily eating time of reindeer. Journal of Applied Ecology 18:63–81.

Tulp, I., H. Schekkerman, L. W. Bruinzeel, J. Jukema, G. H. Visser, and T. Piersma. 2009. Energetic demands during incubation and chick rearing in a uniparental and a biparental shorebird breeding in the high Arctic. Auk 126:155–164.

Tyler, N. J. C., P. Gregorini, K. L. Parker, and D. G. Hazlerigg. 2020. Animal responses to environmental variation: physiological mechanisms in ecological models of performance in deer (Cervidae). Animal Production Science 60:1248–1270.

Vagnoni, D. B., R. A. Garrott, J. G. Cook, and P. J. White. 1996. Urinary allantoin:creatinine ratios as a dietary index for elk. Journal of Wildlife Management 60:728–734.

Valentini, A., F. Pompanon, and P. Taberlet. 2008. DNA barcoding for ecologists. Trends in Ecology and Evolution 24:110–117.

Van Vuren, D., and B. E. Coblentz. 1985. Kidney weight variation and the kidney fat index: an evaluation. Journal of Wildlife Management 49:177–179.

Vleck, C. M., and T. L. Bucher. 1998. Energy metabolism, gas exchange, and ventilation. Pages 89–116 in J. M. Starck and R. E. Ricklefs, editors. Avian growth and development—evolution within the altricial–precocial spectrum. Oxford University Press, New York, New York, USA.

Walker, A. B. D., K. L. Parker, and M. P. Gillingham. 2006. Behaviour, habitat associations and intrasexual differences of female Stone's sheep. Canadian Journal of Zoology 84:1187–1201.

White, R. G. 1983. Foraging patterns and their multiplier effects on productivity of northern ungulates. Oikos 40:377–384.

White, R. G., D. E. Russell, and C. J. Daniel. 2014. Simulation of maintenance, growth and reproduction of caribou and reindeer as influenced by ecological aspects of nutrition, climate change and industrial development using an energy-protein model. Rangifer 34, Special Issue No. 22:1–125.

Wickstrom, M. L., C. T. Robbins, T. A. Hanley, D. E. Spalinger, and S. M. Parish. 1984. Food intake and foraging energetics of elk and mule deer. Journal of Wildlife Management 48:1285–1301.

Williams, T. M., W. A. Friedl, M. L. Fong, R. M. Yamada, P. Sedivy, and J. E. Haun. 1992. Travel at low energetic cost by swimming and wave-riding bottlenose dolphins. Nature 355:821–823.

Williams, T. M., L. Wolfe, T. Davis, T. Kendall, B. Richter, Y. Wang, C. Bryce, G. Hugh Elkaim, and C. C. Wilmers. 2014. Instantaneous energetics of puma kills reveal advantage of felid sneak attacks. Science 346:81–85.

Worker, S. B., Kielland, K., and P. S. Barboza. 2015. Effects of geophagy on food intake, body mass, and nutrient dynamics of snowshoe hares (*Lepus americanus*). Canadian Journal of Zoology 93:323–329.

12 PLANT–ANIMAL INTERACTIONS

KELLEY M. STEWART

INTRODUCTION

In the broadest context, plant–animal interactions are classified as any relationship that occurs between the kingdoms Animalia and Plantae (Anderson 2012). Nearly every ecosystem on the planet is affected by interactions between animals and plants, especially those mediated by pollinators or by herbivory. These interactions are common in virtually every environment, including terrestrial, marine, and freshwater biomes (Anderson 2012). Nearly all terrestrial animals depend on primary production from plants in some manner because more than 90% of the energy in terrestrial ecosystems is fixed by green plants (Price 2002). Additionally, most plant species are influenced by animal-mediated pollination or dispersal of seeds. Therefore plant distributions often define the location and abundance of animals (Price 2002). Regal (1977) argued convincingly the case for dominance of flowering plants based on mutualistic relationships between plants, pollinators, and seed dispersers; because the basis for wide-ranging occupation of animals around the world is dependent on the resources provided by plants (Price 2002). Even species that do not directly consume plant materials depend on plants for cover, nest or bed sites, foraging sites, or in a secondary way by preying upon or parasitizing herbivores (Price 2002). Interactions between plants and animals have structured the ecology of natural communities and the evolution of interacting species worldwide. Those interactions are important influencers of ecological processes and the evolution and maintenance of biodiversity (Becerra 2007, Steele et al. 2018, Hamann et al. 2021).

Much of the research on plant–animal interactions describes relationships between plants and insects, including insect pollinators, plant–ant mutualisms, insect herbivory, and others. Additionally, researchers have focused on interactions between plants and animals in marine or freshwater ecosystems. This chapter, however, in the context of wildlife ecology, conservation, and management, is more specific to, and will focus primarily on, interactions between vertebrate animals and plants in terrestrial ecosystems, except for when it is also important to discuss interactions between plants and insects. Therefore this discussion includes some information on pollination by bees (Hymenoptera) and mutualisms between ants (Formicidae) and plants.

Animals are often defined by what they eat: omnivores eat a variety of foods, including plant material and herbivores eat plants (Danell and Bergström 2002). Herbivores forage on a variety of plants or plant parts, but this description can be further broken down into specific plant types or parts. For instance, moose (*Alces alces*) and mule deer (*Odocoileus hemionus*) that feed on woody plants and forbs are termed browsers. Bison (*Bison bison*) and Cape buffalo (*Syncerus caffer*) are grazers that primarily consume grasses and some other herbaceous plants. Mixed feeders like North American elk (*Cervus canadensis*) feed on a variety of forbs, grasses, and shrubs. Some herbivores, like Botta's pocket gophers (*Thomomys bottae*), focus on belowground tissues, primarily roots and storage organs. Many rodents and birds, like kangaroo rats (*Dipodomys spp.*) and white-winged crossbills (*Loxia leucoptera*), are termed granivores, which primarily consume seeds rather than other plant tissues. Frugivores forage specifically on fruits; for example, trumpeter hornbills (*Bycanistes bucinator*) are the largest obligate frugivores in South Africa (Mueller et al. 2014). Terrestrial animals that feed directly on plants also include nectarivores that consume nectar, pollen feeders, folivores that consume leaves, and exudativores like the pygmy marmoset (*Cebuella pygmaea*) that consume exudates or secretions that ooze or diffuse from plant tissues (Price 2002).

Herbivores can be highly specialized, such as giant pandas (*Ailuropoda melanoleuca*) that feed exclusively on bamboo (Bambusoideae) or hummingbirds (Apodiformes) that feed on nectar. Conversely, some animals can be unselective generalists like coyotes (*Canis latrans*) that consume a variety of foods including animal material, insects, and different types of fruits. Vertebrate herbivores range in body size from tiny hummingbirds, like the 3.5-cm bee hummingbird (*Mellisuga helenae*), small rodents (Rodentia), and shrews (*Sorex* spp.) that eat the highest-quality plant parts, to large-bodied species like Cape buffalo that are relatively unspecific in species of grasses that they consume (Anderson 2012). Cape buffalo often consume several species of grasses in a single bite. Specialization in food choices can be extreme in its manifestation. The kakapo (*Strigops habroptilus*) is an endangered, flightless parrot from New Zealand that feeds on a variety of forages. Kakapos only produce young, however, when the rimu tree (*Dacrydium cupressinum*) is full of fruit, which only happens every 3–5 years (Fidler et al. 2008). The fruit of the rimu trees contains dietary phytochemicals that are hypothesized to stimulate ovarian follicles to develop to ovulation and egg laying in female kakapos (Fidler et al. 2008).

The ultimate value for interactions between animals and plants is for each species to increase their own reproductive success or fitness. Moreover, those interactions among plants and animals may be antagonistic, neutral, or facilitative (Bronstein 2009, Traveset and Richardson 2014). Antagonistic interactions include competition, where both partners are negatively affected, and predation (including herbivory) and parasitism, where one player benefits but the other is harmed or killed (Anderson 2012). Facilitative interactions are primarily mutualisms, like pollination, seed dispersal, and fertilization (Bronstein 2009, Chamberlain and Holland 2009, Klinger and Rejmanek 2010, Vander Wall et al. 2017). Mutualisms are facilitative relationships where both partners benefit from the interaction. Commensalism is a neutral interaction where one partner benefits with no effect on the other. Animals disperse reproductive parts of plants by mechanisms with rewards, like nectar, which would be a mutualism, or by deception without using a reward, such as using stickers or hooks on the seeds, which could be classed as commensalism if the carrier is unaffected or may even be parasitism (Pacini et al. 2008). Anyone

who has had to remove cockleburs (*Xanthium* spp.) or grass (*Poa* spp.) seeds from their pet's fur, eyes, or feet would consider the animal to be harmed by the experience, so in that instance the relationship is more antagonistic, like parasitism, rather than neutral.

Commensal interactions are generally neutral, but demonstrating a neutral effect on one player can be difficult (Anderson 2012). Anderson (2012) described an excellent example of a bird nesting in a tree (Fig. 12.1); the tree provides shelter and nesting habitat for the bird. The tree provides benefits to the bird, but the nesting bird may not have any effect on the tree, which would be a commensal relationship. If, however, the bird eats insects that feed on and harm the tree, then the relationship would be more of a mutualism because the tree benefits from the presence of the bird. Conversely, if the presence of the nest reduces photosynthesis by removal of leaves by the bird or blocking sunlight, it may have a cost or negative effect on the tree, which would be an antagonistic interaction, such as parasitism, where the bird benefits at the expense of the tree (Fig. 12.1). Therefore an experiment is needed to determine whether the relationship between the bird and the tree is truly commensal (Anderson 2012).

Surprisingly, the negative interactions including competition, parasitism, and predation have received more attention from ecologists than positive interactions like facilitation. Positive relationships—including pollination, seed dispersal, fertilization, and cycling of nutrients—are probably every bit as important as negative interactions in structure and function of communities and ecosystems (Bruno et al. 2003, Kiers et al. 2010, Traveset and Richardson 2014). Pollination is obviously

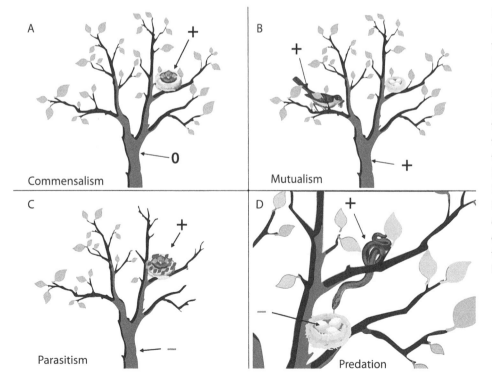

Figure 12.1. Examples of interactions among species. (*A*) Commensalism, where the bird benefits from having its nest in the tree and the tree is unaffected by the presence of the nest. (*B*) Mutualism, where the bird benefits by having the nest in the tree and the tree benefits because the bird removes insects that harm the tree. (*C*) Parasitism, where the bird benefits from having the nest in the tree but the tree is harmed because the bird is removing leaves, decreasing photosynthesis. (*D*) Predation, where the snake benefits from consuming the eggs, which are killed in the process.

an essential service to functioning of ecosystems because animals pollinate more than 90% of flowering plants worldwide (Ollerton et al. 2011, Traveset and Richardson 2014). Additionally, seed dispersal by animals occurs in most species of plants, including 90% of plant species in tropical, and 60% of plant species in temperate regions (Farwig and Berens 2012, Traveset and Richardson 2014). Therefore positive and negative interactions between animals and plants are important in both the structure and functioning of ecosystems and maintenance of biodiversity worldwide.

COEVOLUTION

Coevolution is a central theme in studies of species interactions, especially those between animals and plants. This process may lead to complex interactions among species within communities and ecosystems. Coevolution is defined as the reciprocal evolutionary change between interacting species that is influenced by natural selection (Mode 1958, Thompson 1989, 2005). Darwin used the term *coadaptation* to describe the same processes using the phrase *reciprocal change*. "Thus I can understand how a flower and a bee might slowly become, either simultaneously or one after the other, modified and adapted in the most perfect manner to each other by continued preservation of individuals presenting mutual and slightly favourable deviations of structure" (Darwin 1859:95, Thompson 1989). Coevolution is the idea that interactions among species can result in the evolution of coordinating traits from gene pools that do not mix (Thompson 2005). For example, coevolution occurs between bees and flowers, but the bee genome does not mix directly with the flower genome. Ehrlich and Raven (1964) used this concept of reciprocal change in a broad sense to link adaptation and speciation between butterflies and flowering plants. The central idea in studies of coevolution is that ecological and genetic conditions allow interacting, but unrelated, species to undergo repeated genetic change specifically because of their interactions over time (Thompson 1989).

Demonstrating coevolution is often not easily accomplished. Coevolution includes a variety of scenarios, including "evolutionary arms races" described for interactions like herbivory, plant–parasite interactions, divergence of traits among competing species, and convergence of traits in mutualisms (Ehrlich and Raven 1964, Dawkins and Krebs 1979, Thompson 1989). Coevolution can also lead to highly specialized relationships between species, such as between ants and plants, plants and pollinators, or seed dispersers that become highly specific to one another. This increased specialization, however, can make species vulnerable to extinction as ecosystems change. For example, if the pollinator or host plant becomes rare or extinct, the other becomes more vulnerable to extinction as well. When the dodo (*Raphus cucullatus*) became extinct on Mauritius, there was concern that the Tambalaoque, or dodo tree (*Sideroxylon grandiflorum*), would go extinct as well. Dodos were large, flightless, pigeon-type birds that ate the fruits of the Tambalaoque (Owadally and Temple 1979, Hershey 2004) and then dispersed the seeds. Tambalaoque seeds were believed to require passage through the digestive tract of a dodo to germinate, which would have caused the extinction of the tree concurrent with the loss of the dodo (Temple 1977, Owadally and Temple 1979). Seeds of the Tambalaoque germinate more effectively if the fleshy fruit is removed from the seed, which lowers the incidence of fungus or bacteria that impeded germination growing on the seeds as the fruit rots. Thus passage through the gut of the dodo was probably important for cleaning the fruit off the seeds, which prevented fungal growth. The dodo was probably the primary seed disperser of the Tambaloque, but other now extinct species on Mauritius, including the broad-billed parrot (*Lophopsittacus mauritianus*), Mauritian giant skink (*Leiolopisma mauritiana*), and the Aldabra giant tortoises (*Aldabrachelys gigantea*), also consumed Tambalaoque fruits and dispersed the seeds (Hershey 2004, Catling 2001). The Tambaloque survived the extinction of its seed dispersers, but Tambaloque trees suffered widespread deforestation that destroyed more than 95% of forests on Mauritius (Baider and Florens 2006). The continued decline of the Tambaloque tree appears to result from seed predation from introduced mammals and competition with invasive plants (Baider and Florens 2006).

Coevolution may occur among groups or guilds of species, like bees or hummingbirds, rather than just among pairs of species (Thompson 1989). Examples of this kind of coevolution could include some of the forms of mimicry or interactions among frugivorous birds and plants with fleshy fruits, ants and plants, or pollinators and flowering plants (Thompson 1989). This form of interaction may be broader than just a pair of species, however, because multiple species of bees or hummingbirds obtain nectar and pollinate many types of flowering plants. The original interaction may only be a pair of species, but other unrelated species may become part of the interaction when viewed in a broader context. This incorporation of other species is most likely to occur in facilitative rather than antagonistic interactions, especially those in which a resource, like nectar, is provided that would attract multiple species to pollinate the host plants (Thompson 1989).

Many species have specific functional roles in ecological communities that no other species can provide (Bender et al. 2018). These specialists are especially important for maintaining functional interactions with ecosystems (Mello et al. 2015, Sebastián-González et al. 2017, Bender et al. 2018). Matching traits have been identified in mutualistic interactions between plants and animals, like flower tube length and the bill of a hummingbird, which generally indicates coevolution between them (Wheelwright 1985, Jordano 1987, Stang et al. 2009, Dehling et al. 2014, Maglianesi et al. 2015, Bender et al. 2018). Size matching between plant and animals also leads to corresponding traits like size of fruit and gape of the animal seed disperser's mouth (Wheelwright 1985, Stang et al. 2009, Garibaldi et al. 2015, Sebastián-González et al. 2017, Bender et al. 2018). Another group of matching traits are related to foraging behavior of consumers and to mobility of animals, for example, wing shape of birds and location food in their habitats. Birds with rounded wings are more effective at moving through forested habitats than those with pointed wings (Moermond and Denslow 1985, Schleuning et al. 2011). Bender et al. (2018) examined matching traits between fruiting plants and avian frugivores in the Andes. They found that fruit size was strongly correlated with beak size in birds. Additionally, plant height above the ground was correlated with wing shape of birds in interactions between plants that produce large crops of fruit like the gaque plant (*Clusia elliptica*) and the frugivorous blue-banded toucanet (*Aulacorhynchus coeruleicinctis*; Bender et al. 2018).

ANTAGONISTIC INTERACTIONS

Antagonistic interactions between herbivores and plants as well as defense against natural enemies can be described in two primary categories: resistance and tolerance. Resistance and tolerance appear to be alternative strategies of victims (plants) responding to attacks by enemies (herbivores), although some species invest in both strategies (Van der Meijden et al. 1988, Strauss and Agrawal 1999, Núñez-Farfán et al. 2007, Anderson 2012, Moore and Johnson 2017, Svensson and Råberg 2010). Resistance is the decrease in susceptibility to damage, which relies on physical deterrents (thorns, spines) or chemical compounds that reduce or prevent damage by herbivores (Pilson 2000). Species use resistance to minimize both the number and success of attacks from natural enemies; chemical or physical defenses used by plants to reduce herbivory are good examples of resistance (Svensson and Råberg 2010). Conversely, tolerance—the ability to regrow following damage by herbivory—is used to minimize the effects of the attack on reproductive fitness of plants (Pilson 2000, Strauss and Zangerl 2002, Svensson and Råberg 2010). Both of these strategies may be costly to the victim because of energetic expense of rebuilding lost tissue or investing in physical or chemical defenses, resulting in a decrease in the fitness of the plant. Importantly, resistance has a direct and negative cost to the herbivore, either by reduced nutrition or fitness from avoiding eating the plant or being injured by physical or chemical defenses. Tolerance, however, only has a cost to the plant that has to replace lost tissues, because it does not appear to have any negative effects on the attackers or their population dynamics (Rosenthal and Kotanen 1994, Fay et al. 1996, Roy and Kirchner 2000, Tiffen 2000, Restif and Koella 2003, Fornoni et al. 2004a). Tolerance may allow the herbivore to benefit nutritionally from the plant that is consumed, thereby increasing the nutritional condition and fitness of the herbivore. Therefore tolerance can limit antagonistic coevolution between victims and enemies, but resistance would prolong coevolutionary dynamics because the "evolutionary arms race" would continue (Rauscher 2001, Fornoni et al. 2004b).

Tolerance and resistance are probably alternative strategies in response to damage by herbivores (Strauss and Zangerl 2002). A plant that invests in resistance would be unattractive to herbivores and be avoided, likely because of low nutritional value or physical defenses (Agrawal et al. 1999). Conversely, a plant that uses a tolerance strategy may be preferentially consumed, especially if it is a high-quality food and supports high fitness of herbivores (Agrawal et al. 1999). The ability to regrow after damage is determined by the amount of tissue removed by the herbivore, and plants should invest either in defenses to reduce attacks (resistance) or in the ability to regrow after an attack (tolerance; Van der Meijden et al. 1988). Which strategy is most appropriate for a plant species in different environments depends on which has the higher cost, production of resistance structures or regrowth following herbivory (Fornoni et al. 2003). Most results indicated that plants often use both tolerance and resistance strategies, which supports the idea that intermediate levels of each strategy are favored by natural selection (Valverde et al. 2001, Fornoni et al. 2004a). The annual plant jimsonweed (*Datura stramonium*) uses both tolerance and resistance strategies by varying density of trichomes, fine hairlike structures on the surface of leaves and stems of plants. Some populations used high densities of trichomes on leaves as a resistance strategy against their insect herbivores (Valverde et al. 2001, Fornoni et al. 2004a). Other populations of jimsonweed employed a tolerance strategy and rebuilt tissues following herbivory rather than using trichomes for defense (Valverde et al. 2001, Fornoi et al. 2004a). Surprisingly, more investigation has been directed toward the role of resistance than tolerance in the animal literature (Svensson and Råberg 2010). Whether resistance, tolerance, or some aspect of each strategy is most effective in defense against enemies is largely determined by the cost of each strategy for the defending species in their specific environment.

A small group of plants are carnivorous, and consume animals rather than vice versa. Carnivory in plants is uncommon and usually has evolved as an adaptation to living in soils that lack certain nutrients, especially nitrogen and phosphorus (Juniper et al. 1989, Porembski and Barthlott 2006, Pacini et al. 2008, Chin et al. 2010). Some speculation exists that carnivory by plants is a useful but not necessarily indispensable means of obtaining nutrients because most plants survive even when they fail to catch prey (Adamec 1997, Pacini et al. 2008). Nevertheless, minerals obtained from digestion of animals tend to increase the success of sexual over vegetative reproduction (Adamec 1997, Pacini et al. 2008). Carnivorous plants have evolved an impressive array of strategies to capture animals, usually arthropods, including snap traps and sticky traps used by Venus flytraps (*Dionaea muscipula*), vacuum traps by bladderworts (*Utricularia* spp.), and pitfall traps used by tropical pitcher-plants (*Nepenthes* spp.; Lloyd 1942, Juniper et al. 1989, Chin et al. 2010). Most carnivorous plants specialize in invertebrate prey, primarily insects and other arthropods; some larger species of pitcher-plants are big enough to capture vertebrate prey (Chin et al. 2010). There is no empirical evidence, however, that vertebrate prey plays any significant part of the nutritional requirements of pitcher-plants, because nearly all of the prey found in the pitchers were arthropods, primarily ants (Chin et al. 2010).

Herbivory

Herbivory is the most common mode of antagonistic interactions between animals and plants. Herbivory is the primary method for transferring energy from sunlight into animal biomass (Anderson 2012). Herbivores began consuming plants shortly after land plants evolved about 475 million years ago (Wellman et al. 2003, Hamann et al. 2021). Herbivores shape plant communities in many ways, through foraging and movements, including reduction of vegetation density and biomass, creating gaps in woody vegetation, facilitating coexistence among plant species, dispersing seeds, suppressing sensitive species, reducing the potential for fire through removal of plant tissue, and accelerating cycling of nutrients through deposition of urine and feces (Hobbs 1996, Knapp et al. 1999, Gill 2006, Hester et al. 2006, Johnson 2009).

Most herbivores do not kill their food plants because most plants have parts that are low value as food and are largely unused (Danell and Bergström 2002). Mortality of plants can and sometimes does occur, however, especially at high levels of herbivory, or when herbivores feed on small plants or seedlings. Winter browsing or bark stripping also can cause mortality of plants, particularly during hard winters when deer, snowshoe hares (*Lepus americanus*), and other mammals remove bark from trees or shrubs. Additionally, some species cause mortality of adult trees, like beavers (*Castor canadensis*) building dams or African elephants (*Loxodonta africana*) bark stripping or removing entire trees (Danell and Bergström 2002). Much of this activity by both beavers and elephants maintains or improves habitat for other herbivores, even if it causes some mortality of plants. Grazing by high densities of snow geese (*Chens caerulescens*) on their summer ranges in the Arctic has led to loss of coastal vegetation and increased salinity of soils, in turn leading to high plant mortality, which has resulted in extensive areas devoid of vegetation (Abraham et al. 2005).

Traditionally, populations of animals have been seen as products of plant communities, but plants were observed as inputs to animal populations (Hobbs 1996). More recently, that perspective has shifted such that herbivores, especially ungulates, are also viewed as important regulators of ecosystems processes, and have substantial effects on primary productivity and successional patterns in plant communities (Pastor et al. 1993, Hobbs 1996, Stewart et al. 2004). Large-bodied species have greater capacity to use lower-quality (high fiber or structural carbohydrates) vegetation than small-bodied species, so they consume greater proportions of plant tissues (Owen-Smith 1988, Johnson 2009). African elephants maintain open savannah by removal and suppression of woody plants and create grassy openings in forests and thickets (Owen-Smith 1988, Dublin et al. 1990, Johnson 2009). Similarly, grazing by white rhinos (*Ceratotherium simum*) maintains shortgrass areas within tussock grasses and thickets (Owen-Smith 1988, Waldram et al. 2008, Johnson 2009). Extensive browsing on deciduous trees by high densities of moose on Isle Royal drove successional changes from communities dominated by deciduous trees to those dominated by conifers, like white spruce (*Picea glauca*), which also slowed rates of decomposition and nitrogen cycling in soils (Pastor et al. 1993). Foraging by bison helps to maintain high species diversity of plants in tallgrass prairie (Knapp et al. 1999, Johnson 2009, Tarleton and Lamb 2021).

Herbivores are estimated to remove ~10% of the net primary production in terrestrial environments (Crawley 1983, Danell and Bergström 2002). Data available on production or standing crop of plants, however, do not necessarily reflect the availability of food for herbivores; not every plant part is edible because of low nutritional value, toxicity, or locations outside the reach of herbivores (Danell and Bergström 2002). For example, leaves on the top of acacia (*Acacia* spp.) trees in Africa are available to a giraffe (*Giraffa camelopardalis*), but not to a Thompson's gazelle (*Eudorcas thomsonii*) or a black rhino (*Diceros bicornis*; Danell and Bergström 2002). Availability and quality of food plants vary on both temporal and spatial scales, which usually results from variation of light, temperature, and soil nutrients (Danell and Bergström 2002). Herbivores are regularly exposed to spatial and temporal variation in both quality and abundance of food. Seasonal changes in abiotic conditions affect plant phenology by changing timing of growth, amount of structural carbohydrates, lignification, and wilting, which affects feeding by herbivores (Danell and Bergström 2002). Additionally, abiotic conditions, including variation in weather, affect timing and rate of plant growth, which in turn affects the quality (energy and protein) and quantity of leaves, stems, and reproductive parts of plants that are consumed by herbivores.

Herbivores select different plants and plant parts to extract the needed or required nutrients while also processing those components of the plant that reduce forage quality, including cell walls and secondary compounds. In general, the parts of plants that are most commonly consumed are those with high nutritional value (protein and energy) to herbivores, which corresponds to those parts that have the most reserves (sugars, lipids, proteins, water), are metabolically active, are most tender or accessible, and are without many structural components (Pacini et al. 2008). In general, plants have low nutritional value compared with non-herbivorous diets because they are generally low in nitrogen, must be digested slowly, and often require symbiosis with microorganisms to digest (Danell and Bergström 2002). To digest plant cell walls, vertebrates rely on symbiotic microbes to ferment and break down these tissues. Microbes are located in specialized organs in the foregut or rumen, the hindgut, or the cecea or large intestine. Plant cells are divided into those parts that are metabolically active and those that provide structural support to the plant. The metabolically active portion of the cell, or the cell contents, consist of protein, sugars, starches, and fats that are rapidly digested or fermented in the digestive track (Laca and Demment 1991). Cell contents also may contain secondary metabolites that range from low digestibility to toxic (Danell and Bergström 2002). Additionally, the soluble components of plants also contain water, minerals, and vitamins that increase the quality of the food for herbivores (Danell and Bergström 2002). Plants have cell walls that provide structural support. These plant cells consist of cellulose, which is relatively digestible; hemicellulose, which is moderately digestible; and lignin and cutin, which are mostly indigestible (Danell and Bergström 2002). As plants age, the cell walls become thicker and more lignified, which results in them being more difficult to digest. In response, herbivores undergo many specific adaptations and coevolution with microorganisms to aid in digestion of plant materials. Fermentation is less efficient than digestion for extracting nutrients, however, and greater proportions of cells walls in plants tends to reduce forage quality and slows passage rate of food through the digestive tract (Chapter 11).

Plant Defenses against Herbivory

Plants also have evolved complex defenses, both chemical and mechanical, to reduce or avoid herbivory, and herbivores have simultaneously evolved ways to circumvent those defenses (Hamann et al. 2021). Plants that are tolerant to herbivory have fast growth rates and can reallocate stored carbohydrates from roots or other remaining tissues to defoliated stems so they can regrow relatively rapidly (Anderson 2012). Additionally, plants that are tolerant of herbivory have architecture that protects storage organs by keeping them out of the reach of most herbivores (Anderson 2012). Plants can respond to herbivory with compensatory growth, in which photosynthetic tissue is replaced relatively rapidly following loss to herbivory (Vicari and Bazely 1993). Grasses, for example, have morphological characteristics, like basal meristems, that make them better able than most plants to recover from herbivory. Nevertheless, tolerance does not necessarily require additional growth. Tolerance to herbivory may result in a change in efficiency of existing structures, so partial defoliation by herbivores may result in more effective use of remaining leaves than production of new ones (Danell and Bergström 2002). Experimental removal of leaves from several biennial plants—including mullein (*Verbascum* spp.), groundsel (*Senecio* spp.), thistle (*Cirsium* spp.), bugloss (*Echium* spp.), and hound's tongue (*Cynoglossum* spp.)—indicated that the ability to regrow following herbivory was correlated with the severity of damage, which varied among plant species (Van der Meijden et al. 1988).

Plants that are resistant to herbivory employ structural or chemical defenses to deter or harm the herbivores that feed on them. Plants may not need to produce defenses to herbivory at all times; induced defenses are those that are activated after a plant is browsed or grazed by herbivores. Constitutive defenses are chemicals that are circulating within plants, including when plants are not under threat of herbivory. The level of constitutive defense may be low to nothing, where the plant is undefended, and increase when the plant comes under attack. Some plant species, however, may have low or even high levels of physical or chemical defenses even in the absence of herbivores (Strauss and Zangerl 2002).

Physical Defenses

Plants produce cell walls and fibrous tissues, which provide structural support and are composed of cellulose, hemicellulose, and lignin that are difficult for herbivores to chew and digest. Specialized structures to reduce herbivory include thorns, spines, hooks, trichomes or hairs, and tough leaves that protect the plant, especially the photosynthetic tissues (Anderson 2012). Thorns and other physical defenses are probably more effective against mammals than insects that may be small enough to move between them (Strauss and Zangerl 2002). For example, the thorns of the honey locust (*Gleditsia triacanthos*) tree may be 20 cm in length and probably are good deterrents to vertebrate, but not insect, herbivores (Strauss and Zangerl 2002). Grasses, one of the most widespread families in the plant kingdom, use silica for structural rigidity and in part to defend against insect and vertebrate herbivores. Grasses actively accumulate silica as monosilicic acid from soils. Silica content of most forage grasses is about 2% to 3% of total dry weight, which is about an order of magnitude greater silica content than observed in most flowering plants (Vicari and Bazely 1993). These silica molecules are incorporated throughout the plant and are sequestered in cell walls (Vicari and Bazely 1993). Silica increases toughness and rigidity of stems, which increases resistance to insects, but silica also protects the plant from vertebrate herbivores through increased tooth wear and reduced digestibility, especially to rumen microbes (Van Soest and Jones 1968, Vicari and Bazely 1993). Ungulate herbivores characteristically have evolved hypsodont, or high-crowned, teeth, and many rodents have ever-growing teeth that wear more slowly than those characteristic of non-herbivorous mammals. In some plants, silica particles create edges on plant leaves that resemble files and saws, which are defensive weapons aimed at abrading the teeth of grazing animals (Herrera 1982) and probably have been an important selective factor in dental patterns of mammalian herbivores (Gutherie 1971).

Thorns and spines are used by plants as protection against herbivores, and to navigate these structures, herbivores must carefully manipulate the plant with their mouths, which reduces the foraging rate. Herbivores must spend more time plucking and chewing a plant part, and the thorns and spines generally require the herbivore to bite smaller pieces. This foraging strategy also reduces the amount of food that the herbivore can acquire per unit time (Belovsky and Schmitz 1991). Spines and thorns are most prevalent, with exceptions of course, in arid environments with low primary productivity, which makes investment in those structures important for reducing removal of tissues by herbivores (Belovsky and Schmitz 1991). Plant growth and defense are dependent on resource acquisition, however, and plants must have the resources, like silica, to produce those defenses (Zuest and Agrawal 2017).

Chemical Defenses

Chemical defenses that are employed by plants to reduce herbivory are referred to as secondary compounds or secondary metabolites. These secondary compounds, also known as allelochemicals, are most likely intended to impede or damage other organisms (Pacini et al. 2008). These compounds are considered secondary because few of them have any known primary metabolic functions in the plant (Robbins 1993). Secondary compounds most likely evolved as defenses against herbivores, but some may have been retained from another purpose employed by plants in the past. In general, plant secondary compounds are more likely to be found in flowering plants, like forbs, trees, and shrubs, than grasses or graminoids (Mithofer and Boland 2012, Felton et al. 2018, Wu et al. 2020). Chemical defenses of plants are generally bitter-tasting, poisonous, smell offensive, or have some form of anti-nutritional effects on herbivores, and the effects range from avoidance to impaired digestion to toxicity (Harborne 1991, Robbins 1993, Danell and Bergström 2002, Dearing et al. 2005, Sorensen et al. 2005). Secondary compounds may act as toxins or feeding deterrents that protect the plant by reducing herbivory through reduced palatability, toxicity, or digestibility, thereby decreasing the nutritional quality of the plant (Robbins et al. 1987, Iason 2005, DeGabriel et al. 2009, Felton et al. 2018). Plant secondary compounds are used by humans for a variety of purposes, including cooking spices, stimulants like caffeine or nicotine, narcotics including opioids, and medicines like aspirin obtained from willow (*Salix* spp.) bark or taxol from the Pacific yew tree (*Taxus brevifolia*), which is used to treat cancer (Fig. 12.2; Anderson 2012). Although there are extensive types of secondary compounds produced by plants, they generally fall into three primary classes: phenolics, nitrogenous compounds, and terpenes (Fig. 12.2; Harborne 1991, Danell and Bergström 2002). The most common are soluble phenolics, alkaloids, and terpenoids (Robbins 1993, Pacini et al. 2008). Many of those compounds are also toxic to the plants, and therefore toxins are usually held in the vacuoles of plant cells away from organelles or plant tissues to avoid damage (Robbins 1993). Other compounds in plants that are toxic to herbivores include quinones, cyanogenic glycosides, resins, and others (Dearing et al. 2005, Kohl et al. 2014, Matocq et al. 2020).

Concentrations of secondary compounds in plants may vary in space, time, or both, and they also may vary among individual plants in a single location or even among structures of a single plant (Mattson 1980, Robbins 1993). Secondary compounds in woody plants may be induced with herbivory and may increase with browsing pressure (Robbins 1993). The response of plants to browsing and production of secondary compounds may depend on the physiological state, age, degree of damage, or availability of resources for the plant (Bryant et al. 1991, Robbins 1993). Juvenile forms of plants often have higher amounts of secondary compounds than adults, but excessive browsing has been shown to cause adult plants to revert to a juvenile form with higher amounts of secondary metabolites than usually occur in this age class (Bryant et al. 1991). Captive deer preferred to forage on plants with high energy content and avoided those with high levels of phenolic compounds even if forage quality was higher (McArthur et al. 1993, Felton et al. 2018).

Recent research has noted some functions of secondary compounds that produce positive benefits for plants beyond reducing or avoiding herbivory, including attracting pollinators or seed dispersers, protecting plants from ultraviolet radiation, and defending plants and even sometimes herbivores from oxidative stress, disease, and pathogens (Villalba et al. 2017). Additionally, some of these compounds may provide anti-parasitic properties to herbivores. Domestic sheep and goats have been reported to select foods that contain condensed tannins when burdened by parasitic infections in the gut (Villalba et al. 2010, Juhnke et al. 2012, Amit et al. 2013, Villalba et al. 2017). Interestingly, when parasitic burdens subside, sheep reduce their preference for tannin-containing foods and return to food of higher nutritional quality, because continued ingestion of those tannin-rich foods reduces reproductive performance (Villalba et al. 2017). Although secondary compounds, in this case condensed tannins, have negative effects on reproduction, the positive effect of reducing parasite loads with short-term foraging on plants that contain those second-

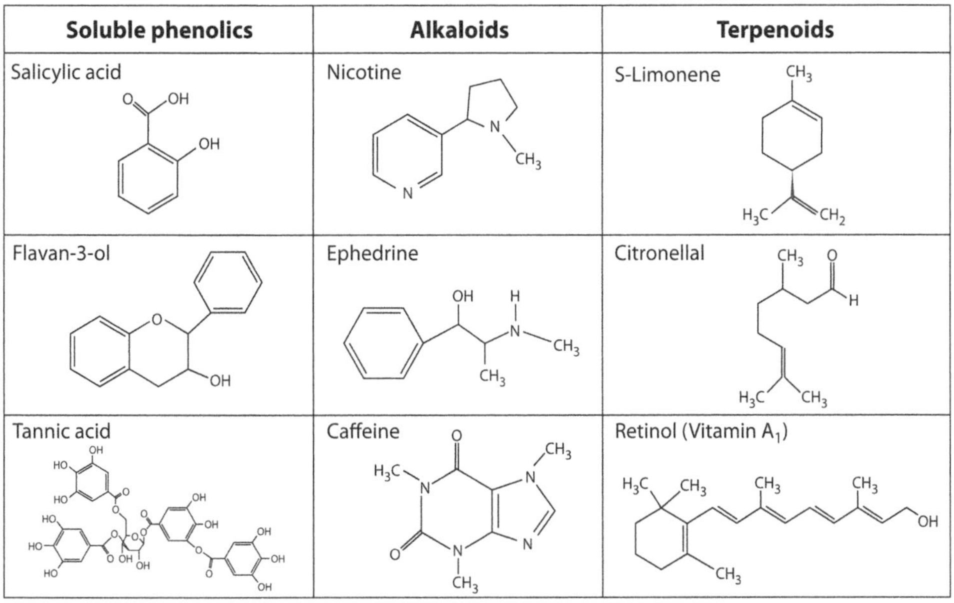

Soluble phenolics	Alkaloids	Terpenoids
Salicylic acid	Nicotine	S-Limonene
Flavan-3-ol	Ephedrine	Citronellal
Tannic acid	Caffeine	Retinol (Vitamin A₁)

Figure 12.2. Chemical structures of some selected secondary compounds found in plants.

ary metabolites has important implications for the health of herbivores, including that of free-ranging species of wildlife.

Tannins and flavonoids are phenolic compounds that include one or more hydroxyl groups bonded directly to a hydrocarbon group (Fig. 12.2). Tannins are water-soluble phenolics that can precipitate proteins from aqueous solution. The name tannin comes from the ability of these compounds to tan animal hides into leather (Robbins 1993). Tannins are more often observed in woody compared with herbaceous plants, but there are a few herbaceous plants like sumac (*Rhus* spp.) that also contain them. Tannins are the most widespread of plant defensive compounds and occur in 14% of herbaceous perennial plants, 79% of deciduous shrubs, and 87% of evergreen woody plants (Rhoades and Cates 1976, Robbins 1993). Research on tannins in plant–herbivore interactions usually focuses on leaf tissue, but tannins may be common in many plant tissues, including twigs, bark, roots, seeds, and fruit (Barbehenn and Constabel 2011). Because tannins are not specific in precipitating proteins, they are effective in preventing fungal, viral, and bacterial attacks on the plant in addition to reducing herbivory (Robbins 1993). Tannin concentration is highly variable among and within plant tissues, and has been shown to vary with genotype, developmental stage of tissues, and environmental conditions (Barbehenn and Constabel 2011).

Tannins can be classified into two major groups: hydrolyzable and condensed (Barbehenn and Constabel 2011). Hydrolyzable tannins have a carbohydrate in the base molecule and may be broken down by acids or enzymes, or hydrolyzed into simple acids. Condensed tannins are not readily hydrolyzed and do not contain a carbohydrate molecule. Flavan-3-ol is a soluble condensed tannin found in tea and cocoa (Fig. 12.2). In oak trees (*Quercus* spp.), observed levels of condensed tannins increased with maturity of leaves, but levels of hydrolyzable tannins declined (Salminen et al. 2004, Barbehenn and Constabel 2011). Additionally, in trembling aspen (*Populus tremuloides*), condensed tannin concentrations in leaves varies from 2% to 25% even among individuals growing in the same conditions or leaves on the same tree, and condensed tannins are twice as high in mature compared with developing leaves (Donaldson et al. 2006, Rehill et al. 2006, Barbehenn and Constabel 2011).

For some mammalian herbivores, higher levels of tannins in forages are correlated with declines in use of proteins (Robbins et al. 1987, McArt et al. 2009, Barbehenn and Constabel 2011). DeGabriel et al. (2009) studied the effects of tannins on protein digestibility of brushtail possums (*Trichosurus vulpecula*) that consume eucalyptus (*Eucalyptus* spp.) leaves and showed that tannins reduced digestibility of proteins, which also was correlated with reduced reproductive fitness. Moreover, tannins from faba beans (*Vicia faba*) decreased the digestibility of crude protein between 7% and 9% in the small intestine of pigs (*Sus scrofa*; Jansman et al. 1995). The effects of tannins are not always negative on mammalian herbivores; moderate concentrations of condensed tannins reduced foaming of protein-rich forage in the rumen and increased availability of amino acids in cattle (Barbehenn and Constabel 2011).

High concentrations of tannins resulted in decreased consumption, digestion, and growth rates of cattle and other ruminants (Aerts et al. 1999, McSweeney et al. 2001, Barbehenn and Constabel 2011).

Many mammalian herbivores produce tannin-binding proteins in their saliva that have much higher affinity for tannins than other types of proteins (Shimada 2006, Barbehenn and Constabel 2011, Schmitt et al. 2020). By binding the tannins, those salivary proteins prevent tannins from interacting with other proteins during digestion and microbial fermentation as they move through the digestive system (Shimada 2006). Those tannin-binding salivary proteins have been observed in a wide variety of mammalian species, including megaherbivores like giraffe and black rhinos (Schmitt et al. 2020), browsers like deer and moose, and even some grazers and omnivores (Barbehenn and Constabel 2011, Schmitt et al. 2020). Nevertheless, not all animals that consume tannin-rich forages produce salivary proteins that bind tannins. Despite feeding on eucalyptus, brushtail possums do not produce salivary proteins, which may explain why tannins decreased digestibility of protein in possum diets and also reduced their reproductive fitness (Shimada 2006, DeGabriel et al. 2009, Barbehenn and Constabel 2011). Nevertheless, reproductive fitness of brushtail possums was strongly correlated with dietary protein in foliage, but not tannin or protein concentration alone (DeGabriel et al. 2009, Barbehenn and Constabel 2011).

Flavonoids are another class of soluble phenolic compounds. Flavonoids give color to flowers, fruits, and leaves, but they can also inhibit digestion and may hinder reproduction in mammals (Robbins 1993, Danell and Bergström 2002). In ruminant mammals the microorganisms in the foregut, or rumen, usually degrade flavonoids into water-soluble products that are readily excreted (Simpson et al. 1969). Conversely, some isoflavonoids, like phytoestrogens, can cause abortions, sterility, or liver damage (Simpson et al. 1969, Robbins 1993). How much these estrogenic isoflavonoids affect free-ranging wildlife is unknown, but health problems in captive animals have occurred with soybean-based or clover feeds (Robbins 1993). Positive effects of flavonoids also have been recently reported. Giant pandas are highly specialized consumers, and more than 99% of their diet is bamboo (Schaller et al. 1985, Nie et al. 2015). Glycoside flavonoids are common in bamboo, and the flavonoid content seems to be positively correlated with migratory movements of pandas (Wang et al. 2020). Consumption of different species of bamboo occurs at different elevations and times of year, as pandas appear to be selecting for high levels of flavonoids in their bamboo foods (Wang et al. 2020). Additionally, some studies indicate that bamboo leaf extract has antioxidant (Hu et al. 2000, Nie et al. 2015), antimicrobial (Tao et al. 2018), and anti-inflammatory (Wedler et al. 2014) properties, which likely provide benefits to pandas in addition to food (Wang et al. 2020).

Alkaloids have one or more nitrogen atoms in a heterocyclic ring (Fig. 12.2; Robbins 1993). Alkaloids tend to occur in plants in nutrient-rich environments where they can take advantage of excess nitrogen (Mattson 1980, Robbins 1993,

Danell and Bergström 2002). Alkaloids occur in ~20% of flowering plants, and more than 10,000 different alkaloid compounds have been identified (Harborne 1991, Robbins 1993). Many of the plant-derived pharmacological drugs used by humans are alkaloid compounds, including nicotine, morphine, caffeine, ephedrine, and many others (Robbins 1993). Terpenoids are a large class of plant secondary compounds that are best known for insecticidal, antimicrobial, and toxic properties (Robbins 1993). Examples of common terpenes are Vitamin A, carotenes, and the volatile oils in sagebrush (*Artemisia tridentata*) and conifers (gymnosperms; Robbins 1993). Eucalyptol of eucalyptus and oils obtained from peels of citrus fruits also are terpenes. Small doses of terpenes can be tolerated by microbial organisms in the gut, but bacteria in the rumen that consume cellulose are especially susceptible to toxicity from terpenoids (Schwartz et al. 1980, Robbins 1993).

While foraging, animals must balance the acquisition of nutrients while minimizing exposure to toxins or digestion inhibitors from plant secondary compounds. Some animals are able to avoid ingesting plants with secondary compounds, but others use myriad means to avoid the toxic effects or reduced digestibility of those compounds in food. Many small-bodied species must manage toxins in the plants that they eat (Dearing et al. 2005, Matocq et al. 2020). Many herbivores, especially small mammals, cache food before consumption, and often secondary compounds degrade during storage, so by storing the foods prior to ingestion, the animals can avoid consuming those compounds. Therefore animals may avoid or reduce the effects of secondary compounds prior to ingestion of the food, thereby reducing the energetic cost of dealing with those metabolites or toxins (Dearing et al. 2005). This behavioral mechanism for reducing secondary compounds may enhance diet breadth by allowing herbivores to consume plants that they may not be able to process otherwise, which also allows them to expand their dietary niche (Dearing et al. 2005).

In some instances, vertebrate herbivores may be able to tolerate high doses of toxic secondary compounds in their foods, without lost body mass, impaired growth, or reduced survival or reproduction (Kohl et al. 2014). The ability to tolerate toxins in their plant food has important implications for determining dietary niche of many species of herbivores (Dearing et al. 2000, Moore and Foley 2005, Kohl et al. 2014, Kohl and Dearing 2016). Kohl et al. (2014) reported that gut microbes in the desert woodrat (*Neotoma lepida*) were tolerant of toxins in resin of creosote bush (*Larrea tridentata*), which allowed for expansion of the dietary niche. In their experiment, Kohl et al. (2014) transplanted microbiota from woodrats that had experience eating creosote to naive recipients, which resulted in the ability of the recipients to feed on creosote without deleterious effects of ingesting the toxin. Naive individuals usually lost body mass when fed diets of 2% resin from creosote (Mangione et al. 2000, Kohl et al. 2014). Therefore, although adaptations of the host to the microbiota also likely played a role in allowing the woodrats to consume such toxic food, their results strongly indicated that the microbiota in the gut substantially improve the ability of those herbivores to cope with and overcome the toxic resins in their foods (Kohl et al. 2014).

Matocq et al. (2020) studied two species of woodrats (Bryant's wood rat [*Neotoma branti*], desert woodrat) occupying different habitats that were relatively close together, and each ingested a highly toxic plant as their primary food. Bryant's woodrat (*Neotoma bryanti*) occupied relatively mesic Sierra Nevada Mountains, hill habitat, and consumed California coffeeberry (*Frangula californica*) as their primary food. California coffeeberry has high concentrations of anthraquinones, which when ingested have laxative effects as a result of damage to the lining of the intestine (Surh et al. 2013, Matocq et al. 2020). In the nearby Kelso Valley, in the Mojave Desert, the desert woodrat consumed desert almond (*Prunus californica*) as their primary food source. Desert almond contains high concentrations of cyanogenic glycosides, which when chewed and consumed are transformed into hydrogen cyanide, with toxic consequences for the woodrats that consume them (Santamour 1998, Matocq et al. 2020). Analysis of feces indicated that each species had unique metabolites that were not seen in the other species, meaning that each species used a different strategy to detoxify the toxic compounds in their primary food. Hydrogen cyanide is at least partly detoxified by the enzyme rhodanese in the liver (Dooley et al. 1995, Matocq et al. 2020). Anthraquinones may partially be detoxified by cytochrome P450 1A2 also in the liver (Mueller et al. 1998, Matocq et al. 2020). Matocq et al. (2020) also considered that the gut microbiomes of each species of woodrat may help to detoxify those compounds. Additionally, most of the individuals of Bryant's woodrat that consumed California coffeeberry also consumed sulphur buckwheat (*Epilobium umbellatum*), and those researchers hypothesized a possible interaction of the two plants to add nutrition or possibly as protection from the toxin (Caesar and Cech 2019, Matocq et al. 2020).

Herbivore Optimization

The idea of herbivore optimization is that at intermediate levels of herbivory, net primary productivity of plants is increased. This hypothesis predicts increased plant production with moderate levels of herbivory compared to heavily grazed or ungrazed areas (McNaughton 1979, Hik and Jefferies 1990, Stewart et al. 2004). This effect has been documented in clipping experiments (Seagle et al. 1992) and in field studies involving large mammals, including moose (Pastor and Naiman 1992), elk (Augustine and Frank 2001, Frank et al. 2002, Stewart et al. 2004), black brant (*Branta bernicla*; Person et al. 2003), and snow geese (Hik and Jefferies 1990). In African savannas with multiple species of ungulates, grazing increased palatability of forages through high nitrogen content of aboveground biomass or by shifting the phenology of plants toward younger, actively growing individuals (McNaughton 1979, Bryant et al. 1983, Hik and Jefferies 1990). At intermediate levels of herbivory by snow geese, higher net aboveground primary productivity (NAPP) was reported relative to heavily grazed or ungrazed areas (Hik and Jefferies 1990). Grazing by black brant on Ramenski's sedge (*Carex ramenskii*) and Hoppner's sedge

Figure 12.3. Grazing by black brant on the Yukon-Kuskokwim delta in Alaska, where grazing lawns of sedges are formed as a result of moderate intensities of herbivory. Courtesy of Jim Sedinger

(*C. subpathacea*) led to formation of grazing lawns (Fig. 12.3), which are grassland communities with prostrate growth forms that have high nutritional quality, on the Yukon-Kuskokwim Delta, Alaska (Person et al. 2003). Following exclusion of herbivory, however, the sedges returned to their upright growth forms (Fig. 12.3; Person et al. 2003). A grazing lawn is defined as an area of vegetation maintained by herbivory that is dominated by high densities of grazing-tolerant and rapidly growing plants with high nutrient concentrations (Person et al. 2003, Stewart et al. 2004). Grazing lawns have been observed in grassland-type ecosystems that are influenced by herbivore species and densities, and they can range from a few square meters (Person et al. 2003) to many square kilometers (McNaughton 1983, 1985; Stewart et al. 2004).

Several authors have questioned the validity of the herbivore optimization hypothesis and reported that the effects of grazing were either neutral or negative, but not positive (Belsky 1986, 1987, Painter and Belsky 1993, Stewart et al. 2004). Nonetheless, Milchunas and Lauenroth (1993) reviewed 236 studies on the effects of grazing on primary production. They found that 17% of studies reported increased production with herbivory, but most reported that the effects of herbivory were neutral or negative. Availability of nutrients and moisture is likely a major factor influencing how plants respond to herbivory and whether a grazing lawn will be formed. Areas with inadequate nitrogen or water likely prevent plants from responding with compensatory growth to herbivory. Areas where this hypothesis has been supported, like Yellowstone National Park, the Serengeti savannah in East Africa, or arctic floodplains, are characterized by high soil fertility and mois-

ture or occur during seasons when nutrients and moisture were high (Stewart et al. 2004).

Whether plants respond to herbivory by increasing or decreasing production is strongly related to intensity of grazing; increased growth results from foraging by high densities of herbivores, but usually for short periods. When herbivores are at high density and exhibit intense foraging for long periods, plant populations decline in NAPP and nutrient cycling (Ruess et al. 1998, Jefferies and Rockwell 2002, Person et al. 2003, Schoenecker et al. 2004, Stewart et al. 2004, Abraham et al. 2005). As plants decline in production and nutrient cycling slows down, there are feedbacks on the quality and abundance of forage for herbivores, which results in animals having poor body condition and reduced fecundity (Stewart et al. 2004, Abraham et al. 2005). Concurrent with those density-dependent feedbacks on animal populations, negative effects on the plant communities result in changes in successional pathways or lead to degradation of plant communities through trampling vegetation, compacting soil (Packer 1953, 1963), denuding vegetation (Jefferies and Rockwell 2002), or lowering plant species diversity (Nicholson et al. 2006), which results in communities dominated by fewer plant species that are unpalatable to herbivores (Stewart et al. 2004) or loss of vegetation entirely (Abraham et al. 2005). Low or intermediate levels of herbivory resulting from populations at low density, however, can generate positive feedbacks on the plant community with increases in NAPP and enhanced nutrient cycling (Hik and Jefferies 1990, Frank et al. 1994, Stewart et al. 2004).

Large, herbivorous mammals often act as keystone species in ecosystems that they inhabit, and depending on their population density relative to ecological carrying capacity, they may increase or decrease plant diversity (Stewart et al. 2009). When populations of large mammals are close to ecological carrying capacity, the corresponding high levels of herbivory lead to declines in plant diversity and loss of highly palatable species from the plant community (Olff and Ritchie 1998, Rooney and Waller 2003, McShea 2005, Stewart et al. 2009). Conversely, low to moderate levels of herbivory from herbivore populations at low density have led to positive effects on ecosystems, including increased NAPP and nutrient cycling because of inputs of nitrogen from urine and feces (McNaughton 1979, Ruess and McNaughton 1987, Hik and Jefferies 1990, Frank and McNaughton 1993, Stewart et al. 2004). In addition, low to moderate levels of herbivory may upset competitive interactions among plants and prevent highly competitive species from dominating the plant community, resulting in higher diversity (Pastor and Cohen 1997, Jacobs and Naiman 2008, Stewart et al. 2009). Thus, at low population density of herbivores, grazing and browsing reduce biomass and canopy cover of competitively dominant plants, leading to greater spatial heterogeneity, which allows a greater number of plant species to coexist (Olff and Ritchie 1998, Jacobs and Naiman 2008, Stewart et al. 2009). Conversely, high levels of herbivory lead to declines in species diversity as some plant species are eliminated or reduced by herbivory or trampling, and the resulting plant community becomes dominated by fewer plant species that are resistant to herbivory or trampling (Olff and Ritchie 1998, Vellend et al. 2003, Vellend 2004, McShea 2005, Nicholson et al. 2006, Stewart et al. 2009). Stewart et al. (2009) observed an indirect relationship between herbivory and plant species diversity. In areas where herbivory was low, NAPP was increased compared with areas of high or no herbivory. Then in areas where NAPP was highest, species diversity of plants also was highest, and low levels of herbivory stimulated NAPP, which then resulted in higher diversity of plants in those areas (Stewart et al. 2009). This relationship is probably at least partly dependent on an evolutionary history between plants and herbivores, such that most of the plants in the community were tolerant of herbivory. Communities that are more resilient to herbivory are more likely to respond quickly to changes in densities of herbivores, and thus levels of herbivory, through changes in plant productivity rather than rapid changes in species composition (Cingolani et al. 2005, Stewart et al. 2009).

Animal Movements and Resource Waves

Movements of animals are largely influenced by plants because animals move in some manner in response to variation in food resources. Plant phenology is the key factor that shapes use of landscapes by herbivores because abundance and quality of plant food supplies vary seasonally, leading to substantial variation in movements of animals. Links between soil fertility and plant defenses, especially in ecosystems with patchy distribution of resources, have been suggested to produce similarly patchy distributions of animals (McKey et al. 1978, Oates et al. 1990, Cork 1992, Ganzhorn 1992, Bryant 2003, DeGabriel et al.

2009). Migratory movements of large, herbivorous mammals and birds is an adaptive response to seasonal variation in food supplies, which is largely influenced by changing phenology of plants (Danell and Bergström 2002). Migration is defined as the back-and-forth movements between seasonal ranges that are not used during other times of year (Berger 2004). In that context, migration allows animals to exploit more abundant plant forage resources in one seasonal range while simultaneously avoiding deficits of plant food on the other (Alerstam et al. 2003, Berger 2004, Sawyer and Kauffman 2011, Blum et al. 2015, Aikens et al. 2017).

Resource waves are described as natural gradients that create pulses of resources that propagate over space and time. Spring green-up, or early growth of vegetation, and peaks in flower or fruit production are a good examples of resource waves (Armstrong et al. 2016). Spring migrations of ungulates in northern latitudes are timed to these resource waves. This phenomenon of animal movements coinciding with spring green-up is known as surfing the green wave, as ungulates move back to summer range following the patterns of plant growth (Bischoff et al. 2012, Aikens et al. 2017). During spring, emergence of green plants across the landscape appears to move as a wave with increasing elevation or latitude because plant emergence starts at low area and moves up in elevation or latitude, and movements of migratory animals follow that wave (Aikens et al. 2016, Armstrong et al. 2016). Migratory ungulates, including red deer (*Cervus elaphus*; Bischof et al. 2012, Rivrud et al. 2016), mule deer (Sawyer and Kauffman 2011, Blum et al. 2015, Aikens et al. 2017), and pronghorn (*Antilocapra americana*; Berger 2004, Sawyer et al. 2005), have been shown to track green-up of vegetation as they move north in latitude to summer range during spring. Barnacle geese (*Branta leucopsis*) also appear to track waves of plant green-up as they move north in latitude during spring migration (Shariatinajafabadi et al. 2014, Kölzsch et al. 2015). Migrations of African ungulates, including wildebeest (*Connochaetes taurinus*), are influenced more by wet and dry seasons than warm and cold as observed in northern species, but those migrations are influenced by availability and quality of plant food resources. Wildebeest migrate from open grasslands used during the wet season to wooded grasslands in higher-rainfall areas used during the dry season (Fryxell et al. 1988). Migratory wildebeest appear to be regulated by the amount of green grass available during the dry season, and food abundance directly influences survival of adults and young (Sinclair et al. 1985, Fryxell et al. 1988). In the Serengeti ecosystem, food plants of wildebeest have a grazing-free period that enables them to recover from high herbivory during seasons when wildebeest were present. Although predation has been suggested as a reason for migratory movements of animals, it is clear that changes in forage quality and availability—meaning changes in plant abundance, age, and structural composition—ultimately influence migratory movements of animals. Plants also have an indirect effect on movements of predators, many of which move in response to movements of herbivores.

Other resource waves are usually described as some type of color, as in the green wave of plants. Grizzly bears (*Ursus arctos*) are said to track the brown wave because they track spatial and temporal variation in the protein content of plant roots (Armstrong et al. 2016). Plants store protein and energy in their roots over winter and then transfer those materials back into aboveground parts during the growing season (Armstrong et al. 2016). Grizzly bears appear to track that brown wave at the end of the growing season and into the fall prior to entering dens for the winter, thereby exploiting those high-protein roots (Hamer et al. 1991, Coogan et al. 2012, Armstrong et al. 2016). Asiatic black bears (*Ursus thibetanus*) track fruit development of yamazakura (*Prunus jamasakura*) when the fruits peak in sugar content, which lasts about four days. Those bears track altitudinal gradients in fruit phenology and are able to exploit that resource wave for much longer, about a month (Koike et al. 2008, Armstrong et al. 2016).

FACILITATIVE INTERACTIONS

Mutualism

Mutualisms between plants and animals usually result from some exchange of resources that benefit the animal for a service to the plant, such as dispersal of seeds (Vander Wall and Moore 2016). Because mutualisms typically involve provision of resources, this type of species interaction likely has a greater potential to foster higher species diversity than antagonistic interactions (Vander Wall and Moore 2016). Pollination and seed dispersal are important examples of mutualisms between vertebrates and plants. The most common reward that plants use to pay for animal services is nectar or pollen, although often with seed dispersal some of the seeds also are the reward (Del-Claro et al. 2016). Nectar is an aqueous solution that is rich in carbohydrates, sucrose, fructose, or both, and it sometimes also contains lipids and amino acids (Koptur 1994, Blüthgen et al. 2004, Gonzalez-Teuber and Heil 2009, Del-Claro et al. 2016). Nectar from flowers is most commonly associated with pollination (Faegri and Van der Pijl 1976, Del-Claro et al. 2016). Not all nectar comes from flowers; however, nectar can be secreted onto most aboveground structures of plants, termed extrafloral nectar (Del-Claro et al. 2016). Extrafloral nectar is provided to insects, especially ants, in another mutualistic relationship in exchange for protection from herbivory (Rosumek et al. 2009, Zhang et al. 2015, Del-Claro et al. 2016).

An interesting and unusual example of a mutualism is between several species of pitcher-plants and tree shrews (*Tupaia montana*) in the cloud forests of Borneo. The largest pitcher-plants produce pitchers with high capacity, up to 2 L, for holding liquid in the pitcher (Clarke 1997, Phillipps et al. 2009, Robinson et al. 2009, Chin et al. 2010). Researchers hypothesized that large pitchers facilitated consumption of larger, perhaps even vertebrate, prey. Chin et al. (2010) recorded the types of prey documented in 36 individual pitchers of Rajah Brooke's pitcher-plant (*Nepenthes rajah*) and reported that ~58% of the contents of the pitchers consisted of the feces of tree shrews. Rajah Brooke's pitcher-plant and Low's pitcher-plant (*N. lowii*) produce sugar-based exudates or nectar that tree shrews feed upon. While feeding, the shrews defecate into the pitchers, providing an important source of nitrogen and phosphorus for the plants (Clarke et al. 2009, Chin et al. 2010). Between 57% and 100% of foliar nitrogen in Low's pitcher-plant is derived from animal feces, mostly that of tree shrews (Clarke et al. 2009). Additionally, these toilet pitchers have a concave lid for tree shews to sit on while feeding, which facilitates capture of feces. Although most of the pitcher-plants also capture and consume arthropods, the pitchers of Low's pitcher-plants are excellent and probably specialized for catching feces of tree shrews, and they are relatively ineffective for capturing arthropods (Clarke et al. 2009, Chin et al. 2010). Finally, the distance from the front of the pitcher opening to the inner surface of the lid correlates almost exactly to the head to body length of tree shrews, so the conclusion as to the large size of the pitchers was to facilitate collecting feces of tree shrews, not for consumption of vertebrate prey (Chin et al. 2010). This relationship is an interesting example of a mutualism; the shrew receives a food source that is rich in carbon-based sugars, and the plant receives a food source that is rich in nitrogen and phosphorus.

Ant–Plant Mutualisms

Probably no other group of animals interacts with plants in such diverse ways as do ants (Formicidae). Ants often protect plants from herbivores or competitors, and they alter both the physical environment and community dynamics by selective weeding, farming plants and fungus, altering nutrient availability, pollinating flowers, and dispersing or harvesting seeds (Beattie and Hughes 2002). Leaf cutter ants (*Atta* spp.) are the primary consumers in New World terrestrial ecosystems, and they have been estimated to remove between 12% and 17% of the total leaf production (Cherrett 1986, Beattie and Hughes 2002). The Philidris ant (*Philidris nagasau*) cultivates the epiphytic plant Squamellaria (*Squamellaria* spp.) for nesting and floral rewards (Chomicki et al. 2020). When Squamellaria was cultivated in full sun by the ants, the floral food reward increased 7.5-fold compared with shade-cultivated plants, and the higher reward was correlated with higher levels of protection from the ants (Chomicki et al. 2020).

Coevolution between ants and plants has resulted in complex mutualisms that involve rewards and services, known as ant-guard systems (Beattie and Hughes 2002). In this mutualism the rewards are provision of nest sites and extrafloral nectar provided to the ants via specialized food bodies; the service is protection of the plants from herbivory. Extrafloral nectar is provided to ants from tissues other than from within flowers and is not associated with pollination. Those food bodies are located on leaves, twigs, or occasionally on external surfaces of reproductive parts; they contain nectar and amino acids and lipids used by ants (Beattie and Hughes 2002). Investment in food bodies for ants can be high (e.g., Bahasa plant [*Macaranga triloba*] invests ~9% of the cost of growing aboveground tissue into food bodies; Heil et al. 1997, Beattie and Hughes 2002). Additionally, in the absence of ants, some plants cease producing food bodies. The ant piper tree (*Piper cenocladum*) stops producing food bodies in the absence of the ant (*Pheidole bicornis*) but immediately begins producing them when the ants return (Risch and Rickson 1981). The location of those food bodies can influence plant fitness; placement on the outside of reproductive structures resulted in foraging ants protecting the plant from seed-eating herbivores and thus higher plant fitness (Del-Claro et al. 1996, Calixto et al. 2020). These mutualisms can range in strength; for example, if ants only attack some herbivores and not others, the mutualism is weak. Conversely, strong mutualisms occur if rewards from the plant result in highly effective services that increase the fitness of both players (Beattie and Hughes 2002). Damage to pauterra (*Qualea parviflora*) by sucking insects led to increased nectar production and aggressiveness by the ants (Raupp et al. 2020). Additionally, plants subjected to higher levels of leaf damage produced more extrafloral nectar and with higher calories, which attracted more ants (Calixto et al. 2020).

Myrmecochory is seed dispersal by ants as part of an ant–plant mutualism (Vander Wall and Moore 2016, Konečná et al. 2018). Plants whose seeds are dispersed via myremecochory occur in 77 plant families and have been observed on all continents except Antarctica (Lengyel et al. 2010, Konečná et al. 2018). Generally, seeds that are subject to myrmecochory have an elaisome, which is a structure attached to the seed that attracts the ants (Vander Wall and Moore 2016). Ants bring the diaspore, seed plus elaisome, into the nest, and the elaisome is separated from the seed and given to ant larvae and workers (Fischer et al. 2005, Caut et al. 2013, Detrain and Bologna 2019). The seed is then discarded and can germinate outside the nest (Detrain and Bologna 2019). Therefore elaisomes are food resources for the ants, and the seed is the energy source for the germination and establishment of the young plant (Konečná et al. 2018). Elaisomes contain more easily digestible metabolites than do seeds, and elaisomes can originate from seed, fruit, or floral tissues (Ciccarelli et al. 2005, Mayer et al. 2005, Konečná et al. 2018). This mutualism provides a nutrient-rich food source for ants and facilitates dispersal of seeds for plants.

Pollination

Pollination occurs when pollen grains are transferred from the male anther of a flower to the female stigma, which enables fertilization and the production of seeds. All of the genetic information needed to produce new plants is contained in the seeds. Some plants can self-pollinate or fertilize, whereas others need an animal, the wind, or occasionally water to transfer pollen between flowers of the same species. Many plants can self-pollinate in the short term, but long-term self-pollination is a losing strategy similar to inbreeding, and many plants are incompatible with self-pollination. Pollination of plants by animals allows many plants to reproduce because animal-mediated pollination is often required for sexual reproduction in flowering plants or angiosperms, the most di-

Figure 12.4. Examples of pollinator species. (*top*) Little spiderhunter sunbird (*Arachnothera longirostra*) on hanging lobster (*Heliconia rostrate*) influorescence and honeybee (*Apis mellifera*) covered with pollen. (*bottom*) Anna's hummingbird and Jamaican fruit-eating bat (*Artibeus jamaicensis*), also covered with pollen. By Noicherrybeans, Julia Salgado, Daniel Prudek, and Megan Stewart

verse group of land plants (Kearns et al. 1998, Bao et al. 2019, Li et al. 2019). Plant pollinators include those that feed on nectar or pollen and then transfer pollen to other plants (Fig. 12.4). Insects, including bees, beetles, and flies, are the largest group of pollinators, with between 25,000 and 30,000 species, of which all are flower pollinators (Kearns et al. 1998). Vertebrate pollinators include birds and bats, but they also include rodents, monkeys, and lizards.

Many of the food plants and other products upon which humans depend would disappear without animal-mediated pollination. Although food crops represent <1% of flowering plants, 75% of the food crops from around the world produce more fruits or seeds when they are pollinated by animals (Klein 2007, Ollerton et al. 2011). There are an estimated 352,000 species of angiosperms (Paton et al. 2008, Ollerton et al. 2011), and about 85% of them are likely pollinated by animals, mostly insects (Kearns et al. 1998, Ollerton et al. 2011). Tropical ecosystems tend to support the highest proportion of animal-pollinated plants; about 94% of them are pollinated by animals. That proportion of animal-pollinated plants drops to 78%, however, in plant communities located in the temperate zone (Ollerton et al. 2011). Most angiosperms are located in the tropics, so it makes sense that more animal-pollinated species exist in that ecosystem, but the proportion of animal-pollinated plants is similar across ecosystems (Fig. 12.5). Additionally, in tropical ecosystems a greater proportion of plants have specialized pollinator groups or species of pollinators, such as birds, bats, hawkmoths, primates, or others (Ollerton et al. 2006, 2011).

Although pollination is a mutualism between plants and animals, each group is focused on their own reproductive

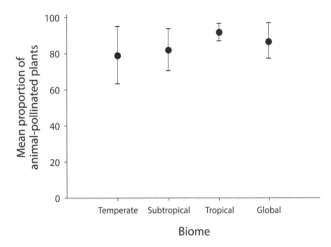

Figure 12.5. Mean proportion of animal-pollinated plant trial communities at different latitudes. The global average accounts for the relative distribution of plants at different latitudes. Redrawn from Ollerton et al. (2011)

fitness. Pollination is how plants reproduce, and animals are gathering food, so the original interactions were probably antagonistic and then evolved to be mutualistic (Kearns et al. 1998). Although many interactions between pollinators and plants involve specialized adaptations with single or few pollinators, others are fairly general, where plants may be pollinated by multiple species. The flowering shrub broadleaved lavender (*Lavandula latifolia*), which is endemic to southern Spain, is visited by 54 different species of insects, although those species vary in their effectiveness as pollinators (Herrera 1987, 1988, 1989; Kearns et al. 1998). In general, pollination tends to occur more as interaction webs that may shift in space and time depending on the associated landscapes (Kearns et al. 1998). There are exceptions, however, where pollination is specific and obligate between the plant and the pollinator.

Plants have evolved many methods to attract pollinators, including visual and scent cues, provision of food, mimicry, and even entrapment. Nectar that is not consumed may be reabsorbed by plants, which is an important adaptation because it is costly for the plant to produce (Pyke 1991, Nepi et al. 1996, Pacini and Nepi 2007, Pacini et al. 2008). Flower traits consist of complex signals, including structure, size, color, and scent that evolved to attract pollinators (Leonard et al. 2012, Leonard and Masek 2014, Bergamo et al. 2019). For example, plants that are pollinated by hummingbirds tend to be red in color, grow pendant flowers, have a tubular shape, have little to no odor, and provide nectar rich in sucrose (Tadey and Aizen 2001).

Coevolution between plants and their pollinators has resulted in many morphological structures that are specific to groups of pollinators. Curved flowers, for example, have been shown to be a hallmark of plant adaptations to pollination by sunbirds (Nectariniidae) and honeyeaters (Meliphagidae) that perch while feeding on nectar (Goldblatt et al. 1999, Goldblatt and Manning 2006, Cronk and Ojeda 2008, Johnson et al.

2020). Sunbirds and honeyeaters perch while feeding and tend to have curved bills, whereas hummingbirds have straight bills, tend to hover while feeding, and prefer straight-tubed flowers (Fig. 12.5; Patton and Collins 1989, Johnson et al. 2020). Johnson et al. (2020) showed that sunbirds preferred flowers that curved downward because it took the least time to insert their bills into the flower structure and begin feeding. Gill and Wolf (1978) showed that species with curved bills could extract nectar more quickly from curved flowers than species with straight bills, and the opposite was seen for straight-tubed flowers and species with straight bills. Flower curvature, flower orientation, and availability of perches had strong influence in feeding position and time to feeding of sunbirds (Johnson et al. 2020). Conversely, many hummingbird-pollinated flowers are pendant shaped, but metabolic studies indicate that it is more energetically costly for hummingbirds to feed from pendant flowers than from horizontal flowers (Sapir and Dudley 2013, Johnson et al. 2020). Pendant flowers, however, may prevent dilution of nectar from rainfall, which may be a stronger selective factor toward evolution of pendant-structured flowers for pollination by hummingbirds, although dilution of nectar by rainfall has so far been difficult to document (Aizen 2003, Johnson et al. 2020).

In addition to flowers, plants produce other structures, such as bracts or modified leaves located close to the flower, to attract pollinators and mutualists (Bergamo et al. 2019). Bracts appear to function as honest signaling of reward production and protection of flowers for hummingbird pollinators (Borges et al. 2003, Sun et al. 2008, Pélabon et al. 2012, Bergamo et al. 2019). Bract colors have been shown to be important and reliable visual cues in attracting hummingbird pollinators for several species of plants (Armbruster 1996, Herrera 1997, Borges et al. 2003, Sun et al. 2008, Bergamo et al. 2019). Interestingly, bract and petal colors of plants that are pollinated by hummingbirds provide a reliable cue for hummingbird pollinators but not for bees (Bergamo et al. 2019). Additionally, long-tubed flowers are effectively pollinated by hummingbirds, and bees often act as nectar robbers in these types of flowers rather than pollinators (Bergamo et al. 2019). Therefore long-tubed flowers pollinated by hummingbirds may use bracts to avoid visitation by bees and encourage pollination by hummingbirds.

Traits such as color, luminance, pattern, size, and scent are considered honest displays that convey accurate information about quality or quantity of rewards available to pollinators or seed dispersers. Yet the mechanisms of how pollinators respond to those displays is still not well understood (Leonard and Francis 2017). Importantly, this idea of communication between plants and animals is a popular theme of research, relating to the mechanisms of signal production, including the biochemistry of scent and colors (Leonard et al. 2012, Leonard and Masek 2014, Leonard and Francis 2017). This communication between plants and animals can be remarkably complex. For instance, the same plant may have different pollinators than seed dispersers and may be consumed by yet another animal. In addition, animals may respond differently to the same flower or fruit based on the color or scent, because different mechanisms are used to respond to olfactory versus visual cues (Leonard and Francis 2017). That information is used by animals to guide selection of nectar or fruit resources among individual plants (Knauer and Schiestl 2015, Leonard and Francis 2017, Bergamo et al. 2019). Quality and quantity of nectar rewards from plants are not consistent across species, largely as a result of their varying life histories. For example, no single trait of flower or fruits signals the same value of rewards to different species of animals (Leonard and Francis 2017). Some species may respond better to color, including birds and primates, but others may respond more to scents. For example, the scent of the guarana (*Paullinia cupana*) plant in the Amazon attracts its nocturnal bee (*Megalopta* spp.) pollinator (Krug et al. 2018).

Seed Dispersal

Dispersal of seeds is an important process for reproduction and recruitment of plants. Seed dispersal is defined as transfer of seeds away from the parent to the place where they germinate (Forget and Milleron 1991). Dispersal and predation are the primary processes that determine survival of seeds in many ecosystems (Fig. 12.6). There are seven general types of seed dispersal, collectively known as syndromes because multiple species use each strategy (Vander Wall et al. 2017). Those seven methods of seed dispersal include (1) frugivory, where fleshy fruit is consumed and viable seeds are excreted through feces or regurgitation; (2) scatter-hoarding, where seeds are buried by animals for future use and those that are forgotten will germinate; (3) myrmecochory, or seed dispersal by ants; (4) ectozoochory, which facilitates dispersal of seeds by hooking onto the exterior of an animals through morphological structures including hooks, burs, and barbs; (5) wind dispersal; (6) water dispersal; and (7) ballistic dispersal, where seeds are propelled by pressurized fruit or some trigger mechanism (Fig. 12.6; Vander Wall and Moore 2016, Vander Wall et al. 2017). The first four of these methods involve animal-mediated dispersal, and they are generally described as mutualisms, possibly with the exception of ectozoochory, because animals usually receive some sort of reward for the service (Bronstein 1994, Vander Wall and Moore 2016, Vander Wall et al. 2017).

Seed dispersal is an important mutualism between animals and plants. Plants produce fruits to facilitate the dispersal of seeds by providing a food reward to animal dispersers (Tweksbury 2002), and animal-mediated dispersal of seeds is an important component of population dynamics of plants and influences their community assemblages (Howe and Miriti 2000, Nathan and Muller-Landau 2000, Wang and Smith 2002, Russo et al. 2006, Cousens et al. 2010, McConkey et al. 2012, Tarszisz et al. 2018). For many plants, long-distance dispersal of seeds, especially large seeds, would not be possible without animal vectors (Anderson 2012). Birds, carnivores, and ungulates move at large spatial scales and therefore may disperse seeds very long distances, often many kilometers (Pacini et al. 2008). Seed dispersal provides a mechanism for plants to reduce intraspecific competition, among conspecifics, or interspecific competition, with seeds of other species. Seed dispersal also facilitates colonization of bare ground or vacant sites, helps to maintain gene flow and genetic diversity, and influences adaptation to changing environments (Traveset et al. 2013, Traveset and Richardson 2014). Pires et al. (2017) estimated that seed dispersal by now extinct megafauna were up to 10 times greater than long-distance seed dispersal by smaller-sized extant mammals. Long-distance dispersal of seeds in South America contracted by at least two-thirds after the megafauna assemblages went extinct (Pires et al. 2017). In the Brazilian Atlantic Forest, where agoutis (*Dasyprrocta leporine*) were reintroduced, dispersal of the seeds of the vulnerable

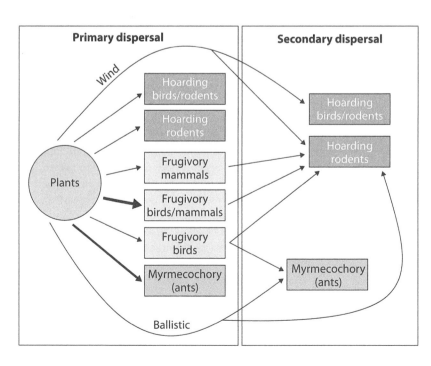

Figure 12.6. Dispersal pathways for plants with seed-dispersal mutualisms in North America. Width of the arrows is proportional to the number of records of mutualists. The different modes of seed dispersal and animal vectors for dispersal are in shaded boxes. Myrmecochory is seed dispersal by ants, which is an ecological important ant–plant interaction with worldwide distribution. Redrawn from Vander Wall and Moore (2016)

arara nut-tree (*Joannesia princeps*) was increased, and those seeds dispersed and buried by agoutis had higher probability of survival away from the adult tree (Mittelman et al. 2020).

While grains of pollen that are dispersed by animals are usually attached to the body of an animal, seeds may be dispersed by attachment to the body or by ingestion. Seed dispersal following ingestion is referred to as endozoochory, or internal dispersal of seeds. Endozoochory, which is facilitated by frugivory, is one of the most widely studied dispersal mechanisms of plants (Baltzinger et al. 2019). Feer (1995) showed that duikers (*Cephalophinae*) in African rainforests dispersed seeds with hard shells by consuming the fruit and then spitting out the seeds during rumination. Germination of fruits of pokeweed (*Phytolacca amaericana*) was accelerated, without impaired viability of seeds, by transit through the gastrointestinal tract of two species of birds (Orrock 2005, Pacini et al. 2008). Germination rates of acacia seeds swallowed and defecated by large ungulates, increased with body size, probably because slower passage rate of seeds through the digestive tract of bigger animals resulted in greater removal of the seed coat, which facilitated germination (Rohner and Ward 1999, Bodmer and Ward 2006). Forbs that rely on large herbivores that consume small fruit have seeds that are probably less likely to be destroyed during consumption by species that take larger and generally less discriminate mouthfuls (Janzen 1984). Additionally, large-bodied species consume greater volumes of fruit, which also facilitates movement of greater numbers of seeds over longer distances (Guimarães et al. 2008a, b).

Carnivores also often act as seed dispersers. In temperate rainforests in Spain, red foxes (*Vulpes vulpes*), martens (*Martes* spp.), badgers (*Meles meles*), and weasels (*Mustela* spp.) were important dispersers of the seeds of fleshy-fruit plants like hawthorn (*Crataegus monogyna*) and wild rose (*Rosa* spp.; Peredo et al. 2013). Moreover, seeds were only occasionally damaged when ingested by those carnivores, and most seeds found in their scats were viable (Peredo et al. 2013). Foxes and badgers dispersed seeds of bramble (*Rubus* spp.) almost exclusively, whereas martens largely dispersed the seeds of wild rose (Matías et al. 2010, López-Bao and González-Varo 2011, Peredo et al. 2013). Additionally, compared with bird-dispersed seeds that were largely trees and in forested areas, mammal-dispersed seeds were primarily shrubs and were dispersed in deforested areas. In Japan the seeds of the liana *Actinidia arguta* had higher rates of germination following ingestion by Japanese martens (*Martes melanpus*) compared with those from fruits that were not consumed (Tsuji et al. 2020).

Dispersal of seeds by birds or mammals through endozoochory may be followed by further movement of seeds by rodents or insects, where seeds are obtained from the feces or following regurgitation (Fig. 12.6). This two-stage process of seed dispersal is termed diplochory, which involves endozoochory followed by movement of seeds by animals, such as dung beetles or rodents, that are obtained from feces without ingestion, termed epizoochory (Blanco et al. 2020). Fecal deposits can be unfavorable to seedling establishment if they are deposited in inhospitable locations (Vander Wall et al. 2005)

or in large clumps, resulting in higher competition among individual plant propagules (Howe 1989). Therefore further movement of seeds by other animals may facilitate seedling establishment into the soil (Vander Wall et al. 2005). Dung beetles (Scarabaeidae) and some species of scatter-hoarding rodents have been shown to remove seeds from mammal feces and bury them in the soil, usually at depths that allow germination if they are not retrieved (Estrada and Coates-Estrada 1991; Forget and Milleron 1991; Wenny 1999; Andresen 2001, 2002; Feer and Forget 2002; Vander Wall et al. 2005). Vander Wall et al. (2005) created artificial bird feces with radiolabeled seeds to test whether rodents in the Sierra Nevada Mountains, Nevada, would remove and cache those seeds. They found 38 caches containing 52 seeds (1.4 seeds/cache) from the experimental pseudo-feces, which accounted for about 27% of the large seeds taken from their test feces. Although seeds are dispersed at greater distances by birds or large mammals compared with rodents, diplochory favored establishment of seedlings and the overall effectiveness of seed dispersal (Vander Wall and Longland 2004, Vander Wall et al. 2005). This two-phased dispersal of seeds may be more common than previously realized, which also may help explain patterns of species diversity and species interactions in temperate communities (Vander Wall et al. 2005).

Seed dispersal is an important ecosystem service provided by primates in tropical ecosystems (Chapman and Russo 2006, Gómez and Verdú 2012). Primates consume fruit fairly regularly, and interactions between primates and plants can be mutualistic or antagonistic, depending on whether the seeds are fully digested. In primates, 64% of the extant species are frugivores, but about 96% of primate species, occasionally at least, consume fleshy fruits and may act as seed dispersers. The establishment of mutualistic relationships between plants and primates resulted in larger geographic ranges for plants, which led to an increase in speciation rate and a decrease in probability of extinction. The diversification rate of primate species with mutualistic interactions with plants, in this instance seed dispersal, was higher than among those with antagonistic relationships, like herbivory (Gómez and Verdú 2012).

Fertilization

There are multiple examples of fertilization of plants by animals, especially colonial species like seabirds and bats that provide nutrient-rich guano into ecosystems. Additionally, foraging by animals also provides fertilization to plants, such as grizzly bears bringing salmon (*Oncorhynchus* spp.) carcasses into terrestrial ecosystems or animal latrine sites that have been shown to provide nutrient enrichment to plants (Ben-David et al. 1998b, Levi et al. 2020). Grazing ungulates also provide inputs of nitrogen through urine and feces that stimulate plant growth in multiple ecosystems, including on grazing lawns (McNaughton 1979, Hik and Jefferies 1990).

Seabirds consume fish and other marine organisms that tend to excrete nutrient-rich guano or feces on islands, thereby moving marine-derived nutrients into terrestrial ecosystems. For example, guano deposited on islands in the Gulf of California averaged about 9% nitrogen and 11% of the phosphorus on those islands, which enhances plant growth. In a comparison of islands with and without seabird colonies on nutrient-poor islands in the Gulf of California, those with colonial seabirds had higher concentrations of nitrogen and phosphorus in soils, with enhanced plant growth compared to islands without seabirds (Anderson and Polis 1998). Those marine-derived nutrients in some island plants, including saltbrush (*Atriplex barclayana*) and buckhorn cholla (*Opuntia acanthocarpa*), were 1.6- to 2.4-fold greater on islands with colonial seabirds (Anderson and Polis 1998). Penguin colonies depositing guano on Marion Island represented about 85% of all organic matter deposited on the island and was a major source of nitrogen and phosphorous on the coastal plain (Burger et al. 1978). Otero et al. (2018) reported that the concentrations of nitrogen and phosphorus in seabird colonies were among the highest measured on the earth's surface, but those effects are relatively short term and limited to the breeding season. Seabirds disperse during the nonbreeding season, which results in small effect on nutrient cycling during that time (Otero et al. 2018). Migratory snow geese on winter range in on the Bosque del Apache Wildlife Refuge in New Mexico provided ~40% of the nitrogen and 75% of the phosphorus during winter (Kitchell et al. 1999). Geese provided a major source of nutrients to those wetlands during winter (Kitchell et al. 1999).

Predators and scavengers often contribute to the redistribution of nutrients derived from spawning Pacific salmon in riparian ecosystems and their predators (Ben-David et al. 1998b, O'Keefe and Edwards 2002, Wilkinson et al. 2005). River otters (*Lutra canadensis*), brown bears, and bald eagles (*Haliaeetus leucocephalus*) moved salmon carcasses away from stream environments in southeast Alaska, thereby contributing to fertilization of plants from the salmon carcasses beyond just waterways and into riparian ecosystems (Ben-David et al. 1998b). When salmon abundance is high, bears selectively feed on the energy-rich parts of the fish and distribute the remaining carcass up to several hundred meters from rivers and streams (Levi et al. 2020). Those carcasses provide nutrients, phosphorus and nitrogen, for plants in the immediate vicinity, but extensive movement of salmon carcasses by bears can provide large nutrient subsidies for plants in those ecosystems (Levi et al. 2020). Bears can transfer <50% of salmon carcasses away from streams (Levi et al. 2020). Additionally, urine and feces from the bears distributed salmon-derived nutrients to plants (Hildebrand et al. 1999, Wilkinson et al. 2005, Levi et al. 2020). Finally, those input of salmon nutrients by bears increased productivity of riparian plants, increased tree growth, and shifted community structure to plant species that were adapted to nutrient rich sites (Levi et al. 2020).

River otters that inhabit coastal environments forage in the intertidal and subtidal zones for marine fish but also mark specific sites along the coast, known as latrine sites (Larsen 1984; Bowyer et al. 1994, 1995; Testa et al. 1994; Ben-David et al. 1998a). Those latrine sites may have a radius of 5–20 m and are generally between 25 and 300 m apart (Ben-David et al. 1998a). The latrine sites fertilized the surrounding vegetation, forming an interesting example of effects of carnivores on plants without herbivory (Ben-David et al. 1998a). Galapagos sea lions (*Zalophus wollebaecki*) also moved marine-derived nutrients into shoreline ecosystems in the Galapagos where those nutrients were documented in soils and shoreline plants, but only in locations where sea lions occurred (Fariña et al. 2003).

CONSERVATION

Recent evidence for declines of pollinator abundance and diversity has generated concern worldwide (Potts et al. 2010, Ollerton et al. 2011). Declines of pollinators and disruption of pollination services have been reported on every continent on the planet except Antarctica (Kearns et al. 1998). Pollinators are important components of biodiversity and provide important ecosystem services to wild plants and crops globally (Benadi et al. 2013). Declines in pollinator populations and seed dispersers can result in loss of ecological processes and ecosystem services, which will result in significant loss of biodiversity and ecosystem stability. Moreover, because most agricultural crop plants depend on animal pollination, loss of pollinators also will have substantial economic ramifications on human communities (Benadi et al. 2013).

Changing climate with increasing temperatures and increasing concentrations of CO_2 has important implications for herbivores and plants. Changing conditions may lead to shifts in plant phenology, or species composition, which may reduce the availability of preferred foods for animals or result in trophic mismatches that could easily reduce recruitment in animal populations (Parmesan and Hanley 2015). Elevated temperatures and concentrations of atmospheric CO_2 are predicted to modify plant chemistry and cause increases in secondary compounds in plants, potentially with toxic results for herbivores (Zvereva and Kozlov 2006, Forbey et al. 2013, Beale et al. 2018, Matocq et al. 2020). Zvereva and Kozlov (2006) reported that under increasing temperature and concentrations of CO_2, both woody and herbaceous angiosperms had decreased nitrogen concentrations, although the signal was stronger in woody plants.

Giant pandas are an important symbol of international conservation, but they are in decline partly because of declines of their food plants and loss of their habitats. Human societies are continuing to grow and will impose stronger selective pressures on plants and both their herbivores and mutualists as environments change rapidly. Habitat loss and fragmentation, overharvest of animals and plants, and climate change are causing catastrophic declines in plant and animal populations, leading to increasing rates of extinction worldwide (Baranosky et al. 2001, Dirzo et al. 2007, Mittelman et al. 2020). As extinctions increase, ecosystems are becoming more impoverished, and loss of these important community interactions continues to hinder ecosystem function as important ecological processes like pollination and seed dispersal disappear (Kurten 2013, Galetti et al. 2017, Mittelman et al. 2020). Restoration of ecosystems and reintroduction of species in areas of local extirpation are steps toward reversing

HARRIET LAWRENCE HEMENWAY (1858–1960)

The progressive conservation crusade of the early 20th century was greatly advanced by women who used social and political activism to advance nature protection. Harriet Hemenway was a leader in the movement to protect birds against the ravages of a plume industry influenced by fashion trends of America's gilded age. Every year, thousands of herons, egrets, woodpeckers, owls, bluebirds, warblers, and many other birds were killed to decorate women's hats and clothing. Drawing upon her social standing and assisted by her cousin Minna B. Hall, Hemenway urged the wealthy women of Boston to reject feathered fashions and to join other forces that advocated for bird protection. The two cousins organized boycotts of stores and companies that dealt in feathered fashions. They organized meetings between the socialites and the leading ornithologists of New England, leading to the founding of the Massachusetts Audubon Society. More than 100 women joined the new society, serving as officers and providing leadership to local chapters. The society was influential in the 1897 passage of a Massachusetts law that prohibited trade in wild bird feathers. These efforts paved the way for the passage of key federal laws, such as the Lacey Act in 1900, which prohibited interstate commerce of protected species taken in violation of state laws. The plume trade and other forms of market hunting ended with the 1913 passage of the Weeks-McLean Law, sponsored by Massachusetts Representative John Weeks and Connecticut Senator George McLean. Also known as the Migratory Bird Act, the law was a milestone in the early conservation movement and remains a legacy of visionary activists such as Harriet Lawrence Hemenway.

Prepared by Winifred Kessler

Courtesy of the Massachusetts Audubon Society

these losses, but prevention of loss of plants and animals their interactions is an important focus for maintaining biodiversity and thus stability in ecological communities, but it is an uphill battle.

SUMMARY

Nearly every ecosystem on the planet is affected by interactions between animals and plants, especially those mediated by pollinators or by herbivory. Interactions may be neutral, antagonistic, or facilitative. Coevolution between plants and animals is an important evolutionary process that shapes communities and ecosystems, and leads to highly complex interactions among multiple species. Plants can strongly affect locations of animals and how animals move across the landscape or in local areas, and animals have strong effects on functioning and structure of plant communities. Herbivory is the most prevalent antagonistic interaction between animals and plants, and coevolution has led to an evolutionary arms race where plants evolve to reduce effects of herbivory through tolerance or resistance, and animals seek to overcome those obstacles to foraging. Plants have evolved complex forms of resistance to herbivores, including elaborate chemical and physical defenses. Tolerance to herbivory has led some species of plants to increase productivity at moderate levels of herbivory, if enough resources are available to them. Mutualisms also have resulted from coevolution into elaborate methods of pollination, seed dispersal, and fertilization that benefits both plants and animals. Plant–animal interactions are highly complex and lead to higher biodiversity, stability, and resilience of ecosystems.

Literature Cited

Abraham, K. F., R. L. Jefferies, and R. T. Alisauskas. 2005. The dynamics of landscape change and snow geese in mid-continent North America. Global Change Biology 11:841–855.

Adamec, L. 1997. Mineral nutrition of carnivorous plants: a review. Botanical Review 63:273–299.

Aerts, R. J., T. N. Barry, and W. C. McNabb. 1999. Polyphenols and agriculture: beneficial effects of proanthocyanidins in forages. Agriculture, Ecosystems, and Environment 75:1–12.

Agrawal, A. A., S. Y. Strauss, and M. J. Stout. 1999. Costs of induced responses and tolerance to herbivory in male and female fitness components of wild radish. Evolution 54:1093–1104.

Aikens, E. O., M. J. Kauffman, J. A. Merkle, S. P. H. Dwinnell, G. L. Fralick, and K. L. Monteith. 2017. The greenscape shapes surfing of resource waves in a large migratory herbivore. Ecology Letters 20:741–750.

Aizen, M. A. 2003. Down-facing flowers, hummingbirds and rain. Taxon 52:675–680.

Alerstam, T., A. Hedenstrom, and S. Akesson. 2003. Ecology of long-distance movements: migration and orientation. Oikos 103:247–260.

Amit, M., I. Cohen, A. Marcovics, H. Muklada, T. A. Glasser, E. D. Ungar, and S. Y. Landau. 2013. Self-medication with tannin-rich browse in goats infected with gastro-intestinal nematodes. Veterinary Parasitology 198:305–311.

Anderson, T. M. 2012. Plant-animal interactions. Pages 298–301 in J. C. Nagle, B. Pardy, O. J. Schmitz et al., editors. Ecosystem management and sustainability. Vol. 5 of the Berkshire encyclopedia of sustainability. Berkshire, Great Barrington, Massachusetts, USA.

Anderson, W. B., and G. A. Polis. 1998. Marine subsidies of island communities in the Gulf of California: evidence from stable carbon and nitrogen isotopes. Oikos 81:75–80.

Andresen, E. 2001. Effects of dung presence, dung amount and secondary dispersal by dung beetles on the fate of Micropholis guyanensis (Sapotaceae) seeds in central Amazonia. Journal of Tropical Ecology 17:61–78.

Andresen, E. 2002. Dung beetles in a central Amazonian rainforest and their ecological role as secondary seed dispersers. Ecological Entomology 27:257–270.

Armbruster, W. S. 1996. Evolution of floral morphology and function: an integrative approach to adaptation, constraint, and compromise in *Dalechampia* (Euphorbiaceae). Pages 241–272 *in* D. G. Lloyd and S. C. H. Barret, editors. Floral biology. Springer, Boston, Massachusetts, USA.

Armstrong, J. B., G. Takimoto, D. E. Schindler, M. M. Hayes, and M. J. Kauffman. 2016. Resource waves: phenological diversity enhances foraging opportunities for mobile consumers. Ecology 97:1099–1112.

Augustine, D. J., and D. A. Frank. 2001. Effects of migratory grazers on spatial heterogeneity of soil nitrogen properties in a grassland ecosystem. Ecology 82:3149–3162.

Baider, C., and F. B. V. Florens. 2006. Current decline of the "Dodo Tree": a case of brown-down interactions with extinct species or the result of new interactions with alien invaders? Pages 199–214 *in* W. F. Laurance and C. A. Peres, editors. Emerging threats to tropical forests. University of Chicago Press, Chicago, Illinois, USA.

Baltzinger, C., S. Karimi, and U. Shukla. 2019. Plants on the move: hitchhiking with ungulates distributes diaspores across landscapes. Frontiers in Ecology and Evolution 7:38.

Bao, T. B. Wang, J. Li, and D. Dilcher. 2019. Pollination of Cretaceous flowers. Proceedings of the National Academy of Sciences 116:24707–24711.

Baranosky, A. D., N. Matzke, S. Tomiya, G. O. U. Wogan, B. Swartz, et al. 2011. Has the earth's sixth mass extinction already arrived? Nature 471:51–57.

Barbehenn, R. V., and C. P. Constabel. 2011. Tannins in plant-herbivore interactions. Phytochemistry 72:1551–1565.

Beale, P. K., K. J. Marsh, W. J. Foley, and B. D. Moore. 2018. A hot lunch for herbivores: physiological effects of elevated temperatures on mammalian feeding ecology. Biological Reviews 93:674–692.

Beattie, A. J., and L. Hughes. 2002. Ant-plant interactions. Pages 211–235 *in* C. M. Herrera and O. Pellmyr, editors. Plant-animal interactions: an evolutionary approach. Blackwell, Malden, Massachusetts, USA.

Becerra, J. X. 2007. The impact of herbivore-plant coevolution on plant community structure. Proceedings of the National Academy of Sciences USA 104:7483–7488.

Belovsky, G. E., and O. J. Schmitz. 1991. Mammalian herbivore optimal foraging and the role of plant defense. Pages 1–28 *in* R. T. Palo and C. T. Robbins, editors. Plant defenses against mammalian herbivory. CRC Press, Boca Raton, Florida, USA.

Belsky, A. J. 1986. Does herbivory benefit plants? A review of the evidence. American Naturalist 127:870–892.

Belsky, A. J. 1987. The effects of grazing: confounding of ecosystem, community, and organism scales. American Naturalist 129:777–783.

Benadi, G., N. Bluthgen, T. Hövestadt, and H.-J. Poethke. 2013. When can plant-pollinator interactions promote plant diversity? American Naturalist 182:131–146.

Ben-David, M., R. T. Bowyer, L. K. Duffy, D. D. Roby, and D. M. Schell. 1998*a*. Social behavior and ecosystem processes: river otter latrines and nutrient dynamics of terrestrial vegetation. Ecology 79:2567–2571.

Ben-David, M., T. A. Hanley, and D. M. Schell. 1998*b*. Fertilization of terrestrial vegetation by spawning Pacific salmon: the role of flooding and predator activity. Oikos 834:47–55.

Bender, I. M. A., W. D. Kissling, P. G. Blendinger, K. Böhning-Gaese, I. Hensen, et al. 2018. Morphological trait matching shapes plant-frugivore networks across the Andes. Ecography 41:1910–1919.

Bergamo, P. J., M. Wolowski, F. J. Telles, V. Lourenco, G. DeBrito, I. G. Varassin, and M. Sazima. 2019. Bracts and long-tube flowers of hummingbird-pollinated plants are conspicuous to hummingbirds but not to bees. Biological Journal of the Linnean Society 126:533–544.

Berger, J. 2004. The last mile: how to sustain long-distance migration in mammals. Conservation Biology 18:320–331.

Bischoff, R., L. E. Loe, E. L. Meisingset, B. Zimmermann, B. Van Moorter, and A. Mysterud. 2012. A migratory northern ungulate in the pursuit of spring: jumping or surfing the green wave? American Naturalist 180:407–424.

Blanco, G., C. Bravo, D. Chamorro, A. Lovas-Kiss, F. Hiraldo, and J. L. Tella. 2020. Herb endozoochory by cockatoos: is "foliage the fruit"? Austral Ecology 45:122–126.

Blum, M. E., K. M. Stewart, and C. Schroeder. 2015. Effects of large-scale gold mining on migratory behavior of a large herbivore. Ecosphere 6:74.

Blüthgen, N., N. E. Stork, and K. Fiedler. 2004. Bottom-up control and co-occurrence in complex communities: honeydew and nectar determine a rainforest ant mosaic. Oikos 106:344–358.

Bodmer, R., and D. Ward. 2006. Frugivory in large mammalian herbivores. Pages 232–260 *in* K. Danell, R. Bergström, P. Duncan, et al., editors. Large herbivore ecology, ecosystem dynamics, and conservation. Conservation Biology 11. Cambridge University Press, Cambridge, UK.

Borges, R. M., V. Gowda, and M. Zacharias. 2003. Butterfly pollination and high-contrast visual signals in a low-density distylous plant. Oecologia 136:571–573.

Bowyer, R. T., W. J. Testa, and J. B. Faro. 1995. Habitat selection and home ranges of river otters in a marine environment: effect of the *Exxon Valdez* oil spill. Journal of Mammalogy 76:1–11.

Bowyer, R. T., W. J. Testa, J. B. Faro, C. C. Schwartz, and J. B. Browning. 1994. Changes in diets of river otters in Prince William Sound, Alaska: effects of the *Exxon Valdez* oil spill. Canadian Journal of Zoology 72:970–976.

Bronstein, J. L. 1994. Conditional outcomes in mutualistic interactions. Trends in Ecology and Evolution 9:214–217.

Bronstein, J. L. 2009. The evolution of facilitation and mutualism. Journal of Ecology 97:1160–1170.

Bruno, J. F., J. J. Stachowicz, and M. D. Bertness. 2003. Inclusion of facilitation in ecological theory. Trends in Ecology and Evolution 18:119–125.

Bryant, J. P. 2003. Winter browsing on Alaska feltleaf willow twigs improves leaf nutritional value for snowshoe hares in summer. Oikos 102:25–32.

Bryant, J. P., F. S. Chapin, and D. R. Klein. 1983. Carbon/nutrient balance of boreal plants in relation to herbivory. Oikos 40:357–368.

Bryant, J. P., P. J. Kuropat, P. B. Reichardt, and T. P. Clausen. 1991. Controls over allocation of resources by woody plants to chemical anti-herbivore defense. Pages 83–102 *in* R. T. Palo and C. T. Robbins, editors. Plant defenses against mammalian herbivory. CRC Press, Boca Raton, Florida, USA.

Burger, A. E., H. J. Lindeboom, and A. J. Williams. 1978. The mineral and energy contributions of guano of selected species of birds to the Marion Island terrestrial ecosystem. South African Journal of Science 8:59–70.

Caesar, L. K., and N. B. Cech. 2019. Synergy and antagonism in natural product extracts: when 1 + 1 does not equal 2. Natural Products Reports 36:869–888.

Calixto, E. S., D. Lange, J. Bronstein, H. M. Torezan-Silingardi, and K. Del-Claro. 2020. Optimal defense theory in an ant-plant mutualism: extrafloral nectar as an induced defence is maximized in the most valuable plant structures. Journal of Ecology 109:167–178.

Catling, P. M. 2001. Extinction and the importance of history and dependence in conservation. Biodiversity 2:2–13.

Caut, S., M. J. Jowers, X. Cerda, and R. R. Boulay. 2013. Questioning the mutual benefits of myrmecochory: a stable isotope-based experimental approach. Ecological entomology 38:390–399.

Chamberlain, S. A., and J. N. Holland. 2009. Quantitative synthesis of context dependency in ant-plant protection mutualisms. Ecology 90:2384–2392.

Chapman, C. A., and S. E. Russo. 2006. Primate seed dispersal: linking behavioral ecology with forest community structure. Pages 510–526 in C. J. Campbell, A. F. Fuentes, K. C. MacKinnon, et al., editors. Primates in perspective. Oxford University Press, Oxford, UK.

Cherrett, J. M. 1986. History of the leaf-cutting ant problem. Pages 10–17 in C. S. Lofgren and R. K. Vander Meer, editors. Fire ants and leaf-cutting ants: biology and management. Westview Press, Boulder, Colorado, USA.

Chin, L., J. A. Moran, and C. Clarke. 2010. Trap geometry in three giant montane pitcher plant species from Borneo is a function of tree shrew body size. New Phytologist 186:461–470.

Chomicki, G., G. Kadereit, S. S. Renner, and E. T. Kiers. 2020. Tradeoffs in the evolution of plant farming by ants. Proceedings of the National Academy of Sciences 117:2535–2543.

Ciccarelli, D., A. C. Andeucci, A. M. Pagni, and F. Garbari. 2005. Structure and development of the elaiosome in Myrtus communis L. (Myrtaceae) seeds. Flora 200:326–331.

Cingolani, A. M., I. Noy-Neir, and S. Diaz. 2005. Grazing effects on rangeland diversity: a synthesis of contemporary models. Ecological Applications 15:757–773.

Clarke, C. M. 1997. Nepenthes of Borneo, Kota Kinabalu, Sabh, Malaysia. Natural History Publications, Borneo.

Clarke, C. M., U. Bauer, C. C. Lee, A. A. Tuen, K. Rembold, and J. A. Moran. 2009. Tree shrew lavatories: a novel nitrogen sequestration strategy in tropical pitcher plants. Biology Letters 5:632–635.

Coogan, S. C. P., S. E. Nielsen, and G. B. Stenhouse. 2012. Spatial and temporal heterogeneity creates a "brown tide" in root phenology and nutrition. ISRN Ecology 618257:1–10

Cork, S. J. 1992. Polyphenols and the distribution of arboreal, folivorous marsupials in Eucalyptus forests of Australia. Pages 653–663 in R. W. Hemingway and P. E. Laks, editors. Plant polyphenols: synthesis, properties, significance. Plenum Press, New York, New York. USA.

Cousens, R. D., J. Hill, K. French, and I. D. Bishop. 2010. Towards better prediction of seed dispersal by animals. Functional Ecology 24:1163–1170.

Crawley, M. J. 1983. Benevolent herbivores? Trends in Ecology and Evolution 2:167–168.

Cronk, Q., and I. Ojeda. 2008. Bird-pollinated flowers in an evolutionary and molecular context. Journal of Experimental Botany 59:715–727.

Danell, K., and R. Bergström. 2002. Mammalian herbivory in terrestrial environments. Pages 107–131 in C. M. Herrera and O. Pellmyr, editors. Plant-animal interactions: an evolutionary approach. Blackwell, Malden, Massachusetts, USA.

Darwin, C. 1859. On the origin of species by means of natural selection or the preservation of favoured races in the struggle for life. John Murray, London, UK.

Dawkins, R., and J. R. Krebs. 1979. Arms races between and within species. Proceedings of the Royal Society B 205:489–511.

Dearing, M. D., W. J. Foley, and S. McLean. 2005. The influence of plant secondary metabolites on the nutritional ecology of herbivorous terrestrial vertebrates. Annual Review of Ecology and Systematics 36:169–189.

Dearing, M. D., A. M. Mangione, and W. H. Karasov. 2000. Diet breadth of mammalian herbivores: nutrient versus detoxification constraints. Oecologia 123:397–405.

DeGabriel, J. L., B. D. Moore, W. J. Foley, and C. N. Johnson. 2009. The effects of plant defensive chemistry on nutrient availability predict reproductive success in a mammal. Ecology 90:711–719.

Dehling, D. M., T. Töpfer, H. M. Schaefer, P. Jordano, K. Böhning-Gaese, and M. Schleuning. 2014. Functional relationships beyond species richness patterns: trait matching in plant-bird mutualisms across scales. Global Ecology and Biogeography 23:1085–1093.

Del-Claro, K., V. Berto, and W. Reu. 1996. Effect of herbivore deterrence by ants on fruit set of an extrafloral nectary plant, Qualea multiflora (Vochysiaceae). Journal of Tropical Ecology 12:887.

Del-Claro, K., V. Rico-Gray, H. M. Torezan-Silingardi, E. Alves-Silva, R. Fagundes, D. Lange, W. Dáttilo, A. A. Vilela, A. Aguirre, and D. Rodriguez-Morales. 2016. Loss and gains in ant-plant interactions mediated by extrafloral nectar: fidelity, cheats, and lies. Insect Sociology 63:207–221.

Detrain, C., and A. Bologna. 2019. Impact of seed abundance on seed processing and dispersal by the red ant Myrmica rubra. Ecological Entomology 44:380–388.

Dirzo, R., E. Mendoza, and P. Ortiz. 2007. Size-related differential seed predation in a heavily defaunated neotropical rain forest. Biotropica 39:355–362.

Donaldson, J. R., M. T. Stevens, H. R. Barnhill, and R. L. Lindroth. 2006. Age-related shifts in leaf chemistry of clonal aspen (Populus tremuloides). Journal of Chemical Ecology 32:1415–1429.

Dooley, T. P., S. K. Nair, R. E. Garcia IV, and B. C. Courtney. 1995. Mouse rhodanese gene (Tst)—cdna cloning, sequencing, and recombinant protein expression. Biochemistry and Biophysics Research Communications 216:1101–1109.

Dublin, H. T., A. R. E. Sinclair, and J. McGlade. 1990. Elephants and fire as causes of multiple stable states in the Serengeti-Mara woodlands. Journal of Animal Ecology 59:1147–1164.

Ehrlich, P. R., and P. H. Raven. 1964. Butterflies and plants: a study in coevolution. Evolution 18:586–608.

Estrada, A., and R. Coates-Estrada. 1991. Howler monkey (Alouatta palliate), dung beetles (Scarabaeidae) and seed dispersal: ecological interactions in the tropical rain forest of Los Tuxtlas, Mexico. Journal of Tropical Ecology 7:459–474.

Faegri, K., and L. Van der Pijl. 1976. The principles of pollination ecology. Pergamon Press, Oxford, UK.

Fariňa, J. M., S. Salazar, K. P. Wallem, J. D. Witman, and J. C. Ellis. 2003. Nutrient exchanges between marine and terrestrial ecosystems: the case of the Galapagos sea lion Zalophus wollebaecki. Journal of Animal Ecology 72:873–887.

Farwig, N., and D. G. Berens. 2012. Imagine a world without seed dispersers: a review of threats, consequences, and future directions. Basic and Applied Ecology 13:109–115.

Fay, P. A., D. C. Hartnett, and A. K. Knapp. 1996. Plant tolerance of gall-insect attack and gall-insect performance. Ecology 77:521–534.

Feer, R. 1995. Seed dispersal in African forest ruminants. Journal of Tropical Ecology 11:683–689.

Feer, R., and P.-M. Forget. 2002. Spatio-temporal variation in post-dispersal seed fate. Biotropica 34:555–566.

Felton, A. M., H. K. Wam, C. Stolter, K. M. Mathisen, and M. Wallgren. 2018. The complexity of interacting nutritional drivers behind food selection, a review of northern cervids. Ecosphere 9(5):e2.2230. doi:10.1002/ecs2.2230.

Fidler, A. E., S. B. Lawrence, and K. P. McNatty. 2008. An hypothesis to explain the linkage between kakapo (*Strigops habroptilus*) breeding and the mast fruiting of their food trees. Wildlife Research 35:1–7.

Fischer, R. C., S. M. Ölzant, W. Wanek, and V. Mayer. 2005. The fate of *Corydalis cava* elaiosomes within an ant colony of *Myrmica rubra*: elaiosomes are preferentially fed to larvae. Insectes Sociax 52:55–62.

Forbey, J. S., N. L. Wiggins, G. G. Frye, and J. W. Connelly. 2013. Hungry grouse in a warming world: emerging risks from plant chemical defenses and climate change. Wildlife Biology 19:374–381.

Forget, P.-M., and T. Milleron. 1991. Evidence for secondary seed dispersal by rodents in Panama. Oecologia 87:596–599.

Fornoni, J., J. Núñez-Farfán, P. L. Valverde, and M. D. Rausher. 2004*b*. Evolution of mixed strategies of plant defense allocation against natural enemies. Evolution 58:1685–1695.

Fornoni, J., P. L. Valverde, and J. Núñez-Farfán. 2003. Quantitative genetics of plant tolerance and resistance against natural enemies of two natural populations of *Datura stramonium*. Evolutionary Ecology Research 5:1049–1065.

Fornoni, J., P. L. Valverde, and J. Núñez-Farfán. 2004*a*. Population variation in the cost and benefit of tolerance and resistance against herbivory in *Datura stramonium*. Evolution 58:1696–1704.

Frank, D. A., R. S. Inouye, N. Huntly, G. W. Minshall, and J. E. Anderson. 1994. The biogeochemistry of a north-temperate grassland with native ungulates—nitrogen dynamics in Yellowstone National Park. Biogeochemistry 26:163–188.

Frank, D. A., M. M. Kuns, and D. R. Guido. 2002. Consumer control of grassland plant production. Ecology 83:602–606.

Frank, D. A., and S. J. McNaughton. 1993. Evidence for the promotion of aboveground grassland production by native large herbivores in Yellowstone National Park. Oecologia 96:157–161.

Fryxell, J. M., J. Greever, and A. R. E. Sinclair. 1988. Why are migratory ungulates so abundant? American Naturalist 131:781–798.

Galetti, M. A. S. Pires, P. H. S. Brancalion, and F. A. S. Fernandez. 2017. Reversing defaunation by trophic rewilding in empty forests. Biotropica 49:5–8.

Ganzhorn, J. U. 1992. Leaf chemistry and the biomass of folivorous primates in tropical forests. test of a hypothesis. Oecologia 91:540–547.

Garibaldi, L. A., I. Bartomeus, R. Bommarco, A. M. Klein, S. A. Cunningham, et al. 2015. Trait matching of flower visitors and crops predicts fruit set better than trait diversity. Journal of Applied Ecology 52:1436–1444.

Gill, F. B., and L. L. Wolf. 1978. Comparative foraging efficiencies of some montane sunbirds in Kenya. Condor 80:391–400.

Gill, R. 2006. The influence of large herbivores on tree recruitment and forest dynamics. Pages 170–202 *in* K. Danell, R. Bergström, P. Duncan, et al., editors. Large herbivore ecology, ecosystem dynamics, and conservation. Cambridge University Press, Cambridge, UK.

Goldblatt, P., and J. C. Manning. 2006. Radiation of pollination systems in the Iridaceae of sub-Saharan Africa. Annals of Botany 97:317–344.

Goldblatt, P., J. C. Manning, and P. Bernhardt. 1999. Evidence of bird pollination in Iridaceae of southern Africa. Adansonia 21:25–40

Gómez, J. M., and M. Verdú. 2012. Mutualism with plants drives primate diversification. Systematic Biology 61:567–577.

Gonzalez-Teuber, M., and M. Heil. 2009. Nectar chemistry is tailored for both attraction of mutualists and protection from exploiters. Plant Signaling and Behavior 4:809–813.

Guimarães, P. R., M. Galetti, and P. Jordano. 2008*a*. Seed dispersal anachronisms: rethinking the fruits extinct megafauna ate. PLoS One 3:e1745.

Guimarães, P. R., V. Rico-Gray, S. Furtado dos Reis, and J. N. Thompson. 2008*b*. Asymmetries in specialization in ant-plant mutualistic networks. Proceedings of the Royal Society B 273:2041–2047.

Gutherie, R. D. 1971. Factors regulating the coevolution of microtine tooth complexity. Zeitschrift für Säugetierkunde 36:37–54.

Hamann, E., C. Blevins, S. J. Franks, M. I. Jameel, and J. T. Anderson. 2021. Climate change alters plant-herbivore interactions. New Phytologist 229:1894–1910.

Hamer, D., S. Herrero, and K. Brady. 1991. Food and habitat used by grizzly bears, Ursus arctos, along the continental divide in Waterton Lakes National Park, Alberta. Canadian Field-Naturalist 105:325–329.

Harborne, J. B. 1991. The chemical basis of plant defense. Pages 45–59 *in* R. T. Palo and C. T. Robbins, editors. Plant defenses against mammalian herbivory. CRC Press, Boca Raton, Florida, USA.

Heil, M., B. Fiala, K. E. Linsenmair, G. Zotz, and P. Menke. 1997. Food body production in *Macaranga triloba* (Euphorbiaceae): a plant investment in anti-herbivore defense via symbiotic ant partners. Journal of Ecology 85:847–861.

Herrera, C. M. 1982. Grasses, grazers, mutualism, and coevolution: a comment. Oikos 38:254–258.

Herrera, C. M. 1987. Components of pollinator "quality": comparative analysis of a diverse insect assemblage. Oikos 50:79–90.

Herrera, C. M. 1988. Variation in mutualisms: the spatio-temporal mosaic of a pollinator assemblage. Biological Journal of the Linnean Society 35:95–125.

Herrera, C. M. 1989. Pollinator abundance, morphology, and flower visitation rate: analysis of the "quantity" component in a plant-pollinator system. Oecologia 80:241–248.

Herrera, J. 1997. The role of colored accessory bracts in the reproductive biology of Lavandula stoechas. Ecology 78:494–504.

Hershey, D. R. 2004. The widespread misconception that the tambalacoque absolutely required the dodo for its seeds to germinate. Plant Science Bulletin 50:105–108.

Hester, A., M. Bergman, G. R. Iason, and J. Moen. 2006. Impacts of large herbivores on plant community structure and dynamics. Pages 97–141 *in* K. Danell, R. Bergström, P. Duncan, et al, editors. Large herbivore ecology, ecosystem dynamics, and conservation. Cambridge University Press, Cambridge, UK.

Hik, D. S., and R. L. Jefferies. 1990. Increases in the net above-ground primary production of a salt-marsh forage grass: a test of the predictions of the herbivore-optimization model. Journal of Ecology 78:180–195.

Hildebrand, G. V., T. A. Hanley, C. T. Robbins, and C. C. Schwartz. 1999. Role of brown bears (*Ursus arctos*) in the flow of marine nitrogen into a terrestrial ecosystem. Oecologia 121:546–550.

Hobbs, N. T. 1996. Modification of ecosystems by ungulates. Journal of Wildlife Management 60:695–713.

Howe, H. F. 1989. Scatter and clump-dispersal and seedling demography: hypothesis and implications. Oecologia 79:417–426.

Howe, H. F., and M. N. Miriti. 2000. No question: seed dispersal matters. Trends in Ecology and Evolution 15:434–436.

Hu, C., Y. Zhang, and D. D. Kitts. 2000. Evaluation of antioxidant and prooxidant activities of bamboo *Phyllostachys nigra* var. *Henonis* leaf extract in vitro. Journal of Agriculture and Food Chemistry. 48:3170–3176.

Iason, G. 2005. The role of plant secondary metabolites in mammalian herbivory: ecological perspectives. Proceedings of the Nutrition Society 64:123–131.

Jacobs, S. M., and R. J. Naiman. 2008. Large African herbivores decrease herbaceous plant biomass while increasing plant species richness in a semi-arid savanna toposequence. Journal of Arid Environments 72:889–901.

Jansman, A. J. M., M. W. A. Verstegen, J. Huisman, and J. W. O. Van den Berg. 1995. Effects of hulls of faba beans (*Vicia faba* L.) with low or high content of condensed tannins on the apparent ileal and fecal digestibility of nutrients and the excretion of endogenous protein in ileal digest and feces of pigs. Journal of Animal Science 73:118–127.

Janzen, D. H. 1984. Dispersal of small seeds by big herbivores: foliage is the fruit. American Naturalist 123:338–353.

Jefferies, R. L., and R. F. Rockwell. 2002. Foraging geese, vegetation loss, and soil degradation in an arctic salt marsh. Applied Vegetation Sciences 5:7–16.

Johnson, C. N. 2009. Ecological consequences of Late Quaternary extinctions of megafauna. Proceedings of the Royal Society B 276:2509–2519.

Johnson, S. D., I. Kiepiel, and A. W. Robertson. 2020. Functional consequences of flower curvature, orientation, and perch position for nectar feeding by sunbirds. Biological Journal of the Linnean Society 131:822–834.

Jordano, P. 1987. Patterns of mutualistic interactions in pollination and seed dispersal: connectance, dependence asymmetries, and coevolution. American Naturalist 129:657–677.

Juhnke, J., J. Miller, J. O. Hall, F. D. Provenza, and J. J. Villalba. 2012. Preference for condensed tannins by sheep in response to challenge infection with *Hemonchus contortus*. Veterinary Parasitology 188:104–114.

Juniper, B. E., R. J. Robins, and D. Joel. 1989. The carnivorous plants. Academic Press, London, UK.

Kearns, C. A., D. W. Inouye, and N. M. Waser. 1998. Endangered mutualisms: the conservation of plant-pollinator interactions. Annual Review of Ecology and Systematics 29:83–112.

Kiers, E. T., T. M. Palmer, A. R. Ives, J. F. Bruno, and J. L. Bronstein. 2010. Mutualisms in a changing world: an evolutionary perspective. Ecology Letters 13:1459–1474.

Kitchell, J. F., D. E. Schindler, B. R. Herwig, D. M. Post, and M. H. Olson. 1999. Nutrient cycling at the landscape scale: the role of diel foraging migrations by geese at the Bosque del Apache National Wildlife Refuge, New Mexico. Limnology and Oceanography 44:828–836.

Klein, A. M. 2007. Importance of pollinators in changing landscapes for world corps. Proceedings of the Royal Society B 274:303–313.

Klinger, R., and M. Rejmanek. 2010. A strong conditional mutualism limits and enhances seed dispersal and germination of a tropical palm. Oecologia 162:951–963.

Knapp, A. K., J. M. Blair, J. M. Briggs, S. L. Collins, D. C. Hartnett, L. C. Johnson, and E. G. Towne. 1999. The keystone role of bison in North American tall grass prairie—bison increase habitat heterogeneity and alter a broad array of plant, community, and ecosystem processes. Bioscience 49:39–50.

Knauer, A. C., and F. P. Schiestl. 2015. Bees use honest floral signals as indicators of reward when visiting flowers. Ecology Letters 18:135–143.

Kohl, K. D., and M. D. Dearing. 2016. The woodrat gut microbiota as an experimental system for understanding microbial metabolism of dietary toxins. Frontiers in Microbiology 7:1165.

Kohl, K. D., R. B. Weiss, J. Cox, C. Dale, and M. D. Dearing. 2014. Gut microbes of mammalian herbivores facilitate intake of plant toxins. Ecology Letters 17:1238–1246.

Koike, S., S. Kasai, K. Yamazaki, and K. Furubayashi. 2008. Fruit phenology of *Prunus jamasakura* and the feeding habit of the Asiatic black bear as a seed disperser. Ecological Research 23:385–392.

Kölzsch, A., S. Bauer, R. Boer, L. Griffin, D. Cabot, K. M. Exo, H. P. Jeugd, and B. A. Nolet. 2015. Forecasting spring from afar? Timing of migration and predictability of phenology along different migration routes of an avian herbivore. Journal of Animal Ecology 84:272–283.

Konečná, M., M. Moos, H. Zahradníčková, P. Šimek, and J. Lepš. 2018. Tasty reward for ants: differences in elaisome and seed metabolite profiles are consistent across species and reflect taxonomic relatedness. Oecologia 188:753–764.

Koptur, S. 1994. Floral and extrafloral nectars of neotropical Inga trees: a comparison of their constituents and composition. Biotropica 26:276–284.

Krug, C., G. D. Cordeiro, I. Schaffler, C. I. Silva, R. Oliveira, C. Schlindwein, S. Dotterl, and I. Alves-dos-Santos. 2018. Nocturnal bee pollinators are attracted to Guarana flowers by their scents. Frontiers in Plant Science 9:1072.

Kurten, E. L. 2013. Cascading effects of contemporaneous defaunation on tropical forest communities. Biological Conservation 163:22–32.

Laca, E. A., and M. Demment. 1991. Herbivory: the dilemma for foraging in a spatially heterogeneous food environment. Pages 29–44 in R. T. Palo and C. T. Robbins, editors. Plant defenses against mammalian herbivory. CRC Press, Boca Raton, Florida, USA.

Larsen, D. N. 1984. Feeding habits of river otters in coastal southeastern Alaska. Journal of Wildlife Management 48:1446–1452.

Lengyel, S., A. D. Gove, A. M. Latimer, J. D. Majer, and R. R. Dunn. 2010. Convergent evolution of seed dispersal by ants, and phylogeny and biogeography in flowering plants: a global survey. Perspectives in Plant Ecology, Evolution, and Systematics 12:43–55.

Leonard, A. S., A. Dornhaus, and D. R. Papaj. 2012. Why are floral signals complex? An outline of functional hypotheses. Pages 261–282 in S. Patiny, editor. Evolution of plant-pollinator relationships. Cambridge University Press, Cambridge, UK.

Leonard, A. S., and J. S. Francis. 2017. Plant-animal communication: past, present, and future. Evolutionary Ecology 31:143–151.

Leonard, A. S., and P. Masek. 2014. Multisensory integration of colors and scents: insights from bees and flowers. Journal of Comparative Physiology A 200:463–474.

Levi, T., G. V. Hildebrand, M. D. Hocking, T. P. Quinn, K. S. White, et al. 2020. Community ecology and conservation of bear-salmon ecosystems. Frontiers in Ecology and Evolution 8:513304.

Li, H. T., T. S. Yi, L. M. Gao, P. F. Ma, T. Zhang, et al. Origin of angiosperms and the puzzle of the Jurassic gap. Nature Plants 5:461–470.

Lloyd, F. E. 1942. The carnivorous plants. *In* Chronica Botanica. New York, New York, USA.

López-Bao, J. V., and J. P. González-Varo. 2011. Frugivory and spatial patterns of seed deposition by carnivorous mammals in anthropogenic landscapes: a multi-scale approach PLoS ONE 6:e14569.

Maglianesi, M. A., K. Böhning-Gaese, and M. Schleuning. 2015. Different foraging preferences of hummingbirds on artificial and natural flowers reveal mechanisms structuring plant-pollinator interactions. Journal of Animal Ecology 84:655–664.

Mangione, A. M., M. D. Dearing, and W. H. Karasov. 2000. Interpopulation differences in tolerance to creosote bush resin in desert woodrats (*Neotoma lepida*). Ecology 81:2067–2076.

Matías, L., R. Zamora, I. Mendoza, and J. A. Hódar. 2010. Seed dispersal patterns by large frugivorous mammals in a degraded mosaic landscape. Restoration Ecology 18:619–627.

Matocq, M. D., K. M. Ochsenrider, C. S. Jeffrey, D. P. Nielsen, and L. A. Richards. 2020. Fine-scale differentiation in diet and metabolomics of small mammals across a sharp ecological transition. Frontiers in Ecology and Evolution 8:282.

Mattson, W. J., Jr. 1980. Herbivory in relation to plant nitrogen content. Annual Review of Ecology and Systematics 11:119–161.

Mayer, V., S. Olzant, and R. C. Fisher. 2005. Pages 175–196 in P. M. Forget, J. E. Lambert, P. E. Hulme, et al., editors. Seed fate predation, dispersal and seedling establishment. CABI, Wallingford, UK.

McArt, S. H., D. E. Spalinger, W. B. Collins, E. R. Schoen, T. Stevenson, and M. Bucho. 2009. Summer dietary nitrogen availability as a potential bottom-up constraint on moose in south-central Alaska. Ecology 90:1400–1411.

McArthur, C., C. T. Robbins, A. E. Hagerman, and T. A. Hanley. 1993. Diet selection by a ruminant generalist in relation to plant chemistry. Canadian Journal of Zoology–Revue Canadienne De Zoologie 71:2236–2243.

McConkey, K. R., S. Prasad, R. T. Corlett, A. Campos-Arceiz, J. F. Brodie, H. Rogers, and L. Santamaria. 2012. Seed dispersal in changing landscapes. Biological Conservation 146:1–13.

McKey, D., P. G. Waterman, C. N. Mbi, J. S. Gartlan, and T. T. Struhsaker. 1978. Phenolic content of vegetation in two African rain forests: ecological implications. Science 202:61–63.

McNaughton, S. J. 1979. Grazing as an optimization process: grass-ungulate relationships in the Serengeti. American Naturalist 113:691–703.

McNaughton, S. J. 1983. Serengeti grassland ecology: the role of composite environmental factors and contingency in community organization. Ecological Monographs 53:291–320.

McNaughton, S. J. 1985. Ecology of a grazing ecosystem: the Serengeti. Ecological Monographs 55:259–294.

McShea, W. J. 2005. Forest ecosystems without carnivores: when ungulates rule the world. Pages 138–153 in J. C. Ray, K. H. Redford, R. S. Steneck, et al., editors. Large carnivores and biodiversity conservation. Island Press, Covelo, California, USA.

McSweeney, C. S., B. Palmer, D. M. McNeill, and D. O. Krause. 2001. Microbial interactions with tannins: nutritional consequences for ruminants. Animal Feed Science and Technology 91:83–93.

Mello, M. A. R., F. A. Rodrigues, L. D. F. Costa, W. D. Kissling, C. H. Şekercioğlu, F. M. D. Marquitti, and E. K. V. Kalko. 2015. Keystone species in seed dispersal networks are mainly determined by dietary specialization. Oikos 124:1031–1039.

Milchunas, D. G., and W. K. Lauenroth. 1993. Quantitative effects of grazing on vegetation and soils over a global range of environments. Ecological Monographs 63:327–366.

Mithofer, A., and W. Boland. 2012. Plant defense against herbivores: chemical aspects. Annual Review of Plant Biology 63:431–450.

Mittelman, P., C. Kreischer, A. S. Pires, and F. A. S. Fernandez. 2020. Agouti reintroduction recovers seed dispersal of a large-seeded tropical tree. Biotropica 52:766–774.

Mode, C. J. 1958. A mathematical model for the co-evolution of obligate parasites and their hosts. Evolution 12:158–165.

Moermond, T. C., and J. S. Denslow. 1985. Neotropical avian frugivores: patterns of behavior, morphology, and nutrition, with consequences for fruit selection. Ornithological Monographs 36:865–897.

Moore, B. D., and W. J. Foley. 2005. Tree use by koalas in a chemically complex landscape. Nature 435:488–490

Moore, B. D., and S. N. Johnson. 2017. Get tough, get toxic, or get a bodyguard: identifying candidate traits conferring belowground resistance to herbivores in grasses. Frontiers in Plant Science 7:1925. doi:10.3389/fpls.2016.01925.

Mueller, T., J. Lenz, T. Caprano, W. Fiedler, and K. Böhning-Gaese. 2014. Large frugivorous birds facilitate functional connectivity of fragmented landscapes. Journal of Applied Ecology 51:684–692.

Mueller, S. O., H. Stopper, and W. Dekant. 1998. Biotransformation of the anthraquinones emodin and chrysophanol by cytochrome p450 enzymes—bioactivation to genotoxic metabolites. Drug Metabolism and Disposition 26:540–546.

Nathan, R., and H. C. Muller-Landau. 2000. Spatial patterns of seed dispersal, their determinants and consequences for recruitment. Trends in Ecology and Evolution 15:278–285.

Nepi, M., E. Pacini, and M. P. M. Willemse. 1996. Nectary biology of Cucurbita pepo: ecophysiological aspects. Acta Botanica Neerl 45:41–54.

Nicholson, M. C., R. T. Bowyer, and J. G. Kie. 2006. Forage selection by mule deer: does niche breadth increase with population density? Journal of Zoology London 269:39–49.

Nie, Y. G., Z. J. Zhang, D. Raubenheimer, J. J. Elser, W. Wei, and F. W. Wei. 2015. Obligate herbivory in an ancestrally carnivorous lineage: the giant panda and bamboo form the perspective of nutritional geometry. Functional Ecology 29:26–34.

Núñez-Farfán, J., J. Fornoni, and P. L. Valverde. 2007. The evolution of resistance and tolerance to herbivores. Annual Review of Ecology, Evolution, and Systematics 38:541–566.

Oates, J. F., G. H. Whitesides, A. G. Davies, P. G. Waterman, S. M. Green, G. L. Dasilva, and S. Mole. 1990. Determinants of variation in tropical forest primate biomass: new evidence from West Africa. Ecology 71:328–343.

O'Keefe, T. C., and R. T. Edwards. 2002. Evidence for hyporheic transfer and removal of marine-derived nutrients in sockeye streams in southwest Alaska. American Fisheries Society Symposium 33:99–107.

Olff, H., and M. E. Ritchie. 1998. Effects of herbivores on grassland plant diversity. Trends in Ecology and Evolution 13:261–265.

Ollerton, J., S. D. Johnson, and A. B. Hingston. 2006. Graphical variation in diversity and specificity of pollination systems. Pages 283–308 in N. M. Waser and J. Ollerton, editors. Plant-pollinator interactions from specialization to generalization. University of Chicago Press, Chicago, USA.

Ollerton, J., R. Winfree, and S. Tarrant. 2011. How many flowering plants are pollinated by animals? Oikos 120:321–326.

Orrock, J. L. 2005. The effect of gut passage by two species of avian frugivore on seeds of pokeweed, Phytolacca Americana. Canadian Journal of Botany 83:427–431.

Otero, X. L., S. De La Peňa-Lastra, A. Pérez-Alberti, T. O. Ferreira, and M. A. Huerta-Diaz. 2018. Seabird colonies as important global drivers in nitrogen and phosphorus cycles. Nature Communications 9:246.

Owadally, A. W., and S. A. Temple. 1979. The dodo and the tambalacoque tree. Science 203(4387):1363–1364.

Owen-Smith, R. N. 1988. Megaherbivores: the influence of very large body size on ecology. Cambridge University Press, Cambridge, UK.

Pacini, E., and M. Nepi. 2007. Nectar production and presentation. Pages 166–214 in S. W. Nicholson, M. Nepi, and E Pacini, editors. Nectary and nectar. Springer, Dordrecht, Netherlands.

Pacini, E., L. Viegi, and G. G. Franchi. 2008. Types, evolution and significance of plant-animal interactions. Rendiconti Lincei 19:75–101.

Packer, P. E. 1953. Effects of trampling disturbance on watershed conditions, runoff, and erosion. Journal of Forestry 51:28–31.

Packer, P. E. 1963. Soil stability requirements for the Gallatin elk winter range. Journal of Wildlife Management 27:401–410.

Painter, E. L., and A. J. Belsky. 1993. Application of herbivore optimization theory to rangelands of the western United States. Ecological Applications 3:2–9.

Parmesan, C., and M. E. Hanley. 2015. Plants and climate change: complexities and surprises. Annals of Botany 116:849–864.

Pastor, J., and Y. Cohen. 1997. Herbivores, the functional diversity of plant species, and the cycling of nutrients in ecosystems. Theoretical Population Biology 51:165–179.

Pastor, J., B. Dewey, R. J. Naiman, P. F. McInnes, and Y. Cohen. 1993. Moose browsing and soil fertility in the boreal forests of Isle Royale National Park. Ecology 74:467–480.

Pastor, J., and R. J. Naiman. 1992. Selective foraging and ecosystem processes in boreal forests. American Naturalist 139:691–705.

Paton, A. J., N. Brummitt, R. Govaerts, K. Harman, S. Hinchcliffe, B. Allkin, and E. N. Lughadha. 2008. Towards target 1 of the global strategy for plant conservation: a working list of all known plant species—progress and prospects. Taxon 57:602–611.

Patton, D. C., and B. G. Collins. 1989. Bills and tongues of nectar-feeding birds—a review of morphology, function, and performance, with intercontinental comparisons. Australian Journal of Ecology 14:473–506.

Pélabon, C., P. Thone, T. F. Hansen, and W. S. Armbruster. 2012. Signal honesty and cost of pollinator rewards in Dalechampia scadens (Euphorbiaceae). Annals of Botany 109:1331–1339.

Peredo, A., D. Martínez, J. Rodríguez-Pérez, and D. García. 2013. Mammalian seed dispersal in Cantabrian woodland pastures: network structure and response to forest loss. Basic and Applied Ecology 14:378–386.

Person, B. T., M. P. Herzog, R. W. Ruess, J. S. Sedinger, R. M. Anthony, and C. A. Babcock. 2003. Feedback dynamics of grazing lawns: coupling vegetation change with animal growth. Oecologia 135:583–592.

Phillipps, A., A. Lamb, and C. C. Lee. 2009. Pitcher plants of Bornea. 2nd edition. Natural History Publications, Borneo; Kota Kinabalu, Sabah, Malaysia.

Pilson, D. 2000. The evolution of plant response to herbivory: simultaneously considering resistance and tolerance in brassica rapa. Evolutionary Ecology 14:457–489.

Pires, M. M., P. R. Guimarães Jr., M. Galetti, and P. Jordano. 2017. Pleistocene megafaunal extinctions and the functional loss of long-distance seed-dispersal services. Ecography 41:153–163.

Porembski, S., and W. Barthlott. 2006. Advances in carnivorous plant research. Plant Biology 8:737–739.

Potts, S. G., J. C. Biesmeijer, C. Kremen, P. Neumann, O. Schweiger, and W. E. Kunin. 2010. Global pollinator declines: trends, impacts, and drivers. Trends in Ecology and Evolution 25:345–353.

Price, P. W. 2002. Species interactions and the evolution of biodiversity. Pages 3–25 in C. M. Herrera and O. Pellmyr, editors. Plant-animal interactions: an evolutionary approach. Blackwell, Malden, Massachusetts, USA.

Pyke, G. H. 1991. What does it cost a plant to produce floral nectar? Nature 350:58–59.

Raupp, P. R., R. V. Gonçalves, E. S. Calixto, and D. V. Anjos. 2020. Contrasting effects of herbivore damage type on extrafloral nectar production and ant attendance. Acta Oecologica 108:103638.

Rauscher, M. D. 2001. Co-evolution and plant resistance to natural enemies. Nature 411:857–864.

Regal, P. J. 1977. Ecology and evolution of flowering plant dominance. Science 196:622–629.

Rehill, B. J., T. G. Whitham, G. D. Martinsen, J. A. Schweitzer, J. K. Bailey, and R. L. Lindroth. 2006. Developmental trajectories in cottonwood phytochemistry. Journal of Chemical Ecology 32:2269–2285.

Restif, O., and J. C. Koella. 2003. Shared control of epidemiological traits in a coevolutionary model of host-parasite interactions. American Naturalist 161:827–836.

Rhoades, D. F, and R. G. Cates. 1976. Toward a general theory of plant antiherbivore chemistry. Pages 168–213 in J. W. Wallace and R. L. Mansell, editors. Recent advances in phytochemistry. Vol. 10: Biochemical reactions between plants and insects. Plenum, New York, New York, USA.

Risch, S. J., and F. R. Rickson. 1981. Mutualism in which ants must be present before plants produce food bodies. Nature 14:149–150.

Rivrud, I. M., M. Heurich, P. Krupczynski, J. Müller, and A. Mysterud. 2016. Green wave tracking by large herbivores: an experimental approach. Ecology 97:3547–3553.

Robbins, C. T. 1993. Wildlife feeding and nutrition. 2nd ed. Academic Press, San Diego, California, USA.

Robbins, C. T., T. A. Hanley, A. E. Hagerman, O. Hjeljord, D. L. Baker, C. C. Schwartz, and W. W. Mautz. 1987. Role of tannins in defending plants against ruminants—reduction in protein availability. Ecology 68:98–107.

Robinson, A. S., A. S. Fleischmann, S. R. McPherson, V. B. Heinrich, E. P. Gironella, and C. Q. Peña. 2009. A spectacular new species of Nepenthes L. (Nepenthaceae) pitcher plant from central Palawan, Philippines. Botanical Journal of the Linnean Society 159:195–202.

Rohner, C., and D. Ward. 1999. Large mammalian herbivores and the conservation of arid Acacia stands in the Middle East. Conservation Biology 13:1162–1171.

Rooney, T. P., and D. M. Waller. 2003. Direct and indirect effects of white-tailed deer in forest ecosystems. Forest Ecology and Management 181:165–176.

Rosenthal, J. P., and P. M. Kotanen. 1994. Terrestrial plant tolerance to herbivory. Trends in Ecology and Evolution 9:145–148.

Rosumek, F. B., F. A. O. Silveira, F. S. Neves, N. P. Barbosa, L. Diniz, Y. Oki, F. Pezzini, W. G. Fernandez, and T. Cornelissen. 2009. Ants on plants: a meta-analysis of the role of ants as plant biotic defenses. Oecologia 160:537–549.

Roy, B. A., and J. W. Kirchner. 2000. Evolutionary dynamics of pathogen resistance and tolerance. Evolution 54:51–63.

Ruess, R. W., R. L. Hendrick, and J. P. Bryant. 1998. Regulation of fine root dynamics by mammalian browsers in early successional Alaskan taiga forests. Ecology 79:2706–2720.

Ruess, R. W., and S. J. McNaughton. 1987. Grazing and the dynamics of nutrient and energy regulated microbial processes in the Serengeti grasslands. Oikos 49:101–110.

Russo, S. E., S. Portnoy, and C. K. Augspurger. 2006. Incorporating animal behavior into seed dispersal models: implications for seed shadows. Ecology 87:3160–3174.

Salminen, J. P., T. Roslin, M. Karonen, J. Sinkkonen, K. Pihlaja, and P. Pulkkinen. 2004. Seasonal variation in the content of hydrolyzable tannin, flavonoid glycosides, and proanthocyanidins in oak leaves. Journal of Chemical Ecology 30:1693–1711.

Santamour, F. S. 1998. Amygdalin I Prunus leaves. Phytochemistry 47:1537–1538.

Sapir, N., and R. Dudley. 2013. Implications of floral orientation for flight kinematics and metabolic expenditure of hover-feeding hummingbirds. Functional Ecology 27:227–235.

Sawyer, H., and M. J. Kauffman. 2011. Stopover ecology of a migratory ungulate. Journal of Animal Ecology 80:1078–1087.

Sawyer, H, F. Lindzey, and D. McWhirter. 2005. Mule deer and pronghorn migration in western Wyoming. Wildlife Society Bulletin 33:1266–1273.

Schaller, G. B., J. C. Hu, W. S. Pan, and J. Zhu. 1985. The Giant panda of Wolong. University of Chicago Press, Chicago, Illinois, USA.

Schleuning, M., N. Blüthgen, M. Flörchinger, J. Braun, H. M. Schaefer, and K. Böhning-Gaese. 2011. Specialization and interaction strength in a tropical plant-frugivore network differ among forest strata. Ecology 92:26–36.

Schmitt, M. H., A. M. Shrader, and D. Ward. 2020. Megaherbivore browsers vs. tannins: is being big enough? Oecologia 194:383–390.

Schoenecker, K. A., F. J. Singer, L. C. Zeigenfuss, D. Binkley, and R. S. C. Menezes. 2004. Effects of elk herbivory on vegetation and nitrogen processes. Journal of Wildlife Management 68:837–849.

Schwartz, C. C., W. L. Regelin, and J. G. Nagy. 1980. Deer preference for juniper forage and volatile oil treated foods. Journal of Wildlife Management 44:114–120.

Seagle, S. W., S. J. McNaughton, and R. W. Ruess. 1992. Simulated effects of grazing on soil nitrogen and mineralization in contrasting Serengeti grasslands. Ecology 73:1105–1123.

Sebastián-González, E., M. M. Pires, C. I. Donatti, P. R. Guimãres Jr., and R. Dirzo. 2017. Species traits and interaction rules shape a species-rich seed-dispersal interaction network. Ecology and Evolution 7:4496–4506.

Shariatinajafabadi, M., T. J. Wang, A. K. Skidmore, A. G. Toxopeus, A. Kozasch, B. A. Nolet, K.-M. Exo, L. Griffin, J. Stahl, and D. Cabot. 2014. Migratory herbivorous waterfowl track satellite-derived green wave index. PLoS ONE 9:e108331.

Shimada, T. 2006. Salivary proteins as a defense against dietary tannins. Journal of Chemical Ecology 32:1149–1163.

Simpson, F. J., G. A. Jones, and E. A. Wolin. 1969. Anaerobic degradation of some bioflavonoids by microflora of the rumen. Canadian Journal of Microbiology 15:972–974.

Sinclair, A. R. E., H. Dublin, and M. Borner. 1985. Population regulation of Serengeti wildebeest: a test of the food hypothesis. Oecologia 65:266–268.

Sorensen, J. S., J. D. McLister, and M. D. Dearing. 2005. Novel plant secondary metabolites impact dietary specialists more than generalists (Neotoma spp.). Ecology 86:140–154.

Stang, M., P. G. L. Klinkhamer, N. M. Waser, I. Stang, and E. van der Meijden. 2009. Size-specific interactions patterns and size matching in a plant-pollinator interaction web. Annals of Botany 103:1459–1469.

Steele, M. A., X. Yi, and H. Zhang. 2018. Plant-animal interactions: patterns and mechanisms in terrestrial ecosystems. Integrative Zoology 13:225–227.

Stewart, K. M., R. T. Bowyer, J. G. Kie, B. L. Dick, and R. W. Ruess. 2009. Population density of North American elk: effects on plant diversity. Oecologia 161:303–312.

Stewart, K. M., R. T. Bowyer, R. W. Ruess, B. L. Dick, and J. G. Kie. 2004. Herbivore optimization by North American elk: consequences for theory and management. Wildlife Monographs 167:1–24.

Strauss, S. Y., and A. A. Agrawal. 1999. The ecology and evolution of plant tolerance to herbivory. Trends in Ecology and Evolution 14:179–185.

Strauss, S. Y., and A. R. Zangerl. 2002. Plant-insect interactions in terrestrial ecosystems. Pages 77–106 in C. M. Herrera and O. Pellmyr, editors. Plant-animal interactions: an evolutionary approach. Blackwell, Malden, Massachusetts, USA.

Surh, I., A. Brix, J. E. French, B. J. Collins, J. M. Sanders, M. Vallant, and J. K. Dunnick. 2013. Toxicology and carcinogenesis study of senna in C3B6.129F1-Trp53^tm1Brd N12 haploinsufficient mice. Toxicology and Pathology 41:770–778.

Sun, J., Y. Gong, S. S. Renner, and S. Huang. 2008. Multifunctional bracts in the dove tree Davidia involucrate (Nyssaceae: Cornales): rain protection and pollinator attraction. American Naturalist 171:119–124.

Svensson, E. I., and L. Råberg. 2010. Resistance and tolerance in animal enemy-victim coevolution. Trends in Ecology and Evolution 25:267–274.

Tadey, M., and M. A. Aizen. 2001. Why do flowers of a hummingbird-pollinated mistletoe face down? Functional Ecology 15:782–790.

Tao, C., J. Wu, Y. Liu, M. Liu, R. P. Yang, and Z. L. Lv. 2018. Antimicrobial activities of bamboo (Phyllostachys heterocycle cv. Pubescens) leaf essential oil and its major components. European Food Research and Technology 244:881–891.

Tarleton, P., and E. G. Lamb. 2021. Modification of plant communities by bison in Riding Mountain National Park. Ecoscience 28:67–80.

Tarszisz, E., S. Tomlinson, M. E. Harrison, H. C. Morrogh-Bernard, and A. J. Munn. 2018. An ecophysiologically informed model of seed dispersal by orangutans: linking animal movement with gut passage across time and space. Conservation Physiology 6:coy013.

Temple, S. A. 1977. Plant-animal mutualism: coevolution with the dodo leads to near extinction of plant. Science 197(4306):885–886.

Testa, J. W., D. F. Hollman, R. T. Bowyer, and J. B. Faro. 1994. Estimating populations of marine river otters in Prince William Sound, Alaska, using radiotracer implants. Journal of Mammalogy 75:1021–1032.

Tewksbury, J. J. 2002. Fruits, frugivores, and the evolutionary arms race. New Phytologist 156:137–144.

Thompson, J. N. 1988. Variation in interspecific interactions. Annual Review of Ecology and Systematics 19:65–87.

Thompson, J. N. 1989. Concepts of coevolution. Trends in Ecology and Evolution 4:179–183.

Thompson, J. N. 2005. The geographic mosaic of coevolution. University of Chicago Press, Chicago, Illinois, USA.

Tiffen, P. 2000. Are tolerance, avoidance, and antibiosis evolutionary and ecologically equivalent responses of plants to herbivores? American Naturalist 155:128–138.

Traveset, A., R. Heleno, and M. Nogales. 2013. The ecology of seed dispersal. Pages 62–93 in R. S. Gallaguer, editor. Seeds: the ecology of regeneration in plant communities. 3rd edition. CABI, Oxfordshire, UK.

Traveset, A., and D. M. Richardson. 2014. Mutualistic interactions and biological invasions. Annual Review of Ecology, Evolution, and Systematics 45:89–113.

Tsuji, Y. T. Konta, M. A. Akbar, and M. Hayashida. 2020. Effects of Japanese marten (Martes melampus) gut passage on germination of Actinidia argute (Actinidiaceae): implications for seed dispersal. Acta Oecologica 105:103578.

Valverde, P. L., J. Fornoni, and J. Nunez-Farfan. 2001. Defensive role of leaf trichomes in resistance to herbivorous insects in Datura stramonium. Journal of Evolutionary Biology 14:424–432.

Van der Meijden, M. Wijn, and H. J. Verkaar. 1988. Defence and regrowth, alternative plant strategies in the struggle against herbivores. Oikos 51:355–363.

Vander Wall, S. B., S. C. Barga, and A. E. Seaman. 2017. The geographic distribution of seed-dispersal mutualisms in North America. Evolutionary Ecology 31:725–740.

Vander Wall, S. B., K. Kuhn, and J. Gworek. 2005. Two-phase seed dispersal: linking the effects of frugivorous birds and seed-caching rodents. Oecologia 145:281–286.

Vander Wall, S. B., and W. S. Longland. 2004. Diplochory: are two seed dispersers better than one? Trends in Ecology and Evolution 19:155–161.

Vander Wall, S. B., and C. M. Moore. 2016. Interaction diversity of North American seed-dispersal mutualisms. Global Ecology and Biogeography 25:1377–1386.

Van Soest, P. J., and L. H. P. Jones. 1968. Effect of silica in forages upon digestibility. Journal of Dairy Sciences 51:1644–1648.

Vellend, M. 2004. Parallel effects of land-use history on species diversity and genetic diversity of forest herbs. Ecology 85:3043–3055.

Vellend, M., J. A. Myers, S. Gardescu, and P. L. Marks. 2003. Dispersal of trillium seeds by deer: implications for long-distance migration of forest herbs. Ecology 84:1067–1072.

Vicari, M., and D. R. Bazely. 1993. Do grasses fight back—the case for antiherbivore defenses. Trends in Ecology and Evolution 8:137–141.

Villalba, J. J., M. Costes-Thire, and C. Ginane. 2017. Phytochemicals in animal health: diet selection and trade-offs between costs and benefits. Proceedings of the Nutrition Society 76:113–121.

Villalba, J. J., F. D. Provenza, and J. O. Hall. 2010. Selection of tannins by sheep in response to gastrointestinal nematode infection. Journal of Animal Science 88:2189–2198.

Waldram, M., W. Bond, and W. Stock. 2008. Ecological engineering by a mega-grazer: white rhino impacts on a South African savanna. Ecosystems 11:101–112.

Wang, B. C., and T. B. Smith. 2002. Closing the seed dispersal loop. Trends in Ecology and Evolution 17:379–386.

Wang, L., S. Yuan, Y. Nie, J. Zhao, X. Cao, Y. Dai, Z. Shang, and F. Wei. 2020. Dietary flavonoids and the altitudinal preference of wild giant pandas in Fopint National Nature Reserve, China. Global Ecology and Conservation 22:e00981.

Wedler, J., T. Daubitz, G. Schlotterbeck, and V. Butterweck. 2014. In vitro anti-inflammatory and wound-healing potential of a Phyllostachys edulis leaf extract identification of isoorientin as an active compound. Planta Medica 80:1678–1684.

Wellman, C. H., P. L. Osterloff, and U. Mohiuddin. 2003. Fragments of the earliest land plants. Nature 425:282–285.

Wenny, D. G. 1999. Two-state dispersal of *Guarea glabra* and *G. kunthiana* (Meliaceae) in Monteverde, Costa Rica. Journal of Tropical Ecology 15:481–496.

Wheelwright, N. T. 1985. Fruit-size, gape width, and the diets of fruit-eating birds. Ecology 66:808–818.

Wilkinson, C., M. Hocking, and T. Reimchen. 2005. Uptake of salmon-derived nitrogen by mosses and liverworts in coastal British Columbia. Oikos 108:85–98.

Wu, D., X.-W. Wang, S.-Q. Zu, C.-J. Chen, R. Mao, and X.-Y. Liu. 2020. Plant phenols contents and their changes with nitrogen availability in peatlands of northeastern China. Journal of Plant Ecology 13:713–721.

Zhang, S., Y. Zhang, and M. A. Keming. 2015. The equal effectiveness of different defensive strategies. Scientific Report 5:13049.

Zuest, T., and A. A. Agrawal. 2017. Tradeoffs between plant growth and defense against insect herbivory: an emerging mechanistic synthesis. Annual Review of Plant Biology 68:513–534.

Zvereva, E. L., and M. V. Kozlov. 2006. Consequences of simultaneous elevation of carbon dioxide and temperature for plant-herbivore interactions: a metaanalysis. Global Change Biology 12:27–41.

WATER AND OTHER WELFARE FACTORS

JAMES W. CAIN III, PAUL R. KRAUSMAN, SCOTT T. BOYLE, AND STEVEN S. ROSENSTOCK

INTRODUCTION

Wildlife agencies indirectly manage populations by manipulating resources upon which animals depend for survival. These resources are called welfare factors and include food, water, cover, space, and special factors (e.g., essential minerals, dust baths). Because food is discussed more thoroughly in Chapter 11, this chapter will focus on water, cover, and special factors. We use water as a model to show how managers use required resources to manipulate populations. We restrict the discussion of water to its role as a resource consumed by animals to meet physiological requirements rather than as a component of habitat (e.g., for waterfowl).

WATER

Life processes evolved around water (Louw 1993). The thermal capacity, high heat of vaporization, and action as a solvent make water crucial for all organisms (Edney 1977). Water constitutes the majority of molecules within an animal's body and plays a critical role in physiological processes (e.g., hydrolytic reactions, joint lubrication, digestion, thermoregulation, transport of nutrients, excretion of waste products; Robinson 1957).

Animals obtain water from three sources: free water in the environment (e.g., lakes, streams, rain, snow, and dew), preformed water in food, and metabolic water produced during oxidation of organic compounds containing hydrogen. Preformed water content of plants varies by species, season, and among plant parts, ranging from under 10% in dried seeds and senescent grasses (Jarman 1973), to 30% to 60% in browse, and more than 70% in succulent plants (cacti; Cain et al. 2008a, 2017); water content of animal tissues consumed by predators commonly exceeds 70% (Robbins 1983). Metabolic water is derived from the oxidation of fats, proteins, and carbohydrates (Robbins 1983). Oxidation of 1 g of fat, protein, and carbohydrate produces 1.07, 0.4, and 0.56 g of water, respectively (Gill 1994).

WATER REQUIREMENTS OF WILDLIFE

Water requirements vary among species and depend on thermal load (i.e., ambient temperature, solar radiation, thermal radiation, vapor pressure deficits), activity patterns, morphology, diet, metabolic processes, reproductive state, and adaptations for water conservation. Ascertaining water requirements requires measurement of intake from free, preformed, and metabolic water (Robbins 1983) and, when possible, should be determined from free-ranging animals. Birds and mammals with diets composed mainly of fruit, nectar, insects, or animal tissue often meet their water requirements from preformed and metabolic water (Bartholomew and Cade 1963, Schmidt-Nielsen 1979, Golightly and Ohmart 1984, Nagy and Gruchacz 1994). Similarly, many large herbivores may meet their water requirements during all or part of the year from preformed and metabolic water (Zervanos and Day 1977, Jhala et al. 1992, Fox et al. 2000, Gedir et al. 2016). When unable to meet their water requirements from these sources, animals must have access to free water.

OVERVIEW OF WATER BALANCE AND THERMOREGULATION

Wildlife that inhabit arid and semiarid environments must contend with factors affecting thermoregulation and water balance, including high solar radiation, high ambient temperatures, and limited water and food resources (Cain et al. 2006). These species cope by either avoiding or tolerating environmental conditions that cause heat stress and or dehydration (Schmidt-Nielsen 1979, Louw and Seely 1982). Animals that inhabit these environments have evolved a variety of behavioral, morphological, and physiological adaptations. Behavioral adaptations typically facilitate avoidance of stressful environmental conditions, whereas tolerance involves morphological and physiological adaptations (Louw and Seely 1982, Louw 1984). Evaporative cooling is one mechanism by which animals maintain homeothermy. Because high temperatures and scarce free water often characterize arid environments, animals face conflicting needs to maintain body temperature within physiologically acceptable limits and to minimize water loss. Use of multiple adaptations simultaneously allows animals to avoid heat stress and dehydration (Cain et al. 2006, Kihwele et al. 2020). We briefly review some of the physiological, morphological, and behavioral mechanisms involved in water balance and thermoregulation.

Physiological Mechanisms and Water Balance

The total body water (i.e., the amount of water in the body) of animals has four compartments: intracellular fluid (i.e., fluid within cells), extracellular fluid within the digestive tract (e.g., rumen), blood plasma, and interstitial fluid. The distribution of water among these compartments varies by species and with the state of hydration (Louw 1993). Species able to withstand severe dehydration typically maintain blood plasma volume and prevent circulatory failure by moving fluid from other compartments (Carmi et al. 1993, Silanikove 1994).

Water turnover rate (i.e., the time required for all water molecules within an animal's body to be replaced) varies among species and is lower in animals adapted to arid environments (Nagy and Peterson 1988, Tieleman et al. 2002). In normally hydrated animals, water turnover rate increases with increasing temperature (Longhurst et al. 1970, Degen et al. 1983). When dehydrated, water turnover rates will decrease and continue to do so regardless of ambient temperature (McNabb 1969, Maloiy 1973). Water turnover rates are affected by reproductive status and are 40% to 50% higher in pregnant and or lactating mammals (Maloiy et al. 1979) and increase during egg production in birds (Bartholomew and Cade 1963).

There are four primary routes of water loss in animals: feces, urine, cutaneous evaporation, and pulmonary evaporation. Mammalian females also experience additional water loss during lactation. Desert-adapted species evolved physiological adaptations that reduce the amount of water lost through each of these routes or are able to tolerate substantial amounts of water loss, or both (Dawson 1984, McNab 2002).

Water Loss in Feces and Urine

Animals can minimize water loss by reducing water content of feces, reducing urine volume, and by increasing urine concentration (osmolality; Maloiy et al. 1979, Maclean 1996). A major difference between mammals and birds is the excretion of nitrogenous waste as urea and uric acid, respectively. Mammals are less efficient than birds and require 20–40 times more water to excrete an equivalent amount of urea (Bartholomew and Cade 1963, Robbins 1983).

Fecal moisture content in normally hydrated animals varies widely among species. Arid-adapted ungulates such as gemsbok (*Oryx gazella*) typically range from 40% to 50% moisture content, compared to 70% to 80% for ungulates adapted to mesic areas, such as waterbuck (*Kobus ellipsiprymnus*; Maloiy et al. 1979, Woodall and Skinner 1993). When dehydrated, some ungulates have been reported to reduce fecal moisture content by 17% to 50% (Maloiy and Hopcraft 1971, Turner 1973). Similarly, arid-adapted birds typically have lower fecal moisture content than species adapted to mesic environments (Gill 1994), and dehydrated birds can reduce moisture content

of feces substantially (Ohmart and Smith 1971). Arid-adapted mammals and birds typically have reduced urine output and greater urine concentration than do species adapted to temperate or mesic environments (Louw and Seely 1982, Williams et al. 1991).

Thermoregulation

Thermoregulation is the maintenance of body temperature within specific boundaries under varying environmental temperatures. The environmental heat load plus the metabolic heat produced by the animal comprise the overall heat load (Porter and Gates 1969). Environmental heat load is a function of thermal radiation absorbed by the animal, air temperature, wind speed, and vapor pressure deficit.

When body temperature exceeds ambient temperature, animals passively lose heat to the environment by radiation, convection, and conduction (Porter and Gates 1969, Mitchell 1977), and these processes may be intensified by behavioral or morphological mechanisms (see below). Once the total heat load exceeds the levels that can be dissipated by nonevaporative means, animals must use other mechanisms to remove heat from the body, or body temperature will increase.

Evaporative Heat Loss

Evaporative heat loss (i.e., evaporative cooling) occurs when heat is transferred from the animal to the environment via evaporation of water from body surfaces. Evaporative heat loss is accomplished passively via diffusion of water through the skin and respiratory tract, or actively by sweating, panting, and gular fluttering (in birds). The magnitude of heat loss by evaporation is related to rates of sweating or panting, wind speed, and the vapor pressure gradient.

Evaporative Heat Loss in Mammals

The density and distribution of sweat glands and rate of sweating vary among species (Robertshaw and Taylor 1969, Sokolov 1982). Rates of cutaneous water loss are affected by season and the age, nutritional status, and state of hydration of the animal (Taylor 1970a, Maloiy and Hopcraft 1971, Parker and Robbins 1984). To minimize water loss, some desert-adapted species use facultative cutaneous evaporation; sweating rates increase with increasing body temperature when hydrated, but they decrease when dehydrated (Maloiy 1970, Taylor 1970b, Finch and Robertshaw 1979, Baker 1989, Ostrowski et al. 2006). The body temperature at which dehydrated animals begin to sweat is often greater than when hydrated (Schmidt-Nielsen et al. 1957, Taylor 1970a).

The respiratory tract also performs a thermoregulatory function. As body temperature increases, mammals increase respiratory rate and volume to maximize the movement of air across the evaporative surfaces of the upper respiratory tract, increasing evaporative heat loss (Finch 1972). In normally hydrated mammals, respiratory rates increase with increasing ambient temperature. Dehydrated mammals minimize respiratory water loss by maintaining lower respiratory rates and by initiating panting at higher temperatures than do normally hydrated animals (Taylor 1969a, Maloiy and Hopcraft 1971, Finch 1972).

Whether or not animals rely primarily on cutaneous or respiratory evaporation is related to body size. Small mammals (≤40 kg) typically pant. Larger mammals rely more on sweating because they have low rates of nonevaporative heat loss (because of low surface area:volume ratio; see below) and respiratory evaporation is less efficient than cutaneous evaporation in reducing body temperature; panting is often insufficient to dissipate heat generated during activity in large-bodied animals (Robertshaw and Taylor 1969, Maloiy et al. 1979).

Evaporative Heat Loss in Birds

Birds lose excess body heat by evaporation of water from the respiratory tract and skin (Dawson 1982, 1984; Williams and Tieleman 2002). Panting and gular fluttering combined with evaporative water loss from the nasal, buccal, tracheal, and upper pharyngeal regions are commonly used to cope with excessive environmental heat loads (Dawson 1984, McKechnie and Wolf 2019). Rates of panting and gular fluttering in

KNUT SCHMIDT-NIELSEN (1915–2007)

Knut Schmidt-Nielsen was born in Trondheim, Norway, in 1915. He was educated in Oslo and Copenhagen before coming to the United States, studying at Swarthmore College, Stanford University, and the University of Cincinnati College of Medicine. He became a US citizen and a professor at Duke University in 1952. He conducted pioneering work in comparative physiology and ecophysiology. His research included work on thermoregulation and water balance of camels, kangaroo rats, and birds, avian respiration, energetics of locomotion, and allometric relationships, among other topics. He published more than 270 scientific publications and 5 books: *How Animals Work* (1972), *Desert Animals: Physiological Problems of Heat and Water* (1979), *Scaling: Why Is Animal Size So Important?* (1984), *Animal Physiology: Adaptation and Environment* (1997), and the autobiography *The Camel's Nose: Memoirs of a Curious Scientist* (1998). He was a member of the US National Academy of Sciences, the Royal Society of London, and the French Academy of Sciences, among others, and he received numerous awards, including the International Prize for Biology.

From Hoppeler and Weibel (2005)

birds tend to increase with ambient temperature (Weathers 1972). Similar to mammals, dehydrated birds begin panting at a higher body temperature compared with normally hydrated birds (Kleinhaus et al. 1985). Gular fluttering provides some bird species with the ability to dissipate head loads with limited increases in metabolic rates (McKechnie and Wolf 2019).

Despite lacking sweat glands, passive cutaneous water loss represents a large proportion (e.g., 40% to 70%) of total evaporative water loss in birds (Dawson 1982, Williams and Tieleman 2005). The relative contribution of respiratory and cutaneous water loss in birds varies with species and environmental heat load (Dawson 1982, Wolf and Walsberg 1996). For example, in passerines, cutaneous water loss, as a proportion of total evaporative water loss, decreases with increasing temperature, whereas in pigeons (*Columba livia*) and doves (*Zenaida* spp.), cutaneous water loss increases with temperature (Wolf and Walsberg 1996, Tieleman et al. 2003). State of hydration can also influence the rate of cutaneous water loss in birds. For example, cutaneous evaporation increased with increasing temperature in hydrated pigeons and decreased in dehydrated birds that relied on panting as their main evaporative cooling mechanism (Arad et al. 1987).

Physiological Adaptations, Thermoregulation, and Water Balance

Metabolic Rate

Reduction in metabolic rate can decrease metabolic heat gain, thereby reducing the need for evaporative cooling. After accounting for body size, species in arid climates typically have lower metabolic rates than those in mesic environments (Macfarlane et al. 1971; Nagy 1987; Tieleman et al. 2002, 2003; Ostrowski et al. 2006; White et al. 2007). The lower metabolic rates of desert-adapted species are associated with lower water turnover rates in ungulates (Maloiy et al. 1979) and lower total evaporative water loss in birds (Tieleman et al. 2002, 2003). Furthermore, dehydrated ungulates have lower metabolic rates than when hydrated normally (Schmidt-Nielsen et al. 1967, Ostrowski et al. 2006).

Adaptive Heterothermy

Adaptive heterothermy can minimize water loss from evaporative cooling. Body temperature rises during the day, reducing the thermal gradient and heat gain from the environment; excess heat is then passively lost at night, when ambient temperature falls below the body temperature (Taylor and Lyman 1972, Dawson 1984). Similarly, antelope ground squirrels (*Ammospermophilus leucurus*) allow their body temperature to rise when foraging during the day in summer and then periodically return to their cooler burrows to dissipate excess body heat (Chappell and Bartholomew 1981).

Selective brain cooling is a commonly cited mechanism to maintain brain temperature below critical values when body temperature increases (Brinnel et al. 1987). Some investigators have questioned the use of selective brain cooling to protect the brain when body temperature rises (Fuller et al. 2005).

Recent research suggests that the rise in core body temperature observed in thermal studies may have resulted from studying captive animals that were prevented from using behavioral or other thermoregulatory mechanisms (Mitchell et al. 2002, Fuller et al. 2005). Studies of free-ranging ungulates generally have not shown evidence of adaptive heterothermy, suggesting that increased daily fluctuations in body temperature in dehydrated animals were actually dehydration-induced hyperthermia (Mitchell et al. 2002). By lowering brain temperature, selective brain cooling allows body temperatures to rise above those that would normally trigger evaporative cooling mechanisms. In short, dehydration-induced hyperthermia and selective brain cooling increase the temperature at which animals begin to thermoregulate via evaporative cooling, thus conserving water (Mitchell et al. 2002, Fuller et al. 2005). The magnitude of selective brain cooling is increased in both dehydrated and food-restricted ruminants (Ostrowski et al. 2006, Fuller et al. 2007, Hetem et al. 2012a, Strauss et al. 2015).

Behavioral Regulation of Body Temperature and Water Balance

Behavioral adaptations function in combination with physiological and morphological mechanisms and are important for the maintenance of body temperature and water balance (Bartholomew 1964). Diet selection, timing of activity, use of microhabitats, and body orientation are common behaviors used by animals that aid in thermoregulation (Berry et al. 1984, Gill 1994, Sargeant et al. 1994).

Diet and Water Balance

Preformed water provides a significant portion of total water intake and allows some species (e.g., heteromyid rodents) to survive long periods without access to free water (Taylor 1968, 1969a, Schemnitz 1994). Also, several ungulate species can survive extended periods without free water, including Arabian oryx (*Oryx leucoryx*; Ostrowski et al. 2002), Grant's gazelle (*Gazella granti*; Taylor 1968), springbok (*Antidorcas marsupialis*; Hofmeyr and Louw 1987), Cape eland (*Taurotragus oryx*; Taylor 1969a), dik-dik (*Rhynchotragus kirki*; Hoppe 1977), and fringe-eared oryx (*Oryx beisa callotis*; Taylor 1969b, Lewis 1977). Lower moisture content of forage used by grazing animals when compared with browsing animals makes the former more dependent on free water (Maloiy 1973, Kay 1997). During drought, mixed feeders shifted diets to include a higher proportion of browse species in Kruger National Park, South Africa (Abraham et al. 2019). Furthermore, foraging on succulent plants (e.g., cacti), underground storage organs (e.g., tubers), and browse with higher moisture content reduces the amount of free water needed to maintain water balance (Taylor 1969b, Williamson 1987, Gedir et al. 2016). Air-dried seeds typically have very low moisture content (e.g., <10%); as a result, many granivorous birds must supplement their diet with insects and succulent vegetation to meet their water requirements by preformed and metabolic water alone (Goldstein and Nagy 1985, Guthery and Koerth 1992).

Timing of Activity

Animals can reduce heat loads by adjusting the duration or timing of activities. During thermally stressful periods, desert-adapted animals spend less time being active (Hetem et al. 2012a, b). For example, Arabian oryx in Saudi Arabia increased resting time in the shade when temperatures exceeded 40°C at the expense of foraging time; some individuals more than halved their daily foraging time (Sneddon and Ismail 2002). Arid-adapted species often shift to more crepuscular or nocturnal activity patterns, foraging and moving during the cooler periods of the day (Carmi-Winkler et al. 1987, Alderman et al. 1989). Nocturnal feeding occurs during summer for largely diurnal species (Miller et al. 1984). Nocturnal feeding may also increase the intake of preformed water. For example, in areas where relative humidity increases at night, water content of dry vegetation may increase because of the hygroscopic uptake of moisture from the air (Taylor 1968).

Use of Microclimates

Animals commonly spend inactive periods in areas characterized by cooler microclimates (Wolf et al. 1996; Tull et al. 2001; Hetem et al. 2012a, b) or forage in shaded areas (Owen-Smith 1998, Giotto et al. 2013). Use of cooler microclimates reduces environmental heat loads and helps maintain a temperature gradient that facilitates non-evaporative heat loss. Bedding down on cool substrates can increase conductive heat loss. Small mammals like kangaroo rats (*Dipodomys* spp.) and mid-sized carnivores living in arid environments, including badger (*Taxidea taxus*) and kit fox (*Vulpes macrotis*), create their own microclimates by burrowing into the soil, and mountain lions (*Puma concolor*) establish natal dens in dense vegetation to ameliorate heat gain by offspring (Bleich et al. 1996). Species largely incapable of creating their own microclimate (e.g., large mammals, and birds), however, seek shade provided by vegetation or other environmental features during the midday heat (Goldstein 1984, Wolf et al. 1996, van Beest et al. 2012, Rakowski et al. 2019, Gedir et al. 2020). Bats commonly use caves, as do some birds, small mammals, and large mammals, including desert bighorn sheep, mountain lions, and mule deer (*Odocoileus hemionus*; Krausman 1979, Abeloe and Hardy 1997, Cain et al. 2008b).

Body Orientation

Animals active during hotter periods of the day or in areas that lack shade reduce absorption of solar radiation and increase convective heat loss by adjusting body position relative to the sun and wind. Ungulates stand with the long axis of the body parallel to the sun or wind (Jarman 1977, Berry et al. 1984, Maloney et al. 2005). When drinking water was available, black wildebeest (*Connochaetes gnou*) spent less time with their body oriented parallel to the sun's direction (Maloney et al. 2005). Birds likewise minimize exposure to solar radiation by similarly positioning their bodies and increase convective heat loss by lowering their legs during flight (Luskick et al. 1978, Martineau and Larochelle 1988). Small mammals like the antelope ground squirrel actually use their tail like an umbrella to shade their body and reduce thermal loading while foraging during the day (Chappell and Bartholomew 1981).

Morphological Adaptations for Thermoregulation and Water Balance

Desert-adapted species possess a variety of morphological adaptations that aid reduction of heat load and minimize water loss. These include body size and shape, and characteristics of pelage, feathers, or horns.

Body Size and Shape

Large-bodied animals have a lower surface area:volume ratio, which reduces the proportion of the body exposed to solar radiation, resulting in lower rates of heat gain from the environment (Phillips and Heath 1995). This characteristic can be disadvantageous because it also reduces the rate of heat loss. Shaded microclimates of size sufficient to benefit large animals are more limited where vegetation is sparse and other types of cover are unavailable. Conversely, smaller species, with their larger surface area:volume ratio, gain heat from the environment at a faster rate but lose excess body heat at a faster rate and can have greater access to shaded microclimates. The shapes of the body and appendages influence the rates of heat transfer between the animal and the environment; long and narrow appendages minimize radiant heat gain and maximize convective heat loss. Desert-adapted species characteristically have longer, thinner appendages with a higher surface area:volume ratio, thereby facilitating heat loss when compared to similar species inhabiting cooler environments (Phillips and Heath 1995).

Pelage, Plumage, and Horn Characteristics

The thickness and color of pelage and plumage affect heat transfer between animals and their environment. Thick pelage provides increased insulation but limits effectiveness of evaporative and convective cooling. Conversely, thin pelage maximizes heat loss but provides little insulation. Pelage thickness tends to decrease as body size increases in desert-adapted ungulates, facilitating heat loss (Hofmeyr 1985). Sparse pelage or plumage on ventral surfaces may facilitate non-evaporative and evaporative heat loss. Erection of pelage and feathers can also facilitate heat transfer to the environment.

Desert-adapted ungulates typically have glossy, light-colored pelage, which reflects more radiation than does dark colored pelage. Dark-colored pelage or plumage may absorb more solar radiation than light-colored pelage or plumage, however, reducing the amount of radiation that reaches the skin (Walsberg et al. 1978, Walsberg 1983). Increased absorption of solar radiation by dark-colored pelage lowers the thermal gradient between the environment and the surface of the pelage, reducing heat gain (Louw 1993). Highly vascularized horns of bovids, including ossicones of giraffe (*Giraffa giraffa*), may contribute to thermoregulation, and species from arid areas have relatively larger horn cores and thinner keratin sheaths than temperate species, which facilitates heat loss through the horns (Picard et al. 1999, Mitchell and Skinner 2004). The

beaks of birds may have a similar thermoregulatory function and allow passive loss of body heat when ambient temperatures are less than body temperature (Tattersall et al. 2017).

WATER MANAGEMENT FOR WILDLIFE

Liebig's law of the minimum states that growth is determined by the scarcest resource and not by the total amount of all resources. Every population needs a variety of resources (e.g., food, water, cover, space) to grow. When availability of these resources exceeds needs, populations can increase until, at some point, a resource is exhausted or depleted and becomes a "limiting factor." In arid regions, water is often assumed to be a primary factor that limits the distribution and productivity of desert ungulates and upland game birds. Beginning in the 1940s, resource management agencies and sportsmen's organizations began investing in the construction and maintenance of water sources in areas where natural sources were scarce or unreliable. These water developments were first constructed to benefit quail (*Callipepla* spp.; Glading 1943) and, later, for the benefit of chukar (*Alectoris chukar*), mule deer, bighorn sheep (*Ovis canadensis*), pronghorn (*Antilocapra americana*), and other game species. Water developments also have been used mitigate the loss or inaccessibility of naturally occurring water sources, direct losses of habitat, and other effects from urban, agricultural, transportation, and industrial development (Rosenstock et al. 1999, Krausman et al. 2006).

Water developments are widely used by wildlife managers in arid regions. At the end of the 20th century, 10 of 11 western state wildlife agencies had ongoing programs that included ~7,000 water developments (e.g., catchments, guzzlers, modified tinajas, developed springs, and wells; see next section) that provided perennial water for wildlife (Simpson et al. 2011). Overall, more than $1 million per year was allocated to these efforts (Rosenstock et al. 1999). Until relatively recently, few resources have been dedicated toward monitoring the influence of water developments on wildlife distribution, abundance, and demography.

Some investigators have questioned the efficacy of water developments (Broyles and Cutler 1999, Rosenstock et al. 2001), and their use is controversial, especially in some protected areas (Czech and Krausman 1999), including wilderness areas (e.g., the Arizona Desert Wilderness Act of 1990, or P. L. 101-628, and the California Desert Protection Act of 1994, or P. L. 103-433) as well as national parks and monuments (e.g., Death Valley, Joshua Tree, Agua Fria, Grand Canyon–Parashant, Sonoran Desert). The question of whether water developments are effective in meeting population objectives largely reflects the lack of experimental studies assessing their efficacy. Some populations of desert ungulates and upland game birds occupy areas without perennial sources of free water (Krausman and Leopold 1986, Alderman et al. 1989), suggesting that perennial water sources might not be necessary, and some studies have produced conflicting results (Mendoza 1976, Leslie and Douglas 1979, Deblinger and Alldredge 1991). Furthermore, some segments of the public have argued that the presence and maintenance of man-made water sources compromise wilderness values of remote, protected areas, creating unacceptable signs of human intrusion and altering some presumed balance of undisturbed nature or some otherwise intangible benefits of wilderness (Bleich et al. 2019). In response, wildlife managers contend that the failure to build or maintain these facilities puts vulnerable species and other important wildlife at risk (Bleich 2005).

TYPES OF WILDLIFE WATER DEVELOPMENTS

Precipitation Catchments

Precipitation catchments harvest rain or snow, storing water for future use by wildlife. These systems can be particularly effective in desert environments, where net annual precipitation is low but often occurs as brief, intense thunderstorms that generate substantial runoff from impervious surfaces. A brief overview of major designs is presented here; additional information can be found in Halloran and Deming (1958), Lesicka and Hervert (1995), Bleich et al. (2020), and references therein.

Guzzlers

Guzzlers are the most common precipitation catchment used to supply water for wildlife. Numerous designs have been developed over more than 50 years of use in western North America, with management agencies adapting them to accommodate varying project objectives and site characteristics (e.g., accessibility for construction equipment, annual rainfall, topography). Guzzlers capture rainwater on a natural surface or constructed apron, then route the water to a storage tank and drinking trough (Fig. 13.1).

Aboveground storage tanks require a float valve to control flow of water to the drinking trough (Bleich et al. 2020). When placed below ground, water can be piped into the trough without a float valve, reducing maintenance needs and mechanical failures (Lesicka and Hervert 1995, Bleich et al. 2020). Small guzzlers for game birds may hold only a few hundred liters; units for large ungulates can have capacities of more than 60,000 liters. To reduce the potential for animals becoming trapped and drowning, drinking troughs include a ramp that allows animals to escape. Concerns over visual effects, susceptibility to vandalism, and high costs associated with maintenance and water hauling have fostered new guzzler designs that have few aboveground components and a minimal disturbance footprint (Lesicka and Hervert 1995; Fig. 13.2.).

Figure 13.1. A 1960s-era guzzler near Yuma, Arizona. System has a metal collection apron and a buried concrete storage tank/trough. By S. S. Rosenstock

Figure 13.2. (A) Modern, low-maintenance / visual impact guzzler, Kofa National Wildlife Refuge, Arizona. All components except collection points (see B) are below ground, with a trough in the foreground at lower right. The disturbed area has been recontoured and revegetated with native plants. This water development was the subject of litigation asserting adverse impacts on wilderness values. (B) Collection point for buried guzzler system. Water flowing in the shallow rill (right foreground) is diverted by the concrete and rock diversion dam, then piped into buried storage tank. By S. S. Rosenstock

Tinajas

Tinajas occur in many desert mountain ranges and can represent the primary source of free water for wildlife. These natural rock tanks collect and hold water for varying lengths of time after rainstorms; capacity ranges from small amounts to more than 100,000 liters (Bleich et al. 2020). Because many tinajas have insufficient capacity to provide perennial water, many have been modified to increase storage capacity and reduce evaporation. An impermeable dam is commonly built on the downstream side to increase the capacity (Fig. 13.3) and sunshades added to minimize evaporation (Halloran and Deming 1958). Rock gabions can be constructed upstream, trapping sediment, reducing water velocity, and increasing flow into the tinaja. Inclusion of an escape ramp can minimize risk of mortality in steep-sided tinajas (Mensch 1969, Bleich et al. 2020).

Adits

Adits are short (e.g., 4–5 m) tunnels 2–3 m in diameter with a downward sloping floor, constructed into solid rock, typically located immediately adjacent to a wash (Fig. 13.4; Bleich et al. 2020). A diversion dam, perforated by pipes, is often constructed near the entrance to prevent the adit from filling with sand and other debris while allowing water to enter, and gabions can be used for the same purpose (Bleich et al. 2020).

Retention Dams and Sand Tanks

Retention dams with sand tanks or sand dams (Fig. 13.5) were a common method to harvest and store runoff in desert washes (Bleich and Weaver 1983); however, their construction is now less common and has largely been replaced by recent catchment designs. Concrete retention dams are usually built across narrow washes, and sand and gravel then fill the upstream side of the dam during flooding. During subsequent flood events, water seeps into the accumulated material stored behind the dam. The addition of rock gabions upstream of the dam slows the flow of floodwater, increasing infiltration into the sand tank. Pipes extending through the bottom of the dam transport water to a drinking trough or tinaja.

Figure 13.4. Buck Peak Tank, an adit located in the Cabeza Prieta Mountains, Cabeza Prieta National Wildlife Refuge, Arizona. By J. W. Cain III

Figure 13.3. Heart Tank, a modified tinaja located in the Sierra Pinta, Cabeza Prieta National Wildlife Refuge, Arizona. By J. W. Cain III

Figure 13.5. Drainage dam with sand tank, Kofa National Wildlife Refuge, Arizona. Subsurface water stored behind the dam is flowing into the tinaja immediately below. By S. S. Rosenstock

Figure 13.6. Former livestock water converted for use by wildlife, Kofa National Wildlife Refuge, Arizona. The aboveground metal trough was replaced with a buried concrete trough with escape ramp, placed under new shade cover. The water is fenced to exclude feral ass (*Equinus asinus*). By S. S. Rosenstock

Wells and Windmills

In areas where precipitation catchments are not practical, wells equipped with windmills can provide water for wildlife. To avoid increasing cover of vegetation and predation risk near the well, overflow from the trough should be redirected back down the casing. Many wells originally constructed for livestock have been converted for use by wildlife, typically by installing a more wildlife-friendly drinking trough (Fig. 13.6).

Spring and Seep Development

Making water more readily available from existing springs or seeps may be as simple as removing phreatophytes (Bleich et al. 2020). In some situations, extensive modifications may be required to provide a reliable water source, including fencing to exclude domestic livestock and feral equids, construction of access ramps, basins, or pools, or piping water to a nearby drinking trough.

Horizontal Wells

Horizontal wells are a viable option for providing water for wildlife in arid regions and overcome many disadvantages of spring development (Bleich et al. 1982, 2020). Horizontal wells have been successfully developed in areas with historical springs or seeps and the presence of an appropriate geologic formation, such as an impervious dike. Wells are drilled through the impervious formation into the water table. Water flow can then be controlled using a float valve and water distributed to a nearby drinking trough.

Livestock Water Developments

Many livestock waters are used by mule deer, pronghorn, elk (*Cervus canadensis*), small mammals, birds, and breeding amphibians, including some species that are of conservation concern (e.g., ranid frogs; Rosen and Schwalbe 1998). The northern leopard frog (*Rana pipiens*) has been extirpated from most of its historical range in Arizona. On the central Mogollon Rim, where the last known remaining metapopulation occurs, 23 of 25 currently known breeding sites are earthen tanks developed for livestock (S. MacVean, Arizona Game and Fish Department, unpublished data). The location of livestock watering facilities, however, sometimes renders them inaccessible to some species (e.g., desert bighorn sheep, pronghorn; Ockenfels et al. 1994). Steep-walled drinkers and storage tanks may be unavailable to smaller species and present a risk of drowning. Wire fencing, water surface area, and other features of livestock waters affect use by bats (Tuttle et al. 2006). Taylor and Tuttle (2007) provide guidelines for making livestock water developments more wildlife friendly.

POTENTIAL BENEFITS OF PROVIDING WATER SOURCES FOR WILDLIFE

Wildlife management agencies believed that augmentation of water sources would increase survival and recruitment, expand animal distribution (Rosenstock et al. 1999, Simpson et al. 2011), and buffer populations during drought periods. Anticipated benefits of water developments typically encompass one or more interrelated population management objectives, including increasing abundance, increasing available habitat (thereby increasing population size and distribution), improving habitat quality, and benefitting species of conservation concern (e.g., sensitive, threatened, or endangered).

Increase Animal Abundance

The abundance of some populations has been associated with the availability of water sources, with populations having higher abundance in areas with perennial water sources. For example, the presence of water developments can affect the population densities of pronghorn, elk, and mule deer. The high densities of some pronghorn populations are associated with water sources (Yoakum 1994). Water developments, however, are not the sole determinant of pronghorn abundance (Deblinger and Alldredge 1991), and some researchers of desert bighorn sheep did not report an influence of water sources on population size (Krausman and Etchberger 1995). Further, the change in ungulate density around water developments could be related to changes in population abundance or could simply be a shift in the distribution of animals, resulting in increased density near water sources with no corresponding change in the actual size of the population.

Increase Available Habitat

If all necessary resources with the exception of water are available, construction of new water sources may increase animal distribution and abundance by increasing available habitat. For example, following development of water sources, mule deer (Wright 1959) and desert bighorn sheep (*O. c. nelsoni*; Leslie and Douglas 1979) began year-round occupation of what were previously seasonal ranges. Elk (Rosenstock et al. 1999) and mule deer (Marshal et al. 2006a) expanded their distribution in arid habitats, possibly in association with water developments. The addition of water sources may increase abundance, but density may remain unchanged because of the increase in available habitat (Bleich 2008, Bleich et al. 2010). Increases in distribution have not been reported in some instances, however (e.g., desert bighorn sheep, Krausman and Leopold 1986; adult pronghorn, Deblinger and Alldredge 1991, Ockenfels et al. 1994), and the availability of water did not influence habitat use of adult pronghorns in some regions (Deblinger and Alldredge 1991, Ockenfels et al. 1994).

Improve Habitat Quality

Developed water sources may also improve habitat quality. Survival and reproductive success of some wildlife populations in arid areas fluctuate widely from year to year. If this variability arises from lack of reliable water sources, then water developments could reduce variability in survival and reproductive rates. For example, growth rates influence survival and recruitment of neonatal ungulates (Cook et al. 2004), which in turn are influenced by the quantity and quality of milk, which can be adversely affected by water deprivation (Hossaini-Hilali et al. 1994), thereby influencing juvenile survival. In areas where forage resources are sufficient but in which water is limiting, wildlife water developments may result in improved reproductive success and juvenile recruitment.

Recovery and Management of Special Status Species

Although most water developments are constructed to benefit specific game species, a wide variety of other species also use them, including species of conservation concern (Kuenzi 2001, Lynn et al. 2006, O'Brien et al. 2006). The Sonoran pronghorn (*A. a. sonoriensis*) was listed as endangered in 1967, with its distribution in the United States currently limited to southwestern Arizona. Declines in forage quality during recurring droughts reduce fawn recruitment and adult survival and are believed to limit recovery efforts (Hervert and Bright 2005). During dry periods, >40% of Sonoran pronghorn's diet consist of chain fruit cholla (*Cylindropuntia fulgida*) and other succulent vegetation (Hughes and Smith 1990). Areas with abundant chain fruit cholla, however, also had higher predation risk (Hervert and Bright 2005, Hervert et al. 2005). Chain fruit cholla and other cacti have high moisture content (e.g., 85%; Hughes and Smith 1990) but otherwise low nutritional value (e.g., <4% protein); therefore pronghorn are selecting forage based on moisture rather than nutrient content when water is limiting. Constructing water sources in areas with low risk of predation and characterized by higher-quality forage may be a positive tradeoff despite lower availability of chain fruit cholla (Morgart et al. 2005).

POTENTIAL NEGATIVE EFFECTS OF WATER DEVELOPMENTS

Critics of wildlife water developments have cited numerous potential adverse effects to wildlife, wildlife habitat, and other resources. Among these concerns are disease transmission, water quality, increases in predation or competition, direct mortality, and changes to plant communities.

Increased Potential for Disease Transmission

Wildlife water developments receive heavy use by numerous wildlife species, and some critics have suggested that water developments may contribute directly or indirectly to the spread of wildlife diseases (Broyles 1995). One pathogen of concern for direct transmission is the protozoan parasite *Trichomonas gallinae*, which has been found in urban birdbaths and causes epizootic outbreaks in doves and other birds. Rosenstock et al. (2004) collected water samples from a variety of water development types in Arizona and cultured them for *Trichomonas*; all samples were negative. Water samples from other developments proximate to a trichomoniasis outbreak also were also negative. Rosenstock et al. (2004) suggested that exposure to ultraviolet radiation and predation by other microorganisms may limit persistence of *Trichomonas* in water developments. This hypothesis was supported by a follow-up experiment in which replicated "mini-exclosures" in a catchment trough were inoculated with live *Trichomonas* and then sampled over time for presence of the protozoan. Within 24 hours, all samples were negative (S. Rosenstock, Arizona Game and Fish Department, unpublished data). Another water-mediated disease is botulism, caused by toxins produced by the bacterium *Clostridium botulinum*. There is one published occurrence of botulism-related mortality at a wildlife water development in California (Swift et al. 2000). Desert bighorn seeking water became entrapped in the storage tank, drowned, and decomposed, and sheep later consumed the contaminated water. At least 45 animals died during this event, which was subsequently attributed to type C botulism (Swift et al. 2000). Following this event, Bleich (2003) provided substantial evidence that conditions suitable for the production of botulinum toxin at natural water sources used by mountain sheep were not uncommon, and that conditions suitable for the production of botulinum toxin, and subsequent outbreaks of botulism, are not unique to anthropogenic water developments.

Broyles (1995) also suggested that water developments provided nurseries for hematophagous gnat (genus *Culicoides*) larvae, the adults of which are vectors of viral hemorrhagic diseases (bluetongue, epizootic hemorrhagic disease) affecting ungulates (e.g., mule deer) that also use these developments. Rosenstock et al. (2004) trapped *Culicoides* at sites adjacent to wildlife water developments and at unwatered controls. Several species of *Culicoides*, including the known vector (*C. sonorensis*), were widely distributed and common at both watered and control sites. Those authors also sampled potential substrates (sand, mud, gravel) potentially used by larval *Culicoides* at natural and modified tinajas (these materials frequently are absent in troughs at other types of constructed waters). Larval *C. sonorensis* were absent, except for a few individuals found at one site. Wildlife water developments typically do not provide fine silt or mud at the margins of water that is brackish or heavily enriched with animal manure and that provide optimal conditions for larval *Culicoides* (Mullens 1991).

Interactions between conspecifics and other species are common at wildlife water developments (O'Brien et al. 2006), creating the possibility of animal–animal pathogen transmission. While this topic has not yet been studied, it seems unlikely that animal interactions at wildlife water developments would be different from those occurring at natural water sources.

Water Quality

During hot summer months, wildlife water developments in desert environments can have high water temperatures and evaporation rates, inputs of organic material—commonly drowned honeybees (*Apis mellifera*)—and infrequent flushing, characteristics that are also common to natural surface water sources. Nevertheless, some have suggested that poor water quality in developed waters can adversely affect health and or survival of wildlife (Kubly 1990, Broyles 1995). Parameters and constituents of potential concern include dissolved solids, pH, heavy metals, toxic chemicals, and toxins produced by blue-green algae. Rosenstock et al. (2005) and Bleich et al. (2006) examined water quality at water developments in Arizona and California, respectively, at sites that included several types of precipitation catchments, improved springs, modified and natural tinajas, and wells. At sites in both states, water-quality parameters were typically within established guidelines for domestic livestock (standards for wildlife have not been published). Elevated levels of pH, alkalinity, and fluoride occurred at some sites but were deemed of negligible effect to wildlife health (Rosenstock et al. 2005, Bleich et al. 2006). Griffis-Kyle et al. (2014) monitored 10 tinajas and 17 water catchments in southwestern Arizona and reported that ammonia concentration at 27% of catchments was >50 mg/L N-NH₃. While ammonia concentrations were lower in tinajas, species diversity of tadpoles was negatively correlated with ammonia concentration. Arizona waters also were screened for cyanobacterial toxins, and all tests were negative (Rosenstock et al. 2004). Similarly, laboratory analyses eliminated cyanobacterial toxins as a factor in the deaths of numerous bighorn sheep at a wildlife water development in California (Swift et al. 2000). Wildlife water developments do not generally appear to provide appropriate conditions for toxic algal blooms, which typically occur on lakes and other large water bodies (Schwimmer and Schwimmer 1968). Nevertheless, more than 100 elk were found dead in a 1 km² area in northern New Mexico in 2013, and the cause of death was attributed to neurotoxin produced by the cyanobacteria *Anabaena* in an earthen livestock tank used by elk (New Mexico Department of Game and Fish 2013).

Predation

Avian and mammalian predators are regular visitors to water developments and are assumed by some to expand their home ranges to include them. Despite a paucity of evidence, it has been suggested that these facilities increase predation rates on desert bighorn sheep, mule deer, and potentially other species (Broyles 1995, Harris et al. 2020). This speculation may reflect observations in African savannas, where ungulates and predation events can be concentrated around waterholes during the dry season (de Boer et al. 2010, Davidson et al. 2013). Evidence supporting a similar pattern at water developments in North American deserts is scant (Rosenstock et al. 1999, Simpson et al. 2011). DeStefano et al. (2000) reported more predator sign (feces, tracks) around water developments than at unwatered comparison sites, but little evidence of predation events. They suggested that water developments could decrease predation pressure by reducing concentration of potential prey around the few existing natural water sources. In New Mexico, increased water source visitation by pronghorn, mule deer, and elk was associated with periods with higher temperatures and less long-term precipitation, whereas visitation rates of bobcat (*Lynx rufus*), coyote (*Canis latrans*), and mountain lion were associated with higher temperature and/or visitation rates of their primary prey species (Harris et al. 2015). O'Brien et al. (2006) documented only eight attempted or successful predation events in more than 37,000 hours of video observations at three water developments in Arizona, the majority involving bobcats or raptors taking birds or small mammals. Camera monitoring of water developments on the Sevilleta National Wildlife Refuge and Armendaris Ranch, New Mexico, has documented interactions between coyote and pronghorn and mountain lion and mule deer (J. Erz, US Fish and Wildlife Service, unpublished data; C. Prude, Armendaris Ranch, unpublished data), but documentation of predation events at water sources is relatively rare. Prude (2020) evaluated the proximity of mountain lion kills to wildlife drinkers on five study areas spanning a rainfall and temperature gradient across New Mexico and Arizona. Of 1,276 ungulates killed by lions, only 1.2% were within 100 m of a wildlife water development, and only 3% were within 250 m. There has been some speculation that the presence of reliable water sources in the form of water developments can allow some predators to occupy arid areas that would otherwise be inhospitable, thus exacerbating predation on ungulates and other game species. To date, this topic has been subject to little research.

Competition

Differences in habitat and diet selection resulting in niche partitioning between herbivores should serve to minimize competition between species. Expansion of areas of sympatry in association with water developments may, however, increase competition for forage between species with broadly similar diets. For example, water sources constructed along the base of mountains occupied by desert bighorn sheep could increase competition for forage with desert mule deer in the less rugged areas of bajadas that are used by both species. Increased availably of water in arid and semiarid regions may allow water-dependent, feral equids to occupy areas that would otherwise be inhospitable on a seasonal or year-round basis and could contribute to increased competition (Broyles 1995,

Simpson et al. 2011, Larsen et al. 2012). Feral horse (*Equus caballus*) and feral burro (*E. asinus*) are widely distributed across much of the western United States, particularly in the southwestern deserts and Great Basin (McKnight 1958, Stoner et al. 2021), and these non-native equids represent unique agents of disturbance in semiarid ecosystems (Beever 2003). These feral equids affect native ungulates directly by competing for forage or water (Bleich et al. 2010, Marshal et al. 2012, Hall et al. 2018*a*, *b*); indirectly by altering the availability of food, water, or habitat—and thereby reducing resource quality (Weaver 1959, Eldridge et al. 2020); or through interference competition (Berger 1985, Ostermann-Kelm 2008, Hennig et al. 2021). Feral burro activity can change the structure and composition of vegetation communities and may reduce availability of forage species preferred by native ungulates (Sanchez 1974, Hanley and Brady 1977). During periods of reduced forage production (e.g., drought), overlapping habitat use and diet between feral burros and desert bighorn sheep and/or mule deer could increase competition (Marshal et al. 2008, 2012). The potential for competition can be reduced if water developments are placed in spatially disjunct areas occupied by desert bighorn sheep and mule deer, and water sources can be fenced to exclude feral equids (Andrew et al. 1997).

The potential for competitive interactions between predator species can be exacerbated around limited resources such as water developments (Arjo et al. 2007, Atwood et al. 2011). An evaluation of water development use by bobcat, coyote, and gray fox (*Urocyon cinereoargenteus*) in West Texas reported temporal partitioning at water sources, resulting in few encounters among those species. Most visits to water were during the night, but coyotes and bobcats regularly visited water sources during daylight; gray fox rarely visited water during the day (Atwood et al. 2011). Relative activity of gray fox was higher at water sites with lower relative activity of coyotes (Atwood et al. 2011). Conversely, spatial or temporal separation in visits to water developments by kit fox and coyote was not detected in the Mojave and Great Basin deserts (Hall et al. 2013).

Interference competition at water sources also has been documented. In Colorado, feral horses were frequently observed preventing elk from accessing an isolated natural water source at Mesa Verde National Park (Perry et al. 2015). Desert bighorn in California avoided water sources when feral horses and feral burros were present (Dunn and Douglas 1982, Ostermann-Kelm et al. 2008). Pronghorn were more vigilant and spent less time drinking in the presence of feral hoses at water sources (Gooch et al. 2017). Pronghorn shifted their visitation time to water at locations where feral horses were common, and pronghorn and mule deer avoided water sources with high use by feral horses (Hall et al. 2018*a*, *b*). Hennig et al. (2021) did not document interference competition between feral horses, livestock, and native pronghorn in Wyoming, a result likely related to the large size of the water sources that were monitored. They concluded, however, that interference competition between horses and other native ungulates is a potential concern.

Direct Mortality

Wildlife water developments have caused direct mortality of wildlife based on anecdotal reports of drowned animals found at these facilities (Baber 1983) and mortalities at natural tinajas (Mensch 1969). One species of particular concern is the desert tortoise (*Gopherus agassizii*), a federally listed endangered species. Andrew et al. (2001) identified faunal remains removed from catchments in the Sonoran Desert in California. They found remains, primarily of small mammals, birds, and reptiles, at six sites, but none of desert tortoise. Most skeletal materials showed evidence of predation and likely originated from pellets cast by raptors or scats deposited by mammalian predators that had been flushed into the catchments by runoff. O'Brien et al. (2006) video-documented another source of remains found in or at water developments, those brought to the site by predators or scavengers. During more than 600 visits to various types of wildlife water developments in Arizona, Rosenstock et al. (2004) documented 19 presumed drowning incidents that included several mule deer, a coyote, a desert bighorn sheep, and a variety of small mammals and passerine birds. Overall, the information in hand does not support the notion that developed waters are an important cause of mortality for wildlife. Further, wildlife water developments often have escape ramps to minimize drowning of wildlife (Simpson et al. 2011, Bleich et al. 2020).

Detrimental Effects on Vegetation

Some investigators have documented gradients of decreasing forage use and plant community change with increasing distance from developed water sources, with a zone of severe impact (*piosphere*) immediately around the water source (Thrash 1998, Parker and Witkowski 1999). Concentrations of animals at water sources can result in trampling and foraging pressure that affect surrounding plant communities, thereby reducing forage abundance and species composition, increasing cover of invasive species, and altering soil nutrient concentrations (Tolsma et al. 1987, Brooks et al. 2006, Jawuoro et al. 2017). Piospheres are prevalent in areas supporting high abundances of herbivores, including native ungulates and livestock. Krausman and Czech (1998) recognized the potential for similar effects around water developments in North American deserts. Although piospheres are regularly observed in African savanna systems with high diversity and abundances of native ungulates, they are more commonly associated with water sources used by domestic livestock or feral equids (Welles and Welles 1961, Andrew 1988) in the southwestern United States owing to the relatively low densities of native herbivores. Marshal et al. (2006*b*) sampled forage biomass and mule deer use in paired desert washes with and without precipitation catchments. During the hot, dry season, deer sign was more abundant in washes with catchments and increased with proximity to water in washes with catchments. Those authors reported no differences in forage biomass attributable to foraging by deer, likely because desert ungulates occur at relatively low densities and do not linger around catchments after watering (O'Brien et al. 2006, Waddell et al. 2007).

INFLUENCE OF WATER DEVELOPMENTS ON SELECTED WILDLIFE SPECIES

Extensive literature reviews on the effects of wildlife water developments on game and nongame species were conducted by Payne and Bryant (1998) and Rosenstock et al. (1999), and were later updated by Krausman et al. (2006) and Cain et al. (2013). Key findings of these reviews are presented here, along with subsequent research on the influence of water sources on game and nongame wildlife species. Collectively, these data suggest that water developments may have resulted in increased wildlife distribution and abundance in some instances, but not in others.

Mammals

Desert Bighorn Sheep

Development of water sources is central to management of desert bighorn sheep and can interact with other factors that influence sheep populations, such as forage quality and availability, escape terrain, thermal cover, and fire history (Bleich et al. 2010). Desert bighorn sheep require approximately 4% of their body weight in water per day (Turner 1973); the proportion met by preformed water in forage is unknown but likely varies seasonally and among populations depending on availability of succulent forage (Turner 1973, Gedir et al. 2016). Many observational studies documented that desert bighorn use water developments when available (Blong and Pollard 1968, Wilson 1971), and use is concentrated during summer months (Bleich et al. 1997, Harris et al. 2020). Desert bighorn reportedly use water every two to three days (Bradley 1963, Knudsen 1963) and consume 2–4 L of water per visit (Knudsen 1963, Hailey 1967). Others did not document use of newly constructed water sources for at least three years following construction (Krausman and Etchberger 1995). Krausman et al. (1985) reported that two females did not visit water sources for 10 or more consecutive days during the summer, although they were observed frequently consuming barrel cactus (*Ferocactus* spp.). The use of water sources is likely related to preformed moisture content in forage and seasonal water demands (Turner 1973, Krausman et al. 1985, Warrick and Krausman 1989).

Perennial water sources are a key habitat component for desert bighorn sheep (Douglas and Leslie 1999, Longshore et al. 2009), and their presence on the landscape is commonly used to define habitat quality for the species (Bleich et al. 2010). Desert bighorn commonly occur less than 4 km from water sources during the hottest and driest times of the year (Wilson 1971, Turner et al. 2004). Because of their smaller body size and the increased water demands during lactation, females with lambs generally occur closer to water sources than do males or females without lambs (Bleich et al. 1997, Krausman 2002). Although numerous investigators have reported that the summer distribution of desert bighorn sheep was associated with water sources, other populations occupy ranges devoid of perennial water sources (Watts 1979, Kraus-

man et al. 1985, Alderman et al. 1989) and may obtain sufficient water from ephemeral sources, metabolic and preformed water from cacti, and other succulent vegetation (Turner 1973, Warrick and Krausman 1989, Gedir et al. 2016). Distribution and habitat use of desert bighorn are not limited by the availability of perennial water sources in all cases (Krausman and Leopold 1986, Krausman and Etchberger 1995).

Although the influence of water sources on seasonal distribution of desert bighorn has been reported for numerous populations across the desert Southwest, documentation of population-level responses (i.e., increases in abundance or survival) is limited. In some cases, population increases were attributed to addition of water sources (e.g., Leslie and Douglas 1979) and population declines (Douglas 1988), and mortalities have been attributed to loss of water sources (Allen 1980). Others have concluded that addition of water sources does not always increase survival or productivity (Ballard et al. 1998). Nevertheless, wildlife water developments play a potentially important role in maintenance of connectivity among populations because they (1) allow bighorn sheep to make use of otherwise suitable ranges that lack reliable sources of surface water; (2) increase the probability of pioneering individuals encountering surface water in areas that otherwise provide habitat; (3) enhance the likelihood of immigrants encountering conspecifics; or (4) may increase survival rates during periods of thermal stress or drought (Bleich 2009).

The equivocal results in the scientific literature regarding influence of water sources on desert bighorn sheep may in part reflect the fact that most studies were either observational, anecdotal, or derived from research designed to address other objectives (Rosenstock et al. 1999, Krausman 2002). Few investigators have examined the influence of water developments on desert bighorn sheep experimentally. Cain et al. (2008a) conducted a manipulative study that documented the responses of desert bighorn sheep to removal of man-made water sources in southwestern Arizona; response variables included diet selection, home range size, movement, survival, and recruitment. Water sources were maintained in two mountain ranges during 2002–2003, then drained in one mountain range and maintained in the other during 2004–2005. Water removal did not result in the predicted changes in the response variables. When water sources were maintained in both mountain ranges, the general area experienced the most severe drought on record, and high adult mortality was observed. In this situation, provision of water sources during drought may have done little to prevent the adverse consequences of scarce or poor-quality forage, but it is unknown whether mortality would have been higher in the absence of water sources during the drought. Conversely, increased precipitation during the posttreatment period resulted in better forage conditions, increased forage moisture content, and availability of ephemeral water sources. The lack of significant changes after water removal suggested that during years with above-normal precipitation, water might not be limiting to desert bighorn sheep. Because of the climatic conditions observed during this study, it is unknown what effect the removal of water catchments would have had during periods with average or below-average precipitation.

Nonetheless, it is important to recognize that provision of water sources is unlikely to enhance carrying capacity (i.e., population density) of bighorn sheep habitat if either forage availability or forage quality is the limiting factor. Wildlife water developments in specific mountain ranges can help expand the distribution of bighorn sheep into water-depauperate areas within those ranges (Weaver 1973) by providing bighorn sheep with opportunities to expand into areas with otherwise suitable conditions that previously were available only on a seasonal basis because of an absence of water. As a result, bighorn sheep can have a wider annual distribution in those ranges (Bleich 2008). Although wildlife water developments likely do not affect absolute density (i.e., sheep/km²) when nutrient availability per individual defines carrying capacity, they can affect overall numbers of bighorn sheep in a given subpopulation and may enhance the total number of animals because sheep may be able to forage across a greater proportion of the mountain range than in the absence of the water (Bleich 2008, Bleich et al. 2010).

Additional experimental studies are needed to determine whether and under what conditions water developments have population-level influences on desert bighorn sheep. Broader questions also remain about climate change and how it will affect individual populations and metapopulation structure (Epps et al. 2004). In the southwestern United States, average annual temperatures are predicted to increase, while changes in precipitation are predicted to be more variable (i.e., increasing in some areas and decreasing in others). Declines in precipitation will result in lower snowpack levels and decreased surface runoff (Seager et al. 2007). Declining river flows are likely to result in more reliance on groundwater sources to meet agricultural and municipal needs (Alexander et al. 2011). Increasing reliance on groundwater and declining groundwater recharge rates will result in lower water tables and loss of natural surface water sources (e.g., springs) used by wildlife (Zektser et al. 2005). Epps et al. (2004) reported that populations of desert bighorn sheep in Southern California occupying higher-elevation desert mountain ranges with more precipitation and the presence of natural springs were less likely to go extinct than populations in lower-elevation mountain ranges without perennial springs. It seems likely that water developments will play an increasingly important role in the conservation and management of desert bighorn sheep and other wildlife in the future.

Mule Deer

Anthropogenic water sources are commonly managed to benefit desert mule deer. Mule deer often use water developments year-round in arid areas, but use increases during the summer months and during periods of reduced precipitation (Harris et al. 2015). Mule deer tend to use natural water sources when available (Krausman 2002), and newly constructed water sources may receive relatively little initial use compared with well-established water sources. Water sources available for longer than three years exhibited higher use by mule deer than newly constructed waters (Marshal et al. 2006a, b). Mule deer visit water at all times, but most visits occur at sunset or at night (Hervert and Krausman 1986, Hazam and Krausman 1988, Shields et al. 2012). Mule deer visit water sources every one to four days; the frequency of water use is higher for females than males (Hervert and Krausman 1986, Hazam and Krausman 1988, Shields et al. 2012). Water consumption rates range from 1 to 6 L per visit but vary seasonally (Hazam and Krausman 1988). In the Sonoran Desert, females consume an average of 3.3 L per visit in early summer and 4.2 L per visit in late summer (Hazam and Krausman 1988); consumption by males averaged 2.7 and 3.6 L per visit during early and late summer, respectively (Hazam and Krausman 1988).

Mule deer tend to be located closer to water during the summer dry season than other periods of the year (Bowyer 1986, Ordway and Krausman 1986, McKee et al. 2015) and may travel outside their normal home range when water becomes unavailable (Hervert and Krausman 1986). Some populations of desert mule deer may migrate during the dry season to areas with perennial water sources (Rautenstrauch and Krausman 1989). Krausman and Etchberger (1995) reported that during summer, females were located closer to water than males; however, all deer were typically within 5 km of water. In addition to influencing the seasonal distribution of individual mule deer, seasonal ranges became occupied year-round after construction of water developments (Wright 1959).

The addition of water sources has also been associated with increases in abundance in some areas. Mule deer numbers increased for at least a five-year period following the construction of water sources near Fort Stanton, New Mexico (Wood et al. 1970). The increase, however, may have been associated with a redistribution of deer to areas near water sources rather than an increase in population size. There is little conclusive information on population-level influences of water developments on mule deer (Heffelfinger 2006:118). McKee (2012) and Bush (2015) were unable to detect effects of anthropogenic waters on mule deer body condition, twinning rates, or survival in the Mojave Desert. Body condition, twinning rate, and fawn survival were more strongly associated with precipitation than provisioning of water (McKee 2012, Bush 2015, Heffelfinger et al. 2018). Mule deer populations may benefit if water availability allows access to areas that are not forage limited and simultaneously provide lower risk of predation but are unlikely to compensate for scarce or poor-quality forage (Heffelfinger et al. 2018). If forage is the primary limiting factor, simply adding water sources is unlikely to result in population increases for mule deer.

White-Tailed Deer

White-tailed deer (*Odocoileus virginianus*) are typically found only in association with water sources (Krausman and Ables 1981, Maghini and Smith 1990) and appear dependent on free water in the southwestern United States (Heffelfinger 2006). Coues white-tailed deer (*O. v. couesi*) in Arizona select areas within 0.4 km of water sources and avoid areas located more than 1.2 km from water sources (Rosenstock et al. 1999). At the western edge of the distribution of Coues white-tailed deer, animals drank every one to four days (Henry and Sowls 1980). White-tailed deer subjected to water restriction decrease forage intake and lose more weight than those with access to water (Lautier et al. 1988). When forage moisture is low in arid environments, water developments may influence fawn survival and recruitment (Ockenfels et al. 1991).

Elk

Elk habitat is commonly restricted to areas with natural surface water. In the spring and summer, elk tend to be distributed within 0.4–0.8 km from water (Jeffrey 1963, Marcum 1975, Nelson and Burnell 1976, Delgiudice and Rodiek 1984). In North Dakota, however, elk movements and landscape use were independent of perennial sources of water (Sargeant et al. 2014). In New Mexico, elk generally visited water during nocturnal or crepuscular periods (Harris et al. 2015). Water needs of elk most likely vary seasonally and annually, depending on precipitation, forage conditions, and reproductive status (Mackie 1970, Marcum 1975, McCorquodale et al. 1986). In many areas of the western United States, elk inhabit mesic areas with relatively abundant natural sources of surface water (e.g., streams), and water is not a limiting factor, making water developments unnecessary. Development of water sources, however, has most likely contributed to the expansion of elk populations in arid and semiarid areas (Strohmeyer and Peek 1996). In arid regions with unpredictable surface water availability, elk may benefit from water developments, particularly during calving, lactation, or during dry periods (Delgiudice and Rodiek 1984, McCorquodale et al. 1986). Like mule deer and desert bighorn sheep, definitive data on population-level effects of water developments are lacking for elk.

Pronghorn

Much of the research on the dependence of pronghorn on free water is equivocal (Yoakum 2004). Some researchers contend that pronghorn need drinking water to sustain healthy populations (Yoakum 1978), whereas others suggest that some populations obtain sufficient moisture from succulent forage (Beale and Smith 1970). When succulent vegetation is unavailable, water developments are heavily used by pronghorn (O'Gara and Yoakum 1992, Yoakum 1994), and many pronghorn populations use water sources, particularly during dry periods (Sundstrom 1968, Beale and Smith 1970, Yoakum 1994, Harris et al. 2015). Frequency of use ranges from daily to every several days or weeks in areas with succulent forage (Buechner 1950). In western Utah, pronghorn did not use water when moisture content of forage exceeded 75% (Beale and Smith 1970). In New Mexico, pronghorn visits to water sources occurred primarily during the day (Harris et al. 2015). When forage moisture content is inadequate to meet water needs, consumption of free water ranges from 0.95 to 3.8 L/day (Sundstrom 1968, Beale and Smith 1970, Beale and Holmgren 1975). Some populations of Sonoran pronghorn (*A. a. sonoriensis*) previously thought not to use free water sub-

sequently have been confirmed to make use of that resource (Morgart et al. 2005).

When forage moisture content declines, pronghorn in Wyoming were reported within 5–6.5 km of water sources (Sundstrom 1968). In Arizona, Hughes and Smith (1990) reported that Sonoran pronghorn were typically within 6 km of water sources, whereas deVos and Miller (2005) reported that Sonoran pronghorn preferentially used areas within 2 km of water sources. During fawning, pronghorn in areas with limited surface water availability may occur closer to water, particularly females with fawns. Fawns in Arizona selected bed sites within 0.4–0.8 km of water sources (Ticer and Miller 1994), and yearling pronghorn in New Mexico tended to be closer to water sources than adults (Clemente et al. 1995).

The influence of water sources on population density of pronghorn is less clear. High-density populations are often associated with free water (Yoakum 1994). In Wyoming, pronghorn densities are highest in areas with water sources (Sundstrom 1968, Boyle and Alldredge 1984), but Deblinger and Alldredge (1991) reported that the distribution of pronghorn did not change when water became unavailable. Whether these results reflect an influence of water sources on population abundance rather than simply an influence on the distribution is unclear.

Water sources are not the sole determinant of the distribution or abundance of pronghorn (Deblinger and Alldredge 1991). Fawn:female ratios in Arizona were more strongly related to previous winter precipitation than to water availability, suggesting that forage availability was more important for fawn recruitment (Bristow et al. 2006). Based on 27 years of aerial survey and precipitation data in the Trans-Pecos area of western Texas, fawn production was strongly correlated with precipitation in the year (August–July) prior to surveys, whereas abundance was more strongly correlated with precipitation indices (e.g., Palmer Drought Severity Index) incorporating other climatic variables (Simpson et al. 2007). These researchers concluded that fawn production was related to immediate precipitation conditions, and that abundance was influenced more by long-term climatic trends. (Simpson et al. 2007).

Mammalian Predators

Research results regarding free water requirements of mammalian predators have been equivocal (Rosenstock et al. 1999). Many predators have been documented using water developments (Cutler 1996, O'Brien et al. 2006). Occupancy of American badger, bobcat, coyote, and gray fox was higher in areas near water developments in the Mojave Desert compared to unwatered control sites (Rich et al. 2019). Most predators obtain sufficient moisture from their prey, and some, like the ringtail (*Bassariscus astutus*) and kit fox, are believed to be independent of free water (Chevalier 1984, Golightly and Ohmart 1984). Use of man-made water sources by kit fox, however, has been documented in the Sonoran Desert in Arizona (O'Brien et al. 2006) and the Mojave Desert in Utah (Hall et al. 2013). A manipulative experiment that reduced the availability of water in the Great Basin Desert did not change kit fox survival or abundance but did result in a reduction in coyote use of the areas with reduced water availability (Kluever and Gese 2017, Kluever et al. 2017). Water removal did not influence home range size, affect territory boundaries, or result in mortalities of coyotes (Kluever and Gese 2016). Although predators commonly use water developments, their need to drink free water for survival or reproduction has not been thoroughly examined (Ballard et al. 1998, DeStefano et al. 2000).

Small Mammals
Many small mammals inhabiting arid environments typically employ physiological or behavioral mechanisms that minimize the need for free water and obtain sufficient water from preformed water in food and metabolic water (Mares 1983, Walsberg 2000, Nagy 2004). Some small mammals have been documented at water developments, including black-tailed jackrabbit (*Lepus californicus*), desert cottontail (*Sylvilagus audubonii*), and ground squirrels (O'Brien et al. 2006, Rich et al. 2019). Rich et al. (2019) reported that occupancy rates of black-tailed jackrabbits and desert cottontail were greater in areas near water developments in the Mojave Desert. Cutler (1996) captured 10 rodent species near water developments in southwestern Arizona and reported drinking only by the round-tailed ground squirrel (*Spermophilus tereticaudus*). Greater abundance of rodents near water developments has been reported (Burkett and Thompson 1994, Switalski and Bateman 2017), whereas others did not find differences in rodent abundance in areas with and without water developments (Cutler and Morrison 1998). Kluever et al. (2016, 2017) implemented a before-and-after-control-impact water removal experiment in Utah and did not detect a change in rodent abundance or relative abundance of black-tailed jackrabbits following removal of water; they reported rodent abundance was more strongly associated with vegetation and precipitation.

Bats commonly use water developments for drinking and as foraging areas (Kuenzi 2001, Rabe and Rosenstock 2005). Schmidt and DeStefano (1999) documented greater bat activity at desert water developments compared with control sites; bat visitation rates of more than 1,000 passes per hour were documented at several water developments. Some bat species are dependent upon surface water, and water developments may influence the distribution of these species (Schmidt and Dalton 1995). Researchers using isotopically labeled water sources in southwestern Arizona reported that developments provided 6% to 43% ($\bar{x} = 12\%$) of the total body water pool for eight species (Wolf 2010). Water developments are important tools for conservation of some bat species, particularly in areas where natural water sources no longer are available (Tuttle et al. 2006, Taylor and Tuttle 2007).

Birds
Upland game birds received the initial focus of wildlife water developments (Rosenstock et al. 1999). Water developments have been constructed for dove, quail, chukar, and wild turkey (*Meleagris gallopavo*). The responses of game birds and other avian species to water sources have varied.

Upland Game Birds
White-winged (*Zenaida asiatica*) and mourning dove (*Z. macroura*) frequently use water developments, particularly during the summer (O'Brien et al. 2006). When deprived of water, mourning doves lose almost 5% of body weight per day and therefore require access to drinking water (Bartholomew and MacMillen 1960). Both species likely have benefited from water developments in arid areas (Rosenstock et al. 1999). Hyde (2011) used stable-isotope techniques to assess avian use of water developments in southwestern Arizona. After accounting for preformed and metabolic water, more than 60% of the total body water pool of mourning and white-winged doves came from water developments (Hyde 2011). Movements and habitat use data in Idaho suggest that construction of water sources may enhance mourning dove populations (Howe and Flake 1988). Because doves can travel long distances to water (>11 km; Howe and Flake 1988), water might not be limiting in many areas. In the Mojave Desert of California, occupancy of mourning dove was strongly associated with wildlife water sources (Rich et al. 2019).

Studies on the influence of water developments on quail populations have yielded equivocal results. Gambel's (*C. gambellii*), scaled (*C. squamata*), and northern bobwhite quail (*C. virginianus*) use water developments during summer, particularly when succulent vegetation is not available (Hungerford 1962, Campbell et al. 1973, Guthery and Koerth 1992). Occupancy of Gambel's quail in the Mojave Desert is greater in areas near water developments (Rich et al. 2019). Scaled quail and northern bobwhite selected areas closer to water sources in Oklahoma (Tanner et al. 2015). Many quail populations obtain sufficient moisture from succulent forage, but quail populations in areas that commonly experience drought during the breeding season may benefit from the addition of water developments (Campbell 1961). Chronic water deprivation can lead to reproductive failure of northern bobwhite quail (Giuliano et al. 1994). Hyde (2011) reported that more than 43% of the total body water pool of Gambel's quail came from water developments, yet Tanner et al. (2015) did not find evidence that water sources affected scaled quail survival or nest success in Oklahoma.

Development of water sources is the primary habitat management activity for introduced chukar in Idaho, Nevada, and Utah (Benolkin and Benolkin 1994). In Nevada, addition of water sources increased existing chukar populations and facilitated new populations (Benolkin and Benolkin 1994). The use of water sources by chukar is inversely related to forage moisture content (Larsen et al. 2007). During the cooler seasons, chukar may meet water requirements from preformed water in forage but may need drinking water when succulent vegetation is unavailable (Degen et al. 1984). Distribution of chukar during summer is largely dependent on the availability of water sources but is also affected by forage moisture content (Christensen 1996). For example, the summer distribution of chukar in western Utah was associated with water sources in two areas when forage moisture content was under 45%, but not when forage moisture content exceeded 55% (Larsen et al. 2010). Chukar do not use all water sources built specifically for them; guzzlers in areas without adequate hiding cover were less likely to be used (Larsen et al. 2007).

Some consider water sources to be an essential habitat component for Merriam's turkey (*M. g. merriami*) in the Southwest. In areas where natural sources of water are lacking or are seasonally unavailable, development of water sources may benefit turkey populations (Shaw and Mollohan 1992). Management guidelines for Merriam's turkeys suggest that water sources should be regularly available throughout turkey habitat (Hoffman et al. 1993). Distribution of turkey in semiarid regions is associated with water, particularly during drought periods (Collier et al. 2017).

Nongame Birds

A variety of nongame birds, including passerines, raptors, shorebirds, and waterfowl, make use of water developments (Lynn et al. 2006, 2008; O'Brien et al. 2006). In Arizona, Cutler (1996) documented more than 150 bird species near water developments, 60 of which were observed drinking. The use of water developments by birds tends to be highest during the summer, but use has been observed during spring and autumn migration, suggesting that these water sources may be important as stopover locations, either as a source of drinking water or because associated vegetation served as foraging and resting areas during migrations (Rosenstock et al. 1999). Lynn et al. (2006, 2008) documented relatively low use by migrants, however, and suggested that the benefits to nongame birds were primarily to resident species. These results were supported by Hyde (2011), who reported that most neotropical migrants derived less than 10% of their body water pool from water developments.

Resident birds may benefit from water sources to a greater extent than migratory species, and they use them most often during the summer (O'Brien et al. 2006). Cutler (1996) reported that avian abundance and species richness were inversely related to the distance from water development in Arizona, but other investigators reported no difference in species richness or abundance at water developments compared with control sites (Burkett and Thompson 1994). In southwestern Arizona, the proportion of body water among resident species (excluding doves and quail) derived from water developments ranged from 10% to 58%; 9 of 15 species derived more than 25% of their body water pool from the developed waters, and 4 species derived more than 45% from these sources (Hyde 2011). The influence of water developments on survival or reproductive success of most passerines is unknown. Availability of supplemental water has been experimentally shown to increase clutch size of black-throated sparrows (*Amphispiza bilineata*), although it did not influence nest survival rates or the probability of fledging (Coe and Rotenberry 2003).

Raptors commonly use water developments in arid environments (O'Brien et al. 2006, Lynn et al. 2008), with evidence of use being greater at water developments than at control sites (Burkett and Thompson 1994, Schmidt and DeStefano 1996). Raptors use water sources for drinking and bathing, and may also capture birds and or small mammals gathered near those developments (O'Brien et al. 2006). Harris's hawks (*Parabuteo unicinctus*) require surface water during the breeding season and may have expanded their distribution following development of water sources (Dawson and Mannan 1991).

Reptiles and Amphibians

Desert-dwelling amphibians may also benefit from water developments. In Arizona and New Mexico, researchers have observed the Colorado River toad (*Incilius alvarius*), Great Plains toad (*Anaxyrus cognatus*), red-spotted toad (*A. punctatus*), green toad (*A. debilis*), leopard frog, Couch's spadefoot toad (*Scaphiopus couchii*), plains spadefoot toad (*Spea bombifrons*), Mexican spadefoot toad (*S. multiplicata*), and Sonoran tiger salamander (*Ambystoma mavortium*) breeding in various types of water developments, including tinajas, precipitation catchments, and developed springs (Griffis-Kyle et al. 2011, Boeing et al. 2014, Harings and Boeing 2014), and the geographic distribution of Woodhouse's toad (*Anaxyrus woodhousii*) in California has expanded with the development of canals (Bleich 2020). Some amphibians use stock tanks and other water developments as breeding habitat, including some species of conservation concern (e.g., leopard frogs in stock tanks; Rosen and Schwalbe 1998). Some water catchment designs result in inadequate flushing events, leading to accumulation of organic material; decomposition of organic materials increases ammonia concentration, which negatively influences physiological processes of anurans and could impair conservation efforts for these species (Griffis-Kyle et al. 2014). Reptiles that inhabit arid environments occasionally use water sources (O'Brien et al. 2006), although it is believed that most reptiles do not require drinking water (Mayhew 1968).

COVER

Wildlife species select habitat conditions that enhance fitness within the constraints imposed by the physical landscape, thermal conditions, and intra- and interspecific interactions. Thus far in the chapter, we have discussed the importance of water and how it influences the life history, ecology, and distribution of wildlife species. Species must also balance the availability of water with areas that provide them with adequate cover. Cover includes any landscape (e.g., vegetative, topographic, or anthropogenic) feature that conceals an individual, and protects them against harsh environmental conditions or minimizes adverse interspecific interactions. Concealment includes stalking cover for predators and hiding cover for prey. Vegetation cover can include vertical (i.e., canopy) and horizontal (i.e., ground) cover (Nudds 1977, Mysterud and Østbye 1999). Topographic features such as cliff overhangs, elevation gradients, and caves offer concealment and protection (Mysterud and Østbye 1999, Cain et al. 2008b). Anthropogenic features can have mixed effects on wildlife populations, as they increase the cover for some species while degrading habitat conditions for others. Common functional descriptions for cover include winter, thermal, loafing, nesting (Jones

and Hungerford 1972, Sveum et al. 1998), residual (Prose et al. 2002, Vodehnal et al. 2020), refuge, escape, and hiding cover (Gionfriddo and Krausman 1986, Mysterud and Østbye 1999). These generalized categories are sometimes useful but frequently overlook the complexity of the general and specific cover needs of wildlife.

Predation

Predation is often cited as being a primary limiting factor for prey populations (Boonstra et al. 1998, Cresswell 2011, Lukacs et al. 2018). Costs of predation can range from direct mortality to nonlethal effects, including changes in activity budgets, increased vigilance resulting in less efficient foraging, and shifts in habitat selection and or movements, all of which can contribute to increased energetic costs and reduced nutritional condition (Lima and Dill 1990, Creel et al. 2007, Cresswell 2008). Prey species should reduce these risks by occupying areas with adequate concealment or escape cover. Concealment cover can include any landscape feature that reduces detection by predators and is frequently used by individuals during times of inactivity (i.e., resting; Mysterud and Ostbye 1999, Camp et al. 2012). Escape cover includes features (e.g., rugged terrain) that allow prey to out-maneuver predators (Gionfriddo and Krausman 1986).

Species are faced with the challenge of selecting areas that offer protective cover while also providing access to forage and other resources. For example, desert bighorn sheep frequently select areas with intermediate vegetation biomass, suggest-

ing a tradeoff between selecting areas with low vegetation biomass and high visibility, allowing detection of predators, and avoiding areas with reduced foraging opportunities (Gedir et al. 2020). Small mammals have reduced mobility and smaller home ranges, and are vulnerable to a wider variety of predators; therefore they need to forage close to cover. Fox squirrels (*Sciurus niger*) will use open grass fields during autumn and spring when grass is taller but avoid these areas during summer and winter when grass is short and detection by predators is higher (McCleery et al. 2007). White-footed mice (*Peromyscus leucopus*) select travel paths that reduce both their visual and auditory detection (Barnum et al. 1992). When traveling in areas with abundant leaf litter (e.g., during autumn), mice will use downed logs, presumably to reduce the sounds of their movements (i.e., auditory cover), and they avoid areas with high grass density, which shows visual movement (Barnum et al. 1992). Waterfowl select wetlands that are composed of a mosaic of vegetation and water that provide areas for foraging and reduce the probability of detection by predators (Webb et al. 2010).

Life stage also plays an important role in cover selection, as individuals experience higher predation risk shortly after hatching or birth. Some ungulates, like desert bighorn sheep, select parturition sites with low vegetation cover and high visibility that aids in predator detection, but they sacrifice nutrient availability by doing so. Other ungulate species select areas with higher concealing cover for neonates, which may also be associated with a reduction in nutrient availability for the mother (Bleich et al. 1997, Bowyer et al. 1999, Pitman et al. 2014). Habitat selection in birds can influence brood survival (Signorell et al. 2010) through access to sufficient food resources and protection from both terrestrial and aerial nest predators (Sveum et al. 1998, Thompson 2007, Draycott et al. 2008). Sharp-tailed grouse (*Tympanuchus phasianellus*) rely on residual cover for nesting sites because it provides areas of dense, herbaceous cover even in periods of drought and heavy grazing (Vodehnal et al. 2020).

The availability and proximity of cover influence predation risk and affect species behavior. Species that rely on concealing cover as their primary means to avoid predation often increase vigilance, resulting in reduced foraging efficiency when they are farther from protective cover. Black-tailed jackrabbits reduce their food consumption (up to 50%) the farther they are from cover (Longland 1991). Larger species tend to be in larger groups and select habitat that increases their ability to detect predators and evade them (Jarman 1974, Dehn 1990, Hebblewhite et al. 2005). Foraging efficiency is influenced when they move to areas without cover and are forced to spend more time being vigilant (Lima and Dill 1990, Fortin et al. 2004, Creel et al. 2014).

Fleeing from predators requires energy and decreases foraging time (Ydenberg and Dill 1986). Animals need to find a balance between fleeing too early or too late. Flight distance depends primarily on the prey's ability to detect the predator or know when it has been detected by the predator, but it can also be influenced by availability of concealment cover, distance to nearest refuge, temperature (e.g., ectotherms), and frequency of encounters with predators (Ydenberg and Dill 1986, Cooper 2003, Broom and Ruxton 2005, Camp et al. 2012). Following a decision to flee, animals rely on topography and vegetation to aid in their escape. The choice of escape behavior varies among species. For example, white-tailed deer will use hiding cover until it is detected by a predator and then will use established paths within the vegetation in an effort to outrun their predator (Geist 1981). Mule deer escape predators by efficiently ascending steep hills and placing obstacles (e.g., downed logs, shrubs, rocks) between them and the predator (Geist 1981). Likewise, black-tailed jackrabbits will use vegetation cover (e.g., shrubs) to help conceal their movements and slow pursuing predators (Longland 1991). Bighorn sheep use steep slopes and rough terrain to aid in escaping predators (Gionfriddo and Krausman 1986, Bangs et al. 2005). While concealing cover provides protection to prey, it can also function as stalking cover for predators. Thus prey species must balance use of areas with sufficient concealment cover while also not inhibiting their ability to detect or evade predators (Embar et al. 2011, Camp et al. 2012).

Climate

Wildlife adapt their behaviors to offset the negative influences of weather and climate. Thermal cover consists of vegetation or other features that provide protection from snow and cold in winter and from the sun and heat during summer. In some cases, cover might not always be viewed in the traditional sense

of vegetation that provides shade or blocks wind and snow, as an open field can also provide thermal cover. For example, ungulates may use open fields at night during summer to help dissipate excess heat (Elmore et al. 2017) or to minimize heat loss and obtain radiant heat from the sun during the day in winter. Desert bighorn sheep select locations with lower solar radiation and north-facing slopes in summer and south-facing slopes in cooler seasons (Gedir et al. 2020). Therefore use of thermal cover often will vary seasonally depending on the species and their current life stage.

Overhead cover for ground-nesting birds reduces detection by predators but also provides cooler microhabitat for nestlings and increases fledgling survival (Carroll et al. 2015, 2016; Grisham et al. 2016). Prolonged exposure of eggs to temperatures above their critical temperature threshold influences body mass, tarsus length, and time of fledging in common fiscals (*Lanius collaris*), which has cascading effects on their future fitness (Cunningham et al. 2013). Bats also select roost sites based on temperature and select maternal roosts (e.g., tree cavity, cave, rock crevices) that offer warmer daily temperatures to allow for passive rewarming (i.e., use of the sun to rewarm them after torpor) during the day, thereby reducing their metabolic costs (Chruszcz and Barclay 2002, Law and Chidel 2007). Moose (*Alces alces*) use forest stands with greater tree density and taller canopies to reduce heat stress (Olson et al. 2014, Borowik et al. 2020). Forests with dense canopies can also have lower snow depth and increased forage availability for moose during winter (Dussault et al. 2005). Some species use topographic features (e.g., cliff overhangs, burrows, rock outcropping) to reduce thermal stress. American pika (*Ochotona princeps*) inhabit alpine tundra ecotones, a rapidly changing climatic environment. Pikas select rock crevices on talus slopes with internal temperatures ranging from 2°C to 5°C cooler than the surface; therefore talus slopes provide a buffer against warming temperatures and reduce metabolic requirements of pikas (Hall et al. 2016).

Anthropogenic Influences on Cover

Human use of landscapes has a direct influence on cover for wildlife. Changes in availability and quality of cover can occur via forest alterations through burning and logging or agriculture, infrastructure, or other anthropogenic developments. Although most of these activities can result in a drastic reduction in cover, when strategically implemented, some management actions can increase availability or quality of cover for wildlife. Several bird species are relatively successful in a managed forest if blocks of forest cover are 10 × 10 km or larger (Trzcinski et al. 1999). In other cases, however, anthropogenic activities result in a loss of beneficial cover. For example, grasslands that are not grazed by livestock are more productive sites for birds than those that are grazed, and the ungrazed sites support a higher abundance of granivorous birds (Bock and Bock 1999). Grazing management systems that explicitly consider the needs of local wildlife species and involve rotational grazing plans rather than continuous year-round grazing are more likely to contribute to wildlife conservation and management objectives (Krausman et al. 2009). Likewise, agriculture alters cover availability for small mammals. Some rodents are more abundant with dense nesting cover, have intermediate densities in areas where haying is delayed and confined to rights-of-way, and are at their lowest numbers in idle pastures where cover is minimal (Pasitschniak-Arts and Messier 1998).

Management of Cover

Effective management of cover depends on understanding successional processes and manipulating successional trajectories to meet management goals. Post-disturbance structural changes in plant communities via secondary succession depend on the severity and type of disturbance, the species composition of the area, and the duration of the post-disturbance recovery period. Under the classical view of succession, stages in grasslands go from bare soil, followed by annual grasses and forbs, perennial forbs, and short-lived perennial grasses, until they are replaced with sod-forming grass or bunch grass communities (Coupland 1992). Similarly, shrubland and woodland vegetation types advance from bare soil to grasses and forbs, followed by dominant shrubs and trees. This classical framework of succession can be misleading and often ignores the complexity that is occurring, as it assumes that pioneer plants facilitate the growth of the next successional stage, up to the climax community (Johnson and Miyanishi 2008). Although this sequence of vegetation regrowth may occasionally be observed, successional trajectories are often more associated with individual plant life-history characteristics and the surrounding ecosystem (Gleason 1926, Johnson and Miyanishi 2008). Specifically, the rate at which vegetation regenerates varies with soil type, moisture, shade tolerance, climate, topography, slope, and severity of interference (e.g., autogenic forces caused by flora and fauna or allogenic forces caused by such outside influences as fire and wind).

All successional stages represented via various patch shapes, sizes, and distributions, which in turn enhance species diversity and richness. These successional stages can be advanced or delayed by natural or human means (e.g., mechanical, fire, wind, insects, grazing, irrigation, planting, fertilizing; Thomas et al. 1979). Leopold (1933) referred to management practices such as planting, fencing against livestock, and protecting areas from fire as tools that aided in speeding up succession, whereas plowing, burning, grazing, and cutting were tools to set back succession. It is important to note that these tools are not unique to wildlife management nor do they only influence wildlife, as each management action has wide-ranging influences. Therefore, when attempting to manage cover, one must coordinate activities with other land management organizations (e.g., US Forest Service, Bureau of Land Management) that influence habitat conditions for wildlife.

Managing for cover becomes even more complex because resource needs differ among species. For example, the structural diversity of mixed-age forests provides cover for species that use the edge of forests and forest interiors. Species with higher densities along ecotones (e.g., the forest edge) could benefit from some degree of fragmentation, but that increase would be at the expense of species more dependent on the forest interior or those that have more homogeneous cover requirements (e.g., salamanders; Herbeck and Larsen 1999). Reduction of old-growth forest reduces microhabitat availability; by cutting forests, habitat for edge-dependent species could be enhanced, reduced for species depending on old growth interiors, or would likely have minimal influences on generalists (Whitcomb et al. 1981).

In some situations, considerable expenditures of time, effort, and money are allocated to the management of cover for ungulates and other wildlife. Over the past six decades, wildlife biologists managed vegetation for ungulates as a mechanism to enhance survival in cold climates. "Thermal cover has been credited widely with moderating the effects of harsh weather, and therefore, may improve overall performance of populations (i.e., survival and reproduction) by reducing energy expenditures required for thermostasis" (Cook et al. 1998:6). This concern for adequate winter cover led to the widespread belief that winter thermal cover was a key component of ungulate habitat in the western United States. As a result, winter thermal cover was incorporated into the development of large-scale national forest plans, and management agencies made numerous site-specific, case-by-case decisions as to how to harvest forests or to prescribe fire based on the perceived view that winter cover was limiting for ungulates (Edge et al. 1990, Cook et al. 1998). In reality, winter thermal cover had little influence on the herd productivity and demographics of elk, as determined from a series of controlled experiments investigating how cover is used by wildlife (Cook et al. 1998). Prior to discovering the true extent of wildlife use of winter thermal cover, it was managed for elk on millions of hectares and at a high cost. More effort could be placed on forage quality and quantity, and the ability to evaluate forage conditions across the landscape, instead of managing for winter thermal cover for elk (Cook et al. 1996, 1998).

Land management agencies have started to shift away from a strict fire suppression paradigm to fire management. Managing fires for resource benefits can increase understory growth, change stand structures in forest and chaparral systems, and reduce fuel loads (Donovan and Brown 2005, Holl et al. 2012). The effectiveness of using fire for management is species dependent (especially in small mammals), and the benefits vary depending on prefire habitat conditions and the postfire recovery time (Converse et al. 2006, Zwolak 2009). Researchers in Oregon reported that prescribed fire reduced Wyoming big sagebrush (*Artemisia tridentata wyomingensis*) by 95%, reducing escape and nesting cover for sage grouse (*Centrocercus urophasianus*) for several years (Rhodes et al. 2010). Fires can also improve habitat conditions for other species. Fires increase habitat quality for bighorn sheep in chaparral and coastal sage systems in Southern California (Bleich et al. 2019). Prescribed fire is associated with mule deer habitat selection in New Mexico, where mule deer displayed stronger selection for more recent prescribed burns (Roerick et al. 2019). When using fire to manage cover and other habitat conditions, it is critical to consider habitat needs of multiple wildlife species and the short- and long-term effects on the vegetation community.

SPECIAL FACTORS

Welfare factors that are important to a particular species, usually in small quantities and for short periods, have been called special factors (Leopold 1933). Well-known special factors include "gravel for gallinaceous birds and waterfowl, salt licks for herbivores and some birds, mineral springs for pigeons, dust baths for various birds, mud baths and hibernation places for bear, caves or dense shade for desert bighorn sheep and quail to reduce water loss during the heat of the day in arid climates, open wind-swept parks or deep water for the relief of moose and deer in fly season, and sandy knolls for booming grounds of prairie chickens" (Leopold 1933:27). Examples of special factors include lambing areas with adequate protection for desert bighorn sheep, lekking areas for grouse, breeding knolls for swamp deer (*Cervus duvauceli*), and mineral licks that provide critically important micronutrients and vitamins. Distribution of these and additional but yet unrecognized special factors likely has a profound influence on the geographic distribution of wildlife species (Leopold 1933).

SUMMARY

Water and cover are essential for all wildlife species, and many species also require special factors. Wildlife that inhabit arid areas use a combination of physiological, morphological, and behavioral adaptations to avoid or cope with limited water availability. Animals obtain water from three sources: preformed, metabolic, and free water. When preformed and metabolic water are insufficient to meet water demands, animals must have access to free water, either naturally occurring or developed. Wildlife managers have devoted substantial time and resources to developing wildlife water sources in areas where natural sources are scarce or unreliable. The use of water developments likely varies among species, and construction of these water sources has not always resulted in the predicted outcomes. Some species have likely benefitted from water developments, whereas questions remain for others. But information on population-level responses to water developments remains inadequate.

Vegetation and topographic features provide cover for wildlife. Hiding cover, nesting cover, roosting cover, and escape cover are important to the life history of most species. In addition, winter thermal cover provides protection from the snow and cold, whereas summer thermal cover (e.g., shade) provides protection from high summer temperatures and solar radiation, particularly in deserts. Management agencies have spent considerable time, effort, and money attempting to manage or enhance cover for ungulates and other wildlife, and, as noted with respect to water development, the efficacy of those efforts requires further evaluation.

Special factors are welfare factors that are important for wildlife but typically are important in small quantities and or for short periods. Well-known special factors include gravel for gallinaceous birds and waterfowl, salt licks for herbivores, parturition sites, and hibernacula.

The assumption that water developments are essential to wildlife in arid regions needs to be assessed rigorously using experimental studies, especially as funds for environmental enhancement or conservation become scarce. Studies that exemplify the importance of water and cover or identify special factors necessary for wildlife populations will contribute to management and wise use of available resources aimed at sustaining wildlife populations.

Literature Cited

Abeloe, T. N., and P. C. Hardy. 1997. Western screech-owls diurnally roosting in a cave. Southwestern Naturalist 42:349–351.

Abraham, J. O., G. P. Hempson, and A. C. Staver. 2019. Drought-response strategies of savanna herbivores. Ecology and Evolution 9:7047–7056.

Alderman, J. A., P. R. Krausman, and B. D. Leopold. 1989. Diel activity of female desert bighorn sheep in western Arizona. Journal of Wildlife Management 53:264–271.

Alexander, P., L. Brekke, G. Davis, S. Gangopadhyay, K. Grantz, et al. 2011. SECURE Water Act Section 9503(c)—reclamation climate change and water. Report to Congress. US Bureau of Reclamation, Denver, Colorado, USA.

Allen, R. W. 1980. Natural mortality and debility. Pages 172–185 in G. Monson and L. Sumner, editors. The desert bighorn. University of Arizona Press, Tucson, USA.

Andrew, M. H. 1988. Grazing impact in relation to livestock watering points. Trends in Ecology and Evolution 12:336–339.

Andrew, N. G., V. C. Bleich, A. D. Morrison, L. M. Lesicka, and P. J. Cooley. 2001. Wildlife mortalities associated with artificial water sources. Wildlife Society Bulletin 29:175–280.

Andrew, N. G., L. M. Lesicka, and V. C. Bleich. 1997. An improved fence design to protect water sources for native ungulates. Wildlife Society Bulletin 25:823–825.

Arad, Z., I. Gavrieli-Levin, U. Eylath, and G. Marder. 1987. Effect of dehydration on cutaneous water evaporation in heat-exposed pigeons (*Columba livia*). Physiological Zoology 60:623–630.

Arjo, W. M., E. M. Gese, T. J. Bennett, and A. J. Kozlowski. 2007. Changes in kit fox-coyote prey relationships in the Great Basin Desert, Utah. Western North American Naturalist 67:389–401.

Atwood, T. C., T. L. Fry, and B. R. Leland. 2011. Partitioning of anthropogenic watering sites by desert carnivores. Journal of Wildlife Management 75:1609–1615.

Baber, D. W. 1983. Mortality in California mule deer at a drying reservoir: the problem of siltation at water catchments. California Fish and Game 70:248–251.

Baker, M. A. 1989. Effects of dehydration and rehydration on thermoregulatory sweating in goats. Journal of Physiology 417:421–435.

Ballard, W. B., S. S. Rosenstock, and J. C. DeVos Jr. 1998. The effects of artificial water developments on ungulates and large carnivores in the Southwest. Pages 64–105 in Proceedings of a symposium on environmental, economic, and legal issues related to rangeland water developments, 13–15 November 1997, Tempe, Arizona. Center for Law, Science, and Technology, Arizona State University, Tempe, USA.

Bangs, P. D., P. R. Krausman, K. E. Kunkel, and Z. D. Parsons. 2005. Habitat use by desert bighorn sheep during lambing. European Journal of Wildlife Research 51:178–184.

Barnum, S. A., C. J. Manville, J. R. Tester, and W. J. Carmen. 1992. Path selection by *Peromyscus leucopus* in the presence and absence of vegetative cover. Journal of Mammalogy 73:797–801.

Bartholomew, G. A., 1964. The roles of physiology and behavior in the maintenance of homeostasis in desert environments. Symposium of the Society of Experimental Biology 18:7–29.

Bartholomew, G. A., and T. J. Cade. 1963. The water economy of land birds. Auk 80:504–539.

Bartholomew, G. A., and R. E. MacMillen. 1960. The water requirements of mourning doves and their use of sea water and NaCl solutions. Physiological Journal 33:171–178.

Beale, D. M., and R. C. Holmgren. 1975. Water requirements for pronghorn antelope fawn survival and growth. Utah Division of Wildlife Resources, Salt Lake City, USA.

Beale, D. M., and A. D. Smith. 1970. Forage use, water consumption, and productivity of pronghorn antelope in western Utah. Journal of Wildlife Management 34:570–582.

Beever, E. 2003. Management implications of the ecology of free-roaming horses in semi-arid ecosystems of the western United States. Wildlife Society Bulletin 31:887–895.

Benolkin, P. J., and A. C. Benolkin. 1994. Determination of a cost–benefit relationship between chukar populations, hunter utilization and the cost of artificial watering devices. Nevada Department of Wildlife, Reno, USA.

Berger, J. 1985. Interspecific interactions and dominance among wild Great Basin ungulates. Journal of Mammalogy 66:571–573.

Berry, H. H., W. R. Siegfried, and T. M. Crowe. 1984. Orientation of wildebeest in relation to sun angle and wind direction. Madoqua 13:297–301.

Bleich, V. C. 2003. The potential for botulism in desert-dwelling mountain sheep. Desert Bighorn Council Transactions 47:2–8.

Bleich, V. C. 2005. Politics, promises, and illogical legislation confound wildlife conservation. Wildlife Society Bulletin 33:66–73.

Bleich, V. C. 2008. Reprovisioning wildlife water developments: considerations for determining priorities to transport water. ESCAPE Technical Report 2008-01. Eastern Sierra Center for Applied Population Ecology, Bismarck, North Dakota, USA.

Bleich, V. C. 2009. Factors to consider when reprovisioning water developments used by mountain sheep. California Fish and Game 95:153–159.

Bleich, V. C. 2020. Locality records for Woodhouse's toad: have wet washes in a dry desert led to extralimital occurrences of an adaptable anuran? California Fish and Wildlife Journal 106:258–266.

Bleich, V. C., N. G. Andrew, M. J. Martin, G. P. Mulcahy, A. M. Pauli, and S. S. Rosenstock. 2006. Quality of water available to wildlife in desert environments: comparisons among anthropogenic and natural sources in southeastern California. Wildlife Society Bulletin 34:625–630.

Bleich, V. C., M. E. Blum, K. T. Shoemaker, D. Sustaita, and S. A. Holl. 2019. Habitat selection by bighorn sheep in a mesic ecosystem: the San Rafael Mountains, California, USA. California Fish and Game 105:205–224.

Bleich, V. C., R. T. Bowyer, and J. D. Wehausen. 1997. Sexual segregation in mountain sheep: resources or predation? Wildlife Monographs 134:3–50.

Bleich, V. C., J. Coombes, and J. H. Davis. 1982. Horizontal wells as a wildlife habitat improvement technique. Wildlife Society Bulletin 10:324–328.

Bleich, V. C., J. P. Marshal, and N. G. Andrew. 2010. Habitat use by a desert ungulate: predicting effects of water availability on mountain sheep. Journal of Arid Environments 74:638–645.

Bleich, V. C., M. W. Oehler, and J. G. Kie. 2020. Managing rangelands for wildlife. Pages 126–148 *in* N. J. Silvy, editor. The wildlife management techniques manual. Volume 2: management. 8th edition. Johns Hopkins University Press, Baltimore, Maryland, USA.

Bleich, V. C., B. M. Pierce, J. L. Davis, and V. L. Davis. 1996. Thermal characteristics of mountain lion dens. Great Basin Naturalist 56:276–278.

Bleich, V. C., and R. A. Weaver. 1983. "Improved" sand dams for wildlife habitat management. Journal of Range Management 36:133.

Blong, B., and W. Pollard. 1968. Summer water requirements of desert bighorn in the Santa Rosa Mountains, California, in 1965. California Fish and Game Journal 54:289–296.

Bock, C. E., and J. H. Bock. 1999. Response of winter birds to drought and short-duration grazing in southeastern Arizona. Conservation Biology 13:1117–1123.

Boeing, W. J., K. L. Griffis-Kyle, and J. M. Jungels. 2014. Anuran habitat associations in the northern Chihuahuan Desert, USA. Journal of Herpetology 48:103–110.

Boonstra, R., C. J. Krebs, and N. C. Stenseth. 1998. Population cycles in small mammals: the problem of explaining the low phase. Ecology 79:1479–1488.

Borowik, T., M. Ratkiewicz, W. Maślanko, N. Duda, and R. Kowalczyk. 2020. Too hot to handle: summer space use shift in a cold-adapted ungulate at the edge of its range. Landscape Ecology 35:1341–1351.

Bowyer, R. T. 1986. Habitat selection by southern mule deer. California Fish and Game 72:153–169.

Bowyer, R. T., V. Van Ballenberghe, J. G. Kie, and J. A. Maier. 1999. Birth-site selection by Alaskan moose: maternal strategies for coping with a risky environment. Journal of Mammalogy 80:1070–1083.

Boyle, S. A., and A. W. Alldredge. 1984. Pronghorn summer distribution and water availability in the Red Desert, Wyoming. Proceedings of the Biennial Pronghorn Antelope Workshop 11:103–104.

Bradley, W. G. 1963. Water metabolism in desert mammals with special reference to desert bighorn sheep. Desert Bighorn Council Transactions 7:26–39.

Brinnel, H., M. Cabanac, and J. R. S. Hales. 1987. Critical upper levels of body temperature, tissue thermosensitivity, and selective brain cooling in hyperthermia. Pages 209–240 *in* J. R. S. Hales, and D. A. B. Richards, editors. Heat stress: physical exertion and environment. Excerpta Medica, Amsterdam, Netherlands.

Bristow, K. D., S. A. Dubay, and R. A. Okenfels. 2006. Correlation between free water availability and pronghorn recruitment. Pages 55–62 *in* J. W. Cain III and P. R. Krausman, editors. Managing wildlife in the Southwest. The Wildlife Society, Tucson, Arizona, USA.

Brooks, M. L., J. R. Matchett, and K. H. Berry. 2006. Effects of livestock water sites on alien and native plants in the Mojave Desert, USA. Journal of Arid Environments 67 (supplement):125–147.

Broom, M., and G. D. Ruxton. 2005. You can run—or you can hide: optimal strategies for cryptic prey against pursuit predators. Behavioral Ecology 16:534–540.

Broyles, B. 1995. Desert wildlife water developments: questioning use in the Southwest. Wildlife Society Bulletin 23:663–675.

Broyles, B., and T. L. Cutler. 1999. Effect of surface water on desert bighorn sheep in the Cabeza Prieta National Wildlife Refuge, southwestern Arizona. Wildlife Society Bulletin 27:1082–1088.

Buechner, H. K. 1950. Life history, ecology, and range use of the pronghorn antelope in Trans-Pecos, Texas. American Midland Naturalist 43:257–354.

Burkett, D. W., and B. C. Thompson. 1994. Wildlife association with human-altered water sources in semiarid vegetation communities. Conservation Biology 8:682–690.

Bush, A. P. 2015. Mule deer demographics and parturition site selection: assessing responses to provision of water. MS thesis. University of Nevada, Reno, USA.

Cain, J. W., III, J. V. Gedir, J. P. Marshal, P. R. Krausman, J. D. Allen, G. C. Duff, B. Jansen, and J. R. Morgart. 2017. Extreme precipitation variability, forage quality and large herbivore diet selection in arid environments. Oikos 126:1459–1471.

Cain, J. W., III, P. R. Krausman, J. R. Morgart, B. D. Jansen, and M. P. Pepper. 2008a. Responses of desert bighorn sheep to removal of water sources. Wildlife Monographs 171:1–32.

Cain, J. W., III, B. D. Jansen, R. R. Wilson, and P. R. Krausman. 2008b. Potential thermoregulatory advantages of shade use by desert bighorn sheep. Journal of Arid Environments 72:1518–1525.

Cain, J. W., III, P. R. Krausman, J. R. Morgart, B. D. Jansen, and M. P. Pepper. 2008a. Responses of desert bighorn sheep to removal of water sources. Wildlife Monographs 171:1–32.

Cain, J. W., III, P. R. Krausman, and S. S. Rosenstock. 2013. Water and other welfare factors. Pages 174–194 in P. R. Krausman and J. W. Cain III, editors. Wildlife management: contemporary principles and practices. The Wildlife Society and Johns Hopkins University Press, Baltimore, Maryland, USA.

Cain, J. W., III, P. R. Krausman, S. S. Rosenstock, and J. C. Turner. 2006. Mechanisms of thermoregulation and water balance in desert ungulates. Wildlife Society Bulletin 34:570–581.

Camp, M. J., J. L. Rachlow, B. A. Woods, T. R. Johnson, and L. A. Shipley. 2012. When to run and when to hide: the influence of concealment, visibility, and proximity to refugia on perceptions of risk. Ethology 118:1010–1017.

Campbell, H. 1961. An evaluation of gallinaceous guzzlers for quail in New Mexico. Journal of Wildlife Management 24:21–26.

Campbell, H., D. K. Martin, P. E. Ferkovich, and B. K. Harris. 1973. Effects of hunting and some other environmental factors on scaled quail in New Mexico. Wildlife Monographs 34:1–49.

Carmi, N., B. Pinshow, M. Horowitz, and M. H. Bernstein. 1993. Birds conserve plasma volume during thermal and flight-incurred dehydration. Physiological Zoology 66:829–846.

Carmi-Winkler, N., A. A. Degen, and B. Pinshow. 1987. Seasonal time-energy budgets of free-living chukars in the Negev Desert. Condor 89:594–601.

Carroll, J. M., C. A. Davis, R. D. Elmore, and S. D. Fuhlendorf. 2015. A ground-nesting galliform's response to thermal heterogeneity: implications for ground-dwelling birds. PLoS ONE 10:1–20.

Carroll, J. M., C. A Davis, S. D. Fuhlendorf, and R. D. Elmore. 2016. Landscape pattern is critical for the moderation of thermal extremes. Ecosphere 7:1–16.

Chappell, M. A., and G. A. Bartholomew. 1981. Standard operative temperatures and thermal energetics of the antelope ground squirrel Ammospermophilus leucurus. Physiological Zoology 54:81–93.

Chevalier, C. D. 1984. Water requirements of free-ranging and captive ringtail cats in the Sonoran Desert. PhD thesis, Arizona State University, Tempe, USA.

Christensen, G. C. 1996. Chukar (Alectoris chukar). The birds of North America. No. 258. Academy of Natural Sciences, Philadelphia, Pennsylvania, and the American Ornithologists Union, Washington, DC, USA.

Chruszcz, B. J., and R. M. R Barclay. 2002. Thermoregulatory ecology of a solitary bat, Myotis evotis, roosting in rock crevices. Functional Ecology 16:18–26.

Clemente, F., R. Valdez, J. L. Holechek, P. J. Zwank, and M. Cardenas. 1995. Pronghorn home range relative to permanent water in southern New Mexico. Southwestern Naturalist 40:38–41.

Coe, S. J., and J. T. Rotenberry. 2003. Water availability affects clutch size in a desert sparrow. Ecology 84:3240–3249.

Collier, B. A., J. D. Guthrie, J. B. Hardin, and K. L. Skow. 2017. Movements and habitat selection of male Rio Grande wild turkey during drought in south Texas. Journal of the Southeast Association of Fish and Wildlife Agencies 4:94–99.

Converse, S. J., G. C. White, K. L. Farris, and S. Zack. 2006. Small mammals and forest fuel reduction: national-scale responses to fire and fire surrogates. Ecological Applications 16:1717–1729.

Cook, J. G., L. L. Irwin, L. D. Bryant, R. A. Riggs, and J. W. Thomas. 1998. Relations of forest cover and condition of elk: a test of the thermal cover hypothesis in summer and winter. Wildlife Monographs 141:1–61.

Cook, J. G., B. K. Johnson, R. C. Cook, R. A. Riggs, T. Delcurto, L. D. Bryant, and L. L. Irwin. 2004. Effects of summer-autumn nutrition and parturition date on reproduction and survival of elk. Wildlife Monographs 155:1–61.

Cook, J. G., L. J. Quinlan, L. L. Irwin, L. D. Bryant, R. A. Riggs, and J. W. Thomas. 1996. Nutrition–growth relations of elk calves during late summer and fall. Journal of Wildlife Management 60:528–541.

Cooper, W. E., Jr. 2003. Risk factors affecting escape behavior by the desert iguana, Dipsosaurus dorsalis: speed and directness of predator approach, degree of cover, direction of turning by a predator, and temperature. Canadian Journal of Zoology 81:979–984.

Coupland, R. T., editor. 1992. Ecosystems of the world: 8A. Natural grasslands—introduction and Western Hemisphere. Elsevier, New York, New York, USA.

Creel, S., D. Christianson, S. Liley, and J. A. Winnie. 2007. Predation risk affects reproductive physiology and demography of elk. Science 315:960.

Creel, S., P. Schuette, and D. Christianson. 2014. Effects of predation risk on group size, vigilance, and foraging behavior in an African ungulate community. Behavioral Ecology 25:773–784.

Cresswell, W. 2008. Non-lethal effects of predation in birds. Ibis 150:3–17.

Cresswell, W. 2011. Predation in bird populations. Journal of Ornithology 152:251–263.

Cunningham, S. J., R. O. Martin, C. L. Hojem, and P. A. Hockey. 2013. Temperatures in excess of critical thresholds threaten nestling growth and survival in a rapidly-warming arid savanna: a study of common fiscals. PLoS One 8:e74613.

Cutler, T. L. 1996. Water use of two artificial water developments on the Cabeza Prieta National Wildlife Refuge, southwestern Arizona. MS thesis, University of Arizona, Tucson, USA.

Cutler, T. L., and M. L. Morrison. 1998. Habitat use by small vertebrates at two water developments in southwestern Arizona. Southwestern Naturalist 43:155–162.

Czech, B., and P. R. Krausman. 1999. Controversial wildlife management issues in southwestern U.S. wilderness. International Journal of Wilderness 5:22–28.

Davidson, Z., M. Valeix, F. Van Kesteren, A. J. Loveridge, J. E. Hunt, F. Murindagomo, and D. W. Macdonald. 2013. Seasonal diet and prey preference of the African lion in a waterhole-driven semi-arid savanna. PLoS One 8:e0055182.

Dawson, W. R. 1982. Evaporative losses of water by birds. Comparative Biochemistry and Physiology A 71:495–509.

Dawson, W. R. 1984. Physiological studies of desert birds: present and future considerations. Journal of Arid Environments 7:133–155.

Dawson, J. W., and R. W. Mannan. 1991. The role of territoriality in the social organization of Harris' hawks. Auk 108:661–672.

Deblinger, R. D., and A. W. Alldredge. 1991. Influence of free water on pronghorn distribution in a sagebrush/steppe grassland. Wildlife Society Bulletin 19:321–326.

de Boer, W. F., M. J. Vis, H. J. De Knegt, C. Rowles, E. M. Kohi, F. Van Langevelde, and S. E. Van Wieren. 2010. Spatial distribution of lion kills determined by the water dependency of prey species. Journal of Mammalogy 91:1280–1286.

Degen, A. A., B. Pinshow, and P. U. Alkon. 1983. Summer water turnover rates in free-living chukars and sand partridges in the Negev Desert. Condor 85:333–337.

Degen, A. A., B. Pinshow, and P. J. Shaw. 1984. Must desert chukars (*Alectoris chukar sinaica*) drink water? Water influx and body mass changes in response to dietary water content. Auk 101:47–52.

Dehn, M. M. 1990. Vigilance for predators: detection and dilution effects. Behavioral Ecology and Sociobiology 26:337–342.

Delgiudice, G. D., and J. E. Rodiek. 1984. Do elk need free water in Arizona? Wildlife Society Bulletin 12:142–146.

DeStefano, S., S. L. Schmidt, and J. C. deVos Jr. 2000. Observations of predator activity at wildlife water developments in southern Arizona. Journal of Range Management 53:255–258.

deVos, J. C., Jr., and W. H. Miller. 2005. Habitat use and survival of Sonoran pronghorn in years with above-average precipitation. Wildlife Society Bulletin 33:35–42.

Donovan, G. H., and T. C. Brown. 2005. An alternative incentive structure for wildfire management on national forest land. Forest Science 51:387–395.

Douglas, C. L. 1988. Decline of desert bighorn sheep in the Black Mountains of Death Valley. Desert Bighorn Council Transactions 25:36–38.

Douglas, C. L., and D. M. Leslie Jr. 1999. Management of bighorn sheep. Pages 238–262 in R. Valdez and P. R. Krausman, editors. Mountain sheep of North America. University of Arizona Press, Tucson, USA.

Draycott, R. A., A. N. Hoodless, M. I. Woodburn, and R. B. Sage. 2008. Nest predation of common pheasants *Phasianus colchicus*. Ibis 150:37–44.

Dunn, W. C., and C. L. Douglas. 1982. Interactions between desert bighorn sheep and feral 601 burros at spring areas in Death Valley. Desert Bighorn Council Transactions 26:87–96.

Dussault, C., J. P. Ouellet, R. Courtois, J. Huot, L. Breton, and H. Jolicoeur. 2005. Linking moose habitat selection to limiting factors. Ecography 28:619–628.

Edge, W. D., S. L. Olson-Edge, and L. L. Irwin. 1990. Planning for wildlife in national forests: elk and mule deer habitats as an example. Wildlife Society Bulletin 18:87–98.

Edney, E. B. 1977. Water balance in land arthropods. Springer-Verlag, Berlin, Germany.

Eldridge, D. J., J. Ding, and S. K. Travers. 2020. Feral horse activity reduces environmental quality in ecosystems globally. Biological Conservation 241:108367.

Elmore, R. D., J. M. Carroll, E. P. Tanner, T. J. Hovick, B. A. Grisham, S. D. Fuhlendorf, and S. K. Windels. 2017. Implications of the thermal environment for terrestrial wildlife management. Wildlife Society Bulletin 41:183–193.

Embar, K., B. P. Kotler, and S. Mukherjee. 2011. Risk management in optimal foragers: the effect of sightlines and predator type on patch use, time allocation, and vigilance in gerbils. Oikos 120:1657–1666.

Epps, C. W., D. R. McCullough, J. D. Wehausen, V. C. Bleich, and J. L. Rechel. 2004. Effects of climate change on population persistence of desert-dwelling mountain sheep in California. Conservation Biology 18:102–113.

Finch, V. A. 1972. Thermoregulation and heat balance of the East African eland and hartebeest. American Journal of Physiology 222:1374–1379.

Finch, V. A., and D. Robertshaw. 1979. Effect of dehydration on thermoregulation in eland and hartebeest. American Journal of Physiology 6:R192–196.

Fortin, D., M. S. Boyce, E. H. Merrill, and J. M. Fryxell. 2004. Foraging costs of vigilance in large mammalian herbivores. Oikos 107:172–180.

Fox, L. M., P. R. Krausman, M. L. Morrison, and R. M. Kattnig. 2000. Water and nutrient content of forage in Sonoran pronghorn habitat, Arizona. California Fish and Game 86:216–232.

Fuller, A., P. R. Kamerman, S. K. Maloney, A. Matthee, G. Mitchell, and D. Mitchell. 2005. A year in the thermal life of a free-ranging herd of springbok *Antidorcas marsupialis*. Journal of Experimental Biology 208:2855–2864.

Fuller A., L. C. R. Meyer, D. Mitchell, and S. K. Maloney. 2007. Dehydration increases the magnitude of selective brain cooling independently of core temperature in sheep. American Journal of Physiology Regulatory, Integrative and Comparative Physiology 293:438–46.

Gedir, J. V., J. W. Cain III, P. R. Krausman, J. D. Allen, G. C. Duff, and J. R. Morgart. 2016. Potential foraging decisions by a desert ungulate to balance water and nutrient intake in a water-stressed environment. PLoS One 11:e0148795.

Gedir, J. V., J. W. Cain III, T. L. Swetnam, P. R. Krausman, and J. R. Morgart. 2020. Extreme drought and adaptive resource selection by a desert mammal. Ecosphere 11:e03175.

Geist, V. 1981. Behavior: adaptive strategies in mule deer. Pages 156–223 in O. C. Wallmo, editor. Mule and black-tailed deer of North America. Lincoln: University of Nebraska Press.

Gill, F. B. 1994. Ornithology. 2nd edition. W. H. Freeman, New York, New York, USA.

Gionfriddo, J. P., and P. R. Krausman. 1986. Summer habitat use by mountain sheep. Journal of Wildlife Management 50:331–336.

Giotto, N., D. Picot, M.-L. Maublanc, and J.-F. Gerard. 2013. Effects of seasonal heat on the activity rhythm, habitat use, and space use of the beira antelope in southern Djibouti. Journal of Arid Environments 89:5–12.

Giuliano, W. M., R. S. Lutz, and R. Patiño. 1994. Physiological responses of northern bobwhite (*Colinus virginianus*) to chronic water deprivation. Physiological Zoology 68:262–276.

Glading, B. 1943. A self-filling quail watering device. California Fish and Game Journal 29:157–164.

Gleason, H. A. 1926. The individualistic concept of the plant association. Bulletin of the Torrey botanical club 53:7–26.

Goldstein, D. L. 1984. The thermal environment and its constraint on activity of desert quail in summer. Auk 101:542–550.

Goldstein, D. L., and K. A. Nagy. 1985. Resource utilization by desert quail: time and energy, food and water. Ecology 66:378–387.

Golightly, R. T., and R. D. Ohmart. 1984. Water economy of two desert canids: coyote and kit fox. Journal of Mammalogy 65:51–58.

Gooch, A. M. J., S. L. Petersen, G. H. Collins, T. S. Smith, B. R McMillan, and D. L. Eggert. 2017. The impacts of feral horses on pronghorn behavior at water sources. Journal of Arid Environments 138:38–43.

Griffis-Kyle, K. L., J. J. Kovatch, and C. Bradatan. 2014. Water quality: a hidden danger in anthropogenic desert catchments. Wildlife Society Bulletin 38:148–151.

Griffis-Kyle, K. L., S. Kyle, and J. Jungels. 2011. Use of breeding sites by arid-land toads in rangelands: landscape level factors. Southwestern Naturalist 56:251–255.

Grisham, B. A., A. J. Godar, C. W. Boal, and D. A Haukos. 2016. Interactive effects between nest microclimate and nest vegetation structure confirm microclimate thresholds for lesser prairie-chicken nest survival. Condor 118:728–746.

Guthery, F. S., and N. E. Koerth. 1992. Substandard water intake and inhibition of bobwhite reproduction during drought. Journal of Wildlife Management 56:760–768.

Hailey, T. L. 1967. Reproduction and water utilization of Texas transplanted desert bighorn sheep. Desert Bighorn Council Transactions 11:53–58.

Hall, L. E., A. D. Chalfoun, E. A. Beever, and A. E. Loosen. 2016. Microrefuges and the occurrence of thermal specialists: implications for wildlife persistence amidst changing temperatures. Climate Change Responses 3:1–12.

Hall, L. K., R. T. Larsen, R. N. Knight, K. D. Bunnell, and B. R. McMillan. 2013. Water developments and canids in two North American Deserts: A test of the indirect effect of water hypothesis. PLoS One 8:e67800.

Hall, L. K., R. T. Larsen, R. N. Knight, and B. R. McMillan. 2018a. Feral horses influence both spatial and temporal patterns of water use by native ungulates in a semi-arid environment. Ecosphere 9:e02096.

Hall, L. K., R. T. Larsen, M. D. Westover, C. C. Day, R. N. Knight, and B. R. McMillan. 2018b. Influence of exotic horses on the use of water by communities of native wildlife in a semi-arid environment. Journal of Arid Environments 127:100–105.

Halloran, A. F., and O. V. Deming. 1958. Water development for desert bighorn sheep. Journal of Wildlife Management 22:1–9.

Hanley, T. A., and W. W. Brady. 1977. Feral burro impact on a Sonoran Desert range. Journal of Range Management 30:374–377.

Harings, N. M., and W. J. Boeing. 2014. Desert anuran occurrence and detection in artificial breeding habitats. Herpetologica 70:123–134.

Harris, G., J. G. Sanderson, J. Erz, S. E. Lehnen, and M. J. Butler. 2015. Weather and prey predict mammals' visitation to water. PLoS One 10:e0141355.

Harris, G. M., D. R. Stewart, D. Brown, L. Johnson, J. Sanderson, A. Alvidrez, T. Waddell, and R. Thompson. 2020. Year-round water management for desert bighorn sheep corresponds with visits by predators not bighorn sheep. PLoS One15:e0241131.

Hazam, J. E., and P. R. Krausman. 1988. Measuring water consumption of desert mule deer. Journal of Wildlife Management 52:528–534.

Hebblewhite, M., E. H. Merrill, and T. L. McDonald. 2005. Spatial decomposition of predation risk using resource selection functions: an example in a wolf–elk predator–prey system. Oikos 111:101–111.

Heffelfinger, J. 2006. Deer of the Southwest. Texas A&M University Press, College Station, USA.

Heffelfinger, L. J., K. M Stewart, A. P. Bush, J. S. Sedinger, N. W. Darby, and V. C. Bleich. 2018. Timing of precipitation in an arid environment: effects on population performance of a large herbivore. Ecology and Evolution 8:3354–3366.

Hennig, J. D., J. L. Beck, C. J. Gray, and J. D. Scasta. 2021. Temporal overlap among feral horses, cattle and native ungulates at water sources. Journal of Wildlife Management 85:1084–1090.

Henry, R. S., and L. K. Sowls. 1980. White-tailed deer of the Organ Pipe Cactus National Monument, Arizona. Technical Report No. 6. National Park Service and University of Arizona, Tucson, USA.

Herbeck, L. A., and D. A. Larsen. 1999. Plethodontid salamander response to silvicultural practices in Missouri Ozark forests. Conservation Biology 13:623–632.

Hervert, J. J., and J. L. Bright. 2005. Adult and fawn mortality of Sonoran pronghorn. Wildlife Society Bulletin 22:43–50.

Hervert, J. J., J. L. Bright, R. S. Henry, L. A. Piest, and M. T. Brown. 2005. Home range and habitat-use patterns of Sonoran pronghorn in Arizona. Wildlife Society Bulletin 33:8–15.

Hervert, J. J., and P. R. Krausman. 1986. Desert mule deer use of water developments in Arizona. Journal of Wildlife Management 50:670–676.

Hetem, R. S., W. M. Strauss, L. G. Fick, S. K. Maloney, L. C. R. Meyer, M. Shobrak, A. Fuller, and D. Mitchell. 2012a. Activity re-assignment and microclimate selection of free-living Arabian oryx: responses that could minimize the effects of climate change on homeostasis? Zoology 115:411–416.

Hetem, R. S., W. M. Strauss, L. G. Fick, S. K. Maloney, L. C. R. Meyer, M. Shobrak, A. Fuller, and D. Mitchell. 2012b. Does size matter? Comparison of body temperature and activity of free-living Arabian oryx (Oryx leucoryx) and the smaller Arabian sand gazelle (Gazella subgut-turosa marica) in the Saudi desert. Journal of Comparative Physiology B 182:437–449.

Hoffman, R. W., H. G. Shaw, M. A. Rumble, B. F. Wakeling, C. M. Mollohan, S. D. Schmnitz, R. Engel-Willson, and D. A. Hengel. 1993. Management guidelines for Merriam's wild turkeys. Wildlife Report 18. Colorado Division of Wildlife, Fort Collins, USA.

Hofmeyr, M. D. 1985. Thermal properties of the pelages of selected African ungulates. South African Journal of Zoology 20:179–189.

Hofmeyr, M. D., and G. N. Louw. 1987. Thermoregulation, pelage conductance and renal function in the desert adapted springbok, Antidorcas marsupialis. Journal of Arid Environments 13:137–151.

Holl, S. A., V. C. Bleich, B. W. Callenberger, and B. Bahro. 2012. Simulated effects of two fire regimes on bighorn sheep: the San Gabriel Mountains, California, USA. Fire Ecology 8:88–103.

Hoppe, P. P. 1977. How to survive heat and aridity: ecophysiology of the dik-dik antelope. Veterinary Medicine Review 8:77–86.

Hoppeler, H., and E. R. Weibel. 2005. Scaling functions to body size: theories and facts. Journal of Experimental Biology 208:1573–1574.

Hossaini-Hilali, J., S. Benlamlih, and K. Dahlborn. 1994. Effects of dehydration, rehydration, and hyperhydration in the lactating and nonlactating black Moroccan goat. Comparative Biochemistry and Physiology A 109:1017–1026.

Howe, F. P., and L. D. Flake. 1988. Mourning dove movements during the reproductive season in southeastern Idaho. Journal of Wildlife Management 52:477–480.

Hughes, K. S., and N. S. Smith. 1990. Sonoran pronghorn use of habitat in southwest Arizona. Final Report 14-16-009-1564 RWO #6. Arizona Cooperative Fish and Wildlife Research Unit, Tucson, USA.

Hungerford, C. R. 1962. Adaptation shown in selection of food by Gambel's quail. Condor 64:213–219.

Hyde, T. C. 2011. Stable isotopes provide insight into the use of wildlife water developments by resident and migrant birds in the Sonoran Desert of Arizona. MS thesis, University of New Mexico, Albuquerque, USA.

Jarman, P. J. 1973. The free water intake of impala in relation to the water content of their food. East African Agricultural and Forestry Journal 38:343–351.

Jarman, P. J. 1974. The social organisation of antelope in relation to their ecology. Behaviour 48:215–267

Jarman, P. J. 1977. Behaviour of topi in a shadeless environment. Zoologica Africana 12:101–111.

Jawuoro, S. O., O. K. Koech, G. N., Karuku, and J. S. Mbau. 2017. Plant species composition and diversity depending on p100pheres and seasonality in the southern rangelands of Kenya. Ecological Processes 6:16.

Jeffrey, D. E. 1963. Factors influencing elk and cattle distribution on the Willow Creek summer range, Utah. MS thesis, Utah State University, Logan, USA.

Jhala, Y. V., R. H. Giles Jr., and A. M. Bhagwat. 1992. Water in the ecophysiology in blackbuck. Journal of Arid Environments 22:261–269.

Johnson, E. A., and K. Miyanishi. 2008. Testing the assumptions of chronosequences in succession. Ecology Letters 11:419–431.

Jones, R. E., and K. E. Hungerford. 1972. Evaluation of nesting cover as protection from magpie predation. Journal of Wildlife Management 36:727–732.

Kay, R. N. B. 1997. Responses of African livestock and wild herbivores to drought. Journal of Arid Environments 37:683–694.

Kihwele, E. S., V. Mchomvu, N. Owen-Smith, R. S. Hetem, M. C. Hutchinson, A. B. Potter, H. Olff, and M. P. Veldhuis. 2020. Quantifying water requirements of African ungulates through a combination of functional traits. Ecological Monographs 90:e01404.

Kleinhaus, S., B. Pinshow, M. H. Bernstein, and A. A. Degen. 1985. Brain temperature in heat-stressed, water-deprived desert phasianids: sand partridge (*Ammoperdix heyi*) and chukar (*Alectoris chukar sinaica*). Physiological Zoology 58:105–116.

Kluever, B. M., and E. M. Gese. 2016. Spatial response of coyotes to removal of water availability at anthropogenic water sites. Journal of Arid Environments 130:68–75.

Kluever, B. M., and E. M. Gese. 2017. Evaluating influence of water developments on the demography and spatial ecology of a rare, desert-adapted carnivore: the kit fox (*Vulpes macrotis*). Journal of Mammalogy 98:815–826.

Kluever, B. M., E. M. Gese, and S. J. Dempsey. 2016. The influence of wildlife water developments and vegetation on rodent abundance in the Great Basin Desert. Journal of Mammalogy 97:1209–1218.

Kluever, B. M., E. M. Gese, and S. J. Dempsey. 2017. Influence of free water availability on a desert carnivore and herbivore. Current Zoology 63:121–129.

Knudsen, M. F. 1963. A summer waterhole study at Carrizo Spring, Santa Rosa Mountains of Southern California. Desert Bighorn Council Transactions 7:185–192.

Krausman, P. R. 1979. Use of caves by white-tailed deer. Southwestern Naturalist 24:203.

Krausman, P. R. 2002. Introduction to wildlife management: the basics. Prentice Hall, Upper Saddle River, New Jersey, USA.

Krausman, P. R., and E. D. Ables. 1981. Ecology of the Carmen Mountains white-tailed deer. Scientific Monograph Series No. 15. US Department of the Interior, National Park Service, Washington, DC, USA.

Krausman, P. R., and B. Czech. 1998. Water developments and desert ungulates. Pages 138–154 in Proceedings of a symposium on environmental, economic, and legal issues related to rangeland water developments, 13–15 November 1997, Tempe, Arizona. Center for Law, Science, and Technology, Arizona State University, Tempe, USA.

Krausman, P. R., and R. C. Etchberger. 1995. Response of desert ungulates to a water project in Arizona. Journal of Wildlife Management 59:292–300.

Krausman, P. R., and B. D. Leopold. 1986. Habitat components for desert bighorn sheep in the Harquahala Mountains, Arizona. Journal of Wildlife Management 50:504–508.

Krausman, P. R., D. E. Naugle, M. R. Frisina, R. Northrup, V. C. Bleich, W. M. Block, M. C. Wallace, and J. D. Wright. 2009. Livestock grazing, wildlife habitat, and rangeland values. Rangelands 31:15–19.

Krausman, P. R., S. S. Rosenstock, and J. W. Cain III. 2006. Developed waters for wildlife: science, perception, values, and controversy. Wildlife Society Bulletin 34:563–569.

Krausman, P. R., S. Torres, L. L. Ordway, J. J. Hervert, and M. Brown. 1985. Diel activity of ewes in the Little Harquahala Mountains, Arizona. Desert Bighorn Council Transactions 29:24–26.

Kubly, D. M. 1990. Limnological features of desert mountain rock pools. Pages 103–120 in G. K. Tsukamoto and S. J. Stiver, editors. Proceedings of the wildlife water development symposium. Nevada Chapter, The Wildlife Society, US Department of the Interior, Bureau of Land Management, Washington, DC, and Nevada Department of Wildlife, Reno, USA.

Kuenzi, A. J. 2001. Spatial and temporal patterns of bat use of water developments in southern Arizona. Dissertation, University of Arizona, Tucson, USA.

Larsen, R. T., J. A. Bissonette, J. T. Flinders, M. B. Hooten, and T. L. Wilson. 2010. Summer spatial patterning of chukars in relation to free water in western Utah. Landscape Ecology 25:135–145.

Larsen, R. T., J. A. Bissonette, J. T. Flinders, and J. C. Whiting. 2012.

Framework for understanding the influences of wildlife water developments in the western United States. California Fish and Game 98:148–163.

Larsen, R. T., J. T. Flinders, D. L. Mitchell, E. R. Perkins, and D. G. Whiting. 2007. Chukar watering patterns and water site selection. Journal of Range Management 60:559–565.

Lautier, J. K., T. V. Dailey, and R. D. Brown. 1988. Effect of water restriction on feed intake of white-tailed deer. Journal of Wildlife Management 52:602–606.

Law, B. S., and M. Chidel. 2007. Bats under a hot tin roof: comparing the microclimate of eastern cave bat (*Vespadelus troughtoni*) roosts in a shed and cave overhangs. Australian Journal of Zoology 55:49–55.

Leopold, A. 1933. Game management. Charles Scribner's Sons, New York, New York, USA.

Lesicka, L. M., and J. J. Hervert. 1995. Low maintenance water developments for arid environments: concepts, materials, and techniques. Pages 52–57 in D. P. Young Jr., R. Vinzant, and M. D. Strickland, editors. Wildlife water development. Water for Wildlife Foundation, Lander, Wyoming, USA.

Leslie, D. M., Jr., and C. L. Douglas. 1979. Desert bighorn sheep of the River Mountains, Nevada. Wildlife Monographs 66:1–56.

Lewis, J. G. 1977. Game domestication for animal production in Kenya: activity patterns in eland, oryx, buffalo, and zebu cattle. Journal of Agricultural Science 89:551–563.

Lima, S. L., and L. M. Dill. 1990. Behavioral decisions made under the risk of predation: a review and prospectus. Canadian Journal of Zoology 68:619–640.

Longhurst, W. M., N. F. Baker, G. E. Connolly, and R. A. Fisk. 1970. Total body water and water turnover in sheep and deer. American Journal of Veterinary Research 31:673–677.

Longland, W. S. 1991. Risk of predation and food consumption by black-tailed jackrabbits. Rangeland Ecology and Management / Journal of Range Management Archives 44:447–450.

Longshore, K. M., C. Lowrey, and D. B. Thompson. 2009. Compensating for diminishing natural water: predicting the impacts of water developments on summer habitat of desert bighorn sheep. Journal of Arid Environments 73:280–286.

Louw, G. N. 1984. Water deprivation in herbivores under arid conditions. Pages 106–126 in F. M. C. Gilchrist and R. I. Mackie, editors. Herbivore nutrition in the subtropics and tropics. Science Press, Craighall, South Africa.

Louw, G. N. 1993. Physiological animal ecology. Longman Scientific and Technical, Burnt Mill, UK.

Louw, G. N., and M. Seely. 1982. Ecology of desert organisms. Longman Group, Burnt Mill, UK.

Lukacs, P. M., M. S. Mitchell, M. Hebblewhite, B. K. Johnson, H. Johnson, et al. 2018. Factors influencing elk recruitment across ecotypes in the Western United States. Journal of Wildlife Management 82:698–710.

Luskick, S., B. Battersby, and M. Kelty. 1978. Behavioral thermoregulation: orientation toward the sun in herring gulls. Science 200:81–83.

Lynn, J. C., C. L. Chambers, and S. S. Rosenstock. 2006. Use of wildlife water developments by birds in southwest Arizona during migration. Wildlife Society Bulletin 34:592–601.

Lynn, J. C., S. S. Rosenstock, and C. L. Chambers. 2008. Avian use of desert wildlife water developments as determined by remote videography. Western North American Naturalist 68:107–112.

Macfarlane, W. V., B. Howard, H. Haines, P. J. Kennedy, and C. M. Sharpe. 1971. Hierarchy of water and energy turnover of desert mammals. Nature 234:483–484.

Mackie, R. J. 1970. Range ecology and relations of mule deer, elk, and cattle in the Missouri River breaks, Montana. Wildlife Monographs 20:1–79.

Maclean, G. L. 1996. Ecophysiology of desert birds. Springer-Verlag, Berlin, Germany.

Maghini, M. T., and N. S. Smith. 1990. Water use and diurnal seasonal ranges of Coues white-tailed deer. Pages 21–34 in P. R. Krausman and N. S. Smith, editors. Managing wildlife in the Southwest: a symposium. Arizona Cooperative Wildlife Research Unit and School of Renewable Natural Resources, University of Arizona, Tucson, USA.

Maloiy, G. M. O. 1970. Water economy of the Somali donkey. American Journal of Physiology 219:1522–1527.

Maloiy, G. M. O. 1973. Water metabolism of East African ruminants in arid and semi-arid regions. Journal of Animal Breeding and Genetics 90:219–228.

Maloiy, G. M. O., and D. Hopcraft. 1971. Thermoregulation and water relations of two East African antelopes: the hartebeest and impala. Comparative Biochemistry and Physiology A 38:525–534.

Maloiy, G. M. O., W. V. Macfarlane, and A. Shkolnik. 1979. Mammalian herbivores. Pages 185–209 in G. M. O. Maloiy, editor. Comparative physiology of osmoregulation in animals. Volume 2. Academic, London, UK.

Maloney, S. K., G. Moss, and D. Mitchell. 2005. Orientation to solar radiation in black wildebeest (Connchaetes gnou). Journal of Comparative Physiology A 191:1065–1077.

Marcum, C. L. 1975. Summer-fall habitat selection and use by a western Montana elk herd. PhD dissertation, University of Montana, Missoula, USA.

Mares, M. A. 1983. Desert rodent adaptation and community structure. Great Basin Naturalist Memoirs 7:30–43.

Marshal, J. P., V. C. Bleich, and N. G. Andrew, 2008. Evidence for interspecific competition between feral ass Equus asinus and mountain sheep Ovis canadensis in a desert environment. Wildlife Biology 14:228–236.

Marshal, J. P., V. C. Bleich, P. R. Krausman, M. L. Reed, and N. G. Andrew. 2006a. Factors affecting habitat use and distribution of desert mule deer in an arid environment. Wildlife Society Bulletin 34:609–619.

Marshal, J. P., V. C. Bleich, P. R. Krausman, M. Reed, and A. Neibergs. 2012. Overlap in diet and habitat between the mule deer (Odocoileus hemionus) and feral ass (Equus asinus) in the Sonoran Desert. Southwestern Naturalist 51:16–25.

Marshal, J. P., P. R. Krausman, V. C. Bleich, S. S. Rosenstock, and W. B. Ballard. 2006b. Gradients of forage biomass and ungulate use near wildlife water developments. Wildlife Society Bulletin 34:620–626.

Martineau, L., and J. Larochelle. 1988. The cooling power of pigeon legs. Journal of Experimental Biology 136:193–208.

Mayhew, W. W. 1968. Biology of desert amphibians and reptiles. Pages 195–365 in G. W. Brown, editor. Desert biology. Academic Press, New York, New York, USA.

McCleery, R. A., R. R. Lopez, N. J. Silvy, and S. N. Kahlick. 2007. Habitat use of fox squirrels in an urban environment. Journal of Wildlife Management 71:1149–1157.

McCorquodale, S. M., K. J. Raedeke, and R. D. Taber. 1986. Elk habitat use patterns in the shrub-steppe of Washington. Journal of Wildlife Management 50:664–669.

McKechnie, A. E., and B. O. Wolf. 2019. The physiology of heat tolerance in small endotherms. Physiology 34:302–319.

McKee, C. J. 2012. Spatial patterns and population performance of mule deer: responses to water provisioning in Mojave National Preserve, California. MS thesis. University of Nevada, Reno, USA.

McKee, C. J., K. M. Stewart, J. S. Sedinger, A. P. Bush, N. W. Darby, D. L. Hughson, and V. C. Bleich. 2015. Spatial distributions and resource selection by mule deer in an arid environment: responses to provision of water. Journal of Arid Environments 122:76–84.

McKnight, T. L. 1958. The feral burro in the United States: distribution and problems. Journal of Wildlife Management 22:163–179.

McNab, B. K. 2002. The physiological ecology of vertebrates. Cornell University Press, Ithaca, New York, USA.

McNabb, F. M. A. 1969. A comparative study of water balance in three species of quail. I. water turnover in the absence of temperature stress. Comparative Biochemistry and Physiology 28:1045–1058.

Mendoza, V. J. 1976. The bighorn sheep of the state of Sonora. Desert Bighorn Council Transactions 20:25–26.

Mensch, J. L. 1969. Desert bighorn sheep (Ovis canadensis nelsoni) losses in a natural trap tank. California Fish and Game 55:237–238.

Miller, G. D., M. H. Cochran, and E. L. Smith. 1984. Nighttime activity of desert bighorn sheep. Desert Bighorn Council Transactions 28:23–25.

Mitchell, D. 1977. Physical basis of thermoregulation. Pages 1–27 in D. Robertshaw, editor. International review of physiology. Volume 15: environmental physiology II. University Park Press, Baltimore, Maryland, USA.

Mitchell, D., S. K. Maloney, C. Jessen, H. P. Laburn, P. R. Kamerman, G. Mitchell, and A. Fuller. 2002. Adaptive heterothermy and selective brain cooling in arid-zone mammals. Comparative Biochemistry and Physiology B 131:571–585.

Mitchell, G., and J. D. Skinner. 2004. Giraffe thermoregulation: a review. Transactions of the Royal Society of South Africa 59:109–118.

Morgart, J. R., J. J. Hervert, P. R. Krausman, J. L. Bright, and R. S. Henry. 2005. Sonoran pronghorn use of anthropogenic and natural water sources. Wildlife Society Bulletin 33:51–60.

Mullens, B. A. 1991. Integrated management of Culicoides variipennis: a problem of applied ecology. Pages 896–905 in T. E. Walton and B. I. Osburn, editors. Bluetongue, African horse sickness, and related orbiviruses. CRC Press, Boca Raton, Florida, USA.

Mysterud, A., and E. Østbye. 1999. Cover as a habitat element for temperate ungulates: effects on habitat selection and demography. Wildlife Society Bulletin 27:385–394.

Nagy, K. A. 1987. Field metabolic rate and food requirement scaling in mammals and birds. Ecological Monographs 57:111–128.

Nagy, K. A. 2004. Water economy of free-living desert animals. Pages 291–297 in Animals and environments. International Congress Series 1275. Elsevier, Amsterdam, Netherlands.

Nagy, K. A., and M. J. Gruchacz. 1994. Seasonal water and energy metabolism of the desert-dwelling kangaroo rat (Dipodomys merriami). Physiological Ecology 67:1461–1478.

Nagy, K. A., and C. C. Peterson. 1988. Scaling of water flux rates in animals. University of California Press, Berkeley, USA.

Nelson, J. R., and D. G. Burnell. 1976. Elk-cattle competition in central Washington. Pages 71–83 in B. F. Roche, editor. Range multiple use management. University of Idaho, Moscow, USA.

New Mexico Department of Game and Fish. 2013. Toxic algae cause of 100 elk deaths in northeastern NM. News Release. New Mexico Department of Game and Fish. http://www.wildlife.state.nm.us /legacy/publications/press_releases/documents/2013/102213elk finding.html.

Nudds, T. D. 1977. Quantifying the vegetative structure of wildlife cover. Wildlife Society Bulletin 5:113–117.

O'Brien, C. S., R. B. Waddell, S. S. Rosenstock, and M. J. Rabe. 2006. Wildlife use of water catchments in southwestern Arizona. Wildlife Society Bulletin 34:582–591.

Ockenfels, R. A., A. Alexander, C. L. D. Ticer, and W. K. Carrel. 1994. Home ranges, movement patterns, and habitat selection of pronghorn in central Arizona. Technical Report No. 13. Arizona Game and Fish Department, Phoenix, USA.

Ockenfels, R. A., D. E. Brooks, and C. H. Lewis. 1991. General ecology of Coues white-tailed deer in the Santa Rita Mountains. Technical Report No. 6. Arizona Game and Fish Department, Phoenix, USA.

O'Gara, B. W., and J. D. Yoakum. 1992. Pronghorn management guidelines: a compendium of biological and management principles and practices to sustain pronghorn populations and habitat from Canada to Mexico. Proceedings of the Biennial Pronghorn Antelope Workshop 15:1–101.

Ohmart, R. D., and E. L. Smith. 1971. Water deprivation and use of sodium chloride solutions by Vesper sparrows (*Pooecetes gramineus*). Condor 73:364–366.

Olson, B., S. K. Windels, M. Fulton, and R. Moen. 2014. Fine-scale temperature patterns in the southern boreal forest: implications for the cold-adapted moose. Alces 50:105–120.

Ordway, L. L., and P. R. Krausman. 1986. Habitat use by desert mule deer. Journal of Wildlife Management 50:677–683.

Orians, G. H., and J. F. Wittenberger. 1991. Spatial and temporal scales in habitat selection. American Naturalist 137:S29–S49.

Ostermann-Kelm, S., E. R. Atwill, E. S. Rubin, M. C. Jorgensen, and W. M. Boyce. 2008. Interactions between feral horses and desert bighorn sheep at water. Journal of Mammalogy 89:459–466.

Ostrowski, S., J. B. Williams, E. Bedin, and K. Ismail. 2002. Water influx and food consumption of free-living oryxes (*Oryx leucoryx*) in the Arabian Desert summer. Journal of Mammalogy 83:665–673.

Ostrowski, S., J. B. Williams, P. Mésochina, and H. Sauerwein. 2006. Physiological acclimation of a desert antelope, Arabian oryx (*Oryx leucoryx*), to long-term food and water restriction. Journal of Comparative Physiology B 176:191–201.

Owen-Smith, N. 1998. How high ambient temperature affects the daily activity and foraging time of a subtropical ungulate, the greater kudu (*Tragelaphus strepsiceros*). Journal of Zoology 246:183–192.

Parker, A. H., and E. T. F. Witkowski. 1999. Long-term impacts of abundant perennial water provision for game on herbaceous vegetation in a semi-arid African savannah woodland. Journal of Arid Environments 41:309–321.

Parker, K. L., and C. T. Robbins. 1984. Thermoregulation in mule deer and elk. Canadian Journal of Zoology 62:1409–1422.

Pasitschniak-Arts, M., and F. Messier. 1998. Effects of edges and habitats on small mammals in a prairie ecosystem. Canadian Journal of Zoology 76:2020–2025.

Payne, N. F., and F. C. Bryant. 1998. Wildlife habitat management of forestlands, rangelands, and farmlands. Krieger, Malabar, Florida, USA.

Perry, N. D., P. Morey, and G. San Miguel. 2015. Dominance of a natural water source by feral horses. Southwestern Naturalist 60:390–393.

Phillips, P. K., and J. E. Heath. 1995. Dependency of surface temperature regulation on body size in terrestrial mammals. Journal of Thermal Biology 20:281–289.

Picard, K., D. W. Thomas, M. Festa-Bianchet, F. Belleville, and A. Laneville. 1999. Differences in the thermal conductance of tropical and temperate bovid horns. Ecoscience 6:148–158.

Pitman, J. W., J. W. Cain III, S. G. Liley, W. R. Gould, N. T. Quintana, and W. B. Ballard. 2014. Post-parturition habitat selection by elk calves and adult female elk in New Mexico. Journal of Wildlife Management 78:1216–1227.

Porter, W. P., and D. M. Gates. 1969. Thermodynamic equilibria of animals with environment. Ecological Monographs 39:227–244.

Prose, B. L., B. S. Cade, and D. Hein. 2002. Selection of nesting habitat by sharp-tailed grouse in the Nebraska Sandhills. Prairie Naturalist 34:85–105.

Prude, C. H. 2020. Influence of habitat heterogeneity and water sources on kill site locations and puma prey composition. MS thesis. New Mexico State University, Las Cruces, USA.

Rabe, M. J., and S. S. Rosenstock. 2005. Effects of water size and type on bat captures in the lower Sonoran Desert. Western North American Naturalist 65:87–90.

Rakowski, A. E., R. D. Elmore, C. A. Davis, S. D. Fuhlendorf, and J. M. Carrol. 2019. Thermal refuge affects space use and movement of a large-bodied galliform. Journal of Thermal Biology 80:37–44.

Rautenstrauch, K. R., and P. R. Krausman. 1989. Influence of water availability on movements of desert mule deer. Journal of Mammalogy 70:197–201.

Rhodes, E. C., J. D. Bates, R. N. Sharp, and K. W. Davies. 2010. Fire effects on cover and dietary resources of sage-grouse habitat. Journal of Wildlife Management 74:755–764.

Rich, L. N., S. R. Beissinger, J. S. Brashares, and B. J. Furnas. 2019. Artificial water catchments influence wildlife distribution in the Mojave Desert. Journal of Wildlife Management 83:855–865.

Robbins, C. T. 1983. Wildlife feeding and nutrition. Academic Press, Orlando, Florida, USA.

Robertshaw, D., and C. R. Taylor. 1969. A comparison of sweat gland activity in eight species of East African bovids. Journal of Physiology 203:135–143.

Robinson, J. R. 1957. Functions of water in the body. Proceedings of the Nutritional Society 16:108–112.

Roerick, T. M., J. W. Cain III, and J. V. Gedir. 2019. Forest restoration, wildfire, and habitat selection by female mule deer. Forest Ecology and Management 447:169–179.

Rosen, P. C., and C. R. Schwalbe. 1998. Using managed waters for conservation of threatened frogs. Pages 180–202 in Proceedings of a symposium on environmental, economic, and legal issues related to rangeland water developments, 13–15 November 1997, Tempe, Arizona. Center for Law, Science, and Technology, Arizona State University, Tempe, USA.

Rosenstock, S. S., W. B. Ballard, and J. C. deVos. 1999. Benefits and impacts of wildlife water developments. Journal of Range Management 52:302–311.

Rosenstock, S. S., V. C. Bleich, M. J. Rabe, and C. Reggiardo. 2005. Water quality and wildlife water sources in the Sonoran Desert, United States. Rangeland Ecology and Management 58:623–627.

Rosenstock, S. S., J. J. Hervert, V. C. Bleich, and P. R. Krausman. 2001. Muddying the water with poor science: a reply to Broyles and Cutler. Wildlife Society Bulletin 29:734–743.

Rosenstock, S. S., M. J. Rabe, C. S. O'Brien, and R. B. Waddell. 2004. Studies of wildlife water developments in southwestern Arizona: wildlife use, water quality, wildlife diseases, wildlife mortalities, and influences in native pollinators. Technical Guidance Bulletin No. 8. Arizona Game and Fish Department, Phoenix, USA.

Sanchez, P. G. 1974. Impact of feral burros on the Death Valley ecosystem. California-Nevada Wildlife Transactions 21–34.

Sargeant, G. A., L. E. Eberhardt, and J. M. Peek. 1994. Thermoregulation by mule deer (*Odocoileus hemionus*) in arid rangelands of southcentral Washington. Journal of Mammalogy 75:536–544.

Sargeant, G. A., M. W. Oehler, and C. L. Sexton. 2014. Use of water developments by female elk at Theodor Roosevelt National Park, North Dakota. California Fish and Game 100:538–549.

Schemnitz, S. D. 1994. Scaled quail. Birds of North America 106. Academy of Natural Sciences, Philadelphia, Pennsylvania, and the American Ornithologists Union, Washington, DC, USA.

Schmidt, S. L., and D. C. Dalton. 1995. Bats of the Madrean Archipelago (Sky Islands): current knowledge, future directions. Pages 274–287 in

L. F. Debano, G. J. Gottfried, R. H. Hamre, et al., technical coordinators. Biodiversity of the Madrean Archipelago: Sky Islands of the southwestern United States and northwestern Mexico. General Technical Report RM-264. US Department of Agriculture Forest Service, Fort Collins, Colorado, USA.

Schmidt, S. L., and S. DeStefano. 1999. Use of water developments by nongame wildlife in the Sonoran Desert of Arizona. Arizona Cooperative Fish and Wildlife Research Unit, Tucson, USA.

Schmidt-Nielsen, K. 1979. Desert animals: physiological problems of heat and water. Dover, New York, New York, USA.

Schmidt-Nielsen, K., E. C. Crawford, A. E. Newsome, K. S. Rawson, and H. T. Hammel. 1967. Metabolic rate of camels: effect of body temperature and dehydration. American Journal of Physiology 212:341–346.

Schmidt-Nielsen, K., B. Schmidt-Nielsen, S. Jarnum, and T. R. Houpt. 1957. Body temperature of the camel and its relation to water economy. American Journal of Physiology 188:103–112.

Schneider, K. J. 1984. Dominance, predation, and optimal foraging in white-throated sparrow flocks. Ecology 65:1820–1827.

Schwimmer, M., and D. Schwimmer. 1968. Medical aspects of phycology. Pages 279–358 in D. F. Jackson, editor. Algae, man, and the environment. Syracuse University Press, Syracuse, New York, USA.

Seager, R., M. Ting, I. Held, Y. Kushnir, J. Lu, et al. 2007. Model predictions of an imminent transition to a more arid climate in southwestern North America. Science 316:1181–1184.

Sneddon, P. J., and K. Ismail. 2002. Influence of ambient temperature on diurnal activity of Arabian oryx: implications for reintroduction site selection. Oryx 36:50–55.

Shaw, H. G., and C. Mallohan. 1992. Merriam's turkey. Pages 331–349 in J. O. Dickson, editor. The wild turkey: biology and its management. Stackpole Books, Harrisburg, Pennsylvania, USA.

Shields, A. V., R. T. Larsen, and J. C. Whiting. 2012. Summer watering patterns of mule deer in the Great Basin Desert, USA: implications of differential use by individuals and the sexes for management of water resources. Scientific World Journal 9:846218.

Signorell, N., S. Wirthner, P. Patthey, R. Schranz, L. Rotelli, and R. Arlettaz. 2010. Concealment from predators drives foraging habitat selection in brood-rearing Alpine black grouse Tetrao tetrix hens: habitat management implications. Wildlife Biology 16:249–257.

Silanikove, N. 1994. The struggle to maintain hydration and osmoregulation in animals experiencing severe dehydration and rapid rehydration: the story of ruminants. Experimental Physiology 79:281–300.

Simpson, D. C., L. A. Harveson, C. E. Brewer, R. E. Walser, and A. R. Sides. 2007. Influence of precipitation on pronghorn demography in Texas. Journal of Wildlife Management 71:906–910.

Simpson, N. O., K. M. Stewart, and V. C. Bleich. 2011. What have we learned about water developments for wildlife? Not enough! California Fish and Game 97:190–209.

Sokolov, V. E. 1982. Mammal skin. University of California Press, Berkeley, USA.

Stoner, D. C., M. T. Anderson, C. A. Schroeder, C. A. Bleke, and E. T. Thacker. 2021. Distribution of competition potential between native ungulates and free-roaming equids on western rangelands. Journal of Wildlife Management. doi:10.1002/jwmg.21993.

Strauss, W. M., R. S. Hetem, D. Mitchell, S. K. Maloney, L. C. R. Meyer, and A. Fuller. 2015. Selective brain cooling reduces water turnover in dehydrated sheep. PLoS ONE 10:1–18.

Strohmeyer, D. C., and J. M. Peek. 1996. Wapiti home range and movement patterns in a sagebrush desert. Northwest Science 70:79–87.

Sundstrom, C. 1968. Water consumption by pronghorn antelope and distribution related to water in Wyoming's Red Desert. Biennial Pronghorn Antelope States Workshop 3:39–46.

Sveum, C. M., W. D. Edge, and J. A. Crawford. 1998. Nesting habitat selection by sage grouse in south-central Washington. Rangeland Ecology and Management / Journal of Range Management Archives 51:265–269.

Swift, P. K., J. D. Wehausen, H. B. Ernest, R. S. Singer, A. M. Pauli, H. Kinde, T. E. Rocke, and V. C. Bleich. 2000. Desert bighorn sheep mortality due to presumptive type C botulism in California. Journal of Wildlife Diseases 36:184–189.

Switalski, A. B., and H. L. Bateman. 2017. Anthropogenic water sources and the effects on Sonoran Desert small mammal communities. PeerJ 5:e4003.

Tanner, E. P., R. D. Elmore, S. D. Fuhlendorf, C. A. Davis, E. T. Thacker, and D. K. Dahlgren. 2015. Behavioral responses at distribution extremes: how artificial surface water can affect quail movement patterns. Rangeland Ecology and Management 68:476–484.

Tattersall, G. J., B. Arnaout, and M. R. E. Symonds. 2017. The evolution of the avian bill as a thermoregulatory organ. Biological Reviews 92:1630–1656.

Taylor, C. R. 1968. Hygroscopic food: a source of water for desert antelopes. Nature 219:181–182.

Taylor, C. R. 1969a. Metabolism, respiratory changes and water balance of an antelope, the eland. American Journal of Physiology 217:317–320.

Taylor, C. R. 1969b. The eland and the oryx. Scientific American 220:88–95.

Taylor, C. R. 1970a. Dehydration and heat: effects on temperature regulation of East African ungulates. American Journal Physiology 219:1136–1139.

Taylor, C. R. 1970b. Strategies of temperature regulation: effects on evaporation in East African ungulates. American Journal Physiology 219:1131–1135.

Taylor, C. R., and C. P. Lyman. 1972. Heat storage in running antelopes: independence of brain and body temperatures. American Journal of Physiology 222:114–117.

Taylor, D. A. R., and M. D. Tuttle. 2007. Water for wildlife: a handbook for ranchers and range managers. Bat Conservation International, Austin, Texas, USA.

Thomas, J. W., R. J. Miller, C. Maser, R. G. Anderson, and B. E. Carter. 1979. Plant communities and successional states. Pages 22–39 in J. W. Thomas, editor. Wildlife habitats in managed forests: the Blue Mountains of Oregon and Washington. Agricultural Handbook 533. US Department of Agriculture Forest Service, Washington, DC, USA.

Thompson, F. R., III. 2007. Factors affecting nest predation on forest songbirds in North America. Ibis 149:98–109.

Thrash, I. 1998. Impact of water provision on herbaceous vegetation in Kruger National Park, South Africa. Journal of Arid Environments 38:437–450.

Ticer, C. L. D., and W. H. Miller. 1994. Pronghorn fawn bed site selection in a semidesert grassland community of central Arizona. Pronghorn Workshop Proceedings 6:86–103.

Tieleman, I. B., J. B. Williams, and P. Bloomer. 2003. Adaptation of metabolism and evaporative water loss along an aridity gradient. Proceedings of the Royal Society B 270:207–214.

Tieleman, I. B., J. B. Williams, and M. E. Buschur. 2002. Physiological adjustments to arid and mesic environments in larks (Alaudidae). Physiological and Biochemical Zoology 75:305–313.

Tolsma, D. J., W. H. O. Ernst, and R. A. Verwey. 1987. Nutrients in soil and vegetation around two artificial water points in eastern Botswana. Journal of Applied Ecology 24:991–100.

Trzcinski, M. K., L. Fahrig, and G. Merriam. 1999. Independent effects of forest cover and fragmentation on the distribution of forest breeding birds. Ecological Applications 9:586–593.

Tull, J. C., P. R. Krausman, and R. J. Steidl. 2001. Bed-site selection by desert mule deer in southern Arizona. Southwestern Naturalist 46:354–357.

Turner, J. C., Jr. 1973. Water, energy, and electrolyte balance in the desert bighorn sheep, *Ovis canadensis*. PhD dissertation, University of California, Riverside, USA.

Turner, J. C., C. L. Douglas, C. R. Hallum, P. R. Krausman, and R. R. Ramey. 2004. Determination of critical habitat for the endangered Nelson's bighorn sheep in southern California. Wildlife Society Bulletin 32:427–448.

Tuttle, S. R., C. L. Chambers, and T. L. Theimer. 2006. Effects of livestock water trough modifications on bat use in northern Arizona. Wildlife Society Bulletin 34:602–608.

van Beest, F. M., B. Van Moorter, and J. M. Milner. 2012. Temperature-mediated habitat use and selection by a heat-sensitive northern ungulate. Animal Behaviour 84:723–735.

Vodehnal, W. L., G. L. Schenbeck, and D. W. Uresk. 2020. Sharp-tailed grouse in the Nebraska Sandhills select residual cover patches for nest sites. Wildlife Society Bulletin 44:232–239.

Waddell, R. B., C. S. O'Brien, and S. S. Rosenstock. 2007. Bighorn use of a developed water in southwestern Arizona. Desert Bighorn Council Transactions 49:8–17.

Walsberg, G. E. 1983. Coat color and solar heat gain in animals. BioScience 33:88–91.

Walsberg, G. E. 2000. Small mammals in hot deserts: some generalizations revisited. BioScience 50:109–120.

Walsberg, G. E., G. S. Campbell, and J. R. King. 1978. Animal coat color and radiative heat gain: a re-evaluation. Journal of Comparative Physiology 126:211–222.

Warrick, G. D., and P. R. Krausman. 1989. Barrel cactus consumption by desert bighorn sheep. Southwestern Naturalist 34:483–486.

Watts, T. J. 1979. Status of the Big Hatchet desert sheep population, New Mexico. Desert Bighorn Council Transactions 23:92–94.

Weathers, W. W. 1972. Thermal panting in domestic pigeons, *Columba livia*, and the barn owl, *Tylo alba*. Journal of Comparative Physiology A 79:79–84.

Weaver, R. A. 1959. Effects of wild burros on desert water supplies. Desert Bighorn Council Transactions 3:1–3.

Weaver, R. A. 1973. California's bighorn management plan. Desert Bighorn Council Transactions 17:22–42.

Webb, E. B., L. M. Smith, M. P. Vrtiska, and T. G. Lagrange. 2010. Effects of local and landscape variables on wetland bird habitat use during migration through the Rainwater Basin. Journal of Wildlife Management 74:109–119.

Welles, R. E., and F. B. Welles. 1961. The bighorn of Death Valley. Fauna Series No. 6. US National Park Service, Washington, DC, USA.

Whitcomb, R. F., C. S. Robbins, J. F. Lynch, B. L. Whitcomb, M. K. Klimbiewicz, and D. Bystrak. 1981. Effects of forest fragmentation on avifauna of the eastern deciduous forest. Pages 125–205 *in* R. L. Burgess and D. M. Sharpe, editors. Forest island dynamics in man-dominated landscapes. Springer-Verlag, New York, New York, USA.

White, C. R., T. M. Blackburn, G. R. Martin, and P. J. Butler. 2007. Basal metabolic rate of birds is associated with habitat temperature and precipitation, not primary productivity. Proceedings of the Royal Society B 274:287–293.

Williams, J. B., M. M. Pacelli, and E. J. Braun. 1991. The effect of water deprivation on renal function in conscious unrestrained Gambel's quail (*Callipepla gambelii*). Physiological Zoology 64:1200–1216.

Williams, J. B., and B. I. Tieleman. 2002. Ecological and evolutionary physiology of desert birds: a progress report. Integrative and Comparative Biology 42:68–75.

Williams, J. B., and B. I. Tieleman. 2005. Physiological adaptation in desert birds. BioScience 55:416–425.

Williamson, D. T. 1987. Plant underground storage organs as a source of moisture for Kalahari wildlife. African Journal of Ecology 25:63–64.

Wilson, L. O. 1971. The effect of free water on desert bighorn home range. Desert Bighorn Council Transactions 15:82–89.

Wolf, B. O. 2010. The use of water developments by the bird and bat communities on the Kofa National Wildlife Refuge, Arizona. Report to Arizona Game and Fish Department. University of New Mexico, Albuquerque, USA.

Wolf, B. O., and G. E. Walsberg. 1996. Respiratory and cutaneous evaporative water loss at high environmental temperatures in a small bird. Journal of Experimental Biology 199:451–457.

Wolf, B. O., K. M. Wooden, and G. E. Walsberg. 1996. The use of thermal refugia by two small desert birds. Condor 98:424–428.

Wood, J. E., T. S. Bickle, W. Evans, J. C. Germany, and V. W. Howard Jr. 1970. The Fort Stanton mule deer herd: some ecological and life history characteristics with special emphasis on the use of water. Agricultural Experiment Station Bulletin 567. New Mexico State University, Las Cruces, USA.

Woodall, P. F., and J. D. Skinner. 1993. Dimensions of the intestine, diet, and faecal water loss in some African antelope. Journal of Zoology 229:457–471.

Wright, J. T. 1959. Desert wildlife. Wildlife Bulletin No. 6. Arizona Game and Fish Department, Phoenix, USA.

Ydenberg, R. C., and L. M. Dill. 1986. The economics of fleeing from predators. Advances in the Study of Behavior 16:229–249.

Yoakum, J. D. 1978. Pronghorn. Pages 103–121 *in* J. L. Schmidt and D. L. Gilbert, editors. Big game of North America. Stackpole Books, Harrisburg, Pennsylvania, USA.

Yoakum, J. D. 1994. Water requirements for pronghorn. Proceedings of the Pronghorn Antelope Workshop 16:143–157.

Yoakum, J. D. 2004. Habitat characteristics and requirements. Pages 409–445 *in* B. W. O'Gara and J. D. Yoakum, editors. Pronghorn ecology and management. University Press of Colorado, Boulder, and Wildlife Management Institute, Washington, DC, USA.

Zektser, S., H. A. Loáiciga, and J. T. Wolf. 2005. Environmental impacts of groundwater overdraft: selected case studies in the southwestern United States. Environmental Geology 47:396–404.

Zervanos, S. M., and G. I. Day. 1977. Water and energy requirements of captive and free-living collared peccaries. Journal of Wildlife Management 41:527–532.

Zwolak, R. 2009. A meta-analysis of the effects of wildfire, clearcutting, and partial harvest on the abundance of North American small mammals. Forest Ecology and Management 258:539–545.

14 PREDATOR–PREY RELATIONSHIPS AND MANAGEMENT

CLINT W. BOAL, BRENT D. BIBLES, AND JOHN M. TOMEČEK

INTRODUCTION

Few topics in the study of wildlife ecology are as viscerally captivating as the dramatic life-and-death interactions of predators and their prey. These interactions, presented in nature documentaries and popular articles, can elicit awe, wonder, and empathy for both predators and prey. This can lead to public perceptions of wildlife ecology that may be based more on emotion than an ecological understanding. Complicating this are perspectives of some user groups that perceive predators as a direct conflict with desires for more harvestable game animals rather than as important components of intact ecological communities. Alternatively, there are opposing perspectives that predators should not be managed regardless of their effect on other species. These societal perspectives can make predator–prey management challenging for wildlife managers. For example, predator control was once widely applied as a default management action to increase game species; today, few contemporary wildlife management actions are as controversial as predator control.

In this chapter we provide a general overview of predator–prey relationships in the context of wildlife management. We begin with an introduction to predator ecology and behavior, transition to modeling approaches used to better understand predation as an interaction between species, and then discuss aspects of predator management. We focus on mammalian and avian predators because these are the primary focus of most predator- and prey-related wildlife management efforts.

PREDATORS

The complex interactions among predators and their prey are a fascinating aspect of wildlife ecology. The secretive and elusive behaviors of most predators, and their often low population densities, make it challenging to study their role in ecological communities. It is uncommon to witness actual predation events (Fig. 14.1); much of the data on prey use are compiled indirectly by analyzing prey remains at kill sites, feces (scat), by observing prey delivered to nests or dens (Fig. 14.2), and more recently by isotopic analysis of predator tissues. Such approaches can be subject to substantial biases, and the interpretation of predator influences on prey at the population level may be limited. Predators are also logistically challenging (and expensive) to study, so few studies examine the range of prey animals that a given predator will attempt to capture, or the success and failure rates (Toland 1986, Temple 1987, Roth et al. 2006). Hence understanding the role of predation in ecosystem processes has often been elusive and incomplete. Furthermore, as the complexity of predator–prey relationships increases from single predator–single prey to multi-predator–multi-prey, our capacity to understand them decreases. In such cases, ecologists often resort to complex mathematical models (Boyce 2000, Eberhardt et al. 2003). Although potentially limited in representing what actually occurs in nature, such models can be valuable in helping wildlife biologists make sense of complex systems (Taylor 1984, Starfield and Bleloch 1991).

Before we continue, we need to clarify what we mean by predator. A straightforward definition could be that predation occurs when one organism kills another for food or when individuals of one species eat living individuals of another (Taylor 1984). These definitions focus on the act rather than the consequences. This can result in identification of a predatory animal being a matter of perspective and situation. For example, if you are concerned more about a field of winter wheat (living organisms), then elk (*Cervus elaphus*) may be perceived as a problematic predator (Brelsford et al. 1998, Hegel et al. 2009). If, however, you are concerned about elk, then wolves (*Canis lupus*) and cougars (*Felis concolor*) may be perceived as important predators (Eacker et al. 2016, Horne et al. 2019). Although depredation of crops by wildlife is an important management issue, the type of predation we are addressing is that of *carnivory*, in which one animal kills and consumes another. Even that restriction, however, can be confusing when discussing predators because many animals will engage in opportunistic carnivory (Fig. 14.3). For example, red squirrels (*Tamasciurus hudsonicus*) primarily forage on pine cones but also regularly consume eggs and nestlings of songbirds (Bayne and Hobson 2002, Siepielski 2006), and woodcock

Figure 14.1. A bobcat (*Lynx rufus*) is photographed (*left*) pouncing on and (*right*) capturing a cotton rat (*Sigmodon hispidus*). The elusive nature of most predatory animals makes observations such as this an uncommon occurrence. Courtesy of Jeffry Scott

(*Scolopax minor*) feed extensively on earthworms (Lumbricina et al. 1985). By definition, these are acts of predation, but that does not make red squirrels and woodcock predators in the conventional sense.

When we discuss predatory animals in a wildlife management context, we are usually concerned with those species that capture and consume other vertebrates as an obligatory part of their diet (although there are exceptions we discuss below). Among terrestrial mammals, these are members of the order Carnivora, including Canidae (e.g., wolf, golden jackal [*C. aureus*]), Felidae (e.g., cougar, leopard [*Panthera pardus*]), Mustelidae (e.g., marten [*Martes americana*], river otter [*Lontra canadensis*]); and Ursidae (e.g., polar bear [*Ursus maritimus*], sun bear [*Hearctos malayanus*]) and marine mammals such as seals and sea lions (i.e., Pinnipediformes). Often overlooked are the marsupial carnivores, primarily in the order Dasyuromorphia and family Dasyuridae (e.g., quolls [*Dasyurus maculatus*], Tasmanian devil [*Sarcophilus harrisii*]). Although also in the order Artiodactyla, the toothed whales (i.e., Odontoceti) are the largest predators of marine ecosystems. Among birds, those in the orders Accipitriforms (e.g., kites, harriers, hawks, and eagles), Strigiformes (e.g., owls), and Falconiformes (i.e., falcons and caracaras) are most readily recognized as predators. Other bird species, such as the skuas and jaegers (*Stercorarius* spp.), herons and egrets (Ardeidae), and crows and ravens (Corvidae), will prey upon both terrestrial and aquatic prey, and most sea birds are piscivorous. Additionally, several song birds such as loggerhead shrike (*Lanius lucovicianus*), gray butcherbird (*Cracticus torquatus*), and greater roadrunner (*Geococcyx californianus*) also forage extensively on vertebrate prey. Among the reptiles, virtually every snake and lizard (Squamata) is predatory.

Predator Behavior

One way in which predators respond to prey population fluctuations is by increasing or decreasing their own numbers (i.e., numerical response discussed further below.). An important consideration, however, is that territoriality (e.g., intraspecific

Figure 14.2. Video recording of prey deliveries to nests, such as this American kestrel (*Falco sparverius*) delivering a juvenile cottontail rabbit (*Sylvilagus* spp.), can provide excellent diet data but is limited to the nesting period. Courtesy of Clint Boal

tolerance) can keep predator numbers below what the local prey population may be capable of supporting. For example, tawny owls (*Strix aluco*), dependent on small mammals for successful reproduction, do not increase their own nesting density in response to prey population increases owing to their own territorial behavior (Jedrzejewski and Jedrzejewski 1996).

Mammalian predators are typically territorial year-round, but this behavior varies among species. Wolves are social, with packs strongly defending defined territorial boundaries, whereas cougars are solitary predators, with males holding a large territory that can encompass multiple female territories. In contrast, territoriality among avian predators is primarily limited to the breeding season. When multiple predator species occur in a given area, it can result in interspecific competition for prey (Gilg et al. 2006). For example, foraging success of herring gulls (*Larus argentatus*) was negatively correlated with the presence and abundance of northern gannets (*Morus bassanus*; Furness et al. 1992).

Figure 14.3. Many non-carnivorous animals will engage in opportunistic predation and carnivory, such as this round-tailed ground squirrel (*Xerospermophilus tereticaudus*) that captured and is eating a Gambel's quail (*Callipepla gambelii*) chick. Courtesy of Yessica Wheeler

Predators use a combination of visual, auditory, and olfactory cues to locate prey, and this has led to some highly evolved senses and distinctive traits. For example, diurnal raptors are highly dependent on vision and have evolved eyes so large that they occupy about 2/3 of the skull; the greater abundance of rod and cone cells (i.e., the photoreceptor tissues in the eye) allows visual image resolution estimated as 5 to 10 times that of humans. Even more impressive are the auditory capacities of owls. This group has evolved disproportionately large cochlea (i.e., the sensitive inner ears) compared to other birds, and a structural asymmetry of their ears such that they can triangulate prey locations by sound alone (Norberg 1978). In addition to both vision and hearing, most terrestrial mammalian predators also depend heavily on olfaction and have evolved highly developed senses of smell. Some species have evolved highly specialized abilities for locating prey. Probably the most unique are the abilities of pit vipers (Crotalinae) to detect prey by infrared radiation (i.e., body heat) or of bats (Chiroptera) and dolphins (Delphinidae) to detect prey by using sonar (i.e., echolocation).

Once prey has been located, predators employ different tactics to capture them. Cougars, Cooper's hawks (*Accipiter cooperii*), and rattlesnakes (*Crotalus* spp.) are classic examples of stalking or ambush predators; they wait for prey to approach or stealthily approach their prey, and then make a rapid, surprise attack. Species such as wolves, cheetahs (*Acinonyx jubatus*), and gyrfalcons (*Falco rusticolus*) are pursuit hunters, built for endurance and speed to chase down prey. Others systematically search areas until they locate prey: a rat snake (*Elaphe* spp.) or raccoon (*Procyon lotor*) use both visual and olfactory

cues to locate bird nests, whereas a red fox (*Vulpes vulpes*) can use hearing to detect voles moving under snow. Finally, most predators are solitary hunters, but some, such as wolves, lions (*Panthera leo*), and Harris's hawks (*Parabuteo unicinctus*), hunt cooperatively in groups, whereas others, such as brown boobies (*Sula leucogaster*) and northern gannets, will forage in free-for-all aggregations when chasing schools of fish.

A variety of factors influence prey selection, including the presence of alternate prey, size of prey populations, age- and sex-specific vulnerabilities of prey, specializations of the predator, and environmental conditions that influence the vulnerability of prey or effectiveness of the predator. Predation usually occurs between trophic levels, such as when sharp-shinned hawks (*Accipiter striatus*) prey upon small songbirds (Passeriformes). Within-trophic-level predation, however, can occur, and we discuss these predator community interactions below.

Most predators are generalists, regularly capturing a wide variety of prey. In North America, coyotes (*Canis latrans*) and red-tailed hawks (*Buteo jamaicensis*) are mammalian and avian epitomes, respectively, of generalist opportunistic predators. Generalist predators do not necessarily take prey in direct relation to their abundance but in relation to their availability (e.g., competitive exclusion from safe habitat patches) and vulnerability (e.g., younger or older animals, poor physiological condition). Generalist predators are able to switch among prey, focusing on that which becomes more available, and their populations tend to remain relatively stable because they are less likely to be influenced by a decrease in any one prey species. Generalist predators may even suppress prey population cycles if predator densities are unrelated to any single prey species and if the predation rate is density dependent (Hanski and Korpimaki 1995, Thirgood et al. 2000a).

Some predators are specialists, and although they can and do capture other prey, they would likely not persist without their key prey species. A classic North American example is the black-footed ferret (*Mustela nigripes*), which is dependent upon prairie dogs (*Cynomys* spp.; Dobson and Lyles 2000). Another is the Canada lynx (*Lynx canadensis*), whose cyclic relationship with the snowshoe hare (*Lepus americanus*) has been the topic of considerable study for decades (Elton 1924, Stenseth et al. 1998, Vik et al. 2008, Yan et al. 2013). Similarly, the gyrfalcon is considered a specialist upon ptarmigan (*Lagopus* spp.) throughout much of its arctic breeding range (Mossop 2011). Hypothetically, other species can specialize under local circumstances. For example, wolves can capture and consume a variety of prey, but on Isle Royale they specialized on moose (*Alces alces*) because it was the only ungulate species available (Mech and Peterson 2003).

Prey Vulnerability

An important factor in prey selection and capture success is the vulnerability of the individual prey animal. This can be influenced by a suite of factors, including age, physical condition, environmental conditions, activity, and group size. Black bears (*Ursus americanus*) and grizzly bears (*U. arctos*) are a significant cause of mortality among elk, deer (*Odocoileus* spp.), and moose, but predation focuses almost entirely on calves and fawns (Boertje et al. 1988, Ballard and Miller 1990, Zager and Beecham 2006). Bald eagles (*Haliaeetus leucocephalus*) will readily prey on hunter injured waterfowl (Anseriformes), or waterfowl experiencing lead toxicosis from ingesting lead shot (Miller et al. 1998). Wolves sometimes kill in excess of their immediate needs (i.e., surplus killing), a behavior that appears to be linked to environmental (i.e., deep snow hindering escape and food availability for prey) and physiological (i.e., poor nutrition of prey because of lack of food access) conditions during winter (DelGiudice 1998). Further, there can be a temporal component to vulnerability. Smith et al. (2004) found winter had a significant, progressive influence on ungulate vulnerability in the Greater Yellowstone area; wolves' hunting success rose from an estimated 1.6 kills/wolf/month in early winter to 2.2 kills/wolf/month during late winter, presumably because of prey's loss of condition caused by limited food access. Vulnerability can also be associated with specific behaviors of prey. Roth et al. (2006) reported that sharp-shinned hawks had significantly greater success when attacking small birds that were distracted by feeding compared to those that were engaged in other activities.

Management efforts to increase prey populations can result in unintended increased predation. Animals captured and translocated, or reared in captivity and released, can experience higher predation mortality than would normally be expected. In the case of the former, timing of the year in which releases are made can substantively influence survival (Snyder et al. 1999, Letty et al. 2007). With the latter, Perkins et al. (2018) found raptors disproportionately selected for captive reared and released northern bobwhite (*Colinus virginianus*) compared to wild bobwhite. Providing supplemental food may also lead to increased predation risk. Red-tailed hawks (Turner et al. 2008) and bobcats (*Felis rufus*; Godbois et al. 2004) were both detected three and ten times closer, respectively, to quail feeders than would be expected by chance. This is not just restricted to natural landscapes; many urban residents are dismayed to find that attracting songbirds in their backyard feeders can result in predation by both avian and mammalian predators (Dunn and Tessaglia 1994, Hammer et al. 2016). Another unintended consequence is when one prey species is increased such that a predator responds numerically to the point that it negatively influences other prey (see "Prey Relationships," below).

Finally, the potential for research activities to influence prey vulnerability cannot be ignored. A primary concern is that human researchers may leave scent trails that predators will follow to find nests, but studies have found little support of this (Olson and Rohwer 1998, Skagen et al. 1999, Jacobson et al. 2011, Border et al. 2017). Another concern is the potential handicapping of animals when they have markers and radio-global positioning system transmitters attached to them (Millspaugh et al. 2012). More recently, Perkins (2019) determined Harris's hawks disproportionately selected for radio-tagged bobwhite when presented with both tagged and untagged individuals in controlled trials, suggesting an influence of the transmitter.

Changing Views on Predation

Humans have a long evolutionary history with predatory animals. As Quammen (2003:3) soberly noted, one of the "earliest forms of human self-awareness was the awareness of being meat"; no doubt, this was a foundation for instincts that, unchecked, can color our perspectives even today. As societies developed, depredations upon domesticated livestock only increased human antipathy toward predators. More recent in our history, especially since the development of firearms, the conflict with predators has expanded to a sense of competition for game animals. As the discipline of wildlife management developed in the early 1900s, a guiding principle was that higher numbers of game species were desirable, and that predators limited populations of game species. Thus managers undertook regular efforts toward predator control without understanding predator–prey relationships.

It was not until the mid-1930s that researchers began scientifically investigating predator–prey relationships. Leopold (1933) proposed that ecosystem processes produced surpluses of game animals and implicitly referenced density-dependent and density-independent factors in population limitations, all of which are important in understanding predator–prey relationships.

Food and other resources ultimately limit prey populations, and when a prey population nears the maximum size the environment can support (i.e., ecological carrying capacity), *intraspecific competition* for resources can occur (Leopold 1933,

ADOLPH MURIE (1899–1974)

Adolph Murie was born in Minnesota in 1899 and became one of the first professional biologists to promote wildlife management based on ecosystems rather than single species. While an undergraduate at Concordia College, he visited his brother, biologist Olaus Murie, at Mt. McKinley National Park in in 1922. The experience inspired him to earn a doctoral degree in biology from the University of Michigan 1929. Following graduate school, Murie worked for the Museum of Zoology at the University of Michigan, most notably conducting research and publishing on the mammals of Guatemala and Honduras. In 1934 Murie became employed as a biologist in the Wildlife Division of the National Park Service. As a Park Service biologist, Murie was involved in studies of a variety of species. But his study of coyotes in the Yellowstone region was groundbreaking as one of the first science-based examinations of predator ecology in natural ecosystems. Assigned the task of determining the importance of coyote predation as a limiting factor of deer and elk populations, and after detailed food habitat analysis of coyotes, Murie concluded that it was inadequate winter range, not coyote predation, that limited big game species. Recognizing coyotes as important components of ecosystems, his resulting book, *Ecology of the Coyote in Yellowstone*, published in 1937, met with controversy by going against popularly held beliefs. However, it was one of the first science-based arguments against the practice of predator control and eradication efforts commonplace within the National Park Service at the time. Murie was reassigned to Mt. McKinley (now Denali) National Park in 1939 and became first biologist to study wolves in their natural habitat. Murie's detailed field observations revealed that wolves, weather, and disease were all significant influences on declines of the Dall sheep population. Murie recommended wolf control be continued until the sheep population began to recover. Rather than broad-based control, however, Murie led efforts targeted to specific wolves and packs to meet management objectives. His subsequent book, *The Wolves of Mt. McKinley*, published in 1944, is considered a classic and was foundational in changing the commonly held perspectives of predators in natural systems. He was ultimately presented with the National Park Service's Distinguished Service Award and received the John Burroughs Medal for his book *A Naturalist in Alaska*. In recognition of his role in both understanding and the sound management of park ecosystems, the National Park Service dedicated the Murie Science and Learning Center at Denali National Park to Adolph Murie in 2004.

Errington 1946). In these situations, predation can serve as an ecological counterbalance to the reproductive potential of that prey, and as an evolutionary mechanism in the culling of less fit individuals. A key component of this premise was that predation occurred—and could occur at high rates—but could hypothetically have no influence on the subsequent breeding population (Errington 1946). Few studies had directly examined predator–prey relationships, however. Murie's (1940) groundbreaking study of coyotes in Yellowstone was one of the first to provide fact-based justification for termination of the predator eradication program conducted by the National Park Service. Murie's (1944) subsequent research of wolves at Mount McKinley, Alaska, was the first detailed observational study of wolves and again provided data supporting termination of eradication, but not control programs. In a landmark study, Craighead and Craighead (1956) provided strong evidence that raptor predation typically occurred proportional to prey densities. They also reported evidence of situations where predators can regulate prey populations at levels below what could be supported by habitat conditions, such as when prey populations were low in numbers for other reasons; similar results have been reported by Thirgood et al. (2000*a*, *b*).

Simply counting the number of prey animals killed does not provide a satisfactory understanding of predator–prey relationships; rather, one must understand what factors make an individual vulnerable to predation in the first place. To do otherwise can result in ineffectual management actions and a waste of limited financial resources. For example, Peek (1986) noted that predator control was easier to implement and more readily demonstrative of action (even if ineffectual) to a public demanding more game than was addressing the ultimate limitations of habitat loss and greed.

For their part, wildlife professionals spend a substantial amount of time trying to better understand predator–prey dynamics and how to apply that information to better manage wildlife resources. Predator–prey interactions are a primary mechanism of evolutionary change, a perpetual cycle of adaptation; as a predator species adapts to increase its success in capturing prey, prey species correspondingly adapt to avoid capture, leading to new adaptations by predators, and so on. Dawkins and Krebs (1979) likened this process to an evolutionary arms race. Selective pressure is also greater on prey species, and as would be expected, prey adapt to predator strategies more rapidly, through a mechanism known as the life-dinner principle (Dawkins and Krebs 1979). During any given predation attempt, the individual prey's failure results in death. Thus, when predators are successful, they remove vulnerable individuals from a prey species population, thereby influencing the behavior of that prey species (Newton 1998). In contrast, failure by the predator (over the short term) results only in missing a meal (Dawkins and Krebs 1979). It is important to also recognize, however, that repeated failures will lead to starvation, death, and ultimately selection against inferior individuals of the predator populations (Newton 1998).

As we address further below, the function of predator–prey relationships is more complex than just one of direct mortality (Lima and Steury 2005). Predators also influence prey behavior (i.e., the ecology of fear; Brown et al. 2001, Preisser et al. 2005) such that prey must balance predation risk with use of preferred habitat (Hernández and Laundré 2005, Coleman and Hill 2014, Schmidt and Kuijper 2015). Further, these interactions are not limited to just one predator and one prey but can have cascading influences throughout ecological communities, especially when keystone species are, or are not, present (e.g., trophic cascades; Paine 1980, Sinclair and Krebs 2002).

In the context of applied wildlife management, the study of predator–prey relationships usually focuses on the influences of predator populations on game species (e.g., wolves as predators of elk), on species of conservation (e.g., rat snake predation on red-cockaded woodpeckers [*Picoides borealis*]) or economic concern (e.g., tern [Sterninae] predation of salmon [*Oncorhynchus* spp.] smolt), and on depredation issues (e.g., herons at aquaculture facilities). More recently, there has been an increased awareness of the importance of the interactive role of all members in biotic communities, and of efforts for conservation at the ecosystem level rather than just for a particular species of interest. This shift included conservation of both predators and prey as part of functioning ecosystems (Fascione et al. 2004, Reynolds et al. 2006).

PREDATION AS AN INTERACTION

The understanding of predator–prey relationships is complicated by the numerous other interactions in which the predator and prey are involved. A predator is interacting with other members of its population, is subject to competitive interactions with other species, and is dependent upon habitat components other than its food. Similarly, the prey must also interact directly with other members of its population, its own suite of competitors, other predator species, and its habitat components. To understand a given predator–prey relationship, it is necessary to consider the role of each of these other interactions on the predator–prey system.

Modeling Predator–Prey Relationships

Early modeling approaches to understanding predator–prey relationships necessarily simplified these systems to single predatory–single prey relationships that operated in an absence of the interactions with habitat and other species interactions. Early models of predator–prey relationships originated with Lotka (1925) and Volterra (1926), and their approach forms the basis of contemporary models. In general, this approach makes the simplifying assumptions that prey populations experience exponential or logistic growth and mortalities are due to the overall predation rate that is influenced by the functional and numerical response of the predator. Although simple, the development of these models, and the efforts to improve them and make them more complex, have informed our understanding of predator–prey relationships.

In these equations the prey is assumed to grow exponentially based on the size of the prey population (N_p) and its intrinsic rate of growth (r). In the absence of predation, the prey population grows according to the exponential population growth equation, $(dN_p/dt) = rN_p$, where dN_p/dt is the change in prey numbers over a small unit of time. Unlimited growth of the prey population is prevented by loss of individuals due to predation, which is modeled as a function of the capture efficiency or predation rate of the predator (α), the prey population size, and the predator population size (P), expressed as $\alpha N_p P$. This results in the change in size of the prey population under predation pressure being given by the equation $(dN_p/dt) = rN_p - \alpha N_p P$ (Fig. 14.4). The loss of prey individuals to predation influences the cycling of the prey population. The two external influencing factors to the cycling of the prey population are predation rate (i.e., $\alpha = $ *#prey killed* / *#prey*) and size of the predator population (Fig. 14.4). The number

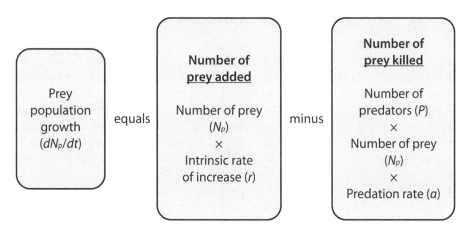

Figure 14.4. Prey population growth components in the Lotka-Volterra model.

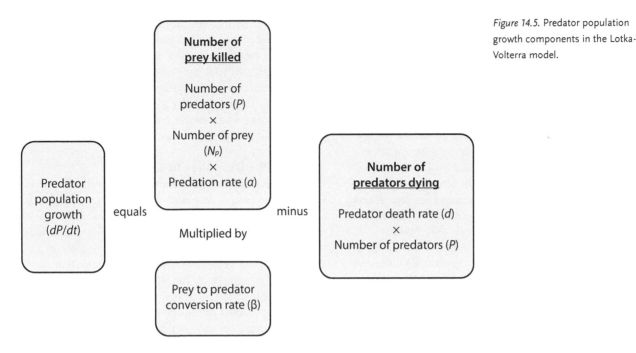

Figure 14.5. Predator population growth components in the Lotka-Volterra model.

of prey killed is proportional to the rate at which predators encounter their prey. When both prey and predator populations are high, high numbers of prey are removed by predation, overwhelming the number added to the prey population via birth rate of prey, and the population declines. At low numbers of prey and predators, fewer prey are removed by predators than are added by population growth, and the prey population grows.

Predator population growth is based on two values, the number of prey consumed from the prey killed portion of the prey equation ($\alpha N_p P$), and a term that describes how efficient predators are at converting a prey animal into an offspring (β = conversion efficiency). Loss of predators from the population is a function of the predator's death rate (d) and the number of predators (P). This results in the equation $dP/dt = \alpha N_p P \beta - dP$, where dP/dt is the change in predator numbers over a small increment of time (Fig. 14.5). Hence predator population growth is influenced by the availability of prey. When prey are abundant, predators are added to the population faster than predators die, but as prey numbers decline to low levels, few individuals are added to the population compared to the number of predators dying.

The Lotka-Volterra equations rely on a series of assumptions that are overly simplified. Of the seven parameters involved in the equations, five remain constant, and only the population sizes of the predator and prey are variable. These equations do produce cyclic patterns of predator and prey population growth (Fig. 14.6), but long-term isolation is dependent on the initial values of the sizes of the predator and prey populations, and the values of the constant parameters. When cycles do occur, a notable concern is that the prey population begins to rebound when it is rare, and the declining predator population is still relatively high (Fig. 14.6). This unrealistic situation, where a prey population can recover from

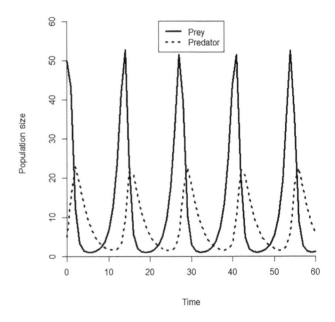

Figure 14.6. Example of the oscillation of predator and prey populations created by Lotka-Volterra equations.

very low numbers instead of moving toward extinction by the more abundant predator population, is called the atto-fox problem (Mollison 1991).

Many laboratory experiments have attempted to demonstrate the cycling exhibited by the Lotka-Volterra equations, notably Gause's (1935) experiments using protozoans, only to find that the typical results were the extinction of the prey, followed by the extinction of the predator, or occasionally extinction of the predator followed by explosive growth of the prey. Creating stable oscillations of predator and prey proved challenging and usually involved introducing other elements to the model. Mathematical ecologists have reframed the basics of the Lotka-Volterra model to focus on isoclines (Rosenzweig and MacArthur 1963) or to view the problem in a ratio-dependent context (Arditi and Ginsburg 1989) as means of improving the models themselves.

Predator Interactions

Predators interact with more than just their prey. You might have noted that in the Lotka-Volterra equations the predator population grows only because of the constant rate process of converting a single species of prey into new predators and a constant (density-independent) death rate. In reality, predators are usually feeding on multiple prey species with a variety of factors influencing how many of a given prey they consume; interactions within their own population can create variations in birth and death rates as their population changes in size, and interactions with other species may influence both population size and prey consumption.

Birth and death rates of predators change as the size of the population varies in relation to the carrying capacity of its habitat, a process known as *density dependence*. Territoriality, a common feature of many predators, can lead to density-dependent patterns of population growth through several mechanisms. The habitat heterogeneity hypothesis (Dhondt et al. 1992) follows an ideal preemptive distribution (Pulliam and Danielson 1991), resulting in lower reproduction. The ideal preemptive distribution hypothesizes that as an area of habitat of varying quality becomes occupied by a species, the highest-quality habitat will be occupied first, and later individuals will occupy lower-quality habitat. Occupancy of lower-quality habitat translates to lower productivity. Consequently, the per capita rates of birth decline, or per capita death rates increase, as the population grows owing to territoriality occurring in a heterogenous environment. Spanish imperial eagles (*Aquila adalberti*) and northern goshawk (*Accipiter gentilis*) were reported to exhibit density-dependent changes in fecundity that followed patterns predicted by the habitat heterogeneity hypothesis (Ferrer and Donazar 1996, Krüger and Lindström 2001). Alternatively, as populations increase, territoriality could lead to increasing interference competition among predators, leading to lower reproduction (Lack 1966). Density dependence influenced by interference competition has been observed in griffon vultures (*Gyps fulvus*; Fernandez et al. 1998). Social structure may be important; solitary species such as the Spanish imperial eagle and northern goshawk may be more influenced by habitat heterogeneity at the territorial scale, whereas grouping or colonial species like the griffon vulture may be more influenced by interference competition (Fernandez et al. 1998). The existence of floating (i.e., nonbreeding) individuals in the population, however, may create another mechanism of interspecific competition that decreases reproductive rates (Brown 1969). Hansen (1987) hypothesized this possibility for bald eagles in southeast Alaska based on the large number of floaters in some areas, and this may influence the very low reproductive rates observed on Prince of Wales Island (Anthony 2001).

Predator Responses to Prey Abundance

Predators do not consume prey at a constant rate regardless of the density of prey available; they alter their feeding behavior in response to prey abundance. These changes create a variety of responses in individual predators and their populations. A *numerical response* is a change in the number of predators owing to a change in the prey population. Because prey are an important component of habitat for a predator, an increase in prey intuitively leads to an increase in the carrying capacity of the habitat for the predator, and we should see the population of predators increase in response given enough time for an increase in reproduction. Reproductive rates in predators are commonly correlated with food availability (Newton 1979). Hansen (1987) found that bald eagle nest territories near abundant prey resources were more likely to be active and produced more fledglings than those without access to abundant prey. This observation of food availability influencing both the likelihood of nesting and number of fledglings produced is common in raptors (Newton 1979). Similarly, reproductive rates in mammalian predators vary with abundance of prey. Both Canada lynx and coyote demonstrate a reduction in pregnancy rates as snowshoe hare populations decline, and kit fox (*Vulpes macrotis*) pregnancy rates were also tied to prey availability (White and Ralls 1993).

With persistent high prey availability, we can expect numerical responses in predators due to reproduction, as predicted by the Lotka-Volterra equations. Movement of predators into areas of temporary high prey density, however, is also a mechanism for short-term numerical response of predators. Korpimaki and Norrdahl (1991) observed that European kestrel (*Falco tinnunculus*), short-eared owl (*Asio flammeus*), and long-eared owl (*Asio otus*) populations respond numerically to rapid changes in the populations of vole (*Microtus* spp.). Based on mark-recapture work, Korpimaki (1988) demonstrated much lower rates of recapture among kestrels in his study area than expected given their annual survival rates and noted that kestrels were moving in and out of the area in response to changes in vole densities. Ephemeral concentrations of prey such as spawning runs of salmon may create a form of rapid numerical response known as an aggregative response, such as that observed in bears and bald eagles in Alaska.

Another way in which predators vary the number of prey consumed is through a functional response. A *functional response* is a change in the number of prey killed by an individual predator as prey density changes. The importance of the functional response in understanding predator–prey cycles was first described by Solomon (1949). Holling (1959a, b, 1965) later described three types of functional responses of predators to prey (Fig. 14.7). The Type I functional response is based on a linear increase in the number of prey consumed by an individual predator as prey density increases. Above a threshold density, however, the predator is limited to a maximum consumption rate owing to satiation. This form of response has been observed in invertebrate predator–prey systems, such as filter-feeders, in which there is no requirement of the predator to process prey and predator–prey encounter is constant

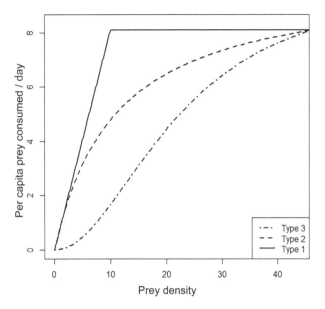

Figure 14.7. Hypothetical Types I, II, and III functional response curves.

and random, but it is based on unrealistic assumptions for the majority of vertebrate predators. For example, other biological activities (e.g., sleep, self-maintenance, reproduction) and handling time (e.g., time required to capture and consume prey when it is encountered) prevent constant searching.

The Type II functional response illustrates a gradual decline in the increase of number of prey consumed per individual per unit time as prey densities increase (Fig. 14.7). When prey are encountered, the predator must engage in attempting to capture it and, if successful, consuming and digesting it. This creates a handling time requirement based on the time it takes to kill, consume, and digest prey that limits the maximum rate of consumption, as opposed to the limit of satiation for a Type I response. The Type II functional response is observed in predator–prey systems characterized by one primary predator species and one primary prey species, and although it predominates in invertebrate systems, it can also apply to vertebrate systems. Korpimaki and Norrdahl (1991) reported that European kestrels exhibited a Type II functional response to vole densities, and Dale et al. (1994) observed a similar relationship between wolves and caribou (*Rangifer tarandus*).

The Type III functional response follows a sigmoid curve similar to a logistic population growth curve (Fig. 14.7). This response illustrates a slow increase in the rate of consumption of a prey as it increases from low densities, followed by a rapid increase until the rate levels at an asymptote. Type III responses appear when predators rely on a variety of prey species, which is most typical of vertebrate systems. A prey species that is more abundant and presumably more available will be encountered and captured more frequently than less abundant prey species. As a result, at low prey densities, predators will switch their focus to more abundant prey species. They will use a search image that focuses on these more abundant prey species, which means that the less abundant species is often missed or passed up. As the less abundant species increases

in density, the predator eventually develops a search image for it and exhibits *prey switching* (Sinclair et al. 2006). Once the predator has switched to a given prey species, the response follows the pattern of a Type II response with a maximum rate determined by search and handling times. Redpath and Thirgood (1999) observed hen harrier (*Circus cyaneus*) predation on red grouse (*Lagopus lagopus*) chicks consistent with a Type III response, while peregrine falcon (*Falco peregrinus*) predation on adult grouse followed a Type II response. Gilg et al. (2006) reported that snowy owls (*Bubo scandiacus*), long-tailed jaegers (*Stercorarius longicaudus*), and Arctic fox (*Vulpes lagopus*) all exhibited a Type III response to changes in the lemming (*Dicrostonyx* spp.) density.

What is the importance of the numerical and functional responses? When the Lotka-Volterra equations are modified to incorporate functional responses, the predator–prey system can demonstrate a stable equilibrium that no longer exhibits unstable cycling patterns. This suggests that capping the rate at which predators consume prey prevents the predator from driving the prey to extinction, with the prey becoming protected as they become rarer. It is counterintuitive, however, that a generalist predator such as a Harris's hawk would forgo capturing an unwary cottontail rabbit (*Sylvilagus* spp.; Fig. 14.8) only because cotton rats (*Sigmodon hispidus*) happen to be locally more abundant. Matter and Mannan (2005) discuss evidence involving the brown tree snake (*Boiga irregularis*) and the Nile perch (*Lates niloticus*), both generalist predators, suggesting that prey switching did not act to protect prey from predation. Rather, prey switching may be best viewed as the observed result of prey being relatively invulnerable to predation rather than the result of the predator changing search images (Matter and Mannan 2005). In addition, prey species subject to Type II functional responses also may be subject to predator-derived Allee effects in which reproductive rates decline, or mortality rates increase, as populations become low, potentially leading to extinction in contradiction of the prediction that functional responses stabilize the predator–prey relationship (Gregory and Courchamp 2010).

It is important to view functional responses as a pattern of changing prey consumption rate with changing prey density. Although it is tempting to categorize observed responses into the three Holling's types, the observed responses of predators exhibit a continuum that often defies such categorization (Denny 2014). Functional responses are dependent on the range of prey densities that the predator (and researcher) encounter. They describe an observed pattern rather than the mechanism that generates the pattern as evidenced by differing explanations for the low-density pattern of the Type III response previously mentioned.

Predator Community Interactions

Most environments possess multiple predator species that not only potentially prey on the same species but may prey on each other. These interactions among predator species can have large effects on the structure and function of the ecological system. Interspecific competition among species

Figure 14.8. Despite a population peak of cotton rats that were heavily preyed upon by raptors, this Harris's hawk (*Parabuteo unicinctus*) did not pass on the opportunity to capture a comparatively less common cottontail rabbit. Courtesy of Clint Boal

that share the same limiting food resource is common. Interspecific competition among predators, however, can have a substantially stronger antagonistic component than competitive interactions among non-predators, including direct killing (Palomares and Caro 1999). Competition among predator species often involves a dominant and subordinate species, and there are three effects the competition may have on the subordinate species (Creel et al. 2001). First, the subordinate predator may alter the time or location in which they occur to avoid the competitor. For example, coyotes may utilize buffer zones between wolf territories, themselves a mechanism for reducing competition among wolf packs, in order to avoid direct conflict with wolves (Fuller and Keith 1981). Second, dominant predators may influence subordinate predators off of their prey kills, a process known as *kleptoparasitism*. While widely documented among mammalian predators of the Serengeti (Creel et al. 2001), kleptoparasitism occurs among raptors. Dekker et al. (2012) reported that wintering bald eagles routinely stole ducks from peregrine falcons, resulting in the falcons shifting to feed on dunlins (*Calidris alpina*) to avoid the kleptoparasitism. Gyrfalcons would then kleptoparasitize the peregrines off their dunlin kills as well. Overall, the presence of eagles and gyrfalcons led to an increased rate of predation on dunlins by the peregrines.

The third effect of interspecific predation among predators involves direct killing of the subordinate competitor, commonly known as *intraguild predation*. In a review of intraguild predation in carnivores, 97 pairwise interactions were documented involving 54 victim species and 27 killer species, with canids and felids being the most prominent killer species (Palomares and Caro 1999). Intraguild predation is not just a form of predator–prey relationship where the prey is another predator; it can also function as a form of interspecific competition. Of the 21 species for which information was

available, 11 species did not consume the victim. Intraguild predation is common among raptors (Mikkola 1976) and may be symmetric such that either species may kill the other. In an experiment with four species of raptors subject to intraguild predation, Lourenço et al. (2011) supported the idea that killing of other predators was to reduce the potential of being predated, known as the predator-removal hypothesis, rather than removal of a potential competitor, the competitor-removal hypothesis.

Interspecific competition among predators can have significant management implications. Loss of a dominant competitor from the system can lead to an increase in the populations of the subordinate competitors, a situation known as *mesopredator release* (Soulé et al. 1988). The loss of apex predators, such as the wolf, in much of North America is correlated with the increase in both range and abundance of mesopredators (Richie and Johnson 2009). Reductions in coyote populations owing to urbanization in Southern California have been associated with increases in raccoons, feral cats, and other mesopredators. Interspecific competition among predators, however, can also create more complex patterns of release. For example, in Minnesota, the presence of wolves suppresses coyote populations and leads to the release of red fox (*Vulpes vulpes*) populations (Levi and Wilmers 2012). The result of mesopredator release is often trophic cascades, as discussed below.

Prey Relationships

Prey interact with more than just predators. They are subject to intra- and interspecific competition and availability of elements of their habitat. The first modification to the initial Lotka-Volterra equations was Volterra's (1931) modification to restrict the exponential growth of the prey species following the logistic growth model, that is, impose density-dependent growth rates on the prey population.

A common result of interspecific competition is declines in abundance of one or more species that are competing. The influence of prey abundance on the total response (the combination of numerical and functional responses) of a predator may create *enemy-mediated apparent competition* among prey species that are not, in fact, competing for resources with each other (Holt 1977). In most cases, this is apparent amensalism as the combined density of the prey species leads to a higher total response on one of the species, rather than both species being negatively affected by the predator population (Chaneton and Bonsall 2000). A particularly novel form of apparent competition that also involves intraguild predation occurred on the Channel Islands of California and involves golden eagles (*Aquila chrysaetos*), island foxes (*Urocyon littoralis*), and feral pigs (*Sus scrofa*). The presence of feral pigs, along with the extirpation of bald eagles, enabled golden eagles to establish on the islands (Roemer et al. 2002). High populations of feral pigs were facilitating presence of golden eagles, which were then increasing predation on the endangered foxes. Management options were complex, as focusing management solely on eradication of feral pigs would likely create increased predation pressure on the foxes. Management recommendations involved simultaneous removal of both pigs and golden eagles to reduce predation on the foxes (Courchamp et al. 2003). In another example of apparent competition influencing an endangered species, cougar predation on endangered Sierra Nevada bighorn sheep (*Ovis canadensis sierrae*) was mediated by overlap of the sheep with abundant mule deer (*Odocoileus hemionus*). Similarly, the expansion of barred owls (*Strix varia*) into the Pacific Northwest may facilitate increased predation on spotted owls (*Strix occidentalis*) by great horned owls (*Bubo virginanus*) and red-tailed hawks (Wiens et al. 2014).

The Role of Habitat in Prey Persistence

A common finding of laboratory experiments to demonstrate predator–prey coexistence based on the Lotka-Volterra models was that additional features needed to be added to the laboratory system to create coexistence. Gause (1935) was able to maintain predator–prey cycles that mimicked those predicted by the equations but only after he introduced two components—existence of refugia and immigration of both predator and prey—to the system. His results conveyed an important understanding of how prey continue to exist, that is, the need for refuges where they are relatively invulnerable to predation. In Gause's experiments the refuge was a sediment in which the prey *Paramecium* was safe from predation by *Didinium*. Later work by Huffaker (1958) reinforced the importance of environmental heterogeneity in prey persistence. His experiments enabled movement of a mite prey (*Eotetranychus sexmaculatus*) and a mite predator (*Typhlodromus occidentalis*) among patches of resources (oranges) in which prey could colonize new patches faster than the predators could; therefore the prey had a temporary spatial refuge from predation. Similarly, Errington's (1946) work on muskrats (*Ondatra zibethicus*) led him to conclude that predated muskrats were those individuals that were forced to emigrate out of refuges because of high

densities. In this view, the refuges define the carrying capacity, and loss of individuals to predation is a form of compensatory mortality that removes individuals in the population in excess of the carrying capacity of the environment. Behavioral or physical anti-predator adaptations of prey such as timing of space use, group living, speed, or size can function to make an individual relatively invulnerable to predation. Habitat features that influence prey vulnerability are important to consider. Gulsby et al. (2017) reported that white-tailed deer (*Odocoileus virginianus*) fawn mortality to coyotes was reduced by the presence of heterogenous landscapes. In this example the presence of habitat features that provide refuge to a vulnerable age class enables these individuals to survive to a size that provides refuge.

Prey population size, in most cases, is a function of the habitat available to prey individuals. As a result, there is an interplay among habitat components, including food, water, and cover. Prey populations serve as the food component that, among other habitat components, determines population size of predators. This interplay among habitat components can make identification of the role of predators in influencing prey numbers difficult to ascertain. Predators can remove large numbers of prey from a population without having a demographic effect on the population size. They can also exert behavioral influences on prey populations that do not involve the numbers killed (see the discussion of ecology of fear, below).

The aforementioned relationship between snowshoe hare and Canada lynx populations is perhaps one of the most widely known predator–prey cycles and has been used as an example of predators driving prey cycling. A series of field experiments have demonstrated that while lynx population cycles are influenced by abundance of their primary prey, the hare, the cycling of hare populations is due to both fluctuations in food availability and a time-lagged numerical response (Krebs et al. 2001). African ungulate communities, while not cycling, exhibit a similar pattern of combined effect of predators and interspecific competition for food among the prey (Sinclair and Arcese 1995). In particular, individuals that were predated were in lower body condition than those that lived, suggesting that poor nutrition led prey to risky behaviors in search of food. Bishop et al. (2009) increased winter food availability by providing supplemental feed to mule deer in large experimental units in Colorado. By comparing survival among treated and untreated areas, they reported evidence that the deer were food limited. The enhanced nutrition substantively increased survival and reduced female and fawn mortality by coyotes and cougars. Their findings suggested coyote predation was likely compensatory in removing fawns in poor condition that would have likely succumbed to other causes. Their results revealed that observed rates of coyote predation alone would not have been sufficient for evaluating effects on the deer population. Similarly, Hurley et al. (2011) conducted a predator removal experiment in Idaho in which both coyotes and cougars were removed. Mule deer population growth was most influenced by winter severity rather than predation. Coyote removal had no effect on fawn ratios or population growth. Although cougar removal led to some increases in fawn ratios and survival rates, they were unable to detect any appreciable change in population growth in response to the removals. Thirgood et al. (2000b) reported that habitat loss was the influencing factor behind long-term declines in red grouse, but that high levels of raptor predation could subsequently limit the cyclic growth of the grouse population. Predation can be a proximal cause for low numbers, but the ultimate cause has usually been reported to be long-term patterns of anthropogenic landscape change and habitat loss. Similarly, Zager and Beecham (2006) indicated the uncertainty of black and brown bear predation as the proximate or ultimate cause of ungulate mortality and concluded that the success of attempts to reduce bear predation likely depends on the prey population's relationship to habitat carrying capacity.

A particularly strong example of the role of predators in defining habitat for prey species is demonstrated by what has been termed the *ecology of fear* (Brown et al. 1999). The ecology of fear focuses on the nonlethal ways in which predators alter the behavior of prey animals, resulting in alteration of habitat use and effectively altering carrying capacity by changing what constitutes habitat for the prey with an emphasis on refugia. Prey are believed to assess predation risk, resulting in avoidance of areas in which risk is high or practicing behaviors such as aggregation to minimize risk, especially when needing to use risky areas. Perhaps the best-known example of the ecology of fear is Ripple and Beschta's (2004) use of wolf reintroduction to Yellowstone National Park and the resulting change in elk distributions within the park. Another example is Laundré's (2010) evaluation of mule deer and cougar space use. He reported that mule deer spent most of their time in low-risk areas (e.g., open areas), and cougars hunted the periphery of these areas.

PREDATION'S INFLUENCE ON ECOLOGICAL SYSTEMS

Cycling of predator and prey has drawn the attention of ecologists since Elton (1924) described the snowshoe hare–lynx cycle and Lotka and Volterra developed their equations. This start was focused on the role of predators in driving these cycles. Researchers since, as described previously, have painted a more complex picture in which predators are important in the process but not the sole influences of the cycling. Furthermore, cycling of prey species is predominantly a feature of high latitudes (Sinclair and Gosline 1997), presumably owing to the simpler systems that derive from a less diverse ecological community at high latitudes. Because most ecological systems consist of more complex systems of generalist and specialist predators, multiple predator species, and multiple prey species, we can expect that regular cycling of predator and prey species is not as common as the interest generated by these cycles when they occur.

Several of the concepts described previously have important implications to understanding predator–prey dynamics and management of predator and prey populations. Some of these management implications have already been presented. Intraguild predation, apparent competition, and the ecology of fear all have the ability to create *trophic cascades*. Trophic cascades occur when changes in populations at high trophic levels (i.e., predators) cause alternating changes to the trophic levels below them. For example, the lack of presence of an apex predator such as the wolf may lead to increase in abundance of the coyote through mesopredator release. The increase in coyotes may then decrease abundance of pronghorn (*Antilocapra americana*). Berger et al. (2008) found this cascade in Yellowstone National Park such that areas with wolves had pronghorn fawn survival four times higher than areas without wolves. Interestingly, the number of resident coyotes did not vary between these areas; the effect appears to arise from the influence of wolves on abundance of transient coyotes. Trophic cascades may extend through multiple trophic levels of predators. Levi and Wilmers (2012) used the restoration of wolves in Minnesota to demonstrate an apparent trophic cascade among three levels of predators. Wolf presence resulted in reduced coyote populations, which led to increases in fox populations. This additional level has management implications for prey owing to the release of the lowest level of predator (e.g., foxes). Following the findings of Crooks and Soulé (1999), songbird diversity might be reduced as the result of reduction of the coyote population in response to wolf presence.

Predator management planning needs to incorporate potential influences on community structure from presence or absence of a keystone predator. For example, Henke and Bryant (1999) found that coyote removal reduced diversity of rodents, with only the Ord's kangaroo rat (*Dipodomys ordii*) being abundant in coyote removal areas. In their area, the Ord's kangaroo rat was the competitively dominant rodent, and the presence of coyotes effectively mediated their competitive dominance, allowing other rodent species to increase in abundance, a form of *predator-mediated competition*.

Effects of trophic cascades extend into the primary producer trophic level. Estes and Duggins (1995) convincingly demonstrated that loss of sea otters (*Enhydra lutris*) in coastal waters resulted in population expansion of their primary prey, sea urchins (*Stronglyocentrotus* spp.), which in turn denuded the sea floor of its kelp forest, negatively affecting the entire ecosystem.

Trophic cascades are influenced by direct influences of predators on lower trophic levels and through the indirect effects of the ecology of fear. Kauffman et al. (2010) discussed behaviorally mediated trophic cascades, in which fear of predation alone can lead to reduction in grazing pressures by ungulates and subsequent increases in plant productivity. For example, debate among ecologists has occurred over the benefits of wolf restoration on riparian woody vegetation and subsequent benefits to neotropical migrant birds in Yellowstone National Park (Berger et al. 2001, Ripple et al. 2001, Kauffman et al. 2010).

Another implication drawing from the total response of predators to increases in prey is the concept of the predator pit.

Predator pits occur when a prey species density declines to low density by some factor other than predation, often a stochastic event such as weather or a human-caused event. Although the prey may be able to sustain higher densities with predation, at low densities they are unable to exhibit positive population growth and are kept by predation at lower than expected densities. Krebs (1996) suggested that collared lemmings (*Dicrostonyx groenlandicus*) were maintained at the low part of their cycle and prevented from cycling by an abundance of generalist predators that were also supported by tundra vole (*Microtus oeconomus*) populations. Messier and Crête (1985) believed that moose populations at low densities in southwestern Quebec were being maintained at levels lower than expected by the combined effect of wolf and bear predation and lack of alternative ungulate prey.

Above we discussed the importance of refugia in persistence of prey in the face of predation, with the typical result of laboratory experiments being extinction of prey. The introduction or expansion of novel predators can also have significant effects on prey species that do not have evolved adaptations and refugia in the face of these new predators. As a result, management of native species in the presence of novel predators is challenging. This has been most apparent on islands (Blackburn et al. 2004) but occurs worldwide. A well-known example of the influence of a novel predator is the brown tree snake on Guam, which was implicated in the extinction or steep (>90%) declines of 17 out of 18 native bird species (Wiles et al. 2003). Salo et al. (2007) conducted a meta-analysis of the effect of novel predators and found that consequences for prey populations were more substantial for novel predators than for native predators and that the effect was greater on mainland populations, although this latter observation may have been because of the large effect of novel predators in Australia. Large changes to native faunas due to introduced predators are issues beyond islands and Australia. Dorcas et al. (2012) documented steep declines in observations of bobcats, raccoons, and opossum (*Didelphis virginiana*), and nearly complete disappearance of cottontail rabbit (*Sylvilagus* spp.), in response to increases in Burmese pythons (*Python molurus bivittatus*) in Florida, indicating direct effects of the introduced predator at multiple trophic levels.

Novel predators are usually the result of intentional or unintentional introduction by humans. Anthropogenic activities, however, can also modify the environment to create opportunities for predators to expand into areas where they are novel. The example of the expansion of the barred owl into the Pacific Northwest discussed above is also speculated to be creating a trophic cascade that may negatively influence prey species shared by both barred and spotted owls, including several already sensitive species (Holm et al. 2016). Anthropogenic activities that result in reduced populations of sea lions and seals are also suspected in the shift of killer whales (*Orcinus orca*) to feeding on sea otters in western Alaska; this has initiated a novel trophic cascade leading to increased urchin populations, reversing the cascade created by reestablishment of the keystone sea otter, discussed above (Estes et al. 1998).

MANAGEMENT OF PREDATION

As noted above, when discussing the changing views on predation, we must first acknowledge the complicated history of humans, both as predators and prey, in the formation of our own interactions with carnivorous wildlife. Human interactions, aside from the fear of death from predators, historically reflect issues of conflict with wildlife, commonly in the form of interference competition as another predator. Although we document interference competition well in nonhuman species today, as wildlife professionals we often lose sight of the underpinning aspects of human ecology and sociality that create negativism toward species that compete with us for resources. Perhaps no more obvious than this is the damage to animal agriculture from predators, which gave rise to the earliest forms of documented wildlife management (Frank and Conover 2015). Needs to efficiently and economically protect livestock from predation in order to ensure a sustainable food supply for humans gave rise to a profession and discipline devoted to this pursuit. It would be foolish and shortsighted to regard such damage and conflict with predators to be a thing of the past. To wit, management of negative human interactions with predators, whether damage to economically valuable resources or human safety, has begun to once again increase, necessitating that the wildlife professional have a clear understanding of their work in managing predators.

In this section we address the management of predators and predation on native wildlife, humans, and their livestock aside. Nevertheless, before attempting to address how one may manage predation from one native species upon another, we must first acknowledge that predation management is a set of value judgements inextricably tied to human goals. As wildlife managers, we strive to conserve wildlife according to the best science available, recognizing that our view of a desirable ecosystem state is shaped by historic accounts, scientific understanding, and human value judgements. Within this context the management of predatory species, and the force of predation, is situated within human goals. In our case, there is a negativistic interaction, whether perceived or real, which evokes a response from humans to ameliorate the causal agent (Messmer 2000). Most often, this takes the form of lethal removal of the species that engages in the predatory action of concern. Lethal management may be warranted, as may be nonlethal actions, habitat modifications, or even no action at all. Mindful wildlife managers avoid conflating management of predation—an ecological force within the system—with management of predatory species as individuals or populations of wildlife. Rather, they seek the health of the system and long-term conservation of native species.

Assessing Sources and Rates of Predation

In order to manage predation as a force, or predators as species in terms of abundance, density, or carrying capacity, one must first be able to identify both the species-specific sources of predation on a population and the rates of loss. From this knowledge, management goals soundly grounded in science

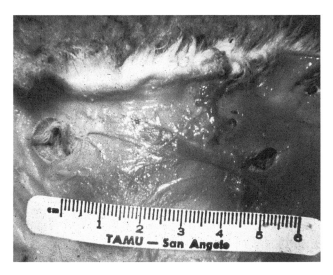

Figure 14.10. The location, spacing, and orientation of the puncture wounds indicate this lamb was killed by a golden eagle (*Aquila chyrsaetos*). Courtesy of John Tomeček

Figure 14.9. Many mammalian predators can be identified by careful examination for diagnostic characteristics of kills. In this case the location of the bite and spacing of canine punctures indicate predation by a cougar (*Felis concolor*). Courtesy of John Tomeček

may be established. Estimating predation rates is difficult because of the inherent mobility of predators and their tendency to avoid humans. Further, it is difficult to know rates of predation-related mortality under most management conditions, given the paucity of necessary data. Wildlife researchers can estimate mammalian predation rates by finding kill sites, by following radio-tagged predators to find killed or cached prey, and most commonly by monitoring survival and causes of mortality of radio-tagged prey animals.

A common feature of many studies on prey animals, particularly those also pursued as game by humans, is to account for sources of predation. Nevertheless, most individuals conducting investigations of kill sites and attributing predation to a specific species have little, if any, formal training in the science of forensic investigation of predation. To improve the science of predation management, it is critical for the wildlife biologist to improve their scientific understanding of the forensic signs of predation by various species (Figs. 14.9 and 14.10). It is also valuable to seek the counsel of those within our profession who have experience with these methods and forensic signs of predation, such as professional wildlife damage managers and carnivore biologists.

Before investigations can begin as to specific species of predator causing losses, the rate of mortality in a given prey population that can be attributed to predation must be considered. This is an important parameter in the monitoring, conservation, and sometimes recovery of a species. When approaching the need to assess predation, the mindful wildlife biologist first considers their need to know (1) rates of predation mortality in a given population and (2) species-specific sources of mortality, or perhaps both items in tandem. Just as we must crawl before we walk, and walk before we run, so too must wildlife biologists carefully address the total contribution of predation to the annual mortality rate of a populations. If

losses to predation mortality do not appear to contribute to decline of a species, or if the population in question is robust enough to sustain assessed rates of predation, then one needs to spend little time considering the precise proportion of take attributed to each predator in the community. But if predation appears to be a limiting factor to the species survival or recovery, one must investigate further.

It is common to use data from monitored prey animals to estimate their survival rates and identify sources of mortality. Radio and global positioning system attachments, however, can predispose some smaller prey animals to predation by negatively influencing their behavior or mobility (Murray and Fuller 2000). Predation rate estimates derived from monitoring tagged smaller prey animals may be biased, and caution is warranted when extrapolating these estimates to prey populations at large (Perkins 2019). For static wildlife resources, such as bird or herpetofauna nests, remotely triggered cameras are often used to determine the abundance and diversity of predators, as well as rates of nest predation (Coates et al. 2008, Vilardell et al. 2012). This approach, however, can introduce bias of the camera itself in some predatory species that avoid novel features in the environment (Harris and Knowlton 2001, Mettler and Shivik 2007).

One must be cautious as to the interpretation of a given predator species' influence on a given prey population, as these can be highly variable, depending largely on the methods used. Based on high-resolution prey use data from video recording at multiple Swainson's hawk (*Buteo swainsoni*) nests, two studies in West Texas found that quail accounted for 0.2% to 0.9% of prey deliveries (Giovanni et al. 2007, Behney et al. 2010). This is a measure of predation rates on bobwhite by a specific raptor species during a specific period (the nesting period of the hawks). In contrast, Carter (1995) reported that avian predators accounted for 25% of bobwhite mortality from February to July during two years in West Texas. This is a measure of cause-specific mortality by a guild across time. Additionally, predation rates are often associated with both

WARREN BAXTER BALLARD JR. (1947–2012)

Warren Ballard was born in Boston, Massachusetts, in 1947, but soon moved and grew up in Albuquerque, New Mexico. After earning a BS in fish and wildlife management from New Mexico State University in 1969 and an MS in environmental biology from Kansas State University in 1971, he went to work for the Alaska Department of Fish and Game as a wildlife biologist and research scientist. For more than 18 years he conducted in depth, groundbreaking research on predator–prey relationships among wolves, bears, and ungulate populations. He went on to earn a PhD from the University of Arizona in 1993 with the dissertation "Demographics, Movements, and Predation Rates of Wolves in Northwest Alaska." His research contributed significantly to the National Research Council report *Wolves, Bears and Their Prey in Alaska: Biological and Social Challenges in Wildlife Management*, published by the National Academies Press in 1997. He then served as director and associate professor with the New Brunswick Cooperative Fish and Wildlife Research Unit at the University of New Brunswick, and then as a research supervisor with the Arizona Game and Fish Department. In 1998 he joined the faculty at Texas Tech University, where he supervised graduate students in his lab as they conducted studies of a variety of species and issues, including white-tailed deer, mule deer, wild turkey, lesser prairie-chickens, and songbirds. But it was his lifelong interest in the ecology and conservation of predators and their relationships with prey that he pursued with passion. This resulted in not only continued study of wolf and bear ecology in Alaska, but also expansion of the research to Mexican wolves and black bears in the arid Southwest, as well as groundbreaking studies of the ecology and interactions of coyote and swift fox. Throughout his career, Warren authored or coauthored more than 200 peer-reviewed journal articles. He twice served as editor-in-chief of the *Wildlife Society Bulletin* and was named a Wildlife Society Fellow by The Wildlife Society in 2005. Ballard received the Outstanding Achievement Award from the Texas Chapter of the Wildlife Society in 2007. He was recognized academically with the Texas Tech University Chancellor's Council Distinguished Research Award in 2002 and the Outstanding Research Award from College of Agricultural Sciences and Natural Resources in 2009, held the Bricker Chair in Wildlife Management, and was named a Horn Professor (the highest honor a faculty member can receive from Texas Tech University) in 2008. But his legacy lies with the more than 60 students that received graduate degrees under his supervision.

prey diversity in a given area and competitive exclusion among predators (Allen et al. 2015, Sivy et al. 2018). For example, in Greenland, where prey diversity is low compared with temperate and tropical zones, Booms and Fuller (2003) reported that gyrfalcons preyed heavily upon rock ptarmigan (*Lagopus muta*); combined with Arctic hares (*Lepus arcticus*), the two prey species accounted for 79% to 91% of the diet.

If one had adequate knowledge of both predator and prey abundance in a given area, it may be possible to estimate rates of take by a species. Kenward (2006) proposed estimates of kill rates as a function of the daily food requirements of a given predator, the edible biomass available from each prey, and the proportion of given prey in the diet. This included material not consumed by the predator, and the amount wasted if a sated predator abandons prey (such as when a scavenger consumes cached prey before the predator retrieves it; Kenward et al. 1981). This estimate of waste is highly variable among predator species and does not account for individual variation, competitive interference, or conspecific interactions. For example, there would be no wastage of fish swallowed whole by great blue herons (*Ardea herodias*), but wastage could occur by osprey (*Pandion haliaetus*), which tear apart their fish prey. A number of predator species will cache prey for later consump-

tion. The degree of waste is difficult to infer, as it is determined by environmental factors (Komar and Beattie 1998). In contrast, wolf packs will gorge themselves and remain close by the kill until it is consumed.

Confounding factors in Kenward's approach include accounting for how daily food requirements vary with environmental conditions and behavior. For example, a male Cooper's hawk hunts for himself and his mate as nesting season approaches, and then for their brood. Additionally, the behavioral influence to hunt and capture food during the breeding period can result in surplus killing. Such situations are difficult to document and, given the difficulty of capturing prey, probably rare; however, they do occur and can result in spoilage (Boal 1997). Such behaviors are not limited to the breeding season. Eurasian pygmy owls (*Glaucidium passerinum*) exhibited a functional response to abundant prey by surplus killing and caching during winter (Solheim 1984). Regardless, such estimates of kill rates are still commonly based on knowledge of the proportion of prey in the diet and understanding the daily food requirements of the given predator species, both of which are often lacking. Furthermore, understanding the kill rate is only one component of more sophisticated models required to estimate the population-level effects of predators on prey (Eberhardt et al. 2003). The field of wildlife science is increasingly aware of single predation events resulting in multi-predator and multitrophic interactions that result in food for a number of species (Wilson and Wolkovich 2011).

A poignant example is the diet of specific raptors, which can be partially understood by identifying prey (or their remains) at nests. This limits data only to the breeding season diets; few quantitative studies have attempted to determine wintering raptor diets (Roth and Lima 2003, Roth et al. 2006, Millsap et al. 2013). Extension of inference from temporally limited data sets may bias estimates of species-specific contributions to overall predation-related mortality of a prey species. Biologists are rarely able to assign a kill to a specific raptor species, much less an individual. In contrast, mammalian predator species, and sometimes specific individuals (e.g., wolf pack, territorial cougar), can often be identified by interpreting evidence (e.g., location, tracks, prey condition) at the kill site. Many mammalian carnivores present sign upon carcasses related to unique methods of killing, handling, and consuming prey. For example, cougars typically kill their prey by bites to the back of the neck and head, usually feed on the shoulders of the prey first, and then cover the remains with leaves, soil, and other debris (Fig. 14.11) until returning to feed again (VerCauteren et al. 2005). In contrast, wolves typically bring down prey by attacking the hind legs and flanks until it is disabled, and they then disembowel it and eat the viscera and hindquarters first (VerCauteren et al. 2005).

It must be further recognized that some predators of concern act as predators upon other predatory species, such as the consumption of a diversity of mesocarnivores by jaguars (*Panthera onca*) and cougars (Cassaigne et al. 2016). Thus one must be careful that in directly manipulating the abundance of a particular species, unintended consequences do not occur, given the potential for interference competition among

carnivores. For example, the presence of wolves may result in coyote avoidance, leading to an increase in some prey items, such as pronghorn fawns (Berger et al. 2008), or increase in abundance of smaller predators often preyed upon by coyotes, such as foxes (Levi and Wilmers 2012).

Assessing Less-Than-Lethal Effects

During the latter 20th and early 21st centuries, wildlife scientists became increasingly interested in the effects of predators on prey populations that were negativistic, but not necessarily lethal. Evaluation of such less-than-lethal effects of predators on prey can be difficult to assess outside the context of properly designed studies. It is incumbent upon the wildlife manager to "do no harm" in terms of management efforts, which requires a thorough understanding of the system. In many cases, native prey have evolved behaviors (Laporte et al. 2010) and behavioral plasticity (Peluc et al. 2008) in response to native predators that, in the presence of sufficient habitat, ameliorates deleterious effects of predation. In the context of wildlife management, then, one must consider our role to be one of the ecosystem manager, to provide a more complete system that fulfills the life history needs of predator and prey, rather than seeking to directly intervene in their interactions. As a final consideration, we must be equally aware that humans are not without their less-than-lethal negative influence on other species as a result of our presence, and thus "nonconsumptive" use of native wildlife (e.g., bird-watching) may also still serve to perturb the system (Tapper 2006).

Predator and Predation Management: What Is the Goal?

Allowing for the difficulty in assessing sources and effects of predation on prey within complex multitrophic systems, the cognizant wildlife biologist would feel somewhat uncertain

Figure 14.11. Cougars will often cover remains of prey and return for subsequent meals. Courtesy of John Tomeček

of how to proceed with managing predators. Doubtless, the management of ecological systems is a lofty goal when we possess an incomplete understanding of the system. The difficulty in understanding predator–prey relationships, and how to manage them, is compounded by human influence on natural systems. Habitat loss, habitat degradation, altered faunal communities, and altered predator–prey ratios contribute to an environment where predation losses could have implications for species conservation (Reynolds and Tapper 1996). Further, lest we forget, we too are an apex predator that influences wildlife behavior, and our consumptive use of wildlife resources (e.g., hunting, fishing, or trapping) introduces a competitive element to wildlife management (Wam et al. 2012).

Perhaps no other wildlife management practice generates more controversy among humans than predator control (Van Ballenberghe 2006). Nevertheless, management of predation is one of the oldest forms of wildlife management, practiced by humans for protection of livestock, crops, and game since antiquity (Frank and Conover 2015). Predator control to conserve species of concern is a more contemporary practice (Sinclair et al. 1998). The advent of this form of management came into its own in the great game restorations of the early 20th century, typified by Aldo Leopold's "Fierce Green Fire" that noted the duality of killing predators to restore prey overharvested by humans (Leopold 1949). Thus we experience the conundrum of the modern wildlife manager: when, and where, is predator management warranted to achieve a conservation goal. When one considers these actions, we must first ask: *Is it the predator, a species itself, or predation the ecological force that we seek to manage?*

When it is determined that the goal is to manage the population of predatory wildlife to achieve some systems-level goal, such as limiting the abundance, density, or geographic range of a species, one must first assess (1) present abundance and range of the species, (2) carrying capacity vis-à-vis prey resources, and (3) other factors that may warrant reduction below carrying capacity, such as restoration of a previously extirpated prey or competitor species. If direct predator management is warranted, it must be clear what one hopes to achieve and what strategy will best achieve those goals. One must also divorce the management of predation from the removal of individual predators or population-level reductions to seek management actions capable of achieving conservation goals.

When considering predation as an ecological force, one must consider whether intervening is biologically justified. Often, such justification comes in the discussion of landscape carrying capacity of either predator or prey, and stochasticity in weather, overuse of vegetative resources by animal agriculture, or both, which may severely limit carrying capacity of the prey in the short term. Thus wildlife managers often justify direct removals of predators as a salve to a damaged plant community. In a system on the brink of habitat or community collapse, one might find themselves removing a few predators, especially those with compensatory reproduction (Minnie et al. 2016, Kilgo et al. 2017) as the lesser of two real evils. There are numerous approaches to managing predation, including nonlethal (e.g., deterrents, translocation) and lethal means. Regardless of method, reliably predicting the effects of a management action on species of interest is problematic. The dynamic nature of natural systems results in changing interactions among biological, environmental, and practical factors (Boertje et al. 2010).

Recognizing that it is physically straightforward to conduct lethal predator management, it is therefore critical for the wildlife professional to work within a schema for making the best decision as to type of management action to employ. Just as with management of habitat, it is easier to carefully remove than have to restore following over-removal. When examining predator control for moose management, Boertje et al. (2010) made recommendations for criteria allowing for virtually any predator control program. When met, these recommended criteria work to ensure that predator control is justified and scientifically defensible. First, it should be established that predators kill substantial numbers of the species of interest that would usually otherwise survive if the predator was removed. Second, that reduced predation can facilitate reliably higher harvests or abundance of the species of interest. Third, that given less predation, habitats can sustain more of, but also be protected from, the species of interest. Fourth, that sustainable populations of predator species will persist in and out of control areas. Ballard et al. (2001) and Ballard (2011) presented criteria for predator reduction programs that might result in increased numbers of mule deer, black-tailed deer (*O. hemionus columbianus*), and white-tailed deer. The researchers emphasized that predator control programs are not likely to influence an ungulate population that is near carrying capacity.

In situations where carrying capacity has not been attained, predator control can have positive influences on prey populations. Keech (2005) concluded that relocation of black and grizzly bears resulted in an increase in moose calf survival in interior Alaska. Similarly, relocation of black bears from Great Smoky Mountains National Park improved calf elk recruitment from 0.306 to 0.544 (Yarkovich et al. 2011). Removal and relocation of predators might be short-term or partial solutions, however: although moose calf proportions increased 5% to 24% in Saskatchewan following bear removal, they later returned to pre-removal levels when bears were no longer removed (Stewart et al. 1985). Brown and Conover (2011) reported that coyote control in Utah and Wyoming resulted in increased pronghorn fawn survival but had no effect on mule deer. Thus it is incumbent upon the wildlife manager to be both thorough in their background research about the desired outcome and ancillary effects of a management action, and to be adaptive when management does not produce the expected result. Another consideration is that predators typically have lower survival once they are translocated. It is therefore prudent to consider whether a management action that reduces predators is economically feasible and ecologically justifiable, and for how long.

An extreme example of a system not at landscape carrying capacity for a native species is a restoration cycle, typified by the reintroduction of otherwise extirpated species. In theory, the goal of restoration is a self-sustaining population that can tolerate predation from native predators. Often, creation and improvement of habitat are a key focus of this work, but predator removals may ameliorate increased rates of predation related to insufficient habitat quantity or quality in the meantime (Reynolds and Tapper 1996, Reynolds et al. 2010). In the course of this conservation action, predator removal can be justified until such a time as monitoring indicates sufficient population size that predator removals are no longer needed to ensure survival. Nevertheless, a successful restoration event must be coupled with adequate habitat for the species to survive and the understanding that predator removal is a means to an end, rather than an end itself.

As the wildlife professional gains more detailed insight into community-level interactions, more unintended consequences of predator control arise. Examples include the increase of undesirable small mammal and lagomorph populations (Henke and Bryant 1999) or unintended predation of lower-level prey by mesocarnivores following removal of top-level predators (Letnic et al. 2009). Clearly, removal of predatory species is not a one-size-fits-all approach, and the goal of such actions must be clear before beginning so as to prepare for ancillary effects that will require additional management from the wildlife biologist. As a living laboratory for unintended consequences of predator removal, the Yellowstone Ecosystem demonstrates that wolf removal resulted in multitrophic changes in the ecosystem, from the loss of woody plants caused by overgrazing by ungulates (Ripple and Beschta 2003) to, as Olechnowski and Debinski (2008) suggested, overgrazing by elk being the ultimate cause for substantially reduced songbird species richness and abundance. Following wolf reintroductions in the 1990s, woody cover has increased (Ripple and Beschta 2012) and, presumably in time, so will avian species. As with all things, we as a profession must be careful not to overextend the results from the Yellowstone Ecosystem beyond its inferential limit: a wise wildlife professional reexamines each community in which they work.

In contrast to mammalian and reptilian predators, predatory birds in the United States are protected at the federal level by the Migratory Bird Treaty Act (MBTA) and, depending on species and status of other federal legislation, they usually receive similar protection at the state level. Avian predators are therefore primarily controlled via nonlethal methods (VerCauteren et al. 2005). Legal lethal control of predatory birds is an exceptional occurrence and requires issuance of permits by the state and federal agencies (e.g., US Fish and Wildlife Service). When lethal control is initiated, it is more often associated with human–wildlife conflict than predator–prey issues. For example, piscivorous birds can have substantial economic effects on aquaculture facilities (Taylor and Strickland 2008). Some form of predator control could potentially be necessary to reduce human–wildlife conflict and for conservation of threatened and endangered species. Although various nonlethal methods have been developed, lethal control of problem species may be allowed when other deterrents are

not available or are ineffective. Notably, such lethal control is often initiated not to reduce abundance of the species broadly, but rather to illicit a predator avoidance response in target species groups, causing them to flee (Lowney 1999). Lethal control continues to be socially controversial, exemplified by the ongoing conflict between fishermen and nonconsumptive wildlife users over control of cormorants (*Phalacrocorax* spp.; Schusler and Decker 2002, Ovegard et al. 2021). Societal and cultural values often influence these debates, and science used to support differing positions can be misunderstood, misinterpreted, or inherently biased (Boertje et al. 2010). The wildlife biologist, a scientist above all else, must first examine what is biologically justifiable and then what is socially acceptable. Aside from social concerns, lethal control can be challenging when the predator is also a species of concern either as game or a declining species. Situations such as this remind the wildlife profession that management of systems, not species, is a higher-level goal that must be considered in our work to advance conservation.

To this end, many today favor development of methods for nonlethal control of predators. Although such approaches sound good in theory, we must delve deeper into practicality of application and longer-term effects. Traditionally, these techniques have been developed for the protection of livestock (Bodenchuk and Hayes 2007), primarily in the form of human-escort, dissuasive devices, fencing, netting, or other exclusion devices. A more contemporary method is the use of dogs (*Canis familiaris*), llamas (*Lama glama*), and donkeys (*Equus asinus*) as livestock guardians (Fig. 14.12; Andelt 2001). Although these approaches work with domestic animals, particularly in closed environments, few effective methods have been developed to reduce predation on non-static wildlife prey species. Some examples include fencing for waterfowl or sea turtle (*Caretta caretta*) nesting grounds (Baskale and Kaska 2005, Malpas et al. 2013). Logically, the wildlife manager experiences difficulty addressing free-roaming wildlife in the same terms as confined livestock. As wildlife management goes, the manager is most commonly concerned about improvement to the quality and quantity of habitat. Habitat management can result in landscapes less attractive to a given predator species, whether or not such a configuration is typical of the native vegetative structure. While this does not result in direct mortality and hence lethal control, mortality may result because of decreased habitat quality or dispersal and competition with individuals already residents in occupied areas. As a consequence, society increasingly favors the concept of translocating predators.

Translocation has become a frequent tool in lieu of lethal control and also for population introduction or enhancement; results have been mixed (Linnell et al. 1997). Before attempting a translocation as a nonlethal control measure, the mindful wildlife manager must ask if this action is truly nonlethal and nondisruptive to the receiving ecosystem. In some cases, translocated animals experience far lower survival when removed from their natural range. In other cases, the influx of new individuals disrupts social dynamics in the receiving ecosys-

Figure 14.12. The use of livestock guardian dogs is an increasingly common and effective practice to reduce livestock depredations. Courtesy of John Tomeček

tem, perhaps increasing predation there. Although translocations may be considered more socially acceptable than lethal control, one cannot quickly characterize them as nonlethal or innocuous; delayed death by starvation due to relocation is still a lethal end point. Nevertheless, we must consider both biological and social implications of translocation as a tool for predator management. In the case of livestock protection, or to provide shorter-term support of restoration efforts, translocations may be both economical and socially justifiable (Armistead et al. 1994).

Translocations of predators have been met with mixed results. Ruth et al. (1998) reported that 2 of 14 translocated cougars returned to their home ranges in New Mexico despite being translocated an average of 477 km away. Eight others headed in a general direction that led the authors to suspect they potentially could have been attempting to return to their original ranges. Additionally, 9 of the 14 died during the study period, but their mortality rate was not different from that of cougars in a reference area. The authors reported that translocations of cougars would be most successful with younger (12- to 27-month-old) individuals. Bradley et al. (2005) translocated wolves with mixed success. Most translocated wolves (67%) never established, or joined, a pack and experienced lower survival than reference animals. The primary cause of mortality was lethal control as a result of more than 25% of translocated wolves preying upon livestock. Translocated wolves also demonstrated a homing tendency, either returning to or moving in the direction of their original area. Bradley et al. (2005) summarized that those individuals that returned tended to be adults, had been translocated comparatively shorter distances, and had gone through hard releases. A hard release is when animals are essentially captured, transported, and released. A soft release is when the animal is relocated to the area but held in an enclosure for a period of time to allow adjustment and acclimation to the new area. Bradley et al. (2005) suggested that translocation efforts would likely be more successful if soft releases were used. Homing behavior is often thought of as a trait of higher, more sophisticated vertebrates (e.g., mam-

mals, birds), which could be a concern in virtually any translocation project. For example, Nowak et al. (2002) reported that translocated western diamondback rattlesnakes (*Crotalus atrox*) experienced decreased survival, and more than 50% returned to their original locations.

Ultimately, little is known about the parameters by which translocations need be successful to prevent return to the original area. For species such as black bears (Armistead et al. 1994, Smith and Clark 1994) or American alligators (*Alligator mississippiensis*; Janes 2004), translocation of "nuisance" animals is a common tool to ameliorate human–wildlife conflicts, but for some species, little is known about the effects of relocations. This represents a clear area for development of this tool and also a point of caution within the North American Model of Wildlife Conservation: science must inform management, and management should not be done without scientific assessment.

Habitat Management as Indirect Predator Management

Unlike management of predation upon livestock, native wildlife coevolved with native predators in a particular ecological context. As discussed earlier in this chapter, a challenge for wildlife managers is the difficulty of identifying the correct action to reduce predation rates, especially in a multi-predator, multi-prey community. A solution to this may lie in a systems-level approach. For many professional wildlife biologists, an overarching goal of their program is the creation, maintenance, improvement, and expansion of "habitat." Incumbent upon the wildlife manager, then, is an appropriate balance of herbivorous species within the plant community. When considering management of predation, understanding that quality habitat can ameliorate higher rates of predation provides an avenue to predator management that exists within systems-focused, rather than species-focused, management. Not only can more prey be supported by increasing the nutritional plane of the area, but they may also be killed by predators at a decreased rate. The type of cover, structure of plant community, linearity of the habitat (Frey and Conover 2006, Eglington et al. 2009), and structure and interspersion of habitat patches may determine more about predation than we have previously believed (Bowyer et al. 1998, Grovenburg et al. 2012). Conversely, structure of habitat can increase the ease with which predators capture prey, sometimes increasing the time predators spent in that area (Grant et al. 2005, Andruskiw et al. 2008). Although many other elements of predator management may confound the wildlife biologist, the creation, improvement, and manipulation of habitat to achieve a conservation goal is among the oldest, and most important, tools to conserve wildlife. So, too, when considering managing predation, habitat may indeed be the answer.

SUMMARY

Predation has long garnered the public's attention and generated numerous, often erroneous, ideas about its role and function. Despite the difficulties involved in the study of predators and their prey, the wildlife profession has made great strides in understanding predator ecology and predator–prey relationships. We have progressed from primarily viewing predators as simply a threat to ourselves, our livestock, and game animals to developing an understanding of the complex interplay of predator–prey interactions and their influence on ecosystems. This increased understanding has led to more nuanced approaches to the management of predators.

There is a diverse array of predators, and it is important for wildlife biologists and managers to consider their level of specialization and hunting techniques. The vulnerability of prey to predation was seen as a critical piece in understanding which prey are consumed and how this effects prey populations, emphasizing the importance of mechanisms that create refugia. Through research, this discipline has gained knowledge as to the influences of density-dependence mechanisms and numerical and functional responses on predator–prey dynamics. These insights have enabled a deeper understanding of how prey populations persist in the face of predation.

As our understanding of simpler predator-prey systems evolved, biologists started focusing on the role of predation from a more complex community structure. This includes the influences of predation that arise from interspecific competition among predators themselves, the role of other prey in facilitating predation on a species, and the effects of predation risk on prey's use of the landscape. The understanding of these interacting influences on ecological systems has led to a better understanding of the effects of management of predators or predation, and when predator control may or may not be justified. Modern predator management options range from lethal to nonlethal approaches and habitat management. When management of predators is considered, it is important that the goals are clear, and that management actions will be likely to achieve their desired outcomes.

Predators and their prey continue to generate intense scientific interest. Undoubtedly, our understanding of these relationships and our options to manage them will continue

to expand rapidly. To conclude, the words of John and Frank Craighead (1956:352), written more than half a century ago, continue to ring true: "predation is a powerful and complex natural force that should be visualized in its ecological entirety."

Literature Cited

Allen, M. L., L. M. Elbroch, C. C. Wilmers, and H. U. Wittmer. 2015. The comparative effects of large carnivores on the acquisition of carrion by scavengers. American Naturalist 185:822–833.

Andelt, W. F. 2001. Effectiveness of livestock guarding animals for reducing predation on livestock. Endangered Species Update 18:182–185.

Andruskiw, M., J. M. Fryxell, I. D. Thompson, and J. A. Baker. 2008. Habitat-mediated variation in predation risk by the American marten. Ecology 89:2273–2280.

Anthony, R. G. 2001. Low productivity of bald eagles on Prince of Wales Island, southeast Alaska. Journal of Raptor Research 35:1–8.

Arditi, R., and L. R. Ginzburg. 1989. Coupling in predator-prey dynamics: ratio-dependence. Journal of Theoretical Biology 139:311–326.

Armistead, A. R., K. Mitchell, and G. E. Connolly. 1994. Bear relocations to avoid bear/sheep conflicts. Proceedings of the Vertebrate Pest Conference 16:31–35.

Ballard, W. B. 2011. Predator–prey relationships. Pages 251–285 in D. G. Hewitt, editor. Biology and management of white-tailed deer. CRC Press, Boca Raton, Florida, USA.

Ballard, W. B., D. Lutz, T. W. Keegan, L. H. Carpenter, and J. C. deVos Jr. 2001. Deer–predator relationships: a review of recent North American studies with emphasis on mule and black-tailed deer. Wildlife Society Bulletin 29:99–115.

Ballard, W. B., and S. D. Miller. 1990. Effects of reducing brown bear density on moose calf survival in south-central Alaska. Alces 26:9–13.

Baskale, E., and Y. Kaska. 2005. Sea turtle nest conservation techniques on southwestern beaches in Turkey. Israel Journal of Zoology 51:13–26.

Bayne, E. M., and K. A. Hobson. 2002. Effects of red squirrel (Tamasciurus hudsonicus) removal on survival of artificial songbird nests in boreal forest fragments. American Midland Naturalist 147:72–79.

Behney, A. C., C. W. Boal, H. A. Whitlaw, and D. R. Lucia. 2010. Prey use by Swainson's hawks in the lesser prairie-chicken range of the Southern High Plains. Journal of Raptor Research 44:317–322.

Berger, J., P. B. Stacey, L. Bellis, and M. P. Johnson. 2001. A mammalian predator-prey imbalance: grizzly and wolf extinction affect avian neotropical migrants. Ecological Applications 11:947–960.

Berger, K. M., E. M. Gese, and J. Berger. 2008. Indirect effects and traditional trophic cascades: a test involving wolves, coyotes, and pronghorn. Ecology 89:818–828.

Bishop, C. J., G. C. White, D. J. Freddy, B. E. Watkins, and T. R. Stephenson. 2009. Effect of enhanced nutrition on mule deer population rate of change. Wildlife Monographs 172:1–28.

Blackburn, T. M., P. Cassey, R. P. Duncan, K. L. Evans, and K. J. Gaston. 2004. Avian extinction and mammalian introductions on oceanic islands. Science 305:1955–1958.

Boal, C. W. 1997. An urban environment as an ecological trap for Cooper's hawks. PhD dissertation, University of Arizona, Tucson, USA.

Bodenchuk, M. J., and D. J. Hayes. 2007. Predation impacts and management strategies for wildlife protection. Pages 221–263 in A. M. T. Elewa, editor. Predation in organisms: a distinct phenomenon. Springer-Verlag, Berlin, Germany.

Boertje, R. D., W. C. Gasaway, D. V. Grangaard, and D. G. Kellyhouse. 1988. Predation on moose and caribou by radio-collared grizzly bears in east central Alaska. Canadian Journal of Zoology 66:2492–2499.

Boertje, R. D., M. A. Keech, and T. F. Paragi. 2010. Science and values influencing predator control for Alaska moose management. Journal of Wildlife Management 74:917–928.

Booms, T. L., and M. R. Fuller. 2003. Gyrfalcon diet in central west Greenland during the nesting period. Condor 105:528–537.

Border, J. A., L. R. Atkison, I. G. Henderson, and I. R. Hartley. 2017. Nest monitoring does not affect nesting success of whinchats Saxicola rubetra. Ibis 160:624–633.

Bowyer, R. T., J. G. Kie, and V. Van Ballenberghe. 1998. Habitat selection by neonatal black-tailed deer: climate, forage, or risk of predation? Journal of Mammalogy 79:415–425.

Boyce, M. S. 2000. Modeling predator–prey dynamics. Pages 253–287 in L. Boitani and T. K. Fuller, editors. Research techniques in animal ecology: controversies and consequences. Columbia University Press, New York, New York, USA.

Bradley, E. H., D. H. Pletscher, E. E. Bangs, K. E. Kunkel, D. W. Smith, C. M. Mack, T. J. Meier, J. A. Fontaine, C. C. Niemeyer, and M. D. Jimenez. 2005. Evaluating wolf translocation as a nonlethal method to reduce livestock conflicts in the northwestern United States. Conservation Biology 19:1498–1508.

Brelsford, M. J., J. M. Peek, and G. A. Murray. 1998. Effects of grazing by wapiti on winter wheat in northern Idaho. Wildlife Society Bulletin 26:203–208.

Brown, D. E., and M. R. Conover. 2011. Effects of large-scale removal of coyotes on pronghorn and mule deer productivity and abundance. Journal of Wildlife Management 75:876–882.

Brown, J. L. 1969. Territorial behavior and population regulation in birds. Wilson Bulletin 81:293–329.

Brown, J. S., B. P. Kotler, and A. Bouskila. 2001. Ecology of fear: foraging games between predators and prey with pulsed resources. Annales Zoologici Fennici 38:71–87.

Brown, J. S., J. W. Laundré, and M. Gurung. 1999. The ecology of fear: optimal foraging, game theory, and trophic interactions. Journal of Mammalogy 80:385–399.

Carter, P. S. 1995. Post-burn ecology of northern bobwhites in West Texas. MS thesis, Angelo State University, San Angelo, Texas, USA.

Cassaigne, I., R. A. Medellín, R. W. Thompson, M. Culver, A. Ochoa, K. Vargas, J. L. Childs, J. Sanderson, R. List, and A. Torres-Gómez. 2016. Diet of pumas (Puma concolor) in Sonora, Mexico, as determined by GPS kill sites and molecular identified scat, with comments on jaguar (Panthera onca) diet. Southwestern Naturalist 61:125–132.

Chaneton, E. J., and M. B. Bonsall. 2000. Enemy-mediated apparent competition: empirical patterns and the evidence. Oikos 88:380–394.

Coates, P. S., J. W. Connelly, and D. J. Delehanty. 2008. Predators of greater sage-grouse nests identified by video monitoring. Journal of Field Ornithology 79:421–428.

Coleman, B. T., and R. A. Hill. 2014. Living in a landscape of fear: the impact of predation, resource availability and habitat structure on primate range use. Animal Behaviour 88:165–173.

Courchamp, F., R. Woodroffe, and G. Roemer. 2003. Removing protected populations to save endangered species. Science 302:1532.

Craighead, J. J., and F. C. Craighead Jr. 1956. Hawks, owls and wildlife. Wildlife Management Institute, Washington, DC, USA.

Creel, S., G. Spong, and N. Creel. 2001. Interspecific competition and the population biology of extinction-prone carnivores. Pages 35–60 in J. L. Gittleman, S. M. Funk, D. Macdonald, et al., editors. Carnivore conservation. Cambridge University Press, Cambridge, UK.

Crooks, K. R., and M. E. Soulé. 1999. Mesopredator release and avifaunal extinctions in a fragmented system. Nature 400:563–566.

Dale, B. W., L. G. Adams, and R. T. Bowyer. 1994. Functional response of wolves preying on barren-ground caribou in a multiple-prey ecosystem. Journal of Animal Ecology 63:644–652.

Dawkins, R., and J. R. Krebs. 1979. Arms races between and within species. Proceedings of the Royal Society B 205:489–511.

Dekker, D., M. Out, M. Tabak, and R. Ydenberg. 2012. The effect of kleptoparasitic bald eagles and gyrfalcons on the kill rate of peregrine falcons hunting dunlins wintering in British Columbia. Condor 114:290–294.

DelGiudice, G. D. 1998. Surplus killing of white-tailed deer by wolves in northcentral Minnesota. Journal of Mammalogy 79:227–235.

Denny, M. 2014. Buzz Holling and the functional response. Bulletin of the Ecological Society of America 95:200–203.

Dhondt, A. A., B. Kempenaers, and F. Adriansen. 1992. Density-dependent clutch size caused by habitat heterogeneity. Journal of Animal Ecology 61:643–648.

Dobson, A., and A. Lyles. 2000. Black-footed ferret recovery. Science 288:985–988.

Dorcas, M. E., J. D. Willson, R. N. Reed, R. W. Snow, M. R. Rochford, et al. 2012. Severe mammal declines coincide with proliferation of invasive Burmese pythons in Everglades National Park. Proceedings of the National Academy of Sciences of the United States of America 109:2418–2422.

Dunn, E. H., and D. L. Tessaglia. 1994. Predation of birds at feeders in winter. Journal of Field Ornithology 65:8–16.

Eacker, D. R., M. Hebblewhite, K. M. Proffitt, B. S. Jimenez, M. S. Mitchell, and H. S. Robinson. 2016. Annual elk calf survival in a multiple carnivore system. Journal of Wildlife Management 80:1345–1359.

Eberhardt, L. L., R. A. Garrott, D. W. Smith, P. J. White, and R. O. Peterson. 2003. Assessing the impact of wolves on ungulate prey. Ecological Applications 13:776–783.

Eglington, S. M., J. A. Gill, M. A. Smart, W. J. Sutherland, A. R. Watkinson, and M. Bolton. 2009. Habitat management and patterns of predation of northern lapwings on wet grasslands: the influence of linear habitat structures at different spatial scales. Biological Conservation 142:314–324.

Elton, C. S. 1924. Periodic fluctuations in the numbers of animals: their causes and effects. Journal of Experimental Biology 2:119–163.

Errington, P. L. 1946. Predation and vertebrate populations. Quarterly Review of Biology 21:144–177, 221–245.

Estes, J. A., and D. O. Duggins. 1995. Sea otters and kelp forests in Alaska: generality and variation in a community ecological paradigm. Ecological Monographs 65:75–100.

Estes, J. A., M. T. Tinker, T. M. Williams, and D. F. Doak. 1998. Killer whale predation on sea otters linking oceanic and nearshore ecosystems. Science 282:473–476.

Fascione, N., A. Delach, and M. E. Smith. 2004. People and predators: from conflict to coexistence. Island Press, Washington, DC, USA.

Fernandez, C., P. Azkona, and J. A. Donazar. 1998. Density-dependent effects on productivity in the griffon vulture Gyps fulvus: the role of interference and habitat heterogeneity. Ibis 140:64–69.

Ferrer, M., and J. A. Donazar. 1996. Density-dependent fecundity by habitat heterogeneity in an increasing population of Spanish imperial eagles. Ecology 77:69–74.

Frank, M. G., and M. R. Conover. 2015. Thank goodness they got all the dragons: wildlife damage management through the ages. Human–Wildlife Interactions 9:156–162.

Frey, S. N., and M. R. Conover. 2006. Habitat use by meso-predators in a corridor environment. Journal of Wildlife Management 70:1111–1118.

Fuller, T. K., and L. B. Keith. 1981. Non-overlapping ranges of coyotes and wolves in northeastern Alberta. Journal of Mammalogy 62:403–405.

Furness, R. W., K. Ensor, and A. V. Hudson. 1992. The use of fishery waste by gull populations around the British Isles. Ardea 80:105–113.

Gause, G. F. 1935. Experimental demonstration of Volterra's periodic oscillation in the numbers of animals. Journal of Experimental Biology 12:44–48.

Gilg, O., B. Sittler, B. Sabard, A. Hurstel, R. Sane, P. Delattre, and I. Hanski. 2006. Functional and numerical responses of four lemming predators in high arctic Greenland. Oikos 113:193–216.

Giovanni, M. D., C. W. Boal, and H. A. Whitlaw. 2007. Prey use and provisioning rates of breeding ferruginous and Swainson's hawks on the Southern Great Plains, USA. Wilson Journal of Ornithology 119:558–569.

Godbois, I. A., L. M. Conner, and R. J. Warren. 2004. Space-use patterns of bobcats relative to supplemental feeding of northern bobwhite. Journal of Wildlife Management 68:514–518.

Grant, J., C. Hopcraft, A. R. E. Sinclair, and C. Packer. 2005. Planning for success: Serengeti lions seek prey accessibility rather than abundance. Journal of Animal Ecology 74:559–566.

Gregory, S. D., and F. Courchamp. 2010. Safety in numbers: extinction arising from predator-driven Allee effects. Journal of Animal Ecology 79:511–514.

Grovenburg, T. W., K. L. Monteith, R. W. Klaver, and J. A. Jenks. 2012. Predator evasion by white-tailed deer fawns. Animal Behaviour 84:59–65.

Gulsby, W. D., J. C. Kilgo, M. Vukovich, and J. A. Martin. 2017. Landscape heterogeneity reduces coyote predation on white-tailed deer fawns. Journal of Wildlife Management 81:601–609.

Hammer, H. J., R. L. Thomas, and M. D. E. Fellowes. 2016. Provision of supplementary food for wild birds may increase risk of local nest predation. Ibis 159:158–167.

Hansen, A. J. 1987. Regulation of bald eagle reproductive rates in southeast Alaska. Ecology 68:1387–1392.

Hanski, I., and E. Korpimaki. 1995. Microtine rodent dynamics in northern Europe: parameterized models for the predator–prey interaction. Ecology 76:840–850.

Harris, C. E., and F. F. Knowlton. 2001. Differential responses of coyotes to novel stimuli in familiar and unfamiliar settings. Canadian Journal of Zoology 79:2005–2013.

Hegel, T. M., C. C. Gates, and D. Eslinger. 2009. The geography of conflict between eld and agricultural values in the Cypress Hills, Canada. Journal of Environmental Management 90:222–235.

Henke, S. E., and F. C. Bryant. 1999. Effects of coyote removal on the faunal community in western Texas. Journal of Wildlife Management 63:1066–1081.

Hernández, L., and J. W. Laundré. 2005. Foraging in the "landscape of fear" and its implications for habitat use and diet quality of elk Cervus elaphus and bison Bison bison. Wildlife Biology 11:215–220.

Holling, C. S. 1959a. The components of predation as revealed by a study of small mammal predation of the European pine sawfly. Canadian Entomology 91:293–320.

Holling, C. S. 1959b. Some characteristics of simple types of predation and parasitism. Canadian Entomology 91:385–398.

Holling, C. S. 1965. The functional response of predators to prey density and its role in mimicry and population regulation. Memoirs of the Entomological Society of Canada 45:5–60.

Holm, S. R., B. R. Noon, J. D. Wiens, and W. J. Ripple. 2016. Potential trophic cascades triggered by the barred owl range expansion. Wildlife Society Bulletin 40:615–624.

Holt, R. D. 1977. Predation, apparent competition, and the structure of prey communities. Theoretical Population Biology 12:197–229.

Horne, J. S., M. A. Hurley, C. G. White, and J. Rachael. 2019. Effects of wolf pack size and winter conditions on elk mortality. Journal of Wildlife Management 83:1103–1116.

Huffaker, C. 1958. Experimental studies on predation: dispersion factors and predator-prey oscillations. Hilgardia 27:343–383.

Hurley, M. A., J. W. Unsworth, P. Zager, M. Hebblewhite, E. O. Garton, D. M. Montgomery, J. R. Skalski, and C. L. Maycock. 2011. Demographic response of mule deer to experimental reduction of coyotes and mountain lions in southeastern Idaho. Wildlife Monographs 178:1–33.

Jacobson, M. D., E. T. Tsakiris, A. M. Long, and W. E. Jensen. 2011. No evidence for observer effects on lark sparrow nest survival. Journal of Field Ornithology 82:184–192.

Janes, D. 2004. A review of nuisance alligator management in the southeastern United States. International Urban Wildlife Symposium 4:182–185.

Jedrzejewski, W., and B. Jedrzejewski. 1996. Tawny owl (Strix aluco) predation in a pristine deciduous forest (Bialowieza National Park, Poland). Journal of Animal Ecology 65:105–120.

Kauffman, M. J., J. F. Brodie, and E. S. Jules. 2010. Are wolves saving Yellowstone's aspen? A landscape-level test of a behaviorally mediated trophic cascade. Ecology 91:2742–2755.

Keech, M. A. 2005. Factors limiting moose at low density in Unit 19D East, and response of moose to wolf control. Federal Aid in Wildlife Restoration Final Research Performance Report, Grants W-27-5 and W-33-1 through W-33-3, Project 1.58. Alaska Department of Fish and Game, Juneau, USA.

Kenward, R. E. 2006. The goshawk. T. and A. D. Poyser, London, UK.

Kenward, R. E., V. Marcstrom, and M. Karlbom. 1981. Goshawk winter ecology in Swedish pheasant habitats. Journal of Wildlife Management 45:397–408.

Kilgo, J. C., C. E. Shaw, M. Vukovich, M. J. Conroy, and C. Ruth. 2017. Reproductive characteristics of a coyote population before and during exploitation. Journal of Wildlife Management 81:1386–1393.

Komar, D., and O. Beattie. 1998. Effects of carcass size on decay rates of shade and sun exposed carrion. Canadian Society of Forensic Science Journal 31:35–43.

Korpimaki, E. 1988. Factors promoting polygyny in European birds of prey—a hypothesis. Oecologia 77:278–285.

Korpimaki, E., and K. Norrdahl. 1991. Numerical and functional responses of kestrels, short-eared owls, and long-eared owls to vole densities. Ecology 72:814–826.

Krebs, C. J. 1996. Population cycles revisited. Journal of Mammalogy 77:8–24.

Krebs, C. J., R. Boonstra, S. Boutin, and A. R. E. Sinclair. 2001. What drives the 10-year cycle of snowshoe hares? Bioscience 51:25–35.

Krüger, O., and J. Lindström. 2001. Habitat heterogeneity affects population growth in goshawk Accipiter gentilis. Journal of Animal Ecology 70:173–181.

Lack, D. 1966. Population studies of birds. Clarendon Press, Oxford, UK.

Laporte, I., T. B. Muhly, J. A. Pitt, M. Alexander, and M. Musiani. 2010. Effects of wolves on elk and cattle behaviors: implications for livestock production and wolf conservation. PLoS ONE. https://doi.org/10.1371/journal.pone.0011954.

Laundré, J. W. 2010. Behavioral response races, predator-prey shell games, ecology of fear, and patch use of pumas and their ungulate prey. Ecology 91:2995–3007.

Leopold, A. 1933. Game management. Charles Scribner's Sons, New York, New York, USA.

Leopold, A. 1949. A Sand County almanac. Oxford University Press, Oxford, UK.

Letnic, M., M. S. Crowther, and F. Koch. 2009. Does a top-predator provide an endangered rodent with refuge from an invasive mesopredator? Animal Conservation 12:302–312.

Letty, J., S. Marchandeau, and J. Aubineau. 2007. Problems encountered by individuals in animal translocations: lessons from field studies. Ecoscience 14:420–431.

Levi, T., and C. C. Wilmers. 2012. Wolves-coyotes-foxes: a cascade among carnivores. Ecology 93:921–929.

Lima, S. L., and T. D. Steury. 2005. Perception of predation risk: the foundation of nonlethal predator–prey interactions. Pages 166–188 in P. Barbosa and I. Castellanos, editors. Ecology of predator–prey interactions. Oxford University Press, Oxford, UK.

Linnell, J. D. C., R. Aanes, J. E. Swenson, J. Odden, and M. E. Smith. 1997. Translocation of carnivores as a method for managing problem animals: a review. Biodiversity and Conservation 6:1245–1257.

Lotka, A. J. 1925. Elements of physical biology. Williams and Wilkins, Baltimore, Maryland, USA.

Lourenço, R., V. Penteriani, M. M. Delgado, M. Marchi-Bartolozzi, and J. E. Rabaça. 2011. Kill before being killed: an experimental approach supports the predator-removal hypothesis as a determinant of intraguild predation in top predators. Behavioral Ecology and Sociobiology 65:1709–1714.

Lowney, M. S. 1999. Damage by black and turkey vultures in Virginia, 1990–1996. Wildlife Society Bulletin 27:715–719.

Malpas, L. R., R. J. Kennerley, G. J. M. Hirons, R. D. Sheldon, M. Ausden, J. C. Gilbert, and J. Smart. 2013. The use of predator-exclusion fencing as a management tool improves the breeding success of waders on lowland wet grassland. Journal for Nature Conservation 21:37–47.

Matter, W. J., and R. W. Mannan. 2005. How do prey persist? Journal of Wildlife Management 69:1315–1320.

Mech, L. D., and R. O. Peterson. 2003. Wolf–prey relations. Pages 131–160 in L. D. Mech and L. Boitani, editors. Wolves: behavior, ecology, and conservation. University of Chicago Press, Chicago, Illinois, USA.

Messier, F., and M. Crête. 1985. Moose-wolf dynamics and the natural regulation of moose populations. Oecologia 65:503–512.

Messmer, T. A. 2000. The emergence of human-wildlife conflict management: turning challenges into opportunities. International Biodeterioration and Biodegradation 45:97–102.

Mettler, A. E., and J. A. Shivik. 2007. Dominance and neophobia in coyote (Canis latrans) breeding pairs. Applied Animal Behaviour Science 102:85–94.

Mikkola, H. 1976. Owls killing and killed by other owls and raptors in Europe. British Birds 69:144–154.

Miller, D. L., and M. K. Causey. 1985. Food preferences of American woodcock wintering in Alabama. Journal of Wildlife Management 49:492–496.

Miller, M. J., M. Restani, A. R. Harmata, G. R. Bortolotti, and M. E. Wayland. 1998. A comparison of blood lead levels in bald eagles from two regions on the great plains of North America. Journal of Wildlife Diseases 34:704–714.

Millsap, B. A., T. F. Breen, and L. M. Phillips. 2013. Ecology of the Cooper's hawk in north Florida. North American Fauna 78:1–58.

Millspaugh, J. J., D. C. Kesler, R. W. Kays, R. A. Gitzen, J. H. Schulz, C. T. Rota, C. M. Bodinof, J. L. Belant, and B. J. Keller. 2012. Wildlife ratiotelemetry and remote monitoring. Pages 258–283 in N. J. Silvy, editor. The wildlife techniques manual. Volume 1, 7th edition. Johns Hopkins University Press, Baltimore, Maryland, USA.

Minnie, L., A. Gaylard, and G. I. H. Kerley. 2016. Compensatory life-history responses of a mesopredator may undermine carnivore management efforts. Journal of Applied Ecology 53:379–387.

Mollison, D. 1991. Dependence of epidemic and population velocities on basic parameters. Mathematical Biosciences 107:255–287.

Mossop, D. H. 2011. Long-term studies of willow ptarmigan and gyrfalcon in the Yukon Territory: a collapsing 10-year cycle and its apparent effect on the top predator. Pages 1–13 in R. T. Watson, T. J. Cade, M. Fuller, et al., editors. Gyrfalcons and ptarmigan in a changing world. Peregrine Fund, Boise, Idaho, USA.

Murie, A. 1940. Ecology of the coyote in the Yellowstone. Fauna of the National Parks of the United States 4. US Department of the Interior, Washington, DC, USA.

Murie, A. 1944. The wolves of Mount McKinley. University of Michigan, Ann Arbor, USA.

Murray, D. L., and M. R. Fuller. 2000. A critical review of the effects of marking on the biology of vertebrates. Pages 15–64 in L. Boitani and T. K. Fuller, editors. Research techniques in animal ecology: controversies and consequences. Columbia University Press, New York, New York, USA.

Newton, I. 1979. Population ecology of raptors. Buteo Books, Vermillion, South Dakota, USA.

Newton, I. 1998. Population limitation in birds. Academic Press, San Diego, California, USA.

Norberg, R. A. 1978. Skull asymmetry, ear structure and function, and auditory localization in Tengmalm's owl, Aegolius funereus (Linne). Philosophical Transactions of the Royal Society B 282:325–410.

Nowak, E. M., T. Hare, and J. McNally. 2002. Management of "nuisance" vipers: effects of translocation on western diamond-backed rattlesnakes (Crotalus atrox). Pages 533–560 in G. W. Schuett, M. Haggren, M. E. Douglas, et al., editors. Biology of the vipers. Eagle Mountain, Eagle Mountain, Utah, USA.

Olechnowski, B. F. M., and D. M. Debinski. 2008. Response of songbirds to riparian willow habitat structure in the Greater Yellowstone Ecosystem. Wilson Journal of Ornithology 120:830–839.

Olson, R., and F. C. Rohwer. 1998. Effects of human disturbance on success of artificial duck nests. Journal of Wildlife Management 62:1142–1146.

Ovegard, M. K., N. Jepsen, M. B. Nord, and E. Petersson. 2021. Cormorant predation effects on fish populations: a global meta-analysis. Fish and Fisheries 2021. https://doi.org/10.1111/faf.12540.

Paine, R. T. 1980. Food webs: linkages, interaction strength, and community infrastructure. Journal of Animal Ecology 49:667–685.

Palomares, F., and T. M. Caro. 1999. Interspecific killing among mammalian carnivores. American Naturalist 153:492–508.

Peek, J. M. 1986. A review of wildlife management. Prentice Hall, Englewood Cliffs, New Jersey, USA.

Peluc, S. I., T. S. Sillett, J. T. Rotenberry, and C. K. Ghalambor. 2008. Adaptive phenotypic plasticity in an island songbird exposed to a novel predation risk. Behavioral Ecology 19:830–835.

Perkins, R. 2019. Impacts of transmitter weight and attachment on raptor agility and survival. PhD dissertation, Texas Tech University, Lubbock, USA.

Perkins, R., C. W. Boal, and C. B. Dabbert. 2018. Raptor selection of captive reared and released Galliform birds. Wildlife Society Bulletin 42:713–715.

Preisser, E. L., D. I. Bolnick, and M. F. Benard. 2005. Scared to death? The effects of intimidation and consumption in predator–prey interactions. Ecology 86:501–509.

Pulliam, H. R., and B. J. Danielson. 1991. Sources, sinks, and habitat selection: a landscape perspective on population dynamics. American Naturalist 137:S50–S66.

Quammen, D. 2003. Monster of god: the man-eating predator in the jungles of history and the mind. W. W. Norton, New York, New York, USA.

Redpath, S. M., and S. J. Thirgood. 1999. Numerical and functional responses in generalist predators: hen harriers and peregrines on Scottish grouse moors. Journal of Animal Ecology 68:879–892.

Reynolds, J. C., C. Stoate, M. H. Brockless, N. J. Aebischer, and S. C. Tapper. 2010. The consequences of predator control for brown hares (*Lepus europaeus*) on UK farmland. European Journal of Wildlife Research 56:541–549.

Reynolds, J. C., and S. C. Tapper. 1996. Control of mammalian predators in game management and conservation. Mammal Review 26:127–155.

Reynolds, R. T., R. T. Graham, and D. A. Boyce Jr. 2006. An ecosystem-based conservation strategy for the northern goshawk. Studies in Avian Biology 31:299–311.

Richie, E. G., and C. N. Johnson. 2009. Predator interactions, mesopredator release and biodiversity conservation. Ecology Letters 12:982–998.

Ripple, W. J., and R. L. Beschta. 2003. Wolf reintroduction, predation risk, and cottonwood recovery in Yellowstone National Park. Forest Ecology and Management 184:299–313.

Ripple, W. J., and R. L. Beschta. 2004. Wolves and the ecology of fear: can predation risk structure ecosystems? BioScience 54:755–766.

Ripple, W. J., and R. L. Beschta. 2012. Trophic cascades in Yellowstone: the first 15 years after wolf reintroduction. Biological Conservation 145:205–213.

Ripple, W. J., E. J. Larsen, R. A. Renkin, and D. W. Smith. 2001. Trophic cascades among wolves, elk and aspen on Yellowstone National Park's northern range. Biological Conservation 102:227–234.

Roemer, G. W., C. J. Donlan, and F. Courchamp. 2002. Golden eagle, feral pigs, and insular carnivores: how exotic species turn native predators into prey. Proceedings of the National Academy of Sciences 99:791–796.

Rosenzweig, M. L., and R. H. MacArthur. 1963. Graphical representation and stabilization conditions of predator–prey interactions. American Naturalist 47:209–223.

Roth, T. C., II, and S. L. Lima. 2003. Hunting behavior and diet of Cooper's hawks: an urban view of the small-bird-in-winter paradigm. Condor 105:474–483.

Roth, T. C., II, S. L. Lima, and W. E. Vetter. 2006. Determinants of predation risk in small wintering birds: the hawk's perspective. Behavioral Ecology and Sociobiology 60:195–204.

Ruth, T. K., K. A. Logan, L. L. Sweanor, M. G. Hornocker, and L. J. Temple. 1998. Evaluating cougar translocation in New Mexico. Journal Wildlife Management 62:1264–1275.

Salo, P., E. Korpimäki, P. B. Banks, M. Nordström, and C. R. Dickman. 2007. Alien predators are more dangerous than native predators to prey populations. Proceedings of the Royal Society B 274:1237–1243.

Schmidt, K., and D. P. J. Kuijper. 2015. A "death trap" in the landscape of fear. Mammal Research 60:275–284.

Schusler, T. M., and D. J. Decker. 2002. Engaging local communities in wildlife management area planning: an evaluation of the Lake Ontario Islands search conference. Wildlife Society Bulletin 30:1226–1237.

Siepielski, A. M. 2006. A possible role for red squirrels in structuring breeding bird communities in lodgepole pine forests. Condor 108:232–238.

Sinclair, A. R. E., and P. Arcese. 1995. Population consequences of predation-sensitive foraging: the Serengeti wildebeest. Ecology 76:882–891.

Sinclair, A. R. E., J. M. Fryxell, and G. Caughley. 2006. Wildlife ecology, conservation, and management. Blackwell, Malden, Massachusetts, USA.

Sinclair, A. R. E., and J. M. Gosline. 1997. Solar activity and mammal cycles in the Northern Hemisphere. American Naturalist 149:776–784.

Sinclair A. R. E., and C. J. Krebs. 2002. Complex numerical responses to top-down and bottom-up processes in vertebrate populations. Philosophical Transactions of the Royal Society B 357:1221–1231.

Sinclair, A. R. E., R. P. Pech, C. R. Dickman, D. Hik, P. Mahon, and A. E. Newsome. 1998. Predicting effects of predation on conservation of endangered prey. Conservation Biology 12:564–575.

Sivy, K. J., C. B. Pozzanghera, K. E. Colson, M. A. Mumma, and L. R. Prugh. 2018. Apex predators and the facilitation of resource partitioning among mesopredators. Oikos 127:607–621.

Skagen, S. K., T. R. Stanley, and M. B. Dillon. 1999. Do mammalian nest predators follow human scent trails in the shortgrass prairie? Wilson Bulletin 111:415–420.

Smith, D. W., T. D. Drummedr, K. M. Murphy, D. S. Guernsey, and S. B. Evans. 2004. Winter prey selection and estimation of wolf kill rates in Yellowstone National Park, 1995–2000. Journal of Wildlife Management 68:153–166.

Smith, K. G., and J. D. Clark. 1994. Black bears in Arkansas: characteristics of a successful translocation. Journal of Mammalogy 75:309–320.

Snyder, J. W., E. C. Pelren, and J. A. Crawford. 1999. Translocation histories of prairie grouse in the United States. Wildlife Society Bulletin 27:428–432.

Solheim, R. 1984. Caching behavior, prey choice and surplus killing by pygmy owls *Glaucidium passerinum* during winter, a functional response of a generalist predator. Annals Zoologica Fennici 21:301–308.

Solomon, M. E. 1949. The natural control of animal populations. Journal of Animal Ecology 18:1–35.

Soulé, M. E., D. T. Bolger, A. C. Alberts, J. Wright, M. Sorice, and S. Hill. 1988. Reconstructed dynamics of rapid extinctions of chaparral-requiring birds in urban habitat islands. Conservation Biology 2:75–92.

Starfield, A. M., and A. L. Bleloch. 1991. Building models for conservation and wildlife management. Interaction Book Company, Edina, Minnesota, USA.

Stenseth, N. C., W. Falck, K.-S. Chan, O. N. Bjornstad, M. O'Donoghue, H. Tong, R. Boonstra, S. Boutin, C. J. Krebs, and N. G. Yoccoz. 1998. From patterns to processes: phase and density dependencies in the Canadian lynx cycle. Proceedings of the National Academy of Science 95:15,430–15,435.

Stewart, R. R., E. H. Kowal, R. Beaulieu, and T. W. Rock. 1985. The impact of black bear removal on moose calf survival in east-central Saskatchewan. Alces 21:403–418.

Tapper, R. 2006. Wildlife watching and tourism: a study on the benefits and risks of a fast growing tourism activity and its impacts on species. United Nations Environment Programme and the Secretariat of the Convention on the Cosnervation of Migratory Species of Wild Animals, Bonn, Germany.

Taylor, J., and B. Strickland. 2008. Effects of roost shooting on double-crested cormorant use of catfish ponds—preliminary results. Proceedings of the Vertebrate Pest Conference 23:98–102.

Taylor, R. J. 1984. Predation. Chapman and Hall, New York, New York, USA.

Temple, S. A. 1987. Do predators always capture substandard individuals disproportionately from prey populations? Ecology 68:669–674.

Thirgood, S. J., S. M. Redpath, D. T. Haydon, P. Rothery, I. Newton, and P. J. Hudson. 2000b. Habitat loss and raptor predation: disentangling long- and short-term causes of red grouse declines. Proceedings of the Royal Society B 267:651–656.

Thirgood, S. J., S. M. Redpath, P. Rothery, and N. J. Aebischer. 2000a. Raptor predation and population limitation in red grouse. Journal of Animal Ecology 69:504–516.

Toland, B. 1986. Hunting success of some Missouri raptors. Wilson Bulletin 98:116–125.

Turner, A. S., L. M. Conner, and R. J. Cooper. 2008. Supplemental feeding of northern bobwhite affects red-tailed hawk spatial distribution. Journal of Wildlife Management 72:428–432.

Van Ballenberghe, B. 2006. Predator control, politics, and wildlife conservation in Alaska. Alces 42:1–11.

VerCauteren, K. C., R. A. Dolbeer, and E. M. Gese. 2005. Identification and management of wildlife damage. Pages 740–778 *in* C. E. Braun, editor. Techniques for wildlife investigations and management. 6th edition. The Wildlife Society, Bethesda, Maryland, USA.

Vik, J. O., C. N. Brinch, S. Boutin, and N. C. Stenseth. 2008. Interlinking hare and lynx dynamics using a century's worth of annual data. Population Ecology 50:267. https://doi.org/10.1007/s10144-008-0088-2.

Vilardell, A., X. Capalleras, J. Budó, and P. Pons. 2012. Predator identification and effects of habitat management and fencing on depredation rates of simulated nests of an endangered population of Hermann's tortoises. European Journal of Wildlife Research 58:707–713.

Volterra, V. 1926. Fluctuation in the abundance of species considered mathematically. Nature 118:558–560.

Volterra, V. 1931. Leçons sur la théorie mathématique de la lutte pour la vie. Gauthiers-Vilars, Paris, France.

Wam, H. K., H. C. Pedersen, and O. Hjeljord. 2012. Balancing hunting regulations and hunter satisfaction: an integrated biosocioeconomic model to aid in sustainable management. Ecological Economics 79:89–96.

White, P. J., and K. Ralls. 1993. Reproduction and spacing patterns of kit foxes relative to changing prey availability. Journal of Wildlife Management 57:861–867.

Wiens, J. D., R. G. Anthony, and E. D. Forsman. 2014. Competitive interactions and resource partitioning between northern spotted owls and barred owls in western Oregon. Wildlife Monographs 185:1–50.

Wiles, G. J., J. Bart, R. E. Beck, and C. F. Aguon. 2003. Impacts of the brown tree snake: patterns of decline and species persistence in Guam's avifauna. Conservation Biology 17:1350–1360.

Wilson, E. E., and E. M. Wolkovich. 2011. Scavenging: how carnivores and carrion structure communities. Trends in Ecology and Evolution 26:129–135.

Yan, C., N. C. Stenseth, C. J. Krebs, and Z. Zhang. 2013. Linking climate change to population cycles of hares and lynx. Global Change Biology 19:3263–3271.

Yarkovich, J., J. D. Clark, and J. L. Morrow. 2011. Effect of black bear relocation on elk calf recruitment at Great Smoky Mountains National Park. Journal of Wildlife Management 75:1145–1154.

Zager, P., and J. Beecham. 2006. The role of American black bears and brown bears as predators of ungulates in North America. Ursus 17:95–108.

ANIMAL BEHAVIOR

JOHN L. KOPROWSKI, W. SUE FAIRBANKS, AND
MELISSA J. MERRICK

INTRODUCTION

Animal behavior is an integral component for developing strategies to conserve and manage wildlife. Leopold (1933:123) recognized this in *Game Management*, stating, "The game manager who observes, appraises, and manipulates these half-known properties of mobility, tolerance, and sex habits of wild creatures, is playing a game of chess with nature. He but dimly sees the board, the men, or the rules. He can be sure of only two things: for intricacy and interest, any other game pales into insignificance; he must win if wild life is to be restored. If any braver challenge inheres in any human vocation, it takes something more than a sportsman to see it."

Why do pronghorn (*Antilocapra americana*) fawns conceal themselves in cover for their initial days of life? Why do eastern gray squirrels (*Sciurus carolinensis*) nest in groups within the cavity of a tree? What are the consequences of sage grouse (*Centrocercus urophasianus*) mating in leks? These are questions that are of importance to a wildlife biologist.

Behavior is often thought of as a conscious or instinctive choice. Innate behaviors are those instinctive actions that have been honed over many generations by the process of natural selection, the differential reproductive success of individuals differing in more than one genetic trait. Recall that for natural selection to operate, three components are required. A behavior must be variable, heritable, and result in differential reproductive success. The popular label for natural selection, survival of the fittest, is a misnomer for natural selection because differential reproductive success is the key to the process. Observed variation in behavior traits, such as alternative reproductive tactics, within wildlife populations is the result of context-dependent differential reproductive success. Some behaviors confer increased reproductive success in certain contexts. For example, in bighorn sheep (*Ovis canadensis*), mule deer (*Odocoileus hemionus*), and white-tailed deer (*O. virginianus*) and other cervids and bovids, smaller, less dominant males employ alternative reproductive tactics than those used by dominant individuals, such as coursing and concurrent courtship to gain access to mates (Airst and Lingle 2020; read more on social systems, below). In a highly variable and chang-

ing world, environmental, community, and social heterogeneity serves to maintain behavioral complexity and diversity in population. Learned behaviors are accrued through individual experience over the course of an individual's lifetime, though there are also genetic components to the ability to learn.

Our objectives in this chapter are to provide an overview of the principles of animal behavior interpreted in a contemporary context that relates behaviors to the fitness of wildlife. Such an approach enables us to make sense of behavior by assessing the costs and benefits. We review common means of studying wildlife behavior as a prelude to examining the diversity of mating systems and how individuals select mates. We provide an assessment of key influences on space and habitat use, including sexual segregation that is common in wildlife, with an emphasis on the need to understand sex-specific needs. Understanding sex-specific resource needs and fitness limiting factors enables us to interpret social systems and processes such as dispersal and migration. Finally, we demonstrate the importance of the characteristics of individuals to conservation and management, an emerging area of study in wildlife biology.

WHAT IS THE NAME OF THE GAME? FITNESS

A useful way to think about behavior is from an economic perspective. Behaviors have certain associated costs (e.g., energetic, lost time that could be spent in other ways, predation risk, lost mating opportunities) but also likely provide benefits (e.g., energetic savings, more time to feed, decreased predation, enhanced mating opportunities, more young surviving to join the population) to the actor. Ultimately, we measure the costs and benefits in terms of fitness. Fitness has a number of connotations in biology. A population geneticist measures the fitness of a trait with respect to the average contribution to the gene pool of the average individual with the trait, whereas a physiologist might measure physical fitness. One must be careful in how the term is used. When examining behavior, we consider fitness in the sense of individual fitness where the propensity to contribute offspring to the next generation is the key measure. An individual with high fitness has a high likelihood of contributing young to the next generation, whereas an individual with low fitness has a low likelihood of contributing offspring to the next generation. As a result, individual fitness is a correlate of lifetime reproductive success, which is often used as a proxy in behavioral research (Clutton-Brock 1988, Brommer et al. 2005). If we think more broadly beyond production of offspring, the fitness of individuals may also benefit from the success of relatives with whom they share genes, the concept of inclusive fitness. Production of offspring and direct descendants, such as grand-offspring, contributes to the direct component of fitness; the indirect component of inclusive fitness comes from helping nondescendant relatives (e.g., offspring of siblings or parents) to produce offspring that would not have survived to join the population without the assistance of the helper. The direct and indirect components collectively result in the inclusive fitness of an individual, which is especially important in social animals.

HOW IS BEHAVIOR STUDIED?

Wildlife behavioral studies are used in numerous ways to inform conservation and management strategies. Direct observation of individuals by an observer is commonly used to assess diet and foraging ecology, habitat use, space use, dominance and social interactions, nesting ecology, reproductive behavior, and anthropogenic influence, to name a few applications. The most common techniques used to sample behavior follow.

1. Focal animal sampling. Following individual animals and recording behaviors over a set period. Often used to create a catalog of all behaviors (i.e., ethogram) or determine how individuals apportion their time to specific behaviors (i.e., activity or time budgets).
2. Scan sampling. Scanning a study population at predetermined intervals and recording all behaviors and locations of individuals. Often used to examine space or habitat use, interindividual distances, and population-wide activity-time budgets.
3. All-occurrence or ad libitum sampling. Recording all observations of a specific behavior or behaviors within a study population. Often used to assess the frequency of rare behaviors such as direct social interactions or matings.

Indirect observations are also used to quantify wildlife behavior. Traditional techniques such as using sign (e.g., tracks, scat, food remains, nests) to provide data on behavior can provide important insight into the ecology of wildlife. Increasingly, technology permits wildlifers to use indirect methods with considerable effectiveness. Radiotelemetry and satellite telemetry permit remote monitoring of individuals, trail cameras record time and location of individuals on the landscape, radioisotopic analyses permit assessment of diet, and molecular genetics provide insight into mating patterns and reproductive success.

MATING SYSTEMS

An integral part of wildlife behavior is the mating system in which individuals must operate to achieve reproductive success. The mating system is integrated with the social system of most species and is profoundly influential to the ecology of wildlife species because of its immediate effect on reproduction. Mating systems are defined by the number of mates garnered by each sex. For example, elk (*Cervus canadensis*) have a polygynous mating system where harems of females are tended by a large male (Fig. 15.1). Males compete for the opportunity to maintain a harem and must be able to attract females and defend the harem from other males. Traits such as large body size and large antlers are the result of sexual selection for success in such competitions. Sexual selection acts solely upon traits that are necessary to maximize access to mates and reproductive success and can result in what may seem like rather bizarre traits such as the massive antlers (record lengths of main beam exceed 1.4 m) and extreme body size of adult male elk (>450 kg). Such traits are also traditionally valued by humans for food and trophies; as a result, management and conservation efforts are influenced by mating systems, and mating systems are defined by the number of mates generated by each sex (Fig. 15.2).

Monogamy
Monogamy occurs when a single male and female are paired. At least 90% of avian species are classified as monogamous, whereas <5% of mammals form pair bonds with a single mate. Serial monogamy is a specific case of monogamy where individuals pair with only a single mate during a breeding season but that mate changes during each season. Many passerine birds are examples of this type of monogamy. Lifelong mo-

Figure 15.1. Male elk tending his harem of female elk. Note the numbered telemetry collars found attached to the male and several females that permit individual identification at a distance. As in nearly all ungulates, males are considerably larger than females and are the only sex adorned with weaponry, antlers in this case. Courtesy of Eric Godoy

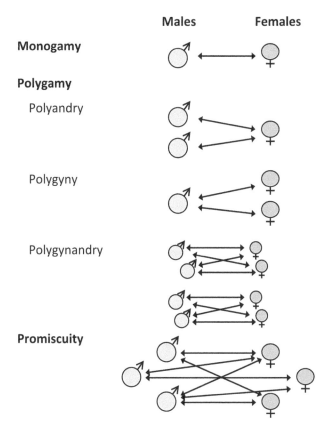

Males   **Females**

Monogamy

Polygamy

  Polyandry

  Polygyny

  Polygynandry

Promiscuity

Figure 15.2. Schematic diagram of wildlife mating systems. Arrows that connect individuals symbolize a mating between the individuals.

Figure 15.3. Sage grouse lek with males displaying for female. Note the proximity of males to each other and the lack of resources at the lekking grounds, a traditional site used only for mating. The exaggerated plumage and yellow esophageal air sacs for sound production are the products of sexual selection to demonstrate vigor in courtship. *(top)* Courtesy of David M. Shumway. *(bottom)* Courtesy of James Ownby

nogamy is the other type of monogamy, where individuals pair with a single mate for that breeding season and for all subsequent breeding seasons when both individuals remain alive. Canada geese (*Branta canadensis*) are an example of this form of monogamy.

Polygamy

Polygamy is a mating system where only one sex has multiple mates. We can further classify polygamy dependent on the sex that obtains multiple mates. Polyandry occurs when one female has multiple male mates. Many sandpipers, jacanas, and raptors practice this type of polygamy. Polygyny is when one male has multiple female mates. Polygyny is common in many mammals, including white-tailed and mule deer. Polygynandry occurs when two or more males have an exclusive relationship with two or more females; the numbers of males and females need not be equal, and in vertebrate species studied so far, the number of males is usually less. Red foxes (*Vulpes vulpes*), bonobos (*Pan paniscus*), and avian species, including alpine accentors (*Prunella collaris*) and acorn woodpeckers (*Melanerpes formicivorus*), are examples of species exhibiting polygynandry.

Polygamy is often further categorized based upon the nature of the resource that is defended. Mate-defense polygamy is a form of polygamy in which the sex that has multiple mates defends the mates from potential competitors. Many

ungulates practice this type of polygamy. Resource-defense polygamy is a form of polygamy in which the sex that has multiple mates defends a resource that attracts mates from potential competitors. A number of pinnipeds, such as elephant seals (*Mirounga angustirostris*), defend protected haul-out areas that females visit. Lek-based polygamy is a form of polygamy where the resource defended is a position at a lek, a traditional site where one sex gathers to display and the other sex visits to select mates. The lek contains no resources (other than potential mates) of value to the visiting sex, and the only value to the displaying sex is that of location in the lek. Central locations often are more successful than peripheral locations. Sage grouse (Fig. 15.3) and greater prairie chickens (*Tympanuchus cupido*) are excellent examples of this mating system.

Promiscuity

A promiscuous mating system is where males and females have multiple mates and do not maintain an exclusive relationship with any individuals. This is a common mating system found in mammals, especially in many small- and medium-sized species such as squirrels (*Sciurus* and *Tamiasciurus* spp.), rabbits (*Sylvilagus* spp.), hares (*Lepus* spp.), and many furbearers.

Through the use of molecular genetics, we know that extra-pair fertilizations (EPFs) are more common than previously believed and that the observable and apparent mating system may not tell the whole story. For instance, within some species classified as monogamous, more than 70% of young are the result of EPFs (Griffith et al. 2002). As a result, it is common to refer to the behavioral or social mating system and the genetic mating system of a species to distinguish between the potential differences that EPFs create.

Mating systems also have an important effect on the ecology of wildlife in subtler ways through the process of sexual selection. Sexual dimorphism, differences between the sexes in one or more traits, is often the result. In species where mates must be actively defended, intrasexual competition (i.e., competition between members of the same sex) results in intrasexual selection for characteristics that enhance success in physical combat such as large body size, horns and antlers, and teeth and claws. Additionally, where one sex actively chooses mates, mate choice can be extremely influential owing to intersexual selection. The bright colors and elaborate plumage of many male birds are results of the response of traits to mate choice by females. Sexual dimorphism also has an effect on how we manage wildlife populations, and characteristics attractive to hunters must be incorporated into our management plans. The differences in physical appearance also permit more refined management because we can plan harvests to target specific sex and age classes. For species without obvious sexual dimorphism (e.g., many small game species such as rabbits, squirrels, doves, and most carnivores), we are not able to target specific segments of the population and are likely to have a reduced ability to manipulate population growth through selective harvest.

SPACE AND HABITAT USE

How animals use space is an important question in wildlife management. The answer is relevant to issues dealing with habitat use, carrying capacity, and even population estimation. When we think about space use by animals, we often classify animals in terms of their fidelity for an area along a continuum (Fig. 15.4). Truly nomadic species with no fidelity to space are not known, likely because knowledge of resources in an area is beneficial. Home ranges result from fidelity to an area without exclusive use. Home ranges are defined as the area traversed by an individual in which their daily needs are met; home ranges are not defended and can be used by other individuals of the same species. Many species maintain nearly exclusive access to core areas within a home range, whereas exclusive use, or territoriality, occurs when the entire area is defended from others by an individual or social group. The key determinant in the development of areas of exclusive use is the concept of economic defendability. Benefits of defending a site include access to resources (e.g., food, limited nutrients,

dens, nests, mates, space). Costs that accrue include lost time, lost opportunities elsewhere, energetic costs, and catastrophic risk of territory loss (Fig. 15.5).

Territoriality is important to wildlife populations in numerous ways. Territoriality often results in a relatively uniform distribution of individuals on the landscape as resources are defended by territory holders. The behaviors of defense limit access to resources and ultimately can control population size in a given area. Such behavior also makes it challenging to translocate individuals to new locations where resident animals already exist. Defense of territories can be through direct interactions that can escalate to conflict between individuals. The weaponry in many male ungulates and carnivores often is attributed to such territorial squabbles. Other subtler behav-

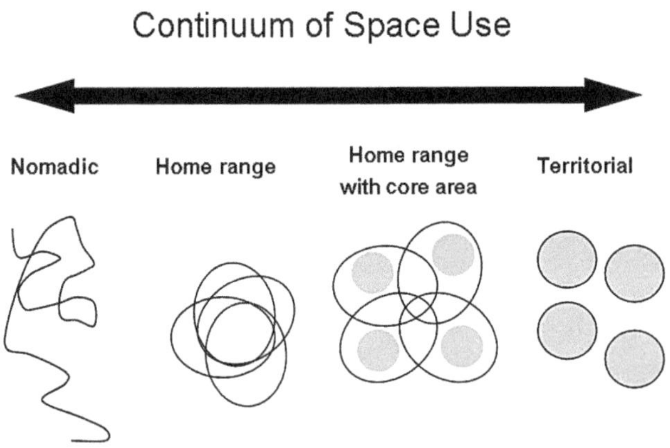

Continuum of Space Use

Nomadic   Home range   Home range with core area   Territorial

Figure 15.4. Patterns of space use based upon affinity for space and level of defense.

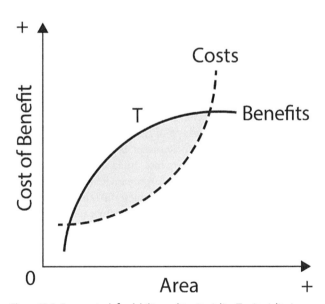

Figure 15.5. Economic defendability and territoriality. Territoriality is possible wherever the benefits exceed the costs (shaded); however, we would expect the most common territory size to be found where the benefits exceed the costs by the greatest amount (point T).

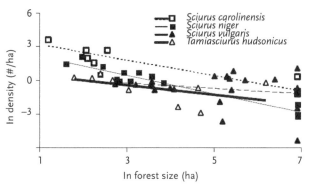

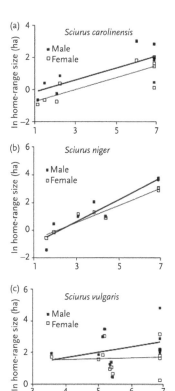

Figure 15.6. Forest fragmentation results in high densities of tree squirrels and a concomitant reduction in home range size as animals are relegated to the remaining fragment of quality of habitat. From Koprowski (2005)

iors, however—such as scent marking by urine, feces, or other body secretions; visual displays by territory owners; and vocalizations—are used to define territory boundaries and settle differences without repeated combat. Wildlife managers capitalize on many of these territorial behaviors to develop indices of the abundance of wildlife populations.

Body size energetics and the economics of energy likely set the size of home ranges and territories (Harestad and Bunnell 1979). Once the effect of body size on home range size is removed from consideration, significant variation remains and is related to resource availability, predation risk, interspecific competition, diet, and climate. Thus we might expect variation in the home range size of a species across habitats that differ in quality. Habitat fragmentation increases population density and can result in severe compaction of home ranges (Koprowski 2005; Fig. 15.6). Some additional patterns are clear among wildlife species. Males often traverse larger areas than females, except in monogamous systems where the ranges of the sexes are typically equivalent. This suggests that mating system is a strong influence on the space use of a majority of mammals and some birds (Clutton-Brock 1989). Other factors such as climate can also have a great effect on the extent of sex differences in home range size and may be the dominant factor in taxa such as the large marsupials of Australia, where large home ranges are in poor-quality habitats in which both males and females must traverse vast areas to find resources (Fisher and Owens 2008). Seasonal variation in the availability of mates, food, cover, climate, and predation risk induces seasonal changes in home range size. Sexual dimorphism in

home range size often is reduced outside of the breeding season when mate searching declines, as males may not need to roam large home ranges (Cudworth and Koprowski 2010).

SEXUAL SEGREGATION

Sexual segregation (i.e., the sexes living apart during certain seasons) is an often overlooked component of space use. During the period of segregation, the sexes may use different habitats (habitat segregation), or they may select similar habitats in different places (spatial segregation). Social segregation, defined as individuals of different sex, age, or size classes living in separate social groups, may or may not be associated with differences in habitat or space use by the social groups (Conradt 2005).

Sexual segregation occurs in many migratory species of birds and marine mammals when one sex migrates farther than the other, or the sexes migrate to different locations for specific needs. For example, male sperm whales (*Physeter macrocephalus*) migrate to high latitudes following breeding, while groups of females and offspring remain closer to tropical breeding grounds (Lyrholm et al. 1999). Among migrating birds, males in numerous species winter at higher latitudes than females (Cristol et al. 1999). Many aquatic reptiles exhibit habitat segregation with respect to foraging habitats (Shine and Wall 2005). In Galapagos marine iguanas (*Amblyrhynchus cristatus*), only the males forage in the sea, diving to feed on algae. Females and smaller males feed at the edges of the ocean (Buttemer and Dawson 1993). While many animal taxa exhibit

CARL B. KOFORD
(1915–1979)

Carl Buckingham Koford was born 3 September 1915 in Oakland, California, and passed away at the age of 64 on 3 December 1979 in Berkeley, California. He completed his undergraduate studies at the University of Washington and his doctoral work at the University of California, Berkeley. At Berkeley, he held appointments with the Department of Forestry and Conservation and the Museum of Vertebrate Zoology. Dr. Koford was a passionate naturalist, rugged outdoorsman, conservationist, and explorer who demonstrated the need to integrate studies on behavior and conservation. Often working in remote locations on some of the most secretive species, he gained renown for his natural history studies in the Americas on vicuña, jaguar, ocelot, prairie dogs, and searched without success for a reported relict population of grizzly bears in Mexico. Dr. Koford is perhaps best known for his pioneering, intensive field work on the behavior of California condors, initiated in March 1939. His graduate studies were interrupted by service in the US Navy during World War II but were continued in 1946. In 1953 he published his seminal work, *The California Condor*, as one of three groundbreaking National Audubon Research Reports (the other two were Tanner's *The Ivory-billed Woodpecker* and Allen's *The Whooping Crane*) in which he alerted the conservation community to a dwindling population of about 60 individuals. The Museum of Vertebrate Zoology at the University of California, Berkeley, established the Carl B. Koford Memorial Fund in 1980 to support field research on vertebrates.

sexual segregation, it has been most studied and is particularly widespread in ungulates (Bowyer 2004).

Why is it important to recognize the differences in grouping or spatial patterns in wildlife species? Studies designed to investigate habitat relationships of particular species commonly focus on only one sex, typically the most important demographic class and the sex that is easiest to observe or fit with radio transmitters. In mammals, this is usually the females; in birds, it is often the males. Depending on the objectives of the study, this sex bias may be defendable. In grazing mammals, females often live in larger groups than males and may have greater effects on the habitat. Alternatively, females may have specific needs for rearing offspring, and thus knowledge of their use of habitat would be critical to the dynamics of the entire population. During the period of sexual segregation, however, the sexes may have different habitat needs, requiring attention to the management of multiple habitats or larger areas that accommodate both sexes. Further, the sexes may respond differently to management activities owing to differences underlying sexual segregation (Stewart et al. 2003, Long et al. 2009). In a review of nonbreeding ecology of 66 North American migratory bird species, sexual segregation had not been investigated or reported for a large proportion of the species (Fig. 15.7), and only 8% of conservation recommendations considered sexual segregation (Bennett et al. 2019). In the case of the golden-winged warbler (*Vermivora chrysoptera*), forest loss was greater on nonbreeding habitat dominated by females than males, and existing conservation focal areas were biased toward male-dominated habitat (Bennett et al. 2019).

Males and females may be exposed to different mortality factors, such as predation or vehicle collisions, or different

rates of mortality during segregation (Bowyer 2004). For example, male bighorn sheep have larger home ranges and move over greater distances, traveling between groups of females. Thus males may sustain higher mortality rates from vehicle collisions than females (Rubin and Bleich 2005). Differences in movement patterns and habitat use may also result in differential exposure of the sexes to disease and parasites (Rubin and Bleich 2005). Counts of lungworm larvae (*Protostrongylus* spp.) were higher in fecal samples from nursery groups (predominantly female) of bighorn sheep than in samples from male groups on Antelope Island in the Great Salt Lake, Utah (Fig. 15.8; Rogerson et al. 2008). Abundance of the intermediate hosts (terrestrial gastropods) of the parasite was higher around water sources, and the only gastropods infected with lungworm were collected near water sources, leading Rogerson et al. (2008) to suggest that the higher larvae counts in female bighorn sheep may arise, in part, from different patterns of water use during sexual segregation. Indeed, not only did males use a larger area than females during segregation, but males also spent less time at water sources than females, and males and females used different water sources during segregation (Fig. 15.9; Whiting et al. 2010). Thus the potential for differential exposure of the sexes to parasites must be considered as an explanation for differential parasite loads or infection rates.

Most of the hypotheses offered to explain sexual segregation result from studies of sexually dimorphic ungulates in which males are larger than females (Bowyer 2004). Thus the following hypotheses may not relate well to other groups of organisms (Wearmouth and Sims 2008). The predation-risk hypothesis (or reproductive strategy hypothesis) is based on

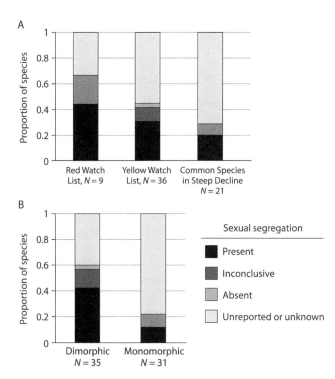

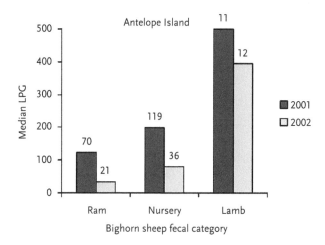

Figure 15.7. Evidence of nonbreeding sexual segregation in 66 North American migratory bird species of conservation concern based on (*A*) Partners in Flight conservation status and (*B*) sexual dimorphism. Reproduced with permission from Bennett et al. (2019)

Figure 15.8. Median lungworm larvae per gram of feces from male bighorn sheep, nursery groups (predominantly female), and lambs on Antelope Island, Great Salt Lake, Utah, May–August 2001–2002. Numbers above the bars indicate sample size (number of pellet groups sampled). Modified with permission from Rogerson et al. (2008)

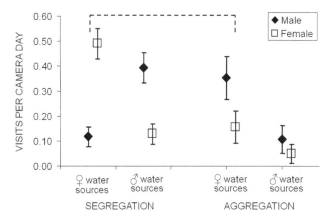

Figure 15.9. Number of visits divided by the number of days in which cameras were set (+ 95% CI) at water sources located in areas used by both sexes of bighorn sheep on Antelope Island State Park, Utah, during 2005 and 2006. Males visited water sources used by females almost three times more often during aggregation compared with when bighorns were segregated (dashed line highlights relationship). Reproduced with permission from Whiting et al. (2010)

the idea that males and females use different strategies to maximize their lifetime reproductive success. Outside the breeding season, females must balance the energy needs required for offspring production with the need to select habitats that minimize predation risk to their offspring or themselves (Rachlow and Bowyer 1998). Males, however, may be less vulnerable to predators during some seasons owing to their size or weapons. The best strategy for males might then be to select habitats offering abundant or high-quality forage, facilitating an increase in body condition prior to the mating season.

Another hypothesis, the forage selection hypothesis (or sexual dimorphism or body-size hypothesis), has several variations. These hypotheses are based on differences in nutritional requirements between the sexes. Larger individuals, which require more food absolutely, are also more efficient at digesting food owing to greater gut capacity. Thus the larger sex should choose habitats with high availability of forage, even if it is of lower quality, while the smaller sex must obtain a high-quality diet to make up for their lower digestive efficiency. The diets of male and female Siberian ibex (*Capra sibirica*) differed in the warm season but overlapped in the cold season (Fig. 15.10), and although male diets exhibited a higher correlation with relative abundance of plant species than did female diets, neither crude protein nor percentage of crude fiber affected diet composition of either sex (Han et al. 2020).

Other hypotheses fall under the category of social factors hypotheses. In the study of Siberian ibex, the forage selection hypothesis did not fully explain sexual segregation, and it was

suggested that differences in the activity budgets of males and females might explain the occurrence of unisex and mixed-sex groups during the cold season (Han et al. 2020). Social factor hypotheses might also be important in other taxa, such as marine mammals (Wearmuth and Sims 2008). As sexual selection is studied in additional groups of animals, other hypotheses may be proposed. For example, the thermal niche–fecundity hypothesis may be particularly relevant to sexual segregation in ectotherms. Under this hypothesis, the sexes segregate to maintain the optimal body temperatures that maximize reproductive output (Sims 2005). While these hypotheses have not received much support as explanations of sexual segregation among ungulates, they may be useful among other taxa, such

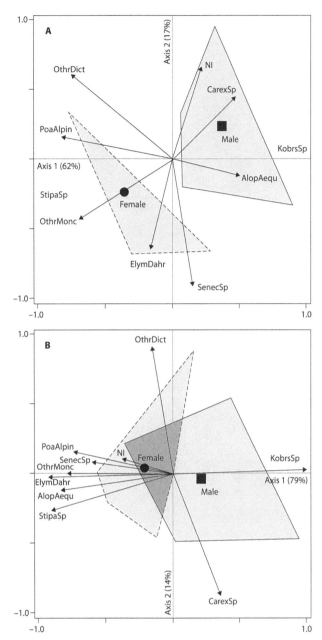

Figure 15.10. Principal components analysis biplot of diet composition of Siberian ibex males (square with surrounding black solid line) and females (circle with surrounding gray dashed line) in the (*A*) warm season and (*B*) cold season. NI, unidentified plant species; CarexSp, sedge (*Carex* spp.); KobrsSp, bog sedge (*Kobresia* spp.); AlopAequ, shortawn foxtail (*Alopecurus aequalis*); SenecSp, groundsel (*Senecio* spp.); ElymDahr, Dahurian wild rye (*Elymas dahuricus*); OthrMonc, other monocots; StipaSp, feather grass (*Stipa* spp.); PoaAlpin, alpine bluegrass (*Poa alpine*); and OthrDict, other dicots. Reproduced with permission from Han et al. (2020)

as marine mammals (Wearmouth and Sims 2008). As sexual segregation is studied in additional groups of animals, other hypotheses may be proposed. For example, the thermal niche-fecundity hypothesis may be particularly relevant to sexual segregation in ectotherms. Under this hypothesis, the sexes

segregate to maintain the optimal body temperatures that maximize reproductive output (Sims 2005).

DISPERSAL AND MIGRATION

Dispersal and migration are two important events in the life history strategies of many wildlife species. They are similar because both involve movement of individuals away from their current home ranges, but the causes and consequences of these behaviors are very different. Dispersal is the one-way movement of an individual from one home range to another location in which it establishes a new home range. Migration is roundtrip movement between different seasonal ranges. Thus dispersal is more or less permanent; migration is seasonal shuttling between two locations. It is important to establish the definitions used in any discussion, as in some fields the definitions are used differently. For example, population geneticists may use the term *migration* to refer to the movement of alleles from one gene pool to another. In an ecological or behavioral sense, this is usually accomplished by dispersal.

Dispersal and migration are behaviors that are relevant to conservation and management. Dispersal is an individual behavior that is at the heart of metapopulation dynamics; dispersal is the mechanism of gene flow that connects discrete subpopulations into a metapopulation and allows recolonization of vacant habitat. Without dispersal, the recolonization of subpopulations within the metapopulation could not occur by natural processes and would require an active management intervention through translocation. The recolonization of subpopulations by dispersers within a metapopulation has been referred to as demographic rescue and increases the likelihood of persistence of the metapopulation as a whole. Dispersal may have significant effects on local demographics of subpopulations and with large numbers of immigrants (dispersers moving in) or emigrants (residents dispersing out) affecting population growth or decline, respectively. Thus dispersal can have major effects on population genetics, evolution, and persistence of metapopulations. Dispersal processes affect predator–prey dynamics and disease transmission and epidemiology, as well. An understanding of dispersal will be important in the field of invasion ecology and prediction or control of invasive species. Dispersal and migration affect species distribution and will be highly relevant to the consequences of land-use change and climate change for wildlife. Migration is closely related to habitat selection and necessitates large-scale conservation and management strategies. Migration may also have important effects on the communities that they move between with respect to what animals may carry with them, contaminants (e.g., DDT or other pesticides), pathogens, and parasites.

Dispersal

Several different types of dispersal are recognized. Two major types are natal dispersal and breeding dispersal. Natal dispersal is the movement of an animal from the home range into which it was born or hatched to the home range in which it first reproduces. Breeding dispersal is movement to a new home range or territory between reproductive events.

Dispersal has been difficult to study in the past because it is a behavior of an individual. New technologies, however, are providing unprecedented opportunities to conduct dispersal studies. Advances in satellite telemetry enable biologists to remotely track movements of individual animals over large distances or inhospitable habitats at all times of the day or night. Indirect genetic methods facilitate study of the genetic consequences of dispersal, and genetic markers may even allow identification of dispersers or distinction between immigrants and residents in a population. While stable isotope methods are becoming widely used for study of migration, they might also be useful in dispersal studies (Caudill 2003).

In birds and mammals, there is a strong bias as to which sex disperses (Greenwood 1980). In birds, females are the primary dispersers; in mammals, it is the males that typically disperse. A smaller proportion of the other sex may also disperse, however. In species in which both sexes disperse, one sex often disperses farther than the other. Extrinsic factors that include interplay among environmental and maternal conditions are also shown to influence offspring behavior and the propensity for offspring dispersal in response to resource availability (Duckworth 2009). Maternal condition (e.g., good, poor) and level of glucocorticoid stress hormones reflect current conditions and resource availability and can influence the proportion of sons and daughters that disperse from the natal area (Merrick and Koprowski 2016a).

Dispersal can be divided into three basic stages: emigration (leaving the current home range), transfer (moving through unfamiliar territory), and immigration (settlement into a new home range). The proximate mechanism that triggers an individual to emigrate may be fixed (genetically hardwired) or condition dependent. If current population or environmental parameters (e.g., sex ratio, population density, habitat quality, body condition, competitive ability) are at least somewhat reliable indicators of future conditions in the natal home range, selection may favor sensitivity to these indicators with respect to whether or not an individual disperses (i.e., condition-dependent dispersal). In an experimental field study of prairie voles (*Microtus ochrogaster*), Lin et al. (2006) manipulated food and cover resources in habitat patches created by mowing intervening vegetation (Fig. 15.11), released a mated pair and their two to three weaning-aged offspring into each patch, and observed dispersal. Males were significantly more likely to disperse overall, but both sexes were significantly less likely to disperse from the highest-quality patches (i.e., those with supplemental food and cover) than from other patches. Voles were more likely than expected to disperse to habitat patches of similar or better quality than the one into which they were released. Sample size of dispersing females was small, but females only dispersed from patches without supplemental food; all females released in patches with supplemental food were philopatric (i.e., remained in the original home range). Thus local conditions affected dispersal by both sexes.

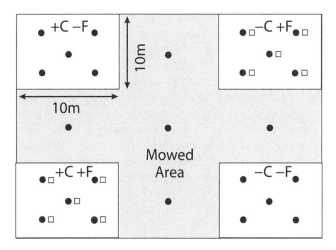

Figure 15.11. Diagram of habitat patches within each of the four 0.1-ha enclosures. Locations of habitats were randomly assigned within each enclosure. The four habitats are coded as (+C +F) supplemental cover, supplemental food; (−C +F) no supplemental cover, supplemental food; (+C −F) supplemental cover, no supplemental food; and (−C −F) no supplemental cover, no supplemental food. Locations of trapping stations (filled circles) and feeding stations (open squares) are indicated. Reproduced with permission from Lin et al. (2006)

Less is known about the transfer stage of dispersal for logistic reasons, but also because interpretation of some behaviors is difficult. Are short excursions from the home range failed dispersal attempts? Or are they fact-finding trips such that when dispersal occurs, the individual already knows where it is going? Evidence suggests that there may be different search strategies employed by short- versus long-distance dispersers within a population during the transfer stage. In Eurasian red squirrels (*Sciurus vulgaris*) and Siberian flying squirrels (*Pteromys volans*), individuals that dispersed short to medium distances from the natal area tended to revisit a potential site several times before eventually settling there, whereas long-distance dispersers performed few revisits (Selonen and Hanski 2009, Hämäläinen et al. 2020). Whether an individual is a long-distance disperser or remains philopatric and near the natal area may also be correlated with individual behavioral differences. In many species across taxonomic groups, bold, aggressive, or highly explorative individuals tend to disperse farther and colonize new places (Merrick and Koprowski 2016a, Duckworth 2009, Cote et al. 2010). For simplicity, in many metapopulations and other types of models, dispersers are often assumed to take a straight path in a random direction. In reality, of course, many factors in the landscape and biological community likely affect the path of dispersal and depend on species' perceptual abilities (i.e., how far away can the animal detect suitable areas) and tolerances to different microhabitats. Another difficult question to answer is whether survival probabilities of dispersers differ from survival of philopatric individuals. If dispersers differ in body condition, competitive ability, and sex, simply comparing survival of dispersing individuals to that of philopatric individuals will not be an appropriate test.

Settling into a new home range (immigration) is largely a function of habitat selection, but for some species the ability to integrate into a new social group may also play a role. Conspecifics already inhabiting the area may act as a social fence, preventing the disperser from establishing a home range. In other cases, the presence of conspecifics may be a factor in the recognition of suitable areas by the disperser (i.e., conspecific attraction). In some species, especially generalist species (Davis 2008), choice of settlement habitat may be influenced by the habitat into which the individual was born (i.e., habitat imprinting or natal habitat preference induction). Dispersers may also respond differently to cues about the quality of the habitats they encounter during transfer, depending on time since leaving their natal home range (Stamps et al. 2007). For example, many insects and some other taxa initially exhibit refractory periods after leaving home, during which they do not respond to cues from even high-quality settlement habitats. Alternatively, dispersers may be limited in terms of time or resources to support dispersal and so may exhibit lower acceptance thresholds for settlement habitat as time since emigration increases (Stamps et al. 2007). In other words, a disperser might be willing to accept a lower-quality habitat the longer it has been searching. A refractory period or declining acceptance thresholds (or both) may explain why dispersers sometimes move through high-quality habitats and later settle in similar or lower-quality habitats. Natal habitat imprinting or natal habitat preference induction (Davis and Stamps 2004) could also influence observed patterns of individuals settling in apparently lower-quality habitat, and this mechanism for post-dispersal habitat selection could have important management consequences. In the case of habitat restoration projects or translocations of individuals to areas identified as high-quality habitat, animals may fail to recruit to these sites despite our best efforts if the cues for settlement differ from those found in the natal area. Further, if animals rely primarily on structural cues similar to the natal area for settlement, individuals may inadvertently settle in areas experiencing rapid environmental change, such as tree death and defoliation, or under invasion by non-native plant species. In such cases, the structural cues may become decoupled from the fitness benefits they once represented, such as nest sites and food, leading to reduced survival and reproduction for individuals—an ecological trap (Merrick and Koprowski 2016b).

Two major hypotheses have been suggested as evolutionary explanations of dispersal: inbreeding avoidance and avoidance of local competition. The inbreeding avoidance hypothesis suggests that dispersal reduces the likelihood of inbreeding depression, the reduced reproductive success that can result from breeding with close relatives. The strong sex bias in avian and mammalian dispersal seems to support this hypothesis. The hypothesis does not predict which sex should disperse, however. The avoidance of local competition hypothesis has two parts and offers suggestions as to why males are the dispersing sex in mammals, and females are the dispersing sex in birds. Local mate competition suggests that young males should disperse to avoid competing for mating opportunities

MARDY MURIE (1902–2003)

Margaret Thomas "Mardy" Murie, known as the grandmother of the conservation movement, was born in Seattle, Washington, on 18 August 1902 and passed away at the age of 101 on 19 October 2003. As a child, Mardy Murie moved to Fairbanks, Alaska, where she became the first female graduate of what was to become the University of Alaska, Fairbanks. There she also met and married Olaus Murie, a biologist and eventual president of The Wildlife Society; their honeymoon was spent following caribou through the Brooks Range of Alaska, one of many exciting wilderness adventures by this wildlife biologist, author, and avid conservationist. After completing the caribou research, the couple moved to Moose, Wyoming, to conduct research on elk. Mardy called Moose her home for more than 75 additional years. Beyond wildlife research, Mardy was a tireless proponent of conservation and collaborated with various agencies and nongovernmental organizations to lobby for legislation that set aside conservation and wilderness lands. Mardy was instrumental in the passage of the Wilderness Act of 1964 and the establishment and enlargement of the Arctic National Wildlife Refuge, among many other similar accomplishments. For her efforts, Mardy received the Audubon Medal, the John Muir Award, and the highest civilian honor awarded by the US government, the Presidential Medal of Freedom. Her life on behalf of conservation is chronicled in a book and documentary titled *Arctic Dance*.

Courtesy of US Fish and Wildlife Service

with their fathers, or other closely related males. Of course, dispersers will likely compete with unrelated males for mates, but by avoiding competition with close relatives, they avoid decreasing their own inclusive fitness. Alternatively, one sex may disperse to avoid competition with close relatives for food or other resources. Polygyny is the primary mating system in mammals, typically resulting in greater competition among males for mates than among females. This may explain the strong bias toward dispersal by males in mammals. The majority of bird species, however, are socially monogamous, and philopatric males may be more likely to obtain a territory, crucial to their reproductive success, in a familiar area. Females may benefit from searching widely for high-quality territories and high-quality mates, and may have a fitness advantage by dispersal.

Migration

Migration represents a critical, repeated phase in the life cycle of migratory animals that has major effects on the physiology and behavior of individuals. Whether or not an animal migrates varies even among closely related species. In fact, within a species, some populations may be migratory, whereas other populations remain in an area as year-round residents. Partial migration occurs when some individuals in a population migrate and others do not. For example, in the elk herd that winters on the National Elk Refuge in northwestern Wyoming, some individuals migrate to summer ranges in southern Yellowstone National Park, while others remain close to the winter range during summer (Boyce 1989, Smith and Robbins 1994). Understanding migration, its requirements, and consequences is important to the management and conservation of these species, but it is difficult owing to the often large geographic range, numerous habitat types, and diverse political and land-use backdrops across which it occurs.

Migration can be studied as four different stages: preparation, movement, stopovers, and arrival (Ramenofsky et al. 1999). During the preparation stage, both physiological and behavioral changes occur to support migration. Changes in metabolism and foraging assist with fuel loading to cover the energetic costs of sustained movement and periods of fasting during migration. Before migration, animals will often exhibit increased activity with an orientation in the direction that they will eventually take when they migrate. The movement stage may occur over land, through the air, or through water, and each mode has its advantages and disadvantages. In general, migration is costlier and more barrier-rich over land than by air or water, so there are many more examples of extreme long-distance migration in birds and marine animals than in terrestrial wildlife. During stopovers, migrants settle in a particular habitat to rest and feed. Some species use many stopovers during migration, some very few. Stopover habitats may be used for a few hours to several weeks, depending on the species. In some species, animals store fuel for the entire journey before the start of migration. In others, feeding may occur during stopovers to replace energy stores used up to that point, which can be necessary if carrying large fuel loads is detrimental (e.g., with respect to predator evasion) or if unpredictable conditions occur during movement that increase the use of stored energy (e.g., storms). Upon arrival in wintering habitat, animals must insert themselves into a community of competitors (and predators) made up of migrants and year-round residents. For migrants, this may mean different diets or foraging strategies, and different social settings (flocks vs. territories), than experienced on their breeding grounds. But the seasonal influx of migratory species may also have profound effects on the existing community of resident species. Upon arrival back on their breeding grounds, migrants may need to fight for and set up exclusive territories, choose mates, and build up energy stores quickly to support reproduction.

In addition to experiencing different competitors and predators in breeding areas, stopover habitats, and nonbreeding ranges, migrants may be exposed to novel parasites and pathogens in these different areas (Altizer et al. 2011). Subsequently, migrating animals may carry parasites and pathogens between their breeding and nonbreeding ranges. Although there has been concern about migratory species causing long-distance spread of zoonotic pathogens (those that can be transmitted from animals to humans), few studies have clearly documented such cases (Altizer et al. 2011). Conversely, migration may result in reduced disease or parasite levels by at least two mechanisms. First, diseased individuals or those with high parasite loads may have decreased ability to survive migration and are removed from the population (Loehle 1995, Altizer et al. 2011). For example, lesser black-backed gulls (*Larus fuscus*) that migrated longer distances exhibited lower seroprevalence of the gull-specific subtype of avian influenza (H16) than those that migrated shorter distances (Fig. 15.12; Arriero et al. 2015). The authors suggested that the difference was likely due to migratory culling because low pathogenic strains of avian influenza have been related to decreased migratory performance or body condition in waterfowl (van Gils et al. 2007, Latorre-Margalef et al. 2009). Second, infective stages of parasites might build up over a season in areas used heavily by wildlife populations. Migration effectively removes the host from that area (migratory escape), potentially decreasing the number of surviving parasites such that migrants return to an environment with lower parasite levels (Loehle 1995, Altizer et al. 2011). Post-calving migration by reindeer (*Rangifer tarandus tarandus*) removes herds from calving grounds, where larval warble fly (*Hypoderma tarandi*) abundance is high, which may explain significantly lower larval abundance in reindeer that migrate than those that remain on or near the calving grounds throughout the summer (Folstad et al. 1991).

Managing and conserving migratory species involve many challenges. In addition to the wide range of habitats used (e.g., winter, breeding, and stopover) and the geographic scale at which their life history plays out, migratory species face nu-

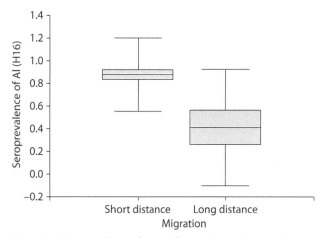

Figure 15.12. Seroprevalence of avian influenza-hemagglutinin subtype H16 in long-distance and short-distance migrants. Mean is indicated by the lines in the center of shaded boxes, standard error is the shaded boxes, and error bars depict the standard deviation. Reproduced with permission from Arriero et al. (2015)

merous threats. As with all species, habitat loss and degradation are a threat, but for migratory species a single population may face these threats in numerous countries or even continents. The political challenges of managing a species that is dependent on regulations and land-use changes in multiple countries can be daunting. Even within a country, terrestrial migrants are in danger of having their migratory routes cut off by development or human activities, such as increased oil and gas extraction efforts (Berger et al. 2006). Climate change has already affected the timing of migration in numerous species (McCarty 2001). For species that rely on invariant cues such as photoperiod to trigger migration, a concern is that changes in habitats and phenology due to climate change will result in arrival at times when the habitat is no longer suitable. Ultimately, the more we can learn about migration, the better we will be able to manage migratory species in the face of rapid changes in ecosystems.

WHY DO ANIMALS LIVE IN GROUPS? SOCIAL SYSTEMS

"Cutting across many of these properties is the habit in many species of forming gregarious units. The existing literature tells which species form coveys, herds, and packs, and which do not, but it seldom suggests what these units consist of . . . a brief summary of what little is known about this question is a necessary basis for an understanding of management technique" (Leopold 1933:48). Aldo Leopold realized the importance of understanding the dynamics of groups and the dearth of knowledge on these behaviors in wildlife species that are often secretive and wary of humans. Although our knowledge has increased dramatically during the nearly 90 years since these words were written, we still have much to learn. Social systems influence how animals use space. Most mammals and birds are not highly social and live either a solitary life or life paired with a mate; however, where resources are dense and aggregated, assemblages may form. Where groups form, more elaborate social interactions and inter-individual bonds may occur. Most species form aggregations at some time during the year, even when they are only ephemeral feeding aggregations at a seasonal food source (Table 15.1).

Social organization in birds and mammals tends to follow a continuum from solitary individuals to increasing levels of overlap and shared duties. The incipient unit of groups is often a pair, or dyad, of individuals. In monogamous species this may be a male and female; however, in birds, males are the fundamental unit of a group, whereas in mammals the fundamental unit of a group is the female-female dyad. The distribution of males in birds and females in mammals is often determined by the distribution of resources owing to their high levels of parental investment. Male birds hold territories in which nest and food resources are found to provision offspring, and female mammals are the only sex able to nurse their young. The distribution of male birds and female mammals often determines the distribution of the opposite sex. For example, male ungulates and pinnipeds are distributed most often in ways that maximize their likelihood of encountering females. Most male carnivores, rodents, and lagomorphs expand their home ranges to overlap female groups. Most birds are monogamous, but in the relatively few species that are polygamous or promiscuous, principally game birds, females settle in male territories or select males by visiting lekking grounds.

Most wildlife species are either solitary or form groups with significant spatial overlap but with no shared duties. White-tailed and mule deer and most ungulates, coyotes (*Canis latrans*), raccoons (*Procyon lotor*), eastern foxes (*Sciurus niger*) and eastern gray (*Sciurus carolinensis*) squirrels, most quail (Family Odontophoridae), turkeys (*Meleagris gallopavo*), and blackbirds (Family Icteridae) are examples (Fig. 15.13). Communal groups share space and duties, most commonly in the care and provisioning of offspring. Examples of this advanced level of sociality include black-tailed prairie dogs (*Cynomys ludovicianus*); gray wolves (*Canis lupus*); a number of primates, banded mongooses (*Mungos mungo*), red-cockaded (*Picoides borealis*), and acorn woodpeckers; and Harris's hawks (*Parabuteo unicinctus*). Eusociality, with extreme division of labor and reproductive castes like the social bees, ants, and termites (Order Hymenoptera), is considered to be the highest level of sociality but is only known from one vertebrate, naked mole rats (*Heterocephalus glaber*). Thus the social system is of little consequence in wildlife management and conservation.

Why do animals live in groups? Weighing costs and benefits is a fruitful approach to understanding sociality (Waterman 1997). Survival and reproduction tend to increase with population size (small populations experience Allee effects owing to reduced ability to find mates), and detection of predators and recruitment of nest mates are other important benefits, while inbreeding and competition for resources could be costly (Stephens and Sutherland 1999). That most wildlife species are solitary suggests that the costs of group living must be substantial in most cases. This is especially evident when considering the lengthy list of potential benefits of group living that have been reported in birds and mammals (Table 15.2). Wildlife species that live in groups tend to gain great benefit from reduced predation risk, location of food, energetic advantage through drafting off or huddling with other individuals, or, when relatives are present, through inclusive fitness gains. Inclusive fitness benefits can be considerable when relatives live in proximity. Hamilton's Rule, named after W. D. Hamilton, the scientist who conceived of the idea, predicts that the inclusive fitness benefits that are accrued by assisting relatives must be devalued by the level of relatedness (r) shared between the two relatives. The simple equation that is Hamilton's Rule predicts that cooperation or assistance should be favored when the adjusted benefits (B) exceed the costs (C):

$$C < rB.$$

Birds and mammals that live in groups often include related individuals, and such inclusive benefits may be significant.

Table 15.1. Names of aggregations of mammals and birds

Birds		Mammals	
Species	Aggregation Name	Species	Aggregation Name
Adult birds	A flight (in air), flock (on ground), volary, brace (gamebirds or waterfowl, referring to a pair killed by a hunter)	Apes	A shrewdness
Bitterns	A sedge	Asses	A pace
Buzzards	A wake	Badgers	A cete
Bobolinks	A chain	Bats	A colony
Chicks	A brood, clutch	Bears	A sloth, sleuth
Coots	A cover	Bison	A herd
Cormorants	A gulp	Buffalo	A gang, an obstinacy
Cranes	A sedge	Cats	A clowder, a pounce. For kittens, a kindle, litter, an intrigue
Crows	A murder, horde	Cattle	A drove, herd
Dotterel	A trip	Deer	A herd, bevy (roe deer)
Doves	A dule, pitying (specific to turtle doves)	Dogs	A litter (young), pack (wild), cowardice (of curs). For hounds, a cry, mute, pack, kennel
Ducks	A brace, flock (in flight), raft (on water) team, paddling (on water), badling	Elephants	A herd
Eagles	A convocation	Elk	A gang
Finches	A charm	Ferrets	A business
Flamingos	A stand	Fox	A leash, skulk, earth
Geese	A flock, gaggle (on the ground), skein (in flight)	Giraffes	A tower
Grouse	A pack (in late season)	Goats	A tribe, trip
Gulls	A colony	Gorillas	A band
Hawks	A cast, kettle (flying in large numbers), boil (two or more spiraling in flight)	Hippopotamuses	A bloat
Herons	A sedge, a siege	Horses	A herd, team, harras, rag (colts), stud. For a group of horses of a single owner, string, ponies
Jays	A party, scold	Hyenas	A cackle
Lapwings	A deceit	Kangaroos	A troop
Larks	An exaltation	Leopards	A leap
Magpies	A tiding, gulp, murder, charm	Lions	A pride
Mallards	A sord (in flight), brace	Martens	A richness
Nightingales	A watch	Moles	A labor
Owls	A parliament	Monkeys	A troop, barrel
Parrots	A company	Mules	A pack, span, barren
Partridge	A covey	Otters	A romp
Peacocks	A muster, an ostentation	Oxen	A team, yoke
Penguins	A colony	Pigs	A drift, drove, litter (young), sounder (of swine), team, passel (of hogs), singular (refers to a group of boars)
Pheasant	A nest, nide (a brood), nye, bouquet	Porcupines	A prickle
Plovers	A congregation, wing (in flight)	Rabbits and hares	A colony, warren, nest, herd (domestic only), litter (young). For hares, a down, husk
Ptarmigans	A covey	Rhinoceroses	A crash
Quail	A bevy, covey	Seals	A pod, herd
Ravens	An unkindness	Sheep	A drove, flock, herd
Rooks	A building	Squirrels	A dray, scurry
Snipe	A walk, a wisp	Tigers	A streak
Sparrows	A host	Whales	A pod, gam, herd
Starlings	A murmuration	Wolves	A pack, rout, or route (when in motion)
Storks	A mustering		
Swallows	A flight		
Swans	A bevy, wedge (in flight)		
Teal	A spring		
Turkeys	A rafter, gang		
Widgeons	A company		
Woodcocks	A fall		

Source: Modified from Fellows (2012).

Figure 15.13. Assemblages of wildlife species. (A) Aggregation of pintails at a pond. (B) Covey of bobwhite quail. (C) Nesting group of eastern gray squirrels. (D) Band of white-nosed coati consisting of adult females and yearling offspring. (E) Group of female pronghorns, likely related, with territorial buck nearby. Photos courtesy of (A) US Fish and Wildlife Service Migratory Bird Program, (B) copyright Joe Coelho, (C) Ken Atkinson, (D) Greg Boreham, and (E) James Ownby

White-tailed deer, elk, gray wolves, eastern gray squirrels, white-nosed coati (*Nasua narica*), California quail (*Callipepla californica*), wood ducks (*Aix sponsa*), and acorn woodpeckers appear to capitalize on relatedness within groups and demonstrate amicable behaviors toward related individuals.

Dominance rank is a key aspect of group living wildlife species. A near-universal truth is that dominant individuals have greater reproductive success than subordinate individuals (i.e., those with low dominance rank), owing to their ability to control access to mates and to resources required by mates. Indeed, low-ranking individuals often adopt alternative reproductive tactics (Gross 1996, Koprowski 2007) to enhance their reproductive success. Subordinate male tree squirrels do not challenge the dominant males in pursuit of an estrous female but adopt a satellite tactic that capitalizes on the female's avoidance behavior and garners some mating success (Koprowski 2007). Similarly, dominant bighorn males tend females and rebuff subordinate males that use a coursing tactic to achieve some reproductive success. Coursing involves initiating a distraction of the tending male to enable the female to race away on a coursing run, during which the subordinate male may successfully breed with the female (Hogg 1984). Subordinate animals are often young or very old, small, in poor health, or unrelated to individuals within the group. The consequences of low rank can include reduced access to resources that results in poor condition, reduced survival, low reproductive success, and overall reduced fitness. High-ranking female elk tend to produce more male young, nurse young more substantially, and wean young that are more dominant than low-ranking females (Clutton-Brock et al. 1986). Dominance among pronghorn females results in high-ranking females having enhanced access to food and higher levels of aggression (Fairbanks 1994). Dominant Harris's hawk females have greater access to food, mates, and nest sites (Dawson and Mannan 1991).

Table 15.2. Generalized costs and benefits of living in groups for wildlife species

Costs	Benefits
Increased competition	Dilution effect: decreased per capita risk of predation
Increased spread of parasites and disease	Confusion effect: predator success rates decrease because of inability to focus on an individual
Increased incidence of cuckoldry	Increased number individuals vigilant
Increased visibility to predators	Increased potential for group defense
Increased reproductive suppression	Increased success in locating food because of group foraging
Increased risk of injury during combat	Increase variety of and proximity to mates
	Aero- or hydrodynamic advantages during locomotion
	Thermodynamic advantage of huddling
	Information transfer between individuals about resources
	Increase in cooperation in social interaction
	Increased inclusive fitness

BEHAVIOR IS AN INDIVIDUAL THING

Although many wildlife management questions focus on characteristics of populations (e.g., fecundity, mortality, growth rate, abundance, emigration, and recruitment), as you can see from examples in this chapter, wildlife behavior is an individual phenomenon. The population parameters we commonly measure are emergent properties of individual choices and behaviors displayed by animals responding to their environment. Natural selection acts upon individuals; therefore observed differences in behavior traits within a population likely provide some fitness benefits in several different contexts (Merrick and Koprowski 2017). Observing changes in behaviors or the variability in and frequency of behavior traits within a population can also provide the first cues about how populations are being affected by and responding to environmental change. Behavioral responses to natural and anthropogenic cues are also important for understanding how to best implement conservation and management plans that could include habitat restoration, targeted removals, repelling or attracting wildlife to particular areas, and conservation translocations (Greggor et al. 2016). Merely documenting changes in wildlife behavior is often not enough to address pressing conservation challenges. Many modern conservation challenges involve issues of attraction and avoidance, such as repelling wildlife from and/or attracting to a particular area as part of a management action or intervention. In such cases it is important to also understand species-specific cognition, or how animals perceive, learn, and remember information about their environment (Greggor et al. 2020), and consider how sensory pollutants such as noise, light, and chemicals can affect perceptions and decision-making (Dominoni et al. 2020).

How Individual Behavior Differences Can Influence Wildlife Research

Consistent differences in inter-individual behavior exist for a variety of behavior traits within wildlife populations. Individually consistent behavior traits such as shy-bold, docile-aggressive, social-antisocial, sedentary-active, and tendency to remain philopatric or disperse are often correlated, and correlated traits are known as behavior syndromes (Sih et al. 2004). Individual behavior differences and behavior syndromes can influence population-level estimates such as population size, structure, and gene flow, and have the potential to influence the outcome of wildlife research (Merrick and Koprowski 2017). Individual behavior differences are also tied to underlying physiological responses to stress, growth and development rates, fecundity and maternal investment, and how individuals tolerate disturbance and novel environments (Dantzer et al. 2013, Lowry et al. 2013, Hinde et al. 2015). As you might imagine, novel environments such as increased urbanization exert extreme selection pressures on wildlife, and behavioral shifts are often observed along urban–rural or high–low disturbance gradients. For example, bold, aggressive individuals are more likely to colonize new sites, do well in urban areas, and tolerate living at higher densities, correlated traits sometimes referred to as urban wildlife syndrome (Parker and Nilon 2012). Novel environments also tend to select for behavioral flexibility, the ability to modify behavior responses with new information and experience (learning), and the ability to apply learned skills to novel problems (innovation; Sol et al. 2013).

The presence of individual behavior differences and shifts in behavior trait frequencies within a population in response to biotic and abiotic environmental conditions can influence wildlife research outcomes and potentially bias the inferences we make from applied studies. It is also important to consider how management actions may affect individuals in a population differently (Merrick and Koprowski 2017). For example, if bold, neophilic individuals tend to be found near urban or disturbed areas, these individuals may be more likely to be involved in human–wildlife conflicts and be detected via visual encounter surveys, trapping designs, and citizen science applications compared to less bold individuals inhabiting undisturbed areas. Bold, neophilic individuals are more likely to investigate live traps and camera traps; thus these individuals may be overrepresented in wildlife studies and be captured in translocation efforts. Similarly, bold, active, less wary individuals are more likely to be found in open areas, making them more susceptible to human harvest. In some taxa, bold and active behavior types are correlated with a fast pace of life syndrome that includes high metabolic rate, faster growth rate, and high fecundity (Biro and Post 2008, Mittelbach et al. 2014, but see Royauté et al. 2018). In fishes, bold, active individuals are likely to be larger, mature faster, and forage in open areas and thus are targeted by selective harvest. Selective harvest of the largest, fastest-growing individuals by commercial fisheries has led to shifts in genetic structure, physiology, and behavior of many populations, resulting in future generations

Table 15.3. Examples of documented behavior trait axes and some commonly used methods to assess individual behavior differences from least to most invasive

	LESS INVASIVE					MORE INVASIVE			
Behavior Trait Axes	Camera Traps	Behavior Observations	Novel Object Tests/ Puzzles	Flight Initiation Distance	Trapability	Open Field Trials / Arena Tests	Mirror Image Stimulation	Struggle Rate, Respiration Rate	Telemetry, Biologging
Sedentary–Active	X	X				X			X
Shy–Bold	X	X	X	X	X		X		
Neophilic–Neophobic	X	X	X						
Docile–Aggressive		X		X	X		X	X	
Social–Antisocial		X					X		X
Philopatric–Dispersive						X			X
Non-innovative–Innovative		X	X						

that are smaller sized, less bold, less active, and grow more slowly—an example of fishery-induced evolution of life history traits (Biro and Post 2008, Mittelbach et al. 2014).

Individual behavior differences can also play a role in mate choice and partner compatibility and may represent honest signals of mate quality. Sexual selection for preferred behavior traits can be context-specific and lead to mate choice variation within a population. For example, in female common lizards (*Zootoca vivipara*), mate preference differs as a function of predation risk (Teyssier et al. 2014). Behavior traits are also correlated to mating strategies and are shown to influence the rate of extra-pair copulations. The presence of behaviorally preferred mates in a population is shown to increase reproductive success in some species and is an important consideration for the recovery of threatened populations and captive breeding programs (Greggor et al. 2016).

Quantifying Individual Behavior Differences
There are many methods for assaying individual behavior differences that range from purely observational to more invasive approaches that require trapping and handling of animals (Table 15.3). If focal individuals can be identified, many behavior traits can be quantified by noninvasive sampling methods that include behavior observations and camera traps. Behavior observations can help determine individual differences in sociality, docility, and conspecific aggression. Behavior observations or cameras can also be used to record individual responses to novel objects and novel problems such as puzzle feeders and toys to document neophilia and innovation (Johnson-Ulrich et al. 2019). Depending upon the experimental design, camera traps can infer an individuals' tendency to explore novel objects, and the number of different camera traps an individual is recorded at (trap diversity) can indicate activity level and tendency to explore new areas. How vigilant individuals are, or how readily they flee upon the approach of a predator or perceived threat, is another useful and less invasive tool to assess individuals along the shy-bold and docile-aggressive behavior trait axes. Flight initiation distance (FID) is recorded as the distance between an approaching observer or

threat and a focal animal at which the individual first becomes alert, and the distance at which the individual leaves or takes flight. Flight initiation distances are often shorter in urban compared to rural populations. Uchida et al. (2019) compared FID in Eurasian red squirrels in response to the approach of humans, predators, and novel objects in both urban and rural sites and found that the alert distance and FIDs were shorter in urban areas across all contexts.

If individuals are part of a study that involves trapping and handling, additional behavioral assays are possible. While an animal is in hand, respiration and struggle rate can be recorded and compared as a metric of docility (Humphries et al. 2012). Open field (OF) and mirror image stimulation (MIS) trials are both commonly used methods to assess activity and exploration in a novel environment (OF) and aggression and sociality (MIS; Humphries et al. 2012, Santicchia et al. 2020). OF and MIS trials are often carried out in a behavior arena into which an individual is released, making this type of trial more feasible for smaller species. The activity level of an individual upon initial entrance into the arena is recorded during the open field portion trial. Next, a mirror is exposed, commencing the mirror image stimulation trial, and the individuals' response to its reflection is recorded (Fig. 15.14). Measures of activity, exploration, sociality, and aggression obtained from OF and MIS trials have been shown to be correlated with natal dispersal distance (Merrick and Koprowski 2016a).

Miniaturization of technology associated with very-high-frequency and satellite telemetry, accelerometers, and biologging devices are expanding the degree to which we can monitor animal movement behavior and physiological responses to environmental stressors in an increasingly diverse array of taxa. Spatial telemetry fixes reveal individual differences in space use, sociality, habitat use, and even perceived risk in response to habitat and predator communities. Accelerometers can further reveal additional individual differences in activity levels and activity budgets across environmental contexts such as habitat type and season (e.g., Hammond et al. 2016). When combined with telemetry, biologgers have the potential to expand the field of sensory ecology (Dominoni et al. 2020)

Figure 15.14. A subadult Mt. Graham red squirrel explores a behavior arena designed to test its response to novel stimuli, including how it responds to a new environment (open field) and its own mirror image (mirror image stimulation).

by documenting direct physiological responses of individuals to acute and chronic stressors in the environment and individual differences therein. For example, Ditmer et al. (2018) fit American black bears (*Ursus americanus*) with global positioning system collars and cardiac biologgers and found that individuals perceived roads as an acute stressor, as evidenced by significant increases in heart rate upon approach and crossing of major roads.

SUMMARY

The field of wildlife management has progressed considerably since Leopold's (1933) observations about the value of understanding animal behavior. Our knowledge has increased with advances in the theory that enrich our understanding of wildlife behavior. Most importantly, the ability to apply acquired knowledge to develop useful management strategies is greatly improved. Yet considerable research needs remain and will ensure a plethora of exciting opportunities for wildlife biologists. Pressing future issues involve the response of wildlife species to habitat fragmentation, increased road development, urban sprawl, alternative energy development, and restoration of predators to ecosystems. What will be the behavioral responses to habitat changes induced by invasive plant and animal species and climate change? These challenges for wildlife management will require scientists with a well-rounded education in the many facets of the wildlife management field, which includes a firm grounding in wildlife behavior.

Literature Cited

Airst, J. I., and S. Lingle. 2020. Male size and alternative mating tactics in white-tailed deer and mule deer. Journal of Mammalogy 101:1231–1243.

Altizer, S., R. Bartel, and B. A. Han. 2011. Animal migration and infectious disease risk. Science 331:296–302.

Arriero, E., I. Müller, R. Juvaste, F. Javier Martínez, and A. Bertolero. 2015. Variation in immune parameters and disease prevalence among lesser black-backed gulls (*Larus fuscus* sp.) with different migratory strategies. PLoS One 10:e0118279.

Bennett, R. E., A. D. Rodewald, and K. V. Rosenberg. 2019. Overlooked sexual segregation of habitat exposes female migratory landbirds to threats. Biological Conservation 240:108266.

Berger, J., S. L. Cain, and K. M. Berger. 2006. Connecting the dots: an invariant migration corridor links the Holocene to the present. Biology Letters 2:528–531.

Biro, P. A., and J. R. Post. 2008. Rapid depletion of genotypes with fast growth and bold personality traits from harvested fish populations. Proceedings of the National Academy of Sciences of the United States of America 105:2919–2922.

Bowyer, R. T. 2004. Sexual segregation in ruminants: definitions, hypotheses, and implications for conservation and management. Journal of Mammalogy 85:1039–1052.

Boyce, M. S. 1989. Elk management in North America: the Jackson herd. Cambridge University Press, Cambridge, UK.

Brommer, J. E., K. Ahola, and T. Karstinen. 2005. The colour of fitness: plumage coloration and lifetime reproductive success. Proceedings of the Royal Society B 272:935–940.

Buttemer, W. A., and W. R. Dawson. 1993. Temporal pattern of foraging and microhabitat use by Galapagos marine iguanas, *Amblyrhynchus cristatus*. Oecologia 96:56–64.

Caudill, C. C. 2003. Measuring dispersal in a metapopulation using stable isotope enrichment: high rates of sex-biased dispersal between patches in a mayfly metapopulation. Oikos 101:624–630.

Clutton-Brock, T. H. 1988. Reproductive success. University of Chicago Press, Chicago, Illinois, USA.

Clutton-Brock, T. H. 1989. Mammalian mating systems. Proceedings of the Royal Society B 236:339–372.

Clutton-Brock, T. H., S. D. Albon, and F. E. Guinness. 1986. Great expectations: dominance, breeding success and offspring sex ratios in red deer. Animal Behaviour 34:460–471.

Conradt, L. 2005. Definitions, hypotheses, models and measures in the study of animal segregation. Pages 11–32 *in* K. E. Ruckstuhl and P. Neuhaus, editors. Sexual segregation in vertebrates: ecology of the two sexes. Cambridge University Press, Cambridge, UK.

Cote, J., J. Clobert, T. Brodin, S. Fogarty, and A. Sih. 2010. Personality-dependent dispersal: characterization, ontogeny and consequences for spatially structured populations. Philosophical Transactions of the Royal Society B 365:4065–4076.

Cristol, D. A., M. B. Baker, and C. Carbone. 1999. Differential migration revisited: latitudinal segregation by age and sex class. Pages 33–88 *in* V. Nolan Jr., E. D. Ketterson, and C. F. Thompson, editors. Current ornithology. Volume 15. Plenum Press, New York, New York, USA.

Cudworth, N. L, and J. L. Koprowski. 2010. Influences of mating strategy on space use of Arizona gray squirrels. Journal of Mammalogy 91:1235–1241.

Dantzer, B., A. E. Newmann, R. Boonstra, R. Palme, M. M. Humphries, and A. G. McAdam. 2013. Density triggers maternal hormones that increase adaptive offspring growth in a wild mammal. Science 340:1215–1217.

Davis, J. 2008. Patterns of variation in the influence of natal experience on habitat choice. Quarterly Review of Biology 83:363–380.

Davis, J. M., and J. A. Stamps. 2004. The effect of natal experience on habitat preferences. Trends in Ecology and Evolution 19:411–416.

Dawson, J. R., and R. W. Mannan. 1991. Dominance hierarchies and helper contributions in Harris' hawks. Auk 108:649–660.

Ditmer, M. A., S. J. Rettler, J. R. Fieberg, P. A. Iaizzo, T. G. Laske, K. V. Noyce, and D. L. Garshelis. 2018. American black bears perceive the risks of crossing roads. Behavioral Ecology 29:667–675.

Dominoni, D. M., W. Halfwerk, E. Baird, R. T. Buxton, E. Fernández-Juricic, et al. 2020. Why conservation biology can benefit from sensory ecology. Nature Ecology and Evolution 4:502–511.

Duckworth, R. 2009. Maternal effects and range expansion: a key factor in a dynamic process? Philosophical Transactions of the Royal Society B 364:1075–1086.

Fairbanks, W. S. 1994. Dominance, age and aggression among female pronghorn, *Antilocapra americana* (Family: Antilocapridae). Ethology 97:278–293.

Fellows, D. 2012. Animal congregations, or what do you call a group of . . .? http://www.npwrc.usgs.gov/about/faqs/animals/names.htm.

Fisher, D. O., and I. P. F. Owens. 2008. Female home range size and the evolution of social organization in macropod marsupials. Journal of Animal Ecology 69:1083–1098.

Folstad, I., A. C. Nilssen, O. Halvorsen, and J. Andersen. 1991. Parasite avoidance: the cause of post-calving migrations in *Rangifer*? Canadian Journal of Zoology 69:2423–2429.

Greenwood, P. J. 1980. Mating systems, philopatry and dispersal in birds and mammals. Animal Behaviour 28:1140–1162.

Greggor, A. L., O. Berger-Tal, and D. T. Blumstein. 2020. The rules of attraction: the necessary role of animal cognition in explaining conservation failures and successes. Annual Review of Ecology, Evolution, and Systematics 51:483–503.

Greggor, A. L., O. Berger-Tal, D. T. Blumstein, L. Angeloni, C. Bessa-Gomes, et al. 2016. Research priorities from animal behaviour for maximising conservation progress. Trends in Ecology and Evolution 2157:1–12.

Griffith, S. C., I. P. F. Owens, and K. A. Thuman. 2002. Extra-pair paternity in birds: a review of interspecific variation and adaptive function. Molecular Ecology 11:2195–2212.

Gross, M. R. 1996. Alternative reproductive strategies and tactics: diversity within the sexes. Trends in Ecology and Evolution 11:92–98.

Hämäläinen, S., K. Fey, and V. Selonen. 2020. Search strategies in rural and urban environment during natal dispersal of the red squirrel. Behavioral Ecology and Sociobiology 74:10.1007/s00265-020-02907-z.

Hammond, T. T., D. Springthorpe, R. E. Walsh, and T. Berg-Kirkpatrick. 2016. Using accelerometers to remotely and automatically characterize behavior in small animals. Journal of Experimental Biology 219:1618–1624.

Han, L., D. Blank, M. Wang, A. Alves da Silva, W. Yang, K. Ruckstuhl, and J. Alves. 2020. Diet differences between males and females in sexually dimorphic ungulates: a case study on Siberian ibex. European Journal of Wildlife Research 66:55.

Harestad, A. S., and F. L. Bunnell. 1979. Home range and body weight—a reevaluation. Ecology 60:389–402.

Hinde, K., A. L. Skibiel, A. B. Foster, L. Del Rosso, S. P. Mendoza, and J. P. Capitanio. 2015. Cortisol in mother's milk across lactation reflects maternal life history and predicts infant temperament. Behavioral Ecology 26:269–281.

Hogg, J. T. 1984. Mating in bighorn sheep: multiple creative male strategies. Science 225:526–529.

Humphries, M. M., S. Boutin, R. W. Taylor, A. K. Boon, B. Dantzer, D. Re, J. C. Gorrell, D. W. Coltman, and A. G. M. C. Adam. 2012. Low heritabilities, but genetic and maternal correlations between red squirrel behaviours. Journal of Evolutionary Biology 25:614–624.

Johnson-Ulrich, L., S. Benson-Amram, and K. E. Holekamp. 2019. Fitness consequences of innovation in spotted hyenas. Frontiers in Ecology and Evolution 7:1–9.

Koprowski, J. L. 2005. The response of tree squirrels to fragmentation: a review and synthesis. Animal Conservation 8:369–376.

Koprowski, J. L. 2007. Reproductive strategies and alternative reproductive tactics of tree squirrels. Pages 86–95 *in* J. Wolff and P. Sherman, editors. Rodent societies: an ecological and evolutionary perspective. University of Chicago Press, Chicago, Illinois, USA.

Latorre-Margalef, N., G. Gunnarsson, V. J. Munster, R. A. M. Fouchier, A. D. M. E. Osterhaus, et al. 2009. Effects of influenza A virus infection on migrating mallard ducks. Proceedings of the Royal Society B 276:1029–1036.

Leopold, A. 1933. Game management. University of Wisconsin Press, Madison, USA.

Lin, Y. K., B. Keane, A. Isenhour, and N. G. Solomon. 2006. Effects of patch quality on dispersal and social organization of prairie voles: an experimental approach. Journal of Mammalogy 87:446–453.

Loehle, C. 1995. Social barriers to pathogen transmission in wild animal populations. Ecology 76:326–335.

Long, R. A., J. L. Rachlow, and J. G. Kie. 2009. Sex-specific responses of North American elk to habitat manipulation. Journal of Mammalogy 90:423–432.

Lowry, H., A. Lill, and B. B. M. Wong. 2013. Behavioural responses of wildlife to urban environments. Biological Reviews 88:537–549.

Lyrholm, T., O. Leimar, B. Johanneson, and U. Gyllensten. 1999. Sex-biased dispersal in sperm whales: contrasting mitochondrial and nuclear genetic structure of global populations. Proceedings of the Royal Society B 266:347–354.

McCarty, J. P. 2001. Ecological consequences of recent climate change. Conservation Biology 15:320–331.

Merrick, M. J., and J. L. Koprowski. 2016a. Altered natal dispersal at the range periphery: the role of behavior, resources, and maternal condition. Ecology and Evolution 7:58–72.

Merrick, M. J., and J. L. Koprowski. 2016b. Evidence of natal habitat preference induction within one habitat type. Proceedings of the Royal Society B 283:1–10.

Merrick, M. J., and J. L. Koprowski. 2017. Should we consider individual behavior differences in applied wildlife conservation studies? Biological Conservation 209:34–44.

Mittelbach, G. G., N. G. Ballew, M. K. Kjelvik, and D. Fraser. 2014. Fish behavioral types and their ecological consequences. Canadian Journal of Fisheries and Aquatic Sciences 71:927–944.

Parker, T. S., and C. H. Nilon. 2012. Urban landscape characteristics correlated with the synurbization of wildlife. Landscape and Urban Planning 106:316–325.

Rachlow, J. L., and R. T. Bowyer. 1998. Habitat selection by Dall's sheep (*Ovis dalli*): maternal trade-offs. Journal of Zoology (London) 245:457–465.

Ramenofsky, M., R. Savard, and M. R. C. Greenwood. 1999. Seasonal and diet transitions in physiology and behavior in the migratory dark-eyed junco. Comparative Biochemistry and Physiology A 122:385–397.

Rogerson, J. D., W. S. Fairbanks, and L. Cornicelli. 2008. Ecology of gastropod and bighorn sheep hosts of lungworm on isolated, semiarid mountain ranges in Utah, USA. Journal of Wildlife Diseases 44:28–44.

Royauté, R., M. A. Berdal, C. R. Garrison, and N. A. Dochtermann. 2018. Paceless life? A meta-analysis of the pace-of-life syndrome hypothesis. Behavioral Ecology and Sociobiology 72:64.

Rubin, E. S., and V. C. Bleich. 2005. Sexual segregation: a necessary consideration in wildlife conservation. Pages 379–391 *in* K. E. Ruckstuhl and P. Neuhaus, editors, Sexual segregation in vertebrates: ecology of the two sexes. Cambridge University Press, Cambridge, UK.

Santicchia, F., S. Van Dongen, A. Martinoli, D. Preatoni, and L. A. Wauters. 2020. Measuring personality traits in Eurasian red squirrels: a critical comparison of different methods. Ethology 127:187–201.

Selonen, V., and I. K. Hanski. 2009. Decision making in dispersing Siberian flying squirrels. Behavioral Ecology 21:219–225.

Shine, R., and M. Wall. 2005. Ecological divergence between the sexes in reptiles. Pages 221–253 in K. E. Ruckstuhl and P. Neuhaus, editors. Sexual segregation in vertebrates: ecology of the two sexes. Cambridge University Press, Cambridge, UK.

Sih, A., A. Bell, and J. C. Johnson. 2004. Behavioral syndromes: an ecological and evolutionary overview. Trends in Ecology and Evolution 19:372–378.

Sims, D. W. 2005. Differences in habitat selection and reproductive strategies of male and female sharks. Pages 127–147 in K. E. Ruckstuhl and P. Neuhaus, editors. Sexual segregation in vertebrates: ecology of the two sexes. Cambridge University Press, Cambridge, UK.

Smith, B. L., and R. L. Robbins. 1994. Migrations and management of the Jackson elk herd. National Biological Survey Resource Publication 199. US Department of the Interior, Washington, DC, USA.

Sol, D., O. Lapiedra, and C. González-Lagos. 2013. Behavioural adjustments for a life in the city. Animal Behaviour 85:1101–1112.

Stamps, J. A., J. M. Davis, S. A. Blozis, and K. L. Boundy-Mills. 2007. Genotypic variation in refractory periods and habitat selection by natal disperser. Animal Behaviour 74:599–610.

Stephens, P. A., and W. J. Sutherland. 1999. Consequences of the Allee effect for behaviour, ecology and conservation. Trends in Ecology and Evolution 14:401–405.

Stewart, K. M., T. E. Fulbright, D. L. Drawe, and R. T. Bowyer. 2003. Sexual segregation in white-tailed deer: responses to habitat manipulations. Wildlife Society Bulletin 31:1210–1217.

Teyssier, A., E. Bestion, M. Richard, and J. Cote. 2014. Partners' personality types and mate preferences: predation risk matters. Behavioral Ecology 25:723–733.

Uchida, K., K. K. Suzuki, T. Shimamoto, H. Yanagawa, and I. Koizumi. 2019. Decreased vigilance or habituation to humans? Mechanisms on increased boldness in urban animals. Behavioral Ecology 30:1583–1590.

van Gils, J. A., V. J. Munster, R. Radersma, D. Liefhebber, R. A. Fouchier, and M. Klaassen. 2007. Hampered foraging and migratory performance in swans infected with low-pathogenic avian influenza A virus. PLoS One 2:e184.

Waterman, J. M. 1997. Why do male Cape ground squirrels live in groups? Animal Behaviour 53:809–817.

Wearmouth, V. J., and D. W. Sims. 2008. Sexual segregation in marine fish, reptiles, birds and mammals: behaviour patterns, mechanisms and conservation implications. Advances in Marine Biology 54:107–170.

Whiting, J. C., R. T. Bowyer, J. T. Flinders, V. C. Bleich, and J. G. Kie. 2010. Sexual segregation and use of water by bighorn sheep implications for conservation. Animal Conservation 13:541–548.

HABITAT

R. WILLIAM MANNAN AND ROBERT J. STEIDL

INTRODUCTION

Managing habitat is fundamental to wildlife conservation because individual animals and the populations they comprise cannot persist without an appropriate place to live. Populations largely decimated by environmental contaminants or overharvest, for example, can recover if adequate habitat is available. In contrast, a reduction in the amount of habitat for a species will lead to permanent decreases in abundance and distribution unless habitat is restored. Manipulating habitat is therefore a primary strategy that biologists use to alter the distribution and abundance of species of conservation or management concern. In this chapter we provide a theoretical and practical overview of the concept of habitat, including defining habitat and reviewing ideas about how animals identify and select habitat. We review how managing habitat can be used to meet conservation objectives; introduce concepts and methods important for studying habitat of terrestrial vertebrates, including issues in the design and analysis of studies to characterize habitat; describe habitat use; and assess habitat selection. We conclude with examples that illustrate strategies for managing habitat.

DEFINITION

Habitat is an area with the combination of resources (e.g., food, cover, water) and environmental conditions (e.g., temperature, precipitation) that promotes residency by individuals of a given species and allows them to survive and reproduce (Morrison et al. 2006). Habitat is inherently a species-specific concept because each species has a unique set of physiological, morphological, and behavioral adaptations that are suited to a particular suite of resources and conditions. Although two or more species may inhabit the same general area, their specific habitat needs are likely to be unique; therefore environmental changes will affect them differently, even if they are sympatric. Qualifiers of habitat—breeding habitat, nonbreeding habitat, and foraging habitat—are useful ways to identify areas or sets of conditions required by an animal during some part of its life cycle or to meet a specific need but should not be misconstrued to represent all the places or resources an animal needs to survive and reproduce.

Use of the term *habitat* can differ markedly from the species-specific concept outlined above (Morrison et al. 2006:10). For example, habitat is sometimes used to denote a particular vegetation community, such as ponderosa pine (*Pinus ponderosa*) habitat. Use of the term in this context probably grew from the phrase *habitat type*, coined by Daubenmire (1976:125) to refer to "land units having approximately the same capacity to produce vegetation." Habitat also is sometimes used to describe a general physical environment, such as riparian habitat or mountain habitat. When biologists refer to an area as suitable habitat or unsuitable habitat, they are most likely noting whether an area will (or will not) support a given species. Because habitat refers implicitly to those areas that are suitable for a given species, the phrase suitable habitat is redundant, and unsuitable habitat is an oxymoron. Although these phrases are commonplace, they dilute the species-specific concept of habitat and can hinder communication, especially when mixed within the same document; we therefore suggest that they be avoided.

HABITAT SELECTION

Inherent in the definition of habitat provided above is the idea that individuals of a species will settle and establish residency in areas that contain the set of physical and biological resources necessary for their survival. The duration of time an individual resides in an area can vary. Animals that do not migrate might, after natal dispersal, remain in the same area throughout their lives, whereas migratory animals might inhabit one area for weeks or months during the breeding period, inhabit another area for weeks or months during the nonbreeding period, and inhabit multiple areas, each for a short time, during migration. Presumably, animals seeking an area to inhabit recognize suitable sites by the presence of environmental cues that are either directly or indirectly associated with resources in an area (Lack 1933, Svardson 1949, Hilden 1965). The link between environmental cues and the resources they indicate is essential to the process of habitat selection (Andersen and Steidl 2020). Over evolutionary time, individuals that recognized and selected areas with the appropriate set of resources had higher survival and reproductive success than individuals that selected areas with fewer resources. Over many generations, individuals with genotypes that allowed them to recognize appropriate areas have come to dominate populations. Habitat selection, therefore, is a behavioral process that has been honed by natural selection to ensure that individuals settle in those areas that contain resources necessary for their survival and reproduction (Jaenike and Holt 1991). The patterns of residency that we observe in nature are manifestations of that long-term evolutionary process. And although we might expect some systematic variation in habitat of a species across its geographical range, the suite of resources necessary for an area to provide habitat for individuals of a species is likely to be reasonably consistent within a species.

Environmental cues that trigger an animal to settle in an area might include those associated with food, nest sites, and cover from predators or inclement weather, but also those associated with the presence of conspecifics (Stamps 1987, Citta and Lindberg 2007) and interspecific competitors (Williams and Batzli 1979). Collectively, cues should indicate areas that have all of the resources needed by an animal and that competitors do not inhabit. In some circumstances, learning can play a role in aspects of habitat selection, such as when an animal's experiences in its natal environment (Davis and Stamps 2004) and prior adult experiences (Baker 2005) influence its selection of an area. Importantly, absence of a single essential resource or its associated cue could prevent an animal from settling in an area. Resources that are absent or in short supply and that limit the number of residents are called *limiting factors*. If biologists can identify these features and manage to increase their abundance, then the number of residents an area supports can be increased. In some instances, adding an essential habitat feature that is missing can even create new areas of habitat (Hamerstrom et al. 1973).

Habitat selection by many animals is likely a complex process that involves a set of behavioral responses to environmental cues that span a range of spatial and temporal scales. One set of cues, for example, might trigger an animal to initiate a settling response, but other cues might be necessary for an animal to remain resident throughout its life cycle, breeding cycle, or nonbreeding period. This process is often considered to be hierarchical because cues expressed at large geographical extents (e.g., presence of a particular plant community) can trigger an animal to initiate a localized search (e.g., for a particular species of nest tree), and more specific cues (e.g., a structural feature of a tree that serves as a place to locate a nest) might narrow the residency response to a particular site. There also might be instances where this hierarchy could be inverted, with animals first identifying important resources at

Figure 16.1. Nest box for wood ducks.

a particular site (e.g., a specific feature that would support a nest) before ensuring the presence of necessary resources at larger spatial extents (e.g., presence of a particular plant community; Flesch and Steidl 2010). For highly mobile animals, the spatial scale of the habitat selection process can span the spectrum from broad regional attributes, such as vegetation zones, to particular resources within a home range (Wiens 1985). The process of selection for less mobile animals also could be hierarchical but confined within smaller spatial extents because of limitations in perception and mobility.

Understanding the behavior of habitat selection is important when providing man-made structures as habitat features. For example, if cavities are a limiting factor for a population of secondary cavity-nesting birds, then providing artificial nest sites might create new areas of habitat and increase the size of an existing population. A nest box nailed to a living tree or affixed to a metal pole, however, does not resemble all aspects of a natural cavity in a dead tree. In these situations, biologists rely on the idea that animals sometimes respond only to a subset of features when identifying some important resources. If a nest box mimics appropriately the critical environmental features associated with a natural nest cavity, such as size of the cavity opening and depth of the cavity, and if the box is positioned in an area that includes the other resources needed by a species, then a cavity-nesting bird might settle in the area and use the box for nesting. For example, wood ducks (*Aix sponsa*) nest in cavities excavated in dead trees by woodpeckers (e.g., pileated woodpeckers, *Dryocopus pileatus*). Like many species, wood ducks were relatively scarce during the early 1900s because of overharvest and habitat destruction, primarily from timber harvesting. Today, wood ducks are among the most abundant and important game birds in the Mississippi Flyway (Yetter et al. 1999). Their recovery was due to a combination of protective game laws that reduced harvests and the willingness of wood ducks to use nest boxes (Fig. 16.1) as substitutes for natural cavities. The erection of thousands of nest boxes (Soulliere 1986, 1988) allowed wood ducks to nest in areas where there were no natural cavities, and to nest in greater densities in areas where natural cavities were limited.

Habitat Quality

Habitat selection can be considered a threshold response (Wiens 1985), where settling behavior is triggered when the necessary resources in an area reach critical levels of abundance (Fig. 16.2). Not all areas with resources above this threshold will be equal, however; plus, abundance of many habitat-related resources can change over time. Variation in levels of important habitat resources among areas is likely to be expressed as variation in survival and reproduction among resident individuals (Johnson 2007). Although it is plausible that some of the observed differences in demographic performance among animals could be due to variation in the quality of the animals themselves—including inherent variation in their ability to recognize suitable places or to survive and reproduce after settling—we favor the idea that every individual of a species has the innate ability to recognize habitat and will settle in locations that meet some minimum threshold of resources. Therefore observed variation in reproduction and survival among areas that provide habitat for a species do not, in our view, often reflect variation in genetic fitness but instead reflect variation in habitat quality. Consequently, demographic performance of a population is likely a good measure of the quality of habitat it occupies (Johnson 2007). We favor this idea because habitat selection is a critical behavioral process and, like all essential behaviors and morphologies, is unlikely to vary markedly among most individuals. Thus areas with sufficient resources to support consistently high densities of residents or high rates of survival and reproduction of individuals can be classified as high-quality habitat. Conversely, areas with resource levels that exceed the threshold for habitat and that trigger settling, but where densities of residents and rates of survival and reproduction are consistently low, can be classified as low-quality habitat. Habitat quality for a given species is therefore a product of the types and abundances of important resources in an area, and how those resources allow members of the species to deal with competitors and preda-

tors. If the suite of resources necessary for a species is well known, then habitat quality could conceivably be determined by measuring the resources themselves, but knowledge of the resources critical to most species is uncommon. Thus habitat quality might best be gauged by assessing demography of the species in question (i.e., the number of individuals that inhabit an area and their relative rates of survival and reproduction; Van Horne 1983, Johnson 2007) and the duration of time an area is inhabited.

Habitat selection behaviors are implicit in theoretical models that describe how animals might select among habitat patches (i.e., discrete and identifiable areas) that vary in quality. Some models, such as the preemptive form of ideal free distribution, suggest that within a heterogeneous landscape, patches of the highest quality will be inhabited first (Brown 1969, Fretwell and Lucas 1970), with patches of lesser quality being inhabited only when higher-quality patches are saturated. Models of this kind depend on animals being able to gauge patch quality and to sample widely across numerous patches before settling. Although the highest-quality habitat for a species often is inhabited consistently, predictions of the ideal free distribution model may not be fully expressed in nature because the number of patches of habitat an animal samples depends on its perceptual ability, its mobility relative to the distribution of patches (Lima and Zollner 1996), the patches it encounters first or most often (Kristan 2003, Krishnan 2007), risks associated with predation, and losing a patch to a conspecific competitor while sampling.

Ecological Traps

The process that animals use to select habitat must be reliable. Humans have altered many natural environments, which can sometimes disconnect the environmental cues that animals use to identify habitat and the resources on which animals rely. An ecological trap is an area with cues that indicate it provides habitat resources, but where animals have relatively low rates of survival, reproduction, or both. An ecological trap in essence tricks animals into settling in areas that are not favorable to them (Schlaepfer et al. 2002, Battin 2004, Robertson and Hutto 2006).

Reduced demographic success in ecological traps can arise at least two ways. First, resources important for survival or reproduction are absent in the area, despite cues indicating their presence. The Arizona cotton rat (*Sigmodon arizonae*), a native rodent common in semidesert grasslands of the southwestern United States, is abundant in areas dominated by the non-native plant Lehmann lovegrass (*Eragrostis lehmanniana*). Survival of cotton rats in patches of Lehmann lovegrass is relatively high, but reproduction is well below that of populations in areas of native grasses (Litt and Steidl 2016). Apparently, some critical resource (e.g., food) required for successful reproduction is reduced in grasslands dominated by the non-native grass. Ecological traps can also reduce demographic success by harboring a novel feature, event, or organism with which a species has not evolved and that reduces habitat quality by increasing mortality or reducing reproduction. Farming

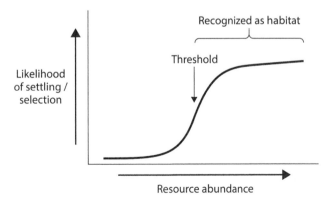

Figure 16.2. Relationship between the likelihood of an animal settling in an area (i.e., selecting it as habitat) and the abundance of resources important for survival.

operations that destroy nests of ground-nesting birds (Roden-house and Best 1983), utility poles that electrocute raptors (Dwyer and Mannan 2007), and cars that kill animals living near roads (Mumme et al. 1999) are examples of events and features that reduce survival, reproduction, or both, and that could create ecological traps (see Chapter 7 for a description of the potential influence that ecological traps might have on populations dynamics). Although ecological traps might exist in natural conditions, we believe they are rare because, over evolutionary time, there has been strong selective pressure against misleading cues. The likelihood of ecological traps is therefore higher in areas where humans have rapidly modified natural landscapes, and where there has been inadequate time for an evolutionary response.

CHARACTERIZING HABITAT

Every wildlife biologist has an intuitive understanding of what constitutes habitat for familiar species, so describing habitat for a species seems like it should be a straightforward exercise. On careful consideration, however, characterizing the suite of biotic and abiotic features that constitute habitat for a species can become surprisingly complex, especially without first constraining the concept of habitat in space or time. The over-arching nature of the habitat concept—which can be defined at multiple spatial and temporal scales, measured in a variety of ways and resolutions, considered for different demographic subsets of a population, and analyzed in what seems like an endless variety of ways—can explain much of the complexity. Consequently, although generally intuitive, the habitat concept comprises a rich and intricate set of topics based on fundamental interactions between animals and their environment, any subset of which might be relevant to a particular set of conservation or management questions. Few topics have generated as much thought and discussion among wildlife scientists as strategies to characterize and evaluate habitat.

The ultimate challenge of characterizing habitat and understanding the process of habitat selection involves identifying the environmental features that trigger an individual's decision to settle in an area and the resources required for successful survival and reproduction. Such identification is sometimes difficult because animals respond to resources at different scales; therefore the scale at which we study habitat can affect our perspective on how habitat is used (Wiens 1989). Further, many species are cryptic and secretive, so we cannot always be certain that we have identified all areas used by a species. Lastly, we often wish to establish the relative importance of resources used by animals compared to the availability of those resources in the environment, which requires that we evaluate resource use relative to some reference; the way in which we establish that reference can influence our conclusions about the relative importance of different resources.

Design Considerations for Habitat Studies
There are many effective strategies for assessing habitat and identifying habitat features important to animals, all of which

involve assumptions and compromises. In general, a strategy for collecting data should reflect the fundamental question of interest, which in turn will dictate the scope of inference (i.e., how broadly the information gained can be applied).

Scale
Scale is a construct we impose on complex systems to help us better identify and understand patterns in nature, including patterns of habitat use by animals (see Chapter 6 for a review of the concept of scale). Because we can consider and evaluate the ways that animals use habitat across a range of spatial scales (we use the term *scale* to refer to extent and not resolution), it is often helpful to consider habitat use as a spatial hierarchy, with choices made by animals at one spatial scale constraining choices at smaller spatial scales. Although habitat for a species can be considered along a continuum of spatial scales, classifying habitat into a three-level spatial hierarchy has proven useful for many wildlife species, with geographical distribution of the species constraining choices at the broadest level, previous choices at the geographical scale constraining choices at the home-range scale, and previous choices at the home-range scale constraining choices of patches for foraging, nesting, and other activities (Johnson 1980). No single spatial scale is correct for all habitat studies because animals respond to resources and other aspects of their environment at multiple scales (Wiens 1989, Levin 1992). Nonetheless, the scale (or scales) we use to evaluate habitat is a critical element of a study design because it constrains the habitat features that we can potentially discern as important to a species. For example, an individual animal may select an area for its home range that is dominated by one vegetation community. If we contrast vegetation at locations used by an individual during its daily activities with the vegetation available within its home range, we could conclude that the vegetation community in which the animal lives out much of its life was not especially important, as it was not used at levels higher than was available. For this reason, the spatial scale of a study can affect its results. The appropriate spatial scale for a study must be based on the scale of the research question and should reflect the scale at which an animal is thought to respond to changes in abundance and distribution of the resources of interest (Wiens 1989). In part because of the uncertainty related to which scale (or scales) is most important to a species, most studies assess habitat use at multiple spatial scales.

In addition to issues of spatial scale, habitat studies must also consider issues of temporal scale because habitat resources important to individuals can vary throughout the day, such as when foraging or resting, throughout the year, especially between breeding and nonbreeding periods, and in response to environmental changes that can occur over longer periods. For example, for species that inhabit dynamic systems that are altered frequently, such as the intertidal zone, habitat features on which these species depend are integrally linked to the periodic disturbances that characterize these systems and govern their structure. Over longer periods, temporal changes in the composition of a plant community that result from the process of ecological succession (see below) have important consequences for the habitat of many animals.

Sampling and Inference

The foundation for inference in habitat studies is similar to any other scientific study where it is usually impractical or impossible to measure every unit in the population we wish to know—inference is based on information gathered from a subset of units selected from all units in the larger population of interest (instances where we study every individual in a population are rare). For example, we might be interested in characterizing habitat features used by grasshopper sparrows (*Ammodramus savannarum*) breeding in a particular geographical region. Because it is unlikely that we would be able to survey all grasslands in the region that might be inhabited by this species, or that we would be able to locate every individual, what we know about this species in this region (i.e., our inference) will be based on the subset of areas we survey and the individuals we locate. When we select the subset of units to study from a larger population through a randomization scheme (e.g., a stratified random sample), then inference is design based, which means we justify inference to the larger population of interest by the approach we used to select sample units (i.e., the sampling design).

In habitat studies, two types of sampling units are common: animals and plots of land. If the sampling unit is an individual animal, such as in radiotelemetry studies or studies based on individual features such as nests or den sites, a subset of units is selected from the larger population of animals or features that is the target of inference. In these studies, individual animals or features are the sampling unit, the collection of individuals selected for study is the sample, and the overall collection of individuals from which the sample is drawn is the population. If the sampling unit is a plot, such as in field surveys where identities and locations of individuals are not known in advance, we use field surveys to assess presence or abundance of individuals on each plot to establish use by animals. When the entire study area will not be surveyed (i.e., in a sampling study), the subset of plots to survey is selected from the larger population of plots that is the target of inference. In these studies, plots are the sampling unit, the collection of plots studied is the sample, and the overall collection of plots from which the sample is drawn (the universe or frame) is the population about which we wish to draw an inference. In studies where plots are the sampling unit, plots are often assigned to different general land cover classes, usually on the basis of the dominant vegetation community or other features thought to influence habitat use by animals.

Common Types of Habitat Studies

Although there are many approaches for studying animal habitat, the two most common types of studies are those that assess *habitat use* and *habitat selection*. Studies of habitat use usually involve gathering information about characteristics of locations used by animals with the goal of characterizing environmental features associated with these locations. Studies of habitat selection usually evaluate the level of use of habitat resources by animals relative to the general availability of those resources in the environment. In contrast to studies of habitat use, studies of habitat selection span a wider range of possible sampling designs and therefore encompass a wider range of analyses that have become increasingly sophisticated.

Imperfect Observations of Animals and Presence-Only Data

Because most animals are impossible to detect with certainty during many field studies, information from studies based solely on descriptions of locations where animals have been observed can be biased when the probability of detecting animals varies across the study area. Specifically, if the probability of detecting an animal is associated or confounded with one or more habitat features, results of habitat studies could be biased. For example, if abundance or presence of a species is relatively consistent across an area, but tree density varies and reduces the ability of surveyors to detect animals that are present, then we might conclude incorrectly that abundance was lower in areas with high tree densities if we failed to adjust for the influence of tree density on detection. Further, if the set of habitat-use locations was not generated from a planned survey, but instead was gathered from incidental observations such as might be available from museum records or Natural Heritage databases, the validity of inferences about habitat use depends on the sample of used locations that represent all locations used by the target species in the area of interest (Pearce and Boyce 2006, Phillips et al. 2009). Contemporary approaches to using presence-only data to model species distributions on the basis of environmental features attempt to overcome these potential biases and to make these inferences more reliable, such as those implemented through MaxEnt software (Phillips et al. 2006, 2009). Simply, the strategy is to develop a model that predicts the distribution of a species based on the contrast between environmental features at use locations and the distribution of those features across the landscape of interest (Elith et al. 2011).

Habitat Use

Understanding the specific resources that are important to habitat of a species is often the first step in developing strategies for conserving or managing animal populations because manipulating habitat is the primary means of influencing the distribution and abundance of animals. Although studies of habitat use often are descriptive, quantitative assessments of important habitat features are critical for developing habitat prescriptions. If all individuals in a population use a particular habitat feature, such as a cavity for nesting, then it seems reasonable to assume that this feature is an important habitat component, regardless of its availability. Further, if all other habitat features required by a species are present in an area where this feature is lacking (i.e., a limiting factor), adding the missing feature could enhance habitat for that species.

All studies designed to identify important habitat features must characterize aspects of the areas used by animals; therefore a fundamental activity of any habitat study is to identify reliably those locations used by the species of interest. This process can focus on locations used by animals when individuals are the sampling unit, whether their identity is known, such as when a bird has been banded or a mammal tagged, or can be inferred, such as when the sample of locations used is determined by observations of unmarked animals. The process can also focus on characterizing plots that have been classified as used by animals, in which case plots, rather than individuals, are the sampling units. The way that use is characterized has implications for analysis, which we explain below.

Habitat Selection

More common than studies to characterize habitat use are studies to assess habitat selection (Thomas and Taylor 2006), where the general goal is to characterize the degree to which animals *use* habitat features relative to their *availability* in the environment. Although these types of studies share a name with the behavioral process of habitat selection—the set of behaviors that animals employ to identify areas with the resources they need to survive and reproduce—habitat selection studies do not often assess this process but instead contrast the characteristics of habitat features in areas used by animals relative to those in locations unused or available to a species in a defined area. In these types of studies, selection implies that a resource is used in greater proportion than is available in the environment.

Even when there is no evidence that a species selects a resource, that resource can still be an essential component of its habitat. In particular, resources that are abundant in the environment are unlikely to be classified as selected even when they are essential to habitat, because selection is not an absolute measure of importance of a resource but a relative measure of use contrasted with availability in the environment. Consequently, resources that animals select in higher proportions than available are not likely to be the only resources important as habitat features (Garshelis 2000). When a habitat resource is essential, such as a cavity for secondary cavity-nesting birds, but is not in limited supply, we will likely find no evidence of selection for that resource, even though it is clearly an essential habitat component. In general, we expect to observe strong evidence of selection for an important habitat resource when the resource is uncommon and to observe no evidence of selection when the feature is common. If evidence of selection for a habitat resource increases as availability increases, however, this resource might be a limiting factor for the species. Therefore evaluating the importance of a resource to a species must be interpreted within the context of how availability is defined, regardless of whether a resource is limiting for a species, and the relative amount of resources in the area considered available to a species.

If absolute use of a particular habitat feature is consistent among individuals, the feature is likely to be important to the species in the region of interest. In contrast, the magnitude of selection can vary among individuals, even when resource use is consistent, because of variation in availability of resources

in the reference area. For example, use of several nest-related resources was consistent across multiple vegetation communities inhabited by cactus ferruginous pygmy owls (*Glaucidium brasilianum cactorum*) in Sonora, Mexico. Because the availability of these resources varied geographically, however, the magnitude of selection of these resources varied with availability (Flesch and Steidl 2010).

Habitat selection is a specific topic within the broader discipline of resource selection, and many of the methods and tools relevant to evaluating resource selection are applicable to studies of habitat selection (Manly et al. 2002). There are multiple sampling designs available for assessing habitat selection (Table 16.1), which have been classified into four main types based on the approaches used to quantify resource use and resource availability (Thomas and Taylor 1990, 2006; Manly et al. 2002). In design I studies, animals are not identified uniquely, resource use is quantified on the basis of information gathered from surveys on plots (often presence or abundance), and availability is characterized at the level of the study area. In design II studies, animals are identified uniquely, resource use is quantified on the basis of information gathered from individuals, and availability is characterized at the level of the study area. Design II studies are common when animals are radio-marked. Resource availability in design I and design II studies can be based on either complete surveys of resources in the study area (e.g., with data from remote sensing) or estimates of resource availability generated by sampling a subset of the entire study area, such as might be gathered through a sample of plots established at random across a study area. In design III studies, animals are identified uniquely, and both resource use and resource availability are quantified for each individual. An example design could involve defining use based on specific locations of a radio-marked or territorial animal and defining availability based on resources measured in the home range of that individual (Aebischer et al. 1993). Lastly, in design IV studies, resource use is defined by locations for each individual animal, and resource availability is defined uniquely for every use location. The idea is to assess the choices an individual makes in light of the resources available in the immediate vicinity; the spatial and temporal limits of immediate are flexible (Erickson et al. 2001).

Habitat Analyses

The type of sampling unit, the type of data collected, and how, when, and where those data are collected (i.e., the sampling design) govern the types of analyses that are appropriate for all scientific studies, including habitat studies. Some common considerations involved with analyzing data from habitat studies relate to collecting multiple observations from the same animal, accounting for our inability to detect individuals with certainty and the breadth of sampling designs available in studies of habitat selection, many of which require careful attention to details of the analysis.

Multiple Observations per Animal

Collecting multiple observations from the same sample unit over time is a common strategy in ecological studies. In

Table 16.1. Characteristics of common statistical methods of resource selection

Characteristics	Neu et al. (1974)	Johnson (1980)	Friedman (1937)	Compositional Analysis: Aebischer et al. (1993)	Logistic Regression: Hosmer and Lemeshow (2000)	Log-Linear Modeling: Knoke and Burke (1980)	Discrete Choice: Hensher et al. (2005)
Use based on unmarked individuals; availability measured at population level (design I)	Yes	No	No	No	Yes	Yes	Yes
Use based on marked individuals; availability measured at population level (design II)	No[a]	Yes	Yes	Yes	Yes	Yes	Yes
Use based on marked individuals; availability measured for each individual (design III)	No[a]	Yes	Yes	Yes	Yes	Yes	Yes
Assumes temporal independence of locations	Yes	No	No	No	Yes/no[b]	Yes/no[b]	No
Assumes independence among animals	Yes	Yes	Yes	Yes	Yes	Yes	Yes
Assumes sample of animals representative of the population; inferences based on average selection in the population	Yes[c]	Yes	Yes	Yes	Yes[d]	Yes[d]	Yes
Allows for categorical covariates (e.g., sex, age class)	No	No	No	Yes	Yes	Yes	Yes
Allows use of continuous covariates (e.g., distance to roads, body mass)	No	No	No	No	Yes	No	Yes

Source: Adapted from McDonald et al. (2005).

[a]Method can be applied after pooling data across individuals, but this is not recommended.

[b]When data are collected for multiple animals, independence among animals is important; animals must be identified as the unit of analysis.

[c]Assumes that the measure of habitat use represents the measure for the population.

[d]Inference to the population is justified if animals are treated as units of analysis (i.e., replicates).

habitat studies involving marked animals, this approach can be an efficient strategy for evaluating the range of locations and habitat features used by an individual during different activities or different time periods (or both). Multiple observations collected from the same individual are not independent (Aebischer et al. 1993, Otis and White 1999), and therefore the approach to data analysis must ensure that the number of animals—not the number of locations—are treated as the primary sampling unit.

In general, for habitat studies where individual animals are the sampling unit and habitat use by individuals is characterized by recording a series of locations over time, there are three common alternatives for analysis: analyze data for each individual separately (alternative 1), analyze data for all individuals combined but disregard the identities of individuals sampled (alternative 2), or analyze data for all individuals combined but include the identity of individuals as a factor in the analysis (alternative 3). If data from individuals are analyzed separately (alternative 1), then inference about the population from which they were selected can be based on evaluating consistency in patterns of habitat use or selection among individuals. Although this approach can be reasonable when the number of individuals studied is small, analyses at the individual level are not as statistically powerful or as scientifically compelling as analyzing data for all individuals combined. We do not recommend combining observations across individuals without also identifying the individual as the sample unit in the analysis (alternative 2), because this approach inflates the true sample size in the study. The appropriate sample size is typi-

cally the number of sample units studied and not the number of locations recorded (Aebischer et al. 1993, Alldredge et al. 1998); using the number of locations inflates sample size and exaggerates precision of estimates. For example, analysis of habitat use recorded for 20 individuals on five occasions should reflect a sample size of 20 sample units, not 100. Further, when the number of observations recorded varies among individuals, individuals with more locations will have a disproportionate influence on results. Combining observations across individuals while also identifying individuals as the sampling unit in the analysis (alternative 3) overcomes the limitations inherent in the other approaches, reflects the underlying two-level hierarchy inherent in the data (individuals and observations), and reflects explicitly the inferential foundation of individual samples representing the larger population from where they were selected. There are several methods for analyzing data in this way, one of which is to treat individual units as a random effect in the model for analysis (Gillies et al. 2006). Identifying individuals as a random effect recognizes that the particular set of individuals studied is a sample of all individuals within a population, and includes a measure of uncertainty that reflects this source of sampling error (i.e., if we had selected a different subset of individuals, the estimates we generated would be slightly different).

Sampling Designs for Habitat Selection
Two common approaches to study habitat selection are to contrast characteristics of habitat features in areas *used* by animals relative to those in locations that are *unused* (use-nonuse

studies) or locations thought to be available to the species (use-availability studies). Use-nonuse designs maximize the contrast between locations used by an individual (habitat) and the reference (nonhabitat); the disadvantage is that it is difficult to classify with certainty areas as unused. Use-availability designs have the advantage of not needing to classify sites as unused, although the contrast between used locations and the reference (availability) is weaker than in use-nonuse studies, because measures of availability usually include areas used by animals, which reduces the relative contrast between used and available locations relative to used and unused locations. Classifying with certainty areas as unused, however, can be difficult in many circumstances, because cryptic species that are present in an area can be easily overlooked during surveys or temporarily absent from an area they routinely inhabit. Contrasting used with unused or available locations while accounting for uncertainty in the process of identifying used locations introduced by imperfect detection is a useful and important strategy for overcoming potential biases in the sample of used locations (Gu and Swihart 2004); these studies require additional effort in the form of multiple visits to some or all of the sampling units (MacKenzie 2006).

For most studies of habitat selection, there is usually more than one reasonable strategy for analyzing data. Strategies can vary appreciably in their flexibility to work for different sampling designs, both in their assumptions and in the type of information they provide to facilitate interpretation of results, and ultimately our understanding of habitat selection. In a classic study of moose (*Alces alces*) habitat selection in Minnesota (Neu et al. 1974), the amount of available habitat was based on the size of the study area classified in each of four general land cover classes; habitat use was based on the number of locations of moose or tracks observed in each class during aerial surveys (Table 16.2). This is an example of a design I study, where use was measured for individual animals that were not identified uniquely, and habitat availability was determined at the level of the study area (Table 16.1). A common analysis for habitat data recorded as counts of animals in different habitat classes is based on contingency tables, which are often used to evaluate the statistical null hypothesis that habitat use is independent of availability (i.e., the null that

moose do not use habitat classes more or less than their availability in the study area). A common test statistic for analysis of contingency tables is Pearson's chi-square (Neu et al. 1974, Byers et al. 1984, Alldredge and Griswold 2006):

$$\chi^2 = \sum_{i=1}^{k} \frac{(O_i - E_i)^2}{E_i},$$

where k is the number of habitat classes, O_i is the observed frequency of use for each habitat class i, and E_i is the expected frequency of use based on the availability of each habitat class i in the study area. For the moose data, the test statistic indicates that we should reject the null hypothesis of independence among habitat use and availability ($\chi^2 = 43.5$, $P < 0.0001$), which suggests that moose use one or more habitat classes disproportional to its availability (Table 16.2).

Although this analysis is appropriate given the study design and the type of data collected, it is not especially informative because it provides no information on which habitat classes were selected, avoided, or used in proportion to availability; additional calculations are necessary to evaluate the specific patterns of habitat selection. This shortcoming is true of all analyses based on contingency tables that yield a chi-square test statistic because they are not designed around models with parameters that link directly to patterns of selection, which should be the primary focus of results of any resource selection study. One parameter that can be computed to facilitate the biological interpretation of these data is a selection ratio, which is available in several forms but is always based on a ratio of habitat use to availability. When use and availability are approximately equal, the basic selection ratio (use:availability) will be near 1.0, indicating no evidence of selection, and when use is higher or lower than availability, the ratio will be appreciably greater than or less than 1.0, respectively, indicating that resource use is more or less than available. In the moose example, the estimated selection ratio for the habitat class out of burn, near periphery, was approximately equal to 2.5 (standard error, or SE = 0.39), suggesting that moose used these areas more frequently than expected based on availability. Confidence intervals can be computed for selection ratios to standardize decisions about whether use

Table 16.2. Habitat classes and habitat-related statistics for moose in Minnesota, USA

Habitat Class	Available		Use		Number of Moose Expected under Null Hypothesis	Selection Ratio	
	Hectares in Study Area	Proportion of Study Area	Number of Moose Observed	Proportion of Use		Estimate	SE[a]
In burn, not near periphery	4,570	0.340	25	0.214	39.8	0.63	0.111
In burn, near periphery	1,355	0.101	22	0.188	11.8	1.86	0.358
Out of burn, near periphery	1,394	0.104	30	0.256	12.2	2.46	0.388
Out of burn, not near periphery	6,128	0.455	40	0.342	53.2	0.75	0.096
Total	13,447		117				

Source: Neu et al. (1974).

[a]SE, standard error.

of a resource differs significantly from availability (McDonald et al. 2005).

When used in combination with selection ratios, analyses based on contingency tables can be informative for studies that evaluate few habitat factors, and the data collected take the form of counts of animals or their sign. When study designs become more complex, however, results from analyses of contingency tables become less informative and are cumbersome to interpret. In most circumstances, we prefer analyses that are designed around models with parameters that facilitate interpretation of results. Among the available alternatives, many biologists have adopted regression-type approaches to analysis, most of which are based on *generalized linear models*, which provide a comprehensive, flexible, and compelling framework for analyzing data from a wide range of resource selection studies (Manly et al. 2002). Generalized linear models link the mean of the response variable to one or more explanatory variables, where the appropriate link function depends on the distribution of the response variable (McCullagh and Nelder 1989). Parameter estimates are straightforward to interpret because they represent the influence of a one-unit change in the level of the explanatory variables on the mean response (on the scale of the link function), which in studies of habitat selection is always some measure of the magnitude of selection. Generalized linear models can be run with nearly all general-purpose statistical software (e.g., SAS, R, JMP).

To illustrate this approach, we assessed the effect of woody plants on habitat selection of two songbirds, verdin (*Auriparus flaviceps*) and eastern meadowlark (*Sturnella magna*), that were surveyed on 40 ten-hectare plots in semidesert grasslands of southern Arizona. We surveyed plots for breeding birds six times between May and August 2006 and estimated the combined density of shrubs and trees on each plot, which we transformed with the natural log to normalize the right-skewed distribution (Table 16.3). We classified a plot as used for breeding by each species if the species was detected at least once during surveys, and assumed that the sample of 40 plots represented availability across the entire study area, as plots were established at random. This, too, is a design I study (Table 16.1).

To model the effect of woody plants on selection for each species, where presence or absence was the response variable and log density of woody plants was the explanatory variable, we used a form of the generalized linear model called *logistic regression*:

$$\text{logit}(\pi) = \beta_0 + \beta_1 X_1 + \ldots + \beta_p X_p + \varepsilon,$$

where π is the average proportion or probability of a positive response (i.e., a species present on a plot), β_0 is the y intercept (which is frequently excluded; Manly et al. 2002:100), $\beta_1 - \beta_p$ are regression coefficients that describe the influence of the explanatory variables $X_1 - X_p$ on the mean response (in these examples, $p = 1$), and ε is the distribution of the errors around the predicted relationship, which is assumed binomial. In logistic regression, the mean of the binary response variable, π, is linked to the regression structure with the logit function,

Table 16.3. Presence or absence of eastern meadowlarks and verdins and natural log-transformed density of shrubs and trees on 40 plots in southern Arizona, 2006

Eastern Meadowlark	Verdin	Log(Density of Woody Plants)
Present	Absent	0.83
Present	Absent	1.19
Absent	Present	1.41
Present	Absent	1.63
Present	Absent	2.02
Present	Absent	2.08
Present	Absent	2.22
Present	Present	2.58
Absent	Absent	2.63
Present	Absent	2.79
Present	Absent	2.85
Absent	Present	2.87
Present	Absent	3.04
Present	Present	3.22
Present	Present	3.25
Present	Present	3.27
Absent	Present	3.27
Absent	Absent	3.29
Present	Absent	3.32
Absent	Present	3.41
Absent	Present	3.42
Present	Present	3.44
Present	Absent	3.47
Present	Present	3.53
Present	Present	3.57
Present	Present	3.61
Present	Present	3.69
Absent	Present	3.75
Absent	Present	3.75
Absent	Present	3.89
Absent	Present	3.91
Absent	Present	4.01
Absent	Present	4.12
Absent	Present	4.20
Absent	Present	4.21
Absent	Present	4.39
Absent	Present	4.42
Present	Present	4.77
Absent	Present	5.05
Absent	Present	5.15

which is the log of the odds, $\log(\pi/1 - \pi)$. Odds represent the ratio of the probability of a positive response (π) to a negative response $(1 - \pi)$ and provide an intuitive metric for interpreting results of these analyses.

Habitat use of both species of songbirds changed markedly in response to variation in the amount of woody vegetation (Fig. 16.3). For verdins, as the amount of woody vegetation increased, the probability of selecting an area for breeding increased ($\chi^2 = 8.79$, $P = 0.003$). We estimated the effect of woody vegetation on presence of verdins to be 2.11 (SE = 0.71) and the y intercept to be –5.91 (SE = 2.27), both on the logit scale. Therefore the relationship between presence of verdins and woody plants based on these data is described by logit(presence) = –5.91 + 2.11 log(woody plant density).

Because the response variable is modeled as the log of the odds of a positive response, we need to back-transform the estimated regression coefficient, $\hat{\beta}_1$, to describe the effect of the explanatory variable on the mean response on the original scale of measure,

$$\text{odds of a positive response} = e^{\hat{\beta}_1},$$

where e is the base of the natural logarithm (2.718 . . .). If an explanatory variable has no effect on the response, $\hat{\beta}_1$ will be near 0, and the estimated odds will be near 1 ($e^0 = 1$). For verdins, we estimated the change in the odds of a positive response for each one-unit increase in the explanatory variable to be $e^{2.11} = 8.3$. That means that for a one-unit increase in the log density of woody plants, the odds of a verdin selecting an area for breeding increased about eight times (Fig. 16.3).

In contrast to verdins, the probability of eastern meadowlarks selecting an area for nesting decreased as the amount of woody vegetation increased ($\chi^2 = 6.11$, $P = 0.013$). We estimated the effect of woody vegetation on presence of meadowlarks to be –1.19 (SE = 0.48), and the change in odds for a one-unit increase in the log density of woody plants to be $e^{-1.19} = 0.30$. That means that for each one-unit increase in log density of woody plants, the odds of a meadowlark selecting an area for breeding decreased about threefold (Fig. 16.3). Given a convenient property of odds, by simply changing the sign of the parameter estimate ($e^{1.19} = 3.3$), we can state equivalently that for each one-unit *decrease* in log density of woody plants, the odds of a meadowlark selecting an area for breeding will increase about three times.

Although the examples above are based only on a single explanatory variable, by modeling the probability of habitat use as a function of different types or levels of one or more resources, these analyses provide examples of *resource selection*

DAVID LAMBERT LACK (1910–1973)

Born in London on 16 July 1910, David Lack was an avid naturalist as a child, studied natural sciences at Magdalene College in Cambridge, and became a well-known evolutionary ecologist. He was influenced early in his professional career by the opportunity to study bird behavior in the Galapagos Islands and by interactions with Ernst Mayr. Lack became director of the Edward Grey Institute of Field Ornithology in Oxford in 1945 and served in that post until his death in 1973. He is perhaps best known for his 1947 book *Darwin's Finches* and for his research on population biology. Lack was among the first researchers to propose that animals use environmental cues to help them find and settle in places with resources needed for their survival and reproduction, and he coined the term *habitat selection* (Lack 1933).

functions (RSFs). This general approach to modeling resource use versus availability or nonuse in a regression-type framework extends readily to many different types of resource selection studies, including those that include both categorical and continuous factors, and where multiple observations have been recorded for each sample unit (Manly et al. 2002, Keating and Cherry 2004, Johnson et al. 2006). Although RSFs are a reliable strategy for analyzing data from resource selection studies, resource units classified as available to animals might include units that were used by animals. When this occurs, estimates from RSFs may not represent the actual probability of selection (Keating and Cherry 2004). Resource selection probability functions (RSPFs) were developed to overcome this potential shortcoming and to provide direct estimates of resource-use probabilities (Johnson et al. 2006, Lele and Keim 2006).

The important framework for analyzing use-availability data within the context of generalized linear models continues to develop to encompass an increasing range of ecological questions expressed at an increasing number of spatial and temporal scales. For example, step-selection analysis (SSA) extends the RSF approach by allowing resource selection to be evaluated as an animal moves through a landscape and local resource availability changes (Fortin et al. 2005). Paths are divided into a series of steps that correspond to straight-line segments between successive locations of a focal individual. Each observed step is then paired with a set of random steps established at the same starting point, but differing in length or direction. Environmental features are contrasted between ob-

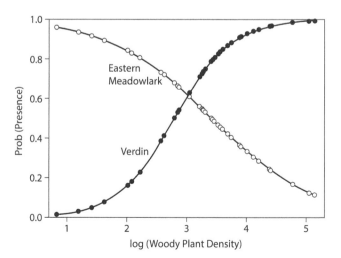

Figure 16.3. Relationship between density of woody vegetation and predicted probability of presence of eastern meadowlarks and verdins in semidesert grasslands of Las Cienegas National Conservation Area, southeastern Arizona, in 2006.

PAUL LESTER ERRINGTON (1902–1962)

Paul Errington was born in Bruce, South Dakota, on 14 June 1902. He grew up working on his family's farm and developed a keen interest in natural history, especially in surrounding marshes and streams, which became evident in his work as a scientist. Errington graduated from South Dakota State College in 1929 and earned a PhD from the University of Wisconsin in 1932. Errington then attended Iowa State University to establish and lead the first cooperative wildlife research unit in the United States, as well as to serve as professor of zoology, which he did through his entire career. Considered a great pioneer of animal ecology, Errington authored more than 200 scholarly papers that focused on predator–prey relationships, population dynamics, population regulation, and the biology and management of muskrats, minks, bobwhite quail, and great horned owls. He also authored four popular books, including *Of Men and Marshes*, that influenced both scientists and the public. In 1961 Errington was named one of ten outstanding naturalists by *Life* magazine, and in 1962 he was awarded the Aldo Leopold Medal, the highest honor bestowed by The Wildlife Society.

By Alfred Eisendstadt; courtesy of the US Fish and Wildlife Service

served and random steps using conditional logistic regression to identify the set of environmental features that influence movements. Therefore SSAs integrate the behaviors of habitat selection and movement (Thurfjell et al. 2014, Prokopenko et al. 2016).

One important assumption in these examples is that presence of a species in the area surveyed was determined without error. That is, no areas where the species was truly present were classified as absent (i.e., no false negatives; false positives can also be a problem, but for easily identifiable species, these errors should be rare). For many vertebrates, this assumption is unlikely to be met. As mentioned previously, if animals cannot be detected with certainty, and the probability of detection is associated with one or more habitat features being evaluated, there is potential for results to be biased and for study conclusions to be incorrect. The design and analysis of studies where the chance of imperfect detections is likely have been explored in many contexts, including habitat selection (Gu and Swihart 2004, MacKenzie 2006).

We offered here a brief introduction to the many issues relevant to design and analysis of habitat selection studies.

Additional coverage of the topic might begin with the book by Manly et al. (2002), papers and book chapters by Erickson et al. (2001) and McDonald et al. (2005), and the special section on resource selection in the *Journal of Wildlife Management* (Strickland and McDonald 2006).

MANAGING HABITAT

Knowledge of the environmental conditions and resources needed to support a species is required before developing a strategy to meet habitat-based conservation or management goals. Frequently, these goals involve maintaining or increasing abundance of a population by protecting existing habitat, restoring degraded habitat, or adding resources that are limiting factors. Occasionally, the goal of management is to decrease abundance of a nuisance population, such as Canada geese (*Branta canadensis*) or white-tailed deer (*Odocoileus virginianus*), by reducing the amount or quality of their habitat (Gosser at al. 1997, Ayers et al. 2010). One of the primary ways biologists can either increase or decrease habitat resources for a species is by manipulating vegetation, and among the most effective ways to change vegetation is to control plant succession.

Habitat Management and Plant Succession

Plant succession describes the natural changes in plant community composition on a site after a disturbance (Fig. 16.4). Patterns of succession vary geographically, but in many areas of the temperate zone, succession begins with annual plants that colonize a site rapidly after a disturbance, then proceeds through a series of different species assemblages called seres or seral stages, and is eventually characterized by large, long-lived tree species (Fig. 16.4). Because requirements of individual species are thought to influence succession (Glenn-Lewin et al. 1992), the assemblage of plant species at any given point along the succession trajectory is a collection of species that thrive in current environmental conditions. Because succession is a continuous process, discrete plant communities often cannot be delineated clearly. Some seral stages, however, are dominated by characteristic plant species that are often named for those species.

Many animal species rely on plants for food and cover. As succession proceeds and the plant species on a site change, animal species that inhabit the site also change. If biologists understand the set of animal species that inhabit each seral stage, they can manipulate vegetation on a site to match the appropriate seral stage in the appropriate configuration to meet the needs of target species. For example, some threatened and endangered species, such as the northern spotted owl (*Strix occidentalis caurina*) and Mount Graham red squirrel (*Tamiasciurus hudsonicus grahamensis*), depend on conditions and resources in mature forests (i.e., late successional forest stages) (Forsman et al. 1984, Smith and Mannan 1994). For these species, and others like them, protecting and maintaining old forests are critical to their persistence (see the example for northern spotted owls, below). Conversely, other species of conservation concern use early stages of succession for critical

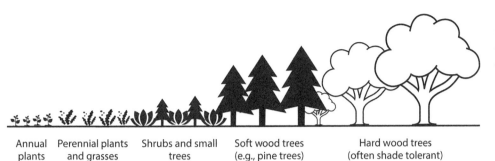

Annual   Perennial plants   Shrubs and small   Soft wood trees   Hard wood trees
plants   and grasses   trees   (e.g., pine trees)   (often shade tolerant)

Time

Figure 16.4. An example of plant succession, or how plant species change over time on a site after disturbance.

aspects of their life cycle. The Kirtland's warbler (*Setophaga kirtlandii*), a migratory species that spends its summers in the northern midwestern states and southern Canada, nests in large stands of dense jack pine trees (*Pinus banksiana*) that are between 5 and 23 years old (Marshall et al. 1998). Historically, frequent fires maintained enough stands in these age classes (i.e., early successional stages) to support the warbler. Suppression of fire in the mid-1900s nearly eliminated the nesting habitat of Kirtland's warblers, and they were among the first species listed as endangered under the Endangered Species Act of 1973. Currently, there are more than 2,300 nesting pairs of Kirtland's warblers, and the species was removed from the list of endangered species in 2019 (US Department of the Interior 2019). Recovery of the Kirtland's warbler occurred, in part, as a result of habitat management, including a carefully scheduled combination of logging, prescribed burning, and tree planting to maintain nesting habitat. Stands of pine are harvested and replanted on a 50-year rotation so that a portion of state and federal lands under management is maintained in a structural condition that provides nesting habitat for this warbler (Marshall et al. 1998).

Other animal species require habitat elements provided by a combination of seral stages. Ruffed grouse (*Bonasa umbellus*), for example, need brushy areas and stands of young aspen (*Populus tremuloides*) that provide food in summer; stands of mature aspen that provide drumming sites in spring and food in fall, winter, and spring; and stands of dense aspen that provide cover for broods in spring (DeStefano et al. 1984). Thus ideal habitat conditions for ruffed grouse are likely to exist in areas with an equal mix of young (<15 years), intermediate-aged (15–30 years), and old (>30–40 years) stands of aspen trees in small patches so that each seral stage can be incorporated into the home range of individual grouse (Natural Resource Conservation Service 2001). A general prescription for managing forests for ruffed grouse habitat might involve harvesting 1-ha patches of aspen from 4-ha blocks every 10 years to maintain a mosaic of different-aged stands in proximity to each other (Natural Resource Conservation Service 2001).

Restoration of Natural Environments

If environments of conservation concern, especially those that provide habitat for many species, are degraded or destroyed, complete restoration can be challenging. If a stand of old-growth timber in western Oregon or Washington is harvested, it would require more than 200 years to regrow naturally, and there are few things biologists can do to speed up the process. Thus, in the short term, the best strategy for maintaining these kinds of environments and the animal species that inhabit them is protection. In contrast, when some other environments are degraded, such as wetlands, they sometimes can be restored relatively quickly.

The Rio Grande is one of the major rivers in the southwestern United States; it flows from Colorado through New Mexico and eventually into the Gulf of Mexico. Before settlement by Europeans, the river flooded periodically in the spring and summer, forming wetlands in its floodplains that supported many migratory species, including sandhill cranes (*Antigone canadensis*) and a variety of other waterbirds and waterfowl. When Europeans settled the area, they installed dams and irrigation ditches to reduce flooding, which eliminated many of the flood-maintained marshes. In the 1930s the Civilian Conservation Corps began to restore the wetlands by diverting water from the river, and an area along the river in central New Mexico was established as the Bosque del Apache National Wildlife Refuge in 1939 (US Fish and Wildlife Service 2020). Today, refuge personnel use gates and ditches to move water from the river to create ponds and flood fields, and the refuge supports thousands of migratory ducks, geese, and cranes during winter.

Habitat Management and Human Activities

Human activities affect wildlife habitat in many ways and at many spatial and temporal scales. Activities that are likely to have adverse effects on wildlife populations can be divided into those that function primarily by altering the physical environment and those that affect an animal's behavior (Steidl and Powell 2006). Agriculture, forestry, and urban development are examples of human activities that alter significantly the local, physical environment. In contrast, climate change—also a result of human activities—alters the physical environment more broadly and in ways that can affect profoundly the quality and amount of habitat for wildlife. Some effects of climate change on wildlife habitat likely will include increases in temperature, which could render currently habitable areas as uninhabitable if the limits of thermal tolerance for the species are exceeded (Ward and Mannan 2011). Changes in temperature or rainfall patterns likely will alter abundance and distribution of a wide array of physical and biological resources, many of which provide critical habitat elements for wildlife species. Examples range from sea ice, an important habitat element for polar bears (*Ursus maritimus*; Stirling et al. 2011), to individual plant species crucial for many species (Kerns et al. 2009).

Although changes in land use and climate affect wildlife habitat by altering the biotic and abiotic features of an area, effects of climate change are likely to be more pervasive than those of other human activities (see Chapter 18 for a discussion of climate change). For example, wildlife habitat in a forest that is eliminated as a result of timber harvest or clearing for agriculture will likely recover, given sufficient time. Conversely, if the dominant plant community in an area changes as a result of shifts in temperature and rainfall regimes, habitat that has been eliminated is unlikely to recover within any reasonable time frame. Although changes in climate present unique challenges to the maintenance of biodiversity, a critical element in managing effects of most human activities—understanding the resources that constitute habitat for a species—also will be a significant factor in predicting and managing the effects of climate change.

Perhaps less obvious in their effects on wildlife habitat are nonconsumptive human activities that do not appreciably alter the physical environment but nonetheless can reduce habitat quality. Examples include recreational activities such as hiking, wildlife viewing, and boating. Many factors influence the magnitude of the effect of nonconsumptive human activities on wildlife, including the type, duration, frequency, magnitude, location, and timing of the disturbance and the particular species of interest. Although effects of these activities are typically of short duration, they can cumulatively affect wildlife populations adversely in both the short and long term (Steidl and Powell 2006). For example, if recreational activities preclude the use of an area by a species, that area could contain all of the resources necessary for a species but remain uninhabited. In this situation, controlling the number of people and their recreational activities is of obvious importance. For the remainder of the chapter, we focus on management activities that affect the resources upon which animals depend.

Human activities that modify natural landscapes inevitably alter habitat resources and lead to declines in abundance and distribution of some species and increases in others. Eliminating these activities is not possible because many provide resources that humans need to exist. But sound management practices can reduce the negative effects of each of these activities on animal populations, sometimes with relatively minor modifications. On lands under federal or state jurisdiction, improving conditions for target wildlife species often can be accomplished relatively easily, if the modifications help meet mandates of federal and state environmental laws. On lands that are privately owned, cooperation of landowners is critical to the implementation of such modifications, and ultimately to maintenance of wildlife populations on those lands. Financial and other incentives can encourage landowners to par-ticipate in habitat management programs that benefit wildlife, such as set-aside programs for agricultural lands (Warner et al. 2005) and strategies to improve the productivity of working landscapes when they also enhance wildlife habitat (Naugle et al. 2019). Habitat management to improve conditions for animals in areas being managed primarily for agriculture, forestry, or livestock production is summarized by Warner et al. (2005), Yahner et al. (2005), and Bleich et al. (2005). Below, we outline briefly how the negative effects of urbanization on animal populations can sometimes be reduced.

Urban environments, which we define broadly as lands developed for human habitation (e.g., towns and cities), support a variety of wildlife species (Adams 1994) but favor those that can live in close association with people. For example, assemblages of bird species in urban areas tend to have a higher total number of individuals and biomass but fewer species than those in more natural areas (Hansen et al. 2005, Chace and Walsh 2006), primarily because non-native species—such as house sparrows (*Passer domesticus*) and European starlings (*Sturnus vulgaris*)—thrive, whereas many native species are eliminated. Land-use changes and introduction of many non-native predators, especially domestic cats and dogs, are examples of challenges to maintaining native species in urban areas. Relatively simple management activities—including using native plant species when landscaping, discouraging open lawns on public and private property, maintaining patches of native vegetation in parks, and maintaining connections, potentially via riparian zones, between urban parks and natural areas outside developed areas—can enhance the presence of some native species of all taxonomic classes in urban areas (Marzluff and Ewing 2001).

The importance of habitat features in maintaining populations of native species in urban environments depends to some extent on how the features are distributed across the landscape (Daniels and Kirkpatrick 2006). Some native species can take advantage of resources even if they are present in small amounts at small scales, such as a single plant or feeding station in a backyard. For these species, actions of individual homeowners can potentially influence their distribution in urban settings (McCaffrey and Mannan 2012). Other species require that resources be distributed more broadly, and habitat management for these species might require actions of neighborhood groups or city and county planners (McCaffrey and Mannan 2012).

Habitat Management on Large Spatial Scales for Single Species

Managing land primarily for the conservation of a single species is costly and usually motivated by the needs of species listed under the Endangered Species Act. Questions that must be answered when developing a strategy to manage land for one species often focus on how much habitat is needed, how it should be distributed, and how it can be maintained over time. Below we describe how the conservation plan for the northern spotted owl addressed these questions.

Conservation Strategy for the Northern Spotted Owl

The original plan to conserve habitat for the northern spotted owl (Thomas et al. 1990) is among the best examples of a management strategy based on the resource needs and behaviors of a single species over a broad geographical range. Specific areas identified for conservation in that plan have been modified to some extent in subsequent recovery plans (US Fish and Wildlife Service 2008, 2011), but the rationale used to develop the plan has remained essentially unchanged and serves as one model for designing habitat reserves for a single species.

The northern spotted owl inhabits coniferous forests from southwestern British Columbia through western Washington, western Oregon, and northwestern California (Gutierrez et al. 1995). It primarily inhabits old-growth forests (Forsman et al. 1984) and was listed as threatened under the Endangered Species Act, primarily because of habitat loss. At that time, old-growth forests were being harvested at an unsustainable rate (Parry et al. 1983). Therefore maintaining northern spotted owls required that some old-growth forests be protected from harvest, but deciding how much to protect and where to protect it was controversial because any protection would result in lost revenues to the timber industry.

How much habitat is needed? The conservation strategy for the northern spotted owl was based on maintaining habitat for roughly 1,500 breeding pairs of owls. A panel of experts convened by the Audubon Society identified this target based on concerns about persistence of spotted owls given demographic and environmental stochasticity. The total protected area would be the amount of habitat capable of supporting about 1,500 breeding pairs (see below).

How big should habitat patches be? The plan called for maintaining patches of forest, called habitat conservation areas (HCAs), that were large enough to support at least 20 pairs of owls. Maintaining large patches was considered an advantage because the potential for internal recruitment enhanced the likelihood of persistence of owls inside an HCA (Thomas et al. 1990). The formula used to determine HCA size was

$$\text{HCA size} = [(\text{median annual home range of pairs}) \times 0.75] \times 20 \text{ pairs.}$$

The size of HCAs varied because the size of annual home ranges of adult owls varied across the geographic range. The multiplier of 0.75 was applied to allow for 25% overlap among home ranges. Current demographic models, developed to be spatially explicit, predict the response of owls to the distribution and quality of habitat and allow for evaluation of different planning strategies (US Fish and Wildlife Service 2011).

How should habitat patches be distributed? Habitat conservation areas were distributed across the entire geographical range of the northern spotted owl, and individual HCAs were situated so that the distance between them was no farther than the distance that dispersing spotted owls were likely to travel. The rationale for this distribution was that resident breeding pairs in an HCA could be replaced by dispersing owls either from within the HCA or from adjacent HCAs.

How should areas between habitat patches be managed? Movement of owls between patches depends not only on the distance between patches, but also on the environmental conditions in areas through which dispersing owls must travel. The plan called for areas between habitat patches (sometimes called *habitat matrix*) to be managed so that half was maintained in forest conditions thought to facilitate movement of dispersing owls.

How can habitat be maintained over time? Old-growth patches identified for conservation could not persist forever, even if vigorously protected. How long patches last would depend on the frequency and intensity of natural disturbances such as fire. Although not described explicitly in the original plan, careful planning would be required to identify, set aside, protect, and allow additional forest stands to develop into old-growth forests. These stands could then be used to replace existing HCAs lost over time. Maintaining habitat reserves for a particular species over long periods will depend in part on the environmental stability of areas selected as reserves. Changes in resources in response to climate change could affect how well an area functions as a habitat reserve, making management of reserves an increasingly complex challenge (Griffith et al. 2009).

Habitat Management on Large Spatial Scales for Multiple Species

Strategies to conserve habitat for single species are based commonly on identifying and maintaining areas that include the specific set of resources required by that species. As the number of species targeted for conservation increases, however, identifying the habitat resources necessary to support all species becomes much more challenging, as does identifying lands that will provide these resources over long time horizons. Therefore strategies to conserve habitat for multiple species tend to focus on identifying lands that provide the full range of environmental conditions thought necessary to support all target species. Further, given the inevitable changes in landscapes we anticipate in response to natural and anthropogenic processes, conservation strategies must include lands that provide redundancy of these environmental conditions so that all species can persist despite natural and human-caused changes in landscape composition over time.

The Sonoran Desert Conservation Plan

Pima County in southern Arizona initiated a comprehensive land-use plan to identify and establish an integrated system of conservation lands, with the goal of maintaining biodiversity while providing a framework to guide future land use. The effort was called the Sonoran Desert Conservation Plan and encompassed an area of nearly 24,000 km^2. The biological foundation for the plan was based on identifying lands of high conservation value that would form a network of conservation lands around which other regional planning needs would be incorporated. Although several ecological targets could have provided the basis for evaluating the potential con-

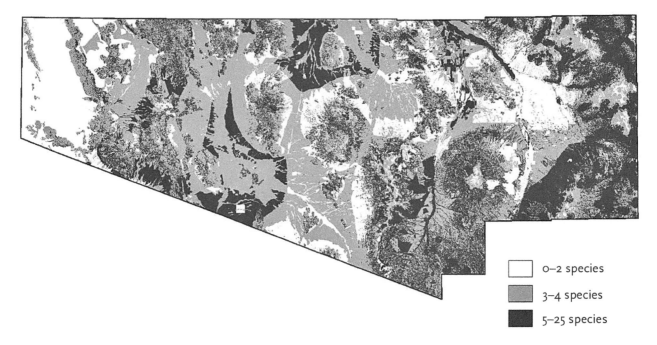

Figure 16.5. Predicted richness of target species across Pima County in southern Arizona, which provided the basis for establishing a network of conservation lands on which the Sonoran Desert Conservation Plan was designed. From Steidl et al. (2009)

servation value of different areas, the strategy was based on understanding habitat needs of 39 target species (nine mammals, eight birds, seven reptiles, two frogs, six fish, and seven plants) selected to represent the range of biological diversity in the region (Steidl et al. 2009).

A detailed account of the habitat requirements of each species was gathered from the literature and from experts, which were then used as the basis for creating a series of spatially explicit models to predict the distribution of potential habitat across the region for each species. Models of potential habitat were used because published distributions tend to be too general in scope, documented locations are uneven in geographic coverage and often biased toward areas that are traveled commonly, and expert opinion has significant limitations in that on-the-ground knowledge is rarely complete (Steidl et al. 2009). Additionally, habitat can often be identified even if the target species is currently absent from an area, which seems especially likely for many species in jeopardy.

Models of potential habitat were based on values established for four major categories of environmental features represented by 130 variables, including vegetation and land cover (60 variables, including mixed broadleaf forest cover, agriculture), hydrology (11 variables, including perennial stream width, groundwater depth), topography and landform (45 variables, including elevation, slope, aspect), and geology (14 variables, including soil type, presence of carbonates). Scores based on the value of these environmental features to each species were combined to produce a habitat suitability surface that represented the distribution of habitat potential for each species across the landscape.

A geographic information system was used to overlay areas of high potential habitat for all species that resulted in a map that portrayed species richness (i.e., number of species in an area where potential habitat for species was classified as high) across the region; this map illustrated the relative importance of different areas to conservation of species across the region (Fig. 16.5). The landscape was then divided into a collection of discrete polygons representing areas with different levels of species richness on which the conservation lands system was built. To establish boundaries for each contiguous collection of landscape units with the same level of species richness (termed a patch), the guidelines on reserve design provided by the literature were followed to maximize conservation benefits of each patch and the overall network of patches. Specifically, the size of each patch was maximized, distances between adjacent patches minimized, contiguity maximized and fragmentation within and among patches minimized, and connectivity maximized among patches to maintain processes that occur at scales larger than individual patches. Additionally, boundaries were adjusted to meet a set of specific conservation objectives established to capture ecological processes that manifest at spatial scales broader than an individual patch, and to capture other elements important to conservation of biodiversity in the region that were too small to be captured in landscape-scale analyses (Steidl et al. 2009).

Ultimately, areas within the conservation lands system are predicted to conserve an average of 75% of potential habitat (range = 28% to 100%) for each of the target species when the region is fully developed. Currently, only about 4% of the area targeted for conservation has been developed, with an additional

4% of these lands predicted to be developed in the future (ESI Corporation 2003). Of the 12,000 km² of conservation lands, 57% is federal, 24% state, 14% private, and 5% county and city combined. With such a high percentage of these lands in public ownership, achieving the conservation objectives that drive this large-scale process seems tenable.

CONTEMPORARY CHALLENGES

As the number of people occupying our planet continues to increase, the challenge of maintaining habitat for animals and all other forms of life will become increasingly difficult. Although humans will make extraordinary efforts to maintain populations of a few individual species (which almost always means maintaining their habitat), maintaining environmental conditions that support the needs of all forms of life on earth will depend increasingly on efforts designed to provide for a great many species over long periods in the face of changes we can anticipate, as well as those that will manifest unexpectedly. Some measure of success in this task can be attained only if biologists continue to strive to understand what constitutes habitat for a wide range of species, and to be innovative and persistent in developing strategies to reduce the adverse effects of anthropocentric activities while enhancing effects that are positive. We suggest that effective strategies include activities that consider the needs of humans and animals across a range of spatial and temporal scales.

SUMMARY

Habitat is a species-specific concept that describes an area with the combination of resources and environmental conditions that promotes residency by individuals of a species and allows those individuals to survive and reproduce. Animals seeking habitat presumably recognize suitable sites by the presence of environmental cues that are directly or indirectly associated with the resources they require. The process of selecting an area to inhabit may span a range of spatial and temporal scales and is likely a hierarchical process for many species. Variation in levels of important resources among habitat patches often is expressed as variation in the demographic performance of resident individuals. Sampling strategies for identifying habitat and important habitat features should be influenced by the research question of interest, which in turn will dictate the scope of inference of the study. The two most common types of habitat studies are those that gather information about characteristics of locations used by animals (i.e., habitat use) and those that evaluate the use of habitat resources by animals relative to the general availability of those resources in the environment (i.e., habitat selection). Sampling designs available for assessing habitat selection have been classified into four main types on the basis of the approaches used to quantify resource use and resource availability, and many biologists have adopted regression-type approaches for analyzing resource selection. Biologists can either increase or decrease the amount and quality of habitat for a species by manipulating the quantity of important resources, setting aside key lands in reserve systems and altering the way humans use land.

Literature Cited

Adams, L. W. 1994. Urban wildlife habitats. University of Minnesota Press, Minneapolis, USA.

Aebischer, N. J., P. A. Robertson, and R. E. Kenward. 1993. Compositional analysis of habitat use from animal radio-tracking data. Ecology 74:1313–1325.

Alldredge, J. R., and J. Griswold. 2006. Design and analysis of resource selection studies for categorical resource variables. Journal of Wildlife Management 70:337–346.

Alldredge, J. R., D. L. Thomas, and L. L. McDonald. 1998. Survey and comparison of methods for study of resource selection. Journal of Agricultural, Biological, and Environmental Statistics 3:237–253.

Andersen, E. M., and R. J. Steidl. 2020. Plant invasions alter settlement patterns of breeding birds. Ecosphere 11:https://doi.org/10.1002/ecs2.3012.

Ayers, C. R., C. E. Moorman, C. S. Deperno, F. H. Yelverton, and H. J. Wang. 2010. Effects of mowing on anthraquinone for deterrence of Canada geese. Journal of Wildlife Management 74:1863–1868.

Baker, M. B. 2005. Experience influences settling behaviour in desert isopods, *Hemilepistus raumuri*. Animal Behaviour 69:1131–1138.

Battin, J. 2004. When good animals love bad habitats: ecological traps and the conservation of animal populations. Conservation Biology 18:1482–1491.

Bleich, V. C., J. G. Kie, E. R. Loft, T. R. Stephenson, M. W. Oehler, and A. L. Medina. 2005. Managing rangelands for wildlife. Pages 873–897 *in* C. Braun, editor. Techniques for wildlife investigations and management. The Wildlife Society, Bethesda, Maryland, USA.

Brown, J. L. 1969. The buffer effect and productivity in tit populations. American Naturalist 103:1313–1325.

Byers, C. R., R. K. Steinhorst, and P. R. Krausman. 1984. Clarification of a technique for analysis of utilization-availability data. Journal of Wildlife Management 48:1050–1053.

Chace, J. F., and J. J. Walsh. 2006. Urban effects on native avifauna: a review. Landscape and Urban Planning 74:46–49.

Citta, J. J., and M. S. Lindberg. 2007. Nest-site selection of passerines: effects of geographic scale and public and personal information. Ecology 88:2034–2046.

Daniels, G. D., and J. B. Kirkpatrick. 2006. Does variation in garden characteristics influence the conservation of birds in suburbia? Biological Conservation 133:326–335.

Daubenmire, R. 1976. The use of vegetation in assessing the productivity of forest lands. Botanical Review 42:115–143.

Davis, J. M., and J. A. Stamps. 2004. The effect of natal experience on habitat preferences. Trends in Ecology and Evolution 19:411–416.

DeStefano, S., S. R. Craven, and R. L. Ruff. 1984. Ecology of the ruffed grouse. University of Wisconsin, Madison, USA.

Dwyer, J. F., and R. W. Mannan. 2007. Preventing raptor electrocutions in an urban environment. Journal of Raptor Research 41:259–267.

Elith, J., S. J. Phillips, T. Hastie, M. Dudik, Y. E. Chee, and C. J. Yates. 2011. A statistical explanation of MaxEnt for ecologists. Diversity and Distributions 17:43–57.

Erickson, W. P., T. L. McDonald, K. G. Gerow, S. Howlin, and J. W. Kern. 2001. Statistical issues in resource selection studies with radio-marked animals. Pages 211–245 *in* J. J. Millspaugh and J. M. Marzluff, editors.

Radio tracking and animal populations. Academic, San Diego, California, USA.

ESI Corporation. 2003. Pima County economic analysis. Report to Pima County. Phoenix, Arizona, USA. http://www.pima.gov/cmo/sdcp/reports.html.

Flesch, A. D., and R. J. Steidl. 2010. Importance of environmental and spatial gradients on patterns and consequences of resource selection. Ecological Applications 20:1021–1039.

Forsman, E. D., E. C. Meslow, and H. M. Wight. 1984. Distribution and biology of the northern spotted owl in Oregon. Wildlife Monographs 87:3–64.

Fortin, D. L. B. Hawthorne, M. S. Boyce, D. W. Smith, T. Duchesne, and J. S. Mao. 2005. Wolves influence elk movements: behavior shapes a trophic cascade in Yellowstone National Park. Ecology 86:1320–1330.

Fretwell, S. D., and H. L. Lucas. 1970. On territorial behaviour and other factors influencing habitat distribution in birds. Acta Biotheoretica 19:16–36.

Friedman, M. 1937. The use of ranks to avoid the assumption of normality implicit in the analysis of variance. Journal of the American Statistical Association 32:675–701.

Garshelis, D. L. 2000. Delusions in habitat evaluation: measuring use, selection, and importance. Pages 111–164 in L. Boitani and T. K. Fuller, editors. Research techniques in animal ecology: controversies and consequences. Columbia University, New York, New York, USA.

Gillies, C. S., M. Hebblewhite, S. E. Nielsen, M A. Krawchuk, C. L. Aldridge, J. L. Frair, D. J. Saher, C. E. Stevens, and C. L. Jerde. 2006. Application of random effects to the study of resource selection by animals. Journal of Animal Ecology 75:887–898.

Glenn-Lewin, D. C., R. K. Peet, and T. T. Veblen, editors. 1992. Plant succession: theory and prediction. Chapman and Hall, New York, New York, USA.

Gosser, A. L., M. R. Conover, and T. A. Messmer. 1997. Managing problems caused by urban Canada geese. Berryman Institute Publication 13. Utah State University, Logan, USA.

Griffith, B., J. M. Scott, R. Adamcik, D. Ashe, B. Czech, R. Fischman, P. Gonzales, J. Lawler, A. D. McGuire, and A. Pidgorna. 2009. Climate change adaptation for the U.S. National Wildlife Refuge System. Environmental Management 44:1043–1052.

Gu, W., and R. K. Swihart. 2004. Absent or undetected? Effects of nondetection of species occurrence on wildlife–habitat models. Biological Conservation 116:195–203.

Gutierrez, R., A. B. Franklin, and W. S. Lahaye. 1995. Spotted owl. Pages 1–28 in A. Poole and F. Gill, editors. The birds of North America. No. 179. Academy of Natural Sciences, Philadelphia, Pennsylvania, and American Ornithologists' Union, Washington, DC, USA.

Hamerstrom, F., F. N. Hamerstrom, and J. Hart. 1973. Nest boxes: an effective management tool for kestrels. Journal of Wildlife Management 37:400–403.

Hansen, A. J., R. L. Knight, J. M. Marzluff, S. Powell, K. Brown, P. H. Gude, and K. Jones. 2005. Effects of exurban development on biodiversity: patterns, mechanisms, and research methods. Ecological Applications 15:1893–1905.

Hensher, D. A., J. M. Rose, and W. H. Greene. 2005. Applied choice analysis: a primer. Cambridge University Press, New York, New York, USA.

Hilden, O. 1965. Habitat selection in birds. Annales Zoologici Fennici 2:53–73.

Hosmer, D. W., and S. Lemeshow. 2000. Applied logistic regression. 2nd edition. John Wiley and Sons, Hoboken, New Jersey, USA.

Jaenike, J., and R. D. Holt. 1991. Genetic variation for habitat preference: evidence and explanations. American Naturalist 137:S67–S90.

Johnson, C. J., S. E. Nielsen, E. H. Merrill, T. L. McDonald, and M. S. Boyce. 2006. Resource selection functions based on use-availability data: theoretical motivation and evaluation methods. Journal of Wildlife Management 70:347–357.

Johnson, D. H. 1980. The comparison of usage and availability measurements for evaluating resource preference. Ecology 61:65–71.

Johnson, M. D. 2007. Measuring habitat quality: a review. Condor 109:489–504.

Keating, K. A., and S. Cherry. 2004. Use and interpretation of logistic regression in habitat selection studies. Journal of Wildlife Management 68:774–789.

Kerns, B. K., B. J. Naylor, M. Buonopane, C. G. Parks, and B. Rogers. 2009. Modeling tamarisk (Tamarisk spp.) habitat and climate change effects in the northwestern United States. Invasive Plant Science and Management 2:200–215.

Knoke, D., and P. J. Burke. 1980. Log-linear models. Sage, Newberry Park, California, USA.

Krishnan, V. V. 2007. Optimal strategy for time-limited sequential search. Computers in Biology and Medicine 37:1042–1049.

Kristan, W. B. I. 2003. The role of habitat selection behaviour in population dynamics: source-sink systems and ecological traps. Oikos 103:457–468.

Lack, D. 1933. Habitat selection in birds with special reference to the effects of afforestation on the Breckland avifauna. Journal of Animal Ecology 2:239–262.

Lele, S. R., and J. L. Keim. 2006. Weighted distributions and estimation of resource selection probability functions. Ecology 87:3021–3028.

Levin, S. A. 1992. The problem of pattern and scale in ecology: the Robert H. MacArthur Award lecture. Ecology 73:1943–1967.

Lima, S. L., and P. A. Zollner. 1996. Towards a behavioural ecology of ecological landscapes. Trends in Ecology and Evolution 11:131–135.

Litt, A. R., and R. J. Steidl. 2016. Complex demographic responses of a small mammal to a plant invasion. Wildlife Research 43:304–312.

MacKenzie, D. W. 2006. Modeling the probability of resource use: the effect of, and dealing with, detecting a species imperfectly. Journal of Wildlife Management 70:367–374.

Manly, B. F. J., L. L. McDonald, D. L. Thomas, T. L. McDonald, and W. P. Erickson. 2002. Resource selection by animals: statistical analysis and design for field studies. 2nd edition. Kluwer, Boston, Massachusetts, USA.

Marshall, E., R. Haight, and F. R Homans. 1998. Incorporating environmental uncertainty into species management decisions: Kirtland's warbler habitat management as a case study. Conservation Biology 12:975–985.

Marzluff, J. M., and K. Ewing. 2001. Fragmented landscapes for the conservation of birds: a general framework and specific recommendations for urbanizing landscapes. Restoration Ecology 9:280–292.

McCaffrey, R. E., and R. W. Mannan. 2012. How scale influences birds' responses to habitat features in urban residential areas. Landscape and Urban Planning 105:274–280.

McCullagh, P., and J. A. Nelder. 1989. Generalized linear models. 2nd edition. Chapman and Hall, Boca Raton, Florida, USA.

McDonald, L. L., J. R. Alldredge, M. S. Boyce, and W. P. Erickson. 2005. Measuring availability and vertebrate use of terrestrial habitats and foods. Pages 465–488 in C. E. Braun, editor. Techniques for wildlife investigations and management. 6th edition. The Wildlife Society, Bethesda, Maryland, USA.

Morrison, M. L., B. G. Marcot, and R. W. Mannan. 2006. Wildlife–habitat relationships. 3rd edition. Island Press, Washington, DC, USA.

Mumme, R. L., S. J. Schoech, G. E. Woolfenden, and J. W. Fitzpatrick. 1999. Life and death in the fast lane: demographic consequences of road mortality in the Florida scrub-jay. Conservation Biology 14:501–512.

Natural Resource Conservation Service. 2001. Ruffed grouse (*Bonasa umbellus*): fish and wildlife habitat management guide sheet. US Department of Agriculture, Washington, DC, USA.

Naugle, D. E., J. D. Maestas, B. W. Allred, C. A. Hagen, M. O. Jones, M. J. Falkowski, B. Randall, and C. A. Rewa. 2019. CEAP quantifies conservation outcomes for wildlife and people. Rangelands 41:211–217.

Neu, C. W., C. R. Byers, and J. M. Peek. 1974. A technique for analysis of utilization-availability data. Journal of Wildlife Management 38:541–545.

Otis, D. L., and G. C. White. 1999. Autocorrelation of location estimates and the analysis of radiotracking data. Journal of Wildlife Management 63:1039–1044.

Parry, B. T., H. J. Vaux, and N. Dennis. 1983. Changing conceptions of yield policy on the national forests. Journal of Forestry 81:150–154.

Pearce, J. L., and M. S. Boyce. 2006. Modelling distribution and abundance with presence-only data. Journal of Applied Ecology 43:405–412.

Phillips, S. J., R. P. Anderson, and R. E. Schapire. 2006. Maximum entropy modeling of species geographic distributions. Ecological Modelling 190:231–259.

Phillips, S. J., M. Dudik, J. Elith, C. H. Graham, A. Lehmann, J. Leathwick, and S. Ferrier. 2009. Sample selection bias and presence-only distribution models: implications for background and pseudo-absence data. Ecological Applications 19:181–197.

Prokopenko, C. M., M. S. Boyce, and T. Avgar. 2016. Characterizing wildlife behavioural responses to roads using integrated step selection analysis. Journal of Applied Ecology 54:470–479.

Robertson, B. A., and R. L. Hutto. 2006. A framework for understanding ecological traps and an evaluation of existing evidence. Ecology 87:1075–1085.

Rodenhouse, N. L., and L. B. Best. 1983. Breeding ecology of Vesper sparrows in corn and soybean fields. American Midland Naturalist 110:265–275.

Schlaepfer, M. A., M. C. Runge, and P. W. Sherman. 2002. Ecological and evolutionary traps. Trends in Ecology and Evolution 17:474–480.

Smith, A. A., and R. W. Mannan. 1994. Distinguishing characteristics of Mount Graham red squirrel midden sites. Journal of Wildlife Management 58:437–445.

Soulliere, G. T. 1986. Cost and significance of a wood duck nest-house program in Wisconsin: an evaluation. Wildlife Society Bulletin 14:391–395.

Soulliere, G. T. 1988. Density of suitable wood duck nest cavities in a northern hardwood Forest. Journal of Wildlife Management 52:86–89.

Stamps, J. A. 1987. Conspecifics as cues to territory quality: a preference of juvenile lizards (*Anolis aeneus*) for previously used territories. American Naturalist 129:629–642.

Steidl, R. J., and B. F. Powell. 2006. Assessing the effects of human activities on wildlife. George Wright Forum 23:50–58.

Steidl, R. J., W. W. Shaw, and P. Fromer. 2009. A science-based approach to regional conservation planning. Pages 217–233 *in* A. X. Esparza and G. R. McPherson, editors. The planner's guide to natural resource conservation: the science of land development beyond the metropolitan fringe. Springer, New York, New York, USA.

Stirling, I., T. L. McDonald, E. S. Richardson, E. V. Regehr, and S. C. Amstrup. 2011. Polar bear population status in the northern Beaufort Sea, Canada 1971–2006. Ecological Applications 21:859–876.

Strickland, M. D., and L. L. McDonald. 2006. Introduction to the special section on resource selection. Journal of Wildlife Management 70:321–323.

Svardson, G. 1949. Competition and habitat selection in birds. Oikos 1:57–74.

Thomas, D. L., and E. J. Taylor. 1990. Study designs and tests for comparing resource use and availability. Journal of Wildlife Management 54:322–330.

Thomas, D. L., and E. J. Taylor. 2006. Study designs and tests for comparing resource use and availability II. Journal of Wildlife Management 70:324–336.

Thomas, J. W., E. D. Forsman, J. B. Lint, E. C. Meslow, B. R. Noon, and J. Verner. 1990. A conservation strategy for the northern spotted owl. Interagency Scientific Committee to Address the Conservation of the Northern Spotted Owl, US Department of Agriculture Forest Service, Department of the Interior, Bureau of Land Management, US Fish and Wildlife Service, and National Park Service, Portland, Oregon, USA.

Thurfjell, H., S. Ciuti, and M. S. Boyce. 2014. Applications of step-selection functions in ecology and conservation. Movement Ecology 2014 2:4.

US Department of the Interior. 2019. Endangered and threatened wildlife and plants; removing the Kirtland's warbler from the federal list of endangered and threatened wildlife. Federal Register 84:54,436–54,463.

US Fish and Wildlife Service. 2008. Recovery plan for the northern spotted owl, *Strix occidentalis caurina*. Portland, Oregon, USA.

US Fish and Wildlife Service. 2011. Revised recovery plan for the northern spotted owl, *Strix occidentalis caurina*. Portland, Oregon, USA.

US Fish and Wildlife Service. 2020. Bosque del Apache. https://www.fws.gov/refuge/bosque_del_apache/.

Van Horne, B. 1983. Density as a misleading indicator of habitat quality. Journal of Wildlife Management 47:893–901.

Ward, M. S., and R. W. Mannan. 2011. Habitat model of urban-nesting Cooper's hawks (*Accipiter cooperii*) in southern Arizona. Southwestern Naturalist 51:17–23.

Warner, R. E., J. W. Walk, and C. L. Hoffman. 2005. Managing farmlands for wildlife. Pages 861–872 *in* C. E. Braun, editor. Techniques for wildlife investigations and management. The Wildlife Society, Bethesda, Maryland, USA.

Wiens, J. A. 1985. Habitat selection in variable environments: shrub-steppe birds. Pages 227–251 *in* M. L. Cody, editor. Habitat selection in birds. Academic, San Diego, California, USA.

Wiens, J. A. 1989. Spatial scaling in ecology. Functional Ecology 3:385–397.

Williams, J. B., and G. O. Batzli. 1979. Competition among bark-foraging birds in central Illinois: experimental evidence. Condor 81:122–132.

Yahner, R. H., C. G. Mahan, and A. D. Rodewald. 2005. Managing forestlands for wildlife. Pages 898–919 *in* Techniques for wildlife investigations and management. The Wildlife Society, Bethesda, Maryland, USA.

Yetter, A. P., S. P. Havera, and C. S. Hine. 1999. Natural-cavity use by nesting wood ducks in Illinois. Journal of Wildlife Management 63:630–638.

WILDLIFE RESTORATION

MICHAEL L. MORRISON

INTRODUCTION

Wildlife management was historically guided through trial and error, where classroom knowledge, practical experience (including hunting), and various opinions were used to try and make a desired change in the distribution and abundance of a target species. Most of the wildlife professional's attention was on consumptive species and species that negatively affected humans directly (e.g., predators) or indirectly (e.g., crop damage, livestock depredation). Over time, the collective experience and the range of species studied expanded and—along with new information from the fields of ecology, physiology, genetics, and others—allowed the wildlife profession to become increasingly rigorous scientifically, leading to the development of management applications based on those data.

The restoration profession has been built on a foundation of plant biology and horticulture. Similar to the wildlife profession, restoration developed into an organized discipline as investigators sought to restore ecologies to previous, more desirable conditions; that is, to manage ecological conditions. Trial and error and projects designed and implemented on the basis of prior experience and expert opinion characterized restoration. Restoration is also moving toward the use of more rigorous study designs for use in developing practical applications of the available data (Palmer et al. 2016). Dickens and Suding (2013) noted that the science–practice gap was frequently cited as a major factor limiting the science and practice of restoration.

In this chapter, I review how the principles of wildlife ecology and restoration ecology can be linked to results in comprehensive planning for the management and conservation of natural resources. Elsewhere in the literature is more detailed coverage of wildlife restoration (Morrison 2009). In a condensed manner, herein I give wildlife and restoration professionals—including biologists, managers, and administrators—a basic outline of the fundamental issues of wildlife populations and habitat relationships as related to restoration. I encourage readers, especially students, to delve deeper into restoration and its applications to wildlife.

Because virtually any restoration activity will change conditions for animal species, restorationists must be aware of how their actions influence the abundance and behavior of animals (if for no other reason than to ensure that the animals do not ruin the restoration measures). Likewise, wildlife ecologists can tap into the vast knowledge of the restoration community on how to manipulate soils, plant species, and vegetative associations to achieve a desired outcome. Restoration plans should be guided by the likely responses of current or desired wildlife species in the project area, which includes data on myriad factors such as the current and historical abundance and distribution of animal species, habitat and food requirements, breeding locations, how plant succession will change species composition through time, space requirements, and necessary links between geographic areas.

CONCEPTUAL FRAMEWORK

Restoration represents a synthesis of many biotic and abiotic concepts, including habitat and niche ecology, populations, genetics, ecosystem dynamics, historical ecology, geology and soils, fire history, and climatic patterns. As such, the practice of restoring ecosystems should be based on an interdisciplinary approach (Halle and Fattorini 2004, Palmer et al. 2016). I first briefly review key concepts and terms that form the foundation of animal ecology and hence restoration. Although such topics as economics, sociology, and politics must be considered for restoration projects to be successful, they are beyond my current scope (see Chapter 4). I focus on restoring wildlife, primarily by condensing and revising the more detailed presentation in Morrison (2009). Although wildlife is frequently limited to terrestrial vertebrates, the concepts and applications outlined herein have broad applicability to other taxa. The terminology and concepts reviewed here serve as the foundation for the remainder of this chapter.

DEFINITIONS AND HISTORICAL CONDITIONS

To restore means to bring back into existence or use; thus restoration is the act of restoring. The simplicity ends here because difficulty arises once we ask follow-up questions. What exactly do we want to restore something *to*? What is our desired condition? As I have noted previously, automobile restoration is simple (Morrison 2009). We usually know the year and often the exact day a vehicle was produced. Detailed engineering drawings might be available for every part of the vehicle, pictures might show what it originally looked like, and an owner's manual explains how to maintain the vehicle. Nature does not provide us with such a detailed description of the past; rather, we are only provided hints and oftentimes well-hidden clues concerning former conditions and dynamics.

Two key terms to differentiate are restoration ecology and ecological restoration. Restoration ecology is the scientific process of developing theory to guide restoration and of using restoration to advance ecology. Ecological restoration is the practice of restoring degraded ecological systems (Palmer et al. 2016). There is a clear and necessary link between the practice of restoration (i.e., ecological restoration) and the concepts that were used to develop the restoration plan (i.e., restoration ecology).

There have been attempts to establish an overall conceptual framework for restoration ecology (Hobbs and Norton 1996, Halle and Fattorini 2004, Palmer et al. 2016). Such conceptual frameworks usually attempt to relate the fields of disturbance ecology and succession, where natural succession and human-induced changes are used to guide development of a system that has been disturbed, including those that are now apart from what is considered normal variation. Although some basic functions and ecosystem processes can be manipulated and changed to a desired condition, it is usually impossible or at least impractical to restore any system to some previous state (Halle and Fattorini 2004).

Historical ecosystems are ecological systems that existed in the past. Historical ecosystems have been a focus of restorationists because they can be used as analogs to guide development of a restoration plan (Egan and Howell 2001). The ecological condition desired as a result of a restoration project is referred to as the reference condition. Reference conditions inform the restoration plan by defining what the original condition was compared to the present, determining what factors caused the degradation, defining what needs to be done to restore the system, and developing criteria for measuring the success of the restoration project (Morrison 2009).

The age of the analog reference condition largely determines the difficulty of describing and then duplicating those conditions and associated processes. If the project is designed to correct a recently damaged area (e.g., natural or human-caused catastrophe), then there should be many readily available reference conditions in the general geographic area. But if the goal is to restore an ecological condition that existed many centuries ago, reference conditions can be highly controversial and also subject to economic and sociological constraints.

First, we must determine what period will serve as the historical reference, and then decide what ecological conditions existed during that period. For example, Noss (1985) concluded that whether American Indians should be considered a natural and beneficial component of the environment cannot be answered conclusively, largely because the effect they had varied by time and location. Many restorationists believe that the arrival of European settlers is the point in time when unnatural conditions began to prevail. Willis and Birks (2006) noted that ecosystems change in response to many factors, including climate variability, invasions of species, and wildfires. A central point developed by Willis and Birks (2006) is the difficulty in defining the natural features in ecosystems through time. For example, there are more than 150 species of plants that have been introduced to the British flora by humans between 500 and 4,000 years ago. Deciding which plants are native would have a substantial effect on efforts to conserve or potentially eradicate many species. Alve et al. (2009) thought that the ecological status of an environment should be evaluated by comparison with local reference conditions, which they defined as the preindustrial ecological status of the 19th century.

In my opinion, however, we have little if anything to gain by attempting to categorize human effects as natural or unnatural. Rather, I believe the role of restorationists is to help quantify the likely consequences of human activities and natural events on the environment so that the public and managers can make informed decisions when developing the desired condition for an area. For example, because most large, terrestrial mammalian predators have been extirpated from most of the contiguous United States and Mexico, ecological systems are missing a dominant selective force. An excellent case study was provided by Berger et al. (2001), who reported how a series of ecological events were triggered by the local extinction of grizzly bears (*Ursus arctos horribilis*) and gray wolves (*Canis lupus*) from the southern Greater Yellowstone Ecosystem in the United States. Their removal allowed a substantial increase in the moose (*Alces alces*) population to occur, which in turn caused substantial degradation of riparian vegetation through moose browsing, followed by a reduction of neotropical migrant birds that used the riparian zone. Similarly, Hebblewhite et al. (2005) reported that wolf exclusion from Banff National Park, Canada, caused increased browsing by elk (*Cervus canadensis*), which in turn resulted in changes in beaver (*Castor canadensis*) and songbird behavior.

Rather than attempting to return a system to some historical state, Palmer et al. (2016) suggested that a more realistic goal was to move a damaged system to an ecological state that is within some acceptable limits relative to a less disturbed system. The key is to define acceptable limits. Similarly, Hobbs et al. (2014) reviewed the debate surrounding how to classify ecosystems and associated restoration goals, and concluded that an effective approach involved moving away from the natural versus unnatural dichotomy toward a more complex characterization of ecosystems into varying states of modification, which then presented a wide array of management challenges and opportunities.

Regardless, it is difficult to achieve a condition that is free from human effects for the following reasons: local plant and animal extinctions, introduced species, migration and dispersal of many plants and animals is no longer possible, and ecological processes have been retarded or prevented. In a practical sense we are left with planning a restoration project based on (1) knowledge of historical conditions, (2) knowledge of current regional conditions, (3) knowledge of species-specific ecological requirements, (4) evaluation of legal requirements, and (5) political reality (Morrison 2009). Although biologists support attempts to place restoration plans in context of historical conditions, ecological reality must guide what can and cannot be achieved.

APPROACHES TO RESTORATION

Planning for restoration and subsequent management of the area requires that managers at least generally understand how the ecosystem in the region was developed and how it functions under differing environmental conditions. Thus we need to examine organisms from different taxonomic levels, along with their interactions with each other and with abiotic conditions and processes. An ecosystem (i.e., an ecological system) can be generally defined as consisting of organisms of various taxonomic designations and levels of biological organization, along with their interactions among each other and among abiotic conditions and processes. Understanding wildlife in an ecosystem context requires understanding of population dynamics; the evolutionary context of organisms, populations, and species; interactions between species that affect their persistence; and the influence of the abiotic environment on the vitality of organisms (Morrison et al. 2006:387). Restoration must similarly be considered within a broad spatial context. Of course, we cannot study an ecosystem and the many species contained therein per se; rather, we must work with the individual animals that comprise the species in an area. Although we should develop restoration plans in a broad-scale context, and some agencies (e.g., US Department of Agriculture Forest Service) are attempting broad-scale projects, most management occurs on small spatial scales (i.e., 1–100 ha).

ANIMAL POPULATIONS AND ASSEMBLAGES

The goal of wildlife restoration is to create conditions that provide for the survival and protection of individual organisms in sufficient numbers and locations to maximize the probability of long-term persistence (Morrison 2009:17). The persistence of a species is influenced by the dynamics of interactions among individuals within a population, by interactions among populations and other species, and by interactions between organisms and their habitats and environments. Wildlife restoration requires knowledge of population dynamics and

behaviors and the processes that regulate population trends. Habitat (Chapter 16; discussed below) by itself does not guarantee long-term fitness and viability of populations (Morrison et al. 2006:61).

The restorationist must understand the spatial and geographic factors that influence habitats and environments, population structure, and fitness of organisms because they relate directly to the size and location of the area needing restoration to ensure survival of a species. For example, providing the proper habitat conditions does not provide for long-term persistence if no allowance has been made for immigration from other geographic locations. I first review the population concept, then discuss the importance of considering exotic species in restoration planning, and finish with an approach for determining which species are likely to occur within a planning area.

Population Concepts

Traditionally, a population has been defined as a collection of organisms of the same species that interbreed. Mills (2013) broadened the concept by defining a population as a collection of individuals of a species in a defined area that might or might not breed with other groups of that species elsewhere. As reviewed by Morrison et al. (2020), however, our understanding of animal ecology is limited if we are unaware of the ways that our data and resulting inferences are being affected by biotic and abiotic interactions not included in the study or occurring outside of our study area boundaries. They thus stressed the need to first determine—to the best of our ability—the biological population of interest, and then sample characteristics of interest from that population, such as features of the environment and resources being used. This allows us to identify the relevant area (spatial extent) that is providing inputs of biotic and abiotic factors (e.g., nutrients, predators, competitors) that are in turn influencing the distribution, abundance, and survivorship of populations. If we are to gather meaningful samples, however, we must understand how the animals are assorted into interacting groups of individuals. Thus better understanding of habitat associations necessitates also studying population viability and demography, including reproduction and survival (Rodewald 2015).

A metapopulation structure can arise when there is partial isolation of individuals among populations. A metapopulation is "a species whose range is composed of more or less geographically isolated patches, interconnected through patterns of gene flow, extinction, and recolonization" and has been termed "a population of populations" (Levins 1970:105, Lande and Barrowclough 1987:106). These component populations have also been referred to as subpopulations (i.e., a portion of a population in a specific geographic location). Unfortunately, the term subpopulation has been used on the basis of nonbiological criteria, often for management or legal reasons (e.g., administrative or political boundaries). Thus metapopulations consist of at least two subpopulations (local populations) of the same species, linked by migration or dispersal, such that organisms occasionally change which subpopulation they

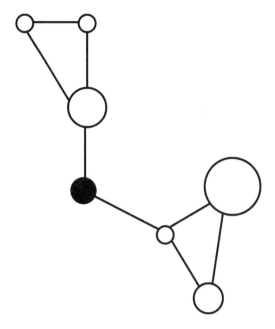

Figure 17.1. Hypothetical arrangement and size of subpopulations (circles) in a metapopulation. The size of the circle reflects population abundance; lines represent dispersal routes. The subpopulation indicated by the solid circle would be a key link between other subpopulations, and thus a top priority site for restoration. From Morrison (2009:fig. 3.2)

are a part of; rates of interaction within subpopulations are much higher than the rates of interaction among subpopulation (Millstein 2010). If the rates of interaction within local groupings of individuals are not significantly higher than the rates of interaction among local groupings, then these would represent a patchy population (and not a metapopulation).

If one or more of the species targeted for restoration are metapopulations, the restoration plan must consider the spatial relationship among areas of habitat so that dispersal among subpopulations can occur. Extinction of the species within one or more subpopulations could occur and become permanent if there is no suitable area of habitat close enough to allow dispersal and recolonization. Subpopulations (within a metapopulation) usually vary in abundance of animals. Note the links between subpopulations (Fig. 17.1), where loss of a subpopulation that is located between other subpopulations could lead to further extinctions because the linkage would be broken. Wildlife restoration requires identification of the structure of the populations of species under consideration, followed by planning of restoration sites to enhance this structure (e.g., by promoting dispersal of individuals between subpopulations, which are sometimes called stepping-stone habitat patches; Morrison 2009). Because of metapopulation structure, not all suitable areas will be occupied at any one time. Even if a habitat patch appears unoccupied in any one year, monitoring wildlife use of habitats should proceed for many years.

Populations are distributed in many patterns, from those in many isolated groups to those that are linked through move-

ments between groups. For multiple isolated populations, each population is susceptible to extinction, usually without the possibility of natural recolonization (Mills 2013). From a restoration perspective, understanding the dynamics of each population and if movement between populations is occurring is a core component of any plan for conserving a species.

Exotic Species

The introduction and subsequent range expansion of exotic species are major challenges for conservation and restoration. Exotic plants and animals can affect the health and ultimately the abundance and distribution of native plants and animals (Office of Technology and Assessment 1993). We usually label a species as exotic if humans accidentally or purposefully introduced it. A species is usually considered native if it resides in its presumed area of evolutionary origin or if it appears because of nonhuman-aided dispersal (Willis and Birks 2006). There are many situations where we are uncertain of the cause of range expansions, however; they might be natural, might have been induced or enhanced by human alteration of environments, or might have begun as a minor introduction by humans. The brown-headed cowbird (*Molothrus ater*), for example, evolved in the Great Plains and then spread throughout most of North America during the 1900s, apparently because of forest clearing and agricultural development (Morrison et al. 1999). Humans did not physically capture and move the cowbirds; they moved as a result of human effects to the environment. Are cowbirds native or exotic within certain portions of their current range? Because cowbirds can negatively affect certain species of (often rare) birds, the answer has a substantial influence on how a restoration project is planned.

Examples of exotic species that have spread throughout the continental United States include the European starling (*Sturnus vulgaris*) and house sparrow (*Passer domesticus*), both introduced from Europe. Both species negatively affect native birds by usurping nest sites and reducing nest productivity. The non-native house mouse (*Mus musculus*) and several species of rats have spread across North America. Morrison et al. (1994) reported that the house mouse was a dominant species of rodent in disturbed areas in Southern California that were scheduled for restoration.

Restorationists must determine the potential effect that exotic species, and native species whose range has expanded (e.g., cowbird), will have on desired species. Starlings and house sparrows, for example, can occupy the majority of nesting cavities (e.g., woodpecker excavations, bird boxes, eaves in buildings) that could be created as part of restoration for native species (e.g., bluebirds, woodpeckers). Cowbirds can cause a high level of nesting failure in songbirds and have been a specific focus in locations undergoing restoration that focus on endangered species (Kus 1999, Kostecke et al. 2005).

There is not a general prescription that calls for removal of all exotic species, however. Native species can become dependent on an exotic species in severely altered environments. For example, channelization of rivers has curtailed flooding and lowered water tables, resulting in a loss of native shrub and tree species. In the southwest United States, for example, the exotic plant saltcedar (*Tamarix* spp.) has become established in many riparian areas following loss of native willow (*Salix* spp.)–cottonwood (*Populus* spp.) woodlands. The endangered southwestern willow flycatcher (*Empidonax traillii extimus*) now nests successfully in salt cedar; removal of salt cedar would eliminate their nesting and foraging locations. And because of the altered hydrology due to the channelization, restoring willow-cottonwood is virtually impossible over large areas. Removal of exotics is not always a preferred restoration strategy, especially when an endangered species is involved. As such, restoration is not necessarily an attempt to restore past species compositions, but rather an attempt to restore specific parts of the previous system. A major activity in restoration planning is prioritizing what can and cannot be restored.

Assembling Groups of Species

A priority step that must be taken in developing a restoration project is determining which species to explicitly incorporate into the plan. As noted above, a host of factors are responsible for the presence of plant and animal species in a location; the scientist's task becomes more difficult as the physical size of the restoration area decreases. One tool for helping plan and organize a comprehensive restoration plan falls under the general ecological concept termed *assemblage* rules. Diamond (1975) attempted to develop general rules that predicted broad patterns of species co-occurrence. He reported that species with similar diets would seldom co-occur; he called this pattern a checkerboard distribution and attributed it to competition between species for limited resources. Assembly rules became controversial, however, because Diamond (1975) and others were largely unsuccessful in showing that competition was the force behind the patterns they reported. Similar patterns as those reported by Diamond (1975) could be created when species were randomly distributed across an area (i.e., a null model); competition was not required (Connor and Simberloff 1979).

Thanks to the pioneering work of Diamond (1975), Connor and Simberloff (1979), and others, the search for assembly rules has gained widespread attention, as evidenced by edited volumes on the subject (Weiher and Keddy 1999, Temperton et al. 2004). Some workers have emphasized biotic interactions (e.g., competition), whereas others have emphasized a more holistic approach that includes interactions of the environment with the organisms. Temperton and Hobbs (2004) concluded, however, that no real consensus existed about which rules should be emphasized. As detailed elsewhere (Morrison 2009), I adhere to a comprehensive view in which biotic and abiotic factors, in combination with other constraints on species, determine the species and their abundance in a specific location. The priority for restoration planning should be identification of the primary factors that can limit the occurrence, abundance, survival, and productivity of a species. The manner in which these limiting factors place boundaries on the distribution of species that can potentially occur in a project area can be captured in the assembly rule process.

Of course, documenting a pattern is not the study of species assembly. Although null models are useful in identifying patterns, they do not identify the mechanisms causing the pattern to exist (Keddy and Weiher 1999). As Temperton and Hobbs (2004) noted, restoration practitioners must try to produce a certain type of ecosystem and need guidelines to follow; identifying the underlying mechanisms substantially enhances what is learned and what can be applied in other locations. Animal ecologists have been able to develop general guidelines for application to large spatial scales and for predicting gross measures of animal performance (e.g., presence–absence). It becomes necessary, however, to identify the mechanisms underlying animal performance as we move into progressively smaller spatial areas. The rules must become more detailed, and more difficult to obtain, as we move to restore more than simple animal presence. Lockwood and Pimm (1999) noted that restoring species composition and diversity was far more difficult than restoring ecosystem function and partial structure.

Terminology

As developed above, an initial evaluation of the likely boundaries of the biological populations of the species of interest should be the starting point for any field study. This is because the assemblage of species (i.e., community) is most likely being influenced by abiotic and abiotic factors occurring outside the project area. With rare exception (e.g., isolated endangered species), remember that the boundaries of a project area are based on some criteria not related to the distribution of the species contained within it, but rather on administrative boundaries (e.g., state park, private landholding). Thus evaluating the species of interest should focus on identifying the filters and constraints (see below) that will modify the species present in an area throughout a successional process, which necessitates looking beyond the boundaries of the project area.

Species Pool

This concept defines a regional species pool as occurring within a biogeographic region and extending over spatial scales of many orders of magnitude larger than those of a local species assemblage. The species pool decreases as we progress to smaller geographic areas, such as moving from the watershed scale down to the stream reach scale (i.e., multiple streams within the larger watershed). Then the local assemblage of species from the larger pool is determined by passing through a series of filters. Van Andel and Grootjans (2006) defined three pools: regional, local, and community. Regional species pools are the set of species occurring in a certain biogeographic or climatic region that are potential members of the target assemblage. Local species pools are the set of species occurring in a subunit of the biogeographic region, such as a valley segment. Community species pools are the set of species present in a site within the target community.

The concept of ecological filters entails a main approach in assembly rule theory. From the total pool of potential colonists, only those that are adapted to the abiotic and biotic

Table 17.1. Filters that influence the assembly of species within a specified geographic area

Abiotic filters
- Climate: Rainfall and temperature gradients
- Substrate: Fertility, soil water availability, toxicity
- Landscape structure: Landscape position, previous land use, patch size, and isolation

Biotic filters
- Competition: With preexisting and potentially invading species and between planted or introduced species
- Predation–trophic interactions: From preexisting and potentially invading species, and predation between reintroduced animal species
- Propagule availability (dispersal): Bird perches, proximity to seed sources, presence of seed banks
- Mutualisms: Mycorrhizae, rhizobia, pollination and dispersal, defense, and so forth
- Disturbance: Presence of previous or new disturbance regimes
- Order of species arrival and successional model: Facilitation, inhibition, and tolerance
- Current and past composition and structure (biological legacy): How much original biodiversity and original biotic and abiotic structure remain

FREDERICK LAW OLMSTED (1822–1903)

As the 20th century approached and Americans migrated to cities from rural areas, it became apparent that cities needed to become more hospitable places. Beautification made cities more aesthetically pleasing and provided restorative areas away from the stress of everyday commerce. Frederick Law Olmsted became a pioneer in the city beautiful movement of the post–Civil War generation. Olmsted is acknowledged as the founder of American landscape architecture even though he had no formal university training and worked a variety of jobs before becoming the superintendent of New York's Central Park, where he became architect-in-chief of construction. He then served as head of the US Sanitary Commission, which was the forerunner of the American Red Cross. Olmsted's goal was to improve American society by designing and promoting recreation in the hearts of cities. He did not envision parks as vast, open spaces but rather as places of harmony—places where people could go to escape city stresses. He wanted these parks to be available to all people no matter their walk of life. He designed trails that curved and flowed with the landscape (see "Designing a Wildlife Restoration Project" for discussion of landscape terminology), screening them with thick plantings along their borders, separating and excluding commercial traffic. He formed his own landscape architecture firm that he ran between 1872 and his retirement in 1895.

conditions present at a site will be able to establish themselves successfully (Hobbs and Norton 2004). A filtering or deletion process takes place that is analogous to a filtering-out of those organisms not adapted to the habitat conditions. This approach focuses on the end product of numerous interactions between a colonist and the ecosystem components. Assembly rules are useful in restoration planning because they allow for a systematic way to determine what factors may be limiting membership in the local pool of species. Some of the rules are obvious and seem trivial (e.g., predators without prey will starve), whereas others are more complicated (e.g., what abundance of prey is necessary for a new predator to enter a community; Temperton and Hobbs 2004). The filters involved that influence the course of system development will determine the approach taken to restore an area (Hobbs and Norton 2004). Filters will also change in type and intensity through space and time. Hobbs and Norton (2004) developed a comprehensive list of potential filters, which Morrison (2009) modified to apply more directly to animals and plants (Table 17.1).

As reviewed elsewhere (Morrison 2009), at least seven figures in Temperton et al. (2004) depicted general pathways from the potential pool of species, through various filters and constraints, to the realized species pool. Morrison (2009:91) synthesized these figures into a single diagram that depicted pathways and filters and how species fit into available space throughout the course of succession. The general diagram (Fig. 17.2) shows how filters change in influence on a species and in actual type, proceeding through succession and development of the species assemblage.

Various abiotic and biotic factors filter the regional species pool, resulting in the local species pool (Fig. 17.2). Species must be physiologically adapted to occupy a given area based on general climatic conditions; additional abiotic fac-

tors will have additional filtering effects. The specific factors that filter or constrain which species will actually occur can change through time, and different limiting factors will affect each species or perhaps groups of similar species. Further, the habitat or niche space available to species will vary through time as succession proceeds through time. Many factors must be considered when trying to predict and guide the actual species that will be present on the target site (e.g., the community species pool of Van Andel and Grootjans 2006). In restoration planning, managers can establish target conditions, such as the structure and floristics of vegetation, and develop lists of desired animal species to be achieved as restoration proceeds. Pools of species change as they proceed through filtering processes that accompany the conditions present within a seral stage (Fig. 17.2). Another interesting feature of this assembly process is that each subsequent community species pool need

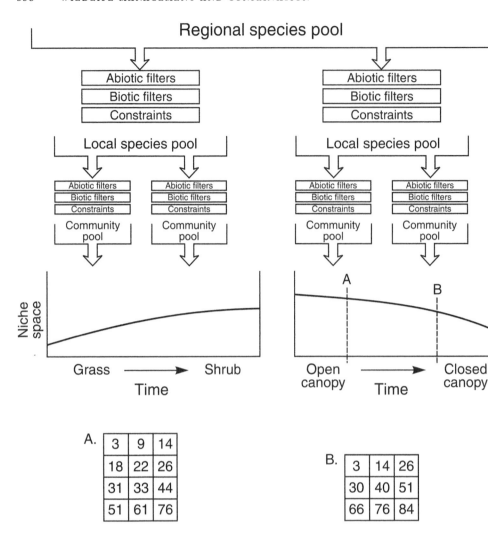

Figure 17.2. The species present in an area (community pool) are those remaining as a result of the filtering process at the local and regional levels. The figure represents two local species pools drawn from the same regional pool and co-occurring in the same two vegetation types. The community pool associated with a seral stage is drawn from the local pool that is specific to the more general vegetation type. The type and number of species present across seral stages will also reflect the size of the target (e.g., project) area and niche space available. The cross-sectional cut depicted as A and B indicates how species (arbitrarily designated by numbers in the squares) change in type and total number. From Morrison (2009:fig. 5.2)

not be a subset of the pool at a previous seral stage; the community pool is under constant reshuffling as a result of the filtering that is occurring through time from the larger local and regional species pools.

Developing a comprehensive restoration plan using the assembly process can be linked directly with monitoring and especially development and implementation of a valid adaptive management plan (Morrison 2009:chap. 19). Knowing what should occur during different stages of the restoration process to the desired condition allows managers to make midcourse corrections and to improve the opportunity for overall project success.

Implications for Restoration

Restoration projects can actively try to modify the effects of filters to allow desired species in and to prevent the establishment of undesired species (Hobbs and Norton 2004). For example, modifying abiotic filters includes providing natural and artificial structures for use as shelter and water sources; modifying biotic filters includes controlling exotic species, predator control, and introducing animals.

HABITAT

Because restoration requires a synthesis of many disciplines, the restorationist must be well schooled in how a diverse array of scientists and managers—including the community of wildlife professionals—communicate. Virtually everyone from scientists to the general public understands that habitat describes where something lives. As reviewed elsewhere for wildlife in general (Hall et al. 1997, Morrison and Hall 2002, Kirk et al. 2018) and for specific application to restoration (Morrison 2009), the term has been used and modified in ways that do not promote understanding and certainly do not easily translate into management actions. This section begins by first providing some brief but critical definitions for key habitat terms (Morrison 2009).

Habitat includes the resources and conditions present in an area that produce occupancy by an animal. If an individual of a species does not occupy a location, then we cannot know for sure that the location is suitable for occupancy. The critical aspect of animal habitat is that it is specific to the presence of a species, population, or individual (animal or plant). Habitat is a synthesis of the specific resources that are needed by the organism. As noted by Miller (2007), an organism-based understanding of habitat is needed to determine appropriate restoration goals.

Habitat and *habitat type* are not synonymous. Daubenmire (1968) developed the latter term to refer only to the potential climax in an area. We can avoid confusion by referring to the vegetation in an area as the vegetation association or vegetation type instead of the confusing term habitat type (Hall et al. 1997, Morrison and Hall 2002). That habitat is organism-specific is a central concept in wildlife ecology and in turn in wildlife restoration, because focusing restoration on vegetation most often will fail to restore the desired assemblage of wildlife.

The definition of habitat has been modified in numerous ways, leading to additional confusion within the wildlife profession (Hall et al. 1997). Several key terms, however, are widely (and mostly appropriately) used to describe how animals use habitat. *Habitat use* is the way an animal uses physical and biological resources in an area. *Habitat selection* refers to the hierarchical and largely innate (as modified by learning) process whereby an animal makes decisions about different scales of the environment (Hutto 1985:458). Johnson (1980) defined selection as a hierarchical process by which an animal chooses which habitat components to use. Habitat preference is restricted to the consequence of the habitat selection process, resulting in the disproportional use of some resources over others.

Habitat availability concerns the ability of an individual to obtain physical and biological components of the environment. In contrast, habitat abundance refers only to the quantity of the resource in the habitat (Wiens 1989:402). Of course, it is challenging to assess resource availability from an animal's perspective (Litvaitis et al. 1994). For example, vegetation that is beyond the reach of an animal is unavailable as a food source.

Lastly, *habitat quality* is measured as an outcome of the survival, reproduction, or persistence of a biological population. Unfortunately, the use of habitat quality in the literature is frequently incorrect and based on habitat conditions absent any link with a measure of demographic performance (e.g., breeding success). Indirect measures of quality, such as animal density, often fail to adequately describe actual animal performance (e.g., survival, productivity; Van Horne 1983). For example, many plants might not flower during short-term drought conditions, which in turn substantially lowers arthropod abundance and in turn negatively affects nest success of insectivorous birds. Yet measurements of plant cover likely remain similar to the previous breeding season. Thus classic measures of habitat (e.g., plant cover) would fail to identify changes in food availability.

Wildlife ecologists frequently use the terms macrohabitat and microhabitat (Johnson 1980). Macrohabitat refers to large spatial extent features such as seral stages or vegetation associations (Block and Brennan 1993), which equates to Johnson's

(1980) first level (order) of habitat selection. Alternatively, microhabitat refers to finer-scaled habitat features, such as would-be important factors in levels 2–4 in Johnson's (1980) hierarchy. Macro- and microhabitat are obviously general categories that have limited usefulness in application, and their use should be minimized except in general conversation.

To advance wildlife ecology and ultimately the transfer of that knowledge to the practice of restoration, we must be sure that the fundamental concepts with which we work are well defined and understood. Peters (1991:76) popularized the term operationalization with regard to ecological concepts, by which concepts such as habitat should have operational definitions that are the practical, measurable specifications of the ranges of specific phenomena the terms represent. But as reviewed above, use of habitat terminology has been imprecise and ambiguous. A lack of explicit definitions leads to ambiguity in what data and results reported in the literature actually mean, which in turn leads to confusion by other workers in the field. As promoted by Hall et al. (1997), Morrison and Hall (2002), Krausman and Morrison (2016), Kirk et al. (2018), and others, there is a strong need for standard definitions or at least a clear understanding of the different uses of the terminology. Without clear definitions of terms, it is unlikely that workers in one field will be able to fully access and understand information generated by those in another; for our purposes, restorationists and wildlife ecologists must understand one another.

Spatial Scale

Ecologists now recognize that animals go through a series of increasingly refined selection decisions: initial selection of a geographic area; selection of specific combinations of elevation, slope, and vegetation type; selection of specific locations to forage, breed, or rest; and selection of specific items to use, such as types and sizes of food (Fig. 17.3). Moving to an increasingly local spatial scale (i.e., from broad to specific), scientists are able to understand a more detailed amount of information about animals (Fig. 17.3). At broad spatial scales, managers can usually quantify such metrics as presence or absence of a species, but they cannot quantify much, if anything, about survival or reproductive success. Data obtained about breeding sites, however, inform us about the number of young produced. For application to restoration, this relationship between spatial scale and information content means we must carefully match the goal of a restoration plan with the appropriate scale of study.

As developed by Lindenmayer and Fischer (2006), landscapes must be analyzed with a clear understanding of the biological entity (organizational issue) and perspective being considered. That is, although humans consider landscape to mean a large viewscape (e.g., a panorama), it is unlikely that many wild animals share the human perspective. With clear operating definitions, we can analyze how individual organisms (or assemblages of similar organisms) respond to variations (planned or not) in the structure of the environment. We are then able to discuss how, say, fragmentation of an area will

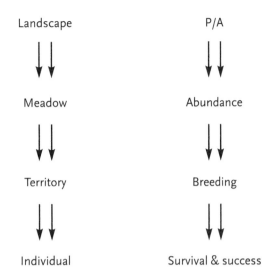

Landscape		P/A	
↓↓		↓↓	
Meadow		Abundance	
↓↓		↓↓	
Territory		Breeding	
↓↓		↓↓	
Individual		Survival & success	

Figure 17.3. Relationship between the measurement of spatial extent and the associated measure of animal performance. From Morrison (2009:fig. 4.1)

affect individual organisms rather than talking broadly and rather meaninglessly about landscape fragmentation. Hodges (2008) noted that determining whether a landscape was connected or fragmented depended on the species of interest and the attributes measured. Thus functional connectivity measurements address movement distances, patterns, and rates, whereas structural connectivity measurements address attributes such as forest density as influencing animal occurrence. What humans view as not fragmented on a large spatial extent (i.e., landscape scale; Chapter 6) might actually be fragmented internally for any number of species.

The Niche

Habitat (in the species-specific sense) provides a general description of the environmental features used by animals and can give initial insight into factors underlying survival and fitness. Other factors, specifically those related to an animals' niche, must be known to understand the mechanisms responsible for animal survival and fitness (Morrison et al. 2006). If we do not know ecological mechanisms, we are bound to misinterpret the phenomenon we observed and, in turn, to develop incomplete and likely misguided restoration and management prescriptions. Morrison et al. (2006) provided a detailed account of how the niche can be used to enhance our understanding of wildlife ecology, and Morrison (2009) translated that account into restoration applications. Here I briefly develop the niche concept and how it can be used to improve our ability to restore habitat.

O'Connor (2002) concluded that only one environmental factor is usually limiting in any particular situation for a species, similar to the limiting factor concept of Leopold (1933). The complication, however, is that the same factor is unlikely to always be limiting, because variation in nature results in a continual shifting of resources, which in turn results in shift-

ing among potentially limiting factors. We must understand a good deal about the niche of a species to manage and restore its populations.

The potential distribution of a species is defined as the physiological or fundamental niche, which encompasses all of the locations a species could occupy if it was uninhibited by other constraining factors. Within the physiological niche resides the realized niche, which is the actual (observed) distribution of the species. The realized niche does not necessarily contain the full range of conditions under which the species does best, because such factors as competitors and predators exclude it from those conditions. Unless we understand the physiological niche, we cannot know the full range of options available for managing a species. Actions like control of predators or manipulation of plant species may allow some species to thrive in areas where they are rarely found under present conditions (Huston 2002).

As developed above for habitat quality, the problem with focusing on habitat alone in restoration is that the features measured (e.g., vegetation cover) can stay the same, while use of important resources (e.g., food) by an animal within that habitat can change. Changes we observe in habitat use by a species through space and time are often caused by changes in the availability of specific resources that we fail to measure; that is, we do not understand the mechanisms that underlie the changes in habitat use. Advancement in knowledge in wildlife ecology and the subsequent translation to restoration entails more than superficial knowledge of natural history and general habitat use. Restoration planning must include these niche factors that have not traditionally been considered because they influence whether an animal occupies a site and how it performs if occupancy is possible.

DESIRED CONDITIONS FOR RESTORING WILDLIFE

An initial step in designing any natural resource restoration endeavor is to decide upon the desired condition. One must first establish the overall project goal and the specific outcomes desired for plants, animals, and the overall environment. As discussed above, restoring an area to mirror a preexisting condition is difficult unless data on historical conditions are available. And, because of changes in the environment, the historical condition may no longer be attainable.

It is difficult, however, to predict the specific species composition of a locality, especially when dealing with small geographic areas. As reviewed herein, the vegetation type of an area does not directly identify the many species-specific habitats of the locality. The best we can usually do is to assemble lists of probable species occurrences based on the available evidence. The more complete our understanding of the filters and constraints that underpin the process of species colonization, the higher the likelihood we are to establish realistic goals. For most species, however, we have either sporadic, often-vague comments on status in the historical literature or data too recent, or too incomplete, to be a basis for reconstructing past communities and changes through time. There are a number of tools available that can help develop a reasonable picture of past environmental conditions and associated species occurrences.

Historical Assessments

The primary methods for assessing historical conditions are existing data sets, museum records, fossils, field notes, and literature. As noted by Morrison (2009), the process of assembling the historical view of an area should be seen as a puzzle whereby one can still identify the picture if some pieces are missing; the number of missing pieces determines the uncertainty in any assessment.

Data Sources

The Biological Resource Division of the US Geological Survey manages the nationwide Breeding Bird Survey (BBS), which began in 1965. The BBS consists of more than 2,000 randomly located 40-km-long permanent survey routes established along secondary roads throughout the United States and southern Canada that are surveyed annually during the breeding season (Robbins et al. 1986). Since 1900 the National Audubon Society coordinated a bird-counting effort across the United States and Canada during winter known as the Christmas Bird Count (CBC), which is a single-day count conducted by volunteers in a 24-km radius of a chosen location (e.g., a city). The CBC is now a valuable database for long-term monitoring of population trends.

The results of numerous natural history surveys beginning in the late 1800s and continuing to date are available in published reports and scientific papers. A thorough literature review will allow reconstruction of the flora and fauna occurring in the general region of a project site prior to or during the advent of intensive human-induced development. Fleishman et al. (2004) presented a good discussion of the use of historical data in understanding faunal distributions, with special reference to the Great Basin of western North America.

Museum Records

Museums that house natural history collections (e.g., skins, skeletons, eggs, plants) are found within private and public universities, federal and state agencies, and various research organizations. They were developed, in part, to record the historical fauna and flora of a region. Of primary value to our discussion here are the data that accompany each specimen, which usually include the date and location of collection, the collector, the species identity, and perhaps a few natural history notes; much of these data are now available electronically. Of course, a museum specimen only indicates that the species was present at the time of collection. The absence of a species cannot be used to conclude that the species did not occur at the time the collecting was underway. Genetic information available from museum specimens is also being used to reconstruct species' ranges and to identify places where they were extirpated. Morrison (2009) discussed other limitations and cautions associated with museum records.

The Fossil and Subfossil Record

Fossils have been used to reconstruct the former ranges of species. A good example comes from Harris (1993), who reconstructed the succession of microtene rodents from the middle to late Wisconsin period of the Pleistocene in New Mexico. Subfossils, which are unmineralized remains, can also be used to reconstruct more recent environments. Subfossils are found in caves and rocky crevasses, mines, and woodrat (*Neotoma* spp.) middens. Analyzing subfossils requires knowledge of vertebrate morphology, although only a good undergraduate training in wildlife science or zoology is usually needed. Restorationists can use these types of information to reconstruct the animals, plants, and successional processes that were occurring in the past (Kay 1998). In Madagascar, Pedrono et al. (2013) proposed using captive Aldabran giant tortoises (*Aldabrachelys gigantean*) to restore missing ecological functions that were previously in place because of extinct giant tortoises. Processes they were hoping to replicate included dispersing large seeds, keeping the understory open, cycling nutrients, and indirectly regulating fire regimes.

Literature

Determining the likely environmental changes that occurred through time requires a temporal baseline against which subsequent records can be compared and evaluated. Our baseline of information is improving through time and has accelerated as we moved through the 20th and into the 21st century. For example, Power (1994) published an example of assembling historical avifaunal records using documents beginning in the 1850s (e.g., from the US Pacific Railroad Survey) and continuing through the early 1900s to reconstruct the distribution and abundance of birds of the coastal islands of California. The National Audubon Society has been publishing a compilation of bird observations submitted by the public since the early 1900s. The field journals and other written records of scientists and amateurs alike are housed in their original form in museums, which can help reconstruct the animals present in the region surveyed.

Uncertainty of Observations

Because historical data records will always be incomplete, we must assess the quality of our historical reconstruction by assigning probabilities of certainty to each data source. Such assignments are relative and qualitative. Factors to consider in assigning uncertainty include age of the data source, recorded distance of the source relative to the project site, the number of records available, presence of an actual specimen versus only a visual observation, and (if known) the reputation of the data source. For each restoration plan, the uncertainty of each conclusion and all assumptions must be clearly stated.

Desired Conditions

Conceptual Model

As developed above, a conceptual model of ecosystem processes, functions, filters, and assemblages of species is the initial step for designing a restoration project. This model is a process of feedback between the desired ecological condition and what is ultimately feasible to implement and maintain given current environmental, budgetary, legal, and political constraints. Likewise, the population structure and movements of the target animal species must be incorporated into the plan. As noted by Morrison (2009), there is no reason to restore habitat for a specific suite of animal species if immigration and emigration are impossible, or if predators or competitors cannot be managed either within the project area or on surrounding lands. For example, if fire is an essential component in maintaining the condition of the vegetation, then either the frequency and intensity of fires must be incorporated into the restoration plan, or alternatives to fire must be developed (e.g., cutting trees and thinning shrubs). The model provides a simple visual representation of the ecological processes necessary to maintain the desired condition.

The success of a project will be based largely on distinguishing between habitat and niche factors, and their relevance to the viability of animals. Because an animal may be absent because the niche factors are inappropriate, models of the presence or viability of animals based on habitat factors often result in poor predictions. Failure to account for niche factors in restoration planning often results in poor success for wildlife diversity and viability. Although evaluation of niche relationships can add substantial time and effort to the planning process, such work substantially improves the efficacy of the final restoration plan and influences anticipated post-restoration management activities. In the previous example of the linkage between wolves, elk, songbirds, beavers, and riparian vegetation (Hebblewhite et al. 2005), failure to incorporate control of elk in the project would likely result in poor plant regeneration and negative responses by the bird assemblage and beavers.

Focal Species

Wildlife management has concentrated time and funds on a restricted set of species, including those regularly hunted and those deemed rare. Many people believe that it is too complicated to address the full potential of the ecological community and that selected species, primarily those of legal concern or consumptive interest, should be the focus of our efforts. Much effort has been expended attempting to simplify how we study and manage species, including a focus on a select list of animal species, which is aptly called a focal species approach.

Lambeck (1997) developed a focal species approach whereby a group of species was selected as a focus of management. These species were the most influenced by specific threatening processes, were area sensitive, and resource and dispersal limited. Related concepts such as indicator, umbrella, and flagship species have generally failed to provide for effective management of multiple species. For example, Lindenmayer et al. (2002) and Nicholson et al. (2013) reviewed and criticized the focal species approach in part on the grounds that it incorporated such troublesome concepts as indicator species. Lambeck (2002) countered Lindenmayer et al. (2002) and

noted that the focal species approach is multifaceted in that it can incorporate multiple species with many different resource requirements and responses to different threatening factors.

Evaluations of the focal species approach have generally concluded that the approach can be a useful starting point for conservation but with major weaknesses, including the fact that the focal species usually do not serve as surrogates for other species, the selection of species often shows high social bias (i.e., large predators), we do not know enough about every species to correctly choose focal species, and empirical testing of the response of species to management actions is minimal (Lindenmayer et al. 2002, Freudenberger and Brooker 2004, Nicholson et al. 2013).

In contrast to the focal species approach, many management agencies have used a coarse filter approach, which entails managing generally defined habitat conditions such that a majority of the associated animal community will be supported (Hunter 1991). Morrison et al. (2006) noted, however, that the coarse filter approach usually fails to protect many native species, because the niche requirements of many species are not usually met by this approach.

The most productive way to design a restoration plan involves developing a conceptual ecological framework that will provide suitable general conditions for many species, along with more specific plans for a set of focal species. There will be a multitude of project-specific reasons for selecting species. What we want to avoid are the extremes; namely, only relying on restoration of general conditions in the hope that the species we desire are supported, versus trying to use only a few species as surrogates of how other species will respond to threatening process and management action. An important aspect of a focal species approach is to document the rationale for selecting species.

Project Success

The Society for Ecological Restoration (SER) has produced and regularly updates recommended practices for all aspects of a restoration project, known as the standards (Gann et al. 2019). The standards established eight principles that underpin ecological restoration, including evaluation of post-project monitoring and modification as needed. The standards emphasize that monitoring should be developed around specific targets and measurable goals and objectives that are identified at the start of the project. They also emphasize that once measures of success are identified, baseline data are collected and milestones determined to gauge whether the rate of progress is on track. They also highlight the role of trigger points that cause corrective actions may be initiated.

Zedler (2007) emphasized the need to develop specific and quantifiable measures of the outcomes, or success, of restoration projects. As reviewed by Wortley et al. (2013), incorporation of measures of success (relative to project goals) has been rapidly increasing as the field of restoration evolves. Most of these evaluations, however, remain focused on measures of vegetation and various other site conditions; little evaluation of animals is being incorporated in most studies.

Implementation Steps

Next, I outline specific steps to implement wildlife restoration that are structured in a spatially explicit context (i.e., from broad spatial extent to local-scale applications; Morrison 2009). There are three major steps.

Step 1: Planning Area

The essential first step in developing a restoration plan is determining the desired ecological condition of the largest planning area (e.g., a valley composed of multiple creeks). Surrounding areas will influence each management unit (e.g., creek) within the overall planning area. Priority steps to conduct include: develop a conceptual ecosystem model of the planning area that identifies major factors that will lead to desired ecological condition, identify key ecological attributes that influence the ecosystem model, identify the suite of species characteristic of the desired condition, identify focal species, and identify the primary constraints and stressors that inhibit proper functioning of the pathways in the ecological model (and thus prevent attainment of the desired condition). By listing constraints and stressors, one quickly identifies those that can be alleviated and those that cannot. Once these limitations have been identified, one can reevaluate the desired condition and associated conceptual model in light of what is possible. Lastly, a monitoring plan that uses the key ecological attributes at the overall project scale must be developed, including establishing quantitative goals, a time frame for attainment, and thresholds for additional action.

Step 2: Project Area

The restoration plan for each project area (e.g., creek) should follow from the desired ecological condition for the overall planning area (e.g., valley). Each project area's current condition allows an investigator to determine how to allocate target levels of each ecosystem component across the specific project area(s) based on the desired condition for the entire planning area. Certain management actions might not be feasible to include because of the location (e.g., cowbird control adjacent to a residential area). Specific steps include: describe the current vegetation and other environmental features, identify the location and amount of special features (e.g., springs, old trees, caves), identify constraints and stressors that occur spatially and temporally, compare special features relative to the larger planning area, develop species list, identify focal species and evaluate the potential of the site to maintain each species and the constraints on maintenance, develop a management plan for all special features and focal species, and develop the preliminary restoration and monitoring plan. Making the best possible effort to identify the location and population boundaries of the focal species—a process that can begin during Step 1—should also be conducted. Understanding whether your project area overlaps the core or the edge of a population should be a central component of project planning. Being at the edge of a population boundary often indicates that the environmental conditions currently present (i.e., before project implementation) are not optimal for continued presence.

Step 3: Adaptive Management Implementation

Implementation and management of a restoration plan should follow the tenets of adaptive management (see Chapter 5 for details). Adaptive management requires the specification, during development of the plan, of the potential actions that could be taken if monitoring thresholds are triggered. The plan specifies values for key variables that must be attained by a certain time following restoration. For example, say that a key project goal is to develop 60% cover of riparian vegetation more than 2 m in height within six years of planting. The plan could incorporate a target (threshold) goal for growth of riparian vegetation at three years post-restoration of 40% cover of more than 1 m height, a threshold that was based on knowledge of riparian vegetation growth rate. Failure to attain this threshold by year three would trigger a specific management action (which was clearly elucidated in the original plan). This process helps eliminate (or at least minimize) the trial-and-error approach followed in most projects. Such scenarios or pathways that the project area is likely to follow are an output of the initial ecosystem model upon which the plan was formulated.

Here we see the value in viewing restoration of one area in an overall context of a larger area, rather than as a series of individual, project-level endeavors, which is essential when dealing with populations structured as a metapopulation. To develop an adaptive management approach, one must develop specific management actions if a threshold is triggered at either the overall project or subproject area scale, implement additional management actions, modify vegetation structure, revise treatment schedules, revise the conceptual model on the basis of new information and changes in management actions, revise thresholds and triggers as indicated by revisions in actions, and continue monitoring at appropriate spatial and temporal scales. Chambers and Miller (2004) presented a succinct example of a hierarchical approach to restoration and management using Great Basin riparian systems. Morrison (2009:chap. 10) provided case studies that developed restoration plans for wildlife.

DESIGNING A WILDLIFE RESTORATION PROJECT

Habitat Heterogeneity

Restoration is perhaps the most integrative of endeavors in the conservation arena. A successful restoration project, regardless of its spatial extent (i.e., how big it is), must consider a plethora of issues that range well outside the physical boundaries of the actual project. Below I review briefly some of the major issues that must be considered in designing a restoration project, including the size and type of patch to be restored, connectivity between patches, fragmentation, and corridors. Broadly, these and other environmental features cause substantial heterogeneity in the habitat available to each animal species.

Habitat heterogeneity is defined as the amount of discontinuity in environmental conditions across a defined space for a particular species. Remember that because animal habitat is a species-specific concept, a particular combination of environmental features may constitute habitat for one species and a barrier for another. Discontinuities in environmental conditions, which lead to heterogeneity, occur naturally with changes in soil type or edges of water bodies, or anthropogenically with agricultural lands or roads. Each species, even those with rather similar general habitat requirements, can show wide differences in tolerance of habitat heterogeneity. After all, at some spatial scale, all locations are heterogeneous (patchy); what may be seen as homogeneous to a large mammal may be heterogeneous at the spatial scale of the amphibian.

Fragmentation

Much has been written, including in the popular press, about fragmentation. Fragmentation is defined as the extent of heterogeneity of (species-specific) habitats across a landscape, with the central qualification (see above) that landscape is viewed as a species-specific concept. The isolation (how far apart) and physical size of resource patches available are used to describe the degree of fragmentation and in turn its effect on specific species (Morrison 2009). Habitat fragmentation appears frequently in the ecological literature and in the past referred (incorrectly) to any type of heterogeneous condition. Similarly, many authors have called it landscape fragmentation, which is strictly incorrect because it is the resources (e.g., habitats for specific species) that become fragmented within landscapes, and not entire landscapes per se. Morrison (2009:119) provided the following truism for designing all restoration projects: "The species-specific concept of habitat selection, which occurs across spatial scales . . . must be the focus of restoration with regard to fragmentation (and the broader issues of habitat heterogeneity). Likewise, the concept of species assembly rules . . . is directly relevant here because fragment size, shape, and the quality of the habitat therein will serve as a filter determining in part if a species can exist in a target area." Morrison et al. (2020) go into more detail on these concepts in reference to the overall field of animal ecology.

Vastly different (and, for specific species, unsuitable) conditions can isolate and surround vegetation patches. Individual animals in these isolated patches can go extinct unless immigrants supplemented their populations. Partial isolation of habitats is also problematic, however, because lowering dispersal rates and thus smaller populations can affect population viability. Other issues involving fragmentation—such as temporal fragmentation and fragmentation caused by subtle discontinuities in environmental conditions within a fragment—are beyond the scope of this chapter but should be reviewed as part of any project (Morrison 2009:119–123).

In planning a restoration project, biologists should consider the following: (1) the absolute loss of habitat area; (2) increased edge; (3) increased distances or permeability for movement of animals between patches; (4) increased penetration of predators, competitors, and nest parasites; and (5) changes

in microclimate with changes in patch area and edge. The species-specific nature of animals' response to habitat heterogeneity can be seen in results of a study by Bolger et al. (1997), who studied the response of birds in a matrix of remaining chaparral vegetation in Southern California. They reported that six species showed negative responses, four species showed positive responses, and ten species showed no apparent response to area size. Morrison (2009) reviewed similar results from other studies, concluding that different aspects of fragmentation affect different species. For highly mobile species, total habitat area is probably more important than the area of single patches. In contrast, less mobile animals will be most affected by the size and proximity of patches.

Corridors

Movements of individuals between populations, especially when a metapopulation is indicated, enhance population viability. Connectivity refers to the extent to which individuals can move among a mosaic of habitat (Hilty et al. 2006). What have become known popularly as corridors are one means of achieving connectivity, which is often recommended when fragmentation of a landscape is involved. As a reminder, however, it is essential that the species-specific nature of fragmentation be incorporated into any planning that seeks to use a corridor (or any type of connection) as a remedy. Clearly, what works for a large, mobile species is likely to be inadequate for a small, less mobile species (e.g., Gilbert-Norton et al. 2010, Ayram et al. 2016).

Because species will show differing responses to patchiness, a restoration plan should include a categorization of species by the potential effects that footpaths, roads, fence lines, canals, structures, changes in vegetation structure, and other features might have on species of interest. Because of the numerous potential natural and human-placed barriers to animal movement, many conservation biologists advocated the retention of natural corridors or development of artificial corridors in an effort to maintain linkages between patches. Although incorporation of corridors into restoration plans has been frequent, their use is controversial.

As reviewed by Hess (1994), in certain situations corridors can increase the chance of metapopulation extinction by promoting the transmission of disease. The probability of disease and other adverse factors (e.g., parasites) should be considered during the planning stages of all reserve networks. The presence of diseases that are known to infect species likely to use corridors should be evaluated prior to linking reserves. Strategies for containing the disease should be developed prior to establishing the linkages. Contingencies for treating epizootics include vaccination, removal of infected individuals, and temporary severing of the linkage. For example, Simberloff and Cox (1987) and Simberloff et al. (1992) cautioned that too little empirical data were available to warrant wholesale adoption of corridors as a conservation action. In addition, corridors can enhance the spread of disease and fires, and they can increase exposure of individuals to predation, domestic animals, and poachers. Although corridors might have great potential, they must be evaluated and planned on a species-specific basis (Cushman et al. 2009). Unfortunately, many restoration plans incorporate corridors with no certitude that they work. If the plan is developed within the context of a true adaptive management plan, then alternatives will be available should corridors fail to provide the necessary linkages between habitat patches to maintain the desired species assemblages. Readers can consult Morrison et al. (2006) and Morrison (2009) for a more thorough review of the empirical evidence for animals dispersing through and otherwise using corridors.

There are many suggested ways to design corridors (Hilty et al. 2006:chap. 4). I describe several of the typical designs for corridors (Fig. 17.4), beginning with the classic depiction of a corridor (Fig. 17.4A) as a continuous passageway of some width that links two or more habitat areas. When a continuous

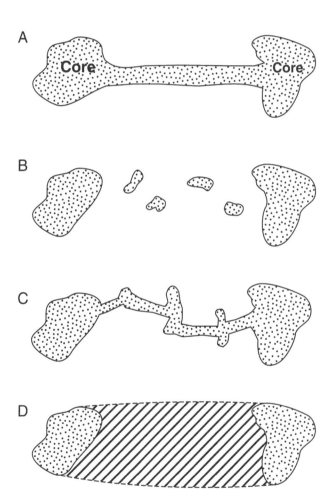

Figure 17.4. Schematic diagrams of general types of movement corridors. (A) The classic design with continuous linkage between core habitat areas. (B) The situation where the linkage is broken into patches of habitat, often called stepping-stones, between core areas. (C) The stepping-stones of Fig. 17.4B have been linked with a narrow but continuous pathway. (D) The core areas are connected by a continuous cover of relatively poor-quality habitat where animals can move, rest, and perhaps feed but not be sustained for prolonged periods. From Morrison (2009:fig. 7.3)

linkage is not possible or perhaps not needed, the stepping-stone concept of corridors (Fig. 17.4B) is available. These stepping-stones have been modified to include physical links between the steps (Fig. 17.4C), where the stepping-stone areas might function as a temporary location for resting and feeding. Lastly, I depict a situation where the corridor is actually a large area of lower-quality habitat relative to the core (and thus higher-quality habitat; Fig. 17.4D); individuals are able to move through but not actually breed or survive indefinitely in the corridor. Other combinations of patches and corridors are possible, such as locating the stepping-stones into the matrix of lower-quality habitat (the corridor). Smallwood (2001) suggested that the area surrounding a corridor be classified according to the functions they individually and collectively serve for specific animal species (e.g., on a scale of low to high). Hilty et al. (2006) provided an in-depth accounting of the factors that must be considered when determining whether corridors are a viable option within a restoration project; I highly recommend that readers review their work.

Restoration Design

The major principles involved with designing areas for the conservation of species have been developed in the context of reserve design and can be summarized as follows: (1) well-distributed species are less susceptible to extinction than species confined to small locations, (2) larger blocks will allow population persistence longer than small blocks, (3) blocks of habitat close together are better than blocks farther apart, (4) habitat in continuous blocks is better than fragmented habitat, (5) interconnected blocks of habitat are better than isolated blocks, and (6) populations that fluctuate are more vulnerable than populations that are stable (Noss et al. 1997).

Guidelines for habitat management were developed by Morrison et al. (2006) and Morrison et al. (2020) and summarized for restoration in Morrison (2009). Below I summarize these guidelines but encourage readers to pursue the literature in greater depth.

Guidelines for Species Richness and Overall Biodiversity

Although most restoration occurs at the within-patch scale, the planner needs to consider the larger spatial area within which the specific project area resides. Design guidelines should be divided into three spatial scales, and they can be based on how population dynamics are likely to vary as a function of habitat complexity at each scale: within habitat patches, or alpha diversity, such as with the observed correlations between foliar height diversity and bird species diversity; between habitat patches, or beta diversity; and among broader geographic areas, or gamma diversity.

This hierarchy links with the filtering concept that I developed above, where the species present in a location are a result of the limitations imposed by abiotic and biotic factors and other constraints. Moderate-disturbance regimes that cause variations across space in vegetation elements, substrates, and abiotic characteristics of the environment usually maintain beta diversity. Of course, beta diversity is reflected at different spatial scales based on body size, size of home range, and movement ability. In contrast, gamma diversity is usually controlled by climate, landform, geographic location, and broad-scale vegetation formations.

Guidelines to Maintain Within-Patch Conditions

The size and the topographic location, proximity of other patches, and the susceptibility to disturbances (e.g., floods, fire, human recreation) usually have a strong effect on within-patch conditions. Further, the context of the patch (conditions in adjacent patches); the type and intensity of natural disturbances; and the type, frequency, and intensity of management activities also affect within-patch conditions. Within-patch conditions must be developed and managed to target specific animal species if the outcome of restoration is to meet project goals.

Guidelines to Maintain a Desired Occupancy Rate of Habitat Patches

Specific animal species often come and go within different geographic locations; occupancy is often not constant. As discussed above, the size and condition of the patches, the proximity and size condition of adjacent patches, the dispersal ability of animals, and other factors determine the dynamics of animal populations within and between patches. Further, the location of patches can change through time because of catastrophic events such as fire, flood, disease, and other disturbances. In restoration planning, habitat patches can often be mapped and links between them identified. Habitat corridors can be retained or developed to provide connections between populations across a landscape.

Planners must consider the range of factors that potentially limit the occurrence of species in a planned restoration area. Such limiting factors and potential ways to mitigate (Table 17.2) these and other factors play a direct role in defining project goals, and in establishing specific management practices for a restoration project. Also, one must keep in mind that these general concepts must be tailored for the suite of species under consideration.

SUMMARY

This chapter reviewed how the principles of wildlife ecology and restoration ecology can be linked to results in comprehensive planning for the management and conservation of natural resources. Restoration plans should be guided by the likely responses of current or desired wildlife species in the project area, including the current and historical abundance and distribution of animal species, habitat and food requirements, breeding locations, how plant succession will change species composition through time, space requirements, and necessary links between geographic areas. There is a necessary connection between the practice of restoration (i.e., ecological restoration) and the concepts that were used to develop the restoration plan (i.e., restoration ecology). Restorationists must

Table 17.2. Potential factors limiting animal populations and potential means of mitigating those factors within a restoration planning context

Limiting Factor	Mitigation Measure
Disturbance caused by human activities	Control human access or timing of access (e.g., no entry into a sensitive area during breeding season)
	Establish buffers around key, sensitive areas (e.g., roosting areas)
Disease	Do not allow animals to concentrate in small areas
	Consider treatment of the environment or treatment of selected animals
Size of area	Functions as a factor limiting occupancy of certain species; consider possibility of linkages (corridors)
Seasonality	Consider availability of water, roosts, and other resources; such resources might need to be artificially established and maintained
Biotic factors, including predation and competition	Consider direct control of exotic, or in some cases native (e.g., cowbirds), animals

Source: Morrison (2009:142).

help quantify the likely consequences of human activities and natural events on the environment so that the public and decision makers can make informed choices when restoring an area. No project should be attempted unless it has a clear conceptual framework with rigor sufficient to allow for continual monitoring and feedback in an adaptive design; monitoring of the path of the restoration project relative to desired outcomes is essential. Restoration projects are part of an ongoing process that will form the building blocks of a strong discipline of wildlife restoration ecology. Although habitat (in the species-specific sense) provides general knowledge on animal requirements, habitat alone seldom identifies the underlying mechanisms determining occupancy, survival, and fecundity. That is, habitat alone provides a limited explanation of the ecology of an animal. Statistical models of habitat are based largely on surrogates of these mechanisms. If restoration is to successfully restore and maintain wildlife populations, then as a discipline it must take advantage of the best that other disciplines, such as wildlife science, have to offer.

Literature Cited

Alve, E., A. Lepland, J. Magnusson, and K. Backer-Owe. 2009. Monitoring strategies for re-establishment of ecological reference conditions: possibilities and limitations. Marine Pollution Bulletin 59:297–310.

Ayram, C. A. C., M. E. Mendoza, A. Etter, and D. R. P. Salicrup. 2016. Habitat connectivity in biodiversity conservation: a review of recent studies and applications. Progress in Physical Geography 40:7–37.

Berger, J., P. B. Stacey, L. Bellis, and M. P. Johnson. 2001. A mammalian predator–prey imbalance: grizzly bear and wolf extinction affect avian neotropical migrants. Ecological Applications 11:947–960.

Block, W. M., and L. A. Brennan. 1993. The habitat concept in ornithology: theory and applications. Current Ornithology 11:35–91.

Bolger, D. T., T. A. Scott, and J. R. Rotenberry. 1997. Breeding bird abundance in an urbanizing landscape in coastal Southern California. Conservation Biology 11:406–421.

Chambers, J. C., and J. R. Miller. 2004. Restoring and maintaining sustainable riparian ecosystems: the Great Basin ecosystem management project. Pages 1–23 in J. C. Chambers and J. R. Miller, editors. Great Basin riparian ecosystems: ecology, management, and restoration. Island Press, Washington, DC, USA.

Connor, E. F., and D. Simberloff. 1979. The assembly of species communities: chance or competition? Ecology 60:1132–1340.

Cushman, S. A., K. S. McKelvey, and M. K. Schwartz. 2009. Use of empirically derived source-destination models to map regional conservation corridors. Conservation Biology 23:368–376.

Daubenmire, R. 1968. Plant communities: a textbook of plant synecology. Harper and Row, New York, New York, USA.

Diamond, J. M. 1975. The assembly of species communities. Pages 342–444 in M. L. Cody and J. M. Diamond, editors. Ecology and evolution of communities. Harvard University Press, Cambridge, Massachusetts, USA.

Dickens, S. M., and K. N. Suding. 2013. Spanning the science-practice divide: why restoration scientists need to be more involved with practice. Ecological Restoration 31:134–140.

Egan, D., and E. A. Howell. 2001. Introduction. Pages 1–23 in D. Egan and E. A. Howell, editors. The historical ecology handbook. Island Press, Washington, DC, USA.

Fleishman, E., J. B. Dunham, D. D. Murphy, and P. F. Brussard. 2004. Explanation, prediction, and maintenance of native species richness and composition. Pages 232–260 in J. C. Chambers and J. R. Miller, editors. Great Basin riparian ecosystems: ecology, management, and restoration. Island Press, Washington, DC, USA.

Freudenberger, D., and L. Brooker. 2004. Development of the focal species approach for biodiversity conservation in the temperate agricultural zones of Australia. Biodiversity and Conservation 13:253–274.

Gann, G. D., T. McDonald, B. Walder, J. Aronson, C. R. Nelson, et al. 2019. International principles and standards for the practice of ecological restoration. 2nd edition. Restoration Ecology 27(S1):S1–S46.

Gilbert-Norton, L., R. Wilson, J. R. Stevens, and K. H. Beard. 2010. A meta-analytic review of corridor effectiveness. Conservation Biology 24:660–668.

Hall, L. S., P. R. Krausman, and M. L. Morrison. 1997. The habitat concept and a plea for standard terminology. Wildlife Society Bulletin 25:173–182.

Halle, S., and M. Fattorini. 2004. Advances in restoration ecology: insights from aquatic and terrestrial ecosystems. Pages 10–33 in V. M. Temperton, R. J. Hobbs, T. Nuttle, and S. Halle, editors. Assembly rules and restoration ecology: bridging the gap between theory and practice. Island Press, Washington, DC, USA.

Harris, A. H. 1993. Wisconsin and pre-pleniglacial biotic changes in southeastern New Mexico. Quaternary Research 40:127–133.

Hebblewhite, M., C. A. White, C. Nietvelt, J. M. McKenzie, T. E. Hurd, J. M. Fryxell, S. Bayley, and P. C. Paquet. 2005. Human activity mediates a trophic cascade caused by wolves. Ecology 86:2135–2144.

Hess, G. R. 1994. Conservation corridors and contagious disease: a cautionary note. Conservation Biology 8:256–262.

Hilty, J. A., W. Z. Lidicker Jr., and A. M. Merenlender. 2006. Corridor ecology: the science and practice of linking landscapes for biodiversity conservation. Island Press, Washington, DC, USA.

Hobbs, R. J., E. Higgs, C. M. Hall et al. 2014. Managing the whole landscape: historical, hybrid, and novel ecosystems. Frontiers in Ecology and the Environment 12:557–564.

Hobbs, R. J., and D. A. Norton. 1996. Towards a conceptual framework for restoration ecology. Restoration Ecology 4:93–110.

Hobbs, R. J., and D. A. Norton. 2004. Ecological filters, thresholds, and gradients in resistance to ecosystem reassembly. Pages 72–95 in V. M. Temperton, R. J. Hobbs, T. Nuttle, and S. Halle, editors. Assembly rules and restoration ecology: bridging the gap between theory and practice. Island Press, Washington, DC, USA.

Hodges, K. E. 2008. Defining the problem: terminology and progress in ecology. Frontiers in Ecology and the Environment 6:35–42.

Hunter, M. L. 1991. Coping with ignorance: the coarse-filter strategy for maintaining biodiversity. Pages 266–281 in K. A. Kohm, editor. Balancing on the brink of extinction. Island Press, Washington, DC, USA.

Huston, M. A. 2002. Introductory essay: critical issues for improving predictions. Pages 7–21 in J. M. Scott, P. J. Heglund, M. L. Morrison, J. B. Haufler, et al., editors. Predicting species occurrences: issues of accuracy and scale. Island Press, Washington, DC, USA.

Hutto, R. L. 1985. Habitat selection by nonbreeding, migratory land birds. Pages 455–476 in M. L. Cody, editor. Habitat selection in birds. Academic Press, Orlando, Florida, USA.

Johnson, D. H. 1980. The comparison of usage and availability measurements for evaluating resource preference. Ecology 61:65–71.

Kay, C. E. 1998. Are ecosystems structured from the top-down or bottom-up? A new look at an old debate. Wildlife Society Bulletin 26:484–498.

Keddy, P., and E. Weiher. 1999. Introduction: the scope and goals of research on assembly rules. Pages 1–20 in E. Weiher and P. Keddy, editors. Ecological assembly rules: perspectives, advances, and retreats. Cambridge University Press, Cambridge, UK.

Kirk, D. A., A. C. Park, A. C. Smith, B. J. Howes, B. K. Prouse, N. G. Kyssa, E. N. Fairhurst, and K. A. Prior. 2018. Our use, misuse, and abandonment of a concept: whither habitat? Ecology and Evolution 8:4197–4208.

Kostecke, R. M., S. G. Summers, G. H. Eckrich, and D. A. Cimprich. 2005. Effects of brown-headed cowbird (Molothrus ater) removal on black-capped vireo (Vireo atricapilla) nest success and population growth on Fort Hood, Texas. Ornithological Monographs 57:28–37.

Krausman, P. R., and M. L. Morrison. 2016. Another plea for standard terminology. Journal of Wildlife Management 80:1143–1144.

Kus, B. E. 1999. Impacts of brown-headed cowbird parasitism on productivity of the endangered Bell's vireo. Studies in Avian Biology 18:160–166.

Lambeck, R. J. 1997. Focal species: a multi-species umbrella for nature conservation. Conservation Biology 11:849–856.

Lambeck, R. J. 2002. Focal species and restoration ecology: response to Lindenmayer et al. Conservation Biology 16:549–551.

Lande, R., and G. F. Barrowclough. 1987. Effective population size, genetic variation, and their use in population management. Pages 87–123 in M. E. Soulé, editor. Viable populations. Cambridge University Press, Cambridge, UK.

Leopold, A. 1933. Game management. Charles Scribner's Sons, New York, New York, USA.

Levins, R. 1970. Extinction. Lectures on Mathematics in the Life Sciences 2:75–107.

Lindenmayer, D. B., and J. Fischer. 2006. Tackling the habitat fragmentation panchreston. Trends in Ecology and Evolution 22:127–132.

Lindenmayer, D. B., A. D. Manning, P. L. Smith, H. P. Possingham, J. Fischer, I. Oliver, and M. A. McCarthy. 2002. The focal-species approach and landscape restoration: a critique. Conservation Biology 16:338–345.

Litvaitis, J. A., K. Titus, and E. M. Anderson. 1994. Measuring vertebrate use of terrestrial habitats and foods. Pages 254–274 in T. A. Bookhout, editor. Research and management techniques for wildlife and habitats. 5th edition. The Wildlife Society, Bethesda, Maryland, USA.

Lockwood, J. L., and S. L. Pimm. 1999. When does restoration succeed? Pages 363–392 in E. Weiher and P. Keddy, editors. Ecological assembly rules: perspectives, advances, and retreats. Cambridge University Press, New York, New York, USA.

Miller, J. R. 2007. Habitat and landscape design: concepts, constraints and opportunities. Pages 81–95 in D. B. Lindenmayer and R. J. Hobbs, editors. Managing and designing landscapes for conservation: moving from perspectives to principles. Island Press, Washington, DC, USA.

Mills, L. S. 2013. Conservation of wildlife populations: demography, genetics, and management. 2nd edition. Wiley-Blackwell, Oxford, UK.

Millstein, R. L. 2010. The concepts of population and metapopulation in evolutionary biology and ecology. Pages 61–86 in M. Bell, D. Futuyma, W. Eanes, and J. Levinton, editors. Evolution since Darwin: the first 150 years. Sinauer Associates, Sunderland, Massachusetts, USA.

Morrison, M. L. 2009. Restoring wildlife: ecological concepts and practical applications. Island Press, Washington, DC, USA.

Morrison, M. L., L. A. Brennan, B. G. Marcot, W. M. Block, and K. S. McKelvey. 2020. Foundations for advancing animal ecology. Johns Hopkins University Press, Baltimore, Maryland, USA.

Morrison, M. L., and L. S. Hall. 2002. Standard terminology: toward a common language to advance ecological understanding and applications. Pages 43–52 in J. M. Scott, P. J. Heglund, M. L. Morrison, et al., editors. Predicting species occurrences: issues of accuracy and scale. Island Press, Washington, DC, USA.

Morrison, M. L., L. S. Hall, S. K. Robinson, S. I. Rothstein, D. C. Hahn, and T. D. Rich. 1999. Research and management of the brown-headed cowbird in western landscapes. Studies in Avian Biology 18:204–217.

Morrison, M. L., B. G. Marcot, and R. W. Mannan. 2006. Wildlife–habitat relationships: concepts and applications. 3rd edition. Island Press, Washington, DC, USA.

Morrison, M. L., T. A. Scott, and T. Tennant. 1994. Wildlife–habitat restoration in an urban park in Southern California. Restoration Ecology 2:17–30.

Nicholson, E., D. B. Lindenmayer, K. Frank, and H. P. Possingham. 2013. Testing the focal species approach to making conservation decisions for species persistence. Diversity and Distributions 19:530–540.

Noss, R. F. 1985. On characterizing presettlement vegetation: how and why. Natural Areas Journal 5:5–19.

Noss, R. F., M. A. O'Connell, and D. M. Murphy. 1997. The science of conservation planning: habitat conservation under the Endangered Species Act. Island Press, Washington, DC, USA.

O'Connor, R. J. 2002. The conceptual basis of species distribution modeling: time for a paradigm shift? Pages 25–33 in J. M. Scott, P. J. Heglund, M. L. Morrison, et al., editors. Predicting species occurrences: issues of accuracy and scale. Island Press, Washington, DC, USA.

Office of Technology and Assessment. 1993. Harmful non-indigenous species in the United States. OTA-F-565. 2 vols. US Congress, Washington, DC, USA.

Palmer, M. A., D. A. Falk, and J. B. Zedler. 2016. Ecological theory and restoration ecology. Pages 3–26 in D. A. Falk, M. A. Palmer, and J. B. Zedler, editors. Foundations of restoration ecology. 2nd edition. Island Press, Washington, DC, USA.

Pedrono, M., O. L. Griffiths, A. Clausen, L. L. Smith, C. J. Griffiths, L. Wilmé, and D. A. Burney. 2013. Using a surviving lineage of Madagascar's vanished megafauna for ecological restoration. Biological Conservation 159:501–506.

Peters, R. H. 1991. A critique for ecology. Cambridge University Press, Cambridge, UK.

Power, D. M. 1994. Avifaunal change on California's coastal islands. Studies in Avian Biology 15:75–90.

Robbins, C. S., D. Bystrak, and P. H. Geissler. 1986. The Breeding Bird Survey: its first fifteen years, 1965–1979. Research Publication 157. US Fish and Wildlife Service, Washington, DC, USA.

Rodewald, A. D. 2015. Demographic consequences of habitat. Pages 19–33 in M. L. Morrison and H. A. Mathewson, editors. Wildlife habitat and conservation: concepts, challenges, and solutions. Johns Hopkins University Press, Baltimore, Maryland, USA.

Simberloff, D., and J. Cox. 1987. Consequences and costs of conservation corridors. Conservation Biology 1:63–71.

Simberloff, D. S., J. A. Farr, J. Cox, and D. W. Mehlman. 1992. Movement corridors: conservation bargains or poor investments. Conservation Biology 6:493–504.

Smallwood, K. S. 2001. Linking habitat restoration to meaningful units of animal demography. Restoration Ecology 9:253–261.

Temperton, V. M., and R. J. Hobbs. 2004. The search for ecological assembly rules and its relevance to restoration ecology. Pages 34–54 in V. M. Temperton, R. J. Hobbs, T. Nuttle, and S. Halle, editors. Assembly rules and restoration ecology: bridging the gap between theory and practice. Island Press, Washington, DC, USA.

Temperton, V. M., R. J. Hobbs, T. Nuttle, and S. Halle, editors. 2004. Assembly rules and restoration ecology: bridging the gap between theory and practice. Island Press, Washington, DC, USA.

Van Andel, J., and A. P. Grootjans. 2006. Concepts in restoration ecology. Pages 16–28 in J. van Andel and J. Aronson, editors. Restoration ecology. Blackwell, Oxford, UK.

Van Horne, B. 1983. Density as a misleading indicator of habitat quality. Journal of Wildlife Management 47:893–901.

Weiher, E., and P. Keddy, editors. 1999. Ecological assembly rules: perspectives, advances, retreats. Cambridge University Press, Cambridge, UK.

Wiens, J. A. 1989. The ecology of bird communities. Volume 1: Foundations and patterns. Cambridge University Press, Cambridge, UK.

Willis, K. J., and H. J. B. Birks. 2006. What is natural? The need for a long-term perspective in biodiversity conservation. Science 314:1261–1265.

Wortley, L., J.-M. Hero, and M. Howes. 2013. Evaluating ecological restoration success: a review of the literature. Restoration Ecology 21:537–543.

Zedler, J. B. 2007. Success: an unclear, subjective descriptor of restoration outcomes. Ecological Restoration 3:162–168.

CLIMATE CHANGE AND WILDLIFE

CHRISTOPHER L. HOVING AND WILLIAM F. PORTER

INTRODUCTION

Climate change is the defining challenge of wildlife conservation for our generation (Bellard et al. 2012, LeDee et al. 2021). The effects of climate change on the wildlife resources that you manage in the future will be profound. Climate change will create opportunities, and it will amplify other threats, like invasive species or habitat fragmentation. Common and valued species will become rare, some rare species will go extinct (Wiens 2016), and some new conservation opportunities will present themselves to those who are ready to act on them (Garcia et al. 2014, Sandifer et al. 2015). Climate change is often portrayed in popular culture as a hopeless crisis. It is certainly a crisis that threatens many of the ecosystem services on which human society depends, but it is not hopeless. During times of crisis, choices of individual conservationists and their professional communities have greater influence. The choices that you make matter; the choices you make in your career may even set the stage for wildlife conservation for decades or centuries to come. You are entering the field at a difficult and pivotal time, a time when your knowledge of the ways the climate is changing and how wildlife respond will determine your effect on what is conserved for future generations.

At one level, climate is simply a statistical description of weather patterns. From a wildlife perspective, climate change is more than just a significant long-term shift of weather patterns, such as averages or variation of temperature or precipitation. Climate is one of the central abiotic components of ecosystems that affects the distribution and abundance of wildlife, and any change in this influence will affect the viability of different wildlife management regimes. Thus wildlife biologists must understand how climate is changing if they hope to be successful in their management programs.

Climate change is not a new phenomenon. As an inherently statistical description of weather, climate can be described across a wide range of scales in space or time. The earth's climate has changed constantly during its 4.5 billion-year history, and each global climatic change has resulted in significant changes in the abundance and distribution of wildlife populations. The climate change that we are currently experiencing is distinguished from past events by at least two factors: the rate and magnitude of change are greater than has been experienced since the advent of civilization, and most of the change is attributable to human activities (International Panel on Climate Change [IPCC] 2014).

We tend to think of climate change as a simple warming of temperatures, which is causing a shift in the distribution of species poleward or higher in elevation. The reality is a more complex suite of changes in global-scale processes that lead to profound effects on ecosystems. As such, climate change has potential to alter the species composition and functioning of ecosystems, affecting wildlife populations worldwide. This potential effect is important to wildlife biologists because much of the management of wildlife habitats and populations is based on existing communities of species and a biogeographic distribution of ecosystems that has been in place for thousands of years. As the climate changes, each species will adapt in different ways, leading to a wholesale reorganization of communities. Wildlife management will need to adapt as well.

The intent of this chapter is to examine the causes of climate change and to address three central questions. What are the biological implications of climate change to wildlife populations? What will this mean to the functioning of ecosystems? What does this mean for the practice of wildlife management?

CAUSES OF CLIMATE CHANGE

In this section we explore the causes of recent climate change. Increases in carbon dioxide (CO_2) and a handful of other greenhouse gases (GHGs) resulting from use of fossil fuels, land cover change, and agricultural practices are now the principal influences of warming in the atmosphere. The increased atmospheric concentrations of four long-lived GHGs have been reported to be the main influence of the recent climate change (IPCC 2014) because they absorb and reemit the outgoing infrared radiation (heat), while at the same time they are transparent to the incoming solar radiation. Those GHGs are CO_2, methane (CH_4), nitrous oxide (N_2O), and halocarbons.

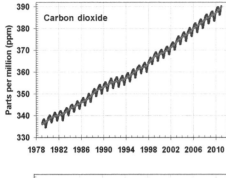

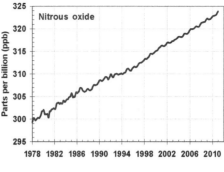

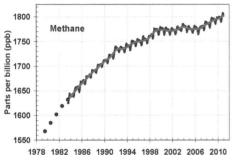

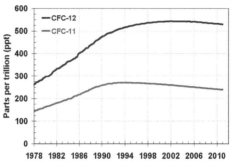

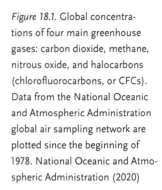

Figure 18.1. Global concentrations of four main greenhouse gases: carbon dioxide, methane, nitrous oxide, and halocarbons (chlorofluorocarbons, or CFCs). Data from the National Oceanic and Atmospheric Administration global air sampling network are plotted since the beginning of 1978. National Oceanic and Atmospheric Administration (2020)

Despite common misconceptions, GHGs do not warm the air directly. Instead, they modify the absorption, scattering, and emission of radiation within the atmosphere, which warms the surfaces of the planet, both land and ocean; the surfaces then transfer some of that heat energy back to the atmosphere (IPCC 2014). This transfer is fast for land and very slow for water, which is causing disproportionately fast heating of the Northern Hemisphere, which has more land area than the Southern Hemisphere.

Burning fossil fuels and land-use changes are the main causes of recent global increases in concentrations of GHGs (IPCC 2014; Figure 18.1). As their name implies, fossil fuels are fossilized remains of dead organisms. Coal is a good example. Because coal is composed of organic matter, it contains high amounts of carbon and some nitrogen, oxygen, and hydrogen. When burned, carbon reacts with oxygen to form CO_2. Two other GHGs, CH_4 and N_2O, are also released while burning fossil fuels. Changes in land cover further contribute to increases in GHGs concentrations. In particular, significant amounts of carbon are released by forest clearing and burning. Agricultural practices are also responsible for increases in GHGs, because CH_4 is a by-product of the fermentative digestion by ruminant livestock (particularly sheep and cattle) and is released from stored manures and inundated rice paddies (Mosier et al. 1998). Soils and manures release N_2O through the microbial transformation of nitrogen (Smith and Conen 2004, Oenema et al. 2005).

Among the GHGs, carbon dioxide has garnered the most attention. The global atmospheric concentration values of CO_2 have increased to 407.4 ± 0.1 ppm in 2019 compared with preindustrial levels of 278 ± 1.2 ppm. Methane concentrations increased almost 2.5-fold since preindustrial times, when CH_4

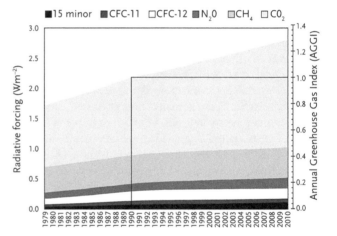

Figure 18.2. The radiative forcing due to increases in the four main greenhouse gases: carbon dioxide (CO_2), methane (CH_4), nitrous oxide (N_2O), and halocarbons (chlorofluorocarbons, or CFCs). Radiative forcing is a measure of the importance of the factor as a potential climate change mechanism. National Oceanic and Atmospheric Administration (2020)

concentration values were 715 ± 15 ppb, compared to concentrations of $1,857.7 \pm 0.8$ ppb in 2018. Nitrous oxide continues to grow at a relatively uniform growth rate; its global concentration increased from preindustrial values of about 270 to 330.9 ± 0.1 ppb in 2018 (Dlugokencky et al. 2019).

Many halocarbons (i.e., mostly man-made chemical compounds in which carbon atoms are linked with halogens such as fluorine, chlorine, bromine, iodine) have also increased from near-zero preindustrial background concentrations. Most of this increase is due to industrial production. Halocarbon concentrations peaked in the late 1990s (Fig. 18.2; National Oceanic and Atmospheric Administration 2020) and are now declining as a response to decreased emissions. These reduced emissions are a product of an international treaty known as the Montreal Protocol, which was signed in 1987. The Montreal Protocol is often cited as an example of a successful attempt to reduce global emissions of harmful gases (Montzka et al. 2011). Unfortunately, the steady rise in CO_2 since 1987 has more than offset the effect of decreasing halocarbons.

Observed Changes in Climate

Climate change is not a uniform effect around the globe. Different parts of the world are warming in different ways, and the effects of climate change are different at local, regional, and global scales. In this section we discuss the environmental consequences of climate change, including rising average temperatures, sea level rise, melting snow and ice cover, and extreme events such as flood, droughts, and heat waves. The principal message of this section is that changes in the abiotic facets of ecosystems have already been observed, and those changes are greater in the Northern Hemisphere than in the Southern Hemisphere.

Temperature

The rate of global temperature increase since 1950 is much faster now than at any other time in the thousands of years for which we can reconstruct decade-scale or finer temperature trends (IPCC 2014). The 100-year linear trend of global temperature increase was 0.74°C/decade from 1906 to 2005. Furthermore, the linear warming trend over the second half of this period (1956–2005) was 0.13°C/decade—nearly twice the rate for the entire 100 years (Fig. 18.3; National Aeronautics and Space Administration 2020). Projected increases in global average temperature depend on when and how fast society shifts energy production away from sources that produce excess pollution of greenhouse gases (IPCC 2014) and

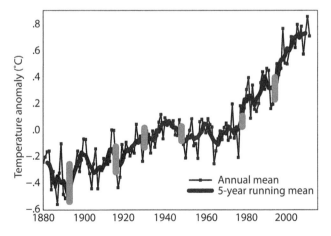

Figure 18.3. Global surface warming for the years 2000–2100, as predicted by different emissions scenarios. National Oceanic and Atmospheric Administration (2020)

the timing of land management (e.g., deforestation, peatland burning) that affects greenhouse gas emissions (Shukla et al. 2019). Many negative effects to essential ecosystem services will occur if global average temperatures warm >1.5°C–2°C (Hoegh-Guldberg et al. 2018).

Despite being a global pattern, the magnitude of temperature increase varies regionally. Higher northern latitudes have been experiencing greater temperature increases. In the Arctic region, temperatures have risen almost twice as quickly over the past 100 years as the global temperatures. The Northern Hemisphere is warming faster than the Southern Hemisphere because it has more land area; land warms faster than ocean (Folland et al. 2002, Sutton et al. 2007, Dong et al. 2009).

Sea Level

Global average sea level increased at an average annual rate of 1.7 mm between 1901 and 2010. The rate of increase was almost twice as high for the decade between 1993 and 2010 and averaged 3.2 mm/yr (Church et al. 2013). The forecasts of the IPCC suggest that the annual trend of 1.7 mm is projected to produce average rise in sea level between 0.26–0.55 and 0.52–0.98 m, depending on emission scenarios, by the end of the century (Church et al. 2013). Regional sea level can be significantly higher than global averages by an order of magnitude because of anthropogenic subsidence, such as that caused by unsustainable groundwater withdrawals. Sea level rise is causing inundation of coastal areas, flooding, erosion, saltwater intrusion, rising water tables and impeded drainage, and habitat loss (Meredith et al. 2019). Sea level rise research has only recently begun to account for increases in the rate of ice flow from the collapse of the marine-based sectors of the Antarctic ice sheet. Therefore the actual sea level rise could exceed the projections, but probably only by several tenths of a meter over this century (Church et al. 2013).

There are several factors that contribute to the sea level rise. Based on observations from a variety of sources, as summarized in Meredith et al. (2019), contributions to sea level rise from 2006 to 2015 were caused primarily by melting ice in Antarctica (13%), Greenland (24%), and other land-based glaciers (19%). Thermal expansion (i.e., the physical increase in volume of water as it warms) of oceans accounts for the remaining 44% of recent sea level rise. Melting sea ice in the Arctic does not contribute to global mean sea level rise when it melts because it is already part of the oceans' mass (Meredith et al. 2019).

Snow Cover, Sea Ice, and Land Ice

The global cryosphere, the part of the planet that is frozen, is changing rapidly relative to changes in the atmosphere (Comiso and Hall 2014, Meredith et al. 2019). Snow cover is changing is complex ways; increased precipitation is partially offset by increased melting as air temperatures increase. In most regions, snow duration is decreasing, although there is significant regional and interannual variation (Brown and Mote 2009). In the Arctic, the minimum sea ice extent has been falling since satellite records became available in the late 1970s (Comiso and Hall 2014). This dynamic is causing high-latitude northern regions to warm faster than the global mean. In the most recent global climate models, the Arctic first becomes ice-free (<1,000 km² of ice) before 2050 under every emissions scenario (Notz et al. 2020).

Global snow cover and arctic sea ice are not the only ice sources threatened by climate change. Glaciers in the tropics and midlatitudes are retreating, and their small size makes them even more vulnerable to climate change. Glaciers outside the polar regions have lost 490 ± 100 kg/m²/yr over 2006–2015, an increase of 30% over 1986-2005 (Hock et al. 2019, Zemp et al. 2019). In the tropics, the Bolivian Chacaltaya Glacier disappeared in 2009, and similar losses are evident in Indonesia and Kenya (Klein and Kincaid 2006, Rostom and Hastenrath 2007). The situation is similar in midlatitudes. The glaciers of the Alps have lost nearly 50% of their surface area and mass since 1850 (Zemp et al. 2006). Nearly 80% of glaciers in Montana's Glacier National Park have retreated since 1850, and predictions indicate that the remaining glaciers in the park will vanish by 2030 (Hall and Fagre 2003, Pederson et al. 2010).

Permafrost is permanently frozen land that is underground. It is a particularly extensive ecosystem, and the Northern Hemisphere has lost 7% to 15% of its total permafrost extent since 1900 (Lemke et al. 2007). Loss of permafrost can destabilize forests, allowing forests to replace open tundra. In addition, melting permafrost releases GHG (e.g., methane) that has been trapped for thousands of years. Because growing forests absorb GHG, the net effect of permafrost melting on future climate change is uncertain and an area of active research (Vincent et al. 2017).

Extreme Events: Heat Waves, Precipitation, and Drought

As climate change intensifies, humans are already experiencing an increase in the number of extreme events such as heat waves, floods, and droughts (Herring et al. 2020). Cold days, cold nights, and frosts have already become less frequent, while hot days and hot nights have become more frequent (IPCC 2014). Heat waves associated with increasing temperatures have become more frequent over most land areas since the 1970s (IPCC 2014).

Increased water vapor and atmospheric moisture have caused higher numbers of precipitation events and increased precipitation intensity, leading to increases in flooding risk in many regions. The risk of flooding is not increasing homogenously around the globe. Regions experiencing increases in the number and intensity of precipitation events are the eastern parts of North and South America, northern Europe, and north-central Asia. Conversely, some parts of the globe have experienced more intense and longer droughts. Regions with increased droughts included the Sahel, the Mediterranean, southern Africa, and parts of southern Asia. At a global scale, precipitation is shifting away from the equator, toward the poles, and such that wet areas become wetter and dry areas become drier (Seager et al. 2010).

WHAT ARE THE EFFECTS OF CLIMATE CHANGE ON WILDLIFE POPULATIONS?

The changes in climatic patterns will have far-reaching consequences for wildlife, with effects already reported for most taxonomic groups (Root et al. 2003, Parmesan 2006, Scheffers et al. 2016). In the following sections, we discuss the observed and predicted effects of climate warming on wildlife. The response of species to long-term environmental changes such as climate change will be determined by their ability (or failure) to adapt (Beever et al. 2016). Adaption will occur through behavior (Beever et al. 2017), phenotypic plasticity (Dunn and Winkler 2010), or rapid evolution (Jezkova and Wiens 2016). Failures of wildlife to adapt, even to the relatively modest climate change experienced to date, have already been documented at the species and local population scales (Wiens 2016).

Changes over Time: Evolutionary Consequences

Beyond the most visible aspects of how wildlife can adapt to climate change, many species and populations could adapt over multiple generations and broader periods of time. Shifts in migration and ranges are often attributed to phenotypic plasticity (i.e., the ability of individuals to modify their behavior, morphology, or physiology to altered environmental conditions; Walther et al. 2002). Many questions remain as to whether these changes have a genetic component that leads to evolutionary consequences for species and populations (Bradshaw and Holzapfel 2006). Generally, we have a limited understanding of the evolutionary responses of wildlife populations to climate change when compared to changes in distribution and phenology, but many of these life history traits (e.g., breeding phenology, dormancy) have a strong genetic component that is climate related.

Past climate cycles have left a strong imprint on the historical structure and patterns of plant and animal communities (Hewitt 1996), and they can exert strong influences on biodiversity and species endemism. For example, by estimating climate change velocity (the geographic rate of climate displacement) since the last glacial period, Sandel et al. (2011) reported that regions with high climate velocities are also characterized by absences of range-restricted amphibians, mammals, and birds. The association between endemism and climate change velocity was weakest in highly mobile birds and strongest in weakly dispersing amphibians. High climate change velocity was also associated with low endemism at regional scales. These results emphasize that areas experiencing relatively low rates of climate change, such as north-south-oriented mountain ranges, will be essential refuges for maintaining biodiversity over broad periods.

Wildlife managers may rightly question the extent to which knowledge of climate-influenced geological events might extrapolate to the current management of wildlife populations. Increasing evidence suggests that life history traits such as breeding phenology, clutch size, and body size are sensitive to long-term trends in climate (Sheldon 2010). Migratory behavior is an excellent example of a trait that has strong heritability and might be particularly sensitive to climate change (Pulido et al. 1996). In a study of blackcaps (*Sylvia atricapilla*), Pulido and Berthold (2010) used a common garden and captive breeding experiment to demonstrate a genetic reduction in migratory activity and evolutionary change in phenotypic plasticity of migration onset. They reported that residency will evolve rapidly in completely migratory bird populations if selection for shorter migration distance persists. These findings suggest that, over time, climate change might favor genotypes that can winter closer to the breeding grounds.

Changes over Time: Phenological Responses

Here we examine the evidence for phenological changes in wildlife populations that result from climate warming: alterations of migration patterns, changes in the timing of breeding and reproductive success, and changes in hibernation patterns. Substantial evidence exists that seasonality of many temperate regions has been altered already (Hansen et al. 2010), leading to changes in phenology of plants and animals. Complicating the problem, the rate and the direction of the phenological change will vary across species or even within populations of the same species (Both 2010). The best evidence that climate warming has already affected wildlife populations comes from long-term studies of butterflies (Parmesan 1996), birds (Thomas and Lennon 1999), and a few examples for mammals (Moyes et al. 2011).

Changes in Migration Patterns

BIRDS

How do we expect birds to respond? Birds may advance their arrival date to breeding sites (either by departing earlier from wintering grounds or by shortening the migration distance) or delay the onset of their fall migration (Lehikoinen and Sparks 2010). Short-distance migrants are more likely to adjust the timing of migration because they rely more on weather conditions, whereas long-distance migrants are cued by photoperiod, thereby being less likely to attune the timing of their migration to increasing temperatures. For this reason, long-distance migrants are expected to suffer higher population losses because they are unable to track the changing climate as closely as short-distance migrants or resident species.

There is mounting evidence that birds have advanced the time of arrival to breeding grounds (Lehikoinen and Sparks 2010, DeLeon et al. 2011, Hurlbert and Liang 2012). Lehikoinen and Sparks (2010) compiled more than 3,800 data series spanning 455 species and 19 countries. Their analysis revealed that 82% of trends in first arrival dates and 76% of trends in mean arrival dates were toward earliness. The recorded advancement was on average 2.8 and 1.8 days/decade for the first and mean dates of arrival, respectively. Hurlbert and Liang (2012) reported shifts in the arrival dates for 18 North American species of 0.8 days earlier for every 1°C of warming of spring temperature. Although these changes are small, when combined with IPCC predictions for the coming decades, climate change will have a profound effect on migratory species.

From a management perspective, the amount of time that a species spends in wintering habitat is important. Migratory waterfowl in eastern North America are triggered to migrate south in the fall by combinations of snowfall and minimum temperatures. These climate triggers are the same across many species of migratory waterfowl, but the thresholds that trigger migration vary from species to species. By the late 21st century, winter severity thresholds are predicted to significantly delay the timing and intensity of migration of waterfowl, which could significantly affect wetlands at the north and south ends of migratory pathways (Notaro et al. 2014).

Where species winter is also important to managers. An increasing number of European long-distance migrants, formerly wintering entirely in tropical and southern Africa, are now wintering in the Mediterranean region (Berthold 2001). Similarly, approximately 33% of short-distance or partial migrants breeding in Germany favor wintering at higher latitudes, while only 10% choose to winter in more southerly latitudes (Fiedler et al. 2004). Some migrants such as Canada goose (*Branta canadensis*) in North America and white stork (*Ciconia ciconia*) in Spain have ceased their migration and stay year-round in breeding or stopover sites (Ferrer et al. 2008).

MAMMALS

Can we expect migratory mammals to respond to climate change in the same way as birds? The evidence suggests that we can. As an example, snow depth triggers seasonal migration of white-tailed deer (*Odocoileus virginianus*) in northern latitudes, and in the absence of snow, deer remain resident on summer range throughout the year (Weiskopf et al. 2019). Moose (*Alces alces*) will start migrating to their winter ranges earlier and start leaving their winter grounds later as future snow conditions are affected by climate change (Ball et al. 1999). Sharma et al. (2009) expected that delays in or incomplete formation of ice in the Arctic will alter the timing of migration of migratory caribou (*Rangifer tarandus*). To ensure safe passage over large water bodies, migratory caribou will either have to advance the spring migration and delay the fall migration or shift their ranges farther north (Sharma et al. 2009).

Changes in Timing of Breeding and Reproductive Success

Changing the timing of breeding is yet another example of phenotypic plasticity that results from climate change. The phenology of reproduction has been widely studied, and evidence for climate warming–induced changes in the timing of breeding is unequivocal (Parmesan 2007). For example, a literature review by Root et al. (2003) revealed that more than 80% of species experienced significant shifts of the phenological events resulting from climate change. The consequences of phenological mismatch include altered trophic interactions.

What is the evidence for climate warming–mediated changes in breeding and reproductive success? A summary of change in laying dates of 68 species of birds from long-term studies reveals that 59% have significantly advanced their laying date, while 79% advanced their laying date in warmer years (Dunn and Winkler 2010). On average, laying dates for these species have advanced by 0.13 days/yr, which amounts to a shift of 2.4 days earlier for every 1°C in temperature increase. The timing of breeding affects other reproductive traits in birds, such as the number and size of clutches, incubation behavior, or recruitment, making them indirectly vulnerable to climate change (Dunn and Winkler 2010).

Studies report similar phenotypic shifts in wild mammals (Reale et al. 2003, Moyes et al. 2011). For example, populations of red squirrel (*Sciurus vulgaris*) in Canada have advanced their breeding by 18 days over a decade (Reale et al. 2003). This trend toward earlier breeding was partly explained by increased food abundance associated with increased temperatures, although selection favoring earlier breeders also could have played a role.

A study by Moyes et al. (2011) is one of the few to provide evidence for phenological advances in a population of an ungulate, red deer (*Cervus elaphus*). Six phenological traits (estrus date and parturition date in females and antler cast date, antler clean date, rut start date, and rut end date in males) advanced between 5 and 12 days over a 28-year study period, and local climate measures explained a significant portion of the variation in these traits.

Hibernation

Hibernation allows animals to conserve energy during winter, and there is evidence for changes in hibernation patterns corresponding to climate change. The timing of hibernation of Arctic ground squirrels (*Spermophilus parryii*) varied between populations living in slightly different environmental conditions in northern Alaska (Sheriff et al. 2011). Squirrels living in slightly warmer conditions (i.e., sites characterized by shorter periods with snow cover) emerged from hibernation nine days earlier and entered hibernation five days later than those living in the cooler region. Such differential response offers evidence that individuals have the ability to respond to variation in environmental conditions.

Yellow-bellied marmots (*Marmota flaviventris*) advanced their emergence from hibernation by 38 days over 25 years (Inouye et al. 2000). The advancement was positively correlated with increasing spring air temperatures and could not be explained by the annual date of snowmelt and plant flowering, because these variables did not change during the study period. Yet other researchers reported the termination of hibernation of dormouse (*Glis glis*) to have advanced eight days per decade (Adamik and Kral 2008) and to be related to rising spring temperatures.

Trophic Mismatch Hypothesis

We have shown that wildlife phenology (e.g., migration patterns, timing of breeding, timing and length of hibernation) will be altered by climate change, with some populations already responding to the warming. How do these adjustments play out at the ecosystem level? The responses of individual species reflect individual traits and are expected to differ across species because of high variability in life history traits and physiological tolerances (Parmesan and Yohe 2003). Species formerly living in synchrony are finding those relationships altered; life cycles of predators and prey, herbivores and host plants, parasitoids and hosts, and symbiotic organisms are being disrupted through the differential effects of climate change (Harrington et al. 1999, Parmesan and Yohe 2003, Visser and Both 2005).

Evidence for altered synchrony in wildlife populations is mounting. A review by Visser and Both (2005) revealed that in seven of eleven cases (nine predator–prey interactions and two insect host–plant interactions), responses of interacting species to climate change were different enough to put them out of synchrony with one another. In another study, pied flycatcher (*Ficedula hypoleuca*), a generalist long-distance migrant, declined significantly in seasonal environments characterized by short food peak but remained stable in less seasonal wetlands where the peak of food availability was longer (Both et al. 2009). In addition, resident and short-distance migratory birds did not decline in any of the habitats, suggesting that it was not the quality of the habitat causing the declines of the long-distance migrant but rather the trophic mismatch that these birds were facing. Sharper declines were also recorded in populations in western Europe, where spring temperatures increased more than in northern Europe (Both et al. 2009).

Post and Forchhammer (2008) provided evidence for a trophic mismatch between migratory caribou and their foraging plants. Migratory herbivores are likely to develop trophic mismatches because photoperiod cues the timing of their spring migration to the breeding grounds, while increasing spring temperatures cue the onset of plants. Caribou lagged climate warming–induced changes in vegetation, and they were not able to keep pace with advancement in plant-growing season on their breeding ranges. Consequences included a fourfold decrease in offspring production and an increase in mortality of those young caribou that were born.

Changes over Space: Distributional Responses

Temporal responses to climate change are often coupled with changes in spatial extents of vegetation and wildlife ranges. Species are likely to respond to changing climate space by moving their range in the direction of changing climatic niche. Let us look at the evidence for latitudinal and elevational shifts in the range boundaries of wildlife and vegetation communities.

Range Boundaries: Latitudinal and Elevational Shifts

Current hypotheses about the location of species distributional boundaries emphasize abiotic and biotic factors (MacArthur 1972, Mittelbach 2012). Range boundaries are set, at least partially, by climatic conditions (Gaston 2003, Cunningham et al. 2009, Geber 2011, Baselga et al. 2012), a pattern first noted more than 200 years ago (von Humboldt and Bonpland 1807). Therefore we expect that species will move their distributions to higher latitudes and higher elevations in response to recent climate warming. Range shifts are leading to changes in community composition because each species is adjusting its distribution in accord with its own ecological tolerances

(Williams and Jackson 2007). Some species have broad ecological tolerances and are expanding their distributions. Local extinctions of species that have narrow tolerances and are unable to effectively track climate change, which cause range contractions, however, are already widespread (Wiens 2016). Ultimately, global extinctions will occur if there is no possibility for all populations of a species to move poleward, to higher elevations, or to adapt in place.

How far can we expect species to move? Predictably, the answer to this question will depend on the studied organism, because the responses will likely vary between different taxonomic groups. Recent studies drawing on Breeding Bird Atlas data reported that songbirds in New York State have shifted their ranges northward an average of 3.6 km between the early 1980s and 2000s (Zuckerberg et al. 2009). Parmesan and Yohe (2003) reported range shifts averaging 6.1 km/decade toward the poles across a range of three different taxa—birds (Aves), butterflies (Rhopalocera), and alpine herbs. Average northward shifts of 31–60 km and elevational shifts of 25 m over a 40-year period were reported for 16 different vertebrate and invertebrate taxonomic groups: dragonflies and damselflies (Odonata), grasshoppers and allies (Orthoptera), lacewings (Neuroptera), butterflies, spiders (Araneae), herptiles (Amphibia and Squamata), freshwater fish (Teleostei), mammals (Mammalia), woodlice (Isopoda), ground beetles (Carabidae), harvestmen (Opiliones), millipedes (Diplopoda), longhorn beetles (Cerambycidae), soldier beetles and allies (Cantharoidea and Buprestoidea), aquatic bugs (Heteroptera), and birds (Hickling et al. 2006). Of 329 species studied by Hickling et al. (2006), 84% showed a significant poleward shift.

Some of the most convincing evidence for elevational range shifts resulting from climate change comes from studies of populations of small mountain mammals. For example, populations of American pika (Ochotona princeps) have been monitored since 1898 throughout the Great Basin ecoregion (Beever et al. 2011). Between 1998 and 2008, the rate of upslope range retraction accelerated 11-fold. The species is now experiencing an upward elevational shift at an average rate of 145 m/decade.

Even though distributional shifts in ranges of large charismatic fauna have rarely been reported, one study deserves mention here. Towns et al. (2010) reported poleward and eastward shifts of spatial distributions of polar bears (Ursus maritimus) in Hudson Bay between 1986 and 2004. The observed distributional changes were a result of three-week advancement of the sea ice breakup since the late 1970s.

Shifts and Reorganization of Vegetation Communities

Habitat suitability is a key determinant of the distribution and abundance of species. Vegetation lies at the heart of the habitat, often influencing wildlife distributions at regional and local scales. Climate change is expected to trigger changes in the species composition of plant communities, leading to changes in overall habitat quality and, ultimately, composition of wildlife species. Plant communities might be altered as changes in temperature and moisture regimes impinge on physiological tolerances of existing plant species, changing the competitive edge to favor invasion by other species (Hellmann et al. 2008, Dukes et al. 2009). An important influence of change in plant communities is occurring as warmer winters allow expansion of pests and pathogens. Dukes et al. (2009) reported that climate change is likely to affect the composition and structure of forests by benefiting several pest species such as hemlock woolly adelgid (Adelges tsugae), forest tent caterpillar (Malacosoma disstria), and pathogens such as Armillaria spp., beech bark (Cryptococcus fagisuga), and invasive plant species such as glossy buckthorn (Frangula alnus) and oriental bittersweet (Celastrus orbiculaturs).

The shifts in competitive advantage among plant species will result in geographic redistribution of major plant biomes (Toot et al. 2020). Most evidence for latitudinal shifts has come from the studies in the boreal biome of the Northern Hemisphere. As an example, Suarez et al. (1999) reported an increase in tree encroachment into Alaskan tundra, resulting in the shift of the forest–tundra biome by approximately 80–100 m over the past 200 years. Similarly, there has been pervasive advancement in the tree line of white spruce (Picea glauca) since 1800 along the forest–tundra border in Alaska (Lloyd and Fastie 2003, Payette 2007). Similar shifts are evident in the vegetation zones of the West African Sahel, where ranges are changing at an annual rate of approximately 500–600 m (Gonzalez 2001): arid species at the northern boundary of the biome are shifting farther north, while mesic species present at the southern boundary are retracting.

The evidence for elevational shifts in the range of biomes is even more compelling. For example, the location and species composition of the northern hardwood–boreal forest biome in Vermont changed significantly over a 40-year study period, experiencing a shift of approximately 90–120 m up in elevation (Beckage et al. 2008). Likewise, the evidence for an advance was found in Yukon, Canada, where white spruce tree lines on south-facing slopes moved up in elevation by 65–85 m over a 30-year period (Danby and Hik 2007). Similar trends have been reported in Eurasia. For example, forest–tundra ecotone in the Polar Urals underwent an upward shift of 20–60 m during the 20th century. This shift was positively correlated with increases in average summer temperatures and precipitation (Devi et al. 2008). In the southern Swedish Scandes, the tree lines of three species—mountain birch (Betula pubescens spp. Czerepanovii), white spruce, and Scots pine (Pinus sylestris)—moved up in elevation by on average 70–90 m during the 20th century (Kullman and Oberg 2009). In the Montseny Mountains in Spain, Penuelas and Boada (2003) reported a replacement of cold-temperature ecosystems by ecosystems associated with warm Mediterranean climates between 1945 and present times.

Ecological communities do not always shift in space; sometimes they reshuffle such that new combinations of species emerge or formerly rare members of the community become abundant. In the paleontological pollen record, there are examples of combinations of species without modern equivalent; they have no modern analog and are referred to as "no-analog communities." No-analog communities in the past have been associated with no-analog climates, which are seasonal patterns of temperature and precipitation without modern equivalent (Williams and Jackson 2007, Blois et al. 2013). The ongoing change in climate is moving many regions of the world toward no-analog climates and no-analog ecological communities. New combinations of species will challenge ideas and metrics for ecosystem integrity and complicate habitat management. Vegetation communities appear more likely to shift in space at higher latitudes, such as in boreal systems, whereas vegetation communities at lower latitudes, such as the tropics, appear more likely to transition to no-analog communities (Garcia et al. 2014).

Population and Community Dynamics

As climate change alters species abundance and distributions, biotic interactions such as predation, parasitism, competition, and mutualism will be subject to disruption, ultimately leading to changes in community composition, biodiversity, and ecosystem functioning. Such indirect effects of climate change, mediated through effects on species relationships, might in fact have far more severe consequences than the direct effects of warming (Bretagnolle and Gillis 2010), especially for ectotherms. In this section we discuss potential consequences of climate change to predator–prey dynamics, parasite–host interactions, and patterns of interspecific competition.

Predator–Prey Dynamics

What mechanisms can lead to alterations in predator–prey interactions? As we saw with trophic mismatches, both predator and prey may experience changes in their phenology or spatial distribution, causing either a temporal or spatial mismatch between the species. As a result, changes in the trophic structure are difficult to predict.

Evidence for mismatches in the predator–prey trophic system is growing. As an example, Both et al. (2009) reported increasing temporal mismatch in a four-level trophic system of plants, caterpillars, four species of passerines that prey on caterpillars, and a raptor that preys on passerines. Climate change resulted in annual advances of 0.17 days in budburst, 0.75 days in caterpillars, and 0.36–0.5 days in passerines hatching dates. Raptors showed no trend toward earliness. Overall, the phenological response of the higher trophic levels (passerines and the raptor) was weaker than that of their prey, resulting in increasing asynchrony between the food demand and food availability.

A study by Vors and Boyce (2009) listed increased predation pressure among the repercussions of climate change for migratory caribou. Currently, predation is not a contributing factor to caribou population regulation because the geographic range of the potential predators does not generally overlap with the distribution of the caribou. If the frontier of the boreal forest shifts poleward, however, spatial overlap between wolf (*Canis lupus*) and caribou will increase, resulting in significant caribou mortality on its winter range.

Evidence is also growing for altered predator–prey dynamics resulting from changes in population cycles. For example, changes in the population dynamics of lemmings (*Lemmus lemmus*) are attributable, in part, to winter weather conditions. Increasingly warm winters cause fewer regular peaks in the lemming cycles, which in turn affect the dynamics of its bird predator (Kausrud et al. 2008).

Changes in rainfall patterns associated with climate warming are likely to produce dramatic effects on predator–prey dynamics. Wild turkeys (*Meleagris gallopavo*) provide a good example. In northern portions of wild turkey range, rainfall in May appears to affect nest predation. Predators using olfactory cues are more effective in finding nests when moisture conditions are high (Roberts and Porter 2001, Fleming and Porter 2007). In contrast, in southwestern portions of the range of wild turkeys, low rainfall reduces insect abundance, resulting in decreases in wild turkey populations (Pattee and Beasom 1979).

Host–Parasite Dynamics

Host–parasite interactions are likely to be affected by climate change through similar mechanisms as predator–prey relationships. These mechanisms include changes in phenology of parasite or host, changes in spatial distribution of parasite or host, changes in the prevalence and intensity of parasitism, and changes in virulence or anti-parasite defenses of hosts (Merino and Moller 2010).

There is compelling evidence for increasing temperatures that affect the timing of emergence of parasites and vectors (Mouritsen and Poulin 2002, Moller 2010). For example, the hippoboscid fly (*Ornitholyia avicularia*), a parasite of a barn swallow (*Hirudo rustica*) in Denmark, has advanced its phenology to emerge during egg laying and early incubation period, rather than during late stages of the swallow breeding cycle (Moller 2010). Consequences of such earlier emergence potentially include lower reproductive success and recruitment of the host. Conversely, brood parasitic cuckoo (*Cuculus canorus*) will likely be affected by climate warming–mediated changes in the phenology of its host, short-distance migratory birds (Saino et al. 2009). Short-distance migrants have advanced their arrival time to breeding grounds on average more than long-distance migratory birds such as cuckoo, and the short-distance migrants are now starting their breeding season prior to the arrival of the cuckoos. This mistiming might ultimately lead to cuckoo population declines because of a lack of nests that they can successfully parasitize.

The prevalence of parasites is also expected to change with increasing temperatures. One of the primary vectors of zoonotic diseases, recently on the rise in Europe, is the castor bean tick (*Ixodes ricinus*). This tick transmits *Borrelia burgdorferi*, tick-borne encephalitis virus, and louping ill virus (Gilbert 2010). *B. burgdorferi* is a spirochete bacteria that is noted for causing Lyme disease. Gilbert (2010) tested whether increasing abundance of castor bean ticks in Scotland is a result of recent climate change. They reported it to be negatively associated with elevation, which they used as proxy for climatic conditions. Given future climate change scenarios, castor bean ticks are likely to become more abundant at higher altitudes. This elevational shift might result in higher prevalence of louping ill virus, because additional competent hosts—red grouse (*Lagopus lagopus scotica*) and mountain hares (*Lepus timidus*)—occur in higher numbers at higher elevations. Similar elevational shifts and increased densities of the castor bean tick were reported over a 15-year study period in Sweden (Lindgren et al. 2000). Relatively mild climatic conditions during the study period in Sweden were thought to be one of the primary reasons for the observed increase of density and geographic range of the castor bean tick. In the United States, winter ticks (*Dermacentor albipictus*) were reported to significantly influence moose population growth (Garner 1994). Mild winters favor higher tick survival, contributing to increased mortality in moose populations (Musante et al. 2010). Climate change is likely to benefit winter tick populations, indirectly contributing to increased mortality in wild ungulates.

Inter- and Intraspecific Competition

Climate change will likely disrupt competitive interactions via similar mechanisms as predator–prey or host–parasite interactions. Climate change may affect competition between migratory and resident birds by allowing short-distance migrants to return earlier to their shared breeding grounds (Forchhammer et al. 2002, Hubalek 2003), and possibly by enhancing overwinter survival of birds wintering in Europe (Lemoine and Bohning-Gaese 2003). The effect of climate change could therefore leave long-distance migrants at a competitive disadvantage. Furthermore, shifts in species geographic ranges might create a spatial overlap between species potentially competing for the same resource.

Ahola et al. (2007) studied interspecific nest-hole competition between resident great tit (*Parus major*) and migrant pied flycatchers (*Ficedula hypoleuca*) in Finland over five decades (1953–2005). Decreasing the interval of the laying date between the species increased the likelihood of pied flycatchers being killed by the great tit after an unsuccessful attempt to take over the tit's nest. Researchers reported that climate change will likely alter the interval of the interspecific laying date, potentially leading to altered competitive interactions between the species. An increasing overlap in the use of nesting boxes was also reported for edible dormouse and some species of birds (Adamik and Kral 2008).

There are similar examples of altered competitive interactions among mammals, as well. Killengreen et al. (2007) reported that populations of Arctic fox (*Vulpes lagopus*) residing at the southern margin of the Arctic tundra have been declining as a result of invasion of red foxes (*Vulpes vulpes*). Red foxes started shifting their ranges poleward as a consequence of increasing temperatures. Other researchers predicted that geographic distributions of two herds of migratory caribou will change following increased temperatures and earlier sea ice breakup (Sharma et al. 2009). These spatial changes will cause an overlap between the herds, leading to increased competition on the calving grounds.

TOOLS FOR STUDYING WILDLIFE RESPONSES TO CLIMATE CHANGE

A wide range of methods exists to study climate change effects on wildlife communities. The most popular methods include bioclimatic envelope models that allow for predicting species distributions under different climate change scenarios, and vulnerability assessments that quantify species exposure, sensitivity, and adaptive capacity to climate change. In this section we discuss the applications, strengths, and limitations of both methods.

Bioclimatic Envelope Modeling

Bioclimatic envelope models (also known as ecological niche models, habitat suitability models, or species distribution models) attempt to define climatic conditions that best describe species range limits by correlating the current species distributions with selected climate variables. Projecting these current relationships to future climate change scenarios, then, can help forecast future species' ranges (Thuiller and Munkemuller 2010).

Different types of algorithms have been used in bioclimatic envelope modeling. The performance of these algorithms is dependent on the availability of different types of species distributional data; therefore it is difficult to judge which method is the best. Factors other than the availability and quality of species distributional data also should be considered while selecting the modeling approach. These factors include the availability of the environmental data, spatial and temporal scale of the study, life history, and the goals of the study (Pearson and Dawson 2003, Hoving et al. 2005).

Because bioclimatic models are simplified versions of the reality, they are necessarily based on assumptions (Pearson and Dawson 2003, Heikkinen et al. 2006, Pearson et al. 2007, Araújo and Peterson 2012). Bioclimatic envelope models have three basic assumptions: species' distributions are determined wholly or partly by climate, species inhabit all areas of suitable climate, and species' climatic envelopes remain constant over time (Araújo and Peterson 2012, Brotons et al. 2012).

Are species distributions determined solely by climate? The existing evidence suggests that species ranges are at least partially influenced by climatic conditions. Land cover, vegetation, and biotic interactions are also important variables in determining habitat suitability, however. Some studies have incorporated land cover in models intended to explain species distributions, and the principal conclusion of these studies is that the value of land cover as a variable in the model is dependent on the geographic scale at which it is measured (Pearson et al. 2004, Luoto et al. 2007). Climatic conditions seem to dictate species distributions at coarse or large geographic scales (e.g., 40–80 km; Luoto et al. 2007). The predictive power of bioclimatic envelope models at large geographic extents is usually not improved by the addition of land cover variables to the model (Pearson et al. 2004, Luoto et al. 2007). Incorporating land cover variables at finer scales (e.g., 10 km), however, can significantly improve predictions of geographic distributions for some species (Pearson et al. 2004, Luoto et al. 2007).

Bioclimatic envelope models are likely to produce some errors because the ecological systems they approximate are complex. Despite their limitations, bioclimatic envelope models are a valuable tool for determining species distributions and the effects of climate warming.

Vulnerability Assessments

In addition to bioclimatic envelope models, another strategy to quantify the effects of climate change is to assess species vulnerability. Vulnerability assessments usually comprise three elements: assessment of sensitivity, adaptive capacity, and potential exposure of individual species to climate change. While traits intrinsic to the species determine sensitivity and adaptive capacity, environmental variables and local microhabitat conditions govern exposure (Williams et al. 2008).

Sensitivity

Factors intrinsic to each species, such as physiological tolerance, behavioral traits, or genetic diversity, will determine sensitivity to climate warming. Not all of these traits are easily characterized, however. As an example, information on reproductive success, the timing of breeding, or the timing of migration is readily available for many taxonomic groups, whereas data on physiological tolerances are more scarce and difficult to obtain. Williams et al. (2008) suggested that where data are lacking for a particular species, it might be reasonable to use information available for a closely related species as a proxy for the responses of the species of interest. For example, thermal tolerances tend to be similar for closely related groups of organisms, and it might suffice to obtain the thermal tolerances only for a number of representatives of the taxonomic groups.

Resilience and Adaptive Capacity

Resilience has many meanings in the context of ecology, climate change, and other fields (Fisichelli et al. 2016). We use the word here interchangeably with adaptive capacity. Resilience and adaptive capacity are the ability of a species to survive or recover from an ecological perturbation. In the case of climate change, those perturbations can be press perturbations (e.g., change in mean air temperature), pulse perturbations (e.g., extreme rainfall), or some combination of the two (e.g., fire risk and drought). Traits such as high reproductive rates, fast life history, or short life span are thought to promote higher resilience and therefore lower risk of extinction (Williams et al. 2008, Thurman et al. 2020). Dispersal abilities are also crucial in maintaining viable populations across the landscapes, especially in the context of climate change, where species will have to track their preferred climatic envelope (Williams et al. 2008).

The adaptive capacity of a species can be expressed either through phenotypic plasticity or genetic change (Beever et al.

2016, Thurman et al. 2020). All organisms have some capacity to adapt to changing conditions. Phenotypic plasticity may include changes in the timing of migration or breeding, or changes in hibernation patterns. Both phenotypic plasticity and evolutionary adaptation have already occurred in a variety of species in response to climate change (Root et al. 2003, Bradshaw and Holzapfel 2006, Parmesan 2007), although a rapid evolutionary response is less likely than phenotypic change to occur for the majority of species (Williams et al. 2008).

Exposure

Two forces influence exposure: the degree of climate change across the geographic range of a species and the availabilities of local microhabitats (Williams et al. 2008). For instance, the availability of thermal refugia or shelters (e.g., small pools, boulder fields, rocks, logs) can buffer a species from the full magnitude of regional climate change. As an example, brushtail possums (*Trichosurus vulpecula*) are buffered from extreme temperature by choosing to den in tree hollows that are 1.6°C cooler than other den locations (Isaac et al. 2008). American pikas are able to tolerate wider temperature ranges in locations with abundant rock-ice features in comparison with sites where those formations are less abundant or absent (Millar and Westfall 2010). Rock-ice features create local environments that are cooler than expected for mean summer temperatures at this elevation and warmer than expected for winter temperatures, reducing pikas' exposure to regional climate warming.

Together, sensitivity, adaptive capacity, and exposure determine the vulnerability of a species to climate change and allow prediction of potential effects. Conservation organizations and management agencies have designed a variety of vulnerability assessment frameworks (Glick and Stein 2011, Rowland et al. 2011, Thurman et al. 2020). For example, NatureServe has created a climate change vulnerability index (Young et al. 2009) that assesses exposure using projection of temperature and precipitation-water balance, sea level rise, and changes in land use. Information on dispersal capabilities, physical tolerances, biotic interactions, and genetics assesses sensitivity. The US Department of Agriculture Forest Service has designed a vulnerability index (Finch et al. 2011, Glick and Stein 2011) that assesses sensitivity and exposure using variables related to four broad categories: habitat, physiology, phenology, and biotic interactions. Finally, the assessment tool developed by the US Environmental Protection Agency evaluates vulnerability of species designated as threatened or endangered under the US Endangered Species Act. The tool specifically determines how species conservation status might be altered by climate change (US Environmental Protection Agency 2009, Glick and Stein 2011, Rowland et al. 2011).

CLIMATE CHANGE ADAPTATION

Climate change adaptation is one of the grand challenges for wildlife management in the 21st century (Mawdsley et al. 2009,

Knutson and Heglund 2012). The ecological consequences of climate change are pervasive (Parmesan 2006, Beever and Belant 2012), and the role of wildlife managers is becoming increasingly important. Wildlife managers do not have primary responsibility to solve climate change; we manage wildlife populations and human interactions with wildlife, not carbon dioxide emissions and energy policy. Our responsibility is to help wildlife adapt to ongoing and foreseeable climate changes by reducing vulnerability to ecological changes wrought by climate change, or by putting in place management regimes that allow managers and wildlife to learn and adapt to novel and unforeseen challenges that inevitably occur when a complex system undergoes systemic change.

Climate Change Uncertainty

For many wildlife managers, a primary concern is that climate change represents a source of environmental uncertainty that is difficult to predict and incorporate into management. Managers are generally comfortable with environmental variation, and its influence on wildlife survival and reproduction, and they use a number of modeling approaches for predicting how vital rates might change in the future (Williams et al. 2002). An assumption of many of these models, however, is that the historic range of environmental variability will remain constant (i.e., stationary). Yet what we know now suggests that climate change and its associated changes in precipitation, snow cover, sea level rises, and extreme weather events will likely alter historical ranges of variability, and perhaps even produce future environmental states with no current analog (Williams et al. 2007b). As such, management and planning should include the biological variables of specific interest (e.g., population size, abundance) and the climatic influences that are expected to change (Conroy et al. 2011, Nichols et al. 2011). Future modeling will require innovative thinking and a decreased dependence on historical data when developing new models to deal with system change.

Decision-making in the face of climate change is complicated because ecological systems are changing, and there is strong uncertainty associated with the magnitude and direction of that change. Two main sources of uncertainty associated with climate change are problems associated with downscaled regional- to local-scale predictions of future climate change and uncertainty associated with predicting the biological responses to these climatic changes (McLennan 2012). Downscaled climate products are being developed for many regions, but reducing the uncertainty associated with biological responses to climate change will only result from further efforts in empirical research and monitoring. A lack of understanding of the biological effects of climate change should not serve as a roadblock; dealing with imperfect knowledge is nothing new to wildlife management. Adaptive management (Williams et al. 2007a) and structured decision-making (Lyons et al. 2008, Martin et al. 2009) are frameworks that have been explicitly designed for accounting for uncertainty (Chapter 5). As an example, McDonald-Madden et al. (2011) demonstrated the ability of dynamic optimization methods and alternative models of system changes to make sound decisions on the benefits and timing associated with relocating species faced with climate change.

The Importance of Monitoring

Climate change has reinvigorated the need for long-term monitoring in wildlife management (Lepetz et al. 2009, Beever and Woodward 2011, Conroy et al. 2011). Monitoring has always been a central component of adaptive management, but climate change will likely test the flexibility of existing monitoring programs. As an example, monitoring efforts often establish survey areas based on historical patterns of where species are more likely to be found during certain times of the year (e.g., migratory corridors, breeding sites). Many species, however, are shifting their distributions and migratory patterns (Root et al. 2003, Walther et al. 2005). Some species of the Northern Hemisphere are exhibiting contractions along southern boundaries and expansions along northern boundaries (Thomas and Lennon 1999, Brommer 2004, Zuckerberg et al. 2009). These geographic shifts will force managers to reconsider their survey strata and boundaries. In addition to spatial considerations, phenological shifts complicate monitoring programs (Penuelas and Filella 2001, Cleland et al. 2007, Visser et al. 2010), which will have to account for species that might be arriving to their breeding grounds earlier, singing or displaying at a different time, or using alternative food resources. These changes will add sources of variation in sampling (e.g., changes in detection probability) when collecting data for monitoring purposes.

Managers should not worry about identifying all sources of variation associated with monitoring but instead understand better how these sources may vary in response to climate change and plan accordingly. There should be a continuous reevaluation of local, state, and regional efforts to develop a more comprehensive network of monitoring sites. Monitoring should be designed in such a way to reflect the seasonally and geographically complex nature of future climate change, capturing a full range of latitudinal, land-use, and seasonal variation. Methods such as occupancy and estimation (MacKenzie et al. 2006) provide enough flexibility to be used for multiple taxa and can account for heterogeneous detectability across species and areas. In addition, many wildlife agencies have a treasure trove of historical surveys and data sets that offer unique opportunities to resurvey sites and to document long-term shifts in species distributions (Tingley and Beissinger 2009, Tingley et al. 2009). These data can be particularly useful for identifying which species are showing the strongest response to climate change.

EFFECTIVELY COMMUNICATING CLIMATE

Climate change is very much a science communication challenge. A growing body of communication science suggests that effective science communication regarding climate change is counterintuitive to most scientists. An initial focus on the gravity of the threat, especially when it is not paired with effective solutions, often leads to lower levels of motivation and sometimes leads to higher levels of denial that climate change is a conservation priority at all (Nisbet 2009, Moser 2010). Poor framing of climate science or climate-related wildlife conservation can have counterintuitive results that may be opposed to the intention of the communicator. Practical guidance for effective climate change communication that is based on sound social science abounds (Moser and Dilling 2007, Jones 2014, Ballantyne 2016). There are at least five common frameworks.

1. Avoid framing threats as overwhelming, without underplaying the threat.
2. Include effective actions or solutions, or point to areas where progress could reveal effective actions or solutions.
3. Avoid engaging in partisan debate in the guise of scientific discourse.
4. Begin communication on partisan topics by establishing common ground and trust with the audience, rather than by appealing to information sources that signal partisan affiliation.
5. Know your perspective and learn to value the perspectives of others who are not from your in-group.

With these recommendations in mind, review the first two paragraphs of this chapter. The authors of this chapter intentionally framed climate change as a challenge and a key ingredient of successful wildlife management. We could have started the chapter with appeals to the IPCC for the reality and gravity of global climate change. By framing the chapter in ways that are engaging to our target audience—students who care a great deal about wildlife and becoming successful professionals—we help them understand the rest of the chapter in ways that were more accurate and less distorted by the anxieties that climate communication sometimes induces. We framed this chapter in this way because we believe that adapting to climate change is one of the defining challenges of the newest generation of conservation practitioners.

SUMMARY

Climate change is occurring. Wildlife species and populations are beginning to respond. There is strong scientific consensus that environmental tipping points are being crossed, and many species are adapting (or failing to adapt) to novel climatic conditions. As wildlife managers, we face the question of how we should adapt. The answer is twofold: we must enhance our abilities to anticipate changes, and we need to be flexible in our responses. Managing wildlife populations under climate change has no single, optimal solution. Instead, both the process of climate change and its resultant ecological consequences are regionally dependent on many variables and difficult to predict. As with any environmental problem, there are multiple stakeholders with disparate values and objectives. Regional and collaborative decision-making is more likely to lead to satisfactory management outcomes in the future. Climate change will require wildlife managers and stakeholders to make decisions despite long-range uncertainty in both the

ecological and human systems. Climate change is a management and communication challenge that can and will be addressed in the coming decades. How you and your colleagues approach this challenge will affect wildlife conservation into the future.

Literature Cited

Adamik, P., and M. Kral. 2008. Climate- and resource-driven long-term changes in dormice populations negatively affect hole-nesting songbirds. Journal of Zoology 275:209–215.

Ahola, M. P., T. Laaksonen, T. Eeva, and E. Lehikoinen. 2007. Climate change can alter competitive relationships between resident and migratory birds. Journal of Animal Ecology 76:1045–1052.

Araújo, M. B., and A. T. Peterson. 2012. Uses and misuses of bioclimatic envelope modelling. Ecology 93:1527–1539.

Ball, J. P., G. Ericsson, and K. Wallin. 1999. Climate changes, moose and their human predators. Ecological Bulletins 47:178–187.

Ballantyne, A. G. 2016. Climate change communication: what can we learn from communication theory? Wiley Interdisciplinary Reviews–Climate Change 7(3):329–344.

Baselga, A., J. M. Lobo, J. C. Svenning, and M. B. Araujo. 2012. Global patterns in the shape of species geographical ranges reveal range determinants. Journal of Biogeography 39:760–771.

Beckage, B., B. Osborne, D. G. Gavin, C. Pucko, T. Siccama, and T. Perkins. 2008. A rapid upward shift of a forest ecotone during 40 years of warming in the Green Mountains of Vermont. Proceedings of the National Academy of Sciences 105:4197–4202.

Beever, E. A., and J. L. Belant. 2012. Ecological consequences of climate change: mechanisms, conservation, and management. CRC Press, Boca Raton, Florida, USA.

Beever, E. A., L. E. Hall, J. Varner, A. E. Loosen, J. B. Dunham, M. K. Gahl, F. A. Smith, and J. J. Lawler. 2017. Behavioral flexibility as a mechanism for coping with climate change. Frontiers in Ecology and the Environment 15(6):299–308.

Beever, E. A., J. O'Leary, C. Mengelt, J. M. West, S. Julius, et al. 2016. Improving conservation outcomes with a new paradigm for understanding species' fundamental and realized adaptive capacity. Conservation Letters 9(2):131–137.

Beever, E. A., C. Ray, J. L. Wilkening, P. F. Brussard, and P. W. Mote. 2011. Contemporary climate change alters the pace and drivers of extinction. Global Change Biology 17:2054–2070.

Beever, E. A., and A. Woodward. 2011. Design of ecoregional monitoring in conservation areas of high-latitude ecosystems under contemporary climate change. Biological Conservation 144:1258–1269.

Bellard, C., C. Bertelsmeier, P. Leadly, W. Thuiller, and F. Courchamp. 2012. Impacts of climate change on the future of biodiversity. Ecology Letters 15(4):365–377.

Berthold, P. 2001. Bird migration: a general survey. 2nd edition. Oxford University Press, Oxford, UK.

Blois, J. L., P. L. Zarnetske, M. C. Fitzpatrick, and S. Finnegan. 2013. Climate change and the past, present, and future of biotic interactions. Science 341(6145):499–504.

Both, C. 2010. Food availability, mistiming, and climatic change. Pages 129–148 in A. P. Moller, W. Fiedler, and P. Berthold, editors. Effects of climate change on birds. Oxford University Press, Oxford, UK.

Both, C., M. van Asch, R. G. Bijlsma, A. B. van den Burg, and M. E. Visser. 2009. Climate change and unequal phenological changes across four trophic levels: constraints or adaptations? Journal of Animal Ecology 78:73–83.

Bradshaw, W. E., and C. M. Holzapfel. 2006. Climate change—evolutionary response to rapid climate change. Science 312:1477–1478.

Bretagnolle, V., and H. Gillis. 2010. Predator–prey interactions and climate change. Pages 227–248 in A. P. Moller, W. Fiedler, and P. Berthold, editors. Effects of climate change on birds. Oxford University Press, Oxford, UK.

Brommer, J. E. 2004. The range margins of northern birds shift polewards. Annales Zoologici Fennici 41:391–397.

Brotons, L., M. De Caceres, A. Fall, and M. J. Fortin. 2012. Modeling bird species distribution change in fire prone Mediterranean landscapes: incorporating species dispersal and landscape dynamics. Ecography 35:458–467.

Brown, R. D., and P. W. Mote. 2009. The responses of Northern Hemisphere snow cover to a changing climate. Journal of Climate 22(8): 2124–2145.

Church, J. A., P. U. Clark, A. Cazenave, J. M. Gregory, S. Jevrejeva, et al. 2013. Sea level change. Pages 1137–1216 in T. F. Stocker, D. Qin, G.-K. Plattner, et al., editors. Climate change 2013: the physical science basis. Contribution of Working Group I to the Fifth Assessment Report of the Intergovernmental Panel on Climate Change. Cambridge University Press, Cambridge, UK.

Cleland, E. E., I. Chuine, A. Menzel, H. A. Mooney, and M. D. Schwartz. 2007. Shifting plant phenology in response to global change. Trends in Ecology and Evolution 22:357–365.

Comiso, J. C., and D. K. Hall. 2014. Climate trends in the Arctic as observed from space. Wiley Interdisciplinary Reviews–Climate Change 5(3):389–409.

Conroy, M. J., M. C. Runge, J. D. Nichols, K. W. Stodola, and R. J. Cooper. 2011. Conservation in the face of climate change: the roles of alternative models, monitoring, and adaptation in confronting and reducing uncertainty. Biological Conservation 144:1204–1213.

Cunningham, H. R., L. J. Rissler, and J. J. Apodaca. 2009. Competition at the range boundary in the slimy salamander: using reciprocal transplants for studies on the role of biotic interactions in spatial distributions. Journal of Animal Ecology 78:52–62.

Danby, R. K., and D. S. Hik. 2007. Variability, contingency and rapid change in recent subarctic alpine tree line dynamics. Journal of Ecology 95:352–363.

DeLeon, R. L., E. E. DeLeon, and G. R. Rising. 2011. Influence of climate change on avian migrants' first arrival dates. Condor 113:915–923.

Devi, N., F. Hagedorn, P. Moiseev, H. Bugmann, S. Shiyatov, V. Mazepa, and A. Rigling. 2008. Expanding forests and changing growth forms of Siberian larch at the Polar Urals treeline during the 20th century. Global Change Biology 14:1581–1591.

Dlugokencky, E. J., B. D. Hall, S. A. Montzka, G. Dutton, J. Mühle, and J. W. Elkins. 2019. Atmospheric composition. Chapter 2 in State of the climate in 2018. Bulletin of the American Meteorological Society 100(9):S48–S50.

Dong, B. W., J. M. Gregory, and R. T. Sutton. 2009. Understanding land-sea warming contrast in response to increasing greenhouse gases. Part I: transient adjustment. Journal of Climate 22:3079–3097.

Dukes, J. S., J. Pontius, D. Orwig, J. R. Garnas, V. L. Rodgers, et al. 2009. Responses of insect pests, pathogens, and invasive plant species to climate change in the forests of northeastern North America: what can we predict? Canadian Journal of Forest Research 39:231–248.

Dunn, P. O., and D. W. Winkler. 2010. Effects of climate change on timing of breeding and reproductive success in birds. Pages 113–128 in A. P. Moller, W. Fiedler, and P. Berthold, editors. Effects of climate change on birds. Oxford University Press, Oxford, UK.

Ferrer, M., I. Newton, and K. Bildstein. 2008. Climatic change and the conservation of migratory birds in Europe: identifying effects and conservation priorities. Convention on the conservation of European wildlife and natural habitats. 1 July 2008, Strasbourg, France.

Fiedler, W., F. Bairlein, and U. Koppen. 2004. Using large-scale data from ringed birds for the investigation of effects of climate change on migrating birds: pitfalls and prospects. Pages 49–67 in A. P. Moller, W. Fielder, and P. Berthold, editors. Birds and climate change. Volume 35 of Advances in ecological research. Academic Press, San Diego, California, USA.

Finch, D. M., M. Friggens, and K. E. Bagne. 2011. Case study 3: species vulnerability assessment for the Middle Rio Grande, New Mexico. Pages 96–103 in P. Glick and B. A. Stein, editors. Scanning the conservation horizon: a guide to climate change vulnerability assessment. National Wildlife Federation, Washington, DC, USA.

Fisichelli, N. A., G. W. Schuurman, and C. H. Hoffman. 2016. Is "resilience" maladaptive? Towards an accurate lexicon for climate change adaptation. Environmental Management 57:753–758.

Fleming, K. K., and W. F. Porter. 2007. Synchrony in a wild turkey population and its relationship to spring weather. Journal of Wildlife Management 71:1192–1196.

Folland, C. K., T. R. Karl, and M. Jim Salinger. 2002. Observed climate variability and change. Weather 57:269–278.

Forchhammer, M. C., E. Post, and N. C. Stenseth. 2002. North Atlantic Oscillation timing of long- and short-distance migration. Journal of Animal Ecology 71:1002–1014.

Garcia, R. A., M. Cabeza, C. Rahbek, and M. B. Araujo. 2014. Multiple dimensions of climate change and their implications for biodiversity. Science 344:1247579.

Garner, D. L. 1994. Population ecology of moose in Algonquin Provincial Park, Ontario, Canada. PhD dissertation, State University of New York, Syracuse, USA.

Gaston, K. J. 2003. The structure and dynamics of geographic ranges. Oxford University Press, Oxford, UK.

Geber, M. A. 2011. Ecological and evolutionary limits to species geographic ranges. American Naturalist 178:S1–S5.

Gilbert, L. 2010. Altitudinal patterns of tick and host abundance: a potential role for climate change in regulating tick-borne diseases? Oecologia 162:217–225.

Glick, P., and B. A. Stein. 2011. Scanning the conservation horizon: a guide to climate change vulnerability assessment. National Wildlife Federation, Washington, DC, USA.

Gonzalez, P. 2001. Desertification and a shift of forest species in the West African Sahel. Climate Research 17:217–228.

Hall, M. H. P., and D. B. Fagre. 2003. Modeled climate-induced glacier change in Glacier National Park, 1850–2100. Bioscience 53:131–140.

Hansen, J., R. Ruedy, M. Sato, and K. Lo. 2010. Global surface temperature change. Reviews of Geophysics 48:RG4004.

Harrington, R., I. Woiwod, and T. Sparks. 1999. Climate change and trophic interactions. Trends in Ecology and Evolution 14:146–150.

Heikkinen, R. K., M. Luoto, M. B. Araujo, R. Virkkala, W. Thuiller, and M. T. Sykes. 2006. Methods and uncertainties in bioclimatic envelope modelling under climate change. Progress in Physical Geography 30:751–777.

Hellmann, J. J., J. E. Byers, B. G. Bierwagen, and J. S. Dukes. 2008. Five potential consequences of climate change for invasive species. Conservation Biology 22(3):534–543.

Herring, S. C., N. Christidis, A. Hoell, M. P. Hoerling, and P. A. Stott. 2020. Explaining extreme events of 2018 from a climate perspective. Bulletin of the American Meteorological Society 101(1):S1–S134.

Hewitt, G. M. 1996. Some genetic consequences of ice ages, and their role in divergence and speciation. Biological Journal of the Linnean Society 58:247–276.

Hickling, R., D. B. Roy, J. K. Hill, R. Fox, and C. D. Thomas. 2006. The distributions of a wide range of taxonomic groups are expanding polewards. Global Change Biology 12:450–455.

Hock, R., G. Rasul, C. Adler, B. Cáceres, S. Gruber, et al. 2019. High mountain areas. Pages 131–202 in H.-O. Pörtner, D. C. Roberts, V. Masson-Delmotte, et al., editors. IPCC special report on the ocean and cryosphere in a changing climate. Intergovernmental Panel on Climate Change, Geneva, Switzerland.

Hoegh-Guldberg, O., D. Jacob, M. Taylor, M. Bindi, S. Brown, et al. 2018. Impacts of 1.5°C global warming on natural and human systems. Pages 175–312 in V. Masson-Delmotte, P. Zhai, H.-O. Pörtner, et al., editors. Global warming of 1.5°C. An IPCC special report on the impacts of global warming of 1.5°C above pre-industrial levels and related global greenhouse gas emission pathways, in the context of strengthening the global response to the threat of climate change, sustainable development, and efforts to eradicate poverty. Intergovernmental Panel on Climate Change, Geneva, Switzerland.

Hoving, C. L., D. J. Harrison, W. B. Krohn, R. A. Joseph, and M. O'Brien. 2005. Broad-scale predictors of Canada lynx occurrence in eastern North America. Journal of Wildlife Management 69:739–751.

Hubalek, Z. 2003. Spring migration of birds in relation to North Atlantic Oscillation. Folia Zoologica 52:287–298.

Hurlbert, A. H., and Z. F. Liang. 2012. Spatiotemporal variation in avian migration phenology: citizen science reveals effects of climate change. PLoS One 7:e31662.

Inouye, D. W., B. Barr, K. B. Armitage, and B. D. Inouye. 2000. Climate change is affecting altitudinal migrants and hibernating species. Proceedings of the National Academy of Sciences 97:1630–1633.

IPCC. Intergovernmental Panel on Climate Change. 2014. Climate change 2014: synthesis report. Contribution of Working Groups I, II and III to the Fifth Assessment Report of the Intergovernmental Panel on Climate Change. R. K. Pachauri and L. A. Meyer, editors. IPCC, Geneva, Switzerland.

Isaac, J. L., J. L. De Gabriel, and B. A. Goodman. 2008. Microclimate of daytime den sites in a tropical possum: implications for the conservation of tropical arboreal marsupials. Animal Conservation 11:281–287.

Jezkova, T., and J. J. Wiens. 2016. Rates of change in climatic niches in plant and animal populations are much slower than projected climate change. Proceedings of the Royal Society B 283:20162104.

Jones, M. D. 2014. Communicating climate change: are stories better than "just the facts"? Policy Studies Journal 42(4):644–673.

Kausrud, K. L., A. Mysterud, H. Steen, J. O. Vik, E. Østbye, et al. 2008. Linking climate change to lemming cycles. Nature 456:93–97.

Killengreen, S. T., R. A. Ims, N. G. Yoccoz, K. A. Brathen, J. A. Henden, and T. Schott. 2007. Structural characteristics of a low Arctic tundra ecosystem and the retreat of the Arctic fox. Biological Conservation 135:459–472.

Klein, A. G., and J. L. Kincaid. 2006. Retreat of glaciers on Puncak Jaya, Irian Jaya, determined from 2000 and 2002 IKONOS satellite images. Journal of Glaciology 52:65–79.

Knutson, M. G., and P. J. Heglund. 2012. Resource managers rise to the challenge of climate change. Pages 261–284 in E. A. Beever and J. L. Belant, editors. Ecological consequences of climate change: mechanisms, conservation, and management. Taylor and Francis, Boca Raton, Florida, USA.

Kullman, L., and L. Oberg. 2009. Post–Little Ice Age tree line rise and climate warming in the Swedish Scandes: a landscape ecological perspective. Journal of Ecology 97:415–429.

LeDee, O. E., S. D. Handler, C. L. Hoving, C. W. Swanston, and B. Zuckerberg. 2021. Preparing wildlife for climate change: how far have we come? Journal of Wildlife Management 85:7–16.

Lehikoinen, E., and T. H. Sparks. 2010. Changes in migration. Pages 89–112 in A. P. Moller, W. Fiedler, and P. Berthold, editors. Effects of climate change on birds. Oxford University Press, Oxford, UK.

Lemke, P., J. Ren, R. B. Alley, I. Allison, J. Carrasco, et al. 2007. Observations: changes in snow, ice and frozen ground. Pages 338–383 in S. Solomon, D. Qin, M. Manning, et al, editors. Climate change 2007: the physical science basis. Contribution of Working Group I to the Fourth Assessment Report of the Intergovernmental Panel on Climate Change. Cambridge University Press, Cambridge, UK.

Lemoine, N., and K. Bohning-Gaese. 2003. Potential impact of global climate change on species richness of long-distance migrants. Conservation Biology 17:577–586.

Lepetz, V., M. Massot, D. S. Schmeller, and J. Clobert. 2009. Biodiversity monitoring: some proposals to adequately study species' responses to climate change. Biodiversity and Conservation 18:3185–3203.

Lindgren, E., L. Talleklint, and T. Polfeldt. 2000. Impact of climatic change on the northern latitude limit and population density of the disease-transmitting European tick Ixodes ricinus. Environmental Health Perspectives 108:119–123.

Lloyd, A. H., and C. L. Fastie. 2003. Recent changes in treeline forest distribution and structure in interior Alaska. Ecoscience 10:176–185.

Luoto, M., R. Virkkala, and R. K. Heikkinen. 2007. The role of land cover in bioclimatic models depends on spatial resolution. Global Ecology and Biogeography 16:34–42.

Lyons, J. E., M. C. Runge, H. P. Laskowski, and W. L. Kendall. 2008. Monitoring in the context of structured decision-making and adaptive management. Journal of Wildlife Management 72:1683–1692.

MacArthur, R. H. 1972. Geographical ecology: patterns in the distribution of species. Harper and Row, New York, New York, USA.

MacKenzie, D. I., J. D. Nichols, J. A. Royle, K. H. Pollock, L. L. Bailey, and J. E. Hines. 2006. Occupancy estimation and modeling: inferring patterns and dynamics of species occurrence. Elsevier, Burlington, Maine, USA.

Martin, J., M. C. Runge, J. D. Nichols, B. C. Lubow, and W. L. Kendall. 2009. Structured decision making as a conceptual framework to identify thresholds for conservation and management. Ecological Applications 19:1079–1090.

Mawdsley, J. R., R. O'Malley, and D. S. Ojima. 2009. A review of climate-change adaptation strategies for wildlife management and biodiversity conservation. Conservation Biology 23:1080–1089.

McDonald-Madden, E., M. C. Runge, H. P. Possingham, and T. G. Martin. 2011. Optimal timing for managed relocation of species faced with climate change. Nature Climate Change 1:261–265.

McLennan, D. 2012. Dealing with uncertainty: managing and monitoring Canada's northern national parks in a rapidly changing world. Pages 209–236 in E. A. Beever and J. L. Belant, editors. Ecological consequences of climate change: mechanisms, conservation, and management. Taylor and Francis, Boca Raton, Florida, USA.

Meredith, M., M. Sommerkorn, S. Cassotta, C. Derksen, A. Ekaykin, et al. 2019. Polar regions. Pages 203–320 in H.-O. Pörtner, D. C. Roberts, V. Masson-Delmotte, et al., editors. IPCC special report on the ocean and cryosphere in a changing climate. Intergovernmental Panel on Climate Change, Geneva, Switzerland.

Merino, S., and A. P. Moller. 2010. Host–parasite interactions and climate change. Pages 213–226 in A. P. Moller, W. Fiedler, and P. Berthold, editors. Effects of climate change on birds. Oxford University Press, Oxford, UK.

Millar, C. I., and R. D. Westfall. 2010. Distribution and climatic relationships of the American pika (Ochotona princeps) in the Sierra Nevada and western Great Basin, USA: periglacial landforms as refugia in warming climates. Reply. Arctic, Antarctic, and Alpine Research 42:493–496.

Mittelbach, G. G. 2012. Community ecology. Sinauer, Sunderland, Massachusetts, USA.

Moller, A. P. 2010. Host–parasite interactions and vectors in the barn swallow in relation to climate change. Global Change Biology 16:1158–1170.

Montzka, S. A., E. J. Dlugokencky, and J. H. Butler. 2011. Non-CO_2 greenhouse gases and climate change. Nature 476:43–50.

Moser, S. C. 2010. Communicating climate change: history, challenges, process, and future directions. Wiley Interdisciplinary Reviews–Climate Change 1(1):31–53.

Moser, S. C., and L. Dilling. 2007. Creating a climate for change: communicating climate change and facilitating social change. Cambridge University Press, Cambridge, UK.

Mosier, A. R., J. M. Duxbury, J. R. Freney, O. Heinemeyer, K. Minami, and D. E. Johnson. 1998. Mitigating agricultural emissions of methane. Climatic Change 40:39–80.

Mouritsen, K. N., and R. Poulin. 2002. Parasitism, climate oscillations and the structure of natural communities. Oikos 97:462–468.

Moyes, K., D. H. Nussey, M. N. Clements, F. E. Guinness, A. Morris, S. Morris, J. M. Pemberton, L. E. B. Kruuk, and T. H. Clutton-Brock. 2011. Advancing breeding phenology in response to environmental change in a wild red deer population. Global Change Biology 17:2455–2469.

Musante, A. R., P. J. Pekins, and D. L. Scarpitti. 2010. Characteristics and dynamics of a regional moose Alces alces population in the northeastern United States. Wildlife Biology 16:185–204.

National Aeronautics and Space Administration. 2020. GISS surface temperature analysis (ver. 4). https://data.giss.nasa.gov/gistemp/graphs_v4/.

National Oceanic and Atmospheric Administration. 2020. The NOAA annual greenhouse gas index. http://www.esrl.noaa.gov/gmd/aggi/.

Nichols, J. D., M. D. Koneff, P. J. Heglund, M. G. Knutson, M. E. Seamans, J. E. Lyons, J. M. Morton, M. T. Jones, G. S. Boomer, and B. K. Williams. 2011. Climate change, uncertainty, and natural resource management. Journal of Wildlife Management 75:6–18.

Nisbet, M. C. 2009. Communicating climate change: why frames matter for public engagement. Environment 51(2):12–23.

Notaro, M., D. Lorenz, C. L. Hoving, and M. Schummer. 2014. Twenty-first-century projections of snowfall and winter severity across central-eastern North America. Journal of Climate 27(17):6526–6550.

Notz, D., J. Dorr, D. A. Bailey, E. Blockey, M. Bushuk, et al. 2020. Arctic sea ice in CMIP6. Geophysical Research Letters 47(10):e2019GL086749.

Oenema, O., N. Wrage, G. L. Velthof, J. W. van Groenigen, J. Dolfing, and P. J. Kuikman. 2005. Trends in global nitrous oxide emissions from animal production systems. Nutrient Cycling in Agroecosystems 72:51–65.

Parmesan, C. 1996. Climate and species' range. Nature 382:765–766.

Parmesan, C. 2006. Ecological and evolutionary responses to recent climate change. Annual Review of Ecology Evolution and Systematics 37:637–669.

Parmesan, C. 2007. Influences of species, latitudes and methodologies on estimates of phenological response to global warming. Global Change Biology 13:1860–1872.

Parmesan, C., and G. Yohe. 2003. A globally coherent fingerprint of climate change impacts across natural systems. Nature 421:37–42.

Pattee, O. H., and S. L. Beasom. 1979. Supplemental feeding to increase wild turkey productivity. Journal of Wildlife Management 43:512–516.

Payette, S. 2007. Contrasted dynamics of northern Labrador tree lines caused by climate change and migrational lag. Ecology 88:770–780.

Pearson, R. G., and T. P. Dawson. 2003. Predicting the impacts of climate change on the distribution of species: are bioclimate envelope models useful? Global Ecology and Biogeography 12:361–371.

Pearson, R. G., T. P. Dawson, and C. Liu. 2004. Modelling species distributions in Britain: a hierarchical integration of climate and land-cover data. Ecography 27:285–298.

Pearson, R. G., C. J. Raxworthy, M. Nakamura, and A. T. Peterson. 2007. Predicting species distributions from small numbers of occurrence records: a test case using cryptic geckos in Madagascar. Journal of Biogeography 34:102–117.

Pederson, G. T., L. J. Graumlich, D. B. Fagre, T. Kipfer, and C. C. Muhlfeld. 2010. A century of climate and ecosystem change in western Montana: what do temperature trends portend? Climatic Change 98:133–154.

Penuelas, J., and M. Boada. 2003. A global change-induced biome shift in the Montseny mountains (NE Spain). Global Change Biology 9:131–140.

Penuelas, J., and I. Filella. 2001. Phenology—responses to a warming world. Science 294:793–795.

Post, E., and M. C. Forchhammer. 2008. Climate change reduces reproductive success of an Arctic herbivore through trophic mismatch. Philosophical Transactions of the Royal Society B 363:2369–2375.

Pulido, F., and P. Berthold. 2010. Current selection for lower migratory activity will drive the evolution of residency in a migratory bird population. Proceedings of the National Academy of Sciences 107:7341–7346.

Pulido, F., P. Berthold, and A. J. vanNoordwijk. 1996. Frequency of migrants and migratory activity are genetically correlated in a bird population: evolutionary implications. Proceedings of the National Academy of Sciences 93:14,642–14,647.

Reale, D., A. G. McAdam, S. Boutin, and D. Berteaux. 2003. Genetic and plastic responses of a northern mammal to climate change. Proceedings of the Royal Society B 270:591–596.

Roberts, S. D., and W. F. Porter. 2001. Annual changes in May rainfall as an index to wild turkey harvest. National Wild Turkey Symposium 8:43–52.

Root, T. L., J. T. Price, K. R. Hall, S. H. Schneider, C. Rosenzweig, and J. A. Pounds. 2003. Fingerprints of global warming on wild animals and plants. Nature 421:57–60.

Rostom, R., and S. Hastenrath. 2007. Variations of Mount Kenya's glaciers 1993–2004. Erdkunde 61:277–283.

Rowland, E. L., J. E. Davison, and L. J. Graumlich. 2011. Approaches to evaluating climate change impacts on species: a guide to initiating the adaptation planning process. Environmental Management 47:322–337.

Saino, N., D. Rubolini, E. Lehikoinen, L. V. Sokolov, A. Bonisoli-Alquati, R. Ambrosini, G. Boncoraglio, and A. P. Moller. 2009. Climate change effects on migration phenology may mismatch brood parasitic cuckoos and their hosts. Biology Letters 5:539–541.

Sandel, B., L. Arge, B. Dalsgaard, R. G. Davies, K. J. Gaston, W. J. Sutherland, and J. C. Svenning. 2011. The influence of late Quaternary climate-change velocity on species endemism. Science 334:660–664.

Sandifer, P. A., A. E. Sutton-Grier, and B. P. Ward. 2015. Exploring connections among nature, biodiversity, ecosystem services, and human health and well-being: opportunities to enhance health and biodiversity conservation. Ecosystem Services 12:1–15.

Scheffers, B. R., L. De Meester, T. C. L. Bridge, A. A. Hoffmann, J. M. Pandolfi, et al. 2016. A globally coherent fingerprint of climate change impacts across natural systems. Science 354:aaf7671.

Seager, R., N. Naik, and G. A. Vecchi. 2010: Thermodynamic and dynamic mechanisms for large-scale changes in the hydrological cycle in response to global warming. Journal of Climate 23:4651–4668.

Sharma, S., S. Couturier, and S. D. Cote. 2009. Impacts of climate change on the seasonal distribution of migratory caribou. Global Change Biology 15:2549–2562.

Sheldon, B. C. 2010. Genetic perspective on the evolutionary consequences of climate change in birds. Pages 149–168 in A. P. Møller, W. Fiedler, and P. Berthold, editors. Effects of climate change on birds. Oxford University Press, Oxford, UK.

Sheriff, M. J., G. J. Kenagy, M. Richter, T. Lee, O. Toien, F. Kohl, C. L. Buck, and B. M. Barnes. 2011. Phenological variation in annual timing of hibernation and breeding in nearby populations of Arctic ground squirrels. Proceedings of the Royal Society B 278:2369–2375.

Shukla, P. R., J. Skea, R. Slade, R. van Diemen, E. Haughey, J. Malley, M. Pathak, and J. Portugal Pereira. 2019. Technical summary. Pages 37–76 in P. R. Shukla, J. Skea, E. Calvo Buendia, et al., editors. Climate change and land: an IPCC special report on climate change, desertification, land degradation, sustainable land management, food security, and greenhouse gas fluxes in terrestrial ecosystems. Intergovernmental Panel on Climate Change, Geneva, Switzerland.

Smith, K. A., and F. Conen. 2004. Impacts of land management on fluxes of trace greenhouse gases. Soil Use and Management 20:255–263.

Suarez, F., D. Binkley, M. W. Kaye, and R. Stottlemyer. 1999. Expansion of forest stands into tundra in the Noatak National Preserve, northwest Alaska. Ecoscience 6:465–470.

Sutton, R. T., B. W. Dong, and J. M. Gregory. 2007. Land/sea warming ratio in response to climate change: IPCC AR4 model results and comparison with observations. Geophysical Research Letters 34:L02701.

Thomas, C. D., and J. J. Lennon. 1999. Birds extend their ranges northwards. Nature 399:213.

Thuiller, W., and T. Munkemuller. 2010. Habitat suitability modeling. Pages 77–85 in A. P. Moller, W. Fiedler, and P. Berthold, editors. Effects of climate change on birds. Oxford University Press, Oxford, UK.

Thurman, L. L., B. A. Stein, E. A. Beever, W. Foden, S. R. Geange, et al. 2020. Persist in place or shift in space? Evaluating the adaptive capacity of species to climate change. Frontiers in Ecology and the Environment 18(9):520–528.

Tingley, M. W., and S. R. Beissinger. 2009. Detecting range shifts from historical species occurrences: new perspectives on old data. Trends in Ecology and Evolution 24:625–633.

Tingley, M. W., W. B. Monahan, S. R. Beissinger, and C. Moritz. 2009. Colloquium papers: birds track their Grinnellian niche through a century of climate change. Proceedings of the National Academy of Sciences 106:19,637–19,643.

Toot, R., L. E. Frelich, E. E. Butler, and P. B. Reich. 2020. Climate-biome envelope shifts create enormous challenges and novel opportunities for conservation. Forests 11(9):1015.

Towns, L., A. E. Derocher, I. Stirling, and N. J. Lunn. 2010. Changes in land distribution of polar bears in western Hudson Bay. Arctic 63:206–212.

US Environmental Protection Agency. 2009. A framework for categorizing the relative vulnerability of threatened and endangered species to climate change. National Center for Environmental Assessment, Washington, DC, USA.

Vincent, W. F., M. Lemay, and M. Allard. 2017. Arctic permafrost landscapes in transition: towards an integrated Earth system approach. Arctic Science 3(2):39–64.

Visser, M. E., and C. Both. 2005. Shifts in phenology due to global climate change: the need for a yardstick. Proceedings of the Royal Society B 272:2561–2569.

Visser, M. E., S. P. Caro, K. van Oers, S. V. Schaper, and B. Helm. 2010. Phenology, seasonal timing and circannual rhythms: towards a unified framework. Philosophical Transactions of the Royal Society B 365:3113–3127.

von Humboldt, A., and A. Bonpland. 1807. An essay on the distribution of plants. Translated in 2009 by S. Romanowski and edited by S. T. Jackson. University of Chicago Press, Chicago, Illinois, USA.

Vors, L. S., and M. S. Boyce. 2009. Global declines of caribou and reindeer. Global Change Biology 15:2626–2633.

Walther, G. R., S. Berger, and M. T. Sykes. 2005. An ecological "footprint" of climate change. Proceedings of the Royal Society B 272:1427–1432.

Walther, G. R., E. Post, P. Convey, A. Menzel, C. Parmesan, T. J. C. Beebee, J. M. Fromentin, O. Hoegh-Guldberg, and F. Bairlein. 2002. Ecological responses to recent climate change. Nature 416:389–395.

Weiskopf, S. R., O. E. Ledee, L. M. Thompson. 2019. Climate change effects on deer and moose in the Midwest. Journal of Wildlife Management 83:769–781.

Wiens, J. J. 2016. Climate-related local extinctions are already widespread among plant and animal species. PLoS Biology 14(12):e2001104.

Williams, B. K., J. D. Nichols, and M. J. Conroy. 2002. Analysis and management of animal populations: modeling, estimation, and decision making. Academic Press, San Diego, California, USA.

Williams, B. K., R. C. Szaro, and C. D. Shapiro. 2007a. Adaptive management: the U.S. Department of the Interior technical guide. US Department of the Interior, Washington, DC, USA.

Williams, J. W., and S. T. Jackson. 2007. Novel climates, no-analog communities, and ecological surprises. Frontiers in Ecology and the Environment 5(9):475–482.

Williams, J. W., S. T. Jackson, and J. E. Kutzbacht. 2007b. Projected distributions of novel and disappearing climates by 2100 AD. Proceedings of the National Academy of Sciences 104:5738–5742.

Williams, S. E., L. P. Shoo, J. L. Isaac, A. A. Hoffmann, and G. Langham. 2008. Towards an integrated framework for assessing the vulnerability of species to climate change. PLoS Biology 6:2621–2626.

Young, B., E. Byers, K. Gravuer, K. Hall, G. Hammerson, and A. Redder. 2009. Guidelines for using the NatureServe climate change vulnerability index, release 1.0. http://www.natureserve.org/prodServices/climatechange/ClimateChange.jsp#v1point2.

Zemp, M., W. Haeberli, M. Hoelzle, and F. Paul. 2006. Alpine glaciers to disappear within decades? Geophysical Research Letters 33:L13504.

Zemp, M., M. Huss, E. Thibert, N. Eckert, R. McNabb, et al. 2019. Global glacier mass changes and their contributions to sea-level rise from 1961 to 2016. Nature 568:382–386.

Zuckerberg, B., A. Woods, and W. F. Porter. 2009. Poleward shifts in breeding bird distributions in New York State. Global Change Biology 15:1866–1883.

19 CONSERVATION PLANNING FOR WILDLIFE AND WILDLIFE HABITAT

ROEL R. LOPEZ, SUSAN P. RUPP, AND ANNA M. MUÑOZ

INTRODUCTION

Aldo Leopold illustrated the importance of careful execution of habitat management by saying, "I have read many definitions of what a conservationist is, and written not a few myself, but I suspect the best one is written not with a pen, but with an axe . . . A conservationist is one who is humbly aware that with each stroke they are writing their signature on the face of the land. Signatures of course differ, whether with axe or pen, and this is as it should be" (Leopold 1949:68). The careful, measured execution of land management plans is the signature on the land of natural resource managers and will be the primary focus of this chapter.

This volume so far has reviewed basic principles in wildlife–habitat relationships, approaches to measuring and evaluating wildlife habitat (Chapter 16), and commonly applied habitat modification practices (Chapter 17). The integration of theory and practice is typically accomplished through conservation planning. This chapter describes the process of conservation planning for wildlife and wildlife habitat and its associated management. Conservation planning is a creative process that requires experience, analysis, intuition, and inspiration. Even though planning can sometimes be unstructured, management plans can add structure to the overall process and are an integral part of conservation (Lopez et al. 2005). Conservation planning is a progression of steps to determine what we have (e.g., animals, plants, physical resources), what we want (e.g., more or fewer white-tailed deer [*Odocoileus virginianus*]), and how to get there (e.g., increase or decrease cover or forage requirements). Although there is no single method or standard approach to developing a conservation plan, there are some general guidelines that can aid the practitioner in successfully engaging in the process and can increase their awareness regarding their signature on the face of the land. We introduce some of these basic concepts in the development and careful execution of habitat management plans.

WHAT IS PLANNING?

A good working definition of planning might include "the deliberate social or organizational activity of developing an op-timal strategy for solving problems and achieving a desired set of objectives" (Yoe and Orth 1996:11). Although this definition oversimplifies an often complex process, it does emphasize a few important planning elements.

1. *Future control.* Planning focuses on the control of future consequences through present actions, suggesting that planning and action together are necessary to ensure and increase predictability in the management of the targeted lands. As such, uncertainties and how to best manage them must be part of the planning process.
2. *Problem-solving.* Planning is a problem-solving approach that addresses a given natural resource issue. Generally, wildlife management problems can include increasing population numbers (e.g., endangered species), decreasing effects from nuisance or invasive species, or maintaining a sustained yield for a game species. The need to address a problem or natural resource issue often influences the reason for conservation planning.
3. *Team effort.* Individuals often do planning in a team environment that considers the opinions and desires of a wide-ranging set of stakeholders (e.g., landowners, state and local governments). This, of course, depends on the spatial extent of the property and landowner (i.e., a plan for small, private property may be done by an individual). Planning should not be done without the consideration of various perspectives, constraints, and approaches. Increasing the level of stakeholder input in conservation planning can ensure the careful review of all viable options.
4. *No single approach.* Plans are uniquely tailored to a specific situation and set of objectives and assumptions. Unfortunately, there is no standard approach in conservation planning. General rules and procedures can be applied, however, and the land manager should be willing to adapt and modify a management plan to the specific situation to improve its overall effectiveness.
5. *Adaptive framework.* Planning should be adaptive and involve feedback loops. Continued monitoring, evaluation, and adjustment are critical components to good manage-

ment plans. Management plans are part of a process. This process typically has a predetermined end point or goal that is reevaluated at the end of the target time line and restarts the management planning process.

6. *Intention to implement.* Planning is done with the intention to implement the strategies outlined in the plan, not just for planning's sake. Management plans require a substantial amount of time, effort, and funding. Land managers should develop management plans with the intention to follow them, and they should not simply put them on a shelf.

Mission and Vision Form the Foundation

Every agency or organization has a specific mission statement and vision, the heart and basic tenet of any conservation planning effort. A clearly stated mission and vision are what clarify the direction in any plan, eliminating circular arguments and allowing a clear development of the desired condition for which goals and objectives become much easier to craft. In short, an agency's mission and vision are the foundation on which conservation planning develops (Fig. 19.1).

The authority for an agency or organization is usually (if not always) stated through a mission statement. A mission statement based on correct principles is like a personal constitution, the basis for making major, life-directing decisions in the midst of the circumstances and challenges that affect an agency or organization. It is the lens through which agencies and organizations see the world. For example, the mission statement for the US Fish and Wildlife Service (USFWS) is "working with others to conserve, protect and enhance fish, wildlife, and plants and their habitats for the continuing benefit of the American people" (https://www.fws.gov/help/about_us.html). This mission statement explains what the organization does, for whom, and the benefit of doing

it, which guides decision-making within the USFWS. It acts as a sieve through which everything is filtered. If a potential decision or action does not fall under the premises outlined in the mission statement, it is not considered to be a priority on which to act. You would not find the USFWS involved with developing a high-rise building in the middle of a major metropolitan area unless it is designed to assist their primary mission in some form or another (e.g., to house employees that assist with the USFWS mission).

Unlike a mission statement that concerns what an organization is all about, a vision statement is what the organization wants to become. Vision statements define the organization's purpose in terms of its values rather than bottom-line measures. It describes how the future will look if the organization achieves its mission. The USFWS vision is to "unite all service programs to lead or support ecosystem level conservation . . . by becoming a more technically capable and culturally diverse organization; through involving stakeholders; through scientific expertise; through land and water management; and, through appropriate regulation" (https://www.fws.gov/help/about_us.html). Once the vision and mission of an agency or organization are established, it becomes much easier to design conservation plans that help to achieve those desired end points.

Goals and Objectives

Our working definition of planning includes a strategy for achieving a desired set of objectives. Management plans should be objective- and goal-oriented. The terms *goals* and *objectives* are often used interchangeably; however, there are some subtle differences between them when used in conservation planning. A goal can be defined as the final or desired end purpose. Goals are overarching statements that establish the overall direction for (and focus of) a conservation plan and define the scope of what the plan should achieve. Conversely, objectives are specific, measurable actions with defined completion dates used to achieve a goal. They outline who will make what change, by how much, where, and by when in order to reach the goal. From a conservation planning perspective, both definitions convey the same basic intent: do the right thing for the resource. But the subtle differences in their definitions also establish a hierarchical structure that suggests we set goals first and then establish objectives to attain them. In other words, goals are overarching outcomes, and objectives are the specific actions—the building blocks—we use to achieve the goals. The best goals are consistent with the mission of the agency or organization. The objectives are then the steppingstones to achieve the goals.

For example, two managers may share the common goal of improving quail (*Colinus* spp.) habitat but differ in how they individually will strive to reach this goal (Table 19.1). Here each land manager has the same goal (i.e., improve quail habitat), yet the objectives are different. In both cases, their individual objectives can allow them to reach their common goal because they are specific, measurable, and have a defined completion date. As Leopold (1949:68) so eloquently said, "sig-

Figure 19.1. The strategic planning pyramid. To be successful in conservation planning, each project should reflect the agency or organization's vision, mission, goals, and objectives.

Table 19.1. Goals and objectives from the perspective of two hypothetical land managers

Land Manager 1	Land Manager 2
Goal: improve quail habitat	Goal: improve quail habitat
Objective 1: apply 16.2-ha prescribed fire in April next year	Objective 1: control predators though active trapping two weeks per year
Objective 2: disc 4.05 ha annually on a rotational basis	Objective 2: plant two 2-ha food plots in spring of next year
Objective 3: maintain brush piles in the northwest corner on annual basis	Objective 3: increase native grasses by 30% across property by end of five-year period

natures of course differ, whether with axe or pen, and this is as it should be."

Types of Planning

There are many different types of plans. Land-use plans maintain the most suitable types of land uses (e.g., developed versus natural areas) and specific activities relative to the management of those lands. Transportation plans may address public transportation strategies to support the projected growth of a region or county. Historic preservation plans might be used in a community to protect a historic area like a Civil War battlefield or the remnants of an old mining community. These are a few examples of the many types of plans used in conservation planning. A common challenge with these types of planning efforts is the lack of coordination and integration among these various processes within the same geographic area. The potential for conflicting approaches and recommendations between transportation and conservation plans, for example, requires a concerted effort by agencies and other resource managers to work collaboratively. Part of this challenge arises in large part from the varying missions of agencies and organizations responsible for the management of natural resources. Consider the multitude of agencies responsible for some form of conservation planning, each with their own varying mission and vision (Table 19.2). Organizations or agencies responsible for conservation planning can range from federal and state agencies to nonprofit organizations, land trusts, and other local or regional partnerships. This wide breadth of organizations and agencies involved in the management of natural resources can be challenging and even daunting at times. Questions regarding who is responsible for what and mission creep (i.e., the expansion of a project or mission beyond its original goals) can make the planning process difficult.

In addition to the multitude of agencies involved, the involvement of the private landowner in successful conservation planning is essential. In many areas of the country, most lands are privately owned (Fig. 19.2). According to the Partners for Fish and Wildlife Act (PL 109-294) passed by Congress in October 2006, roughly 60% of fish and wildlife reside on private lands. In addition to the wildlife present there, private lands also provide many ecosystem services (e.g., clean water, open

space). This makes working with private landowners a critical component of conservation planning. Often, private lands can serve as critical corridors for connecting existing public lands. Because these lands cannot be acquired owing to private ownership or lack of funding to obtain them, private lands have their own stewards (i.e., the landowner) responsible for managing those lands. Even if those lands could be obtained, federal and state budgets are stretched to the breaking point, and public agencies often struggle to manage the properties they already have. The stewardship of natural resources under private ownership in many cases is better than if managed by a federal or state agency. Regardless of ownership, it is important for wildlife managers to work with private landowners to design effective management plans that work across boundaries and can benefit the landowner through strategic planning and careful consideration of the social and economic values of existing lands and supplement essential habitat for wildlife. Despite their importance, however, private lands are often overlooked because agencies tend to think within their own boundaries and may get frustrated because they lack inventories and control of resources on private lands.

Planning also can vary in spatial scale or a particular focus. Conservation plans can range from multistate ecosystem level plans (e.g., America's Longleaf Restoration Initiative) to local habitat conservation plans that focus on a single endangered species (Table 19.3). All levels of planning are important, but each individual plan is likely to be more effective when it integrates with or complements other planning efforts. Including information from a local habitat conservation plan in an ecoregional plan maintains consistency and increases coordination between agencies and conservation management activities. This was the case in the development of The Nature Conservancy's (TNC) Conservation by Design approach. TNC realized funding expenditures would be more effective if they had a long-range plan for land acquisition and restoration based on the best available science at larger regional or global scales. As a result, TNC initiated Conservation by Design (https://www.nature.org/en-us/about-us/who-we-are/our-science/conservation-by-design/), a science-based conservation approach at multiple scales that uses three complementary analytical methods: global habitat assessments, ecoregional assessments, and conservation action planning. Global habitat assessments provide a baseline against which TNC can measure progress toward their mission, identify conservation gaps, and establish priorities for allocating resources on a global scale. Within the global habitat assessments, TNC identifies which specific ecoregions require attention and threats to biodiversity and strategic opportunities that affect one or more major landscapes and demand immediate attention. In some cases, ecoregions span multiple states and countries, and they identify and prioritize conservation land on the current ecology and threat levels within that region. Finally, global and ecoregional priorities are translated into conservation strategies and actions through conservation action planning. This method is used to design and manage conservation projects that advance conservation at any scale, from efforts to con-

Table 19.2. Examples of agencies and organizations responsible for conservation planning

Federal Agencies	State Agencies	Municipal Governments	Nonprofit Organizations	Local Land Trusts	Coalitions and Partnerships
US Forest Service	*Maine* Maine Bureau of Parks and Lands	Conservation Commissions	Trust for Public Lands	Lower Kennebec Regional Land Trust	Mount Agamenticus to the Sea Conservation Initiative
US Fish and Wildlife Service	Maine Forest Service	Town Forest Committees	The Nature Conservancy	Mahoosuc Land Trust	Quabbin to Cardigan Conservation Collaborative
Natural Resource Conservation Service	Maine Department of Inland Fisheries and Wildlife	Open Space Committees	The Conservation Fund	Maine Wilderness Watershed Trust	Great Bay Partnership
Army Corps of Engineers	*Florida* Florida Fish and Wildlife Conservation Commission		Forest Society of Maine	Sebasticook River Watershed Association	Vermont Housing and Conservation Board
Bureau of Land Management	Florida Park Service		Maine Audubon Society	Five Rivers Land Trust	Prairie Coteau Habitat Partnership
Department of Defense	Florida Department of Environmental Conservation		American Farmland Trust	Lakes Region Conservation Trust	
National Park Service	*Texas* Texas Parks and Wildlife Department Texas Forest Service		Society for the Protection of New Hampshire	Monadnock Conservancy	
	Texas Commission on Environmental Quality		Texas Land Trust	Piscataquog River Watershed Association	
	South Dakota South Dakota Department of Game, Fish and Parks		Audubon Vermont	Jericho Underhill Land Trust	
	South Dakota Department of Environment and Natural Resources		Audubon Society of North Carolina	Lake Champlain Land Trust	
	California California Conservation Corps			Middlebury Area Land Trust	
	Department of Boating and Waterways				
	Department of Conservation				
	Department of Fish and Game				
	Department of Forestry and Fire Protection				
	Department of Parks and Recreation				
	Department of Water Resources				

serve species and ecosystems in a single watershed or landscape to efforts to reform regional or multinational policies.

Major Legislation and Policy Directing Wildlife Planning

We have discussed what planning is and have provided examples of several entities typically responsible for or involved in conservation planning. Another question to ask is, Why plan? In addition to some of the aforementioned benefits of conservation planning, there are three basic frameworks that commonly influence conservation planning: agency directives or authorities (e.g., laws or policy that influence an organiza-

tion's mission), environmental compliance (e.g., regulatory laws), and incentive programs (e.g., Farm Bill programs). We provide a brief description of each of these influences, but to understand why planning is necessary, it is important to realize that most conservation funding to support implementation of management activities typically requires a management plan that outlines the activities and desired outcomes the funding will support. A summary of federal and state conservation funding programs illustrates the reasons to develop management plans (Table 19.4). Familiarity with these programs is important for obtaining financial support for conservation or, if working with private landowners, encouraging certain be-

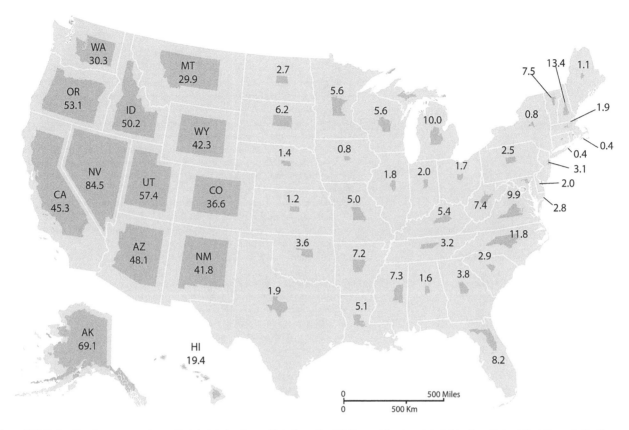

Figure 19.2. Federal land as a percentage of total state land area. Data from the US General Services Administration, Federal Real Property Profile 2004, excluding trust properties

Table 19.3. Examples of conservation planning efforts by focus or scale

Scale or Focus	Example	Description	Source
International	North American Waterfowl Management Plan	The North American Waterfowl Management Plan is an international plan to conserve waterfowl and migratory birds in North America. Established in 1986 by Canada and the United States, it was expanded to include Mexico in 1994.	https://nawmp.org/content/north-american-waterfowl-management-plan
Landscape	America's Longleaf Restoration Initiative	America's Longleaf Restoration Initiative is a voluntary collaborative effort of more than 20 organizations and agencies that seeks to define, catalyze, and support coordinated longleaf pine conservation efforts across a nine-state region.	www.americaslongleaf.org
Regional	The Nature Conservancy's ecoregional assessments	The Nature Conservancy's long-range plan for land acquisition and restoration based on natural ecoregions. Each ecoregional assessment identifies and prioritizes conservation lands on the basis of ecology and threat levels.	http://www.conservationgateway.org/ConservationPlanning/Setting Priorities/EcoregionalAssessment/Pages/ecoregional-assessment.aspx
State	State wildlife action plans	State wildlife action plans take a proactive approach to habitat conservation and species preservation. These plans outline a strategy for protecting priority habitats and species that are at risk but not yet on the endangered species list in conjunction with state wildlife agencies from all 50 states, all US territories, and the District of Columbia.	https://www.fishwildlife.org/afwa-informs/state-wildlife-action-plans
Local	Sonoran Desert Conservation Plan	In 1998, planners and ecologists came together to create a plan that protects critical environmental and cultural resources in Pima County, Arizona. The plan must be updated to remain current, and questions remain about habitat management on these newly protected lands.	https://webcms.pima.gov/government/sustainability_and_conservation/conservation_science/the_sonoran_desert_conservation_plan/
Species-specific	Endangered Species Recovery Plans	Recovery Plans are a critical part of endangered species protection. They outline goals and actions to recover endangered species to a self-sustaining level for one or multiple species.	https://www.fws.gov/endangered/species/recovery-plans.html

Table 19.4. Summary of federal and state conservation funding programs by agency

Acronym	Program	Agency	Description
REPI	Readiness and Environmental Protection Integration	DoD	Cost-sharing program for the acquisition of easements from willing sellers as a way to preserve high-value habitat and limit incompatible development around military ranges and installations.
TE	Transportation Enhancements	DOT	The TE program represents 10% of total Surface Transportation Program (STP) funds. STP is a formula apportionment (maximum 80%) to the states through the federal aid highway program. TE projects include construction, but not maintenance, of various modes of transportation, including the rail trail program, which is funded by an excise tax from the Highway Trust Fund. Projects involve conversion of abandoned railroad corridors into multiuse trails available for recreation.
RTP	Recreational Trails Program	DOT	Apportionments (maximum 80%) to the states to benefit outdoor recreation, including hiking, biking, in-line skating, equestrian use, etc.
LWCF	Land and Water Conservation Fund State Side Program	NPS	State matching grant program (at least 50%) where states request funds from the National Park Service for specific projects.
LWCF	Land and Water Conservation Fund Federal Land Acquisition	NPS	Program to acquire new federal recreation lands in cooperation with the National Park Service.
NERRS	National Estuarine Research Reserves System	NOAA	Grants to coastal states to acquire lands and waters necessary to ensure long-term management of an area as a national estuarine reserve and for operations, construction, and education programs.
CELCP	Coastal and Estuarine Land Conservation Program	NOAA	Competitive state and local grants to acquire property or conservation easements from willing sellers within a state's coastal zone or coastal watershed boundary.
CRP	Conservation Reserve Program	NRCS	Pays farmers to take environmentally sensitive cropland out of production and plant long-term resource-conserving covers (e.g., grasses and trees). The CRP offers 10- to 15-year contracts with annual rental payments, incentive payments for certain activities, and cost-share assistance to establish approved ground cover on eligible cropland in order to reduce erosion on sensitive lands, improve soil and water, and provide significant wildlife habitat.
EQIP	Environmental Quality Incentives Program	NRCS	Technical assistance, cost-share payments, and incentive payments to assist crop and livestock producers with environmental and conservation improvements. The EQIP, established by the 1996 Farm Bill, offers financial, educational, and technical help to install or implement structural, vegetative, and management practices and is one of the several voluntary conservation programs that are part of the US Department of Agriculture Conservation Toolbox.
FPP	Farmland Protection Program	NRCS	State and local grants to help purchase easements that would preclude nonfarm development of productive farmland.
HFRP	Healthy Forests Reserve Program	NRCS	Restores private forest ecosystems to (1) promote the recovery of threatened species, (2) improve biodiversity, and (3) enhance carbon sequestration.
FLP	Forest Legacy Program	USFS	Competitive grant program and direct payments to support land acquisition (fee purchase and easement) to protect important scenic, cultural, fish, wildlife, and recreation resources and riparian areas.
FSP	Forest Stewardship Program	USFS	Provides technical and educational assistance to nonindustrial private forest owners to develop forest management plans.
UCFP	Urban and Community Forestry	USFS	Forest-related technical, financial, research, and educational services to local government, nonprofits, community groups, and educational institutions.
CFOSCP	Community Forest Open Space Conservation Program	USFS	Matching grants (50/50) for local governments, tribes, and nonprofit organizations for full fee purchase of forestlands; differs from the FLP in its community focus and the requirement of fee purchase plus public access. The program is part of the 2008 Farm Bill.
CESCF	Cooperative Endangered Species Conservation Fund	USFWS	Grants to private landowners and groups to implement conservation projects for listed species and at-risk species. Funded activities include developing habitat conservation plans, land acquisition, habitat restoration, research, and wildlife management.
MBCF	Migratory Bird Conservation Fund	USFWS	Land and water acquisition or rental as recommended by the secretary of the interior for the protection of migratory bird species. Funding comes from Duck Stamp revenues, import duties on arms and ammunition, and refuge admission fees.
NCWC	National Coastal Wetlands Conservation Grants	USFWS	Matching grants to states for acquisition, restoration, and enhancement of coastal wetlands. Funding comes from the Sport Fish Restoration and Boating Trust Fund, which is supported by excise taxes on fishing equipment, motorboat and small-engine fuels, and import duties.
NAWCA	North American Wetland Conservation Act Grants	USFWS	Matching grants to organizations and individuals to implement wetlands conservation projects in Canada, Mexico, and the United States. Funding comes from congressional appropriations and fines and penalties collected under the Migratory Bird Treaty Act of 1918, the Sport Fish Restoration and Boating Trust Fund, and interest accrued on the Wildlife Restoration Trust Fund.

Table 19.4. continued

Acronym	Program	Agency	Description
SFR	Sport Fish Restoration Program	USFWS	Apportionments to the states for fishery projects, boating access, and aquatic education. This program is funded by the Sport Fish Restoration and Boating Trust Fund, which is supported by excise taxes on fishing equipment, motorboat and small-engine fuels, and import duties. The annual allocation to the Sport Fish Restoration Program from legislation called the Dingell-Johnson Act or the Wallop-Breaux Act is equal to 57% of the trust fund's receipts (after annual deductions).
WRP	Wildlife Restoration Program	USFWS	Apportionments to the states to restore, conserve, and manage wild birds, mammals, and their habitat. This program is funded by the Wildlife Restoration Trust Fund, which is supported by excise taxes, authorized by the Pittman-Robertson Act, on hunting equipment.
SWGP	State Wildlife Grants Program	USFWS	Grants to plan and implement programs that benefit wildlife and habitats, including species not hunted or fished. Funding comes through appropriations from the Land and Water Conservation Fund.
LIP	Landowner Incentive Program	USFWS	State grants to protect and restore habitats on private lands to benefit at-risk species (including federally listed, proposed, or candidate species).

Note: Agency acronyms are as follows: DoD, Department of Defense; DOT, Department of Transportation; NOAA, National Oceanic and Atmospheric Administration; NPS, National Park Service; NRCS, Natural Resources Conservation Service; USFS, US Forest Service; USFWS, US Fish and Wildlife Service.

haviors that can benefit conservation of wildlife populations and their habitats.

Directives and Agency Guidance

Federal and state agencies work under the authority of laws or policies in support of the organizations' missions. Nongovernmental organizations do not necessarily work under a given authority but may be influenced by some federal or state law (e.g., Endangered Species Act, or ESA), or they may stand to benefit from authorities that provide financial or technical assistance (e.g., the Pittman-Robertson and Dingell-Johnson acts; see Chapter 2). To receive Pittman-Robertson and Dingell-Johnson funds, a state must submit a comprehensive fish and wildlife resource management plan (e.g., state wildlife action plans). The plans must be for at least five years and must be based on long-range projections regarding the desires and needs of the public. In the expenditure of both Pittman-Robertson and Dingell-Johnson funding, management plans are critical components in state conservation agencies that obtain and support their wildlife programs through these funds. These federal acts guide or direct state agencies and their respective state management plans as to how they manage their natural resources.

Environmental Compliance

A second set of laws that influence conservation planning can be labeled under the environmental compliance umbrella. Two landmark pieces of legislation include the National Environmental Policy Act (NEPA) of 1969 and the ESA of 1973. The NEPA was passed during the environmental movement of the 1960s in response to several environmental concerns ranging from clean water and air, pesticide contamination, and declining wildlife species to protection of wilderness areas. The act declares a national policy to "encourage productive and enjoyable harmony between man and his envi-

ronment; to promote efforts which will prevent or eliminate damage to the environment and biosphere and stimulate the health and welfare of man; [and] to enrich the understanding of the ecological systems and natural resources important to the Nation" (42 USC §4321). The NEPA requires federal agencies to identify, analyze, describe (i.e., consider), and publicly disclose the environmental impacts associated with federal actions through publication in the *Federal Register* and a series of public scoping meetings. The act also establishes the Council on Environmental Quality to review government policies and programs for conformity with the NEPA. For actions that may have significant effects to the environment, the development of an environmental impact statement (EIS) or environmental assessment (EA) is required. An EIS includes an examination of environmental impacts and alternatives available to the proposed action, including no action. Prior to preparing an EIS, the agency should coordinate with other state and federal agencies that have expertise on any environmental effect involved (e.g., USFWS and effects to endangered species). Conservation planning conducted by federal agencies typically includes preparation of an EIS when developing management plans. Conversely, an EA is typically a shorter review document that reviews the purpose and need of the proposal, any alternatives, and a brief summary of the affected environment. An EA will either result in a Finding of No Significant Impact (FONSI) or, if significant environmental impacts appear likely, an EIS.

Another example of federal law under the environmental compliance umbrella that influences conservation planning is the ESA. The ESA provides for the conservation of plants and animals considered threatened with or in danger of extinction. The authority for implementing and executing the act is delegated to the USFWS for plants and animals or the National Marine Fisheries Service (NMFS) for marine life and anadromous fishes. But the ESA declares that all federal departments and agencies must also seek to conserve endangered and threatened species and to use their authorities in furtherance of the purposes of the act (16 USC 1531-1544, section 2). The USFWS and NMFS are responsible for determining which species are listed as threatened or endangered and for delineating critical habitats necessary for their conservation. The ESA comprises several sections that provide further guidelines to the aforementioned agencies. These sections offer direction for listing a species as threatened or endangered, designation of critical habitat, and recovery planning (section 4), prohibited actions for listed species (section 9), and penalties and enforcement procedures (section 11). Two sections of the ESA that are noteworthy with regard to conservation planning include sections 7 and 10. The USFWS has developed handbooks (USFWS and NMFS 1998, 2016) to guide applicants and planners through the section 7 and habitat conservation plan (HCP) processes.

Section 7 of the ESA outlines procedures for interagency cooperation to conserve federally listed species and designated critical habitats. Section 7 requires federal agencies to consult with USFWS to ensure that they are not undertaking, funding, permitting, or authorizing actions likely to jeopardize the continued existence of listed species or to destroy or adversely modify designated critical habitat. Section 7 also outlines the processes for consultation to address potential effects to listed species and designated critical habitats, and conferences to address potential effects to species proposed for listing or proposed critical habitat. For projects that are likely to affect adversely a listed species or designated critical habitat, formal consultation is required to ensure the proposed federal actions will not jeopardize the continued existence of a species or destroy or adversely modify designated critical habitat. In these cases, the action agency must develop and submit a biological assessment to USFWS or NMFS that outlines the proposed action and how it may affect listed species, including whether the action may result in the "take" of a species. Take includes the physical removal of an individual animal from the wild or killing thereof, removal of habitat so it can no longer be used by the species, and activities (e.g., recreation, machinery noise) that make an area unsuitable for habitation by the species. Take that is incidental to a proposed federal action may be permitted as long as the agency agrees to comply with the specific measures and conditions set forth by USFWS or NMFS when they issue their biological opinion.

Take can be permitted for scientific purposes and for nonfederal projects (e.g., commercial or residential development on private lands) if the approved plans minimize and mitigate for the incidental take (usually of habitat) of the species. Under section 10 of the ESA, for example, incidental take is authorized through a variety of voluntary agreements to conserve or minimize and mitigate effects upon fish and wildlife, including: candidate conservation agreements, safe harbor agreements, and HCPs with implementation agreements. A brief description of each of these types of agreements and how they influence or are a part of conservation planning follows.

Candidate Conservation Agreements with Assurances

The USFWS (and, to a lesser degree, NMFS) offers nonfederal landowners a policy option of entering into prelisting or candidate conservation agreements with assurances (CCAAs) that provide regulatory assurances to landowners who voluntarily agree to protect habitat for candidate fish and wildlife species before they are listed for protection under the ESA. Under the USFWS policy for CCAAs, successful applicants will receive an enhancement of survival permit if they agree to actions that will provide a conservation benefit to specified candidate species, so listing the species would be unnecessary if other landowners within the range of the species were to manage their land in the same fashion to remove known threats to the species. If species covered under CCAAs are eventually listed for protection, the enhancement of survival permit authorizes incidental take of those species by any action in accordance with the CCAA.

Safe Harbor Agreements

Safe harbor agreements are voluntary arrangements between the USFWS or NMFS and cooperating nonfederal landowners important to the implementation of the various ESA activities. Their purpose is to promote voluntary management for threatened and endangered species on nonfederal property while offering assurances to landowners regarding future regulatory restrictions in the management of these species on their land. The value of safe harbor agreements is that they allow landowners to provide a net conservation benefit and contribute to recovery of a listed species in exchange for the assurance that a return to the baseline habitat conditions at the time of permit inception will not result in liability for unlawful take. For example, if a landowner has five breeding pairs of an endangered species on their property, they may sign a safe harbor agreement. If that landowner were to improve habitat to increase the population to ten breeding pairs, they would not be penalized if, at the end of the permit duration, this number was reduced to the baseline (i.e., five breeding pairs) because of an action (e.g., development of species habitat). As in CCAAs, parties to approved safe harbor agreements will receive an enhancement of survival permit that authorizes incidental take by actions consistent with the terms of the agreement.

Habitat Conservation Plans

Another type of a conservation plan under the ESA is a habitat conservation plan. Section 10(a)(2)(B) of the ESA allows incidental take of listed species resulting from nonfederal actions through the development of an HCP, a planning document required as part of the incidental take permit application and that can cover a single action or a number of similar activities that will occur over a broad area (e.g., expanding commercial and residential development zones or transmission line rights-of-way that cross multiple jurisdictions). HCPs describe the anticipated effects of the proposed taking, how those effects will be minimized and mitigated, and how the plan is to be funded, as explained in section 10(a)(2)(A) of the ESA. Frequently, these plans require habitat protection, restoration, and enhancement in an area in exchange for some lost habitat in another, if it is determined that these actions are required to offset the effects of the taking to the species. In addition to receiving an incidental take permit, an HCP applicant is also provided with long-term regulatory assurances that, if unforeseen circumstances arise, the USFWS will not require additional mitigation from the permittee beyond the level otherwise agreed to in the plan. Programmatic HCPs are developed when a number of similar activities will occur over a broad area, such as when city and county governments want to expand their zone of commercial and residential development, or when a transmission line (and associated right-of-way corridor) will cross numerous governmental jurisdictions. HCPs can be prepared for single species or multiple species and may include unlisted species. In this case, the unlisted species is treated as if it were listed, including the required effects analysis and mitigation. In this way, the permittee can be assured that subsequent listing would not result in additional administrative or conservation requirements. Single-species HCPs are usually easier to develop and implement because they tend to be less complicated, but multispecies HCPs must be developed for areas containing one or more listed or candidate species, and for long-term projects such as phased developments where there is potential for additional listings in the future. Failure to address all listed species that are likely to be incidentally taken by the proposed activities would preclude the USFWS or NMFS from being lawfully able to issue the permit on the grounds that the activities would be in violation of the ESA (e.g., unauthorized take).

Species Status Assessments

The USFWS recently developed the Species Status Assessment (SSA) framework, an analytical tool to communicate scientific information necessary to inform decisions for spe-

cific fish, wildlife, and plants protected under the ESA (US Fish and Wildlife Service 2016). USFWS is confident that focused, repeatable, and rigorous scientific evaluations conducted through the SSA framework will lead to better assessments; clearer and more concise documents; and improved, transparent decision-making. The SSA is a biological risk assessment that provides rigorous scientific information to decision makers about specific species needs. It provides decision makers with a description of species status, focusing on the likelihood that the species will sustain populations under current conditions, and describes uncertainties the species face. Using the SSA framework, USFWS considers what a species needs to maintain viability. In a nutshell, a viable species has highly resilient populations distributed across its range and within areas known to be important for future adaptation. To determine viability, USFWS characterizes the species in terms of its resilience, representation, and redundancy (the three Rs).

- *Resilience* describes the ability of populations to withstand random disturbances.
- *Representation* describes the species' ability to adapt to changing environmental conditions, as characterized by the amount of genetic and environmental diversity within and among populations.
- *Redundancy* is about spreading risk among multiple populations and describes the species' ability to withstand catastrophic events.

The SSA is a living document that follows each plant and animal over the course of its time on the ESA list. The SSA does not directly result in a decision, but it enables decision makers to evaluate each species against procedures outlined in the ESA. Because the SSA informs all decisions made under the ESA (Fig. 19.3), the SSA serves as a baseline for USFWS conservation planning documents such as Recovery Plans, Safe Harbor Agreements, Biological Opinions, Critical Habitat, and Candidate Conservation Agreements, among others.

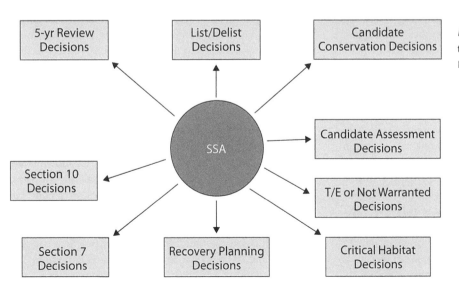

Figure 19.3. Species status assessment relationships in analytical framework used by US Fish and Wildlife Service.

Incentive Programs

The third set of laws that influences conservation planning are incentive programs. Conservation planning on private lands often involves incentivizing private landowners to manage their natural resources to their benefit and the benefit of the public at large. Incentive programs allow federal and state agencies to encourage sustainable use of these ecosystem services by providing financial and technical assistance to landowners.

As mentioned previously, the Partners for Fish and Wildlife Act (PL 109-294) states that roughly 60% of fish and wildlife occur on private lands. The mission of the Partners for Fish and Wildlife Program, which is administered by USFWS, is "to efficiently achieve voluntary habitat restoration on private lands, through financial and technical assistance, for the benefit of Federal Trust Species" (http://www.fws.gov/partners). The Partners Program (as it has come to be called) is active in all 50 states and US territories. It provides technical assistance and delivers on-the-ground restoration projects, particularly to the nation's private landowners, farmers, ranchers, and corporations. In 2007 the USFWS finalized a five-year strategic plan for the Partners Program, consisting of three parts: a vision document that describes the program and its five major goals, regional strategic plans that highlight conservation priorities in each of eight geographic regions, and a national summary document that reflects a national overview of habitat priorities and targets based on the regional strategic plans (US Fish and Wildlife Service 2010). It is important to recognize that even though the Partners Program has its own mission, the program still operates under the overarching mission of the USFWS.

Several state agencies also have incentive programs to assist their landowners with conservation on private lands. Many of these programs are designed to meet specific habitat needs for wildlife species. For example, as part of their wetland and grassland habitat program, wildlife biologists in South Dakota work with landowners and other conservation partners (e.g., nonprofit organizations, USFWS, Natural Resources Conservation Service [NRCS]) to implement wetland and grassland conservation practices that will benefit breeding waterfowl and other wildlife species dependent upon these landscapes, and to meet the needs and management goals of landowners. Up to 100% cost share is provided to the landowner for wetland restoration projects that involve removal of drainage tile or plugging drainage ditches. Other states such as Texas offer wildlife tax reductions to landowners who develop a formal wildlife plan. Furthermore, adjacent landowners can come together and form a wildlife management property association with one wildlife management plan for the group, but every landowner who wants to voluntarily participate is required to sign it. Similarly, landowners in Florida can apply for a conservation exemption in which land dedicated in perpetuity and used exclusively for conservation purposes entitles landowners to a full property tax exemption. We encourage readers to become knowledgeable about the landowner programs available in their own state so they can make the most effective use of all the resources available to them.

A landmark piece of legislation supporting conservation on private lands is the Food, Conservation, and Energy Act of 2018, also known as the 2018 Farm Bill. It is an $867 billion, five-year agricultural policy bill. The 2018 Farm Bill is a continuation of the 2008 Farm Bill and follows a long history of agricultural subsidy in energy, conservation, nutrition, and rural development. One of the three major components of the Farm Bill is providing baseline funding for conservation and working lands programs ($60 billion in 2018). The majority of these programs are managed by the NRCS and the Farm Service Agency (FSA). Originally called the Soil Conservation Service, established by Congress in 1935, the NRCS was refocused and directed to execute the mission of conserving natural resources on private lands in 1994. The NRCS has since expanded to provide landowners technical and financial assistance to improve soil, water, air, plants, and animals that result in productive lands and healthy ecosystems. The more common agency programs outlined below provide a brief description of each program and how funding and technical assistance are delivered (Table 19.5).

Conservation Reserve Program

Generally, Farm Bill conservation programs can be grouped as follows: working lands, land retirement, conservation easements, conservation compliance, and partnership and grants. Land retirement programs authorize the US Department of Agriculture (USDA) to make payments to private landowners to voluntarily retire land from production for less-resource-intensive uses. The primary land retirement program in this category is the Conservation Reserve Program (CRP). The CRP pays producers annual rental payments under 10- to 15-year contracts to set aside previously cropped land that is considered to be marginally productive or highly erodible. In return for establishing and maintaining conservation practices that address soil erosion, water quality, wetland and forest enhancement, and wildlife management, landowners receive annual rental payments, cost-share assistance (not to exceed 50% of the eligible costs), and under certain conditions incentives for enrolling land, undertaking particular practices, and performing certain maintenance practices. Land must meet eligibility requirements to qualify for the CRP. CRP practices include, for example, establishing vegetation cover or trees on erodible cropland, planting native grasses, or placing buffer strips along stream banks to reduce pollution.

Conservation Stewardship Program

A second category of importance in conservation planning is "working lands" conservation programs that allow private land to remain in production while implementing various conservation practices to address natural resource concerns specific to the area. Generally, program participants receive some form of conservation planning and technical assistance to guide decisions on the most appropriate practices to apply, given the natural resource concerns and land condition. The two main working lands programs are the Conservation Stewardship Program (CSP) and the Environmental Quality Incen-

Table 19.5. US Department of Agriculture Natural Resource Conservation Service and Farm Service Agency conservation programs, 2018

Program	Purpose
ACEP	Helps landowners, land trusts, and other entities protect, restore, and enhance wetlands, grasslands, and working farms and ranches through conservation easements.
AMA	Helps agricultural producers manage financial risk through diversification, marketing, or natural resource conservation practices. NRCS administers the conservation provisions, while USDA's Agricultural Marketing Service and Risk Management Agency implement the production diversification and marketing provisions.
CIG	Awards competitive grants that influence innovation and develop the tools, technologies, and strategies for next-generation conservation efforts on working lands. Grantees leverage the federal investment through matching requirements. Through CIG's new On-Farm Trials, partners provide incentive payments to producers to offset the risk of implementing innovative approaches.
CREP	Encourages landowners to sell or lease over the long term to beginning, socially disadvantaged, and veteran farmers and ranchers willing to implement sustainable practices or transition to organic production by providing two years of additional payments for expiring CRP-enrolled land.
CRP	Protects soil, water quality, and habitat by removing highly erodible or environmentally sensitive land from agricultural production through long-term rental agreements.
CSP	Helps agricultural producers maintain and improve their existing conservation systems and adopt additional conservation activities to address priority natural resource concerns. Participants earn CSP payments for conservation performance—the higher the performance, the higher the payment.
EQIP	Provides financial and technical assistance to agricultural producers to address natural resource concerns and deliver environmental benefits, such as improved water and air quality, conserved ground and surface water, reduced soil erosion and sedimentation, and improved or created wildlife habitat.
HFRP	Helps landowners restore, enhance, and protect forestland resources on private lands through easements and financial assistance. Through HFRP, landowners promote the recovery of endangered or threatened species, improve plant and animal biodiversity, and enhance carbon sequestration.
RCPP	Promotes coordination between NRCS and its partners to deliver conservation assistance to producers and landowners. Under partnership agreements, NRCS and its partners leverage and target their respective resources to deliver conservation assistance to producers and landowners to address priority natural resource concerns.
VPA-HIP	Provides state and tribal governments with funding or incentives to expand or improve habitat in existing public access programs.

Note: Most programs require the development of a conservation plan. For further specifics regarding plan requirements and eligible practices, visit http://www.nrcs.usda.gov. Acronyms are as follows: ACEP, Agricultural Conservation Easement Program; AMA, Agricultural Management Assistance Program; CIG, Conservation Innovation Grants; CREP, Conservation Reserve Enhancement Program; CRP, Conservation Reserve Program; CSP, Conservation Stewardship Program; EQIP, Environmental Quality Incentives Program; HFRP, Healthy Forest Reserve Program; RCPP, Regional Conservation Partnership Program; VPA-HIP, Voluntary Public Access and Habitat Incentive Program.

tives Program (EQIP). Combined, both programs account for more than half of all conservation program funding.

The CSP differs from land retirement programs by rewarding farmers and ranchers for undertaking additional conservation activities and improving and maintaining existing conservation systems. Through five-year contracts, the program offers payments to producers who maintain a high level of conservation on their land and who agree to adopt higher levels of stewardship. It provides two possible types of payments. An annual payment is available for installing new conservation activities (e.g., no-till farming) and maintaining existing practices, and a supplemental payment is available to participants who also adopt a resource-conserving crop rotation (e.g., planting soybeans every third year to add nitrogen back to the soil). Eligible lands include cropland, grassland, prairie land, improved pastureland, rangeland, nonindustrial private forest land, and agricultural land under the jurisdiction of an Indian tribe. The NRCS makes the Conservation Stewardship Program available on a nationwide basis through a continuous sign-up process.

Environmental Quality Incentives Program

Similar to the CSP, the EQIP provides technical assistance, incentive payments, and cost sharing to farmers and ranchers to implement, in many cases, longer-term conservation practices on their lands. Contracts can be up to 10 years in duration, and

allowable practices are based on a set of national priorities that are adapted to each state. These priorities range from reduction of point and nonpoint source pollution to watersheds and groundwater to the improvement of wildlife habitat for at-risk species.

MAJOR TYPES OF PLANNING

In this section, we review examples of the types of conservation planning that are conducted by state and federal natural resource agencies to include organizational background, the basic processes followed in the development of that organization's management plan, and funding to support management plan activities. Although the process for conservation planning is specific to each agency, there are some general characteristics that each has in common (Table 19.6). Natural resource students will likely work for one of these agencies in their careers; understanding how agencies operate is fundamental to professional success.

State Wildlife and Forest Action Plans

State Wildlife Action Plans

US laws and policies place the primary responsibility for wildlife management with the appropriate wildlife or natural resource agency in each state (Freyfogle and Goble 2009). State

Table 19.6. Overview of the formal planning process for federal agencies

Step	BLM RMP	NPS GMP	DoD INRMP	USFWS CCP	USACE Master Plans	USFS LRMP	NRCS Conservation Plan
1	Introduction	Identify relevant laws	Description of the installation, its history, and its current mission	Background (introduction, purpose, need)	Specification of the water and related land resource problems and opportunities (relevant to the planning setting) associated with the federal objective and specific state and local concerns.	Introduction	Collection and analysis a. Identify problems and analysis b. Determine objectives c. Inventory resources d. Analyze resource data
2	Purpose statement	Identify issues and concerns (scoping)	Management goals and associated time frames	Refuge overview (location and size, physical resources, biological resources, socioeconomic environment)	Inventory, forecast, and analysis of water and related land resource conditions within the planning area relevant to the identified problems and opportunities.	Desired future conditions, goals, and objectives (forest wide)	Decision support a. Formulate alternatives b. Evaluate alternatives c. Make decisions
3	Authority	Collect data	Projects to be implemented and estimated costs	Plan development (public involvement, planning process, review, and revision)	Formulation of alternative plans	Standards and guidelines	Application and evaluation a. Implement the plan b. Evaluate the plan
4	Organization and scope of an RMP document	Identify alternatives	Review of military mission and training requirements supported	Management direction (vision, goals, objectives, strategies)	Evaluation of the effects of the alternative plans	Desired future conditions, goals, and objectives (management area)	
5	Project history	Prepare draft plan	Legal requirements and biological needs	Plan implementation (proposed projects, funding and personnel, monitoring and adaptive management)	Comparison of alternative plans	Monitoring, evaluation, research, and implementation	
6	Location/setting	Revise and consult	Role of the installation's natural resources in the context of the surrounding ecosystem	Environmental assessment (NEPA)	Selection of a recommended plan based upon the comparison of alternative plans	Appendix and glossary	
7	Overview of public involvement efforts	Approve final plan	Input from the USFWS, state fish and wildlife agency, and the public	Appendices			
8	Overview of consultation efforts	Implement the plan					
9	Management framework						
10	Planning process						
11	Opportunities and constraints						
12	Issues and issue categories						
13	Existing resource inventory						
14	Goals and objectives						
15	Desired future condition						
16	Management action/ direction						
17	Implementation procedures (monitoring, standards and guides, and plan revision or amendment)						

Note: Acronyms are as follows: BLM RMP, Bureau of Land Management Resource Management Plan; DoD INRMP, Department of Defense Integrated National Resources Management Plan; NEPA, National Environmental Policy Act; NPS GMP, National Parks Service General Management Plan; NRCS, Natural Resources Conservation Service; USACE, US Army Corps of Engineers; USFS LRMP, US Forest Service Land and Resource Management Plan; USFWS CCP, US Fish and Wildlife Service Comprehensive Conservation Plan.

fish and wildlife agencies have a long history of success in conserving game species through the support of hunter and angler license fees and federal excise taxes (e.g., the Pittman-Robertson and Dingell-Johnson Acts). How are species that are not hunted or fished (nearly 90%) managed on state and private lands? There is a gap in wildlife conservation funding and the management of nongame species. There are two funding programs used to address this gap in funding for the management of nongame and unlisted species: the State Wildlife Grants Program and Landowner Incentive Program. The State Wildlife Grants Program (created by Congress in 2001) provides federal funding to states to support projects that prevent nongame wildlife from declining to the point of being endangered. Some examples include the restoration of degraded habitat, translocations of native wildlife populations, development of conservation partnerships with private landowners, and data collection on declining species or species of concern. Congress charged each state and territory with developing a statewide wildlife action plan to make the best use of program funding. These plans, technically referred to as comprehensive wildlife conservation strategies, assess the health of each state's wildlife and associated habitat, identify threats, and outline the actions needed to conserve them over the long term. Like the Pittman-Robertson funds, funds annually appropriated by Congress under the State Wildlife Grants Program are based on a predetermined formula that evaluates the state's size and population. State wildlife grants require a nonfederal match, which is also determined by Congress based on annual appropriations, to ensure local ownership and leverage state and private funds to support conservation in each state. A specific format for state wildlife action plans, which comprise eight sections, is required for funding support (Table 19.7).

A complement to the State Wildlife Grants Program is the Landowner Incentive Program, which was designed to benefit at-risk wildlife species and the habitats critical to their survival. The USFWS defines at-risk species as species of greatest conservation need (high priority) in a state's wildlife action plan. Landowner Incentive Program funds are allocated by Congress to USFWS for distribution to state fish and wildlife agencies. Funding for the Landowner Incentive Program is collected from revenues of the Outer Continental Shelf Oil and Gas Leasing Program royalties deposited into the Land and Water Conservation Fund Act of 1965. Funds can go toward development and administration of a dedicated private lands habitat program that provides professional, technical, and financial assistance to private landowners. The Landowner Incentive Program includes two funding tiers. Tier 1 is noncompetitive, and tier 2 is competitive nationally. Under tier 1, each state may receive funding for eligible projects up to $200,000 annually. When funding is available, the program will rank tier 2 grants and make the appropriate awards.

Statewide Forest Resource Assessment and Strategy

The 2008 Farm Bill resulted in the amendment of the Cooperative Forestry Assistance Act (CFAA) of 1973. The purpose of the act is to authorize the US secretary of agriculture to assist in establishing a cooperative federal, state, and local forest stewardship program for management of nonfederal forest lands. Examples of funding programs include the Forestry Incentives Program, Urban and Community Forestry Assistance, Rural Fire Prevention and Control, and direct assistance to state forestry agencies. The State and Private Forestry Organization of the USDA Forest Service manages the majority of these programs. States are required to complete a statewide forest resource assessment and strategy plan to receive funds under CFAA. Like state wildlife action plans, statewide forest assessments provide an analysis of forest conditions and trends in the state and delineate priority rural and urban forest landscape areas. The statewide assessment outlines long-term plans for investing state, federal, and other resources in the most efficient manner. Each statewide forest resource assessment and strategy comprises eight sections that can be used to obtain funding support (Table 19.7).

Federal Planning: An Overview

There are several federal agencies responsible for the management of natural resources directly or indirectly in the United States. For each of these agencies, conservation planning occurs as mandated by agency authorities or directives in meeting their mission. Though each agency uses a different term to describe its management plan, all federal agencies are subject to general requirements as dictated by federal laws (e.g., NEPA) that outline a general process in the development and

Table 19.7. Overview of the formal planning process for state agencies

Step	State Wildlife Action Plans	Statewide Forest Resource Assessments
1	Distribution/abundance of wildlife (particularly low/declining populations)	Description of the priority landscape areas and issues
2	Habitats and community descriptions essential to species conservation	Glossary of terms and acronyms
3	Review of problems and factors adversely affecting species	Investing resources
4	Proposed conservation actions and priorities for identified species and their habitats	List of other plans consulted
5	Monitoring plans for species and their habitats	Listing and description of stakeholder involvement
6	Procedures for plan reassessment and evaluation	Monitoring and reporting
7	Coordination plans with other agencies	Protocol for translating strategies into actions
8	Public participation plans	Strategies to address the priority landscape areas and issues

implementation of an agency's management plan. In general, a federal agency begins its planning process by publishing a notice of intent in the *Federal Register* and in local newspapers to prepare or revise their version of the management plan. Such notices invite the public to identify issues and to submit comments to the agency for consideration during the planning process. This process of public review is called the "scoping" process. Based upon the information gathered (e.g., species data, public input), the agency then prepares a reasonable set of alternatives for managing the proposed public resources within the planning area. Management plan alternatives are designed to address issues identified by the agency during scoping and to comply with applicable laws and agency policy guidance. One alternative typically identified in the management plan is the no-action alternative, which maintains the current management direction. Approaches to the identification of the preferred management alternative may differ between federal agencies, but at some point a given set of alternatives, including the preferred option, are presented to the public as a draft management plan and draft EIS. Public meetings, mass mailings, and other informal discussions are used to solicit input into the prepared draft management plan and EIS. A public review period (e.g., 60–90 days) is usually conducted to receive final public input prior to accepting the final version of the management plan and issuance of the final EIS.

USFWS Comprehensive Conservation Plan

The USFWS mission is to work with others to conserve, protect, and enhance fish, wildlife, plants, and their habitats for the continuing benefit of the American people. Major agency responsibilities include the management of migratory birds, endangered species, freshwater and anadromous fishes, and wetlands. One major aspect of the agency is the management of the National Wildlife Refuge System. Established in 1903 when President Theodore Roosevelt designated Pelican Island as the first national wildlife refuge (NWR), the system has grown to include more than 60.7 million ha comprising 553 national wildlife refuges and 38 wetland management districts. Each refuge requires a management plan to direct activities on USFWS-owned properties. The National Wildlife Refuge System Improvement Act of 1997 (also known as the Refuge Improvement Act) requires USFWS to prepare a comprehensive conservation plan for each NWR. The comprehensive conservation plan is a 15-year refuge management plan that describes the purposes of each refuge; the distribution, migration patterns, and abundance of fish, wildlife, plant populations, and related habitats within the planning unit; significant problems that may adversely affect those populations and the actions necessary to correct or mitigate such problems; the archaeological and cultural values of the planning unit; and opportunities for compatible wildlife-dependent recreational uses (PL 105-57; 9 October 1997). The development of a comprehensive refuge conservation plan generally takes about a year, which does not include the time required for completing the NEPA process. There are five basic steps to the comprehensive con-servation plan process (Table 19.6; see https://www.fws.gov/refuges/planning/ComprehensiveConservationPlans.html).

NRCS Service Conservation Plans

As mentioned previously, NRCS was established with the mission of conserving natural resources on private lands. Because 70% of land is privately owned in the United States (Fig. 19.2), providing support to private landowners is an important aspect of the NRCS mission. The NRCS field agents provide conservation planning expertise and manage incentive programs under the Farm Bill. Management plans by NRCS are outlined in the *National Biology Manual*, which contains policies and procedures for biological resource activities, and the *National Planning Procedures Handbook*, which provides specific policies and procedures for management plan development. The planning process used by NRCS is a three-phase, nine-step process (Fig. 19.4, Table 19.6; see https://www.nrcs.usda.gov/wps/portal/nrcs/main/national/programs/technical/cta/). The planning process is dynamic and not necessarily conducted in a linear or chronological order.

US Army Corps of Engineers Master Plans

The US Army Corps of Engineers (USACE) is a federal agency and US Army command composed of civilian and military personnel involved in a wide range of public works support to the nation and the Department of Defense throughout the world. The USACE mission is to provide public engineering services and to strengthen the nation's security, energize the economy, and reduce risks from disasters (https://www.usace.army.mil/About/Mission-and-Vision/). Some activities of the USACE include planning, designing, building, and operating locks and dams, design and construction of flood protection systems, design and construction management of military facilities for the US Army and Air Force, and environmental regulation and ecosystem restoration. The USACE is the leading provider of outdoor recreation (e.g., boating, fishing, hunting, camping, swimming) and hydropower capacity in the United States, managing 4.9 million ha of public lands and waters at more than 400 lake and river projects in 43 states (USACE 2011). Through its Civil Works Program, the USACE carries out a wide array of projects that provide coastal protection, flood protection, hydropower, navigable waters and ports, recreational opportunities, and water supply. Water conservation planning for USACE initially evolved from the Flood Control Acts (1928, first act) and Rivers and Harbors Acts (1925, first act) to the more modern "308 reports," which are comprehensive studies of the US river basins. USACE (1983) articulates water conservation planning and involves a six-step planning process (Table 19.6; see https://www.nws.usace.army.mil/Home/Master-Plans/).

Bureau of Land Management Resource Management Plans

The US Bureau of Land Management (BLM) was created in 1946 when the Grazing Service was merged with the General Land Office to form the agency. The agency's mission is to

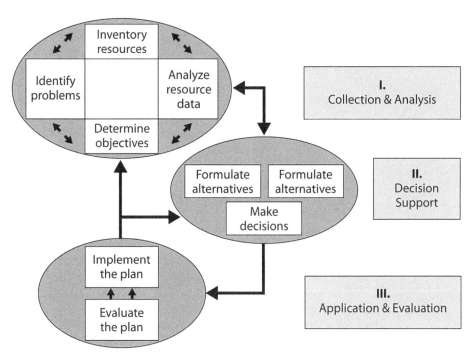

sustain the health, diversity, and productivity of the nation's public lands for the use and enjoyment of present and future generations (BLM 2011). The BLM manages more than 99 million ha of public lands (most of these lands are in the West) and approximately 283 million ha of the subsurface minerals estate of public and private lands. When the BLM was initially created, there were more than 2,000 authorities for managing these public lands. As a result, Congress enacted the Federal Land Policy and Management Act of 1976 (FLPMA; PL 94-579). In it, Congress recognized the value of the remaining public lands by declaring that BLM lands would remain in public ownership and gave the agency the mandate of multiple-use management. Under the FLPMA, the BLM also was required to develop comprehensive land-use plans called "resource management plans." Resource management plans are developed for all BLM-owned lands to include resource areas, national monuments, and national conservation areas. Like other federal agencies, the BLM resource management plans offer opportunities for public input through the NEPA process. There are 17 components of a resource management plan (Table 19.6; see https://www.blm.gov/programs/planning-and-nepa/planning-101/types-of-plans).

National Park Service General Management Planning

The National Park Service (NPS) is responsible for the management of national parks, many national monuments, and other conservation and historical properties. Established through the authority of the National Park Service Organic Act (1916), the NPS oversees 393 units (58 designated as national parks) comprising approximately 34 million ha. The agency's mission is to "conserve the scenery and the natural and historic objects and the wild life therein and to provide

for the enjoyment of the same in such manner and by such means as will leave them unimpaired for the enjoyment of future generations" (16 USC §1.1). The planning process for NPS stems from a base document called a "statement for management" that outlines procedures with the general management planning (GMP) document. NPS units rely on GMP to direct management and development for 10 to 15 years before reevaluation. Each general management plan is a collection of eight action plans that focus on the following key resources: wilderness, wildlife, history, archeology, paleontology, geology, recreation, and access. The NPS prepares for GMP using an eight-step process (Table 19.6; see https://parkplanning.nps.gov/GMPSourceBook.cfm).

US Forest Service Land and Resource Management Plans

The US Forest Service (USFS) is an agency of the USDA that oversees 155 national forests and 20 national grasslands that collectively encompass approximately 78.1 million ha (USFS 2011). Major divisions of the agency include the National Forest System, State and Private Forestry, and Research and Development branches. The Forest Reserve Act (1891) originally authorized withdrawing land from the public domain as forest reserves managed by the Department of Interior. Later, the Transfer Act (1905) transferred the management of forest reserves from the Department of Interior to the Bureau of Forestry, which later became the USFS under USDA.

The National Forest Management Act (1976, 1990; PL 94-588) reorganized and expanded the management of national forest lands. This act requires the US secretary of agriculture to assess forest lands and to develop a resource management plan based on multiple-use, sustained-yield principles for each USFS unit. Land and resource manage-

**CAMILLE PARMESAN
(1961–)**

Dr. Camille Parmesan is
a professor of integrative
biology at the University of
Texas at Austin. She also
holds the National Marine
Aquarium Chair in the Public
Understanding of Oceans
and Human Health at the
University of Plymouth in the
United Kingdom. Parmesan
is currently one of the leading authorities on the
implications of climate change for wildlife populations and natural resources. Her research in the mid-
1990s on Edith's checkerspot butterfly (*Euphydryas
editha*) was the first to demonstrate that species are
shifting their natural ranges in response to climate
change. Her subsequent research demonstrated that
a number of different taxa have already responded to
human-induced climate change.

Parmesan is actively working at the interface of
science, policy, and climate change communication.
She has been asked to present for the White House
and to provide testimony for the US House Select
Committee on Energy Independence and Global
Warming. In addition, Parmesan has been an active
member of the Intergovernmental Panel on Climate
Change (IPCC). As a lead author for the *Third Assessment Report*, she shared the Nobel Peace Prize
awarded to the IPCC in 2007.

Courtesy of Camille Parmesan

ment plans, also known as forest plans, are the product of a
comprehensive notice and comment process established under the National Forest Management Act. Land and resource
management plans provide direction for all future decisions in
the planning area. The secretary must revise and update the
management plans at least once every 15 years. The general
outline for land and resource management plans comprises six
sections (Table 19.6; see https://www.fs.fed.us/emc/nfma/).

Department of Defense Integrated Natural Resources Management Plans

The Department of Defense (DoD) is responsible for coordinating and supervising all agencies and functions related to
national security and the US Armed Forces. The DoD comprises the Office of the Secretary of Defense, the Army, Navy,
Air Force, and other support agencies. The DoD manages
more than 12 million ha of various ecosystems that represent

the major land and climate types in which military personnel may be expected to fight wars. The primary function
of DoD lands is to support the test and training mission for
the US Armed Forces; however, DoD also is responsible for
conserving and protecting biological resources under the authority of the Sikes Act (1960; 16 USC §670) with assistance
from USFWS. In 1997 the Sikes Act was amended to require
military installations to develop and implement mutually
agreed-upon integrated natural resource management plans
through voluntary cooperative agreements between the DoD
installation, the USFWS, and the respective state fish and
wildlife agencies. Integrated natural resource management
plans are planning documents that allow DoD installations
to implement landscape-level management of their natural
resources while coordinating with various stakeholders. The
plans are reviewed and updated annually, and reapproved (to
include the tripartite signatures of the DoD, the USFWS, and
the appropriate state agency) every five years. The integrated
natural resource management plan process also takes into account military mission requirements, installation master planning, environmental planning, and outdoor recreation. The
basic elements on an integrated natural resource management plan include seven steps (Table 19.6; see https://www
.denix.osd.mil/nr/focus-areas/integrated-natural-resource
-management-plans-inrmps/index.html).

Environmental Protection Agency Watershed Protection Plans

The US Environmental Protection Agency (EPA) is responsible for overseeing broad environmental protections to include clean air and water. Watershed protection plans are
implemented by the EPA through state cooperators as part
of the Clean Water Act. The Clean Water Act (1972; 33 USC
§1251) is a US federal law that regulates the discharge of pollutants into the nation's surface waters, including lakes, rivers,
streams, wetlands, and coastal areas. In 2008, EPA published
the *Handbook for Developing Watershed Plans to Restore and
Protect Our Waters* to provide states with a comprehensive resource to develop more effective watershed plans as a means
to improve and protect the nation's water quality. The handbook also provides guidance on how to incorporate the nine
minimum elements from the Clean Water Act section 319
Nonpoint Source Program.

GENERAL CHARACTERISTICS OF PLANNING

From our review of formal plans for state and federal agencies
(Tables 19.6 and 19.7), we exposed some common elements
or themes to management plans. In practice, conservation
planning is somewhat analogous to the scientific method,
which lays out a systematic process for increasing our knowledge and understanding of the world around us. First, you
observe a condition and form a hypothesis. You test your hypothesis in an experiment and compare the results to your
hypothesis. You either confirm your hypothesis or repeat the

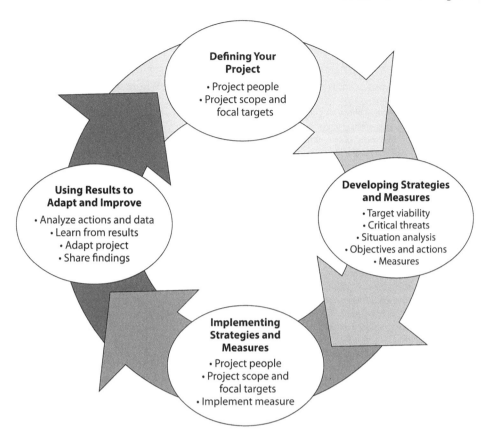

Figure 19.5. Example of the conservation planning process used by The Nature Conservancy.

process with a revised hypothesis. Conservation planning is simply the scientific method modified. For example, the formal conservation planning process used by TNC involves four steps: identify and define the problem, develop strategies and measures, implement or test those measures, and learn and improve from the results you observed (Fig. 19.5). By redefining the problem in their first step with more specific goals and objectives, we propose a five-step process: state goals and objectives, identify and assign tasks, conduct the tasks, evaluate the results, and modify the plan as needed (Fig. 19.6). The remainder of this chapter reviews this generalized approach to better understand the conservation planning process.

Conservation Planning Process

The five steps of the conservation planning process frame the major components of any management plan, regardless of for whom you may work as a wildlife manager. By encompassing these five steps into five distinct sections, we can generalize and define a generic approach in the development of a conservation plan: (1) the introduction and purpose section encompasses goals and objectives, (2) the recommendations section identifies and assigns tasks, (3) the implementation section discusses how tasks will be conducted, (4) the monitoring section discusses how the plan will be evaluated and modified based on its implementation, and (5) the supporting documents section provides supplemental materials (Table 19.8). These major sections describe most management plan formats you are

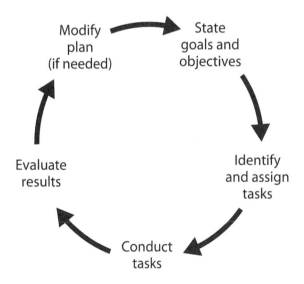

Figure 19.6. The five basic steps of the conservation planning process.

likely to encounter throughout your career. We provide a brief description of each section below.

Management Plan Overview

Introduction and Purpose

The first section of a management plan introduces the reader to the purpose of the plan. It answers the question, Why are

Table 19.8. Example outline used in developing a wildlife management plan for landowners

Major Sections	Plan Elements	Description
Introduction and Purpose	Purpose	Introduces the reader to the purpose of the plan. Answers the question, Why are we doing this plan?
	Owner information	Provides information about property owner or client. Describes the purpose of the property from the landowner's perspective (e.g., recreational use, consumptive).
	Property description	Provides information about property, including general location, vegetation cover maps, and a legal description of area size, historic and current land uses, and other relevant biological descriptions (e.g., soils, topography, cover type, flora, fauna).
	Target species	Identifies species targeted for management and provides relevant life history information. Target species are normally stated within the landowner's objective(s).
	Goals and objectives	Describes the wildlife management goals and objectives for property. A landowner *goal* may include increasing the number of blue birds. A set of *objectives* used to address this goal may include (1) identifying one or more viable management practices and (2) recommending practice(s) to the landowner that would best achieve their goals.
Recommendations	Feasibility assessment	Reviews species requirements in relation to environmental constraints that will serve to frame or "bound" management recommendations. This section describes feasible options for landowner consideration. A "management constraints" section also may be included to identify and discuss any limitations to proposed management activities.
	Alternative recommendations	Outlines proposed management alternatives or options based on a feasibility assessment (i.e., species requirements, constraints, etc.) to support recommendations. A "no management" option is a viable alternative.
	Proposed target areas	Identifies target areas where the proposed management activities are to occur. A map is often included to aid in identifying those areas.
Implementation	Final recommendations	Outlines final management recommendations to the landowner. Identifies how recommendation(s) will support and achieve management goal.
	Time line	Outlines a schedule of events of all proposed activities or necessary steps needed to implement the plan. Typically includes a table or some other calendar that is organized by practice/activity and when that activity should be conducted (e.g., installation of nest box, maintenance, monitoring).
	Budget	Provides a summary of costs for implementing the management plan. Typically includes a table with proposed number of acres treated and cost per treatment organized by management practice. "Matching" or "cost share" is often reflected here as well.
Monitoring	Plan evaluation and modification	Outlines a monitoring protocol based on results that will be observed following implementation of the plan, which is essential in the feedback loop process. Monitoring is essential to determine success and clearly define the metrics that will be monitored to know success (e.g., successfully fledging quail).
Supporting Documents	Appendices, glossary, and literature cited	Includes supporting information relevant to the wildlife management plan. This can include popular literature, information from websites, or scientific publications. Some examples include technical assistance sources and contacts, material/equipment sources, blueprints for construction of nest boxes/platforms, or census and monitoring forms.

we doing this plan? It also describes the current land uses and specific ecological site descriptions for the planning area in question. Ecological sites, defined through the establishment of the *Rangeland Interagency Ecological Site Manual* by the NRCS, USFS, and BLM, are distinctive lands with specific soil and physical characteristics that produce certain kinds and amounts of vegetation that are able to respond similarly to management actions and natural disturbances (Bestelmeyer and Brown 2010). Unlike vegetation classification, ecological site classification uses climate, soil, geomorphology, hydrology, and vegetation information to describe the ecological potential of land areas. Ecological site descriptions are reports that describe the biophysical properties of ecological sites, vegetation and surface soil properties of reference conditions (e.g., pre-European vegetation and historical range of variation or proper functioning condition or potential natural vegetation), state-and-transition model graphics and text, and ecosystem services provided by the ecological site and other interpreta-

tions. Plant-soil inventory data, long-term experimental or management studies, historical reconstructions, and local knowledge go into developing the ecological site descriptions.

In describing current land uses, remote sensing (e.g., aerial photography) or other geographic information system data can be used to physically describe the planning area to include general vegetation cover, area of management units, soil types, property boundaries, hydrology, topography, transportation access (e.g., roads, trails), and other relevant spatial features. This review of the planning area serves to identify any site constraints prior to developing plan alternatives. Maps in this section of the management plan often help tell the story and describe the current state of the area in question. The first part of the management plan also introduces to the reader descriptions of the property owners (i.e., federal agency, private landowner) and their goals and objectives. A brief to extensive review of life history requirements for target wildlife species sometimes is included as background information. This biological background is important, as it will later serve to justify management recommendations based on the biology of the target species. Lastly, the overall goal(s) of the plan, and the more specific objectives that will lead to attaining the goal(s), are specified. In short, the first section of our generic management plan outline describes the current situation for further consideration by the wildlife practitioner.

Recommendations

The second major section of our generic management plan describes the process in developing and identifying feasible alternatives for further development and consideration. This section is an important part of any management plan. The wildlife practitioner, through this process, is made keenly aware of their "signature on the face of the land" (Leopold 1949:68). Two primary factors likely frame the development of management plan alternatives and their feasibility: the biological needs of the species (reference to the life history requirements of the first section) and the physical or site constraints of the planning area (both abiotic and biotic). The development of sound alternatives will have a profound effect on the quality of the management plan's final decision (Lichfield et al. 1973). A good plan cannot be selected from a poor set of alternatives in conservation planning, which is why the recommendations section is so important. During the initial stages of developing reasonable alternatives, a laundry list of possible actions is developed. Alternatives come from people with varied backgrounds and experiences; however, not all alternatives are feasible. A second review of alternatives is typically done on the basis of known constraints and available resources. Feasibility criteria are any factors that may prevent the implementation of the plan (e.g., regulatory constraints, budget or labor constraints) and can be based on cost–benefit analyses, anticipated social or political support, or other predetermined criteria. In addition to the review of feasible alternatives, the recommendations section of the management plan also may identify target areas for proposed management activities.

Implementation

The implementation section of our generic management plan identifies final recommendations for the landowner or manager to consider and apply. In addition to the review of specific management recommendations and supporting management actions or prescriptions, the management plan outlines the proposed time line of events or management plan milestones, costs or budget requirements for implementing the plan, and management guidelines (i.e., best management practices).

Monitoring

The monitoring section addresses what follows implementation of the plan, plan evaluation, and modification. This section is critical because it provides the details needed to successfully evaluate the management plan on the basis of whether desired goals and objectives were achieved. Unfortunately, monitoring is often forgotten in planning activities. Management plans are like road maps. You need to know where you are going, how you will get there (i.e., reach your goals) based on evaluation criteria that are measurable and related to the goals and objectives of the management plan, and whether you reached your destination. If you failed to reach your destination, you must re-create the road map (i.e., plan) using alternatives scenarios to ensure you get there.

Supporting Documents

The last section in our generic management plan consists of supporting materials or documents related to the main body of the plan. Though supporting documents can seem unimportant, they are necessary for full understanding and implementation of the management plan. Appendices, glossary of terms, and literature cited serve to complete your management plan. In more formal management plans for federal agencies, a copy of an EIS or biological opinion may be included in this section. In the case of a management plan for a private landowner, popular literature, how-to publications, equipment lists and suppliers, building plans (e.g., nest boxes), data forms, and other relevant planning documents such as application forms for relevant incentive programs are typically included.

SUMMARY

Management plans are an integral component in managing wildlife habitat and their populations, and developing them is a skill that wildlife practitioners should obtain and refine in the practice and application of wildlife management. The basic elements common to all conservation planning efforts can be encompassed in five sections of a management plan: introduction and purpose, recommendations, implementation, monitoring, and supporting documents (Table 19.8). Even though planning can sometimes be unstructured, management plans add structure to the overall planning process and ultimately improve the management of natural resources.

Conservation planning also begins with the notion that we are dissatisfied with the status quo (i.e., conservation planning begins with problems). A declining endangered population, for example, may provide the reason for plans or actions on the part of the wildlife manager. Because environments are always changing (naturally, human induced, both), one must plan ahead even if currently satisfied. In either case, the decision to address a management problem or maintain the status quo into the future requires some sort of image of a desired state (i.e., goals) and how to get there (i.e., objectives). The conservation planning process begins with a situational diagnosis framed by the clear articulation of goals and objectives. Based on our evaluation of goals and objectives, we then begin to formulate predictions of likely outcomes from the suite of alternatives identified. Next, we conduct a feasibility analysis, where we determine what we ideally desire, any limiting factors that may inhibit implementation of the plan, and reasonable options for further consideration. We evaluate all possible options, and based on our findings that include reality constraints (e.g., budget) and consultation with stakeholders, we select and implement a given set of alternatives in our plan. We monitor and repeat (i.e., conservation planning is an iterative process), because we realize that plans need to be adjusted and revisited from time to time. In essence, this summarizes the conservation planning process.

In closing, conservation planning is an essential component of wildlife management, but it is a process that requires experience, analysis, intuition, and even inspiration. Conservation planning guides management actions in a way that ensures careful evaluation and execution of management plans, which are the basis for conservation planning. We reviewed some formal management plan outlines currently used by state and federal agencies, and the authorities by which these agencies implement conservation planning on private and public land. We also emphasized a few of these common elements and proposed a generic approach that can help develop a management plan. We hope that this process has provided a reasonable road map to determine what you have, what you want, and how you can get there during the conservation planning process. May you be "humbly aware that with each stroke [you are] writing [your] signature on the face of the land" (Leopold 1949:68).

Literature Cited

Bestelmeyer, B. T., and J. R. Brown. 2010. An introduction to the special issue on ecological sites. Rangelands 32:3–4.

BLM. US Bureau of Land Management. 2011. Who we are, what we do. https://www.blm.gov/about/our-mission.

Freyfogle, E. T., and D. D. Goble. 2009. Wildlife law: a primer. Island Press, Washington, DC, USA.

Leopold, A. 1949. A Sand County almanac: sketches here and there. Oxford University Press, New York, New York, USA.

Lichfield, N., P. Kettle, and M. Whitebread. 1973. Evaluation in the planning process. Oxford University Press, Oxford, UK.

Lopez, R. R., B. Hayes, M. W. Wagner, S. L. Locke, R. A. McCleery, and N. J. Silvy. 2005. Integrating land conservation planning in the classroom. Wildlife Society Bulletin 34:223–228.

USACE. US Army Corps of Engineers. 1983. Economic and environmental principles and guidelines for water and related land resources implementation studies. https://planning.erdc.dren.mil/toolbox/library/Guidance/Principles_Guidelines.pdf.

USACE. US Army Corps of Engineers. 2011. USACE website. https://www.usace.army.mil/Missions/Environmental.aspx.

US Fish and Wildlife Service. 2010. Strategic plan the Partners for Fish and Wildlife Program: stewardship of fish and wildlife through voluntary conservation. https://www.fws.gov/northeast/partners/pdf/strategic_plan_national_summary.pdf.

US Fish and Wildlife Service. 2016. USFWS Species Status Assessment Framework: an integrated analytical framework for conservation. Version 3.4.

USFS. US Forest Service. 2011. About us. https://www.fs.usda.gov/about-agency/.

USFWS and NMFS. US Fish and Wildlife Service and National Marine Fisheries Service. 1998. Consultation handbook: procedures for conducting consultation and conference activities under Section 7 of the Endangered Species Act. Washington, DC, USA. https://www.fws.gov/endangered/esa-library/pdf/esa_section7_handbook.pdf.

USFWS and NMFS. US Fish and Wildlife Service and National Marine Fisheries Service. 2016. Habitat conservation planning and incidental take permit processing handbook. Washington, DC, USA. https://www.fws.gov/guidance/sites/default/files/documents/Habitat_Conservation_Planning_and_Incidental_Take_Permit_Processing_Handbook_December%2021%2C%202016.pdf.

Yoe, C. E., and K. D. Orth. 1996. Planning manual. Report 96-R-21. Institute for Water Resources, Alexandria, Virginia, USA.

MANAGING POPULATIONS

WILLIAM P. KUVLESKY JR., SCOTT E. HENKE,
LEONARD A. BRENNAN, BART M. BALLARD, MICHAEL J. CHERRY,
DAVID G. HEWITT, TYLER A. CAMPBELL, RANDALL W. DEYOUNG,
C. JANE ANDERSON, AND FIDEL HERNANDEZ

INTRODUCTION

Managing wildlife populations is not a novel concept that arose in the 1930s when Leopold (1933) introduced the modern philosophy of wildlife management. Indeed, Leopold (1933) recognized this when he formulated his philosophy of wildlife management, and he provides an excellent summary of the history of wildlife management (Chapter 2). He noted that the concept of conserving wildlife emerged when humans were hunters and gathers (Chapter 9). Leopold, and others, suggested that social groups or tribes that survived did so because they regulated their harvest of wild animals in a manner that conserved the resources on which their survival depended. For early Indigenous cultures, there were no laws to regulate or conserve wildlife populations because conserving natural resources was viewed as a cultural obligation and became tradition. As long 10,000 years ago, the Native Americans of the Sierra Nevada of California were sustained by the natural resources of the ecosystem they inhabited, so they intensively managed fish, game, vegetation, and building materials in a manner that had significant ecological and evolutionary consequences (Anderson and Moratto 1996). Many of the Native American cultures that inhabited North America centuries prior to Euro-American exploration and colonization viewed themselves as part of the lands they inhabited (McHugh 1972, Jorgenson 1995, McCorquondale 1997), and because their survival depended on fish and wildlife, many cultures considered numerous fish and wildlife species sacred, even as brothers and sisters (Jorgenson 1995). For example, the Tlingit people of southeastern Alaska (Jorgenson 1995) and the Native Americans of the Great Plains developed and maintained a spiritual link to sea otters (*Enhydra lutris*) and bison (*Bison bison*), respectively, and recognized that to maintain sustainable populations of both species, careful management was necessary.

Leopold (1933) also indicated that the ancient Greeks and Romans had game laws, but justification for the Greeks was that hunting was good preparation for war, and for the Romans, game laws were instituted to protect landowner rights rather than conservation of wildlife. Perhaps the first attempts to manage wildlife for conservation occurred during the 8th century when Charlemagne enacted elaborate game regulations that included bag limits and preserved habitat in the forests of his realm in Europe (Caughley 1985). Another pioneer of wildlife population management was Genghis Kahn, who decreed in the 11th century that the Mongols of his empire could hunt only during the four months of winter to ensure population sustainability (Caughley 1985). Wildlife population management was evidently maintained in the Mongol Empire because Marco Polo wrote that Kublai Kahn in the 14th century prohibited hunting certain wildlife species throughout his empire between March and October to ensure that these species increased and multiplied (Leopold 1933).

Leopold (1933) stated that wildlife population management became better defined about a century after Marco Polo's observations of the Great Kahn's management of wildlife populations in feudal England, where wildlife was managed for the benefit of the aristocracy. During the 14th century, seasons were established for red deer (*Cervus elaphus*) as well as for age and sex restrictions, although these were customs observed by the ruling classes rather than laws decreed by King Henry IV. Hunting was viewed as an elite sport, and the customs established were not really conservation measures but instead political tools to exclude commoners or at least restrict what species, sex, or age class commoners could harvest (MacKenzie 1988). Leopold (1933) indicated that formal written laws were established later, during the 16th century and the reign of Henry VIII, who implemented seasonal waterfowl hunting. Following the lead of Henry VIII, King James I decreed during the late 16th and early 17th centuries that pheasants and partridges would receive seasonal protection. In addition to implementing hunting seasons and bag limits, supplementing wildlife populations via artificial propagation or restocking also apparently first appeared in England during the reign of Henry VIII in the 16th century. Mallard (*Anas platyrhynchos*) propagation was also evident in 17th-century England (Maxwell 1913).

Efforts to conserve and maintain wildlife populations through instituting seasons and bag limits were not the only

CARL MADSEN (1937–)

Born to a Danish immigrant family in Racine, Wisconsin, Carl Madsen received his BS in conservation and biology from the University of Wisconsin (UW), Stevens Point, and his MS in wildlife from Michigan State. He was named Outstanding Alumnus at UW–Stevens Point and was named to the Hall of Fame at Washington Park High School in Racine, Wisconsin.

He served as a wildlife biologist in Wisconsin, Minnesota, and the Dakotas for the US Fish and Wildlife Service from 1967 to 2004. Madsen served in various positions in the service's migratory bird program and wetland habitat protection program, and he was a leader for the Midcontinent Waterfowl Management Project, where he field-tested management techniques and helped develop agricultural programs to benefit wildlife. He served as the habitat coordinator for the North American Waterfowl Management Plan during its early years. He was a visionary and early pioneer of the service's successful Partners for Fish and Wildlife, setting the standard for program accomplishments at the field level in South Dakota.

Mr. Madsen has received many awards for leadership in innovative developments in cooperative wildlife management, including both the Meritorious Service Award and the Distinguished Service Award, the highest recognition given by the US Department of the Interior. He received national awards from Ducks Unlimited and the National Association of Conservation Districts as well as many state and local level awards, including the Professional Award from both the Minnesota and South Dakota Chapters of The Wildlife Society, where he served as president of both chapters.

Carl and his wife of 50 years, Aileen, raised three children and enjoy their family at their rural home in Brookings, South Dakota. He owns and operates a tree nursery, is active in wildlife and community issues, and enjoys gardening, hunting, and fishing.

population management actions introduced in Europe during the early 16th century. Henry VIII placed bounties on several bird species determined to be pests during his reign in the early 16th century, and Queen Elizabeth I retained these bounties and extended them to include numerous additional birds and mammals during her reign in the middle to late 16th century (Leopold 1933). Formal bounties to control wild birds and mammals deemed threats to humans or their livelihoods were first implemented in Finland during the mid-17th century, and this organized persecution was considered an important component of game management (Pohja-Mykra et al. 2005). Caughley (1985) indicated that reducing predator populations in particular was viewed as a way of decreasing nonhunting mortality on game animals. These bounty systems were vigorously pursued and were so effective that many large predator populations were significantly reduced or extirpated. Wolves (*Canis lupus*) and brown bears (*Ursus arctos arctos*) were soon extirpated from England (Caughley 1985) and within a short time were either extirpated or reduced to small, remnant populations in a number of western European countries.

With the exception of the cultural mores that Native Americans used to manage wildlife populations in North America, wildlife populations were exploited for subsistence or to secure safe environments when Europeans began to colonize the continent (Lund 1980). When population management was initiated through hunting, Euro-Americans followed the traditions established in Europe during the 17th century, with the significant exception that recreational hunt-

ing was no longer restricted to the privileged but could be enjoyed by anyone regardless of social standing (Lund 1980). Wildlife was basically available to anyone who wanted to hunt, and because game laws were virtually unknown in the United States until the second half of the 19th century (Caughley 1985), wildlife populations were decimated. For example, unregulated market hunting was a major factor in the extinction Labrador duck (*Camptorhynchus labradorius*) and great auk (*Pinguinus impennis*; Mahoney 2009). The general public's view that wildlife was an inexhaustible resource continued to prevail as American settlers moved west. As these settlers transformed wilderness into farms and settlements, populations of many native bird and mammal species declined significantly, and populations of some species never recovered from unregulated hunting. For example, market hunting and egg collecting have been attributed to the extinction of the passenger pigeon (*Ectopistes migratorius*), which was thought to have been the most abundant bird species on earth during the 18th century; extirpation of this species required only a few decades during the mid-1800s (Askins 2000). The unregulated trapping of the beaver (*Castor canadensis*) and market hunting of American bison almost resulted in the extirpation of two species as well by the middle to late 1800s (Mahoney 2009). Similarly, federal campaigns to eradicate predators during the late 19th and early 20th centuries resulted in the extinction of the plains grizzly (*U. arctos horribilis*) and the extirpation of the Mexican gray wolf (*C. l. baileyi*).

E. CHARLES MESLOW (1937–)

Dr. E. Charles "Chuck" Meslow was born in Waukegan, Illinois. His love of nature and wildlife started during the time he spent on a Wisconsin farm during his early teenage years. After serving three years in the Navy, he received BS and MS degrees in wildlife management at the University of Minnesota and then a PhD in wildlife ecology in 1970 from the University of Wisconsin, where he studied snowshoe hare populations with Dr. L. B. Keith. He started his academic career at North Dakota State University but moved to Oregon State University, where he served as assistant unit leader (1971–1975) and then unit leader of the Oregon Cooperative Wildlife Research Unit and professor of wildlife ecology until he retired in 1994. After his retirement, he served as the northwest regional representative for the Wildlife Management Institute until 1999.

Dr. Meslow has made a significant contribution to the field of wildlife population management through his research on numerous wildlife species that inhabit the forests of the Pacific Northwest. Undoubtedly, the most important contribution he has made to wildlife conservation was the leadership role he provided during the contentious years associated with spotted owl conservation and management. The threatened status enjoyed by spotted owls under the Endangered Species Act might have been much different in the absence of Dr. Meslow's quiet professionalism, enthusiasm, and optimism. As a result of his work on spotted owls, he was a key member of the Forest and Ecosystem Management Assessment Team, which wrote the Northwest Forest Plan that has resulted in the conservation of old forest in the Pacific Northwest. He has authored and coauthored more than 80 peer-reviewed publications and has advised and mentored more than 50 MS and PhD students. He has been involved in The Wildlife Society for almost 50 years, serving on numerous committees and as an officer at the state and national levels. He has received numerous awards from The Wildlife Society during his distinguished career, most notably the organization's most prestigious award, the Aldo Leopold Memorial Award, in 2005.

Dr. Meslow may be retired, but his influence on the field of wildlife population management and other professionals continues today because his advice and counsel continue to be sought by not only his former students and professional associates, many of whom are leaders in the wildlife profession, but also by executives in government and the timber industries who respect his professionalism and honesty.

Although game laws were rare until the second half of the 19th century, a few enlightened states realized that wildlife required protection and subsequently established laws that protected certain species (Chapter 2). For example, the state of Ohio determined that better management of furbearer populations was required and passed laws in 1829 and 1833 that closed seasons on muskrats (*Ondatra zibethicus*), beavers (*Castor canadensis*), mink (*Neovison vison*), and otters (*Lontra canadensis*; Dambach 1948).

In addition, Leopold (1933) noted that New York began managing white-tailed deer (*Odocoileus virginianus*) populations by requiring that hunters purchase a hunting license in 1865, and Iowa began managing greater prairie chickens (*Tympanuchus cupido*) in 1878 by instituting a bag limit of 25 birds/day. Concerted efforts to better manage wildlife populations did not begin until the late 19th century when Theodore Roosevelt, an avid hunter, became concerned with the plight of many game species. As a consequence of this concern, he started the Boone & Crockett Club. Many of the initial members of the club exerted political influence at the federal and state levels to institute hunting seasons and bag limits, and establish protected areas that resulted in population recoveries of many big game species in North America.

The conservation ethic of influential figures such as Theodore Roosevelt laid the foundations upon which Aldo Leopold developed his philosophy of wildlife management during the 1920s and 1930s. Leopold's ideas stimulated the creation of many of the basic wildlife population management principles that were adopted almost universally by many state game and fish agencies in the United States during the 1940s and 1950s. It became clear that effectively managing wildlife populations required a scientific approach, which eventually yielded more accurate and precise surveys and census procedures, population monitoring techniques such as radiotelemetry and satellite telemetry, computer modeling, and modern genetic procedures that have rapidly advanced the science of today's management of wildlife populations.

Clearly, wildlife population management has come a long way since the early efforts to manage wildlife populations practiced by Charlemagne and Genghis Kahn. The efforts of these early conservationists and those that followed indicate that managing wildlife populations is nothing new and has been practiced for centuries. In fact, this abbreviated summary of the history of wildlife population management reveals that three basic themes of wildlife population management have been practiced for the past 800 years: managing to maintain populations, managing to increase populations, and managing to reduce populations. Managing to maintain wildlife populations and perhaps increase them appears to have been the intent of the Genghis and Kublai Kahn and the rulers of feudal Europe, and maintaining wildlife populations is the most common form of wildlife population management that state game and fish agencies try to accomplish on an annual basis today. Closing and establishing hunting seasons and instituting bag limits represented attempts by states in the 1800s to increase wildlife populations, and state and federal natural resource agencies today also attempt to increase endangered species populations via strict protection by suspending harvest or otherwise significantly limiting take, and establishing refuges to protect critical habitats. Finally, the vermin control programs instituted by the rulers of 16th-century England and 17th-century Finland, and the predator eradication programs established by federal agencies in the late 19th and early 20th centuries, resemble what state and federal agencies do today to reduce populations of pest species such as feral hogs (*Sus scrofa*).

The objective of this chapter is to provide an overview of basic wildlife population management practiced by wildlife professionals by selecting some common examples of how populations of prominent wildlife species are managed today. This chapter is organized around the three central themes of wildlife population management that have been practiced throughout the world for centuries: management to increase populations, management to sustain populations, and management to reduce populations. Within each theme we use case histories to illustrate examples of how specific species or classes of wildlife are managed. The case histories for each of the three themes will be organized into three sections that present and discuss the material in a manner that is most informative. These sections include a background section consisting of a brief description of the natural history of the animal or class animals for each case history, a brief overview of techniques used to manage populations of the subject animal, and an overview of the ecological and cultural importance of the animal.

INCREASING WILDLIFE POPULATIONS

Numerous species or classes of wildlife could have been used as case studies for the theme of increasing wildlife populations. The gray wolf (*Canis lupus*), red-cockaded woodpecker (*Picoides borealis*), and Texas horned lizard (*Phrynosoma cornutum*) were selected because each represents a species whose populations had diminished sufficiently that focused efforts were needed to achieve population increases. Moreover, each species was determined to be an important component of their respective ecosystems or important manifestations of national or regional cultures. Considerable research and management have been devoted to increasing populations of these species. Each case study therefore provides a short discussion outlining the reasons, methodology, and cultural and ecological justification for increasing populations.

Gray Wolf

Background

Wolves and people have shared an intricate association through the millennia. It is a relationship in which wolves have been viewed as both friend and foe. The gray wolf, for example, is the ancestral species from which the domestic dog originated. Domestic dogs have provided man with protection, food, fur, and labor. Ironically, the gray wolf also has been the target of human persecution. Gray wolves are carnivores that prey largely on ungulates, including domestic livestock. Consequently, thousands of wolves were killed by poisoning and shooting through government control programs during the turn of the 19th century. Such persecution led to the extermination of wolves in most of the 48 contiguous United States by the late 1950s, with wolf populations being restricted to wilderness areas in northern Minnesota and Isle Royale National Park in Lake Superior (Mech 1995).

Public attitudes toward wolves began to improve in the 1960s and 1970s (Mech 1995). The passage of the Endangered Species Act in 1973 provided wolves legal protection as an endangered species in the lower 48 states, and wolf recovery began in earnest. For the next 20 years, wolf recovery was basically restricted to legal protection, which primarily benefitted wolf populations where they already existed (i.e., in Minnesota, Montana, and Wisconsin). The decision to reintroduce gray wolves in Yellowstone National Park (YNP) and central Idaho during the mid-1990s was the event that prompted wolf recovery. Instituting wolf recovery via reintroduction of wolves in the northern Rocky Mountains (NRM) occurred among considerable controversy that pitted environmental groups against livestock producers and the federal government against state governments. Nevertheless, recovery succeeded in the NRM as a consequence of the reintroductions, and continued federal protection in Great Lake Region (GLR) allowed wolves to recolonize significant portions of their former ranges in Michigan, Minnesota, and Wisconsin.

Wolf populations in the NRM and the GLR have recovered sufficiently, so that wolf populations in at least the NRM are being managed by the state game and fish agencies, and regulated hunting is permitted. Therefore, in parts of the continental United States, increasing wolf populations through management is no longer necessary. Wolf recovery remains ongoing in several other states in the West (California, Colorado, Oregon, and Washington; California Department of Fish and Wildlife 2021, Colorado Parks and Wildlife 2021, Oregon Department of Fish and Wildlife 2021, Washington Department of Fish and Wildlife 2021) and Southwest (Arizona, New Mexico; Arizona Game and Fish Department 2021, New Mexico Game and Fish Department 2021), and discussions continue about recovering wolves in the northeastern United States. The gray wolf therefore continues to represent a case study about increasing wildlife populations, although we can now use strategies that have been applied successfully in the NRM and GLR to enhance recovery efforts in states where recovery is ongoing. The greater challenge is effectively managing the human dimensions associated with wolf recovery and management. The gray wolf reintroduction issue extends beyond basic questions of ecology and reflects core human values (Wilson 1997). Gray wolf reintroduction clearly embodies the complex and often sociopolitical nature of wildlife conservation and management. We use this case study to illustrate the complexity of restoring an endangered species in a modern landscape.

Management Approach

The gray wolf is the largest wild member of the family Canidae (Nowak 1999). It also has the largest natural geographic distribution of living terrestrial mammals besides man. Historically, gray wolves were found throughout the Northern Hemisphere (north of 20°N latitude) and occurred on most landscapes, excluding deserts (Mech 1974). The historical range of the species in North America extended from the Arctic tundra to central Mexico and from the Pacific coast to the Eastern Seaboard (Mech 2017). An apex predator, wolves were important members of the ecosystems they occupied because they helped maintain prey populations (elk [*Cervus canadensis*], deer [*Odocoileus* spp.]) within the carrying capacity of what ecosystems could support. When settlers arrived on wolf ranges, wolves began preying on livestock, bringing them into conflict with humans. Inevitably, human persecution eventually eliminated the species from much of its historical range. It is estimated that gray wolf populations declined in the lower 48 contiguous states from about 2 million to a few hundred by the 1950s (McIntyre 1995).

The wolf situation in the lower 48 states has changed since the 1950s. The increased public acceptance of wolves, the passage of the Endangered Species Act, and the classification of wolves as endangered species in the 1970s created the momentum that resulted in wolf population recovery in portions of their historical range in the lower 48 States. Over the past two decades, wolf reintroduction to the Greater Yellowstone Ecosystem (GYE) captured the most public and scientific interest in gray wolf recovery in the lower 48 states. Gray wolf reintroduction into YNP captured headlines in written and televised media, and in scientific journals.

YNP was established in 1872 (Haines 1977). It was the nation's first national park, and intense control of wolves occurred on the park almost since its establishment. It is estimated that wolves were extirpated from YNP during the mid-1920s. For the next 70 years, wolves were virtually absent from Yellowstone. With the passing of the Endangered Species Act in 1973, gray wolves subsequently received federal protection, and in 1978 the gray wolf became listed as an endangered or threatened species in the lower 48 states.

This federal listing set in motion the development of the Rocky Mountain Wolf Recovery Plan, which was completed in 1987. Its primary objective was the reestablishment wolf populations in the western United States into appropriate habitats. For the NRM region, which included Idaho, Montana, Wyoming, and the GYE, the plan established a recovery goal of 30 breeding pairs (≥10 breeding pairs/area). YNP was designated

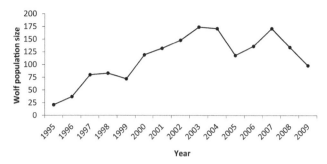

Figure 20.1. Estimated population size of gray wolves in Yellowstone National Park, 1995–2019. Data from the Annual Wolf Project Reports, Yellowstone National Park, USA

as a nonessential experimental population, which represented a compromise to address local concerns and opposition. Such designation permitted flexibility in the management of gray wolf. For example, landowners were permitted to kill wolves on their private property if wolves were caught depredating livestock (US Fish and Wildlife Service [USFWS] 1987).

This conservation and bureaucratic process eventually resulted in the release of 31 gray wolves from southwestern Canada into YNP during 1995–1996 (Bangs et al. 1998). The reintroduced wolf population responded favorably and experienced rapid growth (Fig. 20.1). Within seven years after their reintroduction, wolves had recolonized much of YNP and several adjacent areas (Smith et al. 2003). After reaching a high of about 180 wolves in 2003, wolf numbers began to decline in 2008 and eventually stabilized at around 100 animals (Fig. 20.1), which is close to the level where the population remains today (Smith et al. 2020).

In 2003 gray wolves were downlisted from endangered to threatened status in the eastern and western United States but remained listed as endangered in the Southwest. Such strong population response raised the issue that the estimated 120 breeding pairs outside the GYE, which far exceeded the recovery goal of 30 breeding pairs, was sufficient to grant gray wolves the status of recovered. Environmental groups challenged these estimates in court in the ensuing years. In early 2011 the US Congress delisted the species, which removed the issue from the court system where it had been bogged down for years, and gray wolves in the NRM were delisted and management was transferred to the state wildlife agencies. Wolves in the GLR remained classified as federally endangered for another 10 years.

So, what is the current status of wolves in the lower 48 states? Recently, Mech (2017) summarized the current distributional status of wolves in the lower 48 states and indicated that the wolves that recolonized northwestern Montana and those reintroduced to YNP and central Idaho in the mid-1990s expanded wolf distributions in Idaho, Montana, and Wyoming and have led to wolf recolonization of the Pacific Northwest (Table 20.1). Wolf populations have mixed, and individuals have dispersed to Oregon and Washington, and from Oregon into California. In fact, individual wolves have been reported

Table 20.1. Gray wolf minimum estimates for Wyoming, Idaho, Montana, Minnesota, Michigan, Wisconsin, Oregon, Washington, Utah, California, New Mexico, and Arizona, USA, 1995–2020

Year	AZ	CA	ID	MI	MN	MT	NM	OR	UT	WA	WI	WY
1995				99		30					83	21
1996				113	2,100	59					99	51
1997	4			139	2,445	51	0				148	86
1998	4			169		45	0				178	
1999	9			216		60	6				205	25
2000	15			249		90	7				248	40
2001	21			278		130	4				257	54
2002	34			321		165	7				327	72
2003	42			360	3,020	162	13				335	78
2004	26			405		152	18				373	100
2005	24		512	434		240	18				435	139
2006	25		673	509		325	34				467	173
2007	29		732	520	2,921	430	23				546	180
2008	29		846	577		490	23			5	549	172
2009	27		835	557		510	15	12		14	637	225
2010	29		705	687		550	21	21		18	690	241
2011	32		746	687		650	35	29		37	782	225
2012	37		683	658	2,211	620	43	48		49	815	180
2013	40		659	636	2,423	625	48	63		52	809	195
2014	58		770		2,221	540	54	81		66	660	218
2015	50		786	618	2,278	530	48	110		87	746	270
2016	64				2,865	480	50	113		114	866	296
2017	63		1,000	662	2,655	633	54	123		121	925	347
2018	64					819	67	138		123	905	286
2019	76			662	2,696	850	87	158		145	914	311
2020		14	1,000	695					10	145		

over the past 15 years or so in northeastern Utah (Utah Division of Wildlife Resources 2021), and a pack of six wolves was confirmed to exist in northwestern Colorado in January 2020 (Colorado Parks and Wildlife 2020). Moreover, wolf reintroduction is currently ongoing in New Mexico and Arizona (Mech 2017). Mexican gray wolves in the Mexican Wolf Experimental Area increased from 131 in 2018 to 161 in 2019.

In addition to wolf reintroduction and wolf recolonization of former ranges in the western United States, wolf populations in the GLR have increased substantially over the past 25 years (Table 20.1). The existing wolf population in Minnesota almost tripled since 1975, when there was a little over 1,000 animals, to about 2,600 animals in 2019. The Minnesota population provided dispersing animals to Wisconsin in the mid-1970s that now comprise a population of almost 1,000 wolves. Dispersing wolves from Wisconsin also recolonized the Upper Peninsula of Michigan where currently almost 700 wolves exist (Table 20.1). Wolves have also expanded their ranges within Minnesota and Wisconsin, moving from northern to central portions of these states. Wolves in Michigan may be moving south from the Upper Peninsula. Wolves also have dispersed into Illinois, Indiana, Iowa, Kentucky, Maine, Missouri, Nebraska, New York, North Dakota, and South Dakota, although packs have yet to establish in these states (Mech 2017). Furthermore, there has been interest in restoring wolves to the Adirondacks of New York and remote portions of Maine (Wydeven et al. 1998); however, wolves would need

to be reintroduced by state and federal authorities because insurmountable barriers to dispersal of wolves from neighboring areas in Canada exist.

Gray wolf reintroduction and recolonization of historical ranges in the NRM and GLR have largely been a success owing to the adaptability and productivity of wolves, and almost 40 years of federal protection. Consequently, the USFWS finally removed wolves in the continental United States, with exception of Mexican wolf, from protection under the Endangered Species Act on 4 January 2021. Management of wolves is now under the jurisdiction of the states inhabited by wolves, and wolf conservation and management vary from state to state. For example, wolves were officially federally delisted in Idaho (Idaho Fish and Game 2021), Montana (Inman et al. 2019), and Wyoming (Wyoming Game and Fish Department et al. 2019) several years ago, and wolf hunting and trapping seasons as well as annual bag limits were instituted at the time of delisting. Regulated wolf harvests have been occurring ever since. In addition, wolves in Washington and Oregon were federally delisted in 2011 in the eastern third of the states and are being managed by the respective state agencies, although harvests are restricted to wolves that depredate livestock (Oregon Department of Fish and Wildlife 2021, Washington Department of Fish and Wildlife 2021). A breeding population of wolves has not been established in Utah, but state officials estimate that 10 wolves might exist in the northeastern corner of the State (Utah Division of Wildlife Resources 2021). Wolf management in Utah currently involves harvesting animals that depredate livestock. Wolves have recently recolonized California, where an estimated 15 animals and two breeding packs exist (California Department of Fish and Wildlife 2021). Despite being removed from Endangered Species Act protection, state law prohibits killing wolves in California. Wolves are apparently in the initial stages of recolonizing Colorado, and most Colorado residents support wolf conservation. In November 2020, voters approved legislation to reintroduce and manage wolves in the state (Colorado Parks and Wildlife 2021). Although wolves are no longer receiving federal protection, killing wolves in Colorado is prohibited, as it is in California, because under state law they are considered endangered. Mexican wolves remain protected under the Endangered Species Act because although population recovery in New Mexico and Arizona appears to be succeeding, federal and state wolf biologists believe recovery has not yet been achieved (USFWS 2021).

Regulated wolf hunting had been attempted by state agencies in Minnesota and Wisconsin in the past; however, legal challenges to delisting have thwarted these efforts by the states, and wolves remained listed as endangered despite biological evidence that wolf populations could withstand regulated hunting. Since wolves were removed from the federal endangered list in January 2021, management of GLR wolf populations has been turned over to state natural resource agencies. Currently, legal wolf take in Michigan, Minnesota, and Wisconsin is restricted to wolves that are depredating livestock or pets (Michigan Department of Natural Resources 2021, Minnesota Department of Natural Resources 2021, Wisconsin Department of Natural Resources 2021). Wolf management plans are now being updated in each state, and wolf hunting presumably will be reinstituted when the planning process has been completed. In fact, the Wisconsin Department of Natural Resources anticipates opening a wolf hunting season in fall 2021.

The management approach that succeeded, and is currently succeeding, in reestablishing wolves to their historical ranges is rather simple from a biological perspective. Successful reintroduction requires that an appropriate number of wolves be released in historical range that is sufficiently remote and large, supports sufficient prey populations to maintain a self-sustainable wolf population, and provides wolves federal legal protection until a self-sustainable population is attained that can be managed by state natural resource agencies. Often the more significant challenges to successful wolf reintroduction are ecological and especially cultural.

Ecological and Cultural Context

The controversy surrounding the reintroduction of gray wolf into the United States involved ecological and cultural components. Ecologically, the controversy was not whether wolves should be introduced. Rather, the controversy centered on whether wolf reintroduction would restore ecosystem function via a trophic cascade. The cultural heart of the gray wolf reintroduction controversy was the nature of the relationship between humans and the environment. A brief treatment of the gray wolf controversy from an ecological and cultural perspective will allow for a richer understanding of the complexity of managing a declining population in a society with diverse cultural values.

Community structure and its causes (i.e., top-down or bottom-up processes) have been a source of ecological debate. A top-down influence of community structuring suggests that top predators regulate primary consumers, which in turn allows for a diverse plant community. In contrast, a bottom-up influence proposes that the productivity of plants influences community structure by determining the number of primary consumers, which in turn influences the number of predators. In YNP, the ecological controversy involved the role that reintroduced gray wolves would play in the recovery of declining deciduous trees in the park.

Aspens (*Populus tremuloides*), cottonwoods (*Populus* spp.), and willows (*Salix* spp.) have experienced dramatic declines during much of the 20th century. Intense elk browsing is thought to play a crucial role in the decline of aspen (Ripple and Beschta 2003, Kauffman et al. 2010). Because this decline is coincident with the extirpation of gray wolf in YNP, ecologists have proposed that high levels of herbivory (due to unchecked elk populations) may be responsible for the low recruitment of aspen in YNP (Wagner et al. 1995).

Numerous studies have been conducted that support the theory of top-down trophic cascades. For example, Ripple and Beschta (2012) evaluated the composition of predator guilds and prey densities from 42 studies completed in boreal and temperate ecosystems in North America and Eurasia and reported that sympatric gray wolf and bear (*Ursus* spp.) populations limited prey densities of large browsing herbivores. Cervid densities were six times higher in areas without wolves compared to areas with wolves, suggesting that top-down trophic cascades exist where wolves decrease abundance of large browsing herbivores in forests across the Northern Hemisphere. Painter et al. (2015) proposed that the reintroduction of wolves to YNP contributed to the redistribution of elk (a behavioral response), resulting in the recovery of aspen stands. In addition, Beschta and Ripple (2016) examined 24 studies that assessed woody vegetation riparian communities in YNP during two decades after wolf reintroduction and noted that the majority of these studies demonstrated evidence that riparian woody plant species were recovering simultaneous with a reduction in browsing by elk. They noted that half of the studies they examined compared plant community changes to climatic and hydrologic variables, but the results of these studies were inconsistent. Beschta and Ripple (2016) concluded that a trophic cascade was occurring in northern portions of the park involving an intact large predator guild, elk, and woody plant species.

Evidence of the presence of trophic cascades where wolf have recovered in Wisconsin has also been reported. Flagel et al. (2016) reported evidence of a trophic cascade in northern Wisconsin forests involving wolves, white-tailed deer, and forest vegetation. In areas of high wolf use, deer were 62% less dense, visit durations were reduced by 82%, and deer spent 43% less time foraging compared to low-wolf-use areas. In addition, the proportion of browsing to saplings was sevenfold less in high-wolf-use areas, and average height of maple (*Acer* spp.) sapling height and forb species richness increased 1.37 and 1.17 times, respectively, in high-wolf-use areas compared to low-wolf-use areas. Similarly, Callan et al. (2013) determined that wolves mitigated the effects of deer browsing on understory plant communities in white cedar (*Thuja occidentalis*) wetlands in northern Wisconsin, suggesting that the recolonization of wolves in the area triggered a trophic cascade. They found that forb and shrub species richness at local scales was greater in high-wolf-use areas than in low-wolf-use areas.

Alternative explanations exist for the decline of aspen in YNP, such as fire suppression, climate change, and natural plant–community dynamics. The restoration of ecosystem function following gray wolf reintroduction (i.e., if aspen recruitment would increase following gray wolf reintroduction because of decreased elk browsing) captures the essence of the ecological debate. Preliminary findings suggest that that aspen, cottonwood, and willow populations are recovering in YNP (Ripple and Beschta 2003, 2005). Whether wolves are responsible for these trends and whether these trends are still occurring is still being debated (Kauffman et al. 2010). In addition, the ecological interactions appear to be more complex and involve other factors (e.g., fire) beyond mere predation and herbivory (Mao et al. 2005).

Aspen stem density increases dramatically following fire because of resprouting. Intense herbivory, however, can decrease stem densities to prefire levels within a relatively short time frame (Bartos 1994). Because predators can influence plants via altered herbivore densities or foraging behavior, herbivory, fire, and predation risk appear to play an interactive role in aspen recruitment. The working hypothesis is that the coupling of fire and predation risk may be creating a positive feedback mechanism that results in improved aspen recruitment (Halofsky et al. 2008). Fire creates open areas of increased predation risk. Fire also stimulates aspen resprouting. Elk avoid burned areas because of increased predation risk, which in turn permits the regrowth of aspen following fire and the formation of dense thickets (Halofsky et al. 2008). These dense thickets further increase predator risk because of decreased elk maneuverability, resulting in stronger avoidance. This coupling of fire and predation risk is therefore suggested to be an important link in the apparent recovery of aspen in YNP (Halofsky et al. 2008). The existence of this behaviorally mediated trophic cascade, however, has been challenged (Kauffman et al. 2010). Recent research suggests that aspen recruitment in YNP may not be occurring as suggested or related to this "landscape of fear" created by gray wolf (Kauffman et al. 2010). Thus the debate continues.

The cultural context of the gray wolf reintroduction is just as intricate as the ecological context. Despite the success of wolf recovery in portions of their historic range in the lower 48 states, wolf reintroduction, conservation, and management may be as controversial today as they were prior to wolf recovery. Sociologists have proposed that the controversy surrounding gray wolf reintroduction is not a biological one but rather a sociopolitical one. Sociologists propose that the gray wolf is merely a symbol of a conflict between two opposing social movements: environmentalism and wise use (Wilson 1997, Nie 2001). Environmentalists, who tend to be urban, middle-class citizens (Brick 1995), view the GYE as one of the last remaining wilderness areas that provides important habitat for a large diversity of organisms (Williams 1990). Conversely, the wise use movement, a group composed of people from rural America, perceive the economic viability of the Yellowstone region to be inextricably linked to extractive industries such as grazing, forestry, and mining (Power 1991).

Carlson et al. (2020) recently conducted surveys of US residents and reported that attitudes toward wolves have become more positive over time, but people who lived closer to wolves (NRM, GLS) and were directly affected by them, such as farmers and ranchers, were less tolerant of wolves. Similarly, Meadow et al. (2005) reported that 64% of the registered voters surveyed in Arizona and New Mexico supported restoration of the Mexican gray wolf to the southern Rocky Mountains and that the minority opposed was largely made up of ranchers. Additionally, Williams et al.'s (2002) analysis of 38 quantitative surveys conducted between 1972 and 2000 in the United States and Europe reported that 61% of the general population surveyed expressed positive attitudes toward wolves, and 69% of environmental and wildlife groups had a positive attitude toward wolves. Education and income were correlated with positive attitudes toward wolves, whereas negative attitudes were correlated with age, rural residency, and farming and ranching occupations. Moreover, rural residents of Minnesota who lived both inside and outside of wolf range in the state held similarly negative attitudes toward wolves even though wolves have not occurred in the southern half of the state for more than a century, indicating that cultural attitudes toward wolves among rural residents have not changed (Chavez et al. 2005). The results of these surveys suggest that people who have not had negative experience with wolves because they do not live among them favor wolves and restoration. In contrast, people who share or have shared areas with wolves continue to have negative perceptions of wolves. This is true in western states where there is a long tradition of wise use (Wilson 1999) and in Great Lake states where wolf populations have recently recovered and where hunters in Wisconsin continued to possess negative attitudes toward wolves even during a short period when wolf harvests were legally permitted (Hogberg et al. 2015).

A contrast to the polarized attitudes of environmentalists and wise use proponents is the balanced attitude toward wolves exhibited by Native American tribes. For example, Wilson (1999) details the involvement of the Nez Perce tribe in wolf recovery in Idaho. The Nez Perce agreed to allow and cooperate in the initial release of GYE wolves in the mid-1990s when the state government officials refused. The Nez Perce viewed the land as a source of spiritual and physical sustenance and possessed a cultural attitude of respect for nature and practical realization that natural resource development can provide economic resources for the tribe. In essence, the Nez Perce viewed wolves as respected members of their natural world and recognized that wisely managing wolf populations as an economic resource (harvests) could benefit the tribe. Similarly, Cerulli (2016) reported that the Ojibwe tribe in northern Wisconsin held similar views about wolves as the Nez Perce, not only revering wolves but also recognizing that they were a sustainable resource that could be managed to benefit the tribe. The Ojibwe wanted to recover and maintain wolf populations because they appreciated them as fellow members of an intact, wild, and natural community. The Ojibwe also wanted to manage wolf populations sustainably for the benefit of deer hunting by tribal members, rather than having the authority to reduce overabundant wolf populations forced on them by outsiders.

The balanced attitude of the Nez Perce and Ojibwe tribes toward wolves has unfortunately been almost entirely overlooked or ignored by environmentalist and wise use proponents as well as by the national media. Thus the core underlying issues of gray wolf reintroduction and management are a divergence of social values between the two social movements relative to social power, private property rights, and nature (Wilson 1997). The gray wolf is merely a surrogate for broader cultural issues; namely, preservation versus resource use, recreation versus extraction economies, and rural versus urban values (Primm and Clark 1996, Nie 2001). Wolf populations have been increased and stabilized to the extent that numerous populations are considered recovered, and we know how to manage these wolf populations. Today it is not so much a matter of knowing how to recover a wolf population, but rather how we can successfully manage the human dimensions associated with wolf conservation and management.

Wolf management remains a complex issue because the of the competing interests of environmentalists and wise use proponents, and the realities of the current population status of wolves in the lower 48 states. Clearly, wolf populations have no trouble tolerating annual harvests, as the natural resource agencies of Idaho, Montana, and Wyoming currently demonstrate. In fact, Mech (2017) reports that current wolf harvests in these states have been below the annual reproductive rate of the wolf populations. Therefore managing wolf populations to increase wolves is not necessary in NRM and GLR at this time. Moreover, if the trajectory of wolf populations in California, Colorado, Oregon, and Washington and Mexican gray wolves in Arizona and New Mexico accelerates, there will probably come a time when the emphasis will no longer be on increasing populations but managing populations. In fact, in some situations and some years, wolf harvests may have to be implemented to decrease populations. If the decision is made to reintroduce wolves to the Adirondacks of New York and northern Maine, wolf recovery will probably eventually have to transition to wolf management. The USFWS continues to monitor wolf populations that have recovered in the lower 48 states on an annual basis, and if wolf populations decline below recovery requirements, various environmental organizations may try to pressure the federal government to legally suspend state wolf management programs and again list wolves as threatened.

From a conservation perspective, the reintroduction of gray wolf into NRM and GLR may be hailed as a conservation achievement—an extirpated species has been restored into a significant portion of its former range. Ecologically and culturally, however, the implications and consequences of such reintroduction are not as clearly delineated and likely will be debated for years to come.

Red-Cockaded Woodpecker

Background
The red-cockaded woodpecker is a cardinal-sized woodpecker that was once found throughout the Coastal Plains region of the southeastern United States. Today, the red-cockaded woodpecker is an endangered species whose population abundance is probably less than 3% of what it was before European settlement of the United States.

Red-cockaded woodpeckers are endangered for a simple reason: widespread loss of habitat. The red-cockaded woodpecker is closely linked with the longleaf pine (*Pinus palustris*) forests of the southeastern United States, which today is one of the most endangered ecosystems in the world; less than 1% of the original longleaf pine forests remain (Simberloff 1993). The loss of longleaf pine and the loss of red-cockaded woodpeckers have gone hand in hand.

Today, efforts to sustain, elevate, and recover red-cockaded woodpecker populations rank among the most important wildlife management efforts in the world. More than two dozen leading wildlife science professionals from a variety of agencies and organizations have collaborated to develop a recovery plan for red-cockaded woodpeckers on federal, state, and private lands throughout the southeastern Coastal Plains (USFWS 2003).

The red-cockaded woodpecker was considered endangered and on track for extinction in 1970, three years before passage of the Federal Endangered Species Act. Populations continued to decline through the 1970s and 1980s. During the 1990s, red-cockaded woodpecker recovery efforts began to have a positive effect, and populations started to increase. For example, in 1993, there were about 4,700 known clusters of breeding red-cockaded woodpeckers; this number had increased to 6,100 by 2006 and continues to increase today.

A combination of aggressive habitat management and provisioning of new cavities for nesting and roosting has been the key to recent successful red-cockaded woodpecker conservation efforts. Also, translocations of red-cockaded woodpeckers have helped contribute to the upward population trajectory that we see today.

Management Techniques
As noted above, the devastation of the longleaf pine forest and the decline of red-cockaded woodpecker populations are inextricably linked. Although red-cockaded woodpeckers can

and do use other pines besides longleaf within their range, longleaf pine forest is the critical habitat for this species.

Also, red-cockaded woodpeckers require a specific structure and configuration of habitat within a longleaf pine forest. Suitable habitat must consist of mature and even old-growth trees (longleaf pine >100 years) in an open, parklike configuration and maintained by frequent (every 2–3 years) prescribed fire.

Beginning in the late 1980s and early 1990s, people began to realize the critical importance of the remaining longleaf pine forests and the need to replant longleaf for restoration. For more than a century, longleaf was largely clear-cut and replaced with loblolly pine (*Pinus taeda*), which was preferred by foresters because it initially grew faster and hence produced commercial returns more rapidly than longleaf. The high density of the closed-canopy loblolly plantations, and exclusion of fire, resulted in a habitat that was totally useless for red-cockaded woodpeckers. Thus, as loblolly pine took over the southeastern states, red-cockaded woodpeckers were eliminated.

Red-cockaded woodpeckers need mature and old-growth longleaf pine forests as habitat because they make their nests in living pine trees. It may take red-cockaded woodpecker one or more years to excavate a nest cavity in a living pine tree. The birds typically select old, mature pine trees infected with red heart fungus (*Phellinus pini*), presumably because this makes excavation of the heartwood at the core of the stem somewhat easier once they make it through the outer sapwood zone of the tree.

Living pine trees are used by red-cockaded woodpeckers for nesting and roosting cavities because they maintain an active series of scars or resin wells around the entrance of the cavity, and sometimes around the entire tree. These resin wells allow the pine sap to exude and form a dense, sticky layer around the cavity entrance. This layer of sap makes it difficult to impossible for one of the primary nest predators, the rat snake (*Elaphae* spp.), to gain entry to red-cockaded woodpecker nests. Restoring fire to longleaf pine forests and replanting areas to longleaf pine for restoration are two strategies that are crucial for red-cockaded woodpecker conservation and recovery.

Cavity Provisioning

During the 1980s and 1990s, the development of artificial cavities "revolutionized the management of red-cockaded woodpeckers" (USFWS 2003:81). Cavities for red-cockaded woodpecker nesting and roosting were clearly a limiting factor for the recovery of many populations.

Two different approaches to artificial cavity provisioning were developed: drilling and inserts. The drilling technique (Copeyon 1990) uses a configuration of two different paths to excavate a nest cavity that will be attractive to the birds. The insert technique (Allen 1991) uses prefabricated cavity boxes that are inserted into a section of the tree that is cut out with a chainsaw. Both the drilling and the insert techniques have been highly successful at providing cavities that attract red-cockaded woodpeckers for roosting and ultimately nesting.

Translocations

The Red-Cockaded Woodpecker Recovery Plan (USFWS 2003:94) lists four applications of translocations recovery: augmentation of a population in immediate danger of extirpation, development of a better spatial arrangement of groups to reduce isolation of groups or subpopulations, reintroduction of birds to suitable habitat within their historic range, and management of genetic resources.

Individual red-cockaded woodpeckers are not trapped and translocated to a new area unless there is a specific reason or objective related to doing to. During the past decade or so, translocations have been a successful tool for supplementing small and isolated populations that were clearly on a downward trend. The combination of provisioning artificial cavities in recipient areas (habitat where red-cockaded woodpeckers were translocated to) has been an essential component of this success.

Finally, a key point is that for artificial cavity provisioning and translocation to be successful tactics for red-cockaded woodpecker recovery, suitable habitat for nesting and foraging must be present. Without suitable habitat, which is nearly always mature longleaf pine forest maintained by frequent prescribed fire, no amount of artificial cavity construction or translocation will be successful for recovery.

Cultural and Ecological Context

Red-cockaded woodpeckers are an important part of our natural world. The last decade of the 20th century and first decade of the 21st century represent a breakthrough when it comes to red-cockaded woodpecker conservation and recovery efforts. The first and most critical cultural aspect related to this breakthrough was that people finally realized longleaf pine forests were a critically endangered ecosystem that needed urgent conservation attention to steward what little remained and must be restored wherever possible.

The next two aspects of red-cockaded woodpecker recovery (i.e., artificial cavities, translocation) were largely technical, but they also have a cultural context. Artificial cavities for red-cockaded woodpeckers had been discussed for many years but were not implemented because biologists thought they would not work, would be too expensive, become occupied by other animals, and so on. The devastation from Hurricane Hugo, however, prompted biologists to install artificial cavities because so many natural red-cockaded woodpecker cavities were lost to the storm on the Francis Marion National Forest. To everyone's surprise and pleasure, red-cockaded woodpeckers readily took to the artificial cavities.

Translocation as a red-cockaded woodpeckers population recovery technique had also been discussed and considered for many years. The Safe Harbor approach to endangered species management on private lands, was, in our view, a critical component that helped push translocation from a hypothetical to a practical tool for red-cockaded woodpecker conservation. A Safe Harbor agreement is a contract between a landowner and the federal government that sets a baseline for a population level of an endangered species on a particular property. By setting the baseline in a Safe Harbor agreement to zero, the landowner then does not become legally responsible for a taking under the Endangered Species Act if the population of that species declines or disappears for some reason.

Despite the recent successes in sustaining and elevating red-cockaded woodpecker populations, full recovery and removal of the species from the endangered list is far in the future. The Red-Cockaded Woodpecker Recovery Plan (USFWS 2003) notes that it will most likely take until 2050, or later, for red-cockaded woodpeckers to be sufficiently abundant to be able to be removed from the Endangered Species List.

Texas Horned Lizards

Background

Texas horned lizards are an iconic reptile species of the southwestern United States (Sherbrooke 1981). They can be distinguished from other species of horned lizards by their two sharp occipital spines that protrude from the back of their head, two rows of fringed scales on their sides, pale stripe down the middle of their back, and dark stripes radiating from their eye to their upper lip (Stebbins 1954; Fig. 20.2).

Texas horned lizards generally live in arid and semiarid regions of grasslands, chaparral, and thorn scrub habitats from sea level to approximately 1,850 m elevation (Price 1990, Fair and Henke 1997a). Adult Texas horned lizards range in length from 75 to 125 mm (snout to vent length; Ballinger 1974, Henke and Montemayor 1997) and weigh from 25 to 100 g (Munger 1984a), with the largest specimen on record being 160 mm (Brown and Lucchino 1972). They are active from March until October (Potter and Glass 1931, Fair and Henke 1997b, Henke and Montemayor 1998) and hibernate underground the remainder of the year (Fair and Henke 1999). Hatchlings emerge as fully functional and independent individuals from clutches ranging from 13 to 45 eggs during midsummer (Milne and Milne 1950, Ballinger 1974, Pianka and Parker 1975). Adult Texas horned lizards are considered dietary specialists, with about 70% of their diet consisting of harvester ants (*Pogonomrymex* spp.). Juvenile horned lizards are more opportunistic predators and will consume crickets (Family Gryllidae), grasshoppers (Order Orthoptera), beetles (Order Coleoptera), termites (Order Blattodea), and caterpillars (Order Lepidoptera; Milstead and Tinkle 1969, Munger 1984b). Mortality factors for Texas horned lizards, in order of importance, include predation, traffic accidents, exposure, starvation, and disease (Munger 1986). Owing to numerous and high rates of such mortality factors, annual survival rates of 9% to 54% occur (Fair and Henke 1999). Thus longevity of Texas horned lizards is typically less than five years (Montemayor and Henke 1998).

The distribution of Texas horned lizards is greater than their name implies. Historically, Texas horned lizards could be found throughout much of northeastern Mexico and the south-central United States, including the majority of Texas, Oklahoma, southern and eastern Kansas and New Mexico, the southwestern and northwestern corners of Arkansas, northwestern Louisiana, and southeastern Colorado and Arizona (Sherbrooke 1981; Fig. 20.3). Today, their range has declined, and Texas horned lizards are either absent or rare in Arizona, Arkansas, Colorado, Louisiana, Missouri, and eastern Texas (Donaldson et al. 1994). Texas horned lizards are listed as a threatened species in Texas (Texas Parks and Wildlife Code 31 T.A.C., Section 65.171–65.177); protected in Arizona, Colorado, and Oklahoma; species of concern in Missouri; and nongame or no status in Arkansas, Kansas, Louisiana, and New

Figure 20.2. Texas horned lizard displaying a flattened body, two sharp occipital spines that protrude from the back of their head, two rows of fringed scales on the sides, pale stripe down the middle of the back, and dark stripes radiating from the eye to upper lip.

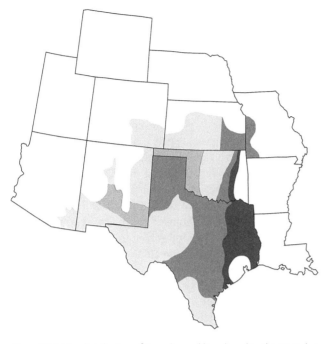

Figure 20.3. The distribution of Texas horned lizards within the United States. Shading highlights the regions where Texas horned lizards are now considered extirpated (black), extremely rare (dark gray), sporadic and isolated populations (light gray), and stable to increasing populations (gray). The distributions within a region were based on state biologists' assessments of populations of Texas horned lizards within their respective state.

Mexico (Price 1990). Today, populations of Texas horned lizards are considered extirpated from Arkansas and Louisiana, decreasing in Texas and Oklahoma, unknown in Missouri and Arizona, and only considered stable in Colorado, Kansas, and New Mexico (Price 1990, Carpenter et al. 1993, Burrow et al. 2001, Henke 2003, Trauth et al. 2004; Fig. 20.3).

Several explanations have been proposed for the decline of Texas horned lizards. The invasive red-imported fire ant (*Solenopsis invicta*) has spread throughout much of Texas (Summerlin and Green 1977) and can negatively influence horned lizards directly by their venomous sting (Webb and Henke 2003) and outcompete native insects, such as harvester ants (*Pogonomrymex* spp.), which is the preferred food of Texas horned lizards (Whitford and Bryant 1979). Widespread and indiscriminate use of broadcast pesticides can further negatively affect the prey base for Texas horned lizards (Donaldson et al. 1994). Texas horned lizards were a popular pet of children prior to 1960 (Donaldson et al. 1994). Children would capture wild lizards and take them home, but because the lizards ate a specialized diet of ants, they rarely survived captivity. Texas horned lizards were over-collected for the pet industry and by children, and they were given as gifts to tourists who stopped at local gas stations (Donaldson et al. 1994). Also, loss of habitat and habitat fragmentation from suburban sprawl, widespread use of exotic grasses that were used to stabilize soil after road construction, and the conversion of

native lands to agricultural crops have been included as potential reasons for the decline of Texas horned lizards (Fair and Henke 1998). Although each suggested reason for the decline has merit, no single proposed reason occurred over their entire distributional range, and Texas horned lizard populations have declined over most of their distribution. Therefore either a combination of these proposed factors has caused the decline, or an unconsidered factor was to blame.

Today, populations of Texas horned lizard are sparse and sporadic across its once vast range (Henke 2003). They are remembered with fondness, especially so by much of today's older generation, who have fond memories of these miniature dinosaurs. Many of the efforts to reverse the declining trends of Texas horned lizards are spawned from this cultural attachment.

Management Techniques

Initially, attempts to repopulate Texas horned lizards to their historical range focused on translocation of adult lizards. Unfortunately, translocation of Texas horned lizards caused a reduction of lizard activity and movements, which potentially affected their ability to acquire sufficient prey, and thus lizards experienced severe weight loss and lost critical mass to survive hibernation (Henke 2013). In addition, the sedentary lifestyle of translocated lizards potentially increased their susceptibility to predators, thus reducing their survival rate to <10% during the initial eight months after translocation (Henke 2013). Therefore translocation of adult Texas horned lizards as a method to repopulate the species to its former distribution and abundance was not considered a feasible option.

Non-native grasses were introduced into the United States from other continents (i.e., Africa, Asia, Australia) with the desire to provide livestock with drought-resistant forage and enhance erosion control along roadways with fast-growing plants (D'Antonio and Vitousek 1992). Unfortunately, non-native grasses are highly competitive with native grasses, and the seeds of non-native grasses have dispersed over large areas, establishing monocultures (Wilcove and Chen 1998). Replanting native vegetation to restore horned lizard habitat to its former vegetative state has been attempted; however, seed banks of non-native grasses are so well established that restored areas must receive constant attention, and non-native plants have to be removed as they grow. Otherwise, non-native grasses will quickly reestablish and outcompete the native plants once again (Gowdy 2020). For example, a 136-ha tract of grassland was restored and maintained free of non-native grasses for 10 years in south-central Texas, at a cost of about $1 million during that time span (Gowdy 2020). Unfortunately, even with such herculean efforts to restore the vegetative community to a native plant habitat, Texas horned lizards did not respond with increased abundance and reproduction (Huerta 2020). Therefore the effort and expense to restore habitat for Texas horned lizards were not practical.

The San Antonio Zoo Center for Conservation and Research is attempting a translocation effort to reestablish populations of Texas horned lizards throughout their former range, but the center's efforts involve captive-reared hatchlings rather than wild-caught adult lizards (Gluesenkamp 2019). Through captive husbandry programs for Texas horned lizards and harvester ants, hatchling lizards are raised in captivity from eggs and then released into preselected habitats. Release sites are screened to ensure that habitat consists of native vegetation with sandy soils and that sufficient prey species such as harvester ants and termites are available. Efforts to control non-native grasses and fire ants on the release sites are performed. Although this program is in its infancy, preliminary results are encouraging.

Cultural and Ecological Context

Texas horned lizards are firmly integrated into the culture of the southwestern United States. Their depictions were common in southwestern tribal art of North America (Sherbrooke 1981), often because several North American tribes regarded horned lizards as sacred. Horned lizards were depicted on ancient pottery and in petroglyphs, representing health and happiness to Native Americans. The days of the Old West with wild horses, cowboys with six-shooters, and stories of Marshall Matt Dillon (okay, you may have to ask your grandparents about these) often included horned lizards as iconic creatures that roamed the western frontier. In fact, in the classic movie *Old Yeller*, the youngest boy, Arliss, captures a Texas horned lizard and puts it in his pocket to play with at a later time. Such events were common, and anecdotes of children with their pet horned lizards were abundant. Today those children are elderly, and populations of Texas horned lizards are sparse and sporadic across its once vast range (Henke 2003). For the generations who grew up with Texas horned lizards, watching horned lizards vanish from the landscape is like watching their childhood die and become a faded memory. It's no wonder the Texas horned lizard was named the official State Reptile of Texas in 1993, and specialty license plates can be purchased, with some of the proceeds going to a state recovery fund for Texas horned lizards. It is hoped that recovery efforts will prove to be successful and that populations of Texas horned lizards can be reestablished throughout their former range. Then perhaps future generations of children can enjoy these amazing lizards.

SUSTAINING WILDLIFE POPULATIONS

Endangered and threatened species populations require human intervention and active management to achieve increase and hopefully population recovery. The goal for managing many wildlife populations is simply to sustain them at specific levels to achieve specific management goals. Moose (*Alces alces*) were selected for the first case study because these large mammals are an important component of the ecosystems they inhabit, and sustainable populations of moose are an important resource to recreational and subsistence hunters. Like moose, waterfowl populations, the subject of the second case study, fulfill important ecological and recreational functions, and managing harvests, a complex annual activity, is integral to

sustaining North American waterfowl populations. The third case study involves prairie dogs (*Cynomys* spp.), which were once treated as vermin, with concerted efforts to reduce populations, but today they are recognized as a keystone species because sustainably maintaining prairie ecosystems in parts of North America requires sustaining prairie dog populations.

Moose

Background

Moose are the largest member of the deer family (*Cervidae*) and are popular with consumptive and nonconsumptive wildlife enthusiasts. Hunter harvest of moose is important in many areas of the species' range because local hunters, including Native Americans, rely on moose meat for sustenance, recreational hunters enjoy the thrill of pursuing such a majestic animal, and local economies benefit from money spent by hunters, photographers, and other people interested in moose. For these reasons, wildlife managers seek to sustain productive moose populations through census, habitat and predator management, and hunting regulations. The phrase "Get your moose yet?" is a common refrain across many rural Alaska towns and villages during moose hunting season.

Knowledge of moose ecology is necessary to implement effective moose management. In North America, moose are distributed in boreal and mixed coniferous-deciduous forests of Alaska, Canada, the Rocky Mountains, and the northern United States. Large predators, particularly wolves and bears, are found throughout most of moose range in North America. Moose have many adaptations that enable them to live in areas that are extremely cold and that accumulate deep snow during winter. The moose's large size and well-insulated winter coat enable them to endure cold temperatures. These adaptations can make moose susceptible to heat stress (Renecker and Hudson 1986, Thompson et al. 2020), and heat can have strong effects on movement and habitat selection (Alston et al. 2020). Large body size and long legs enable moose to move through deep snow and thus live in areas inaccessible to other ungulates during winter. Moose eat aquatic vegetation and leaves, buds, and twigs of woody plants. Buds and twigs can be particularly important because they remain available during winter. Reliance on browse means that moose are dependent on early succession vegetation communities because forage biomass and quality are high in such communities.

Management Techniques

Moose are managed by the appropriate state or provincial wildlife agency. Knowing the size and composition of a moose population, in conjunction with habitat and nutrition information, aids agency management decisions (Boertje et al. 2007). Most moose populations are surveyed along transects or in quadrats using aircraft because large areas can be surveyed quickly, and moose range is dominated by expansive areas with poor road access. Timing of surveys varies from November to March, when snow cover and loss of deciduous canopy cover increase visibility, but in some areas, surveys must be completed before snow accumulation causes moose to use coniferous forests (Timmermann and Buss 1998). Because not all moose within a survey area are observed during surveys, sightability models may be used to correct for unseen animals (Anderson and Lindzey 1996). Moose are classified as calves, cows, and bulls. The calves:100 cows ratio is a common measure of calf production. The sex ratio of bulls:100 cows can index harvest intensity of bulls relative to cows and indicates when composition is sufficiently skewed to influence breeding. Because of their expense, surveys may only be conducted in a given area every two to five years.

While surveys can provide estimates of the abundance of moose, sustaining a moose population requires knowledge of the population size relative to the carrying capacity of vegetation. Population size relative to forage resources can be assessed through a browse survey that measures the proportion of high-quality browse consumed each year (Seaton et al. 2011). Another approach is to use reproductive data, particularly twinning rates, body mass of young moose, and body condition to assess the ability of vegetation to support moose populations (Schwartz and Renecker 1998, Boertje et al. 2007).

Because moose diets are composed primarily of browse, and because forest succession reduces the amount and quality of browse over time, forest manipulation is necessary in many areas to sustain moose populations (Thompson and Stewart 1998). Fire was the primary disturbance that created moose habitat before large-scale forestry operations, and in many areas of the species' range, prescribed fires and wildfires remain important processes to create early successional habitat. Timber harvest can also convert mature forests to an early seral stage useful to moose, with the greatest benefit from cut areas interspersed with mature forest. If cuts are too large, moose will not use the interior until regeneration provides sufficient cover. In areas without marketable timber, mechanical treatment of woody vegetation with some type of drum-roller and cutting blades can stimulate regrowth. An additional form of forest disturbance that can create conditions favorable to moose stems from beavers modifying riparian environments through denning and dam building (Nummi et al. 2019). The restoration of beaver in many portions of the moose's range has likely benefited moose habitat quality.

Moose habitat also has attributes that require protection as opposed to creation. Overstory cover, usually including conifer trees, is important during winter to reduce snow depth and to provide thermal and hiding cover; thus at least some blocks of mature forest should be left undisturbed. Moose use wetlands for feeding and thermoregulation during spring and summer (Alston et al. 2020); such wetlands should be protected, and sufficient cover should remain around them to ensure moose are comfortable using them. Finally, moose have a strong sodium appetite during spring and summer to meet deficiencies acquired during winter and from consumption of large amounts of succulent vegetation during spring. This sodium appetite is met at salt licks and through consumption of aquatic vegetation (Jordan 1987); sites providing minerals should be maintained.

A second management technique that has been applied but is more controversial than habitat management is predator control. In areas where wolf and bear populations are near carrying capacity, moose populations are often lower than predicted by the quality and amount of moose habitat (Gasaway et al. 1992, Hayes et al. 2003). Reducing densities of wolves and bears in such areas may increase the moose harvest several-fold as the moose population increases from low density caused by predation (Boertje et al. 2009, Keech et al. 2011, Severud et al. 2019). Predator control is not necessary to sustain moose populations in such instances but does appear to increase the sustainable harvest.

Another important technique used to manage moose populations is harvest that is structured to provide subsistence and recreational opportunity, to ensure population persistence, and sometimes to promote a change in the moose population's size. Because of long life spans, often low reproductive or recruitment rates, and low population density, moose are susceptible to overharvest. Thus moose hunting opportunities in most areas are allocated through a lottery, which determines when and where a hunter may hunt and the number, sex, and age of moose to be harvested (Timmermann and Buss 1998). Harvest of female moose is manipulated when the goal is to change population size; a larger female harvest causes a decline in population size, and female harvest is restricted when a larger population is desired. In Ontario, Canada, harvest is designated for calf moose in some units with declining moose populations because calf moose are less likely to survive the winter and will be less likely than an adult female to produce offspring the following year (Ontario Ministry of Natural Resources 2011).

Providing opportunities to harvest large male moose is a management objective in some areas. Antler size of moose increases with age, and thus mature males are necessary in such areas. Antler criteria are one means by which to reduce harvest of young and middle-age moose. In some management areas of Alaska, male moose must have antlers with at least a 127-cm spread or have a specific number of brow tines (antler tines arising from the brow palm). In many areas of Alaska, small male moose with either spike or fork antlers are also legal, which fulfills subsistence hunting objectives (Alaska Department of Fish and Game 2011).

Ecological and Cultural Context

The geographic range of moose across large expanses of relatively undeveloped areas of North America raises intriguing and often controversial management issues. First, the presence of large predators, specifically wolves and bears, adds a dimension to ungulate management missing in many areas of North America. While control of predators appears to increase moose density and hunter harvest, attitudes about wolves vary widely (Kellert et al. 1996) such that some people and groups oppose shooting or trapping predators for this purpose (Musiani and Paquet 2004, Decker et al. 2006, Boertje et al. 2010). Furthermore, attempting to increase harvest potential for ungulates through predator control can be ineffective if some members of the predator guild are not eligible for control owing to their conservation status (Peters et al. 2020).

Traditional and subsistence use of moose is part of the management milieu in many parts of moose range. For example, Alaska Department of Fish and Game has a directive to accommodate subsistence needs of Alaska residents before recreational harvest of wildlife. Moose and caribou (*Rangifer tarandus*) meat is an important part of the diet of many rural Alaskans, with residents of some communities consuming 100 kg/person annually (Titus et al. 2009). Many jurisdictions have separate regulations and seasons for traditional and subsistence hunting by Native Americans, some of which date back to treaty obligations in the 1800s and early 1900s (Timmermann and Buss 1998). Incorporating subsistence hunting can be difficult and contentious, as demonstrated by the political and judicial activity at the state and federal level as Alaska's policy developed (Crichton et al. 2007).

Moose management has economic ramifications. Money spent by moose hunters is difficult to quantify, but economic value estimates of greater than $251,600,000 annually as a result of moose hunting in North America during the 1980s and 1990s suggest the value may be substantial, especially in rural economies (Timmermann and Buss 1998). Funds generated from license sales for moose hunting help support state and provincial wildlife agencies and thereby management of many wildlife species. Tourism related to moose hunting generates funds for local communities. Guidelines and regulations on timber harvest to accommodate habitat management objectives for moose also have a cost, albeit rarely quantified.

An emerging issue potentially challenging management for sustainable moose populations is increased mortality along the southern periphery of the species' range. For example, moose populations in northwestern Minnesota declined from more than 3,500 in the 1980s to less than 100 in the 2000s, apparently because a combination of nutritional deficiencies, parasites, and disease reduced adult survival (Murray et al. 2006, Lenarz et al. 2010). Similarly, in northeastern Minnesota, the moose population declined an estimated 65% from 2006 to 2018, likely because of poor health and high susceptibility to predation (Severud et al. 2019). A moose population in New Hampshire appears stable with favorable habitat and conservative harvest. This population's growth rate is apparently limited by heavy infestations of winter ticks, causing low calf survival and poor reproduction by yearling females (Musante et al. 2010). A population in Wyoming had negative population growth rates, most likely because of high adult female mortality and poor calf production (Becker 2008).

Although human-induced habitat changes could help explain some declining or stable moose populations, even in the absence of harvest, mortality is positively correlated to ambient temperature, suggesting climate change may be affecting moose along the southern edge of their range (Murray et al. 2006, Lenarz et al. 2009, Musante et al. 2010). The large size and excellent insulation of moose are adaptations to the long cold winters of northern North America, but these traits could make moose susceptible to warming temperatures.

Evidence for direct mortality caused by heat stress is lacking (Lowe et al. 2000, Murray et al. 2006), but complex interactions among diet quality, foraging behavior, competition with white-tailed deer, parasites, and disease could be influenced by a warming climate (Weiskopf et al. 2019). Furthermore, weather can have a strong effect of moose habitat selection (Alston et al. 2020) and harvest susceptibility (Hasbrouck et al. 2020), ultimately affecting the species value as a consumable resource. Whatever the cause, poor survival or reproduction of moose on the southern edge of their range could result in a shift of moose range to the north and represent a challenge to sustainable management of the world's largest deer species (Lenarz et al. 2010).

Waterfowl

Background

Migratory waterfowl are an important natural resource in North America that provides recreational and economic opportunities across international boundaries. The migratory nature of waterfowl makes them a shared resource among Canada, Mexico, and the United States. These countries have developed treaties and cooperative management plans that acknowledge their shared accountability toward the responsible management and sustainability of North America's migratory birds. Harvest management has been an international cooperative effort since the mid-1900s, acknowledging the shared resources between the United States and Canada. Sport harvest regulations have been developed to ensure the long-term sustainability of populations, while at the same time providing the ability of hunters to harvest migratory birds. The treaties define the general guidelines for sport hunting regulations of waterfowl, including which species are allowed to be harvested and when the hunting seasons can occur. The process of regulating harvest of waterfowl is conducted annually through extensive monitoring of waterfowl populations, breeding habitat conditions, and hunter success (Blohm et al. 2006).

Up until the 1980s, habitat management for waterfowl did not share the same international cooperation as harvest management. Most habitat management for waterfowl was conducted by state and federal agencies or private landowners working independently without any coordinated efforts. In 1986 the signing of the North American Waterfowl Management Plan (NAWMP) between the United States and Canada, and later by Mexico, was a monumental achievement for waterfowl management and helped coordinate management efforts across large regions (US Department of Interior and Environment Canada 1986). One of the most important accomplishments of NAWMP was the creation of joint ventures, or regionally based partnerships composed of state, federal, nonprofit, and private stakeholders. Joint ventures combine knowledge and resources from the diverse group of stakeholders to direct conservation planning for waterfowl habitat at regional levels. Migratory waterfowl rely on habitats across large scales that often include multiple political jurisdictions. Thus the partnership aspect of joint ventures provides the financial and political capacity to successfully develop conservation strategies at large spatial scales. The transformation in philosophy about habitat conservation, from independent small-scale projects to regionally based large-scale planning, has been instrumental in restoring North American waterfowl populations.

Management Techniques

The Breeding Waterfowl Survey in North America is the oldest and most extensive wildlife survey in the world. It covers more than 5.4 million km² of important duck breeding habitat in the northern United States and Canada. The survey estimates the size of breeding populations of waterfowl and evaluates their breeding habitat conditions. The survey was initiated in an experimental phase in 1947 and became operational in 1955. The USFWS, Canadian Wildlife Service, many state and provincial agencies, and tribal organizations cooperate to conduct this intensive survey, which has been conducted each year since its inception (USFWS 2010).

The survey is conducted in May via fixed-wing airplanes flying at low altitude. All species of ducks observed on >89,000 km of transects are recorded. Not all species are readily observed on aerial surveys; therefore biologists on the ground (or using helicopters in some areas) conduct intensive ground-truthing surveys in portions of the survey area to deal with this disparity in detection probabilities. The differences between aerial and ground counts are used to develop visibility correction factors for each species. After the aerial counts are adjusted by the visibility correction factors, they are extrapolated across the survey area to derive estimates of breeding populations for all species of waterfowl observed on the survey. In July, a production survey is conducted on a subset of transects to count the number of broods produced. Breeding habitat conditions (i.e., number and types of ponds) are also surveyed along the transects to predict productivity, which in combination with population estimates are used in models to estimate the fall flight and help set hunting regulations (i.e., season length and species' daily bag limits; USFWS 2010).

Waterfowl biologists across North America also mark individual ducks with aluminum leg bands that contain unique numeric codes. More than 200,000 ducks are banded annually in North America (Blohm et al. 2006). Based on the number of individuals banded, number of bands recovered, and band reporting rates, biologists are able to estimate survival and harvest rates for individual species.

Another assessment of harvest is through surveying hunters. For instance, the USFWS selects a sample of about 60,000 of the roughly 3.5 million migratory bird hunters each year who purchased a hunting license in the United States. (Padding et al. 2006). The participants are asked to keep record of the date, location, and number of ducks and geese they harvest each day afield. The sample is stratified by state and by the hunter's success during the previous hunting season. The responses from the sample of hunters are used to estimate the average number of ducks and geese harvested per hunter within each stratum. Overall harvests of ducks and geese are

then estimated by multiplying the average harvests per hunter by the number of active hunters in each stratum and then adding stratum estimates. Because hunters vary in their ability to identify waterfowl they harvest, another sample of hunters is surveyed in the Waterfowl Parts Collection Survey. These hunters are asked to mail a wing from each duck and a tail fan from each goose they harvest that hunting season to the USFWS, resulting in approximately 90,000 duck wings and 20,000 goose tails annually (Padding et al. 2006). Waterfowl biologists are then able to determine the species, age, and gender based on characteristic of each duck wing, and species and age from goose tail fans. The two harvest surveys together provide species-specific harvest estimates and age and sex ratios of harvested species.

In the United States, adaptive harvest management is used to help the decision-making process on waterfowl hunting regulations each year. In this approach, decisions about waterfowl bag limits and hunting season length are made on the basis of results from statistical models used to select from a fixed set of regulatory alternatives. Each year, information from the annual monitoring programs discussed above is used in the models in an iterative fashion. Subsequent monitoring then assesses the effects of the regulatory decisions on populations. Thus the strength of the decision process is its adaptive quality, which is designed to reduce uncertainty on how the monitoring results relate to harvest (Blohm et al. 2006, Nichols et al. 2007).

Habitat management to sustain waterfowl populations is carried out throughout North America to fulfill the changing requirements of waterfowl across the annual cycle. Habitat management for breeding waterfowl concentrates on providing adequate nesting and brood-rearing habitat that promotes high rates of nest success and brood survival, as these are important vital rates that contribute greatly to changes in population abundance (Hoekman et al. 2002, 2006, Coluccy et al. 2008). Understanding the biological needs of breeding waterfowl, and in particular their specific habitat requirements and the distribution of requisite habitats in the landscape, enables managers to identify areas with currently suitable conditions and those areas that could be made suitable by managing for a missing component (Krainyk et al. 2019). This sort of management occurs through NAWMP joint ventures within conservation planning regions of varying spatial scales. Habitat management for waterfowl on nonbreeding areas focuses on providing adequate food resources because energy is the most limiting factor for waterfowl during the nonbreeding period (Williams et al. 2014). Joint ventures often use bioenergetics models to translate regional waterfowl population objectives into energy requirements by waterfowl during the nonbreeding period within their planning region (Wilson and Esslinger 2002). These energy requirements are then converted to habitat area on the basis of conversion factors of food density in specific habitats. This provides joint ventures a goal in regard to the quantity of specific habitats that are required to support waterfowl populations at desired levels within specific planning regions. This also allows joint ventures to evaluate whether habitat availability within planning regions is adequate to support nonbreeding waterfowl requirements or whether it is at a deficit and in need of increasing to meet those requirements.

Ecological and Cultural Context

Waterfowl are migratory birds that cross international boundaries during migration between breeding and wintering areas. Thus the management of waterfowl in North America is a cooperative effort among Canada, Mexico, and the United States. Treaties among the three countries provide the basis for the cooperative management. Waterfowl hunting regulations are set independently within each country based on the results of the monitoring programs described above. Subsistence harvest is important to the culture of some native tribes in northern Canada and Alaska but is not regulated to the degree of sport harvest. Subsistence harvest of both birds and eggs is legal and is allowed outside of the regulatory framework dates established by the treaties between countries that define the regulations.

Several aspects of waterfowl ecology provide a unique array of management challenges. For instance, there is a large diversity of waterfowl species regulated, each one varying in regard to their abundance, population trend, distribution, use of habitats, and so forth. Harvest regulations that take these differences into account can become complicated. Also, the migratory nature of waterfowl allows them to cross state, provincial, and international boundaries that have varying harvest regulations and often different management interests. Additionally, the geographic distribution of waterfowl in North America necessitates a large monitoring effort and can relegate different populations of a species to varying pressures such as habitat alterations or hunting intensity. To address these challenges, the regulation of the harvest of waterfowl in North America likely represents the most complicated and extensive wildlife management effort in the world (Blohm et al. 2006).

Prairie Dogs

Prairie dogs are an interesting case study because they represent a genus that has received the gambit of wildlife management practices, from population reduction to maintenance to increasing prairie dog populations. Historically, five species of prairie dogs were on the Great Plains and Rocky Mountain regions of North America: the black-tailed prairie dog (*Cynomys ludovicianus*), Gunnison's prairie dog (*C. gunnisoni*), Mexican prairie dog (*C. mexicanus*), white-tailed prairie dog (*C. leucurus*), and the Utah prairie dog (*C. parvidans*; Pizzimenti 1975). Of these, black-tailed prairie dogs had the most extensive range. Their distribution spanned from southern Saskatchewan, Canada, to western Texas and north-central Mexico (Hoogland 2003). Today, all species of prairie dogs are rare, occurring in small remnant colonies dotted throughout their former range of 150 years ago. Populations have become so reduced that the black-tailed prairie dog occupies <2% of its former range (Miller et al. 2000). During the mid-1970s, all

species except the white-tailed prairie dog were protected under the Endangered Species Act. By 2010 the Mexican, Utah, and Gunnison's prairie dogs were listed as endangered, threatened, and candidate species, respectively, while the black-tailed and white-tailed prairie dogs have no federal listing status (USFWS 2011). Ten years later (2020) the Mexican and Utah prairie dogs were federally listed as endangered, while the black-tailed, white-tailed, and Gunnison's prairie dogs have no federal status (USFWS 2020). The International Union for the Conservation of Nature (IUCN) Red List of Threatened Species also classifies the Mexican and Utah prairie dogs as endangered, while the other prairie dog species are species of least concern.

Background

Prairie dogs are colonial and fossorial rodents that belong to the family Sciuridae (Hafner 1984). They are a larger rodent species with an average length of about 360 mm and average weight of 400–1,350 g (Pizzimenti 1975). Although there are five species of prairie dog, the black-tailed prairie dog is the most often referred to because it was the most numerous and had the largest distribution.

Black-tailed prairie dogs live in a territorial, polygynous harem society with family groups called coteries (King 1955). A coterie typically consists of approximately seven individuals: a breeding male, two to three adult females, and three to four offspring (Hoogland 1995), maintaining a territory of 0.05–1.0 ha (King 1955, Hoogland 1995). Number of burrow entrances per coterie range from 5 to 214 (Hoogland 1995). Black-tailed prairie dogs are selectively herbivorous, eating a variety of plants and underground roots (King 1955, Fagerstone et al. 1981), but they also will eat insects (Costello 1970, O'Meilia et al. 1982) and frequently cannibalize conspecific offspring. In addition, black-tailed prairie dogs cut down plants >20 cm within their territory. This mowing behavior is believed to aid predator detection (Hoogland 2003).

Black-tailed prairie dogs are the aerodynamic engineers of the prairies. They apply Bernoulli's principle to keep their burrows well ventilated (Vogel et al. 1973). Prairie dogs always build their burrow system with one opening at ground level and another with a raised entrance. The air over an entrance at ground level moves slower and has greater air pressure over the ground entrance than the air over an entrance in a raised mound, which has increased speed and lower air pressure (Vogel et al. 1973). This type of tunnel construction creates an unbalanced force that passively flows air through the tunnels and out the higher mound entrance.

Historically, prairie dogs were an important food source for certain Native American tribes and early European settlers, presumably because black-tailed prairie dogs do not hibernate and were abundant year-long (Scheffer 1945). As the North American continent became settled, large tracts of prairie dog habitat were converted to farmland. Within the 60-year time span from 1900 to 1960, an estimated 40 million ha of prairie dog habitat had been reduced to less than 600,000 ha (Miller et al. 2000). In addition, ranchers replaced bison on the Great

Plains with domestic livestock. Believing that prairie dogs competed with cattle for forage and that prairie dog burrows would result in injuries (i.e., broken legs) to cattle and horses, ranchers began an intensive eradication campaign. Methods to rid the Great Plains of prairie dogs included shooting, drowning, poisoning, and habitat destruction (Swenk 1915; Randall 1976a, 1976b). In addition, prairie dogs are highly susceptible to bubonic plague (*Yersinia pestis*), a bacterium transmitted most commonly by the bite of certain fleas (Cully et al. 1997). Prairie dogs have little to no immunity to bubonic plague, so once introduced into a colony, it decimates the residents within weeks. Although bubonic plague is considered an accidental introduction into the United States from about 100 years ago via rodent-infested cargo from European ships (Olsen 1981), bubonic plague has been intentionally introduced into prairie dog colonies as an eradication method. The combination of all eradication methods has been successful, resulting in the small remnant populations observed today. Presently, all species of prairie dog are considered rare.

Management Techniques

Public education is the key to prairie dog survival. Ranchers must be reeducated to understand the critical role that prairie dogs play within grassland ecosystems. In addition, several myths about prairie dogs need to be dispelled. Eradication of prairie dogs was advocated because it was believed that prairie dogs compete with livestock for food, destroy grazing habitat, injure livestock with their burrows, and spread disease to livestock and humans. We briefly discuss each of these complaints against prairie dogs.

Prairie dogs consume some plant species that are also consumed by livestock. The majority of the diet of livestock is avoided by prairie dogs, and vice versa (O'Meilia et al. 1982). Prairie dogs are selectively herbivorous, and they have been reported to improve the quality of certain plants; therefore livestock prefer to forage within prairie dog colonies (King 1955, Koford 1958). Because prairie dogs clip tall vegetation within their territory, ranchers have claimed that prairie dogs destroy grazing habitat. In reality, black-tailed prairie dogs prefer to colonize areas of existing low vegetation (Koford 1958). In addition, clipping tall vegetation keeps plants in a young growing state, which can aid digestibility of plants for livestock. Third, prairie dogs do create holes in the ground; the number of burrow entrances can exceed 100/ha (Hoogland 1995). Even in areas with high numbers of burrow entrances, there have been few documented cases where livestock have fractured their legs after stepping into burrows of prairie dogs (Hoogland 1995). It appears that this complaint has been much exaggerated. Lastly, prairie dog colonies harbor disease, particularly bubonic plague. Unfortunately, plague has been used as an eradication tool, which has enhanced its spread across the grassland plains. Vector control programs that treat prairie dog burrows with insecticide can reduce flea populations, which in turn can reduce the probability of bubonic plague outbreaks (Barnes et al. 1972). Therefore the complaints against prairie

dogs can be dismissed. Private landowners must understand these data about prairie dogs and cease eradication efforts for repopulation attempts to have any chance of success.

Other management strategies to help alleviate declines of prairie dog populations include habitat restoration; developing corridors for dispersal between locations of suitable habitat; translocations of young, male prairie dogs from their natal range; and supplemental feeding programs during times when body condition becomes poor. Habitat restoration can be easy for prairie dogs because they colonize areas that have been overgrazed by livestock (Koford 1958, Costello 1970). Natal dispersal is biased toward males (King 1955), and mammalian and avian predators of prairie dogs are numerous (Hoogland 2003). Therefore translocation of juvenile males from their natal range to nearby territories can help reduce predation risks. Overwinter survivorship can be a hardship on prairie dogs; it has been demonstrated to directly vary with body mass (Hoogland 1995). Therefore, during times of drought or when body condition appears poor, a supplemental feeding regime potentially can offset a high winter mortality rate.

Ecological and Cultural Context

Unfortunately, prairie dog eradication was conducted without adequate knowledge of ecosystem function of grasslands. Leopold (1949:190) wrote, "To keep every cog and wheel is the first precaution to intelligent tinkering." Early biologists and wildlife managers could have learned from Leopold's message. Prairie dogs, especially black-tailed prairie dogs, are now considered an essential component of grassland ecosystems.

Prairie dogs meet the definition of a keystone species. Keystone species have a disproportionate effect in line to their abundance, which results in significant effects on ecosystem structure, function, and composition (Paine 1980). Removing prairie dogs from grasslands has resulted in cascading effects, some of which are still being identified. Prairie dogs dig tunnels to protect themselves from weather and predators. Burrow systems / coterie have been as long as 33 m long and 5 m deep (King 1955), removing about 225 kg of soil from each burrow system (Whicker and Detling 1993). Such activity increases soil turnover, soil aeration, and soil macroporosity (Munn 1993). In addition, prairie dogs cache food in their burrows and often defecate within their burrows, which changes soil chemistry, increases soil organic content and soil nitrogen content, and promotes nematode abundance and other invertebrates within soils (Ingham and Detling 1984, Munn 1993, Outwater 1996). Such improvements to the soil result in an improved vegetation community by increasing plant nutritional content, increasing plant digestibility, and increasing the ratio of live to dead plants within the community (Whicker and Detling 1993). This activity coupled with their behavior of cutting down tall vegetation promotes greater diversity of plants. Prairie dogs also have been documented to retard the spread of mesquite (*Prosopis* spp.; List 1997, Weltzin et al. 1997). The enhanced vegetational diversity results in greater species richness, density, and diversity of small mammals, which then cascades into a greater diversity and abundance of mammalian and avian predators (Manzano 1996, Ceballos et al. 1999). Therefore prairie dogs produce significant bottom-up effects on soils, vegetation structure, plant productivity, nutrient cycling, and ecosystem functions where they occur.

To date, about 170 species have been listed that potentially rely on prairie dogs for some aspect of their survival (Miller et al. 2000). Most notable of these are black-footed ferrets (*Mustela nigripes*), whose diet is almost exclusively prairie dogs (Clark et al. 1982), mountain plovers (*Charadrius montana*) that selectively use prairie dog habitat (Knowles et al. 1982), and burrowing owls (*Athene cunicularia*) that rely on prairie dog burrows for homes (Plumpton and Lutz 1993).

The public needs to understand the keystone role that prairie dogs play within grassland ecosystems. The public and policymakers alike need to understand the concept that Leopold espoused concerning intelligent tinkering.

REDUCING POPULATIONS

Just as some wildlife populations are managed to be increased, and some wildlife populations are managed to be sustained on an annual basis, populations of wildlife species determined to interfere with human interests or represent threats to desirable wildlife species or ecosystems are managed with the objectives of reducing population size. Exotic ungulates were selected for the first case study because many species are becoming pervasive in many portions of the United States, but particularly so in the Southwest and especially on Texas rangelands. There is a long tradition among some private landowners in Texas and elsewhere to introduce exotic ungulate species on their properties that they have hunted or encountered in the animal's native habitats in foreign countries. Often the purpose of these introductions is to provide hunting opportunities to the landowner and his friends or even for commercial purposes as way of earning money from individuals who desire an exotic trophy. Unfortunately, many of these exotic ungulates have adapted extremely well to their introduction sites, and populations have increased to an extent that the animals negatively affect native plant communities and the native wildlife species that depend on them. Therefore, in many situations, reducing exotic ungulate populations is desirable.

Feral swine are an example of an exotic ungulate that has become a significant problem in most of North America over the past 40 years. Feral swine were selected for the second case study because they have invaded almost every state in the United States where they have become such significant threats to rural economies and native ecosystems that active population management is ongoing to reduce their numbers.

The final case study involves rose-ringed parakeets (*Psittacula krameri*), one the most invasive bird species worldwide. They are extremely adaptable, so they can successfully invade and become established in wide variety of native plant communities and damage them, facilitating the introduction of exotic invasive plants and the decline of native bird species.

Consequently, like feral swine, active population management is currently ongoing for rose-ringed parakeet populations.

Exotic Ungulates
Background
The archaeological record reveals that humans have a history of introducing non-native species of large mammals for hunting, conservation, and aesthetic purposes (Hofman and Rick 2018). These non-natives are often termed exotics, defined as a species or population living outside its historical distribution (Mungall 2000). While exotics provide some benefits, many also compete with or harm native species. The factors that influence the effect of introduced populations are not well understood, and it is difficult to predict which introductions will become invasive (Sakai et al. 2001). Mounting evidence indicates that non-native species may have long-lasting ecological influences on native plant and animal communities worldwide.

Feral swine are the best-known exotic species of large mammal in North America. Many other introduced species of large mammals have established free-ranging populations, yet few are familiar with the extent and effect of the non-natives. Exotics are often considered local or regional issues, yet they collectively influence plant and animal communities over a broad geographic area. Each species presents a unique set of challenges and influences agriculture, native, and domestic animals.

After feral swine, feral horses (*Equus caballus*) may garner the most recognition because of their status as a symbol of the American West and coverage of their management challenges in the media. Horses evolved in North America but became extinct at the end of the Pleistocene (Grayson 2006). Reintroduced to the continent during European colonization in the 1500s, horses escaped or were released into the wild. Feral horses are presently widespread on public lands throughout much of the western United States, where they cause extensive damage to rangelands and compete with native and domestic species for forage and water resources (Davies and Boyd 2019). Native to East Asia, sika deer (*Cervus nippon*) have been introduced worldwide for sport hunting or as alternative livestock. Few are aware that the largest free-ranging population of sika deer in North America occurs in the northeastern United States. Sika deer were introduced to islands near Delaware and the eastern coasts of Maryland and Virginia in 1916, and have since expanded in number and geographic distribution across the peninsular area shared by the three states. Sika deer are similar in size to native white-tailed deer, and the two species overlap in diet and habitat use, which indicates high potential for competition (Kalb et al. 2018). Free-ranging populations also occur in Florida and Texas (Mungall 2000).

Texas may have the most extensive free-ranging populations of exotic ungulates in the United States largely because of the long history of introductions, well-established market for breeding stock and hunting opportunities, and a regulatory system that considers most species of exotics as livestock.

Common free-ranging exotics include the Aoudad or Barbary sheep, (*Ammotragus lervia*), native to North Africa; axis deer (*Axis axis*), native to India, Nepal, and Sri Lanka; and nilgai antelope (*Boselaphus tragocamelus*), native to India and Pakistan (Traweek and Welch 1992). Aoudad were first introduced into western Texas in the 1950s and have since expanded throughout much of the Trans-Pecos, Llano Estacado, and portions of the Edwards Plateau, where the main concerns are competition with native species and potential for disease transmission to the reintroduced populations of desert bighorn sheep (*Ovis canadensis*; Schmidly and Bradley 2016). Free-ranging populations also occur in New Mexico and California (Mungall 2000). Axis deer were introduced to Texas in the 1930s and are considered the most common free-ranging exotic in Texas (Schmidly and Bradley 2016) after feral swine. Axis deer are common in central Texas, where they are an urban pest, are involved in vehicle collisions, and compete with native species for forage (Faas and Weckerly 2010). Well-established populations of axis also occur in California, Florida, and Hawaii (Mungall 2000).

Nilgai antelope are regarded as the first exotic species in Texas, introduced to south Texas during the 1920s to 1930s as a potential alternative livestock. Nilgai have expanded throughout the southern Gulf Coast of Texas and into Mexico. Nilgai are the focal point of a management challenge that poses a multi-billion-dollar threat to the US beef industry because of disease (e.g., bovine babesiosis). The vector-borne disease is caused by the protozoan parasites *Babesia bovis* and *B. bigemina*. The vector is the cattle fever tick (*Rhipicephalus annulatus* and *R. microplus*), one-host ticks native to the Near and Middle East, Mediterranean, and India (Estrada-Peña et al. 2006). Nilgai can serve as alternative hosts for the ticks and complicate tick eradication efforts owing to their extensive movements and lack of treatment methods (Lohmeyer et al. 2018).

Management Techniques

Nonlethal methods for population control and management are often preferred because they have the greatest acceptance by the public (VerCauteren et al. 2012). Nonlethal methods are most effectively implemented as part of an integrated strategy of management rather than a standalone solution (VerCauteren et al. 2012). Fencing is often effective for the confinement or exclusion of ungulates but may be costly depending on the height, design, and size of area (VerCauteren et al. 2006). The recent deployment of large-scale (hundreds of km) fencing for disease containment in Europe and elsewhere provides a socially appealing solution (Mysterud and Rolandsen 2018), especially when the alternatives are lethal control. The efficacy, cost, and ecological effects of extensive fencing on the landscape are poorly understood.

Disturbance or hazing (Gilsdorf et al. 2002) varies in effectiveness if animals become habituated but can be an effective control measure when combined with other nonlethal methods. The use of dogs to protect crop resources has been successful but presents logistical challenges in the spatial area covered and in training and feeding the dogs; long-term costs

and effectiveness of guard dogs may be similar to fencing (VerCauteren et al. 2014). Capture and translocation can be effective for carnivores but is not often a viable solution for exotic ungulates (VerCauteren et al. 2012). Contraceptives delivered orally or by injection are available and considered safe for use in ungulates (Patton et al. 2007). Contraception is most effective in small or confined populations (VerCauteren et al. 2012) owing to logistical and economic considerations (Fagerstone et al. 2010).

Lethal methods for management include organized hunting by sport hunters, sharpshooting, and trapping. Most examples of organized hunting programs in the United States are for overabundant species of native ungulates in urban settings, such as the white-tailed deer (Williams et al. 2013). Land ownership, parcel size, and distribution may be complicating factors if owners cannot agree on management goals for a population. Sufficient hunters must be available to achieve a meaningful reduction in population density, though recruitment of hunters can be addressed through incentives and advertising (Hansen and Beringer 1997). A review of case studies suggested that hunting was useful in reducing density of urban white-tailed deer so long as the goal was 17–18 deer/km^2; further reductions in density would require the use of alternative methods (Williams et al. 2013). Sharpshooting is preferred for efficiency and where there are safety concerns (e.g., urban areas) but can be more expensive than use of hunters. Exotic ungulates can be captured using a net-gun, cage or corral traps, or chemical immobilization (Schemnitz et al. 2012). Captured animals can then be euthanized or released elsewhere. Logistical problems with capture include holding animals prior to transport, availability of suitable release sites, and carcass disposal (VerCauteren et al. 2012). Some landowners have euthanized animals and sold the meat (where legal) or donated carcasses to charitable organizations.

Feral horses have a unique regulatory status and management because of the Wild and Free-Roaming Horse and Burro Act of 1971. This federal legislation states that wild horses are publicly owned and are to be managed as "part of the natural system of the public lands" (PL 92-195:1). The Bureau of Land Management, a division of the US Department of the Interior, is the lead agency for their management and performs population counts and management actions. Management options are limited in part by agency rules and by strong public opposition to lethal control methods. When populations exceed management goals, horses are rounded up and offered for adoption; most are maintained in private holding facilities for their lifetime, a logistically and financially unsustainable strategy (Garrott and Oli 2013).

Ecological and Cultural Context

Exotic ungulates are valued in large part for their aesthetic appeal and economic potential. The wild horse has a unique status as an iconic symbol of the American frontier, but it has no real economic potential. Most exotic ungulates are perceived as unique and charismatic animals (Jarić et al. 2020).

Many species also offer recreational value for wildlife viewing and for hunters. Several species of exotics are preferred for the quality of their meat, especially axis deer and nilgai antelope (Mungall 2000). Economic opportunities from hunting, meat, or breeding stock are strong motivation for the acquisition and propagation of exotic ungulates other than horses. The propagation of exotic species of ungulates has, in some cases, been a conservation success for populations that are imperiled in their native range. For instance, scimitar-horned oryx (*Oryx dammah*) and Grévy's zebra (*Equus grevyi*) have benefited from privately funded conservation populations in the United States (Mungall 2000).

The legal status of exotic ungulates differs among regions. In the United States, wild horses have a special legal status thanks to the Wild and Free-Roaming Horse and Burro Act (1971) and are perhaps the only species of exotic with a consistent regulatory category. For other species of exotic ungulates, the establishment and growth of populations, and management options, are often a product of differing local and state regulations and land ownership (Mungall 2000). For instance, >97% of land in Texas is privately owned. Most common species of exotic ungulates have a regulatory status much like livestock, where landowners can buy and sell from individuals or at auction. The absence of seasons or regulatory oversight is part of the appeal of exotics, as these animals provide hunting opportunities when seasons for native animals are closed and there are no bag limits or restrictions on method of take. The availability and regulatory status of exotic ungulates combined with strong aesthetic appeal and desire for hunting opportunities has led to a strong hunting industry and the increase of free-ranging populations.

Unfortunately, there is little appreciation for the wider negative effects of exotic species (e.g., competition with native species, habitat alteration and risk of disease transmission). As exemplified by feral horses, the aesthetic appeal often conflicts with the harsh realities of the effects of exotics on native plant and animal communities. Management issues for species of exotic ungulates can be synthesized into mismatches between rates of cultural and ecosystem adaptation, the scale of ecological versus social effects, and the scale of effects versus management actions (Beever et al. 2019). Effective management of exotic ungulates will depend greatly on increased public education, awareness, and involvement (Jarić et al. 2020).

Wild Pigs

Background
One of the greatest emerging threats to wildlife conservation and management in North America are expanding populations of vertebrate invasive species (Pimentel 2011), including wild pigs, feral and free-ranging cats (*Felis catus*), and feral horses and burros (*Equus asinus*). This case history focuses on wild pigs because of their transcontinental distribution, burgeoning populations, game status in some states and provinces, and rich history in North America (VerCauteren et al. 2020). The

cases of feral and free-ranging cats and feral horses and burros are equally interesting and complex and present their own suite of issues to those who are tasked with managing wildlife populations and ecosystems.

Wild pigs are a highly adaptable and mobile species capable of exploiting resources and causing ecological and agricultural damage under a wide range of environmental conditions (Campbell and Long 2009, Strickland et al. 2020). Several biological, ecological, and cultural features of wild pigs contribute to their invasiveness. First, wild pigs are fecund, reaching sexual maturity at a young age, having large litter sizes, and reproducing up to two times per year (Taylor et al. 1998, Delgado-Acevedo et al. 2010, Mayer et al. 2020). Second, wild pigs have no natural predators in North America, and many large predators that do occur are declining or are at too low of population levels to be an effective predator of wild pigs. Third, because wild pigs are prized game that are hunted year-round in many states and provinces, they are often illegally translocated to new locations or to augment existing populations (Sweeney et al. 2003).

Management Techniques
An integrated and comprehensive management program should be used to control or eradicate wild pig populations (Campbell and Long 2009, Ditchkoff and Bodenchuk 2020). Implementation techniques may include lethal and nonlethal approaches. Lethal methods include live-trapping followed by euthanasia (Williams et al. 2011); snaring (Barrett and Birmingham 1994); shooting from the ground by hunters (Engeman et al. 2007) or at night with the aid of night vision equipment and noise suppression devices (Adams et al. 2006); shooting from the air (Campbell et al. 2010a); and shooting with the aid of trained dogs (Katahira et al. 1993). Nonlethal methods primarily include exclusion fencing. For example, researchers found that two strands of electrified fencing reduced wild pig damage to a row crops (Reidy et al. 2008), and 86-cm-high paneling contained wild pigs during simulated depopulations (Lavelle et al. 2011) and excluded wild pigs from a localized resource (Rattan et al. 2010). Each of the above techniques has distinct advantages and disadvantages (Campbell and Long 2009, Ditchkoff and Bodenchuk 2020). These, plus social and legal considerations, should be known and anticipated prior to their deployment. Several techniques aimed at controlling wild pig populations are being developed, including fertility control agents (Campbell and Long 2010, Sanders et al. 2011, Samoylova et al. 2012, Campbell et al. 2017) and toxicants (Cowled et al. 2008, Snow et al. 2017, Poche et al. 2018). For these developing methods to be registered for use, a wild-pig-specific oral delivery system must also be identified (Campbell and Long 2007, Campbell et al. 2011, Lavelle et al. 2018).

Ecological and Cultural Context
Wild pigs are among the 14 mammals noted in a recent list of 100 of the world's worst invasive alien species because of the damage they cause to natural resources, agriculture, livestock

health and production, and human health and safety (Lowe et al. 2000, Ditchkoff and Bodenchuk 2020). Wild pigs are opportunistic omnivores and forage at or below the soil surface in a destructive manner called "rooting" (Gray et al. 2020). Natural resource managers have recognized the effects of rooting and other wild pig behaviors has on sensitive species, habitats, and ecosystems (Kaller and Kelso 2006, Jolley et al. 2010, Gray et al. 2020). Among agricultural producers, wild pig rooting is a major damage complaint (Adams et al. 2005, Mengak and Miller 2020). Wild pigs also pose a major threat to livestock, wildlife, and people because of their ability to transmit diseases (Corn and Yabsley 2020). For example, in southern Texas, wild pigs regularly come into contact with transition domestic swine at facilities with low biosecurity, where disease transmission may occur (Wyckoff et al. 2009). Wild pig threats to human health and safety are both direct (e.g., swine–vehicle collisions or defensive attacks) and indirect (e.g., zoonotic disease transmission). Wild pigs are known to harbor important zoonotic diseases, such as type A influenza virus (Hall et al. 2008) and *Escherichia coli* O157:H7 (Jay et al. 2007), among others.

Several factors complicate wild pig population management. First, similar to other exotic invasive organisms, early detection of their presence is paramount to successful management. Unfortunately, wild pigs are primarily nocturnal (Campbell and Long 2010, Gray et al. 2020), and signs of their presence (including tracks, scats, rooting, wallows, and rubs) often go unnoticed, allowing for populations to become well established before they are detected (Campbell and Long 2009). Second, wild pig populations are difficult to survey using traditional techniques (e.g., spotlight counts, aerial surveys), which hinders population monitoring (Beasley et al. 2020). The recent development of biomarkers, however, may allow for wild pig mark-recapture population analyses relative to control activities (Wiles and Campbell 2006, Reidy et al. 2011, Snow et al. 2018). Third, hunting regulations and wild pig regulations generally vary among states and provinces (Smith et al. 2020). In some states, wild pigs are managed by the wildlife agency as a game animal, in other states it is illegal to hunt wild pigs, and in still other states there are no regulations for wild pigs. Efforts to unify regulations between adjacent states and provinces with wild pigs are needed. Lastly, control or eradication of wild pig populations may not be acceptable to members of society who place high value on wild pig recreational opportunities, who depend upon wild pigs as a source of protein, or who disagree with population control measures altogether (Lowe et al. 2000, Mengak and Miller 2020).

Populations of wild pigs are often managed in a reactionary and piecemeal manner with no cohesive objectives and minimal planning (Campbell and Long 2009). The value of federal and state agencies forging partnerships among agricultural, conservation, industry, public health, and landowner groups cannot be overstated (Hartin et al. 2007). Additional regulatory efforts and enforcement of existing laws are needed to curb the illegal translocation and expansion of wild pig populations (Smith et al. 2020). Decisive actions to fulfill realistic and obtainable objectives, and stable funding to implement management and conduct research toward prioritized data needs, are necessary, and recent successes have been documented (Ditchkoff et al. 2020).

Rose-Ringed Parakeets

Background

Rose-ringed parakeets, also known as ring-necked parakeets, are one of the nearly 400 species within the order Psittaciformes, colloquially known as parrots. Species within this order are native to subtropical and tropical habitats throughout Africa, the Americas, Australia, and Eurasia. Psittaciformes is among the most imperiled bird orders, with 28% of parrot species classified as threatened or endangered under IUCN criteria (Olah et al. 2016). Some parrot species, however, have expanded their ranges and thrived in response to human habitat modification. An estimated 60 parrot species have established populations outside of their native range, predominantly owing to unintentional releases by humans (Menchetti and Mori 2014).

Rose-ringed parakeets are native to equatorial regions of Africa and the Indian subcontinent. They are considered a medium-sized parrot, weighing approximately 110–182 g. They measure around 40 cm in length, about half of which is their long tail (Butler 2003). They are sexually dimorphic, named for the distinct pink-and-black collars of the adult males. They feed on seeds, nuts, buds, berries, vegetables, and fruits.

Rose-ringed parakeets are one of the most invasive bird species worldwide, with observations in over 70 countries. There have been introduced in more than 35 countries in Africa, Asia, Australia, Europe, North America, and South America (Invasive Species Compendium 2009, Menchetti et al. 2016). The species is popular in the pet trade, which has led to intentional and unintentional releases around the world. The ability of this species to thrive in urban and agricultural areas, coupled with its tolerance of a range of climates, has allowed rose-ringed parakeets to become established when released in novel habitats, even with very small founder populations. Balmer et al. (2013) estimated the rose-ringed parakeet breeding range increased by 440 times from 1968 to 2013, making it one of the most rapidly increasing species worldwide.

With the diversity of habitats in which rose-ringed parakeets have been introduced, their demonstrated and potential ecological effects are varied. In Australia, they damage and kill trees by stripping bark (Fletcher and Askew 2007). In Hawaii they consume, and therefore potentially disperse, invasive yellow guava (*Psidium guajava*) and passion fruit (*Passiflora edulis*; Gaudioso et al. 2012, Shiels et al. 2018). In Britain, native songbirds have been reported to decrease feeding rates and increase vigilance in response to rose-ringed parakeet presence (Peck et al. 2014). Perhaps the most detrimental ecological effect is through their nesting behavior. Rose-ringed parakeets are secondary cavity nesters, meaning they rely on cavities that naturally occur or were created by other species. Rose-ringed parakeets are notably antagonistic in claiming these cavities and have reportedly attacked native birds and killed native bats to claim cavities (Hernández-Brito et al. 2018). Non-native rose-ringed parakeets have also been observed attacking, sometimes lethally, native small mammals, squirrels, birds of prey, passerines, and seabirds (Klug et al. 2019).

Rose-ringed parakeets are one of the most economically destructive bird species (Kumschick and Nentwig 2010). In much of their native range, they are regarded as agricultural pests, frequently raiding fruit and grain crops such as sunflower (Khan and Ahmad 1983), corn, sorghum, and citrus (Khan et al. 2011, Klug et al. 2019). In Hawaii, invasive rose-ringed parakeet populations have caused millions of dollars in losses in grain crops, most notably corn and fruit crops such as lychee, mango, and papaya. On the island of Kauai, invasive rose-ringed parakeets cause property damage through excessive droppings from their large communal roosts; these roosts are typically in urban areas and composed of 1,500–6,000 birds (C. J. Anderson, Texas A&M University–Kingsville, personal observation). Beyond property damage, these droppings potentially expose the public to pathogens such as avian flu and psittacosis (Klug et al. 2019). The dense populations of rose-ringed parakeets in urban areas also pose human safety threats through potential bird strikes at airports; at least three rose-ringed parakeet bird strikes have been documented at London's Heathrow airport, costing an average of more than $24,000 in damage each (Fletcher and Askew 2007).

Management Techniques

Nonlethal control of rose-ringed parakeet damage to crops includes modifying the crop and surrounding areas, exclusionary devices, and hazing or frightening devices (Klug et al. 2019). Use of reflective ribbons has demonstrated moderate success in protecting mangos from rose-ringed parakeet damage in Pakistan (Khan et al. 2011). In Kauai, netting is placed over fields of corn to prevent parakeet access. Frightening devices (e.g., lights) appear to be relatively ineffective at roosting sites, as the birds quickly habituate to them. In much of the rose-ringed parakeets' native and introduced range, lethal population management has been deemed necessary to curb population growth and negative effects.

Culling via shotgun can be an effective strategy if it is implemented at a large enough scale and with frequent adaptation. This strategy was used to eradicate an invasive population of rose-ringed parakeets on the island Mahe in the Seychelles, marking the first of only two known eradications of this species. In this program, 548 rose-ringed parakeets were culled, the majority of which were removed using shotguns along the birds' flight lines (Bunbury et al. 2019). This required frequent monitoring and adaptation, as the birds altered flight lines and heights in response to shotgun use (N. Bunbury, Seychelles Island Foundation, personal communication). The final birds removed in this program were through a public reporting bounty campaign (Bunbury et al. 2019). In Kauai, rose-ringed parakeets have been culled using shotguns at corn fields since 2005. Control efforts at the corn fields yielded more than 8,000 culled birds from 2005 to 2019; however, the population remained at more than 10,000 individuals in 2020.

The second documented eradication of an invasive population of rose-ringed parakeets occurred on La Palma Island Biosphere Reserve in the Canary Islands. A total of 175 parakeets were removed from 2015 to 2018. In this program, 120 parakeets were captured via live traps using bait and live decoy birds. An additional 34 parakeets were removed using 5.5-mm caliber air rifles (Saavedra and Medina 2020).

Culling at the roost may be a feasible option for population suppression. As roosts are often in urban areas, this strategy must be conducted in a manner that is safe for the public. Furthermore, rose-ringed parakeets will abandon roosts if they are threatened, making the sustainability of this strategy unclear. Air rifles (25-mm caliber) have been used to cull parakeets that roost in Kauai (C. J. Anderson, personal observation), but the long-term sustainability of this effort is not yet known.

Chemical control, by toxicant or contraceptive, may be a viable option for rose-ringed parakeet control but has not been tested in the natural environment. For example, Starlicide is an avicide used to control starlings (*Sturnus vulgaris*), blackbirds (*Turdus merula*), pigeons (*Columbia* spp.), crows (*Corvus* spp.), ravens (*Corvus corax*), gulls (*Larus* spp.), and magpies (*Pica pica*), but the efficacy on Psittacines has not been tested (Klug et al. 2019). Lambert et al. (2010) reported that the chemical 20,25 diazacholesterol dihydrocholoride (DiazaCon), originally developed as a cholesterol inhibitor for humans, significantly decreased fertility in captive rose-ringed parakeets. Whether this can be used in a natural setting is unclear, as an effective strategy for providing the chemical to rose-ringed parakeets while excluding nontarget species has not been identified.

Ecological and Cultural Context

Management of invasive species can be controversial, especially for those considered charismatic. Studies indicate the public is less supportive of lethal control of charismatic species than those considered less appealing (Verbrugge et al. 2013). Crowley et al. (2019) proposed that many people experience a type of dissonance when observing charismatic invasive species in a novel habitat, finding it to be a surprisingly and therefore pleasurable experience. Despite what may be demonstrated negative ecological or economic effects of an introduced species, local communities may develop emotional attachments to the species or grow to appreciate it as a part of the local culture. In the United Kingdom, culling efforts of invasive monk parakeets (*Myiopsitta monachus*) were halted after local activists voiced opposition (Crowley et al. 2019).

Like monk parakeets, lethal control of rose-ringed parakeets has the potential to draw public controversy, which could lead to a failure of management. The success of eradications in the Seychelles and Canary Islands was largely attributed to extensive public outreach. Therefore management campaigns of rose-ringed parakeets must be not only ecologically informed but also socially conscious.

SUMMARY

There are three important emergent themes from each of the population management case histories outlined in this chapter. The first theme is that an understanding of life history attributes of a species is essential for implementing a successful management program, regardless of whether that program is aimed at increasing, sustaining, or reducing populations.

The second theme is that effective management techniques always use some aspect of an animal's life history to identify a limiting factor and then manipulate that factor to increase, sustain, or reduce populations. The red-cockaded woodpecker is a classic example of managers identifying that nesting and roosting cavities were limiting, and then exploiting this concept by provisioning artificial cavities. This allowed red-cockaded woodpecker recovery efforts to progress at a rate that was far greater than people thought possible.

The third theme is that for any wildlife population management action to be effective, it must be conducted in both cultural and ecological contexts. This not only axiomatic but also as it should be. After all, in the North American Model of Wildlife Management, it is the people who own the wildlife resources in the context of a public trust. As such, logic dictates that all citizens are stakeholders when it comes to issues related to wildlife population management. Attempting to manage wildlife population in the absence of considering interactions with people is folly at best, disaster at worst.

Human activity is essentially the reason that a need exists to increase, sustain, or reduce wildlife populations. Recovering populations of gray wolves and red-cockaded woodpeckers would not be necessary had humans not almost extirpated wolves from North America or removed virtually all of the habitat required to sustain red-cockaded woodpecker populations. Similarly, the recreational and subsistence needs of humans are largely the reasons that moose and waterfowl populations are managed at sustainable levels. Furthermore, feral swine and rose-ringed parakeet populations require reduction because humans introduced these two invasive exotic species to novel habitats. Therefore humans typically create situations with regard to wildlife populations that require further human intervention via population management to rectify. People have a vested interested in managing wildlife populations whether they realize it or not, so encouraging public input should be an integral part of any wildlife population management plan.

Looking into the Future

One topic that was largely ignored in this chapter, at least until this point, is how climate change will influence wildlife population management (Chapter 18). Taking liberal versus conservative politics out of the equation, it is evident that climate change, or global warming, is beginning to have an influence on wildlife populations, especially in the more northern and southern regions of the globe. Examples are legion; shrinking polar ice caps and glaciers, shifts in phenology of plants that break bud and flower earlier than ever, migratory birds that now arrive on their breeding grounds weeks before their usual timing, and so on.

What can be done in a management context for species of wildlife that are likely to be negatively influenced by climate change? This is one of the most challenging—and vexing—questions faced by wildlife scientists and managers today. Should we assume that governmental policies and incentives will result in a reduction in atmospheric carbon and a reversal of the warming trends we are seeing? Probably not. It is probably better to assume that the emerging economies in countries such as India and China will continue to mature and develop at a breakneck pace, and that the atmospheric carbon issue will most likely worsen before it improves.

From the standpoint of wildlife, it is probably safe to assume that some species will be winners and some will be losers if current global warming trends continue. Species such as polar bears (*Ursus maritimus*) and penguins (*Spheniscidae*) are clearly in potential peril from global warming, while other species may simply experience more northward or southward shifts in their geographic distributions. All people, not just wildlife professionals, must understand and cope with these changes as we approach the middle of the 21st century.

Literature Cited

Adams, C. E., B. J. Higginbotham, D. Rollins, R. B. Taylor, R. Skiles, M. Mapston, and S. Turman. 2005. Regional perspectives and opportunities for feral hog management in Texas. Wildlife Society Bulletin 33:1312–1320.

Adams, C. E., K. J. Lindsey, and S. J. Ash. 2006. Urban wildlife management. Taylor and Francis, Boca Raton, Florida, USA.

Alaska Department of Fish and Game. 2011. Identifying a legal moose in antler restricted hunts. http://www.adfg.alaska.gov/static/regulations/wildliferegulations/pdfs/mooseid.pdf.

Allen, D. H. 1991. Constructing artificial red-cockaded woodpecker cavities. Forest Service General Technical Report SE-73. US Department of Agriculture, Washington, DC, USA.

Alston, J. M., M. J. Joyce, J. A. Merkle, and R. A. Moen. 2020. Temperature shapes movement and habitat selection by a heat-sensitive ungulate. Landscape Ecology 35:1961–1973.

Anderson, C. R., Jr., and F. G. Lindzey. 1996. Moose sightability model developed from helicopter surveys. Wildlife Society Bulletin 24:247–259.

Anderson, M. K., and M. J. Moratto. 1996. Native American land-use practices and ecological impacts. Sierra Nevada Ecosystem Project: final report to Congress. Volume 2: Assessments and scientific basis for management options. Center for Water and Wildland Resources, University of California, Davis, USA.

Arizona Game and Fish Department. 2021. Mexican wolf reintroduction and management. https://www.azgfd.com/wildlife/speciesofgreatestconservneed/mexicanwolves/.

Askins, R. A. 2000. Restoring North American birds. Yale University Press, New Haven, Connecticut, USA.

Ballinger, R. E. 1974. Reproduction of the Texas horned lizard, *Phrynosoma cornutum*. Herpetologica 30:321–327.

Balmer, D. E., S. Gillings, B. Caffrey, R. Swann, I. Downie, and R. Fuller. 2013. Bird atlas 2007–11: the breeding and winter birds of Britain and Ireland. British Trust for Ornithology, Thetford, UK.

Bangs, E. E., S. H. Fritts, J. A. Fontaine, D. W. Smith, K. M. Murphy, C. M. Mack, and C. C. Niemeyer. 1998. Status of gray wolf restoration in Montana, Idaho, and Wyoming. Wildlife Society Bulletin 26:785–798.

Barnes, A. M., L. J. Ogden, and E. G. Campos. 1972. Control of the plague vector, *Opisocrostis hirsutis*, by treatment of prairie dog (*Cynomys ludovicianus*) burrows with 2% carbaryl dust. Journal of Medical Entomology 9:330–333.

Barrett, R. H., and G. H. Birmingham. 1994. Wild pigs. Pages D65–D70 *in* S. Hyngstrom, R. Timm, and G. Larsen, editors. Prevention and control of wildlife damage. Cooperative Extension Service, University of Nebraska, Lincoln, USA.

Bartos, D. L. 1994. Twelve years biomass response in aspen communities following fire. Journal of Range Management 47:79–83.

Beasley, J. C., M. J. Lavelle, D. A. Keiter, K. M. Pepin, A. J. Piaggio, J. C. Kilgo, and K. C. VerCauteren. 2020. Research methods for wild pigs. Pages 199–227 *in* K. C. VerCauteren, J. C. Beasley, S. S. Ditchkoff, et al., editors. Invasive wild pigs in North America: ecology, impacts, and management. CRC Press, Boca Raton, Florida, USA.

Becker, S. A. 2008. Habitat selection, condition, and survival of Shiras moose in northwest Wyoming. MS thesis, University of Wyoming, Laramie, USA.

Beever, E. A., D. Simberloff, S. L. Crowley, R. Al-Chokhachy, H. A. Jackson, and S. L. Petersen. 2019. Social–ecological mismatches create conservation challenges in introduced species management. Frontiers in Ecology and the Environment 17:117–125.

Beschta, R. L., and W. J. Ripple. 2016. Riparian vegetation recovery in Yellowstone: the first two decades after wolf reintroduction. Biological Conservation 198:93–103.

Blohm, R. J., D. E. Sharp, P. I. Padding, R. W. Kokel, and K. D. Richkus, 2006. Integrated waterfowl management in North America. Waterbirds around the world. Pages 199–205 *in* G. Boere, C. Galbraith and D. Stroud, editors. A global overview of the conservation, management and research of the world's waterbird flyways. Stationery Office, Edinburgh, UK.

Boertje, R. D., M. A. Keech, and T. F. Paragi. 2010. Science and values influencing predator control for Alaska moose management. Journal of Wildlife Management 74:917–928.

Boertje, R. D., M. A. Keech, D. D. Young, K. A. Kellie, and C. T. Seaton. 2009. Managing for elevated yield of moose in interior Alaska. Journal of Wildlife Management 73:314–327.

Boertje, R. D., K. A. Kellie, C. T. Seaton, M. A. Keech, D. D. Young, B. W. Dale, L. G. Adams, and A. R. Aderman. 2007. Ranking Alaska moose nutrition: signals to begin liberal antlerless harvests. Journal of Wildlife Management 71:1494–1506.

Brick, P. 1995. Taking back the rural West. Pages 61–65 *in* J. A. Baden and R. B. Eby, editors. Let the people judge: wise use and the private property rights movement. Island Press, Washington, DC, USA.

Brown, T. L., and R. V. Lucchino. 1972. A record-sized specimen of the Texas horned lizard (*Phrynosoma cornutum*). Texas Journal of Science 24:353–354.

Bunbury, N., P. Haverson, N. Page, J. Agricole, G. Angell, et al. 2019. Five eradications, three species, three islands: overview, insights and recommendations from invasive bird eradications in the Seychelles. Pages 282–288 *in* C. Veitch, M. Clout, A. Martin, et al., editors. Proceedings of the International Conference on Island Invasives 2017. Occasional paper of the IUCN Species Survival Commission. University of Dundee, Dundee, Scotland.

Burrow, A. L., R. T. Kazmaier, E. C. Hellgren, and D. C. Ruthven III. 2001. Microhabitat selection by Texas horned lizards in southern Texas. Journal of Wildlife Management 65:645–652.

Butler, C. J. 2003. Population biology of the introduced Rose-ringed parakeet *Psittacula krameri* in the UK. Page 312 *in* Biological sciences. University of Oxford, Oxford, UK.

California Department of Fish and Wildlife. 2021. Gray wolf. https://wildlife.ca.gov/conservation/mammals/gray-wolf.

Callan, R., N. P. Nibbelink, T. P. Rooney, J. E. Wiedenhoeft, and A. P. Wydeven. 2013. Recolonizing wolves trigger a trophic cascade in Wisconsin (US). Journal of Ecology 101:837–845.

Campbell, T. A., M. R. Garcia, L. A. Miller, M. A. Ramirez, D. B. Long, J. Marchand, and F. Hill. 2010a. Immunocontraception of male feral swine with a recombinant GnRH vaccine. Journal of Swine Health and Production 18:118–124.

Campbell, T. A., S. J. Lapidge, and D. B. Long. 2006. Using baits to deliver pharmaceuticals to feral swine in southern Texas. Wildlife Society Bulletin 34:1184–1189.

Campbell, T. A., and D. B. Long. 2007. Species-specific visitation and removal of baits for delivery of pharmaceuticals to feral swine. Journal of Wildlife Diseases 43:485–491.

Campbell, T. A., and D. B. Long. 2009. Feral swine damage and damage management in forested ecosystems. Forest Ecology and Management 257:2319–2326.

Campbell, T. A., and D. B. Long. 2010. Activity patterns of wild boar in southern Texas. Southwestern Naturalist 55:564–567.

Campbell, T. A., D. B. Long, and B. R. Leland. 2010b. Feral swine behavior relative to aerial gunning is southern Texas. Journal of Wildlife Management 74:337–341.

Campbell, T. A., D. B. Long, and G. Massei. 2011. Efficacy of the Boar-Operated-System to deliver baits to feral swine. Preventive Veterinary Medicine 98:243–249.

Campbell, S., C. R. Long, B. Pyzyna, M. Westhusin, C. Dyer, and D. Kraemer. 2017. Development and evaluation of an oral contraceptive bait for feral pigs. Reproduction, Fertility and Development 29:189–190.

Carlson, S. B., A. M. Dietsch, K. M. Slagle, and J. M. Bruskotter. 2020. The VIPS of wolf conservation: how values, identity, and place shape attitudes toward wolves in the United States. Frontiers in Ecology and Evolution 8:1–9.

Carpenter, C. C., R. St. Clair, P. Gier, and C. C. Vaughn. 1993. Determination of the distribution and abundance of the Texas horned lizard (*Phrynosoma cornutum*) in Oklahoma. Final Report to Oklahoma Department of Wildlife Conservation, Federal Aid Project E-18. Oklahoma City, Oklahoma, USA.

Caughley, G. 1985. Harvesting of wildlife: past, present, and future. Pages 3–14 *in* S. L. Beasom and S. F. Roberson, editors. Game harvest management. Caesar Kleberg Wildlife Research Institute, Kingsville, Texas, USA.

Ceballos, G., J. Pacheco, and R. List. 1999. Influence of prairie dogs (*Cynomys ludovicianus*) on habitat heterogeneity and mammalian diversity in Mexico. Journal of Arid Environments 41:161–172.

Cerulli, T. 2016. Of wolves, hunters and words: a comparative study of cultural discourses in the western Greta Lakes region. PhD dissertation, University of Massachusetts, Amherst, USA.

Chavez, A. S., E. M. Gese, and R. S. Krannich. 2005. Attitudes of rural landowners toward wolves in northwestern Minnesota. Wildlife Society Bulletin 33:517–527.

Clark, T. W., T. M. Campbell, D. G. Socha, and D. E. Casey. 1982. Prairie dog colony attributes and associated vertebrate species. Great Basin Naturalist 42:572–582.

Colorado Parks and Wildlife. 2020. Colorado Parks and Wildlife officers confirm latest wolf pack sighting in NW Colorado. https://cpw.state.co.us/Lists/News%20Releases/DispForm.aspx?ID=7225.

Colorado Parks and Wildlife. 2021. Wolf management. https://cpw.state.co.us/learn/Pages/CON-Wolf-Management.aspx.

Coluccy, J. M., T. Yerkes, R. Simpson, J. W. Simpson, L. Armstrong, and J. Davis. 2008. Population dynamics of breeding mallards in the Great Lakes states. Journal of Wildlife Management 72:1181–1187.

Copeyon, C. K. 1990. A technique for constructing cavities for the red-cockaded woodpecker. Wildlife Society Bulletin 18:303–311.

Corn, J. L., and M. J. Yabsley. 2020. Diseases and parasites that impact wild pigs and species they contact. Pages 83–126 in K. C. VerCauteren, J. C. Beasley, S. S. Ditchkoff, et al., editors. Invasive wild pigs in North America: ecology, impacts, and management. CRC Press, Boca Raton, Florida, USA.

Costello, D. F. 1970. The world of the prairie dog. Lippincott, Philadelphia, Pennsylvania, USA.

Cowled, B. D., P. Elsworth, and S. J. Lapidge. 2008. Additional toxins for feral pig (Sus scrofa) control: identifying and testing Achilles' heels. Wildlife Research 35:651–662.

Crichton, V. F. J., W. L. Regelin, A. W. Franzmann, and C. C. Schwartz. 2007. The future of moose management and research. Pages 655–663 in A. W. Franzmann and C. C. Schwartz, editors. Ecology and management of the North American moose. 2nd edition. University of Colorado Press, Boulder, USA.

Crowley, S. L., S. Hinchliffe, and R. McDonald. 2019. The parakeet protectors: understanding opposition to introduced species management. Journal of Environmental Management 229:120–132.

Cully, J. F., A. M. Barnes, T. J. Quan, and G. Maupin. 1997. Dynamics of plague in a Gunnison's prairie dog complex from New Mexico. Journal of Wildlife Diseases 33:706–719.

Dambach, C. A. 1948. The relative importance of hunting restrictions and land use in maintaining wildlife populations in Ohio. Ohio Journal of Science 209–229.

D'Antonio, C. M., and P. M. Vitousek. 1992. Biological invasions by exotic grasses, the grass-fire cycle, and global change. Annual Review of Ecology and Systematics 23:63–87.

Davies, K. W., and C. S. Boyd. 2019. Ecological effects of free-roaming horses in North American rangelands. Bioscience 69:558–565.

Decker, D. J., C. A. Jacobson, and T. L. Brown. 2006. Situation-specific "impact dependency" as a determinant of management acceptability: insights from wolf and grizzly bear management in Alaska. Wildlife Society Bulletin 34:426–432.

Delgado-Acevedo, J., A. Zamorano, R W. DeYoung, T. A. Campbell, D. G. Hewitt, and D. B. Long. 2010. Promiscuous mating in feral pigs (Sus scrofa) from Texas. Wildlife Research 37:539–546.

Ditchkoff, S. S., J. C. Beasley, J. J. Mayer, G. J. Roloff, B. K. Strickland, and K. C. VerCauteren. 2020. The future of wild pigs in North America. Pages 465–469 in K. C. VerCauteren, J. C. Beasley, S. S. Ditchkoff, et al., editors. Invasive wild pigs in North America: ecology, impacts, and management. CRC Press, Boca Raton, Florida, USA.

Ditchkoff, S. S., and M. J. Bodenchuk. 2020. Management of wild pigs. Pages 175–197 in K. C. VerCauteren, J. C. Beasley, S. S. Ditchkoff, et al., editors. Invasive wild pigs in North America: ecology, impacts, and management. CRC Press, Boca Raton, Florida, USA.

Donaldson, W., A. H. Price, and J. Morse. 1994. The current status and future prospects of the Texas horned lizard (Phrynosoma cornutum) in Texas. Texas Journal of Science 46:97–113.

Engeman, R. M., A. Stevens, J. Allen, J. Dunlap, M. Daniel, D. Teague, and B. Constantin. 2007. Feral swine management for conservation of an imperiled wetland habitat: Florida's vanishing seepage slopes. Biological Conservation 134:440–446.

Estrada-Peña, A., A. Bouattour, J.-L. Camicas, A. Guglielmone, I. Horak, F. Jongejan, A. Latif, R. Pegram, and A. R. Walker. 2006. The known distribution and ecological preferences of the tick subgenus Boophilus (Acari: Ixodidae) in Africa and Latin America. Experimental and Applied Acarology 38:219–235.

Faas, C. J., and F. W. Weckerly. 2010. Habitat interference by axis deer on white-tailed deer. Journal of Wildlife Management 74:698–706.

Fagerstone, K. A., L. A. Miller, G. Killian, and C. A. Yoder. 2010. Review of issues concerning the use of reproductive inhibitors, with particular emphasis on resolving human-wildlife conflicts in North America. Integrative Zoology 5:15–30.

Fagerstone, K. A., H. P. Tietjen, and O. Williams. 1981. Seasonal variation in the diet of black-tailed prairie dogs. Journal of Mammalogy 62:820–824.

Fair, W. S., and S. E. Henke. 1997a. Effects of habitat manipulations on Texas horned lizards and their prey. Journal of Wildlife Management 61:1366–1370.

Fair, W. S., and S. E. Henke. 1997b. Efficacy of capture methods for a low density population of Phrynosoma cornutum. Herpetological Review 28:135–137.

Fair, W. S., and S. E. Henke. 1998. Habitat use of Texas horned lizards in southern Texas. Texas Journal of Agriculture and Natural Resources 11:73–86.

Fair, W. S., and S. E. Henke. 1999. Movements, home ranges, and survival of Texas horned lizards (Phrynosoma cornutum). Journal of Herpetology 33:517–525.

Flagel, D. G., G. E. Belovsky, and D. E. Beyer Jr. 2016. Natural and experimental tests of trophic cascades: gray wolves and white-tailed deer in Great Lakes forest. Community Ecology 180:1183–1194.

Fletcher, M., and N. Askew. 2007. Review of the status, ecology and likely future spread of the parakeets in England. York Central Science Laboratory, York, UK.

Garrott, R. A., and M. K. Oli. 2013. A critical crossroad for BLM's wild horse program. Science 341:847–848.

Gasaway, W. C., R. D. Boertje, D. V. Grangaard, D. G. Kelleyhouse, R. O. Stephenson, and D. G. Larsen. 1992. The role of predation in limiting moose at low densities in Alaska and Yukon and implications for conservation. Wildlife Monographs 120:1–59.

Gilsdorf, J. M., S. E. Hygnstrom, and K. C. VerCauteren. 2002. Use of frightening devices in wildlife damage management. Integrated Pest Management Reviews 7:29–45.

Gluesenkamp, A. G. 2019. Texas horned lizard reintroduction project, interim report (1 July 2019). San Antonio Zoo Center for Conservation and Research, San Antonio, Texas, USA.

Gowdy, G. G. 2020. Wildlife community response to native-grassland restoration in the Rio Grande Plains. MS thesis, Texas A&M University–Kingsville, Kingsville, USA.

Gray, S. M., G. J. Roloff, R. A. Montgomery, J. C. Beasley, and K. M. Pepin. 2020. Wild pig spatial ecology and behavior. Pages 33–56 in K. C. VerCauteren, J. C. Beasley, S. S. Ditchkoff, et al., editors. Invasive wild pigs in North America: ecology, impacts, and management. CRC Press, Boca Raton, Florida, USA.

Grayson, D. K. 2006. The late Quaternary biogeographic histories of some Great Basin mammals (western USA). Quaternary Science Reviews 25:2964–2991.

Gaudioso, J. M, A. B. Shiels, W. C. Pitt, and W. P. Bukoski. 2012. Rose-ringed parakeet impacts on Hawaii's seed crops on the island of Kauai: population estimate and monitoring movements using radio telemetry. Report QA 1874. US Department of Agriculture, National Wildlife Research Center, Hilo, Hawaii, USA.

Hafner, D. J. 1984. Evolutionary relationships of the Nearctic Sciuridae. Pages 3–23 in J. O. Murie and G. R. Michener, editors. The biology of ground dwelling squirrels. University of Nebraska Press, Lincoln, USA.

Haines, A. L. 1977. The Yellowstone story. Colorado Associated University Press, Boulder, USA.

Hall, J. S., R. B. Minnis, T. A. Campbell, S. Barras, R. W. DeYoung, K. Palilonia, M. L. Avery, H. Sullivan, L. Clark, and R. G. McLean. 2008. Influenza exposure in feral swine from the United States. Journal of Wildlife Diseases 44:362–368.

Halofsky, J. S., W. J. Ripple, and R. L. Beschta. 2008. Recoupling fire and aspen recruitment after wolf reintroduction in Yellowstone National Park, USA. Forest Ecology and Management 256:1004–1008.

Hansen, L., and J. Beringer. 1997. Managed hunts to control white-tailed deer populations on urban public areas in Missouri. Wildlife Society Bulletin 25:484–487.

Hartin, R. E., M. R. Ryan, and T. A. Campbell. 2007. Distribution and disease prevalence of feral hogs in Missouri. Human–Wildlife Conflicts 1:186–191.

Hasbrouck, T. R., T. J. Brinkman, G. Stout, E. Trochim, and K. Kielland. 2020. Quantifying effects of environmental factors on moose harvest in Interior Alaska. Wildlife Biology 2:1–8.

Hayes, R. D., R. Farnell, R. M. P. Ward, J. Carey, M. Dehn, G. W. Kuzyk, A. M. Baer, C. L. Gardner, and M. O'Donoghue. 2003. Experimental reduction of wolves in the Yukon: ungulate responses and management implications. Wildlife Monographs 152:1–35.

Henke, S. E. 2003. Baseline survey of the population demographics of Texas horned lizards, *Phrynosoma cornutum*, in Texas. Southwestern Naturalist 48:278–282.

Henke, S. E. 2013. Coping with invasive invaders: behavioral and morphological modifications of the Texas horned lizard. Pages 159–176 *in* W. I. Lutterschmidt, editor. Reptiles in research: investigations of ecology, physiology, and behavior from desert to sea. Nova Biomedical Press, New York, New York, USA.

Henke, S. E., and M. Montemayor. 1997. *Phrynosoma cornutum* (Texas horned lizard) growth. Herpetological Review 28:152.

Henke, S. E., and M. Montemayor. 1998. Diel and monthly variation in capture success of *Phrynosoma cornutum* via road cruising in southern Texas. Herpetological Review 29:148–150.

Hernández-Brito, D., M. Carrete, C. Ibáñez, J. Juste, and J. L. Tella. 2018. Nest-site competition and killing by invasive parakeets cause the decline of a threatened bat population. Royal Society Open Access 5:1–10.

Hoekman, S. T., T. S. Gabor, M. J. Petrie, R. Maher, H. R. Murkin, and M. S. Lindberg. 2006. Population dynamics of mallards breeding in agricultural environments in eastern Canada. Journal of Wildlife Management 70:121–128.

Hoekman, S. T., L. S. Mills, D. W. Howerter, J. H. Devries, and I. J. Ball. 2002. Sensitivity analyses of the life cycle of midcontinent mallards. Journal of Wildlife Management 66:883–900.

Hofman, C. A., and T. C. Rick. 2018. Ancient biological invasions and island ecosystems: tracking translocations of wild plants and animals. Journal of Archaeological Research 26:65–115.

Hogberg, J., A. Treves, B. Shaw, and L. Naughton-Treves. 2015. Changes in attitudes toward wolves before and after an inaugural public hunting and trapping season: early evidence from Wisconsin's wolf range. Environmental Conservation 43:45–55.

Hoogland, J. L. 1995. The black-tailed prairie dog: social life of a burrowing mammal. University of Chicago Press, Chicago, Illinois, USA.

Hoogland, J. L. 2003. Black-tailed prairie dog. Pages 232–247 *in* G. A. Feldhamer, B. C. Thompson, and J. A. Chapman, editors. Wild mammals of North America: biology, management, and conservation. Johns Hopkins University Press, Baltimore, Maryland, USA.

Huerta, J. O. 2020. Effects of habitat restoration on Texas horned lizards and their prey. MS thesis, Texas A&M University–Kingsville, Kingsville, USA.

Idaho Fish and Game. 2021. Wolves in Idaho. https://idfg.idaho.gov/public/wildlife/wolves.

Ingham, R. E., and J. K. Detling. 1984. Plant herbivore interactions in a North American mixed-grass prairie. III. Soil nematode populations and root biomass on *Cynomys ludovicianus* colonies and adjacent uncolonized areas. Oecologia 63:307–313.

Inman, B., K. Podruzny, T. Smucker, A. Nelson, M. Ross, N. Lance, T. Parks, D. Boyd, and S. Wells. 2019. Montana gray wolf conservation and management 2018 annual report. Montana Fish, Wildlife and Parks, Helena, USA.

International Union for the Conservation of Nature. 2020. https://www.prairiedoghoogland.com/conservation.

Invasive Species Compendium. 2009. *Psittacula krameri* (rose-ringed parakeet). https://www.cabi.org/isc/datasheet/45158.

Jarić, I., F., R. A. Courchamp, S. L. Correia, F. Crowley, A. Essl, et al. 2020. The role of species charisma in biological invasions. Frontiers in Ecology and the Environment 18:345–353.

Jay, M. T., M. Cooley, D. Carychao, G. W. Wiscomb, R. A. Sweitzer, et al. 2007. *Escherichia coli* O157:H7 in feral swine near spinach fields and cattle, central California coast. Emerging Infectious Diseases 13:1908–1911.

Jolley, D. B., S. S. Ditchkoff, B. D. Sparklin, L. B. Hanson, M. S. Mitchell, and J. B. Grand. 2010. Estimate of herpetofauna depredation by a population of wild pigs. Journal of Mammalogy 91:519–524.

Jordan, P. A. 1987. Aquatic foraging and the sodium ecology of moose: a review. Swedish Wildlife Research (Supplement) 1:119–137.

Jorgenson, C. J. 1995. How Native Americans as an indigenous culture consciously maintained a balance between themselves and their natural resources. Pages 31–33 *in* J. A. Bissonette and P. R. Krausman, editors. Integrating people and wildlife for a sustainable future. The Wildlife Society, Bethesda, Maryland, USA.

Kalb, D. M., J. L. Bowman, and R. W. DeYoung. 2018. Dietary resource use and competition between white-tailed deer and introduced sika deer. Wildlife Research 45:457–472.

Kaller, M. D., and W. E. Kelso. 2006. Swine activity alters invertebrate and microbial communities in a coastal plain watershed. American Midland Naturalist 156:163–177.

Katahira, L. K., P. Finnegan, and C. P. Stone. 1993. Eradicating feral pigs in montane mesic habitat at Hawaii Volcanoes National Park. Wildlife Society Bulletin 21:269–274.

Kauffman, M. J., J. F. Brodie, and E. S. Jules. 2010. Are wolves saving Yellowstone's aspen? A landscape-level test of a behaviorally mediated trophic cascade. Ecology 91:2742–2755.

Keech, M. A., M. S. Lindberg, R. D. Boertje, P. Valkenburg, B. D. Taras, T. A. Boudreau, and K. B. Beckmen. 2011. Effects of predator treatments, individual traits, and environment on moose survival in Alaska. Journal of Wildlife Management 75:1361–1380.

Kellert, S. R., M. Black, C. R. Rush, and A. J. Bath. 1996. Human culture and large carnivore conservation in North America. Conservation Biology 10:977–990.

Khan, A. K., and S. Ahmad. 1983. Parakeet damage to sunflower in Pakistan. Pages 191–196 *in* W. B. Jackson and B. Dodd, editors. Proceedings of the Ninth Bird Control Seminar. Center for Environmental Research, Bowling Green State University, Bowling Green, Ohio, USA.

Khan, A. K, S. Ahmad, M. Javed, K. Ahmad, and M. Ishaque. 2011. Comparative effects of some mechanical repellents for management of rose ringed parakeet (*Psittacula krameri*) in citrus, guava and mango orchards. International Journal of Agriculture and Biology 13:396–400.

King, J. A. 1955. Social behavior, social organization, and population dynamics in a black-tailed prairie dog town in the Black Hills of South

Dakota. Contribution No. 67. Laboratory of Vertebrate Biology, University of Michigan, Ann Arbor, USA.

Klug, P. E., W. B. Bukoski, A. B. Shiels, B. M. Kluever, and S. R. Siers. 2019. Critical review of potential control tools for reducing damage by the invasive rose-ringed parakeet (Psittacula krameri) on the Hawaiian Islands. Unpublished Report QA-2836. USDA Animal and Plant Health Inspection Service, Wildlife Services, National Wildlife Research Center, Washington, DC, USA.

Knowles, C. J., C. J. Stoner, and S. P. Gieb. 1982. Selective use of black-tailed prairie dog towns by mountain plovers. Condor 84:71–74.

Koford, C. B. 1958. Prairie dogs, whitefaces, and blue grama. Wildlife Monographs 3:1–78.

Krainyk, A., B. M. Ballard, M. G. Brasher, B. C. Wilson, M. W. Parr, and C. K. Edwards. 2019. Decision support tool: mottled duck habitat management and conservation in the western Gulf Coast. Journal of Environmental Management 230:43–52.

Kumschick, S., and W. Nentwig. 2010. Some alien birds have as severe an impact as the most effectual alien mammals in Europe. Biological Conservation 143:2757–2762.

Lambert, M. S., G. Massei, C. A. Yoder, and D. P. Cowan. 2010. An evaluation of Diazacon as a potential contraceptive in non-native rose-ringed parakeets. Journal of Wildlife Management 74:573–581.

Lavelle, M. J., N. P. Snow, J. M. Halseth, J. C. Kinsey, J. A. Foster, and K. C. VerCauteren. 2018. Development and evaluation of a bait station for selectively dispensing bait to invasive wild pigs. Wildlife Society Bulletin 42:102–110.

Lavelle, M. J., K. C. VerCauteren, J. W. Fischer, G. E. Phillips, T. Hefley, S. E. Hygnstrom, S. R. Swafford, D. B. Long, and T. A. Campbell. 2011. Evaluation of fences for containing motivated feral pigs during depopulations. Journal of Wildlife Management 75:1200–1208.

Lenarz, M. S., J. Fieberg, M. W. Schrage, and A. J. Edwards. 2010. Living on the edge: viability of moose in northeastern Minnesota. Journal of Wildlife Management 74:1013–1023.

Lenarz, M. S., M. E. Nelson, M. W. Schrage, and A. J. Edwards. 2009. Temperature mediated moose survival in northeastern Minnesota. Journal of Wildlife Management 73:503–510.

Leopold, A. 1933. Game management. Charles Scribner's Sons, New York, New York, USA.

Leopold, A. 1949. A Sand County almanac. Oxford University Press, New York, New York, USA.

List, R. 1997. Ecology of kit fox (Vulpes macrotis) and coyote (Canis latrans) and the conservation of the prairie dog ecosystem in northern Mexico. PhD dissertation, University of Oxford, Oxford, UK.

Lohmeyer, K. H., M. A. May, D. B. Thomas, and A. A. Pérez de León. 2018. Implication of nilgai antelope (Artiodactyla: Bovidae) in reinfestations of Rhipicephalus (Boophilus) microplus (Acari: Ixodidae) in south Texas: a review and update. Journal of Medical Entomology 55:515–522.

Lowe, S., M. Browne, S. Boudjelas, and M. De Poorter. 2000. 100 of the world's worst invasive alien species: a selection from the Global Invasive Species Database. Invasive Species Specialist Group of the Species Survival Commission of the World Conservation Union, Auckland, New Zealand.

Lund, T. A. 1980. American wildlife law. University of California Press, Berkley, USA.

MacKenzie, J. M. 1988. The empire of nature. Manchester University Press, New York, New York, USA.

Mahoney, S. P. 2009. Recreational hunting and sustainable wildlife use in North America. Pages 266–281 in B. Dickson, J. Hutton, and W. Adams, editors. Recreational hunting, conservation and rural liveli-

hoods: science and practice. Wiley-Blackwell, Hoboken, New Jersey, USA.

Manzano, P. 1996. Avian communities associated with prairie dog towns in northwestern Mexico. PhD thesis, University of Oxford, Oxford, UK.

Mao, J. S., M. S. Boyce, D. S. Smith, F. J. Singer, D. J. Vales, J. M. Vore, and E. H. Merrill. 2005. Habitat selection by elk before and after wolf reintroduction in Yellowstone National Park. Journal of Wildlife Management 69:1691–1707.

Maxwell, A. E. 1913. Pheasants and covert shooting. Adams and Charles Black, London, UK.

Mayer, J. J., T. J. Smyser, A. J. Piaggio, and S. M. Zervanos. 2020. Wild pig taxonomy, morphology, genetics, and physiology. Pages 7–31 in K. C. VerCauteren, J. C. Beasley, S. S. Ditchkoff, et al., editors. Invasive wild pigs in North America: ecology, impacts, and management. CRC Press, Boca Raton, Florida, USA.

Meadow, R., R. P. Reading, M. Phillips, M. Mehringer, and B. J. Miller. 2005. The influence of persuasive arguments on public attitudes toward a proposed wolf restoration in the southern Rockies. Wildlife Society Bulletin 33:154–163.

McCorquondale, S. M. 1997. Cultural contexts of recreational hunting and native subsistence and ceremonial hunting: their significance for wildlife management. Wildlife Society Bulletin 25:568–573.

McHugh, T. 1972. The time of the buffalo. University of Nebraska Press, Lincoln, USA.

McIntyre, R. 1995. War against the wolf: America's campaign to exterminate the wolf. Voyageur Press, Stillwater, Minnesota, USA.

Mech, L. D. 1974. Canis lupus. Mammalian Species Account No. 37. American Society of Mammalogists, Topeka, Kansas, USA.

Mech, L. D., 1995. The challenge and opportunity of recovering wolf populations. Conservation Biology 9:270–278.

Mech, L. D., 2017. Where can wolves live and how can we live with them? Biological Conservation 210:310–317.

Menchetti, M., and E. Mori. 2014. Worldwide impact of alien parrots (Aves Psittaciformes) on native biodiversity and environment: a review. Ethology, Ecology and Evolution 26:172–194.

Menchetti, M., E. Mori, and F. M. Angelici. 2016. Effects of recent world invasive by ring-necked parakeets Psittacula krameri. Pages 253–266 in F. M. Angelici, editor. Problematic wildlife. Springer, Cham, Switzerland.

Mengak, M. T., and C. A. Miller. 2020. Human dimensions and education associated with wild pigs in North America. Pages 229–243 in K. C. VerCauteren, J. C. Beasley, S. S. Ditchkoff, et al., editors. Invasive wild pigs in North America: ecology, impacts, and management. CRC Press, Boca Raton, Florida, USA.

Michigan Department of Natural Resources. 2021. Wolves in Michigan. https://www.michigan.gov/dnr/0,4570,7-350-79135_79218_79619 -32569—,00.html.

Miller, B., R. Reading, J. Hoogland, T. Clark, G. Ceballos, R. List, S. Forrest, L. Hanebury, P. Manzano, J. Pacheco, and D. Uresk. 2000. The role of prairie dogs as a keystone species: response to Stapp. Conservation Biology 14:318–321.

Milne, L. J., and M. J. Milne. 1950. Notes on the behavior of horned toads. American Midland Naturalist 44:720–741.

Milstead, W. W., and D. W. Tinkle. 1969. Interrelationships of feeding habits in a population of lizards in southwestern Texas. American Midland Naturalist 81:491–499.

Minnesota Department of Natural Resources. 2021. Wolf management. https://www.dnr.state.mn.us/wolves/index.html.

Montemayor, M., and S. E. Henke. 1998. Phrynosoma cornutum (Texas horned lizard) longevity. Herpetological Review 29:157.

Mungall, E. C. 2000. Exotics. Pages 736–764 *in* S. Demarais and P. R. Krausman, editors. Ecology and management of large mammals in North America. Prentice Hall, Upper Saddle River, New Jersey, USA.

Munger, J. C. 1984*a*. Home ranges of horned lizards (*Phyrosoma*): circumscribed and exclusive? Oecologica 62:351–360.

Munger, J. C. 1984*b*. Long-term yield from harvester ant colonies: implications for horned lizard foraging strategy. Ecology 65:1077–1086.

Munger, J. C. 1986. Rate of death due to predation for two species of horned lizard, *Phrynosoma cornutum* and *P. modestum*. Copeia 1986:820–824.

Munn, L. C. 1993. Effects of prairie dogs on physical and chemical properties of soil. Pages 11–17 *in* J. L. Oldmeyer, D. E. Biggins, and B. J. Miller, editors. Management of prairie dog complexes for the reintroduction of the black-footed ferret. US Department of the Interior, Washington, DC, USA.

Murray, D. L., E. W. Cox, W. B. Ballard, H. A. Whitlaw, M. S. Lenarz, T. W. Custer, T. Barnett, and T. K. Fuller. 2006. Pathogens, nutritional deficiency, and climate influences on a declining moose population. Wildlife Monographs 166:1–30.

Musante, A. R., P. J. Pekins, and D. L. Scarpitti. 2010. Characteristics and dynamics of a regional moose *Alces alces* population in the northeastern United States. Wildlife Biology 16:185–204.

Musiani, M., and P. C. Paquet. 2004. The practices of wolf persecution, protection, and restoration in Canada and the United States. BioScience 54:50–60.

Mysterud, A., and C. M. Rolandsen. 2018. Fencing for wildlife disease control. Journal of Applied Ecology 56:519–525.

New Mexico Game and Fish Department. 2021. Mexican gray wolf. http://www.wildlife.state.nm.us/wpfb-file/29-wolves-pdf-2/.

Nichols, J. D., M. C. Runge, F. A. Johnson, and B. K. Williams. 2007. Adaptive harvest management of North American waterfowl populations: a brief history and future prospects. Journal of Ornithology 148:343–349.

Nie, M. A. 2001. The sociopolitical dimensions of wolf management and restoration in the United States. Research in Human Ecology 8:1–12.

Nowak, R. M. 1999. Walker's mammals of the world. 6th edition. Volume 1. Johns Hopkins University Press, Baltimore, Maryland, USA.

Nummi, P., W. Liao, O. Huet, E. Scarpulla, and J. Sundell. 2019. The beaver facilitates species richness and abundance of terrestrial and semiaquatic mammals. Global Ecology and Conservation 20:p.e00701.

Olah, G., S. H. M. Butchart, A. Symes, I. M. Guzmán, R. Cunningham, D. J. Brightsmith, and R. Heinsohn. 2016. Ecological and socioeconomic factors affection extinction risk in parrots. Biodiversity and Conservation 25:205–223.

Olsen, P. F. 1981. Sylvatic plague. Pages 232–243 *in* J. W. Davis, L. H. Karstadt, and D. O. Trainer, editors. Infectious diseases of wild animals. Iowa State University Press, Ames, Iowa, USA.

O'Meila, M. E., F. L. Knopf, and J. C. Lewis. 1982. Some consequences of competition between prairie dogs (*Cynomys ludovicianus*) and beef cattle. Journal of Range Management 35:580–585.

Ontario Ministry of Natural Resources. 2011. Moose regulations. http://www.mnr.gov.on.ca/stdprodconsume/groups/lr/@mnr/@fw/documents/document/239848.pdf.

Oregon Department of Fish and Wildlife. 2021. Wolves in Oregon. https://dfw.state.or.us/wolves/.

Outwater, A. 1996. Water: a natural history. Basic Books, New York, New York. USA

Padding, P. I., J. F. Gobeil, and C. Wentworth. 2006. Estimating waterfowl harvest in North America. Pages 849–852 *in* G. Boere, C. Galbraith, and D. Stroudt, editors. Waterbirds around the world. Stationery Office, Edinburgh, UK.

Paine, R. T. 1980. Food webs: linkage, interaction strength and community infrastructure. Journal of Animal Ecology 49:667–685.

Painter, L. E., R. L. Beschta, E. J. Larsen, and W. J. Ripple. 2015. Recovering aspen follow changing elk dynamics in Yellowstone: evidence of a trophic cascade? Ecology 96:252–263.

Patton, M. L., W. Jöchle, and L. M. Penfold. 2007. Review of contraception in ungulate species. Zoo Biology 26:311–326.

Peck, H. L., H. E. Pringle, H. H. Marshall, I. P. F. Owens, and A. M. Lord. 2014. Experimental evidence of impacts of an invasive parakeet on foraging behavior of native birds. Behavioral Ecology 25(3):582–590.

Peters, R. M., M. J. Cherry, J. C. Kilgo, M. J. Chamberlain, and K. V. Miller. 2020. White-tailed deer population dynamics following Louisiana black bear recovery. Journal of Wildlife Management 84:1473–1482.

Pianka, E. R., and W. S. Parker. 1975. Ecology of horned lizards: a review with special reference to *Phrynosoma platyrhinos*. Copeia 1975:141–162.

Pimentel, D. 2011. Biological invasions. 2nd edition. CRC Press, Boca Raton, Florida, USA.

Pizzimenti, J. J. 1975. Evolution of the prairie dog genus *Cynomys*. Occasional Papers of the Museum of Natural History, University of Kansas 39:1–73.

Plumpton, D. L., and R. S. Lutz. 1993. Nesting habitat use by burrowing owls in Colorado. Journal of Raptor Research 27:175–179.

Poche, R. M., D. Poche, G. Franckowiak, D. J. Somers, L. N. Briley, B. Tseveenjav, and L. Ployakova. 2018. Field evaluation of low-dose warfarin baits to control wild pigs (*Sus scrofa*) in north Texas. PLoS One 13:e0206070.

Pohja-Mykra, M., T. Vuorisalo, and S. Mykra. 2005. Hunting bounties as a key measure of historical wildlife management and game conservation. Oryx 39:284–291.

Potter, G. E., and H. B. Glass. 1931. A study of respiration in hibernating horned lizards, *Phrynosoma cornutum*. Copeia 1931:128–131.

Power, T. M. 1991. Ecosystem preservation and the economy of the Greater Yellowstone Area. Conservation Biology 5:395–404.

Price, A. H. 1990. *Phrynosoma cornutum* (Harlan): Texas horned lizard. Catalogue of American Amphibians and Reptiles 469:1–7.

Primm, S. A., and T. W. Clark. 1996. Making sense of the policy process for carnivore conservation. Conservation Biology 10:1036–1045.

Randall, D. 1976*a*. Poison the damn prairie dogs. Defenders 51:381–383.

Randall, D. 1976*b*. Shoot the damn prairie dogs. Defenders 51:378–381.

Rattan, J. M., B. J. Higginbotham, D. B. Long, and T. A. Campbell. 2010. Exclusion fencing for feral hogs at white-tailed deer feeders. Texas Journal of Agriculture and Natural Resources 23:83–89.

Reidy, M. M., T. A. Campbell, and D. G. Hewitt. 2008. Evaluation of electric fencing to inhibit feral pig movements. Journal of Wildlife Management 72:1012–1018.

Reidy, M. M., T. A. Campbell, and D. G. Hewitt. 2011. A mark-recapture technique for monitoring feral swine populations. Rangeland Ecology and Management 64:316–318.

Renecker, L. A., and R. J. Hudson. 1986. Seasonal energy expenditure and thermoregulatory response of moose. Canadian Journal of Zoology 64:322–327.

Ripple, W. J., and R. L. Beschta. 2003. Wolf reintroduction, predation risk, and cottonwood recovery in YNP. Forest Ecology and Management 184:299–313.

Ripple, W. J., and R. L. Beschta. 2005. Willow thickets protect young aspen from elk browsing after wolf reintroduction. Western North American Naturalist 65:118–122.

Ripple, W. J., and R. L. Beschta. 2012. Trophic cascades in Yellowstone: the first 15 years after wolf reintroduction. Biological Conservation 145:205–213.

Saavedra, S., and F. M. Medina. 2020. Control of invasive ring-necked parakeet (*Psittacula krameri*) in an island biosphere reserve (La Palma, Canary Islands): combining methods and social engagement. Biological Invasions 22:3653–3667.

Sakai, A. K., F. W. Allendorf, J. S. Holt, D. M. Lodge, J. Molofsky, et al. 2001. The population biology of invasive species. Annual Review of Ecology and Systematics 32:305–332.

Samoylova, T. I., A. M. Cochran, A. M. Samoylova, B. Schemera, A. H. Breiteneicher, S. S. Ditchkoff, V. A. Petrenko, and N. R. Cox. 2012. Phage display allows identification of zona pellucida-binding peptides with species-specific properties: novel approach for development of contraceptive vaccines for wildlife. Journal of Biotechnology 162:311–318.

Sanders, D. L., F. Xie, R. E. Mauldin, L. A. Miller, M. R. Garcia, R. W. DeYoung, D. B. Long, and T. A. Campbell. 2011. Efficacy of ERL-4221 as an ovotoxin for feral swine. Wildlife Research 38:168–172.

Scheffer, T. H. 1945. Historical encounter and accounts of the plains prairie dog. Kansas History Quarterly 13:527–537.

Schemnitz, S. D., G. R. Batcheller, M. J. Lovallo, H. Bryant White, and M. W. Fall. 2012. Capturing and handling wild animals. Pages 64–117 in N. J. Silvy, editor. The wildlife techniques manual. Volume 1: Research. 7th edition. Johns Hopkins University Press, Baltimore, Maryland, USA.

Schmidly, D. J., and R. D. Bradley. 2016. The mammals of Texas. 7th edition. University of Texas Press, Austin, USA.

Schwartz, C. C., and L. A. Renecker. 1998. Nutrition and energetics. Pages 441–478 in A. W. Franzmann and C. C. Schwartz, editors. Ecology and management of the North American moose. Smithsonian Institution Press, Washington, DC, USA.

Seaton, C. T., T. F. Paragi, R. D. Boertje, K. Kielland, S. DuBois, and C. L. Fleener. 2011. Browse biomass removal and nutritional condition of Alaska moose *Alces alces*. Wildlife Biology 17:1–12.

Severud, W. J., T. R. Obermoller, G. D. Delgiudice, and J. R. Fieberg. 2019. Survival and cause-specific mortality of moose calves in northeastern Minnesota. Journal of Wildlife Management 83:1131–1142.

Sherbrooke, W. C. 1981. Horned lizards: unique reptiles of the western North America. Southwestern Parks and Monuments Association, Globe, Arizona, USA.

Shiels, A. B., W. B. Bukoski, and S. R. Siers. 2018. Diets of Kauai's invasive rose-ringed parakeet (*Psittacula krameri*): evidence of seed predation and dispersal in human-altered landscape. Biological Invasions 20:1449–1457.

Simberloff, D. 1993. Species-area and fragmentation effects on old growth forests: prospects for longleaf pine communities. Pages 1–14 in S. M. Hermann, editor. The longleaf pine ecosystem: ecology, restoration, and management. Tall Timbers Fire Ecology Conference Proceedings, No. 18. Tall Timbers Research Station, Tallahassee, Florida, USA.

Smith A. L. 2020. Wild pig policy and legislation. Pages 245–273 in K. C. VerCauteren, J. C. Beasley, S. S. Ditchkoff, et al., editors. Invasive wild pigs in North America: ecology, impacts, and management. CRC Press, Boca Raton, Florida, USA.

Smith, D. W., R. O. Peterson, and D. B. Houston. 2003. Yellowstone after wolves. Bioscience 53:330–340.

Smith, D. W., D. R. Stahler, K. A. Cassidy, E. Stahler, M. Metz, et al. 2020. Yellowstone National Park Wolf Project Annual Report 2019. YCR-2020-01. National Park Service, Yellowstone Center for Resources, Yellowstone National Park, Wyoming, USA.

Snow, N. P., J. A. Foster, J. C. Kinsey, S. T. Humphrys, L. D. Staples, D. G. Hewitt, and K. C. VerCauteren. 2017. Development of toxic bait to control invasive wild pigs and reduce damage. Wildlife Society Bulletin 41:256–263.

Snow, N. P., M. J. Lavelle, J. M. Halseth, M. P. Glow, E. H. VanNatta, et al. 2018. Exposure of a population of invasive wild pigs to simulated toxic bait containing a biomarker: implications for population reduction. Pest Management Science 75:1140–1149.

Stebbins, R. C. 1954. Amphibians and reptiles of western North America. McGraw-Hill, New York, New York, USA.

Strickland, B. K., M. D. Smith, and A. L. Smith. 2020. Wild pig damage to resources. Pages 143–174 in K. C. VerCauteren, J. C. Beasley, S. S. Ditchkoff, et al., editors. Invasive wild pigs in North America: ecology, impacts, and management. CRC Press, Boca Raton, Florida, USA.

Summerlin, J. W., and L. R. Green. 1977. Red imported fire ant: a review on invasion, distribution, and control in Texas. Southwestern Entomology 2:94–101.

Sweeney, J. R., J. M. Sweeney, and S. W. Sweeney. 2003. Feral hog. Pages 1164–1179 in G. A. Feldhamer, B. C. Thompson, J. A. Chapman, editors. Wild mammals of North America. Johns Hopkins University Press, Baltimore, Maryland, USA.

Swenk, M. H. 1915. The prairie dog and its control. Nebraska Agricultural Experiment Station Bulletin 154:3–38.

Taylor, R. B., E. C. Hellgren, T. M. Gabor, and L. M. Ilse. 1998. Reproduction of feral pigs in southern Texas. Journal of Mammalogy 79:1325–1331.

Thompson, D. P., J. A. Crouse, S. Jaques, and P. S. Barboza. 2020. Redefining physiological responses of moose (*Alces alces*) to warm environmental conditions. Journal of Thermal Biology 90:102581.

Thompson, I. D., and R. W. Stewart. 1998. Management of moose habitat. Pages 377–401 in A. W. Franzmann and C. C. Schwartz, editors. Ecology and management of the North American moose. Smithsonian Institution Press, Washington, DC, USA.

Timmermann, H. R., and M. E. Buss. 1998. Population and harvest management. Pages 559–615 in A. W. Franzmann and C. C. Schwartz, editors. Ecology and management of the North American moose. Smithsonian Institution Press, Washington, DC, USA.

Titus, K., T. L. Haynes, and T. F. Paragi. 2009. The importance of moose, caribou, deer, and small game in the diet of Alaskans. Pages 137–143 in R. T. Watson, M. Fuller, M. Pokras, and W. G. Hunt, editors. Ingestion of lead from spent ammunition: implications for wildlife and humans. Peregrine Fund, Boise, Idaho, USA.

Traweek, M., and R. Welch. 1992. Exotics in Texas. Texas Parks and Wildlife Department, Austin, USA.

Trauth, S. E., H. W. Robison, and M. V. Plummer. 2004. The amphibians and reptiles of Arkansas. University of Arkansas Press, Fayetteville, USA.

US Department of the Interior and Environment Canada. 1986. North American waterfowl management plan. Washington, DC, USA; Ottawa, Ontario, Canada.

USFWS. US Fish and Wildlife Service. 1987. Northern Rocky Mountain wolf recovery plan. Denver, Colorado, USA.

USFWS. US Fish and Wildlife Service. 1992. Recovery plan for the eastern timber wolf. Twin Cities, Minnesota, USA.

USFWS. US Fish and Wildlife Service. 1994. The reintroduction of gray wolves to Yellowstone National Park and Central Idaho: final environmental impact statement. Helena, Montana, USA.

USFWS. US Fish and Wildlife Service. 1996. Reintroduction of the Mexican wolf within its historic range in the southwestern United States: final environmental impact statement. US Department of the Interior, Albuquerque, New Mexico, USA.

USFWS. US Fish and Wildlife Service. 2003. Recovery plan for the red-cockaded woodpecker (*Picoides borealis*). 2nd revision. Atlanta, Georgia, USA.

USFWS. US Fish and Wildlife Service. 2010. Waterfowl population status, 2010. US Department of the Interior, Washington, DC, USA.

USFWS. US Fish and Wildlife Service. 2011. Endangered species. http://www.fws.gov/endangered/.

USFWS. US Fish and Wildlife Service. 2020. Black-tailed prairie dog. https://www.fws.gov/mountain-prairie/es/blackTailedPrairieDog.php.

USFWS. US Fish and Wildlife Service. 2021. Mexican wolf. https://www.fws.gov/southwest/es/mexicanwolf/.

Utah Division of Wildlife Resources 2021. Federal delisting announcement for gray wolves: fact sheet. https://wildlife.utah.gov/wolves.html.

Verbrugge, L. N. H., R. J. G. Van den Born, and H. J. R. Lenders. 2013. Exploring public perception of non-native species from a visions of nature perspective. Environmental Management 52:1562–1573.

VerCauteren, K. C., J. C. Beasley, S. S. Ditchkoff, J. J. Mayer, G. J. Roloff, and B.K. Strickland. 2020. Invasive wild pigs in North America: ecology, impacts, and management. CRC Press, Boca Raton, Florida, USA.

VerCauteren, K. C., R. A. Dolbeer, and E. M. Gese. 2012. Identification and management of wildlife damage. Pages 232–269 *in* N. J. Silvy, editor. The wildlife techniques manual. Volume 2: Management. 7th edition. Johns Hopkins University Press, Baltimore, Maryland, USA.

VerCauteren, K. C., M. J. Lavelle, and S. Hygnstrom. 2006. Fences and deer-damage management: a review of designs and efficacy. Wildlife Society Bulletin 34:191–200.

VerCauteren, K., M. Lavelle, T. M. Gehring, J. M. Landry, and L. Marker. 2014. Dogs as mediators of conservation conflicts. Pages 211–238 *in* M. E. Gompper, editor. Free-ranging dogs and wildlife conservation. Oxford University Press, Oxford, UK.

Vogel, S., C. P. Ellington, and D. L. Kilgore. 1973. Wind-induced ventilation of the burrow of the prairie dog, *Cynomys ludovicianus*. Journal of Comparative Physiology 85:1–14.

Wagner, F. W., R. Forester, R. B. Gill, D. R. McCullough, M. R. Pelton, W. F. Porter, and H. Salwasser. 1995. Wildlife Policies in US national parks. Island Press, Washington, DC, USA.

Washington Department of Fish and Wildlife. 2021. Gray wolf conservation and management. https://wdfw.wa.gov/species-habitats/at-risk/species-recovery/gray-wolf.

Webb, S. L., and S. E. Henke. 2003. Defensive strategies of Texas horned lizards (*Phrynosoma cornutum*) against red imported fire ants. Herpetological Review 34:327–328.

Weiskopf, S. R., O. E. Ledee, and L. M. Thompson. 2019. Climate change effects on deer and moose in the Midwest. Journal of Wildlife Management 83:769–781.

Weltzin, J. F., S. Archer, and R. K. Heitshmidt. 1997. Small mammal regulation of vegetation structure in a temperate savanna. Ecology 78:751–785.

Whicker, A., and J. K. Detling. 1993. Control of grassland ecosystem processes by prairie dogs. Pages 18–27 *in* J. L. Oldmeyer, D. E. Biggins, and B. J. Miller, editors. Management of prairie dog complexes for the reintroduction of the black-footed ferret. US Department of the Interior, Washington, DC, USA.

Whitford, W. G., and M. Bryant. 1979. Behavior of a predator and its prey: the horned lizard (*Phrynosoma cornutum*) and harvester ants (*Pogonomyrmex* spp.). Ecology 60:686–694.

Wilcove, D. S., and L. Y. Chen. 1998. Management costs for endangered species. Conservation Biology 12:1405–1407.

Wiles, M. C., and T. A. Campbell. 2006. Liquid chromatography–electrospray ionization mass spectrometry for direct identification and quantification of iophenoxic acid in serum. Journal of Chromatography 832:144–157.

Williams, B. L., R. W. Holtfreter, S. S. Ditchkoff, and J. B. Grand. 2011. Trap style influences wild pig behavior and trapping success. Journal of Wildlife Management 75:432–436.

Williams, C. K., B. D. Dugger, M. G. Brasher, H. M. Coluccy, D. M. Cramer, et al. 2014. Estimating habitat carrying capacity for migrating and wintering waterfowl: considerations, pitfalls and improvements. Wildfowl Special Issue 4:407–435.

Williams, C. K., G. Ericsson, and T. A. Heberlein. 2002. A quantitative summary of attitudes towards wolves and their reintroduction. Wildlife Society Bulletin 30:575–584.

Williams, S. C., A. J. Denicola, T. Almendinger, and J. Maddock. 2013. Evaluation of organized hunting as a management technique for overabundant white-tailed deer in suburban landscapes. Wildlife Society Bulletin 37:137–145.

Williams, T. 1990. Waiting for wolves to howl in Yellowstone. Audubon 92:32–41.

Wilson, B. C., and C. G. Esslinger. 2002. North American waterfowl management plan, Gulf Coast joint venture: Texas Mid-Coast Initiative. North American Waterfowl Management Plan, Albuquerque, New Mexico, USA.

Wilson, M. A. 1997. The wolf in Yellowstone: science, symbol, or politics? Deconstructing the conflict between environmentalism and wise use. Society and Natural Resources 10:453–468.

Wilson, P. I. 1999. Wolves, politics, and the Nez Perce: wolf recovery in central Idaho and the role of native tribes. Natural Resources Journal 39:543–564.

Wisconsin Department of Natural Resources. 2021. Wolves in Wisconsin. https://dnr.wisconsin.gov/topic/WildlifeHabitat/wolf/index.html.

Wyckoff, A. C., S. E. Henke, T. A. Campbell, D. G. Hewitt, and K. C. VerCauteren, 2009. Feral swine contact with domestic swine: a serologic survey and assessment of potential for disease transmission. Journal of Wildlife Diseases 45:422–429.

Wydeven, A. P., Fuller, T. K., W. Weber, and K. MacDonald, K., 1998. The potential for wolf recovery in the northeastern United States via dispersal from southeastern Canada. Wildlife Society Bulletin 26:776–784.

Wyoming Game and Fish Department, US Fish and Wildlife Service, National Park Service, US Department of Agriculture, Animal and Plant Health Inspection Service, Wildlife Services, and Eastern Shoshone and Northern Arapahoe Tribal Fish and Game Department. 2019. Wyoming gray wolf monitoring and management 2019 annual report. K. J. Mills and Z. Gregory, editors. Cheyenne, Wyoming, USA.

GLOSSARY

Compiled and revised from various sources and used with permission of Prentice Hall, Upper Saddle River, New Jersey, USA.

abiotic. The nonliving components of the environment (e.g., air, rocks, soil, water; Thomas 1979).

abrupt edge. An edge between stands or communities that is regular (i.e., straight lines or gently sweeping curves), leading to relatively low amounts of edge per unit area (Thomas 1979).

abundance. The number of individuals or of a resource. Abundance usually should be evaluated concurrently with the availability of the resource to specific organisms (Bolen and Robinson 1995).

accident. Death or injury from physical causes alone.

accuracy. The nearness of a measurement to the actual value of the variable being measured.

active management. Direct manipulation of animal populations (e.g., translocation, hunt).

activity budget. Amount of time an animal spends in different activities. Biologists observe the animal for a set period and document behaviors (e.g., foraging, walking, standing, running, grooming) at regular intervals to quantify the behavior of animals over time and across populations.

adapted. The suitability of an organism for a particular condition, usually habitat, which comes from the process by which an organism becomes better suited to its environment or particular functions (Thomas 1979).

adaptive heterothermy. A process by which an animal increases body temperature and reduces water loss via evaporation by storing heat until it can be released to the environment passively via convection or conduction.

adaptive impact management. A variation on adaptive management, adaptive impact management is an approach that regards impacts created by human–wildlife interactions as the focus for fundamental objectives of management (Riley et al. 2003); a method for integrating insights from biological and human dimensions of wildlife management.

adaptive management. The process of implementing a policy decision incrementally, so that changes can be made if the desired results are not being achieved. It is a process similar to a scientific experiment in that predictions and assumptions in management plans are tested, and experience and new scientific findings are used as the basis to improve resources management practices and future planning.

additive mortality. A concept that the effect of one kind of mortality is added to those of other sources of mortality. For example, if predation takes 10% of a population and ice storm takes 20%, the total mortality for the year is 30%. If in the next year predation takes 20% and an ice storm takes 20%, for a total mortality of 40%, the effects of the two factors are said to be additive. In this concept, hunting mortality adds to the total natural mortality rate of a population. See *compensatory mortality* (Bolen and Robinson 1995).

adult. An animal, or age class of animals, that has reached breeding age (Bolen and Robinson 1995).

age ratio. The relative proportions of various age groups in a population. An age ratio may be expressed in several ways: number of juveniles per adult, juveniles per adult female, juveniles per 100 females, or juveniles per pair of adults. Age ratios often are determined from examinations of hunter-killed animals (e.g., wings of woodcock, quail, ducks) and may be used as a measure of breeding success in animal populations (Bolen and Robinson 1995).

aggregation. Coming together of organisms into a group (Anderson 1991).

allopatric. Different, usually used in reference to populations that occupy mutually exclusive (but usually adjacent) geographic areas (Anderson 1991).

allotment. A designated area of land available for livestock grazing. A grazing permit is usually issued, designating a specified number and kind of livestock to be grazed according to direction found in an allotment management plan. An allotment is the basic land unit used in the management of livestock on National Forest System lands and associated lands administered by the US Forest Service.

alluvial soil. Soil deposited by running water, showing practically no horizon development or other modification (Anderson 1991).

alpha diversity. Diversity within a community (Anderson 1991).

ambient. Surrounding.

angiosperm. Any class of plants that is identified by having their seeds enclosed in an ovary (Thomas 1979).

animal community. The species of animals supported by a combination of habitat niches (Thomas 1979).

animal unity (AU). Considered to be one mature cow (455 kg) or the equivalent based upon average daily forage consumption of 11.8 kg of dry matter per day. Therefore 1 AU = 7.7 white-tailed deer or 5.8 mule deer (Bolen and Robinson 1995).

annual. In plants, an annual is a species that completes its life cycle in one growing season. Many common weeds are annuals (Bolen and Robinson 1995).

annulus (plural: annuli). A mark, usually circular, that indicates one year of growth and is therefore a useful measure of age. Accuracy increases or decreases in relation to the distinctiveness of seasonal growth patterns (i.e., north temperature vs. tropical environments). Examples include rings on fish scales, trees, and horns (Bolen and Robinson 1995).

antler. One of paired bony structures protruding from the skulls of deer, elk, moose, and caribou (Cervidae). Antlers are covered with velvet during development, shed annually, and are usually branched. Antlers grow on males only, except caribou (Bolen and Robinson 1995).

apical growth. Growth from the terminal shoot meristem (Thomas 1979).

aquaculture. The rearing of plants or animals in water under controlled conditions (Anderson 1991).

aquatic habitat. Habitat that occurs in free water (Thomas 1979).

arboreal. Living in trees (chiefly said of animals); relating to trees..

area-kill. The annual kill per unit area.

arthropod. Any invertebrate organism of the phylum Arthropoda, which includes the insects, crustaceans, arachnids, and myriapods, having a horny, segmented, external covering and jointed limbs (Thomas 1979).

artificial establishment. A planting of wildlife maintained only through renewed plantings or artificial propagation.

asexual. Having no evident sex or sex organs; sexless; pertaining to or characterizing reproduction involving a single male or female gamete, such as binary fission or budding (Thomas 1979).

aspect. The direction toward which a slope faces (Thomas 1979).

assemblage. A group of species under study.

association. A major unit in community ecology, characterized by essential uniformity of species composition (Anderson 1991).

atrical. Born in helpless states, not able to move or support itself (Anderson 1991).

attitude. A person's expressed favorable or unfavorable evaluation of a person, object, concept, or action is referred to in social psychology as an attitude. Attitudes often are important components in predictions of human behaviors.

autotrophic. Not requiring an exogenous factor for normal metabolism; refers to organisms, usually green plants, that are capable of converting solar energy to chemical energy (sugar) by photosynthesis (Anderson 1991).

availability. With abundance, availability is part of the ecological equation for measuring food and other resources. For example, earthworms often are abundant beneath logs, but under those conditions, such worms are not available as food for robins or woodcock. Similarly, twigs on tall shrubs and small trees, although abundant, may not be available as browse until deep snows bring the upper branches within the reach of deer and rabbits (Bolen and Robinson 1995).

aversive conditioning. Learning that relies on the stimuli of undesirable experiences (i.e., negative reinforcement). For example, some experiments suggest that coyotes fed mutton treated with distasteful substances may avoid killing sheep (Bolen and Robinson 1995).

avifauna. A term referring to all species of birds that occupy a designated area of time. For example, about 487 species of birds make up the current avifauna of Texas (Bolen and Robinson 1995).

backyard management. A part of urban wildlife management in suburban zones; installation of birdhouses, feeders, and layered vegetation are common backyard management techniques (Bolen and Robinson 1995).

bag limit. Number of animals that can be taken in a unit of time, usually a day (Anderson 1991).

band. A loose aggregation of wildlife, sometimes all of one sex.

bare ground. All land surface not covered by vegetation, rock, or litter.

basal area. The cross-sectional area of a tree as measured at breast height (~1.37 m), expressed in square feet per acre.

basal metabolism. The measure of metabolism of the body in the resting state, determined by the amount of oxygen used or of heat produced (Thomas 1979).

bedding. The process of an animal lying down for a rest (Thomas 1979).

behavioral carrying capacity. The maximum population size a given area will support when intrinsic behavioral or physiological mechanisms are the primary force controlling the population.

beliefs. Thoughts or propositions about general classes of objects (e.g., all wildlife) or issues (e.g., climate change) that are believed to be true.

best estimate. The best possible understanding given a combination of available information and understand of a situation (Thomas 1979).

beta diversity. Diversity comparison between similar communities (Anderson 1991).

bias. The difference between the expected value of population estimate and the actual population size; a systematic distortion away from true density.

biennial. A plant living two years, usually flowering the second year; occurring every two years (Anderson 1991).

Big Data. Big Data is defined by the five Vs: volume, variety (heterogeneity of sources, unstructured data), velocity (speed of data generation and collection), veracity (uncertainty and data quality), and value.

big game. Large animals hunted, or potentially hunted, for sport—for example, elk, mule deer, bighorn sheep, pronghorn, black bear (Anderson 1991).

binomial. The latinized name of an organism consisting of two words, the first of which is the genus and the second the species (Thomas 1979).

biological amplification. The process in which organisms higher in the food chain accumulate and retain materials, such as organochlorines, from organisms lower in the food chain (Anderson 1991).

biological clock. An internal mechanism that signals animals that it is time for some activity, such as migration or nesting (Anderson 1991).

biological diversity (biodiversity). The variety of life, typically expressed in terms of species richness but also may be applied to genes and ecosystems. The preservation of biodiversity is the primary goal of conservation biology (Bolen and Robinson 1995).

biological potential. The maximum production of a selected organism that can be attainted under optimum management (Thomas 1979).

biomagnification. The accumulation of matter with each succeeding trophic level, bottom to top, in an ecosystem. Harmless levels of dichlorodiphenyltrichloroethane (DDT) applied at the lower trophic levels were thereby amplified to harmful levels in the bodies of bald eagles and other carnivores (Bolen and Robinson 1995).

biomass. The total quantity of living organisms of one or more species per unit of space, which is called species biomass, or of all the species in a community, which is called community biomass (Thomas 1979).

biome. A complex of communities with a distinct type of vegetation (Anderson 1991).

biosphere. The zone of life on earth, a region usually considered to range from a few hundred meters above ground occupied by high-flying birds to a few meters below ground level, where burrowing animals occur. Species living on the ocean floor extend the biosphere by several kilometers (Bolen and Robinson 1995).

biosphere people. Society that is aware of and uses the biosphere.

biota. All the plants and animals within an area or region (Anderson 1991).

biotic. Life or the act of living (Thomas 1979).

biotic factors. Living entities, or living plants and animals; opposite of abiotic factors (Anderson 1991).

biotic potential. The number of births divided by the population size in a given area in given time (Anderson 1991).

biotrophic levels. The feeding levels in a food chain (Thomas 1979).

birth flow. A breeding system that pertains to populations whose rate of breeding is relatively constant throughout the year.

birth pulse. A breeding system that pertains to populations whose rate of breeding occurs one time of the year.

birth rate. Proportion of a population newly born in a unit of time (Anderson 1991).

body condition. A state, mode, or state of being, generally referring to the level of energy reserves of an animal, which are typically assessed relative to morphology and physiology. The amount of stored energy and nutrients in the body is related to the ability of the animal to survive and to reproduce.

bog. An extremely wet, poorly drained area characterized by a floating spongy mat of vegetation often composed of sphagnum, sedges, and heaths (Anderson 1991).

bottleneck. A term used to describe periods, typically in winter, when food or other resources are limiting, usually to the point of markedly increasing mortality in a wildlife population (Bolen and Robinson 1995). See also *genetic bottleneck*.

breeding (or reproduction) potential. The maximum or unimpeded increase rate of a species in an ideal environment.

brood. A family of young birds from a single mother; sometimes applied to fishes and reptiles (Bolen and Robinson 1995).

brood parasite. See *nest parasitism* (Bolen and Robinson 1995).

browse. Palatable twigs, shoots, leaves, and buds of woody plants. Often used to describe a category of ungulate forage.

brucellosis. A bacterial disease that affects mammals, often causing abortions in cattle and some ungulates (Anderson 1991).

buffer. A species that constitutes food for predators and acts as a buffer to protect primary prey from predators.

buffer strip. A strip of vegetation that is left or managed to reduce the impact of a treatment or action of one area on another (Thomas 1979).

burrow. A hole or tunnel dug in the ground by an animal.

calving area. The areas, usually on spring-fall range, where females give birth to calves and maintain them during their first few days or weeks.

cannibalism. A special case of carnivory where the prey and predator are the same species.

canopy. The more or less continuous cover of branches and foliage formed collectively by the crowns of trees and other woody growth (Thomas 1979).

captive breeding. The practice of breeding animals in captive facility, usually done with engendered species for later release into the wild (Anderson 1991).

carnivore. A flesh-eating animal (Anderson 1991).

carrion. Dead animal flesh (Anderson 1991).

carrying capacity. The maximum rate of animal stocking possible without inducing damage to vegetation or related resources; may vary from year to year because of fluctuation forage production (Thomas 1979).

catastrophic events. Events resulting from a great and sudden calamity or disaster. In the case of forest stands, such events include windstorms, wildlife, floods, snowslides, and insect outbreaks (Thomas 1979).

cave. A natural underground chamber that is open to the surface (Thomas 1979).

cavity. The hollow excavated in snags by birds; used for roosting and reproduction by many birds and mammals (Thomas 1979).

cavity nester. Wildlife species that nest in trees cavities (e.g., woodpeckers; Anderson 1991).

cecum (plural: ceca). A dead-end outpouching of the digestive tract. Prevalent in fishes, birds, and mammals with high-fiber diets (e.g., spruce grouse, with a winter diet of evergreen needles; Bolen and Robinson 1995).

cementum annuli. Layers of the teeth of some animals that can be used for determining age (Anderson 1991).

census. A complete enumeration of an entity.

chaining. A vegetation maintenance technique in which a heavy anchor chain is dragged between two tractors to break off or uproot plants (Anderson 1991).

check-out system. Measuring the number of hunters or their kill by checking them in and out at points of entry and exit.

cherry picking. Failing to report results that did not reach an expected statistical or other threshold (Fraser et al. 2018).

chill factor. The increased chilling effect on an animal, attributable to wind velocity (Thomas 1979).

chlorophyll. A complex of mainly green pigments in the chloroplasts, characteristic of plants whose light-energy-transforming properties permit photosynthesis (Thomas 1979).

chloroplast. The protoplasm body or plastid in plant cells that contains chlorophyll (Thomas 1979).

circadian rhythm. The regular fluctuations in bodily functions (e.g., temperature) and behavior (e.g., sleeping) during a cycle approximating 24 hours (Bolen and Robinson 1995).

clean farming. The elimination of diversity and interspersion on farmland and substitution of a monoculture. Hedgerows, roadside vegetation, and similar areas are eliminated. Instead, fields of single crops are cultivated from border to border without cover for wildlife (Bolen and Robinson 1995).

clear-cut. An area from which all trees have been removed by cutting (Thomas 1979).

climate change. A change in global or regional climate patterns, in particular a change apparent from the middle to late 20th century onward and attributed largely to the increased levels of atmospheric carbon dioxide produced by the use of fossil fuels.

climax. The culminating stage in plant succession for a given site where the vegetation has reached a highly stable condition (Thomas 1979).

climax community. The final or stable biotic community in a developmental series. It is self-perpetuating and in equilibrium with the physical habitat and environment. The climax community is the presumed end point in succession.

clumped distribution. An aggregated distribution pattern, for example, a herd of animals (Anderson 1991).

cluster sampling. Simple random sampling applied to distinct groups of population numbers (Anderson 1991).

clutch. The eggs laid by a birds or reptile in a single nesting attempt (Bolen and Robinson 1995).

coevolution. The reciprocal evolutionary change, influenced by natural selection, between interacting species that are not related.

cohort. A group of individuals in a population born during a particular period, such as a year (Anderson 1991).

colonial nesters. Birds that nest in large groups (Anderson 1991).

colony. A group of the same kind of animals or plants living together (Thomas 1979).

comanagement. Management by two or more entities involving shared control, costs, and responsibility for a situation in wildlife management.

commensalism. The relation between two populations living together when only one receives a benefit; the other population is neither harmed nor benefited (Anderson 1991).

commercial harvest. The cutting and marketing of trees for use (Thomas 1979).

commercial thinning. Any type of thinning that produces merchantable material at least equal to the value of the direct costs of harvesting (Thomas 1979).

commercial timber production. The process of growing wood products for sale or use (Thomas 1979).

community. A group of one or more populations of plants and animals in a common spatial arrangement; an ecological term used in a broad sense to include groups of various sizes and degrees of integration.

community type. An aggregation of all plant communities with similar structure and floristic composition. A unit of vegetation within a classification with no particular successional status implied. A taxonomic unit of vegetation classification that references existing vegetation.

compartment. An organization unit or small subdivision of forest area for the purpose of orientation, administration, and silvicultural operations. An area defined by permanent boundaries, either of natural features or artificially marked, which is not necessarily coincident with stand boundaries.

compensatory mortality. The concept that one kind of mortality largely replaces another kind of mortality in animal population. In simple logic, an animal dying from one cause (e.g., hunting or disease) cannot die from another cause (e.g., predation or starvation), so one source of mortality compensates for the other. Therefore the total mortality rate normally is not greatly influenced by changes for any single cause of death. For example, if predation takes 10% of a population and disease takes 20%, total natural morality for the year is 30%. If in the next year predation takes 25% and disease takes 5%, the effect of predation is said to be compensatory. Of importance, the total annual mortality rate may remain essentially unchanged with or without legal hunting (Bolen and Robinson 1995).

compensatory population growth. Equivalent to density-dependent population growth. That is, growth of a population is rapid when numbers are low and less rapid when numbers are high. Such growth may result from decreased mortality, increased natality, or both (Bolen and Robinson 1995).

competition. The active demand by two or more organisms for a commonly required resource that is limited (Anderson 1991).

competitive exclusion principal (Gause's hypothesis). No two species can occupy the same niche at the same time (Anderson 1991).

compound 1080. Sodium monofluoroacetate, commonly used as a coyote poison before restrictions were imposed in the United States in 1973. Compound 1080 is considered dangerous because of nontarget victims (Bolen and Robinson 1995).

condition index. A measure of an animal's well-being, usually expressed in terms of fat content. Most condition indices are adjusted for size differences among individuals by dividing fat weight by some other physical feature (e.g., wing length for birds). In mammals, fat deposits in bone marrow or around kidneys are common indices (Bolen and Robinson 1995).

conduction. Heat transfer between two solid objects with different temperatures that are in contact with one another.

conifer (coniferous). A cone-bearing plant (Anderson 1991).

coniferous forest. A forest dominated by cone-bearing trees (Thomas 1979).

connectors. Strips or patches of vegetation used by wildlife to move between habitats (Thomas 1979).

conservation. Wise maintenance and use of natural resources (Anderson 1991).

conservation biology. A field of many disciplines united with the common goal of preserving biodiversity. Genetics, physiology, geography, population biology, wildlife management, forestry, and veterinary science are among the basic and applied disciplines contributing to conservation biology; the professional staffs at many zoos also provide crucial expertise (Bolen and Robinson 1995).

consumer. A member of the animal community. Consumers occupy the higher trophic levels in an ecosystem (Bolen and Robinson 1995).

consumptive use. Use of resources that involves removal (e.g., hunting and fishing; Anderson 1991).

contour line. An imaginary line, or its representation on a contour map, joining points of equal elevation (Thomas 1979).

contrast. In wildlife management, contrast is the degree of difference in vegetative destructive structure along edges where plant communities meet or where successional stages or vegetative conditions within plant communities meet (Thomas 1979).

control. In wildlife management, control is the process of managing populations of a species to accomplish an objective; usually used in the sense of depressing population numbers of a pest species to prevent or decrease the impact of the species (Thomas 1979).

convection. Heat transfer between a solid body and a fluid (water or air) or between two fluids that are different temperatures.

coordinated resources management (CRM). The process whereby various user groups are involved in discussion of alternative resources uses and collectively diagnose management problems, establish goals and objectives, and evaluate multiple-use resource management.

coordinated timber-wildlife management. The melding of timber and wildlife management planning and action into one plane so that goals of both timber and wildlife are met (Thomas 1979).

coprophagy. The practice of eating feces. Coprophagy enables rabbits to recover nutrients from their droppings that escape initial digestion (Bolen and Robinson 1995).

corpora lutea. See *corpus luteum* (Bolen and Robinson 1995).

corpus luteum (plural: corpora lutea). A structure formed in the mammalian ovary from the follicle that once contained an ovum (egg). The corpus luteum functions as an endocrine gland by secreting hormones to maintain pregnancy. Corpora lutea, which are visible when an ovary is sectioned, indicated the number of ova (eggs) shed by the ovary, thereby providing data on fertility (Bolen and Robinson 1995).

corridor. A strip or block of habitat connecting otherwise isolated units of suitable habitat that allows the dispersal of organisms and the consequent mixing of genes (Bolen and Robinson 1995).

cover. A general term used to describe vegetation and topography. Vegetative cover is divided into three categories: (1) the overstory of trees; (2) the midstory composed mainly of large shrubs and small trees; and (3) the understory that includes small shrubs, grasses, and forbs. Types of cover include brood cover (low-lying vegetation such as grasses and forbs that afford protection for young game birds, usually quail, turkeys, and grouse), escape cover (shrubby or herbaceous cover, hollow trees, water, rock crevices, or burrows that provide a means of getting away when harassed by predators), and nesting cover (vegetation required by birds or animals to rear their young. Examples are tree dens for squirrels or grassy patches near openings for quail, turkey, grouse, or rabbits), roosting cover (cover required by game birds, ranging from conifers for turkeys to idle fields or sparse timberlands for quail), and wintering cover. All cover required by game to overwinter. (It varies from trees with squirrel dens, to brush cover for quail, and rock outcrops for bear).

cover, canopy. The percentage of ground covered by a particular projection of the outermost perimeter of the natural spread of foliage of plants.

cover, ground. The percentage of material, other than bare ground, that covers the soil surface. It may include organic material such as vegetation basal cover (live and standing dead), mosses and lichens, and litter, and inorganic material such as cobble, gravel, stones, and bedrock. Ground cover plus bare ground will total 100%.

cover patch. A discrete area covered by vegetation that meets either the definition of hiding or thermal cover (Thomas 1979).

cover type. A taxonomic unit of vegetation classification referencing existing vegetation. Cover type is a broad talon based on existing plant species that dominate, usually within the tallest layer.

covert. A geographic unit of game cover.

covey. A small flock of birds that lie.

crash. The period of severe mortality following the peak of a cycle.

critical area. An area that must be treated with special consideration because of inherent site factors, condition, values, or significant potential conflicts among uses.

critical foods. Foods essential or necessary to the welfare of a game species, such as hard mast for gray squirrel.

critical habitat. A legal term that describes the physical or biological features essential to the conservation of a species.

crop. Anatomically, an expandable part of a bird's esophagus used for food storage and perhaps a small amount of digestion. Analysis of crop contents helps determine the diets or birds because foods in the gizzard (stomach) often are ground beyond recognition. Crop also refers to the harvested part of a plant or animal population (Bolen and Robinson 1995).

crop tree. Any tree forming, or selected to form, part of the final crop; generally, a tree selected in a young stand for that purpose (Thomas 1979).

crown. The upper part of a tree or other woody plant that carries the main branch system and foliage, and surmounts at the crown base a more or less clean stem (Thomas 1979).

cruising radius. The distance between locations at which an individual animal is found at various hours of the day or at various seasons, during various years.

culling. Removal of animals, usually individuals at risk, from a population. For example, predators normally remove sick and injured individuals from a prey population. In some cases, culling may apply to the general thinning of a population, as when the density of foxes or raccoons is reduced for the control of rabies (Bolen and Robinson 1995).

cutaneous evaporation. Evaporation of water from the surface of the skin.

cycle. A regular pattern, such as a repetitive change in population size (Anderson 1991).

DDT. Abbreviation for the chemical dichlorodiphenyltrichloroethane, perhaps the most familiar pesticide in the family of chlorinated hydrocarbons, banned in the United States in 1972, partly because of its harmful effects on wildlife (Bolen and Robinson 1995).

dead and down woody material. All woody material, from whatever source, that is dead lying on the forest floor (Thomas 1979).

debris. The scattered remains of something broken or destroyed; ruins, rubble; fragments (Thomas 1979).

deciduous. Pertaining to any plant organ or group of organs that is shed naturally; perennial plants that are leafless for some time during the year (Thomas 1979).

deciduous tree. A tree that drops its leaves each autumn. Usually broadleaved trees, of which maple, oak, birch, and hickory are examples (Bolen and Robinson 1995).

decimating factors. Those that kill directly (e.g., hunting, predation, disease and parasites, accidents, starvation).

decompose. To separate into component parts or elements; to decay or putrefy (Thomas 1979).

decomposer. One of the many organisms, principally bacteria, that reduce animal wastes and the carcasses of complex organisms into elemental components (Bolen and Robinson 1995).

deer yard. An area of heavy cover where deer congregate in winter for food and shelter (Anderson 1991).

Delphi technique. The process of combining expert opinions into a consensus; a method of making predictions (Thomas 1979).

den trees. A rainproof, weather-tight cavity in a tree. Dens take 8 to 30 years to develop following injury, disease, or natural pruning and are aided by periodic enlargement by squirrels or other wildlife.

denning site. A place of shelter for an animal; also where an animal gives birth and raises young (Thomas 1979).

density. The number of individuals per unit area.

density dependent. Having more severe impacts on a population as the population size increases (Anderson 1991).

density-dependent factor. A factor that acts in proportion to the density of animals. Some diseases are density dependent because a higher percentage of the population becomes infected as density increases. Natality and mortality often fluctuate with changes in density (Bolen and Robinson 1995).

density independent. Having impact on a population not related to the population's size (Anderson 1991).

density-independent factor. A factor that acts independently of population density. Weather is often considered density independent. A flood, for example, may kill an entire population regardless of density (Bolen and Robinson 1995).

dependent variable. The variable in a relationship that is influenced by the independent variable (Thomas 1979).

desertion limit. The number of days after incubation starts when normal disturbances of the nest will not cause desertion.

deterministic model. A mathematical model in which all the relationships are fixed and the concept of probability does not enter; a given input produces a predictable output (Anderson 1991).

detritus food chain. A process in which dead organisms are decomposed by other organisms, such as worms, larvae, and bacteria (Anderson 1991).

diameter breast high (dbh). The standard diameter measurement for standing trees, including bark, taken at 1.37 m above the ground (Thomas 1979).

diplochory. A two-step process of seed dispersal. Step 1 is through endozoochory, where seeds are dispersed after ingestion by animals. Step 2 is usually through ectozoochory, where some animals obtain seeds from feces and then move them to another location.

direct habitat improvement. Habitat manipulations primarily for the benefit of wildlife (Thomas 1979).

direct improvement. Land treatment measures or structures installed to benefit game (e.g., seeded wildlife openings, waterholes, or daylighting fruit-producing shrubs).

dispersal. Movement of organisms into unfamiliar locations; behavior usually associated with younger animals upon leaving natal areas (Bolen and Robinson 1995).

dispersion. The pattern of distribution of individuals in an animal population; in the mathematical sense, dispersion describes the probability of occurrence of such individuals in particular places (Thomas 1979). See also *law of dispersion*.

dispersion failure. A planting of wildlife followed by immediate dispersal and disappearance.

distribution. The spread or scatter of an entity within its range.

diversity. The total range of wildlife species, plant species, communities, and habitat features in an area (Anderson 1991).

diversity index. A number that indicates the relative degree of diversity in habitat per unit area. It is expressed mathematically as

$$DI = \frac{TP}{2\sqrt{A \cdot \pi}},$$

where TP is the total perimeter of an area plus any edge within the area in meters or feet, A the area in square meters, and π is 3.1416 (Thomas 1979).

domains of scale. Areas where spatial and temporal patterns do not change or change very little or monotonically with changes in scale. Domains of scale are a way to deal with extrapolation (Wiens 1989).

dominant. Plant species or species groups that by means of their number, coverage, or size have considerable influence over control upon the conditions of existence of associated species. Also, individual animals that determine the behavior of one or more other animals, resulting in the establishment of a social hierarchy (Thomas 1979).

dragging. The use of a log or metal grate behind a tractor to loosen cattle manure in a field (Anderson 1991).

drum. To make a reverberating sound by beating the wings rapidly, as grouse do, or by tapping on a suitable surface, as woodpeckers do (Thomas 1979).

dusting. The process of rolling or exercising vigorously in dust or duff; in birds, dusting has the function of aligning barbules and maintaining feathers (Thomas 1979).

dynamic. Characterized by or tending to produce continuous change (Thomas 1979).

easement. An access area across another's land (Anderson 1991).

ecological characteristics. The basic features of a species related to distribution, habitat, reproduction, growth characteristics and needs, and responses to habitat changes (Anderson 1991).

ecological equivalents. Organisms that occupy the same niche but live in different communities. Examples include bison and wildebeest occupying grazing niches in the plains of North America and Africa, respectively (Bolen and Robinson 1995).

ecological longevity. The average length of life of individuals of a population under stated conditions.

ecological neighborhood. An area defined by three properties: a specific ecological process, a timescale appropriate to that process, and the organism's spatial activity during that period and where an animal's movement defines the neighborhood (Addicott et al. 1987).

ecological niche. The role a particular organism plays in the environment (Thomas 1979).

ecological role. The part or influence of an organism in an ecosystem (Thomas 1979).

ecology. The study of interactions between organisms and their environment. Term coined by German zoologist Ernst Haeckel in 1866 based on the Greek root *oikos*, meaning home, and *logos*, meaning study (Bolen and Robinson 1995).

ecosystem. Living and nonliving components in an environment functioning together (Anderson 1991).

ecosystem management. Using an ecological approach to achieve the multiple-use management of national forests and grasslands by blending the needs of people and environmental values in such a way that national forests and grasslands represent diverse, healthy, productive, and sustainable ecosystems.

ecosystem people. Humans that live directly off the land.

ecotone. The community formed where two other communities meet, sometimes called an edge (Anderson 1991).

ectozoochory. Dispersal of seeds by hooking to the exterior of animals through hooks or barbs that allow the seed to attach to the exterior of an animal.

edaphic. An adjective pertaining to soil (e.g., concerning texture, drainage, fertility). Edaphic factors often are of direct importance to burrowing animals, but they also exert major indirect influences on all wildlife because of ecological links with vegetation (Bolen and Robinson 1995).

edge. The area where two communities (ecotones) meet (Anderson 1991).

edge effect. The ecological result of increasing edges in homogeneous habitats, principally the increased abundance and diversity of species. A benefit in the management of some animals (bobwhite and other edge species), although increased predation or nest parasitism may result in other cases (Bolen and Robinson 1995).

efficiency. Proportion of incoming solar energy converted to chemical energy (Anderson 1991).

elaisome. Fruiting body on seeds that provide a nutrient-rich food source for ants that disperse the seeds.

elk calving habitat. Habitat used by elk for calving, usually located on spring-fall range in areas of gentle slope. Elk calving habitat contains forage areas and hiding and thermal cover close to water (Thomas 1979).

emigration. Movement out of a given area (Anderson 1991).

endangered species. A wildlife species officially designated by the US Fish and Wildlife Service as having its continued existence threatened over its entire range because its habitat is threatened with destruction, drastic modification, or severe curtailment, or because of overexploitation, disease, predation, or other factors (Thomas 1979).

endemic. Native to a region (Anderson 1991).

endozoochory. Fleshy fruits are consumed by animals, and the seeds are then dispersed after traveling through the digestive tract of an animal and discarded in feces.

energy balance (or energy budget). The relation between intake of food and output of work. In positive energy balance, the animal's intake of metabolizable energy exceeds it requirements for basal metabolism, activity, growth, and reproduction, and the body stores extra food as fat. In negative energy balance, energy reserves are depleted and body mass decreases. The relation between intake of food and output of work (e.g., muscular, secretory activity) is positive when the body stores extra food as fats and negative when the body draws on stored fat to provide energy for work.

environment. The sum of all the external conditions that may influence organisms (Thomas 1979).

environmental analysis report. A report on environmental effects of proposed federal actions that require an environmental impact statement under section 102 of the National Environmental Policy Act (PL 91-190, 1970). This is an in-house document that becomes the final document on projects whose effects are so minor as not to require a formal environmental impact statement. Though not formally required by the act, this document is commonly used to determine whether section 102 applies to the contemplated action (Thomas 1979).

environmental factor. Any influence on the combined plant and animal community (Thomas 1979).

environmental impact statement. The final version of the statement required under section 102 of the National Environmental Policy Act (PL 91-190, 1970) for major federal actions that affect the environment; a revision of the draft statement that includes public and governmental agency comments; a formal document that meets legal requirements and used as the basis for judicial decision concerning compliance with the act. An environmental impact statement can also refer to similar statements required by state and local laws patterned after the act (Thomas 1979).

environmental resistance. Factors that act to slow a population's growth (Anderson 1991).

enzootic. The chronic level of disease frequency (i.e., a low but constant occurrence of a disease in a population; Bolen and Robinson 1995).

ephemeral streams. Streams that contain running water only for brief periods (Thomas 1979).

epizootic. An outbreak of disease. Large numbers of animals die in a short period (Bolen and Robinson 1995).

epizootiology. The study of disease ecology. Addresses the how and why of diseases at either enzootic or epizootic levels (Bolen and Robinson 1995).

equilibrium carrying capacity. The density of a population is at a constant equilibrium, but the limiting variables are unknown.

escape cover. Usually vegetation dense enough to hide an animal; used by animals to escape from potential enemies (Thomas 1979).

esophagus. A tube in the digestive system of vertebrates that connects the mouth with the stomach. Expands in some birds for temporary food storage (Bolen and Robinson 1995).

estimate. A result of a statistical sample, often for determining population size. Used instead of a census or index. To obtain an estimate, animals or inanimate objects (e.g., nests or dens) may be counted on one or more sample areas known as plots (Bolen and Robinson 1995).

ethics. The discipline whose focus is formal and rigorous analysis of ethical propositions (Nelson and Vucetich 2012). Whereas other social sciences are concerned with the analysis of descriptive propositions about human values, ethics is concerned with the analysis of prescriptive propositions about human values. Descriptive propositions describe the nature of the world around us, while prescriptive (ethical) propositions are claims about how individuals or society ought to behave, value, or relate to the world.

ethology. The study of animal behavior (Bolen and Robinson 1995).

etiology. The study of the cause of disease.

eury-. A prefix used to identify wide tolerances for specific components of the environment. White-tailed deer are eurythermal because of their wide tolerance of temperature extremes (Bolen and Robinson 1995).

eurytopic. Able to withstand wide variations in environmental conditions (Anderson 1991).

eutrophic. Term describing the enriched nature of some freshwater lakes and rivers. Organic materials and nutrients accumulate, leading to increased biological productivity. The enrichment process is known as eutrophication. Human activities may affect the process abnormally (Bolen and Robinson 1995).

eutrophication. Process of lake succession by addition of nutrients (Anderson 1991).

even-aged management. A system of forest management in which stands are produced or maintained with relativity minor differences in age (Thomas 1979).

evolution. The change in a population's genetic composition over time, leading to adaptions to the environment (Anderson 1991).

existence value. A class of economic values that reflect the benefit people receive from knowing that an environmental resource, such as an endangered species, exists (Ready 2012). Usually measured by a willingness to pay for the existence of that resource.

exotic species. An organism introduced—intentionally or accidentally—from its native range into an area where the species did not previously occur. Asian animals such as pheasants, sika deer, and carp are exotic species in North America (Bolen and Robinson 1995).

exponential growth. Population growth that exceeds the carrying capacity until population numbers saturate the habitat; growth characterized by a progressively increasing, nonlinear relation between population numbers and time (Anderson 1991).

extinction. The complete loss—forever—of a unique constellation of genes known as a species (e.g., passenger pigeon; Bolen and Robinson 1995).

extirpation. The elimination of a species from one or more specific areas, but not from all areas. Extirpation should not be confused with extinction. Bison have been extirpated from most of their former range in North America, but many hundreds still exist on private ranches, zoos, and federal refuges (Bolen and Robinson 1995).

facilitation. The activities of one species enhance conditions for another.

factor. One of the forces reducing the numbers (decimating factors) or retarding the increase rate (welfare factor) of wildlife.

fauna. A term for all animal life.

fawning area. An area, usually on spring-fall range, where females give birth to fawns and maintain them in their first few days or weeks (Thomas 1979).

featured species. The selected wildlife species whose habitat requirements guide wildlife management, including coordination, multiple-use planning, direct habitat improvements, and cooperative programs for a unit of land.

featured-species management. A management policy keyed to a single species, perhaps at the expense of the others. Often issued for endangered species (e.g., forest management for red-cockaded woodpeckers) but also applied to more abundant species (Bolen and Robinson 1995).

fecal material. Material discharged from the bowels; more generally, any discharge from the digestive tract of an organism (Thomas 1979).

fecundity. Actual number of organisms produced.

feedback. The output of a given system that affects the state of that same system (Anderson 1991).

feeding substrate. The surface on which an animal finds its food (Thomas 1979).

feral. Having reverted to a wild state after being domesticated (e.g., feral horses; Anderson 1991).

fertility. The potential capability of an organism to produce young.

fitness. The competence of an organism to pass on its genes to the next generation. All else being equal, some individuals within a population are of greater fitness than others. Fitness may be influenced by many factors, including physical, physiological, and social conditions (Bolen and Robinson 1995).

flagship species. A flagship species is a species selected to act as an ambassador, icon, or symbol for a defined habitat, issue, campaign, or environmental cause. Flagship species are usually relatively large and considered to be charismatic in Western cultures.

flight limit. The maximum distance a bird can traverse at one continuous flight.

flora. A term for all plant life.

fluctuations. Irregular changes.

focal species. Those species focused on that provide an essential ecological function or are indicative of essential habitat conditions. Focal species are linked to particular habitats or ecosystem types and changes in those habitats and ecosystems.

food chain. The energy flow from green plants through consumer organisms at each trophic level (Anderson 1991).

food habits. A term for the diet of wildlife. Some studies of food habits include feeding behavior and food availability (Bolen and Robinson 1995).

food web. A complex food chain (Anderson 1991).

forage. Browse and herbage, which is available and may provide food for grazing animals or be harvested for feeding.

forage areas. Forest stands that do not qualify as either hiding or thermal cover and all-natural and man-made openings (Thomas 1979).

forage (or food) availability. What is available to the animal as food. Food availability changes seasonally and depends on selection by the animal. For carnivores, food availability varies with prey abundance; for herbivores, all herbage is not food because animals only select and consume some forages and some plant parts.

forage medium. The environment in which feeding by a wildlife species occurs (Thomas 1979).

forage (or food) quality. Nutritional value associated primarily with energy and protein, because these are the two nutritional factors most commonly in short supply, and to a lesser extent with minerals and vitamins. Food quality depends on the ability of the animal to digest and then metabolize food components and varies with fiber content and plant biochemical compounds that affect browsers and grazers.

forb. Any herbaceous plant other than grasses or grasslike plants.

forced trailing. Trails that compel animals, particularly big game and domestic livestock, to take a particular route of travel (Thomas 1979).

forest. Generally, an ecosystem characterized by tree cover; more particularly, a plant community predominantly composed of trees and other woody vegetation, growing closely together; an area managed for the production of timber and other forest produce, or maintained in forest cover for such indirect benefits as protection of catchment areas or recreation; an area of land proclaimed to be a forest under a forest act or ordinance (Thomas 1979).

forest management. The application of scientific, economic, and social principles to the management of a forest estate for specified objectives; the branch of forestry concerned with its overall administrative, economic, legal, social, scientific, and technical aspects, especially silviculture, protection, and forest regulation (Thomas 1979).

fossorial species. Animals adapted for living underground (Bolen and Robinson 1995).

free water. Water that is not bound to any surface, particularly soil particles, and is available for transpiration of plants (Thomas 1979).

furbearer. A mammal commonly harvested for its hide (e.g., muskrat, mink; Anderson 1991).

gallinaceous bird. A bird of the order of Galliformes. These are chickenlike birds, including pheasants, quail, grouse, prairie chickens, ptarmigan, and turkeys (Bolen and Robinson 1995).

game. A species of vertebrate wildlife hunted by man for sport (Thomas 1979).

gamma diversity. Diversity comparison of large, heterogenous areas (Anderson 1991).

gene pool. Narrowly, all the genes of a localized interbreeding population; broadly, all the genes of a species throughout its entire range (Anderson 1991).

generality. The applicability of a model to appropriate situations (Anderson 1991).

genetic bottleneck. The temporary reduction of a population to only a few individuals, thereby limiting the gene pool and increasing inbreeding. When such a population later increases in size, it will still have a limited gene pool (Bolen and Robinson 1995).

genetic composition. The total genetic makeup of a population (Anderson 1991).

genotype. The entire genetic constitution of an organism; contrast with *phenotype* (Anderson 1991).

genus. One or a group of related species used in classification of organisms, the first word in binomial or scientific name (Thomas 1979).

geomorphic. Of or like the earth or the configuration or shape of the earth's surface (Thomas 1979).

gestation. The length of time from conception to birth (Anderson 1991).

gizzard. The muscular stomach of birds that grinds food, usually with the aid of grit. Gizzards serve as the functional equivalent of teeth in mammals (Bolen and Robinson 1995).

gleaning. A process of feeding, particularly in birds, in which food items are gathered from the surface of the foraging substrate, usually plants (Anderson 1991).

gradient. The rise or fall of a ground surface expressed in degrees of slope (Thomas 1979).

graminoids. Grasses and grasslike plants, including grasses (Poaceae), sedges (Cyperaceae), and rushes (Juncaceae).

graph theory. A network approach. Mathematically, networks are graphs and are concerned primarily with functional connectivity (Golubski et al. 2016).

grass. Any plant species that is a member of the family Gramineae.

grasslike plant. A plant that vegetatively resembles a true grass of the Gramineae.

gravid. Term describing females with ripening eggs. Major physiological differences aside, the gravid condition in egg-laying species is somewhat analogous to pregnancy in mammals (Bolen and Robinson 1995).

grazing food chain. The movement of energy from green plants to herbivores to carnivores, excluding composition (Anderson 1991).

grazing lawn. Grassland communities with prostrate growth forms with rapidly growing tissues of high nutritional quality for herbivores.

grazing system. A specialization of grazing management that defines systematically recurring periods of grazing and deferment for two or more pastures or management units.

gross primary productivity. The total amount of energy available from the conversion of solar energy to chemical energy during photosynthesis by green plants (Anderson 1991).

ground cover. The percentage of material, other than bare ground, that covers the land surface. It may include live vegetation, standing dead vegetation, litter, cobble, gravel, stones, and bedrock. Ground cover plus bare ground would total 100%.

guild. A group of species that exploits the same class of environment resources in a similar way.

habitat. The resources and conditions present in an area that produce occupancy, including survival and reproduction by a given organism. Habitat is species-specific.

habitat availability: The accessibility and procurability of physical and biological components in a habitat.

habitat avoidance. An oxymoron that should not be used; wherever an animal occurs defines its habitat.

habitat block. An area of land covered by a relatively homogenous plant community, essentially a single successional stage condition (Thomas 1979).

habitat component. A simple part, or a relatively complex entity regarded as a part, of an area or type of environment in which an organism or biological population normally lives or occurs (Thomas 1979).

habitat evaluation procedure (HEP). A method of the US Fish and Wildlife Service that documents the quality and quantity of resources available for selected species of wildlife; involves estimation of a *habitat suitability index (HSI)* and the amount of such habitat available. HEP numerically predicts the effects of altering habitats by development or by natural events (Bolen and Robinson 1995).

habitat niche. The peculiar arrangement of food, cover, and water that meets the requirements of a particular species (Thomas 1979).

habitat preference. Used to describe the relative use of different locations (habitats) by an individual or species.

habitat quality. The ability of the area to provide conditions appropriate for individual and population persistence.

habitat richness. The relative degree of ability of a habitat to produce numbers of species of either plants or animals; the more species produced, the richer the habitat (Thomas 1979).

habitat selection. A hierarchical process involving a series of innate and learned behavioral decisions made by an animal about what habitat it would use at different scales of the environment.

habitat suitability index (HSI). Part of habitat evaluation procedure. A value ranging from 0.0 to 1.0, assigned on the basis of known food and cover requirements of a species. Habitat suitability index is multiplied by the area of habitat to obtain the *habitat units (HUs)* available for each species (Bolen and Robinson 1995).

habitat type. Term referring only to the type of vegetation association in an area or to the potential of vegetation to reach a specified climax stage (Daubenmire 1968:72–73).

habitat unit (HU). Part of habitat evaluation procedure. The product resulting from *habitat suitability index (HSI)* multiplied by the area available with that HSI. Used for assigning a comparative value for habitat of various species (Bolen and Robinson 1995).

habitat use. The way an animal uses (or consumes, in generic sense) a collection of physical and biological entities in a habitat.

habituation. A behavioral tendency wherein animals become accustomed to unnatural components in their environment. Deer that feed at the edges of highways often become habituated to traffic (Bolen and Robinson 1995).

half-shrub. A perennial plant with a woody base; the annually produced stems die each year.

hard snag. A snag composed primarily of sound wood, particularly sound sapwood; generally merchantable (Thomas 1979).

hardwood. The wood of broadleaved trees, and the trees themselves, belonging to the botanical group Angiospermae; distinguished from softwoods by the presence of vessels (Thomas 1979).

HARKing. Hypothesizing after the results are known, rather than before the study, as if it were an a priori hypothesis (Kerr 1998).

harvest. Removal of animals from a population (Anderson 1991).

hawking. The feeding behavior of birds wherein they capture food in flight (Thomas 1979).

herb. Any vascular plant, except those developing persistent woody stems above ground.

herbaceous. Term descriptive of nonwoody plants. Herbaceous vegetation includes forbs and grasses (except bamboo). Also, the nonwoody parts of trees and shrubs (e.g., leaves; Bolen and Robinson 1995).

herbage. The aboveground material of any herbaceous plant.

herbicide. A chemical substance used for killing plants (Thomas 1979).

herbivore. An animal that feeds on plants (Anderson 1991).

herd. Any large aggregation, or detached unit, of hoofed mammals.

herd behavior. The collective behavior exhibited by social groups of ungulates (Thomas 1979).

herptofauna. Term that combined the amphibian and reptilian members of a community. *Herptiles* (or *herps*) is likewise a term that combined amphibians and reptiles (Bolen and Robinson 1995).

heterotroph. An organism that uses chemical energy supplied by autotrophic organisms (Anderson 1991).

heterozygosity. The presence of both forms (known as alleles)—dominant and recessive—of a single gene in an individual or population. Heterozygosity generally conveys an advantage for coping with new situations. An individual with a genetic composition (genotype) represented as AaBbCc is heterozygous, whereas another individual of the same species AabbCC or aaBBCC is homozygous. In a population, heterozygosity provides a variety of genotypes, thereby increasing the likelihood that some individuals are capable of surviving stresses that affect the population and thereby increase fitness (Bolen and Robinson 1995).

hibernacula. Habitat niches where certain animals overwinter (Thomas 1979).

hiding cover for deer. Vegetation capable of hiding 90% of standing adult deer from the view of a human at a distance equal to or less than 61 m; generally, any vegetation used by deer for security or to escape from danger (Thomas 1979).

hiding cover for elk. Vegetation capable of hiding 90% of standing adult elk from the view of a human at a distance equal to or less than 61 m; generally, any vegetation used by elk for security or to escape from danger (Thomas 1979).

hole nesters. Wildlife species that nest in cavities (Thomas 1979).

holistic. Emphasizing the importance of the whole and the interdependence of its parts (Thomas 1979).

home range. The area traversed by an animal during a specified period.

homeostasis. A stable state or the tendency of a system to maintain a stable or balanced state (Anderson 1991).

homeothermic. Term indicating maintenance of stable body temperatures independently of environmental temperatures. Homeotherms—mammals and birds—can produce metabolic heat to meet the stress of low ambient temperatures. Warm-blooded is a population description of homeothermic species (Bolen and Robinson 1995).

homotherm. A warm-blooded animal that can regulate its body temperature physiologically (Anderson 1991).

homozygosity. The limited presence or total absence of either the dominant or recessive form (allele) of a gene in an individual or population. The genes of a homozygous individual may be represented as AabbCC or AABBcc. A population composed mainly of homozygous individuals lacks genetic diversity, thereby reducing the likelihood that some individuals are genetically more capable of surviving stresses than those in a heterozygous population. A homozygous population is therefore generally less able to survive over a long period than a heterozygous population (Bolen and Robinson 1995).

horn. A structure that protrudes from the skulls of goats, sheep, and bovines (antelope, cows, and bison), consisting of a keratin sheath surrounding a core of bone. Usually paired, except in rhinoceros. There is no velvet covering during development. Rarely branched or shed, but pronghorns are notable exceptions. Horns occur in both sexes (Bolen and Robinson 1995).

host. An organism that furnishes food, shelter, or other benefits to another species (Anderson 1991).

human dimensions of wildlife management. Discovery and application of insights about how humans value wildlife (and outcomes and benefits they desire from the wildlife resource), how humans want wildlife to be managed to achieve certain outcomes, and how humans affect or are affected by wildlife and the outcomes of wildlife management (Chapter 4; Decker et al. 2012). Inquiry into human dimensions of wildlife management typically uses stakeholder engagement techniques and an array of social sciences research methods to understand traits of individuals, groups, organizations, social structures, economic activity, cultural systems, communities, and institutions within the wildlife management system.

I-carrying capacity. The population density of the inflection point of the logistic curve.

immature. An animal, or age class, incapable of breeding and not looking like an adult; usually young-of-the-year. *Juvenile* is a synonym (Bolen and Robinson 1995).

immigration. Movement into a given area (Anderson 1991).

impact (population). A change in a population's natality, growth, or survival caused by some disturbance (Anderson 1991).

imprinting. Recognition fixed through a short-interval learning process in animals; young might imprint on parents, or animals might imprint on nesting habitat (Anderson 1991).

inactive management. No direct action is allocated toward the manipulation of wildlife populations.

inbreeding. Breeding among genetically similar individuals in a population that leads to a reduced genetic variability (homozygosity) (Anderson 1991).

inbreeding depression. The undesirable result of repeated matings within a small population of related individuals, typically reducing the gene pool and producing abnormalities and lessened fitness in the offspring (Bolen and Robinson 1995).

inclusion. Areas of less-than-stand size that have an inherently different management type and productivity than that of the stand in which they lie. They can be treated differently than the stand if they are classified as key areas.

independent variable. The variable in a relationship that is judged for its effect on the dependent variable (Thomas 1979).

index. A means of comparing the relative size of an animal population, usually from year to year, but it does not provide a census or estimate of actual numbers. Roadside drumming counts of ruffed grouse provide a useful population index (Bolen and Robinson 1995).

indicator. A condition that is visible and denotes some other condition that is invisible.

indicator species. Species that indicate certain environmental conditions, seral stages, or treatment.

indicator species system. In wildlife, analogous to featured species management; in plant ecology, plant species used to indicate special environmental factors or plant community types (Thomas 1979).

indices. Indicators of population changes through repeated measurements. Indices generally show population trends (Anderson 1991).

indirect habitat improvement. Habitat manipulation done for purposes other than wildlife habitat improvement but exploited to accomplish wildlife management objectives (Thomas 1979).

induced diversity index. A number that indicates the relative degree of induced diversity in habitat per unit area produced by edges formed at the junction of successional stages, or vegetative conditions within plant communities; expressed mathematically as

$$\text{induced DI} = \frac{\text{TE}_S}{2\sqrt{A \cdot \pi}},$$

where TE_s is the total length of edges between successional stages or conditions within plant communities, in meters or feet, A is the area expressed in square meters, and π is 3.1416 (Thomas 1979).

induced edge. An edge that results from the meeting of two successional stages or vegetative conditions within a plant community; can be controlled by management action (Thomas 1979).

influence. An environmental variable that influences a factor.

inherent. In ethology, inherent refers to a behavioral characteristic that is not learned. Innate and instinctive are synonyms (Bolen and Robinson 1995).

inherent diversity index. A number that indicates the relative degree of inherent diversity in habitat per unit area produced by plant community to plant community edges. Expressed mathematically as

$$\text{inherent DI} = \frac{\text{TE}_c}{2\sqrt{A \cdot \pi}},$$

where TE_c is the total edge between plant communities within or on the perimeter of the area, A is the area, and π is 3.1416 (Thomas 1979).

inherent edge. An edge that results from the meeting of two plant community types (Thomas 1979).

innate capacity for increase (r_m). A measure of the rate of increase of a population under ideal conditions (Anderson 1991).

insectivorous. Insect eating (Anderson 1991).

integrated pest management (**IPM**). A combination of chemical, cultural, and biological methods designed to minimize pest damage. For example, IPM recognizes the beneficial role of predaceous insects in the control of herbivorous insects (Bolen and Robinson 1995).

interface. A surface forming a common boundary between two regions or between two things (Thomas 1979).

interference competition. Direct interactions whereby one organism prevents another from accessing resources through threat of or actual use of force or physical aggression.

intermittent stream. A stream that ordinarily goes dry at one or more times during the year but sustains flows for some period (Thomas 1979).

interspecific competition. Competition between individuals of different species (Anderson 1991).

interspersion. The degree to which environmental types are intermingled rangeland.

intraspecific competition. Competition between members of the same species (Anderson 1991).

intrinsic rate of increase. Difference between the birth and death rates in a population (Anderson 1991).

inventory. A detailed list of things in possession, especially a periodic survey of goods and material in stock; the process of survey; the items and the quantity of goods and materials listed (Thomas 1979).

invertebrate. Any animal without a vertebral column (i.e., backbone). Insects and crustaceans are particularly abundant groups of invertebrates (Bolen and Robinson 1995).

irruption. A large, sudden, nonperiodic increase in density, often accompanied by an extension into hitherto unoccupied range.

island biogeography. The study of natural communities on islands, with an emphasis on species diversity as related to an island's area and its distance from the mainland. Principles apply to actual islands and to fragmented forests and other patches of habitat (Bolen and Robinson 1995).

isometric and allometric scaling. The slope of the line representing the relationship between two variables is equal to 1 (i.e., a 45° line). An allometric relationship is disproportionate; one variable increases or decreases more rapidly (West 2017).

juvenile. Immature (Bolen and Robinson 1995).

juxtapose. To situate side by side; to place together (Thomas 1979).

juxtaposition. The act of arranging in space.

K carrying capacity. The maximum number of animals of a given population that can be supported by available resources.

key area. A portion of rangeland selected because of its location, grazing or browsing value, or use. It serves as a monitoring and evaluation point for range condition, trend, or degree of grazing use. Properly selected key areas reflect the overall acceptability of current grazing management over the rangeland. A key area guides the general management of the entire area of which it is a part.

key species. (1) Forage species whose use serves as an indicator to the degree of use of associated species. In many cases, key species include indicator species and species traditionally referenced as increasers, decreasers, desirables, or intermediates. (2) Those species that must, because of their importance, be considered in the management program.

key species management. In wildlife, analogous to featured species management; in range management, the most palatable and common plant species used by livestock; the plants on whose status livestock management decisions are based (Thomas 1979).

keystone species. A species on which other species in an ecosystem largely depend, such that if it were removed the ecosystem would change.

kill. The number of head killed per year from a unit of population.

kill ratio. The proportion or percentage of the game population that can be killed yearly without diminishing subsequent crops. The ratio of the yield to the population.

K-selected species. Species adapted to low birth rates, low death rates, and long life spans. Depressed populations of such species recover slowly and often have specialized habitat needs. As a result, overkill and other disasters easily jeopardize K-selected species. Whales, elephants, and whooping cranes are examples (Bolen and Robinson 1995).

lagomorph. Any member of the mammalian order Lagomorpha, which includes rabbits (cottontails), hares, and pikas. Often mistaken for rodents, lagomorphs have two pairs of upper incisors, one small set directly behind the larger front pair (Bolen and Robinson 1995).

land base. The amount of land with which the land manager has to work (Thomas 1979).

landform. Any physical, recognizable form or feature of the earth's surface having characteristic shape and produced by natural causes.

landscape. A spatially heterogenous area, used to describe features (e.g., stand type, site, soil) of interest.

landscape feature. Widespread or characteristic features within the landscape (e.g., stand type, site, soil, patch).

landscape foreground. Areas managed particularly for their impact on the visual attributes of an area as viewed from a specified point (Thomas 1979).

landscape species. Species that use large, ecologically diverse areas and often influence the structure and function of ecosystems. The landscape species concept is a way of selecting a few species to emphasize in conservation with the belief that meeting their needs will be beneficial to other species and the landscape (Sanderson et al. 2002).

law of dispersion. An ecological theory; the potential density of wildlife species with small home ranges that requires two or more types of habitat and is roughly proportional to the sum of the peripheries of those types (Thomas 1979).

law of interspersion. The number of resident wildlife species that requires two or more types of habitat depends on the degree of interspersions of numerous blocks of such types (Thomas 1979).

law of the minimum. An ecological axiom that states that any factor a population requires that is present in the smallest amount limits the population's growth accordingly (Anderson 1991).

legume. Any of a large group of the pea family that has five pods enclosing the seeds (Anderson 1991).

lek. A site where birds (primarily grouse) traditionally gather for sexual display and courtship (Anderson 1991).

level. Often confused with the term *scale*, as in scale level vs. scale resolution or extent. For precision, consider that the term *level* defines a specific organization in a hierarchy definition (e.g., an organism level or a population level, and is distinguished by differences in the rates, or frequencies, of its characteristic processes; Turner et al. 2001).

life form. A group of wildlife species whose requirements for habitat are satisfied by similar successional stages within given plant communities (Thomas 1979).

life-form association. A group of organisms whose requirements are satisfied by similar successional stages in the development of communities (Bolen and Robinson 1995).

life tables. A table of population data based on a sample (often 1,000 individuals) of the population that shows the age at which each member died. A dynamic or *cohort* life table starts with a group of individuals all born during the same period. A static or time-specific life table has a sample of individuals from each age class in the population (Anderson 1991).

limiting factor. That factor or condition greater than or outweighing other factors in limiting wildlife population growth.

Lincoln index. A ratio based on banding and used for census.

litter. (1) A family of young mammals from a single mother.
(2) Uppermost layer of organic debris on the soil surface, essentially freshly fallen or slightly decomposed vegetative material (Bolen and Robinson 1995).

livestock. Domestic animals, usually ungulates, raised for use, profit, or pleasure (Thomas 1979).

loafing cover. A place that offers shade in summer or sun and wind protection in winter for idling.

logistic growth. Growth of a population that approaches and remains near the carrying capacity (Anderson 1991).

lopping. The chopping of small trees and branches and tops of large trees after felling, so that the resultant slash will lie close to the ground and decay more rapidly or be available for forage (Thomas 1979).

managed forest. A forest that has been brought under management to accomplish specified objectives.

managed stand. A stand subject to silvicultural manipulation planned to give a desired result, usually increased wood production.

management. Manipulation of populations or habitats to achieve desired goals by people (Anderson 1991).

management by objectives. Planning by putting program objectives in priority order (Anderson 1991).

management for species richness. A wildlife management strategy to produce a relatively high number of species per unit area (Thomas 1979).

management systems. A group of interacting, interrelated, or interdependent components that form a unified whole (i.e., systems) concerned with management (Riley and Gregory 2012). Management systems encompass the ecological and human subsystems (such as organizations, institutions, and processes) involved in wildlife management.

marsh. A low, treeless, wet area characterized by sedges, rushes, and cattails (Anderson 1991).

mast. The fruit or nuts of such trees as oaks, beech, walnuts, and hickories. These are commonly referred to as hard mast. Soft mast consists of fruits and nuts of such plants as dogwood, viburnums, blueberry, huckleberry, crataegus, grape, raspberry, and blackberry.

maximum population level. The greatest number of wildlife species that can occur if the constraints of food, cover, and water are removed; the greatest number that can exist without losses causes by social strife (Thomas 1979).

maximum sustained yield. The largest number of fish or wildlife that can be removed without destroying a population's reproduction capability (Anderson 1991).

mean. Average; the total of a series of measurements divided by the number of measurements (Thomas 1979).

mean annual mast yield. Total mast yield divided by the total age.

mesic. A site characterized by a moderate amount of moisture.

metabolic water. Water produced during oxidation of organic compounds containing hydrogen.

metapopulation. Strictly, a system of populations of a given species in a landscape linked by balanced rates of extinction and colonization. More loosely, the term is used for groups of populations of a species, some of which go extinct whereas others are established, but the entire system may not be in equilibrium.

microclimate. The climatic conditions within a small or local habitat that is well defined (Thomas 1979).

micronutrient. One of several minerals necessary for plant and animal growth but required in minute amounts. Primarily functions as a catalyst in enzyme systems. Zinc, copper, cobalt, and iron are examples (Bolen and Robinson 1995).

migration. The periodic movement to and from a breeding area, but the term is often used loosely for other types of movements. Migration recurs each year in individuals of many species, but other species make only one round trip before dying (e.g., Pacific salmon). Other patterns occur (e.g., monarch butterflies) that complicate a universal definition. Altitudinal migration involves changes in elevation (e.g., some grouse), whereas latitudinal migration involves changes in latitude (e.g., the typical north-south migration of geese in North America; Bolen and Robinson 1995).

migration corridor. A belt, band, or stringer of vegetation that provides a completely or partially suitable habitat and that animals follow during migrations (Thomas 1979).

migration homing. The return to a site of previous experience, usually a breeding area. Female ducks of many species, for example, return to the same breeding area where they nested in the previous year. The term is often shortened to *homing* (Bolen and Robinson 1995).

migration route. A travel route used routinely by wildlife in their seasonal movement from one habitat to another (Thomas 1979).

mineral cycling. The cycling of minerals throughout an ecosystem (Anderson 1991).

minimum-impact carrying capacity. The population density that minimizes impact on other wildlife or vegetation without eliminating the population.

mitigation. The replacement, usually by substitution, of wildlife habitat lost to development (Bolen and Robinson 1995).

mobility. The tendency of the individual animal to change location during the day, or between seasons or years.

model. Any formal representation of the real world. A model may be conceptual, diagrammatic, mathematical, or computational.

monitoring. The orderly collection, analysis, and interpretation of resource data to evaluate progress toward meeting management objectives.

monoculture. A term that describes unbroken expanses committed to a single crop, whether in forest or farmland. That is, interspersion is lacking (Bolen and Robinson 1995).

monogamy (monogamous). A type of mating behavior in which one male unites with one female, forming either seasonal (e.g., most ducks) or lifetime (e.g., geese and swans) pair bonds (Bolen and Robinson 1995).

mortality. In wildlife management, mortality is the loss in a population from any cause, including hunter kill, poaching, predation, accidents, or diseases and parasites (Thomas 1979).

mortality rate. The proportion of population dying in a unity of time (death rate; Anderson 1991).

mosaic. The intermingling of plant communities and their successional stages in such a manner as to give the impression of an interwoven design (Thomas 1979).

mosaic edge. An edge between stands or communities that is highly irregular, leading to a relatively large amount of edge per unit area (Thomas 1979).

multiple use. A concept of land management in which a number of products are deliberately produced from the same lad base; in national forests, the simultaneous provision of water, wood, recreation, wildlife is required by the Multiple-Use Sustained Yield Act (PL 86-517, 1960; Thomas 1979).

multiple-use management. A concept of land management that integrates various activities and natural resources. For example, many national forests provide a combination of skiing, hunting, fishing, hiking, canoeing, camping, mining, and grazing in addition to lumbering and watershed protection. Wildlife represents a major component in multiple-use management (Bolen and Robinson 1995).

multiple-use planning. The planning of management activities for a defined area to simultaneously accomplish goals for several distinct purposes, such as production of wood, water, wildlife, recreation, and grazing (PL 86-517, 1960).

mutualism. Mutually beneficial association of different kinds of organisms (Anderson 1991).

myrmecochory. Seed dispersal by ants.

natality. Birth rate. Expressed in several ways, but often as the number of offspring per female per year or per 100 females per year (Bolen and Robinson 1995).

natality rate. The proportion of a population born in a unit of time (birth rate; Anderson 1991).

national forest. In the United States, a federal reservation, forest, range, or other wild land administered by the Forest Service of the US Department of Agriculture under a program of multiple use and sustained yield.

natural regulation. The regulation of a population through naturally occurring biological processes as opposed to regulation through anthropogenic practices.

negative feedback. Feedback that inhibits or stops a system's progress (Anderson 1991).

nest parasitism. The act or result of one species (the parasite) of bird laying its eggs in the nest of another (the host), after which the host raises the parasite's young at the expense of its own offspring. The brown-headed cowbird is a common nest parasite in North America (Bolen and Robinson 1995).

nesting population level. The number of individuals or pairs of species in an area during the breeding season.

net primary productivity. Energy available in a plant following respiration. Gross primary productivity minus respiration equals net primary productivity (Anderson 1991).

net reproductive rate. The average number of female offspring produced by the females in a population (Anderson 1991).

niche. The functional role of an organism within its habitat (Anderson 1991).

nitrogen fixation. The conversion of elemental, atmospheric nitrogen (N_2) to organic combinations or to forms readily usable in biological processes (Anderson 1991).

nocturnal. Active at night (Anderson 1991).

nonconsumptive use. Use of natural resources without removing them (e.g., photography and watching wildlife; Anderson 1991).

nongame wildlife. All wildlife not subject to harvest regulations (Anderson 1991).

nonmanagement. See *inactive management*.

nonpasserine. Term applied collectively to all birds other than those in the order Passeriformes. See *Passeriformes* (Bolen and Robinson 1995).

noxious weeds. Those plant species designated as noxious weeds by federal or state law (it is a legal administrative declaration). Noxious weeds generally possess one or more of the characteristics of being aggressive and difficult to manage, parasitic, a carrier or host of serious insects or disease, and are nonnative, new to, or not common to the United States.

nutrition. A process that animals use to procure and process portions of their external chemical environment for the continued functioning of internal metabolism.

objective. An achievable, quantifiable, and explicit statement of planned results to be achieved within a stated period. The completion of an objective must occur within a stated time frame, and the results must be documented.

obligate. A plant or animal that occurs in narrowly defined habitat (Thomas 1979).

odd area. A farm site where poor drainage or another shortcoming prevents tillage and crop production, thereby providing potential habitat for wildlife (Bolen and Robinson 1995).

old growth stand. A stand that is past full maturity and showing decadence; the last stage in forest succession.

omnivore. An animal that feeds on plants and animals.

open canopy. A canopy condition that allows large amounts of direct sunlight to reach the ground (Thomas 1979).

opening. A break in the forest canopy; the existence of an area essentially bare soil, grasses, forbs, or shrubs in an area dominated by trees (Thomas 1979).

optimum carrying capacity. Wildlife carrying capacity related to human values, which varies from place to place and may change as values change.

optimum yield. The amount of material that, removed from a population, will maximize biomass (or numbers, or profit, or any other type of optimum) on a sustained basis (Anderson 1991).

organic matter in soil. Materials derived from plants or animals, much of it in an advanced state of decomposition (Thomas 1979).

organism. Any living individual of any plant or animal species (Thomas 1979).

outbreeding depression. The undesirable result of crossing between individuals from separate populations with incompatible traits; the offspring lack fitness (e.g., young ibex born in winter instead of spring; Bolen and Robinson 1995).

overgrazing. (1) Severe and frequent grazing during active growth periods that affects the recovery capability of plant species or plant community. Generally, this results in lower plant diversity and low plant community successional levels. (2) A continued overuse, usually by ungulates, that creates a deteriorated range condition (Anderson 1991).

overstory. The upper canopy of plants. Usually refers to trees, tall shrubs, or vines.

palatability. The relative degree of attractiveness of plants to animals as forage.

parameter. Any variable or arbitrary constant that appears in a mathematical expression, the values of which restrict or determine the specific form of the expression (Thomas 1979).

parasite. An organism that benefits while feeding upon, securing shelter from, or otherwise injuring another organism (the host; Anderson 1991).

parasitic. Growing on and deriving nourishment from another organism (Anderson 1991).

parasitism. The interaction of two individuals in which one, the host, serves as a food source for the other, the parasite (Anderson 1991).

Passeriformes. The largest order of birds, commonly known as perching birds or songbirds, which includes such diverse groups as crows, wren, swallows, thrushes, warblers, blackbirds, and sparrows (Bolen and Robinson 1995).

passive management. See *inactive management.*

passive rewarming. Use of the sun or warm surfaces to increase body temperature; typically used by animals after arousing from torpor.

patch. A recognizable area on the surface of earth that contrasts with adjacent areas and has definable boundaries.

pathogen. A disease-causing agent; includes bacteria, viruses, and parasites.

pathology. The anatomic or functional manifestations of diseases (Thomas 1979).

pattern. Type of distribution (random, regular, aggregate; Anderson 1991).

perennial. For plants, species that live two or more growing seasons. Trees and other woody plants are examples, but cattails, sod-forming grasses, and many other herbaceous species also are perennials (Bolen and Robinson 1995).

perennial stream. A stream that ordinarily has running water on a year-round basis (Thomas 1979).

permafrost. Ground that is frozen a few centimeters below the surface all year (Anderson 1991).

pesticide. Any of several chemicals that kill pests. Includes insecticides, herbicides, rodenticides, and fungicides. See *herbicide* (Bolen and Robinson 1995).

P-hacking. Includes several illegitimate practices involving the reporting of statistical significance (e.g., checking statistical significance of results before deciding to collect more data, stopping data collection when results reach statistical significance, excluding data points after checking their impact on statistical significance, and not reporting the impact of the data exclusion; Fraser et al. 2018).

phenology. (1) A branch of science that deals with relations between climate and periodic biological phenomena such as flowering, germination, and growth patterns. (2) Periodic biological phenomena that are correlated with climatic conditions.

phenotype. Expression of the characteristics of an organism as determined by the interaction of its genetic constitution and the environment; contrast with *genotype* (Anderson 1991).

pheromone. A naturally produced chemical secretion for olfactory communication (i.e., smell), usually between individuals of the same species. When synthesized, pheromones acting as sex attractants offer a means of luring insects into lethal traps, thereby reducing the need for toxic insecticides (Bolen and Robinson 1995).

photo point. A permanently identified point from which photographs are taken at periodic intervals. Sometimes called a camera point.

photoperiod. Day length (Anderson 1991).

photoperiodism. Response of plants and animals to the relative duration of light and darkness; e.g., some migration patterns are triggered by day length (Anderson 1991).

photosynthesis. Formation of chemical bond (sugar) from solar energy by green plants (Anderson 1991).

physiognomy. Vegetation classified according to shape and structure, irrespective of the species included.

physiological longevity. Maximum life span of individuals in a population under specified conditions; the organisms die of senescence (Anderson 1991).

phytoplankton. Minute plants that float in an aquatic system; the pant community in marine and freshwater that floats free in the water and contains many species of algae and diatoms (Anderson 1991).

pioneer. The first species of community in succession. On bare soil, annual weeds are usually the pioneer species, but some kinds of woody plants are also pioneers (e.g., jack pine after fires). Together with its associated animal life, such vegetation forms pioneer communities (Bolen and Robinson 1995).

piosphere. The gradient of decreasing forage use and plant community change with increasing distances from a water source.

plant association. A potential natural plant community; definite floristic composition and uniform appearance represented by stands occurring in places with similar environments. A taxonomic unit of vegetation classification.

plant community. An assemblage of plants living and interacting together in a specific location. No particular ecological status is implied. Plant communities may include exotic or cultivated species.

plant vigor. Plant health.

plot. A sampling of an ecosystem or of a site.

poikilothermic. Term indicating the inability to maintain stable body temperatures when environmental temperatures change. Poikilotherms—fishes, amphibians, reptiles, and invertebrates—cannot generate metabolic heat to meet the stress of low ambient temperatures nor cool themselves at high ambient temperatures. *Cold-blooded* is a popular description for poikilothermic species (Bolen and Robinson 1995).

point. A map feature described by a single set or coordinates.

point of resistance. The minimum population or density necessary for recovery of productivity.

polyandry. Mating of a female with two or more males during mating season (Anderson 1991).

polygamy. Mating of a male with two or more females during mating season (Anderson 1991).

population. A group of organisms of a single species that interact and interbreed in a common place (Anderson 1991).

population dynamics. The totality of changes in number, sex, and age that take place during the life of population (Thomas 1979).

positive feedback. Return of output to a system that allows it to continue in its direction; feedback that enhances or promotes a system's progress (Anderson 1991).

precipitation. Any form of atmospheric moisture that reaches the earth's surface, including rain, snow, fog, and hail (Bolen and Robinson 1995).

precision. The closeness to each other of repeated measurements of the same quantity.

precocial. Able to move about at an early age (Anderson 1991).

predation. The act of predators capturing prey (Anderson 1991).

predator. An organism that depends in total or part on killing another animal for its food. Bobcats, owls, and bass are well-known predators, but so are shrews, robins, bullfrogs, and dragonflies. A few species of plants are also predaceous (e.g., Venus flytrap; Bolen and Robinson 1995).

predator lane. A narrow strip of cover in which predators easily can find prey. For example, a predator lane is created when clean farming leaves only a small ring of vegetation around a pond edge; in this way skunks, raccoons, and other predators easily locate and destroy duck nest (Bolen and Robinson 1995).

preformed water. Water contained in food sources.

preregistration and registration. Attempts to correct questionable practices in the sciences. There are two levels of registering a study, including hypotheses, method of data collection, and analyses before the study is conducted (Kupferschmidt 2018).

prescribed burning. The use of fire as a management tool for improving plant and animal habitats. Also known as controlled burning (Bolen and Robinson 1995).

prescribed fire. Fire used as a management tool under specified conditions for burning a defined area (Thomas 1979).

prescription. In silvicultural terms, the formal written plan of action to carry out a silvicultural treatment of a forest stand to achieve specific objectives.

prey. An organism killed and eaten by a predator (Bolen and Robinson 1995).

prey switching. When two prey types are available and one is more abundant than the other. Predators will concentrate on the prey that is more abundant until their numbers are low, at which time they switch to the other prey.

primary association. The relationship between a wildlife species and a habitat condition that reflects a dependence on such habitat; a relationship that is strong and predictable (Thomas 1979).

primary cavity nesters. Wildlife species that excavate cavities in snags or trees.

primary consumer. A trophic level; consists of herbivores (Bolen and Robinson 1995).

primary excavator. A species that digs or chips out cavities in wood to provide itself or its mate with a site for nesting or roosting (Thomas 1979).

primary production. Production by green plants (Anderson 1991).

primitive road. A one-lane unimproved forest road in fair to poor condition that is seldom or never maintained (Thomas 1979).

probability. The frequency, expressed as a proportion or percentage of the total occurrences, that, over a long series of trials, will produce a specified value for a variable (Anderson 1991).

producer. A trophic level; consists of green plants (Bolen and Robinson 1995).

production. Amount of energy (or material) formed by an individual, population, or community in a specified period (Anderson 1991).

productivity. The rate at which mature breeding stock produces other mature stock, or mature removable crop.

promiscuous. Not restricted to one sexual partner (Anderson 1991).

protein. Amino acids linked end to end in a specific order (Anderson 1991).

proventriculus. The glandular prestomach of birds, located between the esophagus and gizzard, where chemical digestion softens foods (Bolen and Robinson 1995).

public trust doctrine. The principle that certain resources are preserved for public benefit, and that the government is required to maintain those resources for the public's reasonable use (Organ and Batcheller 2009). The public trust doctrine applies to wildlife in the United States, where wildlife are considered a common property resource held in trust by state and federal governments for the people, both current and future generations.

pulmonary evaporation. Evaporation of water from mouth, nose, and respiratory surfaces.

put and take. Planting of hatchery fish for removal by fishing enthusiasts or game-farm birds for removal by hunters (Anderson 1991).

quantify. To determine or express the number of some item (Thomas 1979).

quantity. A number or amount of anything, either specific or indefinite (Thomas 1979).

race. A geographic variant of a species; often considered the same as subspecies (Anderson 1991).

radiation heat transfer. Transfer of heat via electromagnetic waves.

radio collar. A collar containing a radio transmitter that is fastened on an animal. Signals from the transmitter are received and used by wildlife biologists to gain information, usually about the position of the animal (Thomas 1979).

rain forest (tropical). A multicanopied forest in humid tropical zones. Although lush, rain forest vegetation is highly susceptible to human disturbances and seldom recovers after being cut. More than half of the world's biota probably occur in tropical rain forest. Rain forests also occur in some temperate zones (e.g., Olympic Peninsula, Washington; Bolen and Robinson 1995).

random distribution. A distribution pattern in which an organism's position is independent of that of others (Anderson 1991).

random sample. A sample in which the selection potential of each individual in the population is known; a sample free from selection bias (Anderson 1991).

randomized sampling. A research method whereby each individual within a population has an equal chance of being selected as a sample, equivalent to picking numbers from a hat (Bolen and Robinson 1995).

range. The limits within which an entity operates or can be found.

range of tolerance. The range of environmental conditions (e.g., temperature) that limits the abundance and distribution of organisms (Bolen and Robinson 1995).

raptor. Any predatory bird (e.g., falcon, hawk, eagle, owl) that has feet with sharp talons or claws adapted for seizing prey and a hooked beak for tearing flesh (Thomas 1979).

recruitment. Increment of new individuals added to a wildlife population by natural reproduction, immigration, or stocking.

reforestation. Reestablishment of a tree crop on forest sites (Thomas 1979).

refugium carrying capacity. The number of animals a habitat will support when the necessary welfare factors to alleviate predation are present in the proper amount.

regenerate. To renew a tree crop through artificial or natural means (Thomas 1979).

regeneration. The renewal of the tree crop by natural or artificial means; also, the young crop (Thomas 1979).

regulate. To control or direct by rule, principle, or method; to adjust to some standard or requirement; to put in desired order (Thomas 1979).

regulating mechanisms. Factors that act to control the density of a population (Anderson 1991).

relative density. The ranking of populations by density (e.g., area A has 40% more mice/ha than area B).

release:kill ratio. The ratio of the number of head game annually released for restocking, and the number killed.

relic. A surviving memorial of something past; an object having interest by reason of its age or association with the past; a surviving trace of something; remaining parts or fragments (Thomas 1979).

remise. A European term for an artificially established game-bird covert. Sometimes includes food.

renesting. A nesting attempt that follows an earlier failure. (Not a second brood following an earlier success.)

replacement rate (net production rate; R_0). The average number of female offspring produced by the females in a population (Anderson 1991).

resident species. The wildlife species commonly found in a specific area (Thomas 1979).

resolution. The smallest spatial scale at which discontinuities in biotic and abiotic factors can be portrayed in map form.

resource. Any biotic and abiotic factor directly used by an organism.

resource abundance. The absolute amount (or size or volume) of an item in an explicitly defined area.

resource availability. A measure of the amount of a resource available to the animal (i.e., the amount exploitable).

resource preference. The likelihood that a resource will be used if offered on an equal basis with others.

resource selection. The process by which an animal chooses a resource.

resource use. A measure of the amount of resource taken directly (e.g., consumed, removed) from an explicitly defined area.

respiration. The breakdown of sugar into usable energy by living organisms (Anderson 1991).

response curve. A curve describing the response in potential use by elk or deer on a land type to changes in the cover:forage area ratio (Thomas 1979).

rest period. The period when an incubating female normally leaves the nest for rest, food, or recreation.

richness. A measure of the relative degree or number of plant or wildlife species or both associated with particular habitat conditions (Thomas 1979).

riparian area. Geographically declinable area with distinctive resource values and characteristics that are composed of the aquatic and riparian ecosystems.

riparian complex. Ecological units that support or may potentially support specified patterns of riparian ecosystems, soils, landforms, and hydrologic characteristics.

riparian ecosystem. A transition between the aquatic ecosystem and the adjacent terrestrial ecosystem: identified by soil characteristics or distinctive vegetation communities that require free or unbound water. Riparian ecosystems often occupy distinctive landforms such as flood plains or alluvial benches.

riparian zone. An area identified by the presence of vegetation that requires free or unbound water or conditions moister than normally found in the area (Thomas 1979).

rookeries. In the United States, colonies of nesting herons, usually great blue (Anderson 1991).

rotation. The planned number of years between the regeneration of a stand and its final cutting at a specified area (Thomas 1979).

r-selected species. Species adapted to high reproductive rates and short life spans, often with wide ranges of tolerance to environmental conditions. Bobwhites, mourning doves, and cottontails are examples (Bolen and Robinson 1995).

rumen. The first of four chambers in the stomach of ruminants (e.g., deer). Food in the rumen undergoes fermentation, after which the contents are returned to the mouth for further chewing (chewing the cud; Bolen and Robinson 1995).

ruminant. A cud-chewing animal with a complex stomach that is often divided into four chambers. Deer, goats, sheep, and bovines (antelopes, cattle, and bison) are examples (Bolen and Robinson 1995).

runoff. Precipitation that is not retained on the site where it fell; natural drainage away from an area (Thomas 1979).

rut. Breeding season of some ungulates (Anderson 1991).

sample. A subset of the total number of units in a population (Anderson 1991).

sampling bias. An error in collecting or analyzing samples representing a population. Fish populations sampled with nets may not include the younger age classes if smaller individuals escape through the mesh, thereby introducing sampling bias based on equipment. Females and fawns usually are underrepresented in the deer harvest simply because of sampling bias introduced by hunter selection for mature males (Bolen and Robinson 1995).

sampling theory. The assumption that samples reflect the same attributes (e.g., age or sex ratios, stomach contents, or parasite loads) as the remainder of the population from which the samples were collected (Bolen and Robinson 1995).

saturation point. The maximum wild density common to widely separated optimum ranges.

savanna. Lowland tropical and subtropical grassland with a scattering of trees and shrubs (Anderson 1991).

scale. The resolution at which patterns are measured, perceived, or represented. Scale can be broken into several components, including grain and extent (Bissonette 1997).

scale of observation. The spatial and temporal scales at which observations are made. Scale of observation has two parts: extent and grain.

scat. Animal fecal matter (Anderson 1991).

scavenger. An organism that feeds on dead organisms. Vultures and hyenas are examples, but many other species also scavenge at times (e.g., crows, golden eagles, and coyotes; Bolen and Robinson 1995).

scientific name. The binomial or two-word latinized name of an organism; the first word describes the genus, the second the species (Thomas 1979).

secondary cavity nester. Wildlife that occupies a cavity in a snag that was excavated by another species (Thomas 1979).

secondary consumer. A trophic level; consists of predators that feed on primary consumers. A bobcat is a secondary consumer (Bolen and Robinson 1995).

secondary poisoning. Unintended contamination of organisms (nontarget species) feeding on previously poisoned animals. Vultures and other scavengers are particularly susceptible (e.g., feeding on the carcasses of poisoned coyotes; Bolen and Robinson 1995).

secondary species. Wildlife species that are less important than the featured species but will be present on a given unit of land. Management activities aimed at the primary species may sometimes benefit the secondary species. At times the secondary species may outnumber and provide more recreation than the primary wildlife species (e.g., deer on an area where bear are featured).

sediment. Material suspended in liquid or air; the deposition of that material onto the surface underlying this liquid or air; usually the deposition of organic and inorganic soil materials by water (Thomas 1979).

selective grazing. The grazing of certain plant species on the range to the exclusion of others.

self-regulation. The process of population regulation in which population increase is prevented by a deterioration in the quality of individuals that make up the population; population regulation by internal adjustments in behavior and physiology within the population rather than by external forces, such as predators (Anderson 1991).

self-sustaining population. A wildlife population of sufficiently large size to ensure its continued existence within the area of concern without introduction of other individuals from outside the area (Thomas 1979).

semiaquatic species. Wildlife species that spend part of their lives on land and part in water (Thomas 1979).

sensitive species. Those plants and animal species identified by a regional forester for which population viability is a concern, as evidenced by (1) significant current or predicated downward trend in population numbers or density, or (2) significant current or predicted downward trend in habitat capability that will reduce species' existing distribution.

seral community. Any community that is not at potential. A relatively transitory community that develops under ecological succession, toward or away from a potential natural community.

seral stage. Successional plant communities are often classified unto quantitative seral stages to depict that relative position on a classical successional pathway.

sere. The stages that follow one another in an ecological succession (Thomas 1979).

sex ratio. The ratio of males to females in animal populations, with the percentage of males expressed first (e.g., in a population with a 45:55 sex ratio, 45% are males). Sex ratios are subdivided by age: primary (fertilization), secondary (birth or hatching), tertiary (juvenile), and quaternary (adult; Bolen and Robinson 1995).

shade-intolerant plants. Plant species that do not germinate or grow well in shade (Thomas 1979).

shade-tolerant plants. Plants that grow well in shade (Thomas 1979).

sheet erosion. Loss or movement of soil in thin layers (Thomas 1979).

shelterbelt. A strip of trees of shrubs planted or left standing in prairie areas to help reduce wind and erosion of topsoil (Anderson 1991).

shelterwood cutting. Any regeneration cutting designed to establish a new tree crop under the protection of remnants of the old stand (Thomas 1979).

shrub. A plant with persistent woody stems, relatively low growth habit, and generally several basal shoots instead of a single bole. It differs from a tree by its low stature and nonarborescent form (Thomas 1979).

sight barrier. Any object that serves to block the vision of an observer (Thomas 1979).

sight distance. The distance at which 90% or more of an adult elk or deer is hidden from the view of a human (Thomas 1979).

sigmoid curve. S-shaped curve (e.g., the logistic curve; Anderson 1991).

silvics. The study of the life history and general characteristics of forest trees and stands, with particular reference to site factors, to provide a basis for the practice of silviculture (Thomas 1979).

single-tree selection. A method of harvesting under an uneven-aged forest management system in which trees are individually selected for harvest (Thomas 1979).

single use. A concept of management in which a single management objective is paramount (Thomas 1979).

sink populations. In a landscape, a population or site that attracts colonists while not supplying migrants to other sites or populations.

site. A single, specific point on the land. The sample point where data measurements are taken.

site type. Classification of an area considered quantitatively in terms of how its environment determines the type, quality, and growth rate of the potential vegetation (Thomas 1979).

slash. The residue left on the ground after trees are felled or accumulated there as a result of storm, fire, or silvicultural treatment (Thomas 1979).

slope. The incline of the land surface measured in degrees from the horizontal or as a percentage, as determined by the number of units change in elevation per 100 of the same measurement units; also characterized by the compass direction in which it faces (Thomas 1979).

snag. A standing dead tree from which the leaves and most of the limbs are missing owing to environmental factors.

source populations. In a landscape, a population or a site that supplies colonists to other patches.

species. The basic taxonomic unit. A population whose members freely interbreed and share a common gene pool. Species are identified scientifically in Latin with a two-part name known as a binomial, which includes both genus and species names (e.g., *Bubo virginianus* is the binomial for the great horned owl; Bolen and Robinson 1995).

species composition. The proportion of plant species or aggregations of species in relation to a total area.

species diversity. A measure combining richness and evenness. Diversity increases when both the number of species increases and the number of individuals of each species are more evenly distributed. See *species evenness* and *species richness* (Bolen and Robinson 1995).

species evenness. The relative abundance of individuals among those species present in a specified area or time. Some species, although present, may have only a few individuals, whereas others are abundant; such conditions diminish evenness. See *species richness* (Bolen and Robinson 1995).

species richness. An indicator of the number of species of plants or animals present in an area. The more species present, the higher the degree of species richness (Anderson 1991).

specific natality rate. The number of individuals produced per unit of time per breeding female.

spring-fall range. An area between summer range at high elevation and winter range at low elevation that is used by deer and elk during spring and fall as they move between summer and winter range (Thomas 1979).

stable isotope. Isotopes are atoms of the same chemical element that contain the same number of protons and the same atomic number but a different number of neutrons. Heavy isotopes have an extra neutron compared to light isotopes and are much less abundant in nature than light isotopes. Stable isotopes are useful in reconstructing diets and movements of species between isotopically distinct areas. Stable isotopes remain stable, unlike radioactive isotopes, for which neutrons are converted to protons during radioactive decay.

stakeholder. A person or group who significantly affects or is significantly affected by wildlife management, normally including future as well as present generations (Decker et al. 2012).

stand. An uninterrupted unit of vegetation, homogenous in composition and of the same age. The vegetation can be of any physiognomic class.

standard deviation. A statistical term; a measure of the dispersion about the mean of a population (i.e., the positive square root of the variance; Thomas 1979).

standard error. A statistical term; the standard deviation of a distribution of means or of any other statistic determined from samples; determined by dividing the standard deviation by the square root of the number of observations (Thomas 1979).

standardized sampling. A research method whereby an individual within a population is selected on a predetermined basis. For example, every twentieth oak tree along a line transect might be measured for acorn production. For comparison, see *randomized sampling* (Bolen and Robinson 1995).

stationary age distribution. The allocation of age classes in a population when the population does not change in size and the age structure is constant over time.

stem. The principal axis of a plant from which buds and shoots develop; with woody species, the term applies to all ages and thickness (Thomas 1979).

steno-. A prefix used to identify narrow tolerances for specific components of the environment. Frogs are stenohaline because their eggs and larvae have almost no tolerance to salt water (Bolen and Robinson 1995).

stenotopic. Having little ability to withstand modification of environmental conditions (Anderson 1991).

steppe. An extensive area of natural, dry grassland; usually refers to grasslands in southwestern Asia and southeastern Europe; equivalent to prairie in North American usage (Anderson 1991).

stochastic. Random or expected by change. Wildfires, storms, and flooding are stochastic events, each with a statistical probability of happening. For example, rain may fall only 10 times per year in a desert, but the date of the next rainfall cannot be predicted accurately, whereas the probability of its occurrence on a given day can be calculated rather precisely. Rainfall is therefore stochastic (e.g., chance variation in a sex ratio or birth rate; Bolen and Robinson 1995).

stochastic model. A mathematical model based on probabilities; the prediction of the model is not a single fixed number but a range of possible numbers; opposite of *deterministic model* (Anderson 1991).

stock. A group of fish or other aquatic animal that can be treated as a single unit for management purposes (Anderson 1991).

straggling failure. A translocation followed by initial thrift but ultimate dwindling and disappearance.

strain. Type of fish that has genetic adaptability to a specific set of physical conditions (e.g., temperature; Anderson 1991).

strata. See *vegetation strata*.

stratified sampling. Sampling by groups: a precision-increasing method (Anderson 1991).

stream management unit. A management zone alongside a stream where the management objective is the protection of the riparian zone or of water quality or quantity or both (Thomas 1979).

stream protection zone. A zone in which management activity is excluded or modified to protect the riparian zone, quality and quantity of water, or both (Thomas 1979).

stringer. Vegetation that is arranged in a long, thin, linear fashion (Thomas 1979).

structure (community). Physical makeup of vegetation in a community (Anderson 1991).

study area. An arbitrary spatial extent chosen by the investigator within which to conduct a study (contrast with site and scale).

study repeatability vs. replicability. Repeatability does not involve use of the same data set and analyses, while reproducibility does (Cassey and Blackburn 2006).

stump. The woody base of a tree left in the ground after felling (Thomas 1979).

subadult. An animal, or age class, that resembles an adult but does not breed because of behavioral or sexual immaturity (Bolen and Robinson 1995).

subspecies. Geographic variant of a species (Anderson 1991).

substitutable area. An area that may serve in the place of another area for a particular purpose selected by the forest manager (Thomas 1979).

substrate. Supporting material. In biology, substrate usually refers to soil or soil-like material as community substrate (Anderson 1991).

success ratio. The ratio of number of hunters to number of game killed.

succession. The process of vegetative and ecological development whereby an area becomes successively occupied by different plant communities.

successional stage. A stage or recognizable condition of a plant community that occurs during its development from bare ground to climax (Thomas 1979).

sucker. A shoot arising from below ground level, either from a rhizome or from a root (Thomas 1979).

summer range. A range, usually at higher elevation, used by deer and elk during the summer; a summer range is usually much more extensive than a winter range (Thomas 1979).

sunning. The process of an animal methodically exposing itself to direct sunlight; also called sunbathing (Thomas 1979).

surrogate species. A close relative of a rare or endangered species; used to determine capture and breeding techniques, rearing procedure, and in physiological tests for endangered species (Anderson 1991).

survivorship curve. Data from column l_x in a life table (individuals alive at the beginning of each interval) plotted on semilog paper (Anderson 1991).

suspended sediment. Soil and other material suspended in water, usually disturbed or moving water; this material drops out of suspension when water movement slows or ceases (Thomas 1979).

sustained yield. Number of animals or pants that can be removed from a population year after year without jeopardizing future yields (Anderson 1991).

swamp. A wet area that usually has standing trees (Anderson 1991).

symbiosis. A relationship between two or more kinds of living organisms where all benefit; sometimes obligatory to one or more of the organisms in the relationship (Thomas 1979).

sympatric. Similar; usually used in reference to populations that occupy the same geographic region (Anderson 1991).

taiga. Boreal forests of the north (Anderson 1991).

talus. The accumulation of broken rocks that occurs at the base of cliffs or other steep slopes (Thomas 1979).

taxon (plural: taxa). Any level of the hierarchy of taxonomy. Species, genus, and family are each a taxon (Bolen and Robinson 1995).

taxonomy. The classification of organisms into units known as *taxa* (singular: *taxon*). Carolus Linnaeus, a Swedish biologist, developed the currently used taxonomic system in 1753 (for plants) and 1758 (for animals). The species is the basic taxonomic unit. After the species level, taxa progress into larger units: genus family, order, class, and phylum. For some plants and animals, scientist identify a smaller taxon, the subspecies (Bolen and Robinson 1995).

terrestrial vertebrates. Animals with backbones that dwell primarily on land (Thomas 1979).

territory. Any area defended by an individual against intrusion by others of the same species; apparently ensures spacing in keeping with availability of food and other resources (the classical explanation). Typical territorial behavior (by males) simultaneously attracts females while warning other males (Bolen and Robinson 1995).

thermal cover. Cover used by animals to ameliorate effects of weather (Thomas 1979).

thermal-neutral zone. An area where the ambient conditions do not trigger a metabolic response on the part of the occupying animal (Thomas 1979).

thermoregulation. Maintenance of body temperature within physiologically acceptable limits.

threatened species. Any species that is likely to become an endangered species within the foreseeable future throughout all or a significant portion of its range and that the appropriate secretary has designated as a threatened species. (Some states also have declared certain species as threatened through their regulations or statutes).

tolerance. The capacity of wildlife to withstand or adjust to any disturbed conditions in the environment.

transect. A linear plot, usually represented by a line, along which are often placed regularly spaced quadrates (plot frames), loops, or other devices.

transition zone. An area of land where two forest types blend. A transition zone occurs between upland and lowland. This band usually contains high-quality wildlife food, including soft and hard mast, forage, and fruiting shrubs.

translocate. To move from one place to another within current or historical habitat (Anderson 1991).

transmutation. Also known as aggregation error, transmutation occurs most often when results qualitatively change as one moves from one hierarchical level to another (O'Neill 1979).

treatment. In experimentation, a stimulus applied to observe its effect on an experimental situation or to compare its effect with the effects of the other treatments. In practice, treatment may refer to anything capable of controlled application according to experimental requirements; the act or manner of treating something (Thomas 1979).

trends. Indications of changes in populations over time (Anderson 1991).

tribe. A subdivision or subfamily based on structure, plumage, habits, or courtship behavior (Anderson 1991).

trophic level. A structure of producer and consumer organisms superimposed on food chains to trace the flow of energy (Anderson 1991).

tundra. Northern, high-elevation biome with few sizable wood plants because of a short growing season (Anderson 1991).

umbrella species. Umbrella species are species selected for making conservation-related decisions, typically because protecting these species indirectly protects the many other species that make up the ecological community of its habitat.

understory. (1) Foliage, consisting of seedlings, shrubs, and herbs, that lies beneath and is shaded by canopy or taller plants (Anderson 1991). (2) In silviculture, trees that grow under the canopy formed by taller trees; in range management, herbaceous and shrub vegetation under a brushwood or tree canopy (Thomas 1979).

uneven-aged management. A forestry practice in which individuals or small group of trees are cut (e.g., single-tree selection cut; group selection cut), thereby producing a forest with trees of different ages and heights as cutting and regrowth continue. For comparison, see *even-aged management* (Bolen and Robinson 1995).

uneven-aged stand. A forest stand managed to maintain an intermingling of trees that differ markedly in age; stands continuously or periodically regenerated, tended, and harvested with no real beginning or end (Thomas 1979).

ungulate. A hooved animal (Anderson 1991).

unit area. A parcel of land in which wildlife needs for food, cover, water, and reproduction must be met.

urbanization. The process or degree of human concentration in relatively small areas (cities), where huge amounts of basic material are consumed from stocks produced elsewhere (e.g., food from farms, ranches, oceans). At its extreme, urbanization covers the soil with a monoculture of buildings and pavement (Bolen and Robinson 1995).

urine osmolality. Measure of urine concentration; large values indicate concentrated urine, and small values indicate dilute urine.

utilization. The available forage by weight consumed or trampled through livestock grazing. Usually expressed as a percentage.

values. Desirable end states, modes of conduct, or qualities of life humans individually or collectively hold dear (Manfredo 2008). Values are general mental constructs that define what is important to people.

vapor pressure deficit. The difference between the amount of moisture in the air and how much moisture the air can hold when saturated.

variable. Generally, any quantity that varies; more precisely, a quantity that may take any one of a set of values; a single influence or one of several measurable influences acting on a particular process (Thomas 1979).

variance. A statistical term; a measure of variability within a finite population or sample; the total of the squared deviations of each observation from the arithmetical mean divided by one less than the total number of observations (Thomas 1979).

vector. An organism that transmits a disease within and between populations, commonly an insect (e.g., mosquito) or other arthropod (e.g., tick). Fleas are vectors for sylvatic plague (Bolen and Robinson 1995).

vegetation classification. The process of analyzing vegetation community data and defining hierarchal entities based on that data. There are two branches of vegetation classification, potential natural and existing. The potential natural vegetation classification hierarchy includes series, subseries, plant associations, and plant association phases. The existing vegetation classification hierarchy covers types and community types.

vegetation management status. The relative degree to which kinds, proportions, and amounts of vegetation in the present plant community resemble the desired plant community chosen for an ecological site.

vegetation strata. The layers of vegetation in a plant community defined by species, age, size of plant, tree layer, shyrb layer, and herb layer. (Thomas 1979).

vegetation structure. The form or appearance of a stand; the arrangement of the canopy; the volume of vegetation in tiers or layers (Thomas 1979).

vegetation type. A kind of existing plant community with distinguishable characteristics described in terms of the present vegetation that dominates the aspect or physiognomy of the area.

velvet. The soft, highly vascularized tissues that cover developing antlers; normally rubbed and sloughed off when antlers reach full growth and harden (Bolen and Robinson 1995).

ventriculus. The muscular, grit-bearing stomach of birds, designed for grinding food; also called gizzard. May be modified in some species (e.g., hummingbirds; Bolen and Robinson 1995).

versatile. Capable of or adapted for survival in several plant communities or successional stages or both (Thomas 1979).

vertebrate. Any animal with a backbone, including cartilaginous fishes (e.g., sharks). A taxon known scientifically as Vertebrata includes fishes, amphibians, reptiles, birds, and mammals. Among the vertebrates, birds and mammals are most often considered in the narrow view of wildlife, although such a limited view is not in keeping with modern wildlife management (Bolen and Robinson 1995).

vertical diversity. The diversity in an area that results from the complexity of the aboveground structure of the vegetation; the more tiers of vegetation or the more diverse the species makeup or both, the higher the degree of vertical diversity (Thomas 1979).

viability. Strictly, the ability to live or grow. In conservation biology, the probability of survival of a population for an extended period (Thomas 1979).

viable population. A population large enough and with adequate habitat to perpetuate itself (Anderson 1991).

vigor. The relative robustness of a plant in comparison to other individuals of the same species. It is reflected primarily by the size of a plant and its parts in relation to its age and the environment in which it is growing.

visual management zone. An area in which the overriding management concern is for an esthetically pleasing appearance; in most forested areas, such zones are managed to give the impression of mature and relatively unbroken forest (Thomas 1979).

vitamins. Organic compounds that occur in food in minute amounts, cannot be synthesized by animals, and are essential for normal life and functioning.

water-holding capacity. A measure of the ability of soil to soak up and hold water (Thomas 1979).

water quality. Determinized by a series of standard parameters: turbidity, temperature, bacterial content, pH, and dissolved oxygen (Thomas 1979).

water quantity. The amount of water that comes from a watershed or drainage (Thomas 1979).

water turnover. Replacement of all water molecules within an animal's body; typically expressed as water turnover rate.

welfare factors. Necessary aspects for survival (e.g., food, water, cover, special factors).

wetland(s). (1) Any area where the water table is near or above the surface of the land during a considerable part of the year (Anderson 1991). (2) Those areas that are inundated by surface or ground water with a frequency sufficient to support, and under normal circumstances do or would support, a prevalence of vegetation or aquatic life that requires saturated or seasonally saturated soil conditions for growth and reproduction. Generally includes swamps, marshes, bogs, and similar areas such as sloughs, potholes, wet meadows, river overflows, mud flats, and natural ponds.

wilderness. Lands designated by law as wilderness; no road building or timber management is allowed on such lands; they are intentionally managed to maintain their primitive character (Thomas 1979).

wildfire. An unplanned fire requiring suppression action, as contrasted with a prescribed fire burning within prepared lines that enclose a designated area under prescribed conditions; a free-burning fire unaffected by fire suppression measures (Thomas 1979).

wildlife. All undomesticated animals in a natural environment (Anderson 1991).

wildlife, consumptive use. Unharvested game or nongame species harvested for sport, food, study, or commerce.

wildlife habitat management. The manipulation or maintenance of vegetation to yield desired results in terms of habitat suitable for designated wildlife species or groups of species (Thomas 1979).

wildlife logs. Logs left in place on the forest floor for wildlife habitat (Thomas 1979).

wildlife management. The scientifically based art of manipulating habitats to produce some level of a desired species or manipulating animal populations to achieve a desired end (Thomas 1979).

wildlife openings. Openings maintained to meet various food or cover needs for wildlife. Wildlife openings may contain native vegetation or planted crops and can be maintained by burning, discing, mowing, planting, fertilizing, grazing, or herbicides.

wildlife range. A range, usually a lower elevation, used by migratory deer and elk during winter months; usually better defined and smaller than summer ranges (Thomas 1979).

wolf tree. A tree of dominant size and position that usurps light and space from smaller understory, preventing its growth (Anderson 1991).

xeric. (1) Deficient in moisture for the support of life, said of a desert environment (Anderson 1991). (2) A site low or deficient in moisture that is available for the support of plant life.

xerophyte. Plant that can grow in dry places (Anderson 1991).

yard. A wintering ground used by deer during deep snow. Paths are trampled down to afford access to browse food.

yield. The sustained kill per unit of area or population.

zooplankton. Animal portion of the plankton; the animal community in marine and freshwater situations that floats free in the water, independent of the shore and the bottom, moving passively with the currents (Anderson 1991).

Literature Cited

Addicott, J. F., J. M. Aho, M. F. Antolin, D. K. Padilla, J. S. Richardson, and D. A. Soluk. 1987. Ecological neighborhoods: scaling environmental patterns. Oikos 49:340–346.

Anderson, S. H. 1991. Managing our wildlife resources. 2nd edition. Prentice Hall, Englewood Cliffs, New Jersey, USA.

Bissonette, J. A. 1997. Scale-sensitive properties: historical context, current meanings. Pages 3–31 in J. A. Bissonette, editor. Wildlife and landscape ecology: effects of pattern and scale. Springer-Verlag, New York, New York, USA.

Bolen, E. G., and W. L. Robinson. 1995. Wildlife ecology and management. 3rd edition. Prentice Hall, Englewood Cliffs, New Jersey, USA.

Cassey, P., and T. M. Blackburn. 2006. Reproducibility and repeatability in ecology. Bioscience 56:958–959.

Daubenmire, R. 1968. Plant communities: a textbook of plant synecology. Harper and Row, New York, New York, USA.

Decker, D. J., S. J. Riley, and W. F. Siemer, editors. 2012. Human dimensions of wildlife management. Johns Hopkins University Press, Baltimore, Maryland, USA.

Fraser, H., T. Parker, S. Nakagawa, A. Barnett, and F. Fidler. 2018. Questionable research practices in ecology and evolution. PLoS One 13:e0200303.

Golubski, A. J., E. E. Westlund, J. Vandermeer, and M. Pascual. 2016. Ecological networks over the edge: hypergraph trait-mediated indirect interaction (TMII) structure. Trends in Ecology and Evolution 31:344–354.

Kerr, N. L. 1998. HARKing: hypothesizing after the results are known. Personality and Social Psychology Review 2:196–217.

Krausman, P. R. 2002. Introduction to wildlife management: the basics. Prentice Hall, Upper Saddle River, New Jersey, USA.

Kupferschmidt, K. 2018. A recipe for rigor. Science 361:1192–1193.

Manfredo, M. J., 2008. Values, ideology, and value orientations. Pages 141–166 in Who cares about wildlife? Springer, New York, New York, USA.

Nelson, M. P., and J. A. Vucetich. 2012. Environmental ethics for wildlife management. Pages 223–237 in D. J. Decker, S. J. Riley, and W. F. Siemer, editors. Human dimensions of wildlife management. Johns Hopkins University Press, Baltimore, Maryland, USA.

O'Neill, R. V. 1979. Transmutations across hierarchical levels. Pages 59–78 in G. S. Innis and R. V. O'Neill, editors. System analysis of ecosystems. International Cooperative Publishing House, Fairland, Maryland, USA.

Organ, J. F., and G. R. Batcheller. 2009. Reviving the public trust doctrine as a foundation for wildlife management in North America. Pages 161–171 in M. J. Manfredo, J. J. Vaske, P. J. Brown, and D. J. Decker, editors. Society and wildlife in the 21st century. Island Press, Washington, DC, USA.

Ready, R. C. 2012. Economic considerations. Page 68 *in* D. J. Decker, S. J. Riley, and W. F. Siemer, editors. Human dimensions of wildlife management. Johns Hopkins University Press, Baltimore, Maryland, USA.

Riley, S. J., and R. S. Gregory. 2012. Decision making in wildlife management. Pages 101–112 *in* D. J. Decker, S. J. Riley, and W. F. Siemer, editors. 2012. Human dimensions of wildlife management. Johns Hopkins University Press, Baltimore, Maryland, USA.

Riley, S. J., W. F. Siemer, D. J. Decker, L. H. Carpenter, J. F. Organ, and L. T. Berchielli. 2003. Adaptive impact management: and integrative approach to wildlife management. Human Dimensions of Wildlife 8:81–95.

Sanderson, E. W., K. H. Redford, A. Vedder, P. B. Coppolillo, and S. E. Ward. 2002. A conceptual model for conservation planning based on landscape species requirements. Landscape and Urban Planning 58:41–56.

Thomas, J. W. 1979. Glossary. Pages 470–494 *in* J. W. Thomas, technical editor. Wildlife habitat in managed forests. Agriculture Handbook 553. US Department of Agriculture Forest Service, Washington, DC, USA.

Turner, M. G., R. H. Gardner, and R. V. O'Neill. 2001. Landscape ecology in theory and practice: pattern and process. Springer, New York, New York, USA.

West, G. 2017. Scale: the universal laws of growth, innovation, sustainability, and the pace of life in organisms, cities, economies, and companies. Penguin Press, New York, New York, USA.

Wiens, J. A. 1989. Spatial scaling in ecology. Functional Ecology 3:385–397.

INDEX

Page numbers followed by *t* indicate tables.